Methods in Enzymology

Volume 131
ENZYME STRUCTURE
Part L

METHODS IN ENZYMOLOGY

EDITORS-IN-CHIEF

Sidney P. Colowick   Nathan O. Kaplan

Methods in Enzymology

Volume 131

Enzyme Structure

Part L

EDITED BY

C. H. W. Hirs

DEPARTMENT OF BIOCHEMISTRY, BIOPHYSICS, AND GENETICS
UNIVERSITY OF COLORADO MEDICAL CENTER
DENVER, COLORADO

Serge N. Timasheff

GRADUATE DEPARTMENT OF BIOCHEMISTRY
BRANDEIS UNIVERSITY
WALTHAM, MASSACHUSETTS

1986

ACADEMIC PRESS, INC.

Harcourt Brace Jovanovich, Publishers

Orlando   San Diego   New York   Austin
Boston   London   Sydney   Tokyo   Toronto

ACADEMIC PRESS, INC.
Orlando, Florida 32887

United Kingdom Edition published by
ACADEMIC PRESS INC. (LONDON) LTD.
24–28 Oval Road, London NW1 7DX

LIBRARY OF CONGRESS CATALOG CARD NUMBER: 54-9110

ISBN 0–12–182031–9

PRINTED IN THE UNITED STATES OF AMERICA

86 87 88 89   9 8 7 6 5 4 3 2 1

Table of Contents

Section I. Unfolding and Refolding of Proteins

v

Contributors to Volume 131

Article numbers are in parentheses following the names of contributors.
Affiliations listed are current.

ROBERT L. BALDWIN (1, 3), *Department of Biochemistry, Stanford University School of Medicine, Stanford, California 94305*

JOHN F. BRANDTS (6), *Department of Chemistry, University of Massachusetts, Amherst, Massachusetts 01003*

THOMAS E. CREIGHTON (5, 9), *Medical Research Council, Laboratory of Molecular Biology, Cambridge CB2 2QH, England*

C. M. DOBSON (18), *Inorganic Chemistry Laboratory, University of Oxford, Oxford OX1 3QR, England*

JOAN J. ENGLANDER (22), *Department of Biochemistry and Biophysics, University of Pennsylvania, School of Medicine, Philadelphia, Pennsylvania 19104*

S. W. ENGLANDER (22), *Department of Biochemistry and Biophysics, University of Pennsylvania, School of Medicine, Philadelphia, Pennsylvania 19104*

ANTHONY L. FINK (10), *Department of Chemistry, University of California, Santa Cruz, Santa Cruz, California 95064*

ROGER B. GREGORY (21), *Chemistry Department, Kent State University, Kent, Ohio 44242*

RAINER JAENICKE (12), *Institut für Biophysik und Physikalische Biochemie, Universität Regensburg, D-8400 Regensburg, Federal Republic of Germany*

MARCEL A. JUILLERAT (11), *Nestec SA, CH-1800 Vevey, Switzerland*

MARTIN KARPLUS (15, 18), *Department of Chemistry, Harvard University, Cambridge, Massachusetts 02138*

PETER S. KIM (8), *Whitehead Institute for Biomedical Research, Cambridge, Massachusetts 02142*

JONATHAN KING (13), *Department of Biology, Massachusetts Institute of Technology, Cambridge, Massachusetts 02139*

A. A. KOSSIAKOFF (20), *Department of Biocatalysis, Genentech, Inc., South San Francisco, California 94080, and Department of Pharmaceutical Chemistry, School of Pharmacy, University of California, San Francisco, San Francisco, California 94143*

A. M. LABHARDT (7), *Zentrale Forschungs-Einheiten (ZFE), F. Hoffmann-La Roche & Co., Aktiengesellschaft, CH-4002 Basel, Switzerland*

JOSEPH R. LAKOWICZ (23), *Department of Biological Chemistry, University of Maryland, School of Medicine, Baltimore, Maryland 21201*

LUNG-NAN LIN (6), *Department of Chemistry, University of Massachusetts, Amherst, Massachusetts 01003*

STANLEY J. OPELLA (17), *Department of Chemistry, University of Pennsylvania, Philadelphia, Pennsylvania 19104-6323*

C. N. PACE (14), *Biochemistry Department, Texas A&M University, College Station, Texas 77843-2128*

FRITZ PARAK (24), *Institut für Physikalische Chemie der Universität Münster, D-4400 Münster, Federal Republic of Germany*

GARY R. PARR (11), *Monsanto Company, St. Louis, Missouri 63167*

GREGORY A. PETSKO (19), *Department of Chemistry, Massachusetts Institute of Technology, Cambridge, Massachusetts 02139*

SERGEY A. POTEKHIN (2), *Institute of Protein Research, Academy of Sciences of the USSR, 142292 Pushchino, Moscow Region, USSR*

PETER L. PRIVALOV (2), *Institute of Protein Research, Academy of Sciences of the USSR, 142292 Pushchino, Moscow Region, USSR*

LOU REINISCH (24), *Physics Department, Northeastern University, Boston, Massachusetts 02115*

DAGMAR RINGE (19), *Department of Chemistry, Massachusetts Institute of Technology, Cambridge, Massachusetts 02139*

JOSE R. ROGERO (22), *Istituto de Pesquisas Energeticas e Nucleares-IPEN, Sao Paulo, Brazil*

ANDREAS ROSENBERG (21), *Department of Laboratory Medicine and Pathology, University of Minnesota Medical School, Minneapolis, Minnesota 55455*

RAINER RUDOLPH (12), *Institut für Biophysik und Physikalische Biochemie, Universität Regensburg, D-8400 Regensburg, Federal Republic of Germany*

FRANZ X. SCHMID (4), *Institut für Biophysik und Physikalische Biochemie, Universität Regensburg, D-8400 Regensburg, Federal Republic of Germany*

HIROSHI TANIUCHI (11), *Laboratory of Chemical Biology, National Institute of Diabetes and Digestive and Kidney Diseases, National Institutes of Health, Bethesda, Maryland 20892*

HIROYASU UTIYAMA (3), *Life Science Group, Faculty of Integrated Arts and Sciences, Hiroshima University, Hiroshima 730, Japan*

GERHARD WAGNER (16), *Institut für Molekularbiologie und Biophysik, Eidgenössische Technische Hochschule, Hönggerberg HPM, CH-8093 Zürich, Switzerland*

KURT WÜTHRICH (16), *Institut für Molekularbiologie und Biophysik, Eidgenössische Technische Hochschule, Hönggerberg HPM, CH-8093 Zürich, Switzerland*

MYEONG-HEE YU (13), *Center for Genetic Engineering, Korea Advanced Institute of Science and Technology, Seoul, Korea*

Preface

Volumes 117 and 130, Parts J and K of Enzyme Structure, were published recently. They were devoted to physical methods. This volume also deals with physical methods. It is hoped that together they will bring up-to-date coverage of techniques currently available for the study of enzyme conformation, interactions, and dynamics.

This volume deals in detail with the questions of protein folding and protein dynamics. Many of the methods presented, even though based on experiments, are conceptual in nature. We hope that the expositions will result in proper guidance to the correct ways of asking questions and analyzing observations in these areas. In those articles which are primarily experimental, techniques have been described in detail and, where necessary, guidance has been provided for assembling the proper equipment if not commercially available.

We wish to acknowledge with pleasure and gratitude the generous cooperation of the contributors to this volume. Their suggestions during its planning and preparation have been particularly valuable. The staff of Academic Press has provided inestimable help in the assembly of this volume. We thank them for their many courtesies.

C. H. W. Hirs
Serge N. Timasheff

METHODS IN ENZYMOLOGY

EDITED BY

Sidney P. Colowick and Nathan O. Kaplan

VANDERBILT UNIVERSITY
SCHOOL OF MEDICINE
NASHVILLE, TENNESSEE

DEPARTMENT OF CHEMISTRY
UNIVERSITY OF CALIFORNIA
AT SAN DIEGO
LA JOLLA, CALIFORNIA

METHODS IN ENZYMOLOGY

EDITORS-IN-CHIEF

Sidney P. Colowick and Nathan O. Kaplan

VOLUME XXXIII. Cumulative Subject Index Volumes I–XXX
Edited by MARTHA G. DENNIS AND EDWARD A. DENNIS

VOLUME XXXIV. Affinity Techniques (Enzyme Purification: Part B)
Edited by WILLIAM B. JAKOBY AND MEIR WILCHEK

VOLUME XXXV. Lipids (Part B)
Edited by JOHN M. LOWENSTEIN

VOLUME XXXVI. Hormone Action (Part A: Steroid Hormones)
Edited by BERT W. O'MALLEY AND JOEL G. HARDMAN

VOLUME XXXVII. Hormone Action (Part B: Peptide Hormones)
Edited by BERT W. O'MALLEY AND JOEL G. HARDMAN

VOLUME XXXVIII. Hormone Action (Part C: Cyclic Nucleotides)
Edited by JOEL G. HARDMAN AND BERT W. O'MALLEY

VOLUME XXXIX. Hormone Action (Part D: Isolated Cells, Tissues, and
Organ Systems)
Edited by JOEL G. HARDMAN AND BERT W. O'MALLEY

VOLUME XL. Hormone Action (Part E: Nuclear Structure and Function)
Edited by BERT W. O'MALLEY AND JOEL G. HARDMAN

VOLUME XLI. Carbohydrate Metabolism (Part B)
Edited by W. A. WOOD

VOLUME XLII. Carbohydrate Metabolism (Part C)
Edited by W. A. WOOD

VOLUME XLIII. Antibiotics
Edited by JOHN H. HASH

VOLUME XLIV. Immobilized Enzymes
Edited by KLAUS MOSBACH

VOLUME XLV. Proteolytic Enzymes (Part B)
Edited by LASZLO LORAND

VOLUME XLVI. Affinity Labeling
Edited by WILLIAM B. JAKOBY AND MEIR WILCHEK

VOLUME XLVII. Enzyme Structure (Part E)
Edited by C. H. W. HIRS AND SERGE N. TIMASHEFF

VOLUME XLVIII. Enzyme Structure (Part F)
Edited by C. H. W. HIRS AND SERGE N. TIMASHEFF

VOLUME XLIX. Enzyme Structure (Part G)
Edited by C. H. W. HIRS AND SERGE N. TIMASHEFF

VOLUME L. Complex Carbohydrates (Part C)
Edited by VICTOR GINSBURG

VOLUME LI. Purine and Pyrimidine Nucleotide Metabolism
Edited by PATRICIA A. HOFFEE AND MARY ELLEN JONES

VOLUME LII. Biomembranes (Part C: Biological Oxidations)
Edited by SIDNEY FLEISCHER AND LESTER PACKER

VOLUME LIII. Biomembranes (Part D: Biological Oxidations)
Edited by SIDNEY FLEISCHER AND LESTER PACKER

VOLUME LIV. Biomembranes (Part E: Biological Oxidations)
Edited by SIDNEY FLEISCHER AND LESTER PACKER

VOLUME LV. Biomembranes (Part F: Bioenergetics)
Edited by SIDNEY FLEISCHER AND LESTER PACKER

VOLUME LVI. Biomembranes (Part G: Bioenergetics)
Edited by SIDNEY FLEISCHER AND LESTER PACKER

VOLUME LVII. Bioluminescence and Chemiluminescence
Edited by MARLENE A. DELUCA

VOLUME LVIII. Cell Culture
Edited by WILLIAM B. JAKOBY AND IRA PASTAN

VOLUME LIX. Nucleic Acids and Protein Synthesis (Part G)
Edited by KIVIE MOLDAVE AND LAWRENCE GROSSMAN

VOLUME LX. Nucleic Acids and Protein Synthesis (Part H)
Edited by KIVIE MOLDAVE AND LAWRENCE GROSSMAN

VOLUME 86. Prostaglandins and Arachidonate Metabolites
Edited by WILLIAM E. M. LANDS AND WILLIAM L. SMITH

VOLUME 87. Enzyme Kinetics and Mechanism (Part C: Intermediates, Stereochemistry, and Rate Studies)
Edited by DANIEL L. PURICH

VOLUME 88. Biomembranes (Part I: Visual Pigments and Purple Membranes, II)
Edited by LESTER PACKER

VOLUME 89. Carbohydrate Metabolism (Part D)
Edited by WILLIS A. WOOD

VOLUME 90. Carbohydrate Metabolism (Part E)
Edited by WILLIS A. WOOD

VOLUME 91. Enzyme Structure (Part I)
Edited by C. H. W. HIRS AND SERGE N. TIMASHEFF

VOLUME 92. Immunochemical Techniques (Part E: Monoclonal Antibodies and General Immunoassay Methods)
Edited by JOHN J. LANGONE AND HELEN VAN VUNAKIS

VOLUME 93. Immunochemical Techniques (Part F: Conventional Antibodies, Fc Receptors, and Cytotoxicity)
Edited by JOHN J. LANGONE AND HELEN VAN VUNAKIS

VOLUME 94. Polyamines
Edited by HERBERT TABOR AND CELIA WHITE TABOR

VOLUME 95. Cumulative Subject Index Volumes 61–74 and 76–80
Edited by EDWARD A. DENNIS AND MARTHA G. DENNIS

VOLUME 96. Biomembranes [Part J: Membrane Biogenesis: Assembly and Targeting (General Methods; Eukaryotes)]
Edited by SIDNEY FLEISCHER AND BECCA FLEISCHER

VOLUME 97. Biomembranes [Part K: Membrane Biogenesis: Assembly and Targeting (Prokaryotes, Mitochondria, and Chloroplasts)]
Edited by SIDNEY FLEISCHER AND BECCA FLEISCHER

VOLUME 98. Biomembranes [Part L: Membrane Biogenesis (Processing and Recycling)]
Edited by SIDNEY FLEISCHER AND BECCA FLEISCHER

VOLUME 99. Hormone Action (Part F: Protein Kinases)
Edited by JACKIE D. CORBIN AND JOEL G. HARDMAN

VOLUME 100. Recombinant DNA (Part B)
Edited by RAY WU, LAWRENCE GROSSMAN, AND KIVIE MOLDAVE

VOLUME 101. Recombinant DNA (Part C)
Edited by RAY WU, LAWRENCE GROSSMAN, AND KIVIE MOLDAVE

VOLUME 102. Hormone Action (Part G: Calmodulin and Calcium-Binding Proteins)
Edited by ANTHONY R. MEANS AND BERT W. O'MALLEY

VOLUME 103. Hormone Action (Part H: Neuroendocrine Peptides)
Edited by P. MICHAEL CONN

VOLUME 104. Enzyme Purification and Related Techniques (Part C)
Edited by WILLIAM B. JAKOBY

VOLUME 105. Oxygen Radicals in Biological Systems
Edited by LESTER PACKER

VOLUME 106. Posttranslational Modifications (Part A)
Edited by FINN WOLD AND KIVIE MOLDAVE

VOLUME 107. Posttranslational Modifications (Part B)
Edited by FINN WOLD AND KIVIE MOLDAVE

VOLUME 108. Immunochemical Techniques (Part G: Separation and Characterization of Lymphoid Cells)
Edited by GIOVANNI DI SABATO, JOHN J. LANGONE, AND HELEN VAN VUNAKIS

VOLUME 109. Hormone Action (Part I: Peptide Hormones)
Edited by LUTZ BIRNBAUMER AND BERT W. O'MALLEY

VOLUME 110. Steroids and Isoprenoids (Part A)
Edited by JOHN H. LAW AND HANS C. RILLING

Section I

Unfolding and Refolding of Proteins

Subeditor

Robert L. Baldwin

[1] Protein Folding: Introductory Comments

By ROBERT L. BALDWIN

In the last decade, experimental work on the pathway and mechanism of protein folding has been, for the most part, a quiet field. The efforts of the small number of workers in this field have been focused largely on development of techniques, in the belief that new methods must be found to speed up progress in solving the folding problem. This volume describes methods which have proved to be widely useful or of basic importance for work on the mechanism of protein folding.

As this volume goes to press, the situation is changing. First, the mechanism of protein folding is now perceived as being a problem of wide interest in molecular biology, and more people are being attracted to work on this problem. The following example illustrates how this has come about. In the study of oncogenes, it has been found that single mutations in *ras* genes can cause cancers in suitable cell lines, and that many cancer-producing mutations result from single amino acid replacements at a single position in one protein, the p21 protein. The amino acid sequence of the p21 protein is known from the DNA sequence of the cloned gene. Next one wants to study the folded structure of the p21 protein and to investigate the changes in its folding brought about by amino acid changes at this crucial position. Since the folding is determined by the amino acid sequence, it should become possible to predict the folding from the sequence. Hence, solving the folding problem appears as a new frontier.

Second, the folding problem is suddenly seen as being soluble through application of the new techniques of genetic engineering. Gene cloning and site-directed mutagenesis now make it possible to produce any desired change in the amino acid sequence of a protein. Given that the sequence encodes the instructions for folding, it appears that only ingenuity, insight, and careful experimentation are needed to solve the folding problem.

Third, powerful new techniques, both genetic and physical, are coming into use. Gene fusion techniques have been developed for cloning large peptides, and folding experiments on long peptides of unique sequence are becoming feasible. For years, improvements in the speed and accuracy of X-ray methods for solving the structures of proteins have appeared at a steady rate; now, area detectors and the use of synchrotron radiation are giving a boost to their development. New methods of two-dimensional (2-D) NMR provide unparalleled sensitivity for the detection

METHODS IN ENZYMOLOGY, VOL. 131

and characterization of folding intermediates. Antibodies are becoming powerful tools in the study of protein folding. Not only can antibodies be made against peptides, but antipeptide antibodies cross-react with the proteins from which the peptides were derived. These and other developments in technique will surely have a major impact in the next decade on work aimed at solving the folding problem.

[2] Scanning Microcalorimetry in Studying Temperature-Induced Changes in Proteins

By Peter L. Privalov and Sergey A. Potekhin

Introduction

The importance of calorimetric studies of changes in the state of proteins induced by various factors stems from the fact that calorimetry is the only method for the direct determination of the enthalpy associated with the process of interest. Calorimetry acquires special significance in studies of temperature-induced changes in the state of the protein, since temperature and enthalpy are coupled extensive and intensive variables. All temperature-induced changes in macroscopic systems always proceed with a corresponding change of enthalpy, i.e., they are accompanied by heat absorption if the process is induced by a temperature increase, or by the evolution of heat if it is caused by a temperature decrease. The functional relation between enthalpy and temperature actually includes all the thermodynamic information on the macroscopic states, accessible within the considered temperature range and this information can be extracted from the enthalpic function by its thermodynamic analysis.

The temperature dependence of the enthalpy can be determined experimentally by calorimetric measurements of the heat capacity of the studied objects over the temperature range of interest. Since the heat capacity determined at constant pressure is a temperature derivative of the enthalpy function

$$C_p = (\partial H / \partial T)_p \tag{1}$$

one can easily estimate the enthalpy function by integration of the heat capacity

$$H(T) = \int_{T_0}^{T} C_p(T)dT + H(T_0) \tag{2}$$

The so-called heat capacity calorimeters are used for heat capacity measurements over a particular temperature range. There are many modifications of this instrument designed for studies of various materials in different aggregate states, over different temperature ranges, and with different accuracies. As a material for calorimetric study protein has some characteristics which distinguish it from other objects.

Among the protein characteristics the size of these molecules should be mentioned first. The molecular weight of a protein is usually greater than 10,000. Thus, these molecules, consisting of many thousands of atoms, can be regarded in themselves as macroscopic systems. In this respect proteins are similar to nucleic acids, the other important representatives of biological macromolecules, the calorimetric studies of which attract no less attention.

The important feature in calorimetric studies of these biological macromolecules is that we are interested in their physical properties not in the isolated state—in vacuum, but in the dispersed state—in solution (particularly in aqueous solution) in which they can be regarded as individual macroscopic systems surrounded by the solvent medium. The solution concentration at which the interaction between macromolecules is sufficiently small to be neglected is of the order of $10^{-4} M$ for average proteins, i.e., of the order of one tenth of the weight percent. In such dilute solutions, however, the macromolecular contribution to the thermal properties of the entire samples should also be small: the protein heat capacity does not exceed 0.03%, while the excess heat capacity at the peak of denaturation is less than 1% of the solution heat capacity. At the same time, even dilute solutions of biological molecules are quite viscous so that they cannot be stirred to achieve rapid thermal equilibration during heating, as is usually done in studies of liquids.

The other important peculiarity of biological molecules is their poor availability and exceptionally high cost due to the difficulty of their isolation and purification. The amount of material which can be used practically in experiments for the most available materials does not exceed a few milligrams.

Thus, heat capacity studies of proteins actually boil down to calorimetric studies of very small heat effects which occur in a few milliliters of a viscous solution heated over a broad temperature range. These experiments cannot be done in any of the known heat capacity calorimeters used for physicochemical studies of nonbiological materials. The realization of these experiments has required the creation of a qualitatively new technique, which is known as heat capacity microcalorimetry.

We will consider in this chapter the experimental technique used in heat capacity studies of individual macromolecules in solution, experiments with these materials, methods of treatment of experimental results, and their analysis.

Experimental Technique

Microcalorimeters for Heat Capacity Studies of Liquids

During the last few years a number of review papers have been devoted to supersensitive and superprecise instruments for heat capacity studies of small volumes of liquids.[1-6] Unfortunately, the technical problems inherent in heat capacity microcalorimetry have been treated rather briefly. Here we also cannot go into technical detail and readers interested in this aspect are referred to the original papers.[7-11] Nevertheless, some general principles of construction of such instruments will be considered here, to make clear what can be expected from this technique and how to work with it.

All contemporary heat capacity microcalorimeters have a number of features in common. First, they do not have a mechanical stirrer as do all the macrocalorimeters for the rapid redistribution of the introduced thermal energy over the sample volume. The stirring of a liquid with a high and variable viscosity produces uncontrollable Joule heat in an amount much greater than the measured heat effect. The elimination of the mechanical stirrer was made possible by the great decrease of the operational volume of the calorimeter. Therefore, a small volume of the calorimetric cell is a key requirement in the construction of heat capacity microcalorimeters.

All microcalorimeters measure heat capacity not in a discrete way at stepwise sample heating by discrete energy increments as do all the classical calorimeters, but continuously with continuous heating or cooling of the sample at a constant rate. In other words, they scan along the temperature scale by measuring continuously small changes in the heat capacity

[1] I. Wadsö, *Q. Rev. Biophys.* **3,** 383 (1970).

[2] J. M. Sturtevant, this series, Vol. 26, p. 227.

[3] J. M. Sturtevant, *Annu. Rev. Biophys. Bioeng.* **3,** 35 (1974).

[4] S. Mabrey and J. M. Sturtevant, *Methods Membr. Biol.* **9,** 237 (1978).

[5] K. C. Krishnan and J. F. Brandts, this series, Vol. 49, p. 3.

[6] P. L. Privalov, *Pure Appl. Chem.* **52,** 479 (1980).

[7] P. L. Privalov, J. R. Monaselidze, G. M. Mrevlishvili, and V. A. Magaldadze, *Zh. Eksp. Teor. Fiz. (USSR)* **47,** 2073 (1964); see also *Sov. Phys. JETP* **20,** 1399 (1965).

[8] P. L. Privalov and J. R. Monaselidze, *Prib. Tek. Eksp. (USSR)* **6,** 174 (1965).

[9] P. L. Privalov, V. V. Plotnikov, and V. V. Filimonov, *J. Chem. Thermodyn.* **7,** 41 (1975).

[10] S. J. Gill and K. Beck, *Rev. Sci. Instrum.* **36,** 274 (1965).

[11] R. Danford, H. Krakauer, and J. M. Sturtevant, *Rev. Sci. Instrum.* **38,** 484 (1967).

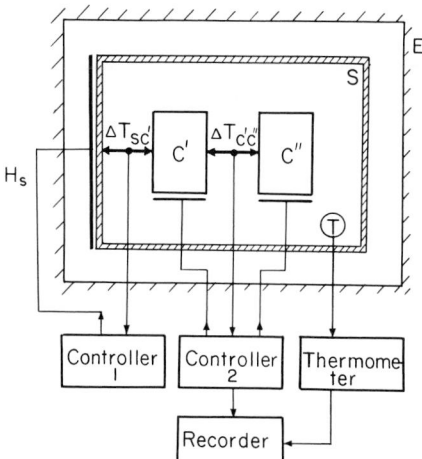

FIG. 1. Differential adiabatic scanning calorimeter with two identical cells (C' and C'') heated at a constant rate and a controlled temperature of the surrounding shell (S).

of the samples. As a result, these instruments are usually called scanning microcalorimeters.

Continuous heating and measurement have great advantages over the discrete procedure: it gives more complete information on the heat capacity function and permits the complete automatization of all the measurement processes. Its disadvantage is that the studied sample is never in complete thermal equilibrium. This sets certain requirements for the samples and for instrument construction: the temperature-induced changes in the samples should not be too sharp and the relaxation process at these changes should be sufficiently fast, while the construction of the calorimetric cell should provide minimal thermal gradients which do not change with heating.

All scanning microcalorimeters measure not the absolute but the difference heat capacity, i.e., they are all differential instruments with two identical calorimetric cells, and measurements consist in a comparison of their heat capacities (Fig. 1). One of the cells is loaded with the solution to be studied and the other one with some standard liquid. Thus, the heat capacity of the studied liquid is determined relative to the chosen standard. In studying dilute solutions, it is convenient to take the solvent as a standard, since, in this case, the measured difference heat capacity will correspond directly to the heat capacity contribution of the molecules dissolved in the solution.

The heat capacity difference of the cells is usually measured by the compensation method: the controller automatically monitors the power in the electric heaters of both cells to maintain identity of their temperatures at heating, and the difference of these powers is recorded as a temperature function. Such a feedback by the heat balance in the cells improves significantly the dynamic characteristics of the instrument which is especially important for a precise registration of the complicated heat capacity functions.

The constant rate of heating over the entire temperature range in scanning microcalorimetry is provided either by the controllable power which compensates for an increasing loss of energy by the cells on heating[12,13] or by the thermoisolation of the cells which prevents the loss of their energy. The first method is simpler to construct. However, the large amplitude of heating power regulation results in excessive noise and an unstable baseline. That is why this method is not used in precision scanning microcalorimetry where the normal practice is to use complete or partial adiabatization of cells, i.e., their controllable thermal isolation from the surrounding thermostat.[6]

The heat exchange of the cells with the thermostat is controlled by enclosing them in thermal shells with a controllable temperature (Fig. 1). The shells completely prevent the heat exchange between the cells and the thermostat if its temperature is maintained equal to that of the cells, while a constant difference between the cell and shell temperatures provides constant positive or negative heat flow into the cells. Such a heat flow is necessary for scanning in both directions along the temperature scale.

The thermal shell is made of metal with high thermal conductivity (silver, copper). Electric heaters insulated by a thin film are uniformly distributed on the outward surface of the shell. Precise instruments usually have several shells to provide a high symmetry and constancy of the thermal field around the twin cells over the entire operational temperature range.

One of the most difficult problems in scanning microcalorimetry is the loading of the cells with equal and definite amounts of the studied and standard materials and of the free volume. It is clear that when measuring heat capacity with an error of less than 10^{-5} J K^{-1}, the error in loading the cells with the sample should not exceed 10^{-6} g. It is practically impossible to load the calorimetric cells with such accuracy by weighing the sample. A calorimeter with extractable cells never gives reproducible results; each replacement of the cells results in a different slope and position of the baseline of the instrument. Therefore, such instruments, being quite sensitive to sharp changes of heat capacity with temperature, cannot be

[12] E. S. Watson, M. J. O'Neil, J. Justin, and N. Brenner, *Anal. Chem.* **36,** 1233 (1964).
[13] E. S. Watson and M. J. O'Neil, U.S. Patent, 3, 263, 484 (1966).

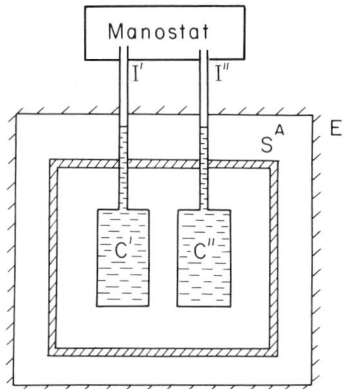

FIG. 2. Undismountable calorimetric block with the cells completely filled with liquid samples through the capillary inlets I′ and I″. The measuring volume of the cells is fixed by the point at which the adiabatic shell (S) contacts with the capillary inlet and shunts the difference heat effect. The manostat is used to apply extra pressure on the liquid for squeezing the bubbles which might be left in the cells and for preventing their appearance upon heating.

used for determining the absolute heat capacity difference and its dependence on temperature, i.e., they cannot be considered as precise instruments for the determination of the difference heat capacity of liquids.

The free volume in the loaded cells raises similar difficulties. It is evident that hermetic cells cannot be filled completely with a liquid sample, since the thermal expansions of the liquid and of the cell are usually different. On the other hand, the free volume in the cell leads also to numerous complications during heating caused by vaporization of the liquid. Although the vapor pressures of the solvent and the dilute solution do not differ greatly, the difference in the heat effect of their vaporization might be quite significant because of the large specific heat of vaporization (about 2 kJ g^{-1} for water).

The problem of loading a scanning microcalorimeter with a sample and that of the free volume was resolved by replacing heat capacity measurements of a sample of definite mass by heat capacity measurements of a sample of definite volume.[6,9] The volume of the studied sample can be fixed by the fixed operational volume of the calorimetric cell. The main requirement which must be fulfilled is that the cell should be filled completely and no microscopic bubbles should be left in it. It is clear that this can be done only by replacing the isolated hermetic cell, which has always been used in heat capacity calorimetry, by an open cell connected with the external vessel (Fig. 2). A thin capillary tube connects the cell with the external vessel so that the cell can be filled with the sample without removing it from the adiabatization system of thermal shells. This permits the liquid expanded on heating to flow from the cell. The operational volume of the cell is determined by the thermal shell which is in thermal contact with the capillary tubes and which plays the role of a thermal shunt, cutting off the influence of the external part of the capillary tube on the cells.

To exclude bubbles from the cells, an excess pressure is applied to the external ends of the capillary tubes by a manostat and all measurements are performed under this constant pressure. A few atmospheres are sufficient to compress all the bubbles in the cells and to prevent their appearance on heating. Since the excess pressure raises the boiling temperature, this extends the operational range of the instrument. For example, 5 atmospheres of excess pressure are enough to heat an aqueous solution to 150°, which is very important for studies of thermostable macromolecules.

When the described cells are heated, the expanding liquid filling the cells is removed through the capillary tube, i.e., the mass of the liquid which is in the cell decreases with an increase of temperature. However, the flow of liquid from both cells is almost identical when studying dilute solutions, since the thermal expansion coefficients of the dilute solution and the solvent are almost the same. Therefore, expansion of the liquid does not affect the measured difference of heat capacity, although a decrease of the mass in the cells should be taken into account when calculating the specific heat capacities.

The principle of fixed volume of liquid heat capacity measurements was used in designing the precision scanning microcalorimeters DASM-1M and DASM-4, manufactured by the Bureau of Biological Instrumentation of the USSR Academy of Sciences. Later this principle was also used in the scanning microcalorimeter Biocal, manufactured by Setarum (France), and in the model MK-2 of the scanning microcalorimeter manufactured by Microcal (United States).

The undismountable calorimetric block of the DASM-1M instrument with the cells filled through the capillary tube and a double adiabatic shell is shown in Fig. 3. The golden 1 ml volume cells are made as flat disks. The capillary tubes with a 1.2 mm inner diameter are made of platinum. The thermal shells are made of silver.

The main difference between calorimeters DASM-4 and DASM-1M is in the construction of the cells. In DASM-4 the cells are made completely from capillary tubes wound into a helix (Fig. 4). Capillary cells have many advantages: they are easily washed and filled without bubbles and they provide a much more homogeneous thermal field with lower temperature gradients in the studied liquids. This permits use of higher heating rates

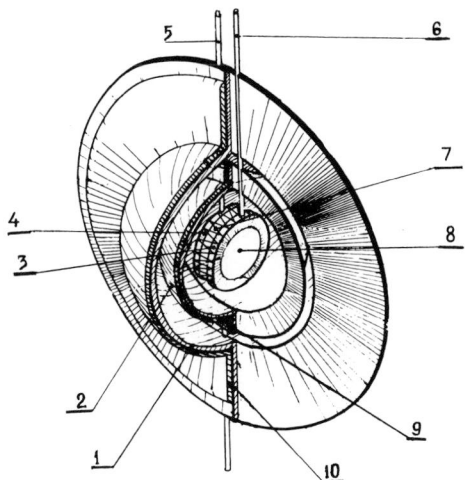

FIG. 3. Calorimetric block of the scanning microcalorimeter DASM-1M. (1,2) Internal and external adiabatic shells with heaters; (3) calorimetric cells; (4) shell thermosensor; (5, 6) capillary inlets; (7) thermopile; (8) cell heater; (9) internal shield rim; (10) external shield rim.

during measurements, and therefore increases the sensitivity of the instrument in heat capacity studies. In the capillary cell there is no thermal convection of the liquid during heating which is one of the sources of artifacts in scanning microcalorimetry. Moreover, thin capillary tubes can withstand much higher pressures than cells of any other shape and this is important for extending the operational temperature range of the instrument (Fig. 5).

Main Characteristics of Scanning Microcalorimeters

Usually the sensitivity of an instrument means the minimal signal which can be detected against the background noise. Therefore, it depends not only on the noise of the instrument, but also on the shape of the signal.

In the case of the scanning microcalorimetry of biological molecules, the spectrum of possible signals varies over a wide range, from fractions of a degree (melting of homopolymers and phospholipid bilayers) to dozens of degress (gradual changes of heat capacity, cooperative transitions with small enthalpies) and the problem of their isolation against the background noise requires detailed studies of noise characteristics, i.e., of the noise spectrum of the instrument. However, this is usually somewhat arbitrarily replaced by the evolution of a mean square deviation of the recording from an ideal line (noise level) and by the evolution of the

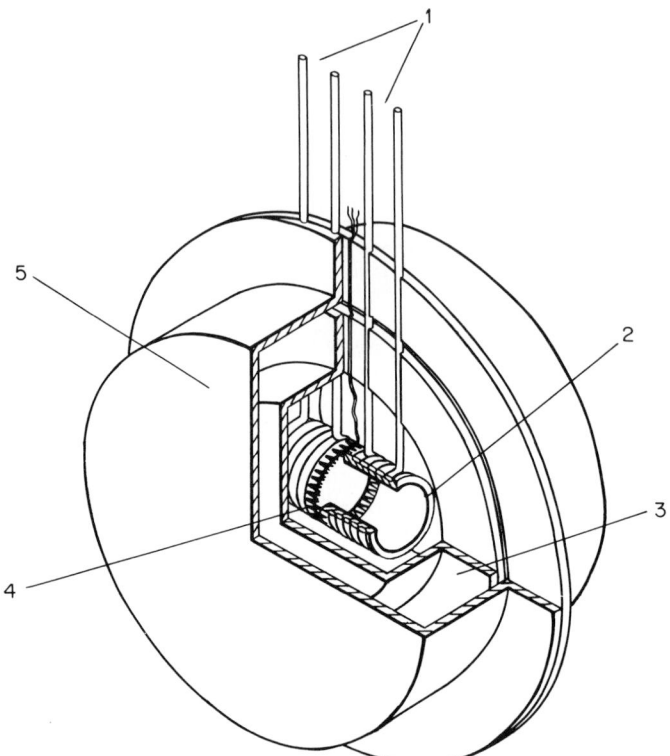

Fig. 4. Calorimetric block of the capillary scanning microcalorimeter DASM-4. (1) Capillary inlets; (2) helical capillary cell with heater; (3) internal shell; (4) thermopile measuring the temperature difference between the capillary cells; (5) external shell.

reproducibility of recordings at consecutive runs of the instrument without refilling the cells with the sample (convergency). But this is far from sufficient for characterizing the ability of the instrument to measure the difference heat capacity. The main parameter showing this ability of a scanning microcalorimeter is the reproducibility of the results after refilling the cells with the same sample. In contrast to the mean noise level and convergency which are usually estimated in power units (watts), the reproducibility of the results of difference heat capacity measurements should be estimated in heat capacity units (J K^{-1}) at the optimal heating rate.[6]

It is evident that the sensitivity of a scanning microcalorimeter depends directly on the amount of the sample studied. The smaller the amount of the sample and, consequently, its heat capacity, the smaller are the heat effects registered on heating. Therefore, one of the widespread

Fig. 5. Capillary scanning microcalorimeter DASM-4, manufactured by the Bureau of Biological Instrumentation of the Academy of Sciences of the USSR.

methods of increasing the apparent sensitivity of a scanning microcalorimeter is reduction of the volume of the calorimetric cells. However, in solution studies it is the sensitivity to the heat effect occurring within the sample of definite volume that is significant, and not the absolute sensitivity to the heat effect, since in solution studies it is the minimal concentrations and not the minimal volumes that are of importance. The characteristics of a scanning microcalorimeter reduced to a unit of volume are denoted as reduced characteristics.[6] It is clear that the reduced characteristics are basic in the assessment of the quality of an instrument designed for solution studies.

Although at present many companies manufacture scanning calorimeters, it is difficult to establish a clear idea of the potentials of these instruments from their descriptive brochures, since usually the descriptions give only one or two most favorable parameters, for example, only the noise level. The table summarizes the basic characteritics determined from the descriptions and illustrations of some manufactured instruments which at present are used most frequently in studies of macromolecules. It is evident that although there is only a 100-fold difference in the noise levels of these instruments, their reduced noise-levels, which per se determine their fitness for studying macromolecules in solution, differ more than 1000-fold, while the relative error in estimating heat capacity varies by two orders of magnitude. This qualitative difference compels us to divide scanning calorimeters into two classes. The dividing line can be the relative error in the estimation of the heat capacity. It is expedient to

Main Characteristics of Commercial Scanning Microcalorimeters

Quality	Characteristic	Unit	Perkin-Elmer DSC-2	Dupont 910 DSC	Daini Seikoshi SSC-50	Acad. Sci. USSR DASM-1M	Acad. Sci. USSR DASM-4	Microcal USA MC-2	Seteram France Biocal
Operational range	Volume of the cell	ml	0.03	0.03	0.07	1.0	0.5	1.3	1.0
	Temperature range	K	100–1000	100–1000	120–400	273–373	250–400	250–390	273–373
Sensitivity	Heating rates	K/min	0.3–320	0.5–100	0.01–5.0	0.1–2.0	0.1–2.0	0.16–1.5	0–1.0
	Noise level	μW	17	4	1.3	0.5	0.2	0.26	1.0
	Reduced noise level	μW/ml	600	150	20	0.5	0.4	0.2	1.0
Precision	Reproducibility without refilling of the cell	μW	—	20	2.0	2.0	0.5	1.3	—
	Reduced reproducibility without refilling of the cell	μW/ml	—	700	30	2.0	1.0	1.0	—
Accuracy	Reproducibility on refilling of the cell	mJ K^{-1}	—	—	—	0.3	0.05	0.13	0.3
	Reduced reproducibility on refilling of the cell	mJ K^{-1}/ml	—	—	—	0.3	0.1	0.1	0.3
	Relative error in the heat capacity determination	%	1.0	1.0	0.5	0.01	0.005	0.005	0.01

consider the instruments for which this error does not exceed 0.1% as precision instruments, as these instruments can be used in quantitative studies and, in particular, studies of macromolecules in solution. The other instruments, which we cannot call precision ones, have a number of advantages: a smaller cell volume and larger ranges of heating rates and working temperatures. In other words, they are more universal. It is evident that precision is the cost for the universality which has ensured their wide usage mainly in applied studies. Such instruments are also used in studies of macromolecules in solution: low sensitivity and accuracy is compensated by using high concentrations and heating rates, but the possibility of their increase is rather limited. The maximally admissible heating rate in studies of macromolecules should be 5 K/min, but in many cases it is much lower.

Calorimetric Experiment

Preparation of Samples

The calorimetric study of a protein usually begins with the choice of appropriate conditions and the preparation of protein solutions. We assume these conditions to be the required purity of the protein, the most suitable solvent (pH, salt content), the concentration, the temperature range, and the scanning rate.

The requirement for protein purity in calorimetric experiments is usually very rigid and not easily realizable, since even the most sensitive instruments require large amounts of material for the experiment. Homogeneity of the studied protein preparation is crucial for a thermodynamic analysis of calorimetric data, especially in studies of complex temperature-induced processes, since an admixture of small amounts of even the same modified protein can alter the result seriously. The homogeneity of the studied preparation must be checked by all possible tests: by electrophoresis of the native protein and of the unfolded one in SDS in the presence of reducing agents, by isoionic focusing, chromatography, ultracentrifugation, and terminal group analysis. None of these tests separately can guarantee the purity and homogeneity of the preparation, while a combination of several tests can give a more or less certain answer. In quantitative studies, the amount of contamination should not exceed 3%.

Arguments for the choice of solvent are as follows: (1) the protein should be soluble in the solvent over the entire studied temperature range, i.e., it should not aggregate on heating; (2) the states of protein to be studied should be realizable in the chosen solvent; and (3) protein transitions between the states should be reversible (at least partly).

Irreversibility of the protein intramolecular transitions is usually caused by various concomitant intermolecular processes, such as aggregation, chemical modification, and intermolecular cross-linking. Therefore, a search for conditions providing maximal reversibility of the considered changes in the state of the protein leads us, first of all, to conditions which prevent direct contact between protein molecules, i.e., a low concentration and maximal electrostatic repulsion between protein molecules, which is achieved by a proper choice of pH and salt content of solution.

In scanning calorimetric experiments pH stabilization is achieved by the use of various buffers which are added in minimally required concentrations to avoid any direct influence of the buffer on the protein. However, one should bear in mind that the pH of buffer solutions is a function of temperature. Since this dependence is proportional to the enthalpy of ionization of the buffering compound, it is preferable to choose buffers with a minimal enthalpy of ionization for scanning calorimetric experiments, so that they should contribute minimally to the calorimetrically measured heat effect. The buffers most preferable for studying conformational transitions of proteins are those which have the same ionization enthalpies as the protein group, since in this case both effects, the heat of protein ionization and the heat of buffer ionization, compensate each other, and we measure directly the net heat effect of the conformational transition. One such buffer is glycine, the carboxylic and amino groups of which are ionized in the pH regions of 2–4 and 8–10, respectively.

In working with proteins which bind specifically ligands other than protons, it is necessary to stabilize not only the pH but also the concentration of these ligands in solution during heating. For example, in studying temperature-induced changes of calcium-binding proteins (calmodulin, torponin C) it is necessary to stabilize the activity of calcium ions in solution with an EDTA/Ca^{2+} buffer. The concentration of free Ca^{2+} in solution $[Ca^{2+}]$ is determined in this case by the equation

$$[Ca^{2+}] \simeq \frac{[Ca]}{[EDTA] - [Ca]} K^{-1} \tag{3}$$

where K is the Ca^{2+} binding constant by EDTA, [Ca] is the concentration of calcium in solution, and [EDTA] is the concentration of EDTA in solution.[14]

The required concentration of protein depends on the sensitivity of the instrument and the experimental goals. The estimation of partial heat capacity requires a quite high concentration of protein (about 2–5 mg/ml

[14] T. N. Tsalkova and P. L. Privalov, *J. Mol. Biol.* **181,** 533 (1984).

for the DASM-4 instrument). The required concentration for studying conformational transitions of proteins depends on the sharpness of the transition and its specific enthalpy: for studying the relatively sharp transitions of globular proteins with a half-width of about 5–10 K and a specific enthalpy of about 20 J g^{-1} the concentration can be about 5 times lower than that indicated above (about 0.4–1 mg/ml). For studying sharp transitions, such as the melting of collagen and phospholipid bilayers, the concentration can be an order lower (about 0.1 mg/ml), while for studying multidomain proteins which melt over a close to 100 K temperature range, the concentration should be high (about 3 mg/ml).

The preparation of a protein solution for calorimetric experiments starts by dissolving the protein in the chosen solvent in amounts required to obtain the desired concentration and volume. Then, the protein solution is carefully dialyzed against the solvent for 12 hr with several changes of buffer to achieve complete equilibrium between the low molecular compounds in the protein solution and the pure solvent. This is an absolutely necessary procedure since the protein solution and solvent which are loaded into the two cells of the differential calorimeter should differ only in the protein content. (The small difference in the solvent composition caused by the Donnan effect is negligible for a dilute protein solution.)

After dialysis, the protein solution should be centrifuged (30 min, 15,000 g) to remove aggregates and dust, and its concentration should be determined by some standard method. For a quantitative thermodynamic analysis of calorimetric results, the error in determining the concentration should not exceed 3%.

In some cases the studied liquids are degassed under vacuum before loading into the calorimetric cells. However, one should bear in mind that this procedure could change significantly the concentrations of the studied solutions. If experiments are done on calorimeters which operate under excess pressure (DASM-1M, DASM-4), no degassing is needed, except for some special purpose, e.g., oxygen release from the solution.

The excess pressure which is used in scanning microcalorimetric experiments to prevent degassing and boiling of the liquids depends on the upper temperature to which the protein solution is to be heated. It is 1 atm for heating to 100°, 1.5 atm for 110°, 3 atm for heating to 130°, and 5 atm for heating to 150°.

When choosing the optimal scanning rate the following considerations should be taken into account: (1) the sensitivity of the scanning calorimeter increases with an increase of the heating rate; (2) an increase of the heating rate leads to a decrease in the uniformity of the temperature field of the sample; this results in smoothing of the studied effects; and (3) the rate of the temperature-induced transition in protein is limited, it is about 10^3 sec^{-1} for the denaturation of globular proteins and only 10^{-2} sec^{-1} for the denaturation of fibrillar proteins.

The main criterion in choosing the optimal heating rate is the lack of dependence of the calorimetric recordings on the heating rate. If the heating rate is too large, the observed transition will appear broader, flatter, and will be shifted to higher temperatures. The optimal rates for calorimeters with capillary cells (DASM-4) are 1–2 K/min for globular proteins, 0.25–1.0 K/min for fibrillar proteins, and 0.1–0.25 K/min for phospholipid bilayers.

Difference Heat Capacity Determination

Although a differential scanning microcalorimeter is designed for the difference heat capacity determination of liquids, this cannot be done by single measurements of the heat capacity of its two cells loaded with the two liquids (solvent and solution). The first thing that should be done for a precise determination of their heat capacity difference is to load both cells with one of the liquids which is considered as a standard (solvent) and to determine the zero or the baseline of the instrument over the entire required temperature region (Fig. 6). This "line" in most cases is neither linear nor horizontal for supersensitive instruments, because it is practically impossible to make the two cells absolutely identical. However, in precise instruments, it is stable and reproducible in repeated runs of the instrument after refilling the cells with new aliquots of the same liquid. The slope of this line can be easily corrected by applying a temperature-controlled power into one of the cells. A more radical correction of the baseline, including its linearization, can be achieved by memorizing the result of the first run of the instrument and subtracting it from the results of all other runs. This can be done by a special electronic corrector with a memory unit or by a computer on line with the scanning microcalorimeter used for further processing of the data.

Once the baseline has been determined, one of the cells is filled with the studied solution. The deviation of the recording for the solution from the baseline should correspond to the heat capacity difference of the same volume of solvent and solution. Therefore, in difference heat capacity determinations we are comparing the heat capacity of the solution with that of the solvent which had been in the same cell previously, while the other cell with the solvent is used only as a reference one. This permits complete exclusion of the influence of nonidentity on the results of measurements.

To estimate the difference heat capacity in heat capacity units, the instrument must be calibrated. This cannot be done with standard liquids

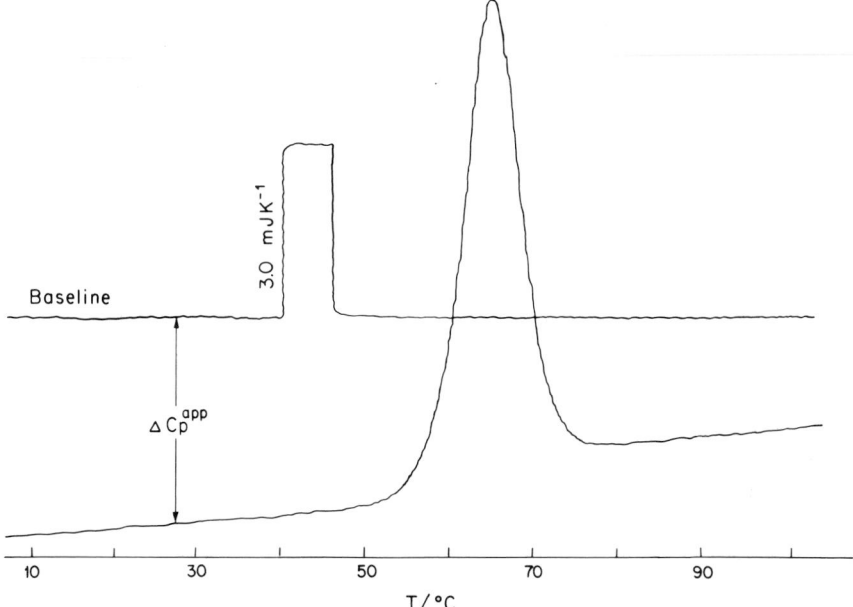

FIG. 6. Microcalorimetric recording of the difference heat capacity of 0.182% lysozyme solution against the solvent—40 mM glycine buffer, pH 2.5. The baseline was obtained on filling both cells with the solvent.

with known heat capacity, since absolute heat capacity has never been determined with the accuracy required for the calibration of scanning microcalorimeters. Therefore, it can be calibrated only electrically by applying some definite power, δW, to one of the cells, which imitates the heat capacity change of this cell. The apparent heat capacity change will be equal to $\delta C_p = \delta W/(dT/dt) = \delta W(dt/dT)$, where dT/dt is the heating rate and dt/dT is the time required for heating the cell by one degree. Dividing δC_p by the observed deviation of the recording from the baseline δl, we obtain the heat capacity value of a unit displacement on heat capacity versus temperature record, $(\delta W/\delta l)(dt/dT)$.

If the calibration power was applied for a duration, δt, the energy released in the cell is $\delta E = \delta W \delta t$ and it corresponds to the δS area of the calibration mark. Therefore $\delta W \delta T/\delta S$ will be the energy value of the square unit on the record.

Determination of Protein Partial Heat Capacity

The difference in heat capacity between a solution of biological macro-molecules and a solvent measured by scanning microcalorimetry is always negative, i.e., the heat capacity of the solution is smaller than the heat capacity of the same volume of solvent (Fig. 6). It follows that the heat capacity of macromolecules in solution is smaller than that of the same volume of the solvent. For the observed difference in heat capacity between the solution and the solvent we have

$$\Delta C_{\mathrm{p}}^{\mathrm{app}}(T)_{\mathrm{pr,sol/solv}} = C_{\mathrm{p}}(T)_{\mathrm{pr}} m(T)_{\mathrm{pr}} - C_{\mathrm{p}}(T)_{\mathrm{solv}} \Delta m(T)_{\mathrm{solv}} \qquad (4)$$

where $C_{\mathrm{p}}(T)_{\mathrm{pr}}$ is the partial specific heat capacity of protein at temperature T, $m(T)_{\mathrm{pr}}$ is the mass of protein which is in the calorimetric cell at temperature T, and $m(T)_{\mathrm{solv}}$ is the mass of the solvent displaced by proteins in solution. The latter equals

$$\Delta m(T)_{\mathrm{solv}} = m(T)_{\mathrm{pr}} \frac{V(T)_{\mathrm{pr}}}{V(T)_{\mathrm{solv}}} \qquad (5)$$

Here $V(T)_{\mathrm{pr}}$ is the partial specific volume of protein at temperature T and $V(T)_{\mathrm{solv}}$ is that of the solvent. From Eqs. (4) and (5) we obtain the partial specific heat capacity of protein:

$$C_{\mathrm{p}}(T)_{\mathrm{pr}} = C_{\mathrm{p}}(T)_{\mathrm{solv}} \frac{V(T)_{\mathrm{pr}}}{V(T)_{\mathrm{solv}}} - \frac{\Delta C_{\mathrm{p}}^{\mathrm{app}}(T)_{\mathrm{pr,sol/solv}}}{m(T)_{\mathrm{pr}}} \qquad (6)$$

It should be noted that the heat capacity of protein determined by the above equation is not strictly the partial heat capacity since the latter is the value which is obtained by extrapolation to an infinitely dilute solution. However, if we take into account that the protein concentration in the solution used for the scanning calorimetric experiment is less than 10^{-4} M and this solution does not show any concentration dependence of the heat capacity it becomes evident that we can consider the value determined by Eq. (6) as the partial specific heat capacity of the protein in solution.

Since partial specific volumes of proteins and solvents do not change significantly with temperature, we can consider their ratio $V(T)_{\mathrm{pr}}/V(T)_{\mathrm{sol}}$ as temperature independent in the first approximation. As for the second term in Eq. (6), it also does not depend significantly on temperature because thermal expansion leads not only to a decrease of the mass of protein in the calorimetric cell (m_{pr}) but also to an increase of the sensitivity of the instrument as a result of the cell heat capacity decrease and these two effects greatly compensate each other. Therefore, if high precision is not required in the partial specific heat capacity determination one can use the values for the specific volumes of protein and solvent and for the mass of protein in the cell which had been estimated for 25°. For many proteins and solvents, the values of the specific volumes at 25° are tabulated and can be found in handbooks. Specific volumes for proteins can be also calculated with quite reasonable accuracy from the known specific volumes of the amino acid residues.

If it is necessary to determine the partial specific heat capacity of a protein with a greater accuracy and over a larger temperature range, the specific volumes of the protein and the solvent should be determined experimentally in the same temperature range using precise densimeters, e.g., the digital vibrational densimeter DMA 02 (Anton Paar, Graz). Specific volumes must be determined in heat capacity studies of phospholipids, since their partial specific volumes change significantly with temperature variation.

In determining the partial specific heat capacity of macromolecules in solution with high accuracy, it is necessary to use precise data on the heat capacity of the solvent, as well as to take into account the change of the mass of macromolecules in the calorimetric cell due to thermal expansion of the solution and the cell.

The specific heat capacity of the solvent can be determined over the desired temperature range with the required accuracy using the same scanning microcalorimeter. To do this, we have to measure the difference heat capacity of the considered solvent in relation to distilled water:

$$\Delta C_p^{app}(T)_{solv/H_2O} = C_p(T)_{H_2O}m(T)_{H_2O} - C_p(T)_{solv}m(T)_{solv}$$

$$= C_p(T)_{H_2O}\frac{v(T)}{V(T)_{H_2O}} - C_p(T)_{solv}\frac{v(T)}{V(T)_{solv}} \qquad (7)$$

where $v(T)$ is the operational volume of a calorimetric cell at temperature T and $V(T)_{H_2O}$ and $V(T)_{sol}$ are the specific volumes of water and solvent at this temperature. From Eq. (7) we have

$$\frac{C_p(T)_{sol}}{V(T)_{sol}} = \frac{C_p(T)_{H_2O}}{V(T)_{H_2O}} - \frac{\Delta C_p^{app}(T)_{sol/H_2O}}{v(T)} \qquad (8)$$

and substituting it in Eq. (6) we get

$$C_p(T)_{pr} = \left[\frac{C_p(T)_{H_2O}}{V(T)_{H_2O}} - \frac{\Delta C_p^{app}(T)_{solv/H_2O}}{v(T)}\right]V(T)_{pr} - \frac{\Delta C_p^{app}(T)_{pr,sol/solv}}{m(T)_{pr}} \qquad (9)$$

The specific volume and the heat capacity of pure water (V_{H_2O} and C_{p,H_2O}) are known with high accuracy over the entire temperature range of its existence in the liquid phase and can be found in handbooks. As for the operational volume of a calorimetric cell $v(T)$ and the mass of protein in the cell $m(T)_{pr}$ at temperature T, they can be calculated from their values at room temperature T_0 and known coefficients of thermal expansion of the solution and the cell material. Bearing in mind that

$$m(T)_{pr} = v(T)\rho(T)_{pr} \qquad (10)$$

where $\rho(T)_{pr}$ is the protein concentration in solution at temperature T, we have

$$\rho(T)_{pr} = \rho(T_0)_{pr} \Big/ \left[1 + \frac{1}{V(T_0)_{solv}} \frac{dV_{sol}}{dT} (T - T_0) \right] \tag{11}$$

$$v(T) \simeq v(T_0)[1 + \beta(T - T_0)] \simeq v(T_0)[1 + 3\alpha(T - T_0)] \tag{12}$$

In Eq. (11) the specific volume of the solution is approximated by the specific volume of the solvent since they do not differ significantly for dilute solutions. In Eq. (12) the volumetric thermal expansion coefficient β is replaced by the linear thermal expansion coefficient α which is equal to 14.3×10^{-6} for gold.

Finally, we obtain the following equation for the partial specific heat capacity of protein in solution

$$C_p(T)_{pr} = \left[\frac{C_p(T)_{H_2O}}{V(T)_{H_2O}} - \frac{\Delta C_p^{app}(T)_{solv/H_2O}}{v(T)} \right] V(T)_{pr}$$

$$- \frac{\Delta C_p^{app}(T)_{pr/sol}}{\rho(T_0)_{pr}v(T_0) \left[1 + \left(3\alpha - \frac{1}{V(T_0)_{sol}} \frac{dV_{sol}}{dT} \right) (T - T_0) \right]} \tag{13}$$

It should be noted that the difference heat capacity of solvent/water could be much larger than the difference heat capacity of protein solution/solvent, since a protein solution is dilute while the solvent can be a quite concentrated solution of a low-molecular-weight compound (salts, alcohol, urea, etc.). Therefore the difference heat capacity of the solvent is usually measured at low sensitivity of the scanning microcalorimeter.

The operational volume of a calorimetric cell $v(T)$ is determined by measuring the difference heat capacity of two liquids with different known specific heat capacities, or just by measuring the heat capacity difference of an empty cell and the cell filled with water.

Changes of Partial Heat Capacity and Enthalpy in Temperature-Induced Changes of a Protein

The partial specific heat capacity of native proteins at 25° varies over the range of 1.2 to 2.3 J K^{-1} g^{-1} for various proteins and increases linearly with a temperature increase with a slope of the order of 10^{-3} J K^{-1} g^{-1}. The heat capacity of a protein in the denatured state is significantly higher than that in the native state (Fig. 7). For various proteins this difference is from 0.30 to 0.70 J K^{-1} g^{-1} and seems to be independent of temperature (for details see Ref. 15). In nucleic acids, the heat capacities of the native

15 P. L. Privalov, *Adv. Protein Chem.* **33**, 179 (1979).
16 V. V. Filimonov and P. L. Privalov, *J. Mol. Biol.* **122**, 465 (1978).

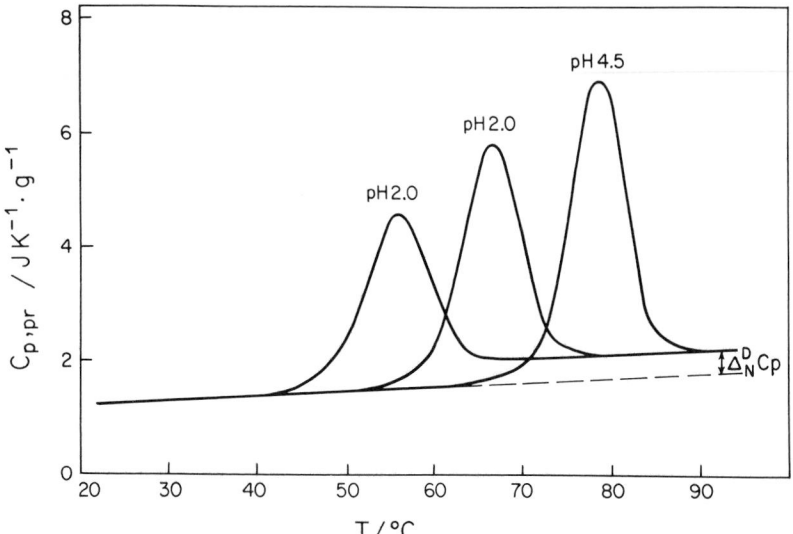

FIG. 7. Partial specific heat capacity of lysozyme in solutions with different pH values.

and denatured states do not differ noticeably, in any case this difference is not greater than 0.06 J K^{-1} g^{-1}.[16]

Since the pure native and denatured states of a protein in any given conditions are realized at different temperatures, it is evident that the heat capacity difference of these states can be estimated either by extrapolation of their heat capacity values to the considered temperature, or by varying solvent conditions if they do not affect the heat capacity of the protein as is true in the case of pH variation (Fig. 7).

If the heat capacity of the protein in the native state is known over a quite extended temperature range, we can calculate the so-called "excess" heat capacity of the protein by extrapolating it to the range in which it cannot be determined by direct measurements. The "excess" heat capacity of the protein relative to some state, considered as its "zero" state, is the difference between the heat capacity of the protein and the heat capcity of this "zero" state at the considered temperature T

$$\langle \Delta C_p(T) \rangle = C_p(T)_{pr} - C_{p,0}(T)_{pr} \tag{14}$$

Correspondingly for the excess enthalpy of protein we have

$$\langle \Delta H(T) \rangle = \int_0^T \langle \Delta C_p(T) \rangle dT \tag{15}$$

Although in the above equation the integration is from absolute zero temperature, in practice it is integrated from some temperature T_0 from which the heat capacity of the protein is distinguished from the heat capacity of the protein in the zero state.

The importance of the excess heat capacity and the excess enthalpy functions follows from the fact that only a part of the observed heat effect, which is accounted by these functions, is related to changes of the macroscopic states of a system. Therefore, these functions actually include all the thermodynamic information on the states realized in the considered temperature range. This is expressed by a very general relation of statistical thermodynamics

$$\langle \Delta H(T) \rangle = RT^2 \frac{d \ln Q}{dT} \tag{16}$$

where $Q = \Sigma K_i$ is the partition function of a system and K_i is the probability of all accessible states.[17,18] Thus, for the partition function of a protein in solution we have

$$Q = \exp \left(\int_0^T \frac{\langle \Delta H(T) \rangle}{RT^2} \, dT \right) = \exp \left\{ \int_{T_0}^T \frac{1}{RT^2} \left[\int_{T_0}^T \langle \Delta C_p(T) \rangle dT \right] dT \right\} \tag{17}$$

Thermodynamic Analysis of Calorimetric Data

Transition between Two States

The simplest temperature-induced process for macromolecules is the transition between two thermodynamically stable states—the native (N) and denatured (D) ones. A general scheme of this process for a biological macromolecule, not necessarily consisting of a single polymer chain, can be presented as

$$N \rightleftharpoons m_1 D_1 + m_2 D_2 + \cdots + m_k D_k$$

assuming that the transition results in the disintegration of the macromolecule into k different subunits (chains) each present in m_i copies. Designating the initial and final states by indices 0 and 1, we have for the equilibrium constant

$$K = \exp \left(-\frac{\Delta_0^1 G}{RT} \right) = \exp \left(\frac{\Delta_0^1 S}{R} - \frac{\Delta_0^1 H}{RT} \right) \tag{18}$$

Let F be the fraction of molecules that have undergone the transition at a given temperature T, and $[N]_0$ be the initial concentration of the preparation in the native state. Then, the concentration of molecules in

[17] E. Freire and R. L. Biltonen, *Biopolymers* **17**, 463 (1978).
[18] R. L. Biltonen and E. Freire, *Crit. Rev. Biochem.* **5**, 85 (1978).

the native state can be presented as

$$[N] = [N]_0(1 - F) \tag{19}$$

The concentration of each subunit in the denatured state will be

$$[D_i] = m_i[N]_0 F \tag{20}$$

and for the equilibrium constant we will have

$$K = \frac{\prod_{i=1}^{k} [D_i]^{m_i}}{[N]} = \frac{\prod_{i=1}^{k} (m_i)^{m_i}[N]_0^{m_i}F^{m_i}}{[N]_0(1 - F)} \tag{21}$$

or

$$K = \frac{[N]_0^{n-1}F^n}{1 - F} \prod_{i=1}^{k} (m_i)^{m_i} \tag{22}$$

where $n = \Sigma_{i=1}^{k} m_i$ is the order of the reaction. Thus, the equilibrium constants of reactions of identical order and various stoichiometries differ only by a constant multiplier. Consequently, the temperature derivatives of the equilibrium constant logarithm will be indistinguishable for such reactions

$$\frac{\partial \ln K}{\partial T} = \frac{\Delta_0^1 H}{RT^2} = (n/F)\frac{dF}{dT} + \frac{1}{1 - F}\frac{dF}{dT} = \frac{n - F(n - 1)}{F(1 - F)}\frac{dF}{dT} \tag{23}$$

Hence we get for the temperature derivative of parameter F

$$\frac{dF}{dT} = \frac{\Delta_0^1 H}{RT^2}\frac{F(1 - F)}{n - F(n - 1)} \tag{24}$$

At the same time for the excess enthalpy of the considered system we have

$$\langle \Delta H \rangle = \Delta_0^1 H F \tag{25}$$

and for its excess heat capacity

$$\langle \Delta C_p \rangle = \frac{d\langle \Delta H \rangle}{dT} = \frac{d(\Delta_0^1 H F)}{dT} = \Delta_0^1 H \frac{dF}{dT} + \Delta_0^1 C_p F \tag{26}$$

Substituting Eq. (24) into Eq. (26) we obtain

$$\langle \Delta C_p \rangle = \frac{(\Delta_0^1 H)^2}{RT^2}\frac{F(1 - F)}{n - F(n - 1)} + \Delta_0^1 C_p F \tag{27}$$

By determining the temperature dependence of F from Eq. (22), one can calculate the excess heat capacity function for a reaction of any order.

For example, for a monomolecular reaction $N \rightleftharpoons D$ for which $n = 1$ and

$$K = \frac{F}{1 - F} \tag{28}$$

we have

$$\langle \Delta C_p \rangle = \frac{(\Delta_0^1 H)^2}{4RT^2} Ch^{-2}(\Delta_0^1 G/2RT)$$

$$+ \frac{\Delta_0^1 C_p}{2} \exp(-\Delta_0^1 G/2RT) Ch^{-1}(\Delta_0^1 G/2RT) \tag{29}$$

This is an analytical expression for the dependence of the excess heat capacity on temperature and the thermodynamic parameters of the transition for a monomolecular reaction.

By comparing the observed excess heat capacity of a macromolecule with the calculated one from this equation, it is possible to find out, using the best fit procedure, whether the studied process indeed represents a monomolecular two-state transition and to determine the enthalpy and entropy of transition which should also satisfy the two basic requirements

$$\frac{\partial \Delta_0^1 H}{\partial T} = \Delta_0^1 C_p \tag{30}$$

and

$$\frac{\partial \Delta_0^1 S}{\partial T} = \frac{\Delta_0^1 C_p}{T} \tag{31}$$

In practice, however, this can be determined by a much simpler analysis of the main characteristics of the heat capacity peak—the area, height, and temperature of the maximum.

Let us estimate the position of the heat capacity peak maximum and its value for an arbitrary process. Differentiating Eq. (27) with respect to temperature and bearing in mind that $(\partial \Delta_0^1 C_p/\partial T)$ is likely to be very small for biopolymers and, therefore, can be neglected,[15] we get

$$\frac{\partial \langle \Delta C_p \rangle}{\partial T} = \frac{F(1 - F)}{1 - kF} \frac{2\Delta_0^1 H(T\Delta_0^1 C_p - \Delta_0^1 H)}{nRT^3} + \frac{(\Delta_0^1 H)^3}{n^2 R^2 T^4}$$

$$+ \frac{F(1 - F)(1 - 2F + kF^2)}{(1 - kF)^3} + \Delta_0^1 C_p \frac{\Delta_0^1 H}{nRT^2} \frac{F(1 - F)}{1 - kF} \tag{32}$$

where $k = (n - 1)/n$. At the extremum this equation is equal to zero. Dividing the above equation by $(\Delta_0^1 H)/(nRT^2)[F(1 - F)]/(1 - kF) \neq 0$, we get at the maximum

$$\frac{2(T\Delta_0^1 C_p - \Delta_0^1 H)}{T} + \frac{(\Delta_0^1 H)^2}{nRT^2} \frac{1 - 2F + kF^2}{(1 - kF)^2} + \Delta_0^1 C_p = 0$$

or

$$\frac{1 - 2F + kF^2}{(1 - kF)^2} = -\frac{nRT(3T\Delta_0^1 C_p - 2\Delta_0^1 H)}{(\Delta_0^1 H)^2} = -nX \tag{33}$$

where

$$X = \frac{RT(3T\Delta_0^1 C_p - 2\Delta_0^1 H)}{(\Delta_0^1 H)^2} \tag{34}$$

Thus, F_{max}, which corresponds to the maximum of the heat absorption peak, must satisfy the equation

$$F_{max}^2 k - 2F_{max} + \frac{1 + nX}{1 + knX} = 0 \tag{35}$$

Since for a first-order reaction $k = 0$, we have two quite different cases for Eq. (35): it is linear for first-order reactions and quadratic for higher order reactions.

Let us consider the first case when

$$F_{max} = \frac{1 + X}{2}$$

and the maximal heat capacity is

$$\langle \Delta C_p \rangle_{max} = \frac{(\Delta_0^1 H)^2}{4RT_{max}^2} (1 - X^2) + \frac{\Delta_0^1 C_p}{2} (1 + X) \tag{36}$$

Equation (36) is exact and can be used for an exact estimation of the enthalpy of the process if the latter represents a "two-state" transition. However, since we can never be certain about the correctness of this assumption, the enthalpy value estimated by this equation based on the van't Hoff relations, is regarded as an effective one and is usually called the van't Hoff enthalpy to distinguish it from the real one measured calorimetrically. Equation (36) can be simplified if the parameter X is sufficiently small relative to unity. In this case, it can be written as

$$\frac{(\Delta_0^1 H)^2}{4RT_{max}^2} = \langle \Delta C_p \rangle_{max} - \frac{\Delta_0^1 C_p}{2} \tag{37}$$

and the value of the van't Hoff enthalpy of the process can be expressed either as

$$\Delta_0^1 H^{vh} = \frac{4RT_{max}^2(\langle \Delta C_p \rangle_{max} - \Delta_0^1 C_p/2)}{\Delta_0^1 H^{cal}} \tag{38}$$

or as

$$\Delta_0^1 H^{vh} = 2T_{max} \sqrt{R\left(\langle \Delta C_p \rangle_{max} - \frac{\Delta_0^1 C_p}{2}\right)} \tag{39}$$

The advantage of the first expression is that the value of the van't Hoff enthalpy $\Delta_0^1 H^{vh}$ in this case does not depend on the absolute values $\langle \Delta C_p \rangle_{max}$ and $\Delta_0^1 H^{cal}$, and, consequently, neither on possible errors in determining the concentration of the studied preparation nor on the calibration of the scanning calorimeter.

The applicability of the simplified Eqs. (37)–(39) can be verified by evaluating the parameter X. As a rule, $3T\Delta_0^1 C_p \gg 2\Delta_0^1 H$ for biological macromolecules. Hence for X we obtain an approximate value

$$X = \frac{3RT_{max}^2 \Delta_0^1 C_p}{(\Delta_0^1 H)^2} = \frac{4RT_{max}^2}{(\Delta_0^1 H)^2} \frac{3\Delta_0^1 C_p}{4} \approx \frac{3}{4} \frac{\Delta_0^1 C_p}{\langle \Delta C_p \rangle_{max} - \frac{\Delta_0^1 C_p}{2}} \tag{40}$$

Let us estimate the error in evaluating the van't Hoff enthalpy using Eq. (37) instead of (36) induced by the nonzero value of the parameter X. From Eq. (36) we have

$$\frac{(\Delta_0^1 H)^2}{4RT_{max}^2} = \frac{\langle \Delta C_p \rangle_{max} - (\Delta_0^1 C_p/2)(1 + X)}{1 - X^2}$$

$$= \left(\langle \Delta C_p \rangle_{max} - \frac{\Delta_0^1 C_p}{2}\right)\left[1 - \frac{\Delta_0^1 C_p}{2(\langle \Delta C_p \rangle_{max} - \Delta_0^1 C_p/2)} X\right] \bigg/ (1 - X^2)$$

$$\approx \left[\left(1 - \frac{2}{3}X^2\right)\left(\langle \Delta C_p \rangle_{max} - \frac{\Delta_0^1 C_p}{2}\right)\right] \bigg/ (1 - X^2)$$

$$\approx (\langle \Delta C_p \rangle_{max} - \Delta_0^1 C_p/2)(1 + X^2/3) \tag{41}$$

Thus

$$\Delta_0^1 H^{vh} = \frac{4RT_{max}^2}{\Delta_0^1 H^{cal}} \left(\langle \Delta C_p \rangle_{max} - \frac{\Delta_0^1 C_p}{2}\right)(1 + X^2/3) \tag{42}$$

In order for the error in the estimate of the van't Hoff enthalpy not to exceed 1%, it is necessary that $X^2 \leq 3 \times 10^{-2}$, and that the heat capacity difference $\Delta_0^1 C_p$ should not exceed approximately 20% of the maximal heat capacity value.

In the case when the transition between two states is a reaction of an order higher than the first one, we have from Eq. (35) for the value of F_{max} at a heat capacity maximum

$$F_{max} = \left(1 - \sqrt{1 - \frac{1 + nX}{1 + knX} k}\right) \Big/ k = \left(1 - \sqrt{\frac{1 - k}{1 + knX}}\right) \Big/ k$$

$$= \frac{1 - \sqrt{1 - k}}{k} + \left(\sqrt{1 - k} - \sqrt{\frac{1 - k}{1 + knX}}\right) \Big/ k$$

$$= \frac{1 - \sqrt{1 - k}}{k} + \frac{\sqrt{1 - k}}{k} \left[1 - \frac{1}{\sqrt{1 + knX}}\right]$$

$$= \frac{n - \sqrt{n}}{n - 1} + \frac{\sqrt{n}}{n - 1} \left[1 - \frac{1}{\sqrt{1 + (n - 1)X}}\right] \tag{43}$$

Taking into consideration the small value of the parameter X, we get from Eq. (43) by expanding it into a series with an accuracy of a quadratic term

$$F_{max} = \frac{n - \sqrt{n}}{n - 1} + \frac{1}{2\sqrt{n}} X - \frac{3(n - 1)}{8n\sqrt{n}} X^2 \tag{44}$$

It is easy to see that even at a very low value of the parameter X the F_{max} value corresponding to the maximal heat capacity differs significantly from 0.5. Indeed, for a bimolecular reaction this value will correspond to approximately 0.59, and in the case of a third-order reaction it will be approximately 0.63.

Substituting the value F_{max} [Eq. (44)] in Eq. (27), we obtain

$$\langle \Delta C_p \rangle_{max} = \frac{(\Delta_0^1 H)^2}{RT_{max}^2 (\sqrt{n} + 1)^2} + \frac{\Delta_0^1 C_p \sqrt{n}}{\sqrt{n} + 1} + \frac{\Delta_0^1 C_p X}{2\sqrt{n}}$$

$$+ \left[\frac{7(\Delta_0^1 H)^2}{4RT_{max}^2} - \frac{3\Delta_0^1 C_p (n - 1)}{2}\right] \frac{X^2}{4n\sqrt{n}} \tag{45}$$

If X is sufficiently small, the van't Hoff enthalpy values are expressed as

$$\Delta_0^1 H^{vh} = \frac{(\sqrt{n} + 1)^2 RT_{max}^2 \left(\langle \Delta C_p \rangle_{max} - \frac{\Delta_0^1 C_p \sqrt{n}}{\sqrt{n} + 1}\right)}{\Delta_0^1 H_{cal}} \tag{46}$$

or

$$\Delta_0^1 H^{vh} = (\sqrt{n} + 1) T_{max} \sqrt{R \left(\langle \Delta C_p \rangle_{max} - \frac{\Delta_0^1 C_p \sqrt{n}}{\sqrt{n} + 1}\right)} \tag{47}$$

The errors related to the deviation of X from zero using Eqs. (46) and (47) are of the same order of magnitude as those for the monomolecular reaction.

The value of the transition entropy can be calculated easily if the temperature corresponding to the heat absorption peak is known. In this case, for the first-order reaction when $F_{max} = (1 + X)/2$ we have

$$K_{max} = \frac{F_{max}}{1 - F_{max}} = \frac{1 + X}{1 - X} \approx 1 + 2X \tag{48}$$

$$\ln K_{max} = \ln(1 + 2X) \approx 2X \tag{49}$$

$$\Delta_0^1 G_{max} = - RT \ln K_{max} = - 2RT_{max}X \tag{50}$$

$$\Delta_0^1 S_{max} = \frac{\Delta_0^1 H_{max} - \Delta_0^1 G_{max}}{T_{max}} = \frac{\Delta_0^1 H_{max} + 2RT_{max} X}{T_{max}}$$

$$= \frac{\Delta_0^1 H_{max}}{T_{max}} + 2RX \tag{51}$$

If we take into account the real values of the parameters included in this expression it becomes evident that the transition entropy at peak temperature amounts to with good accuracy

$$\Delta_0^1 S_{max} = \Delta_0^1 H_{max}/T_{max} \tag{52}$$

For higher order reactions we have

$$K_{max} = \frac{[N]_0^{n-1} F_{max}^n}{1 - F_{max}} \prod_{i=1}^{k} (m_i)^{m_i} \tag{53}$$

$$\ln K_{max} = (n - 1) \ln[N]_0 + \sum_{i=1}^{k} m_i \ln m_i + \ln \frac{F_{max}^n}{1 - F_{max}}$$

$$\ln \frac{F_{max}^n}{1 - F_{max}} = \ln \left\{ \frac{\left(\frac{n - \sqrt{n}}{n - 1} + \frac{1}{2\sqrt{n}} X \right)^n}{1 - \frac{n - \sqrt{n}}{n - 1} - \frac{1}{2\sqrt{n}} X} \right\}$$

$$= \ln \left\{ \frac{\left(\frac{n - \sqrt{n}}{n - 1}\right)^n \left(1 + \frac{n - 1}{2\sqrt{n}\,(n - \sqrt{n})}\,X\right)^n}{\frac{1}{\sqrt{n} - 1} - \frac{1}{2\sqrt{n}}\,X} \right\}$$

$$= \ln \left\{ \frac{\left(1 + \frac{\sqrt{n} - 1}{2n}\,X\right)^n}{1 - \frac{\sqrt{n} - 1}{2\sqrt{n}}\,X} \right\} + \ln \left\{ \left(\frac{\sqrt{n}}{\sqrt{n} - 1}\right)^n (\sqrt{n} - 1) \right\}$$

$$= \ln \left\{ \left(1 + \frac{\sqrt{n} - 1}{2}\,X\right)\left(1 + \frac{\sqrt{n} - 1}{2\sqrt{n}}\,X\right) \right\} + \ln \left\{ \frac{n^{n/2}}{(\sqrt{n} - 1)^{n-1}} \right\}$$

$$\approx \ln \left(1 + \frac{n - 1}{2\sqrt{n}}\,X\right) + \ln \left\{ \frac{n^{n/2}}{(\sqrt{n} - 1)^{n-1}} \right\}$$

$$\approx \ln \frac{n^{n/2}}{(\sqrt{n} - 1)^{n-1}} + \frac{n - 1}{2\sqrt{n}}\,X \tag{54}$$

$$\ln K_{max} = \frac{\Delta_0^1 G_{max}}{RT_{max}} = \ln \frac{[N]_0^{n-1} \prod\limits_{i=1}^{k} (m_i)^{m_i} n^{n/2}}{(\sqrt{n} - 1)^{n-1}} + \frac{n - 1}{2\sqrt{n}}\,X \tag{55}$$

$$\Delta_0^1 S_{max} = \frac{\Delta_0^1 H_{max} - \Delta_0^1 G_{max}}{T_{max}} = \frac{\Delta_0^1 H_{max}}{T_{max}}$$

$$+ R \ln \left\{ \frac{[N]_0^{n-1} \prod\limits_{i=1}^{i} (m_i)^{m_i} n^{n/2}}{(n - 1)^{n-1}} \right\} + RX \frac{n - 1}{2\sqrt{n}} \tag{56}$$

The ratio $\Delta_0^1 H^{cal}/\Delta_0^1 H^{vh}$ can be considered as a measure of the validity of the assumption that the studied process is a two-state transition. For heat denaturation of small compact globular proteins, this ratio is usually very close to 1.0.[15] The deviation from 1.0 in most cases does not exceed 5%. This means that the population of all intermediate states between the native and denatured ones of these proteins is rather low, and denaturation can be considered in the first approximation as an "all-or-none" process.

However, for the denaturation of large proteins, and in some cases even of not very large ones, the ratio $\Delta_0^1 H^{cal}/\Delta_0^1 H^{vh}$ exceeds significantly 1.0, which indicates that the process of disruption of their native structure is not of an "all-or-none" character even in the first approximation, and that in this process several intermediate states are realized.[19]

It should be noted that this ratio can also be smaller than 1.0, but this can occur only in two cases: when the molecular weight of a cooperative unit is determined incorrectly and when the studied process is irreversible.

The first case is found frequently for molecules which exist in solution as dimers and not as monomers, and which denature in the dimeric form. For a reversible process, a value of the ratio $\Delta_0^1 H^{cal}/\Delta_0^1 H^{vh}$ smaller than 1.0 can be considered as an indisputable argument in favor of the existence of a specific complex of several molecules.

Influence of Irreversibility on the Value of the $\Delta_0^1 H^{cal}/\Delta_0^1 H^{vh}$ Ratio

The problem of irreversibility of a process and its influence on the estimated thermodynamic characteristics is rather complicated. In principle, a quantitative thermodynamic analysis is based on equilibrium studies. Nevertheless, in some cases, one can use this analysis for studying processes which are practically irreversible. If the process of protein unfolding is irreversible due to the metastability of the initial state of a molecule, then the equations of equilibrium thermodynamics are inapplicable for its analysis.

However, the observed irreversibility of protein unfolding can be caused also by some ancillary concomitant processes occurring significantly slower and with a significantly smaller enthalpy than the gross conformational changes. For example, this may be the process of proline isomerization, which is rather slow and is practically not accompanied by any enthalpy change, but which hinders sterically the refolding of the polypeptide chain into a compact native structure. This may be also the aggregation of unfolded protein molecules which proceeds much more slowly than the unfolding of the native structure; it depends greatly on concentration and ionic conditions and blocks the refolding of the polypeptide chain into a compact native structure.

Thus, the process of irreversible denaturation can be presented schematically as

$$N \underset{k_2}{\overset{k_1}{\rightleftharpoons}} U \overset{k_3}{\longrightarrow} mD$$

where N is the initial native state, U is the unfolded state, and D is the aggregated state. The reaction $U \rightarrow mD$ results in a decrease of the

[19] P. L. Privalov, *Adv. Protein Chem.* **35**, 1 (1982).

concentration of active molecules participating in the equilibrium $N \rightleftharpoons U$, and the rate of decrease of the active concentration increases with an increase in temperature. This will lead to a sharpening of the observed heat absorption peak and, hence, to an increase of the apparent van't Hoff enthalpy of the transition and to a decrease of the $\Delta_0^1 H^{cal}/\Delta_0^1 H^{vh}$ ratio. It is clear that this change in the van't Hoff enthalpy value is more significant the slower the heating rate; if the reaction $U \rightarrow D$ is not a monomolecular one, the change in the van't Hoff enthalpy will grow with an increase of the sample concentration. Therefore, the influence of irreversibility can be decreased by increasing the heating rate and decreasing the concentration of the studied protein solution.

Since the process of unfolding of the compact protein structure $N \rightarrow U$ is rather fast ($\sim 10^{-3}$ sec^{-1}), the value of $\Delta_0^1 H^{vh}$ obtained upon extrapolation to infinite heating rate and zero concentration should be very close to the real values that could be obtained for the first stage of the process in the absence of the second irreversible stage. This was demonstrated in our laboratory by a comparison of the data obtained on the same protein at different pH values, namely at pH values at which denaturation is completely reversible and, in turn, irreversible.

Analysis of Complex Processes

Under a complex process we mean these cases in which a temperature-induced transformation of a macromolecule proceeds through a number of intermediate macroscopic states. The complex nature of a process can be noticed by the appearance of additional extrema and bends on the melting curve and by the deviation of the calorimetrically measured total melting enthalpy from the effective one estimated by the van't Hoff equation. The main goal of complex process analysis is the determination of its reaction scheme and the thermodynamic description of all the macroscopic states which are realized in the process. A general case, in which the stages of a complex multistage process are not necessarily first-order reactions, is too complicated to be considered here, especially since real cases are usually much simpler: most of their stages are, as a rule, first-order monomolecular reactions. Therefore, here we shall limit our consideration to multistage monomolecular reactions.

It should be noted that the reactions representing a release of ligands can be considered also as monomolecular ones if the concentration of free ligands does not change considerably during the experiments, i.e., if there are free ligands in large excess or if the solution is correspondingly buffered.

Thermodynamic Description of Multistage Processes

Let us assume that a temperature increase induces the macromolecule to pass through distinct macroscopic states. Arranging all these states according to increasing enthalpy values, we can formally describe the equilibrium existing between these states by the following sequential scheme

$$I_0 \underset{}{\overset{K_{0,1}}{\rightleftharpoons}} I_1 \underset{}{\overset{K_{1,2}}{\rightleftharpoons}} I_2 \cdots I_{n-1} \underset{}{\overset{K_{n-1,n}}{\rightleftharpoons}} I_n$$

Denoting the equilibrium constant between the states i and j by $K_{i,j}$, we can write the following thermodynamic relations for this system

$$K_{0,i} = \prod_{j=1}^{i} K_{j-1,j} \tag{57}$$

$$\Delta_0^i G(T) = G_i(T) - G_0(T) = \sum_{j=1}^{i} \Delta_{j-1}^j G(T) \tag{58}$$

$$\Delta_0^i H(T) = H_i(T) - H_0(T) = \sum_{j=1}^{i} \Delta_{j-1}^j H(T) \tag{59}$$

$$\Delta_0^i S(T) = S_i(T) - H_0(T) = \sum_{j=1}^{i} \Delta_{j-1}^j S(T) \tag{60}$$

$$\Delta_0^i C_p(T) = C_{p,i}(T) - C_{p,0}(T) = \sum_{j=1}^{i} \Delta_{j-1}^j C_p(T) \tag{61}$$

The partition function of this system is

$$Q = \sum_{i=0}^{n} K_{0,i} = \sum_{i=0}^{n} \exp(-\Delta_0^i G/RT) \tag{62}$$

and the population of states is

$$F_i = K_{0,i}/Q = \frac{\exp(-\Delta_0^i G/RT)}{Q} \tag{63}$$

The mean excess enthalpy of the system in relation to the initial state can be presented as

$$\langle \Delta H(T) \rangle = \sum_{i=1}^{n} \Delta_0^i H F_i$$

$$= \left[\sum_{i=1}^{n} \Delta_0^i H \exp\left(-\frac{\Delta_0^i H}{RT} + \frac{\Delta_0^i S}{R} \right) \right] \Big/ \left[\sum_{i=1}^{n} \exp\left(-\frac{\Delta_0^i H}{RT} + \frac{\Delta_0^i S}{R} \right) \right] \tag{64}$$

and the excess heat capacity as

$$\langle \Delta C_p(T) \rangle = \frac{d\langle \Delta H \rangle}{dT} = \sum_{i=1}^{n} \Delta_0^i H \frac{dF_i}{dT} + \sum_{i=1}^{n} \Delta_0^i C_p F_i$$

$$= \frac{\langle \Delta H^2 \rangle - \langle \Delta H \rangle^2}{RT^2} + \sum_{i=1}^{n} \Delta_0^i C_p F_i = \langle \delta C_p^{exc}(T) \rangle + \langle \delta C_p^{int}(T) \rangle \quad (65)$$

where

$$\langle \Delta H^2 \rangle = \sum_{i=1}^{n} (\Delta_0^i H)^2 F_i \quad (66)$$

is the mean value of the enthalpy squared.

Our task is to find the number of realizable states and to determine their thermodynamic parameters in such a way that the experimental curve would coincide with the calculated one within the accuracy of the experiment. Since each of the states is specified by three thermodynamic parameters, $(\Delta_0^i H, \Delta_0^i S, \Delta_0^i C_p)$, this task actually comes down to the minimization of the following function

$$\phi(\Delta_0^i H, \Delta_0^i S, \Delta_0^i C_p, n) = \int_0^{\infty} [\langle \Delta C_p(T) \rangle_{theor} - \langle \Delta C_p(T) \rangle_{exp}]^2 dT \quad (67)$$

which depends on $(3n + 1)$ parameters. In the analysis of calorimetric data, the dependence of the transition enthalpy on temperature is usually neglected, since it is weak and does not lead to significant distortion of the $\langle \delta C_p^{exc}(T) \rangle$ term in Eq. (65). However, the second term in this equation, $\langle \delta C_p^{int}(T) \rangle$, should be taken into account and subtracted from $\langle \Delta C_p(T) \rangle$.

The simplest is to approximate $\langle \delta C_p^{int}(T) \rangle$ by a straight line connecting the beginning and the end of the melting curve (Fig. 8). Such an approximation works well in the case of processes which are extended over a broad temperature and which consist of many overlapping broad transitions. A more accurate approximation can be achieved by assuming that the changes of specific enthalpy and specific heat capacity are proportional for all transitions:

$$\frac{\Delta_0^i C_p}{\Delta_0^i H} = \kappa = \frac{\Delta C_p^{tot}}{\Delta H^{tot}}$$

where ΔC_p^{tot} is the heat capacity difference of the final and initial states and ΔH^{tot} is the total denaturation enthalpy. In this case we get the

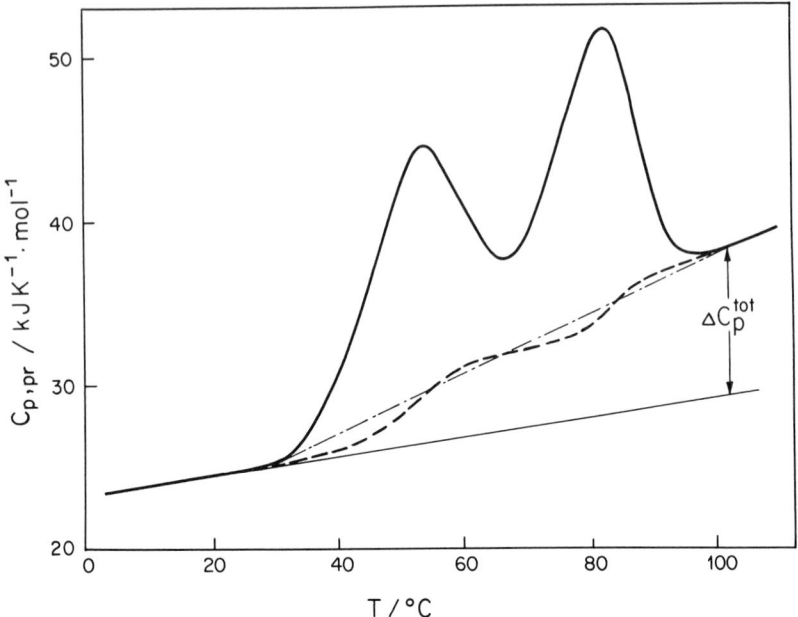

FIG. 8. Evaluation of δC_p^{int} in the transition temperature range, using as example troponin C (10 mM cacodylate buffer, pH 7.25, 10 mM EDTA, 9.5 mM CaCl$_2$) according to Tsalkova and Privalov.[14] Dot-dash line, first approximation of $\delta C_p^{int}(T)$; dashed line, final approximation of $\delta C_p^{int}(T)$ by Eq. (68).

expression

$$\langle \delta C_p^{int}(T) \rangle = \sum_{i=1}^{n} \Delta_0^i C_p F_i = \kappa \sum_{i=1}^{n} \Delta_0^i H F_i = \kappa \langle \Delta H(T) \rangle$$

$$= (\Delta C_p^{tot}/\Delta H^{tot}) \int_0^T \langle \delta C_p^{exc}(\tau) \rangle d\tau \qquad (68)$$

which relates $\langle \delta C_p^{int}(T) \rangle$ to $\langle \delta C_p^{exc}(T) \rangle$. Although this relation is only an approximate one, the error in the $\langle \delta C_p^{exc} \rangle$ value introduced by such an approximation of $\langle \delta C_p^{int} \rangle$ is not great, since $\langle \delta C_p^{exc}(T) \rangle$ is much larger than $\langle \delta C_p^{int}(T) \rangle$.

The determination of $\langle \delta C_p^{int} \rangle$ and its subtraction from $\langle \Delta C_p \rangle$ can be performed in the following way. The straight line connecting the beginning and the end of the excess heat capacity is taken as a zero aproximation of $\langle \delta C_p^{int} \rangle'$. Then $\langle \delta C_p^{exc} \rangle = \langle \Delta C_p \rangle - \langle \delta C_p^{int} \rangle'$ is integrated, and the complete integral is accepted to be ΔH^{tot}. The value of ΔC_p^{tot} is known; it is equal to the difference between the heat capacities of the initial and final states; hence the coefficient κ is determined and the more accurate function $\langle \delta C_p^{int} \rangle''$ can be calculated by Eq. (68). The procedure should be repeated until the values of $\langle \delta C_p^{int} \rangle$ and $\langle \delta C_p^{exc} \rangle$ do not change. The obtained value of $\langle \delta C_p^{exc} \rangle$ is used for further analysis. Thus, we eliminate the parameters $\Delta_0^i C_p$ in Eq. (67) and decrease to $(2n + 1)$ the number of parameters which have to be found.

In using the best fit procedure for this analysis of calorimetric data one should bear in mind that the approximation of the experimental heat capacity curve by the calculated curve improves with an increase in the number of considered states, but this improvement significantly slows down after some definite number of states. This takes place due to a too low population of states added (for details see Ref. 20). Therefore, the states above this critical number can be neglected.

The best fit procedure for approximation of calorimetric data can be simplified significantly in the case when the considered process can be represented as the sum of independent transitions and the system can be considered as consisting of independent subsystems. In this case, the partition function of the system can be represented as the product of the subsystem partition functions

$$Q = \prod_{i=1}^{n} Q_i = \prod_{i=1}^{n} (1 + K_i) \tag{69}$$

The enthalpy and heat capacity of such a system is equal to the sum of the enthalpies and heat capacities of each of the subsystems

$$\langle \Delta H(T) \rangle = RT^2 \frac{\partial \ln Q}{\partial T} = RT^2 \frac{\partial \ln \left(\prod_{i=1}^{n} Q_i \right)}{\partial T}$$

$$= RT^2 \sum_{i=1}^{n} \frac{\partial \ln Q_i}{\partial T} = \sum_{i=1}^{n} \langle \Delta H_i(T) \rangle \tag{70}$$

$$\langle \Delta C_p(T) \rangle = \frac{\partial \langle \Delta H(T) \rangle}{\partial T} = \sum_{i=1}^{n} \langle \Delta C_{pi}(T) \rangle \tag{71}$$

while the heat capacity of each subsystem is described by Eq. (27) for a "two-state" transition. The model of independent subsystems was efficiently used, for example, in studying the melting of transfer RNA.[21]

[20] S. J. Gill, B. Richey, G. Bishop, and J. Wyman, *Biophys. Chem.* **21**, 1 (1985).
[21] P. L. Privalov and V. V. Filimonov, *J. Mol. Biol.* **122**, 447 (1978).

Recommendations for the Best Fit Procedure

What has been said can be summarized by the following recommendations. Determine the excess heat capacity function. Extract from it the effect of the intrinsic heat capacity change $\langle \delta C_p^{int} \rangle$. Minimize the function [Eq. (67)], assuming the most probable number of states. If the experimental curve cannot be approximated by the calculated one, increase the number of states. If the shape of the calculated curve is close to the experimental one, eliminate the states with a low maximal population and repeat the minimization. Obtain the minimal number of well-populated states required for a description of the melting curve. One should bear in mind that the lower the level of the maximal population in the given state, the higher is the error in determining its thermodynamic parameters.

For minimization of function [Eq. (67)] one can use practically any algorithm. However, it should be noted that, since the dependence of this function on variables is of a higher order than quadratic, a change of the second derivatives of this function is significant. Therefore, methods suggesting constant values of second derivatives will give satisfactory results only near the minimum. In regions remote from the minimum the method of steepest descent can be used.

Sequential Analysis of the Excess Heat Capacity Function

An alternative approach to the analysis of the excess heat capacity function was suggested by Freire and Biltonen[17,18] in 1978. The starting point in this approach is that, at sufficiently low temperature, only the first transition contributes noticeably to the observed excess heat capacity. The thermodynamic parameters of the first transition can be determined by using the following function

$$\varphi_1 = \frac{\langle \Delta H \rangle}{1 - F_0} = \left(\sum_{i=1}^{n} \Delta_0^i H F_i \right) \Big/ \left(\sum_{i=1}^{n} F_i \right) \tag{72}$$

For a system which has only two states, this function comes down to $\Delta_0^1 H$, while, for a more complicated system, it is equal to

$$\varphi_1 = \Delta_0^1 H + \frac{\sum_{i=1}^{n} (\Delta_0^i H - \Delta_0^1 H) F_i}{\sum_{i=1}^{n} F_i} = \Delta_0^1 H + \frac{\sum_{i=2}^{n} \Delta_1^i H F_i / F_1}{1 + \sum_{i=2}^{n} F_i / F_1}$$

$$= \Delta_0^1 H + \frac{\sum_{i=2}^{n} \Delta_1^i H \exp(-\Delta_1^i G / RT)}{1 + \sum_{i=2}^{n} \exp(-\Delta_1^i G / RT)} = \Delta_0^1 H + \langle \Delta H \rangle_1 \tag{73}$$

The second term of this function is nothing else than an averaged excess enthalpy for the system consisting of $(n - 1)$ states, and the initial one is considered here as state 1, but not zero. At rather low temperatures at which only the zero state and state 1 are noticeably populated, $\langle \Delta H \rangle_1 = 0$ and φ_1 approaches $\Delta_0^1 H$. Bearing also in mind that $F_0 = 1/Q$ and Q is determined by the excess enthalpy function according to Eq. (17), we get

$$\lim_{T \to 0} \frac{\langle \Delta H \rangle}{1 - F_0} = \lim_{T \to 0} \frac{\langle \Delta H \rangle}{1 - \exp\left(- \int_0^T \frac{\langle \Delta H \rangle}{RT^2} dT\right)} = \Delta_0^1 H \tag{74}$$

The partition function of the new subsystem consisting of $(n - 1)$ states can be estimated from the enthalpy of this subsystem

$$\langle \Delta H \rangle_1 = \varphi_1 - \Delta_0^1 H \tag{75}$$

$$Q_1 = \exp\left(\int_0^T \frac{\langle \Delta H \rangle_1}{RT^2} dT\right) \tag{76}$$

The enthalpy of the second transition can be obtained in the same way as that of the first transition using the function

$$\varphi_2 = \frac{\langle \Delta H \rangle_1}{1 - Q_1^{-1}} = \Delta_1^2 H + \frac{\sum\limits_{i=3}^{n} \Delta_2^i H \exp(-\Delta_2^i G / RT)}{1 + \sum\limits_{i=3}^{n} \exp(-\Delta_2^i G / RT)} = \Delta_1^2 H + \langle \Delta H \rangle_2 \tag{77}$$

for which

$$\lim_{T \to 0} \varphi_2 = \Delta_1^2 H$$

and

$$\langle \Delta H \rangle_2 = \varphi_2 - \Delta_1^2 H$$

Repeating the above procedure and using recurrent relations

$$\varphi_{i+1} = \frac{\langle \Delta H \rangle_i}{1 - Q_i^{-1}} \tag{78}$$

$$Q_i(T) = \sum\limits_{j=i}^{n} \exp(-\Delta_i^j G / RT) = \exp\left(\int_0^T \frac{\langle \Delta H \rangle_i}{RT^2} dT\right) \tag{79}$$

$$\langle \Delta H \rangle_i = \varphi_i - \Delta_{i-1}^i H \tag{80}$$

$$\lim_{T \to 0} \varphi_i = \Delta_{i-1}^i H \tag{81}$$

one can obtain the thermodynamic parameters of all the realizable states. The equilibrium constant for these states, as is clear from Eq. (79), is

$$K_{i-1,i} = \frac{Q_{i-1} - 1}{Q_i} \tag{82}$$

and the population of the ith state, consequently, will be

$$F_i = F_{i-1} K_{i-1,i} = F_{i-1} \frac{Q_{i-1} - 1}{Q_i} \tag{83}$$

The thermodynamic parameters of transitions between the $(i - 1)$th and ith states can be obtained from the condition of equality of their population. Assuming under $T_{t,i}$ a temperature at which $F_{i-1} = F_i$ and the equilibrium constant $K_{i-1,i}$ becomes unity, we can determine $T_{t,i}$ from the condition

$$Q_{i-1}(T_t) - 1 = Q_i(T_t) \tag{84}$$

On the other hand, since

$$K_{i-1,i} = \exp(-\Delta_{i-1}^i H/RT_{t,i} + \Delta_{i-1}^i S/R) = 1$$

we have for the entropy of this transition

$$\Delta_{i-1}^i S = \frac{\Delta_{i-1}^i H}{T_{t,i}} \tag{85}$$

This algorithm is mathematically strict and, as has been shown by Freire and Biltonen, gives the single solution if the function $\langle \Delta C_p \rangle$ and, consequently, $\langle \Delta H \rangle$ are known over the entire temperature range from absolute zero up to temperatures at which the last transition is completed. However in reality the $\langle \Delta C_p \rangle$ value is not known from absolute zero temperature and is estimated with some error.

Actually the $\langle \Delta C_p \rangle$ function is determined only from some temperature T_0 at which the difference $(C_p - C_{p,0})$ begins to exceed the experimental error. Therefore the value of the experimentally determined $\langle \Delta H \rangle_0$ function equals zero at T_0, whereas Q_0 and F_0 are equal to unity. Correspondingly, φ_1 at this point cannot be determined since it tends to $+\infty$ at $T \to T_0$ (Fig. 9). Therefore the function φ_1 approaches the first transition enthalpy value with a reasonable accuracy (of about 7% error) only at

$$T = T_0 + \frac{4RT_{t,1}^2}{\Delta_0^1 H} \approx T_0 + \Delta T_{t,1} \tag{86}$$

i.e., at a distance of half-width of the first transition from T_0. Thus, if we

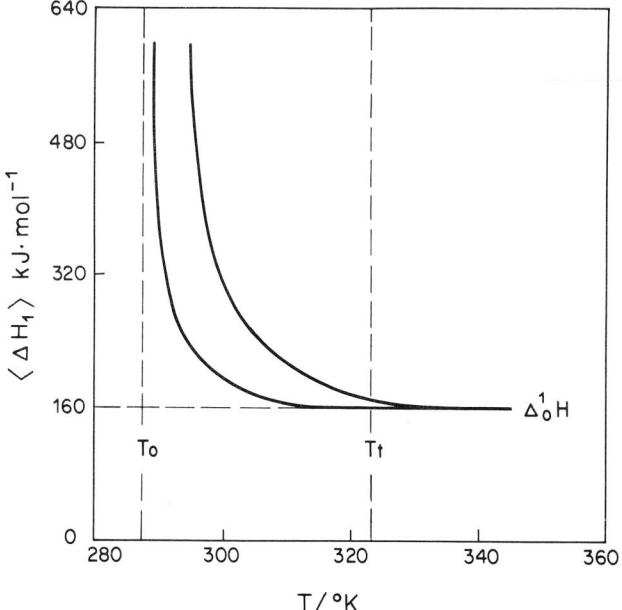

FIG. 9. Temperature dependence of functions $\varphi = \langle \Delta H \rangle / (1 - F_0)$ and $\varphi' = \langle \Delta H^2 \rangle / \langle \Delta H \rangle$ for a two-state transition with $\Delta_0^1 H = 160$ kJ mol^{-1}, $T_t = 323$ K, and $T_0 = 288$ K.

have several transitions, and the first transition is followed by a second one, the determination of its enthalpy value becomes a problem, since the contribution of the subsequent transitions to φ_1 can become noticeable before the value φ_1 approaches $\Delta_0^1 H$. That is why in reality one has to search not for the limit but for the minimum of function φ_1 and to accept this value as the enthalpy of the first transition. Hence, it becomes clear that the extent of transition overlap imposes certain limitations on the applicability of this method to excess heat capacity analysis. A serious obstacle to this method arises also from the accumulation of errors at consecutive steps of the analysis.

In order to decrease the influence of the enumerated factors, the possibility was studied of using for the deconvolution analysis, instead of function φ, some other one which could give also a transition enthalpy at the limit $T \to 0$.[22] In particular, it has been shown that the function $\varphi' = \langle \Delta H^2 \rangle / \langle \Delta H \rangle$ approaches the transition enthalpy value faster than function

[22] V. V. Filimonov, S. A. Potekhin, S. V. Matveyev, and P. L. Privalov, *Mol. Biol. (USSR)* **16**, 551 (1982).

φ (see Fig. 9), and, consequently, is much more useful for sequential analysis of complicated processes. Nevertheless, even with the use of this function, the error accumulation is significant in sequential analysis.

The results of the analysis can be improved by an optimization procedure carried out at each stage of sequential deconvolution.[22] After determining $\Delta_0^1 H$ and $T_{t,1}$, the heat absorption peak corresponding to this parameter is synthesized, and the initial and synthesized curves are compared in the temperature region from T_0 to T_1, where T_1 corresponds to the temperature at which the minimum of the function φ', corresponding to $\Delta_0^1 H$, is achieved. The optimization procedure consists in a search for the minimum of the function

$$\phi(\Delta_0^1 H, T_{t,1}) = \int_{T_0}^{T_1} (\langle \Delta C_p \rangle_{\text{theor}} - \langle \Delta C_p \rangle_{\text{exp}})^2 dT \qquad (87)$$

The refined values of $\Delta_0^1 H$ and $T_{t,1}$ are used to determine $\langle \Delta H \rangle_1$. Then the parameters of the second transition are optimized in the same manner and the parameters of both transitions are optimized in the temperature range from T_0 to T_2 which includes the first two transitions. The same procedure is then repeated. The efficiency of optimization at each stage of the sequential deconvolution analysis of the complex function was demonstrated on synthesized model functions.

Interpretation of the Results of Heat Capacity Function Analysis

The problem which arises after the determination of the number of states of a macromolecule realized in the temperature-induced process and the thermodynamic parameters of the transition which constitutes this process is the interpretation of the found values, their physical and structural meaning.

In many cases, discrete states of macromolecules result in their structural discreteness, i.e., in the subdivision of their structure into cooperative structural blocks, which disrupt on heating in an "all-or-none" way. Thus, each of the realized states represents some definite combinations of blocks in the ordered and disordered states. If there are N such blocks in the molecule, the number of possible states will be 2^N. However, far from all possible states are realized upon heating. The number of such states is determined both by the thermodynamic parameters of stabilization of the blocks and by their interaction.

Let us consider as an example a macromolecule consisting of two cooperative blocks A and B which can be either in state N or D. The scheme of possible transitions in such a molecule is

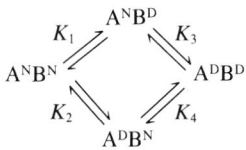

Not all the equilibrium constants in this scheme are independent, since $K_1K_3 = K_2K_4$. It is clear that, if there is no interaction between the transitions, then $K_1 = K_4$ and $K_2 = K_3$.

However, all these transitions can be represented formally by a linear scheme of sequential reactions, as well

$$A^NB^N \underset{}{\overset{K_1^*}{\rightleftharpoons}} A^NB^D \underset{}{\overset{K_2^*}{\rightleftharpoons}} A^DB^N \underset{}{\overset{K_3^*}{\rightleftharpoons}} A^DB^D$$

The equilibrium constants K_i^* are related to the equilibrium constants K_i in the previous scheme in the following way

$$
\begin{array}{lll}
K_1^* = K_1 & & K_1 = K_1^* \\
K_2^* = K_2/K_1 & \text{or} & K_2 = K_1^* K_2^* \\
K_3^* = K_4 & & K_4 = K_3^* \\
& & K_3 = K_2^* K_3^*
\end{array}
\tag{88}
$$

Using these relations, we can obtain the enthalpy and entropy of transition of blocks A and B with the "neighbor" in the ordered or disordered state from the thermodynamic parameters for the linear scheme, which can be determined by the sequential deconvolution analysis of the calorimetric melting curve

$$
\begin{array}{ll}
\Delta_{A^NB^N}^{A^DB^N} H = \Delta H_1^* + \Delta H_2^* & \Delta_{A^NB^N}^{A^DB^N} S = \Delta S_1^* + \Delta S_2^* \\
\Delta_{A^NB^D}^{A^DB^D} H = \Delta H_2^* + \Delta H_3^* & \Delta_{A^NB^D}^{A^DB^D} S = \Delta S_2^* + \Delta S_3^* \\
\Delta_{A^NB^N}^{A^NB^D} H = \Delta H_1^* & \Delta_{A^NB^N}^{A^NB^D} S = \Delta S_1^* \\
\Delta_{A^DB^N}^{A^DB^D} H = \Delta H_3^* & \Delta_{A^DB^N}^{A^DB^D} S = \Delta S_3^*
\end{array}
\tag{89}
$$

Then, for the enthalpy and entropy of interaction between blocks A and B, we have

$$
\begin{aligned}
\delta H_{\text{inter}} &= \Delta_{A^NB^N}^{A^DB^N} H - \Delta_{A^NB^D}^{A^DB^D} H = \Delta H_1^* - \Delta H_3^* \\
\delta S_{\text{inter}} &= \Delta_{A^NB^N}^{A^DB^N} S - \Delta_{A^NB^D}^{A^DB^N} S = \Delta S_1^* - \Delta S_3^*
\end{aligned}
\tag{90}
$$

In real cases, the enthalpy of interaction between the blocks is much lower than the enthalpy of structure stabilization of individual blocks.

Nevertheless, it can be easily determined by measuring the heat effect of melting at conditions with different transition sequences of the blocks (see for example Ref. 23).

Increase of Analysis Reliability

The incorrectness in the determined transition enthalpy values, caused by the inaccuracy of the experimental curve, increases with a decrease of the maximal population of the corresponding state. At the limit when the population of intermediate states tends to zero, the shape of the melting curve becomes indistinguishable from that of the "two-state" transition, and, in this case, the thermodynamic parameters of only the final state can be determined reliably. The population of the intermediate states decreases with an increase of the transition overlap. Therefore, one of the possibilities to increase the reliability of the analysis consists in carrying out experiments at solvent conditions which assure a maximal spread of the studied process over the temperature scale, and, as a result, a minimal overlap of its constituent transitions.

The criterion of reliability of the results of the deconvolution analysis can be the continuity in the changes of the found parameters with variation of environmental conditions. Indeed, variation in the stability of individual states results usually in significant changes in the observed overall melting profile (Fig. 10). However, if the analysis is carried out correctly, the temperature and enthalpy of individual transitions should vary continuously, and their functional dependence on environmental conditions should be interpreted physically (see for example Ref. 24).

When the studied macromolecular system is too complicated, consisting of many subsystems undergoing transitions in the overlapping temperature ranges, it is expedient to simplify it by fragmentation. Studies of melting of a set of fragments permits identification of the found transitions with the melting of a definite part of the molecule, i.e., to carry out actually a structural analysis of the molecule at the domain level (Fig. 11). However, fragmentation of a large molecule into quite definite parts and isolation of highly homogeneous fragments in amounts required for calorimetric experiments by biochemical procedures is much more time consuming than calorimetric measurements and analysis of results.

23 S. V. Matveyev, V. V. Filimonov, and P. L. Privalov, *Mol. Biol. (USSR)* **6**, 1234 (1982).
24 V. V. Novokhatny, S. A. Kudinov, and P. L. Privalov, *J. Mol. Biol.* **179**, 215 (1984).

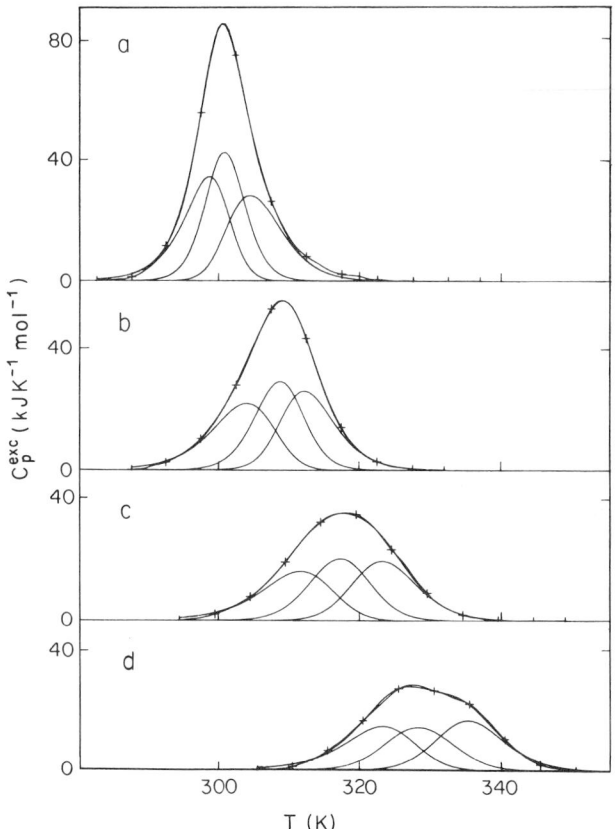

FIG. 10. Deconvolution of the excess heat capacity functions of plasminogen fragment K1-3 in solutions with different pH values: (a) pH 7.4, (b) pH 5.4, (c) pH 4.0, and (d) pH 3.4. Crosses indicate the experimentally determined function to distinguish it from the calculated one using the parameters of the transition estimated by the convolution analysis (see for details Novokhatny *et al.*[24]).

Additional Information Obtained from Melting Curves

Information Obtained from Calorimetric Melting Curves

Carrying out calorimetric experiments at various solvent conditions permits us not only to increase the reliability of the deconvolution analysis, but also to obtain some additional information on the revealed transitions, particularly on the amount of ligands released during the transition.

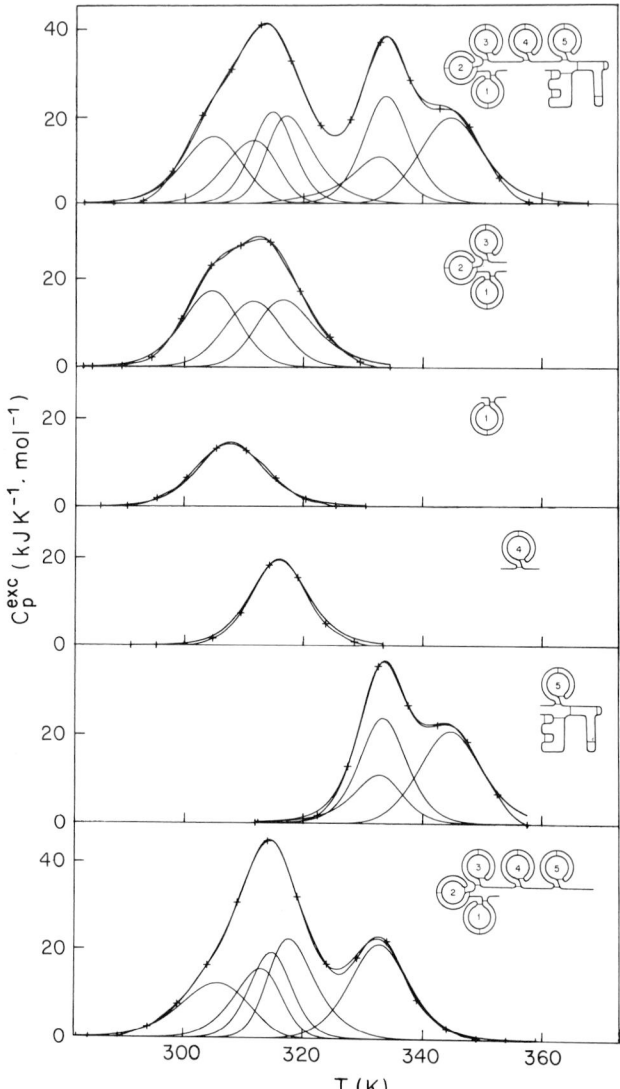

FIG. 11. Deconvolution of the excess heat capacity functions of Lys-plasminogen and its various proteolytic fragments which are given schematically in the upper right-hand corner of each panel (see for details Novokhatny *et al.*[24]).

The equilibrium constant of a reaction which proceeds with ligand binding, such as

$$N \rightleftharpoons D + jL$$

where N and D are the native and denatured states of the macromolecule and L is the ligand released at denaturation, can be expressed as

$$K = \exp\left(-\frac{\Delta G}{RT} + j \ln a_L\right) \tag{91}$$

where a_L is the activity of the ligand in solution. The conditions corresponding to some fixed value of the equilibrium constant, for example, conditions at which the equilibrium constant is equal to unity, i.e., transition conditions, can be found from the requirement

$$\frac{d \ln K}{d \ln a_L} = \left[d\left(-\frac{\Delta_N^D G}{RT} + j \ln a_L\right)\right] \bigg/ (d \ln a_L)$$

$$= -\frac{\Delta_N^D H}{RT_t^2} \frac{dT_t}{d \ln a_L} + j = 0 \tag{92}$$

From this equation, we get immediately

$$\frac{d(1/T_t)}{d \ln a_L} = -\frac{jR}{\Delta_N^D H} \tag{93}$$

Therefore, knowledge of the functional dependence of the reciprocal of the transition temperature on the logarithm of the ligand activity in solution permits us to determine the amount of ligands released during the transition (see for example Refs. 19 and 25).

Information Obtained from Noncalorimetric Melting Curves

In order to carry out a structural interpretation of the stable states that are realized on heating the macromolecule, it is useful to have as many various physicochemical characteristics of these states. However, although obtaining the temperature dependence of some physicochemical parameter characterizing the molecule usually does not present any experimental problems, their deconvolution analysis, intended for the determination of a characteristic specific value for each realizable state, is practically impossible without scanning microcalorimetry data.

Indeed, if a temperature increase induces the molecule to pass through n macrostates which are specified by definite values of the considered parameter $\Delta_0^i A$, the observed overall melting profile of the quantity A will

---

[25] P. L. Privalov, O. B. Ptitsyn, and T. M. Birshtein, *Biopolymers* **8**, 559 (1969).

be

$$\langle \Delta A(T) \rangle = \sum_{i=0}^{n} \Delta_0^i A F_i \tag{94}$$

$$= \left[\sum_{i=0}^{n} \Delta_0^i A \, \exp\left(-\frac{\Delta_0^i H}{RT} + \frac{\Delta_0^i S}{R} \right) \right] \Big/ \left[\sum_{i=0}^{n} \exp\left(-\frac{\Delta_0^i H}{RT} + \frac{\Delta_0^i S}{R} \right) \right]$$

The principal difference between any parameter A and enthalpy, i.e., between Eqs. (94) and (64), is that in the case of enthalpy it figures not only as a coefficient but also in the exponents. Therefore, this equation is solvable only if A is the enthalpy and not any other parameter. The enthalpy function is unique in this respect. For any arbitrary A, this equation can be solved only if we know the thermodynamic characteristics of all the realized states, i.e., their population over the entire considered temperature range. The latter can be determined only by deconvolution analysis of the calorimetric melting curve. If the population of intermediate states, F_i, is already determined from calorimetric experiments all other characteristics of these states can be found from the observed melting curve, $\langle A(T) \rangle$, just by minimization of the function

$$\phi(\Delta_0^i A) = \int_{T_0}^{T} \left[\langle \Delta A(T) \rangle - \sum_{i=1}^{n} \Delta_0^i A F_i(T) \right]^2 dT \tag{95}$$

For this purpose, ϕ is differentiated with respect to parameter $\Delta_0^k A$ and equated to zero, assuming $\Delta_0^k A$ to be independent of temperature.

$$\frac{\partial \phi}{\partial \Delta_0^k A} = 2 \int_{T_0}^{T} \left\{ \langle \Delta A(T) \rangle - \sum_{i=1}^{n} \Delta_0^i A F_i(T) \right\} F_k(T) \, dT$$

$$= 2 \left\{ \int_{T_0}^{T} \langle \Delta A(T) \rangle F_k(T) \, dT \right.$$

$$\left. - \sum_{i=1}^{n} \Delta_0^i A \int_{T_0}^{T} F_i(T) \, F_K(T) \, dT \right\} = 0 \tag{96}$$

Thus, the problem is reduced practically to the solution of the linear system of equations

$$\begin{aligned} L_{11}\Delta_0^1 A + L_{12}\Delta_0^2 A + \cdots + L_{1n}\Delta_0^n A &= M_1 \\ L_{n1}\Delta_0^1 A + L_{n2}\Delta_0^2 A + \cdots + L_{nn}\Delta_0^n A &= M_n \end{aligned} \tag{97}$$

where

$$L_{jk} = L_{kj} = \int_{T_0}^{T} F_j(T) F_k(T) dT, \qquad M_k = \int_{T_0}^{T} \langle A(T) \rangle F_k(T) dT$$

(for details see Ref. 26).

[26] S. V. Matveyev, V. V. Filimonov, and P. L. Privalov, *Mol. Biol. (USSR)* **17**, 172 (1983).

Thermodynamic Description of a Cooperative Unit

Although the enthalpy and entropy of the transition of a cooperative unit between its two realizable states (native and denatured) can be measured calorimetrically only within the transition temperature range, the enthalpy, entropy, and Gibbs energy difference of these states can be estimated over a much broader temperature range if the heat capacities of these states or their differences are known. Indeed, bearing in mind that $(\partial S/\partial T) = C_p/T$ and, at transition temperature, T_t

$$\Delta_N^D G(T_t) = \Delta_N^D H(T_t) - T_t \Delta_N^D S(T_t) = 0 \tag{98}$$

we have for these difference functions at temperature T

$$\Delta_N^D H(T) = \int_T^{T_t} C_p^N(T)dT + \Delta_N^D H(T) + \int_{T_t}^T C_p^D(T)dT$$

$$= \Delta_N^D H(T_t) - \int_T^{T_t} \Delta_N^D C_p(T)dT \tag{99}$$

$$\Delta_N^D S(T) = \int_T^{T_t} \frac{C_p^N(T)}{T} dT + \frac{\Delta_N^D H(T_t)}{T_t} + \int_{T_t}^T \frac{C_p^D(T)}{T} dT$$

$$= \frac{\Delta_N^D H(T_t)}{T_t} - \int_T^{T_t} \Delta_N^D C_p(T)d \ln T \tag{100}$$

$$\Delta_N^D G(T) = \Delta_N^D H(T) - T\Delta_N^D S(T) = \Delta_N^D H(T_t) \frac{T_t - T}{T_t}$$

$$- \int_T^{T_t} \Delta_N^D C_p(T)dT + T \int_T^{T_t} \frac{\Delta_N^D C_p(T)}{T} dT \tag{101}$$

Usually the heat capacity difference between the native and denatured states, $\Delta_N^D C_p(T)$, does not depend significantly on temperature[15] and can be considered in the first approximation as a constant specific for the considered cooperative unit. This permits us to simplify the above equations and to rewrite them in the following way:

$$\Delta_N^D H(T) = \Delta_t H - \Delta_t C_p(T_t - T) \tag{102}$$

$$\Delta_N^D S(T) = \frac{\Delta_t H}{T_t} - \Delta_t C_p \ln \frac{T_t}{T} \tag{103}$$

$$\Delta_N^D G(T) = \Delta_t H \frac{T_t - T}{T_t} - \Delta_t C_p(T_t - T) + \Delta_t C_p T \ln \frac{T_t}{T} \tag{104}$$

where $\Delta_t H \equiv \Delta_N^D H(T_t)$ is the calorimetrically measured transition enthalpy at transition temperature T_t and $\Delta_t C_p \equiv \Delta_N^D C_p(T_t)$ is the heat capac-

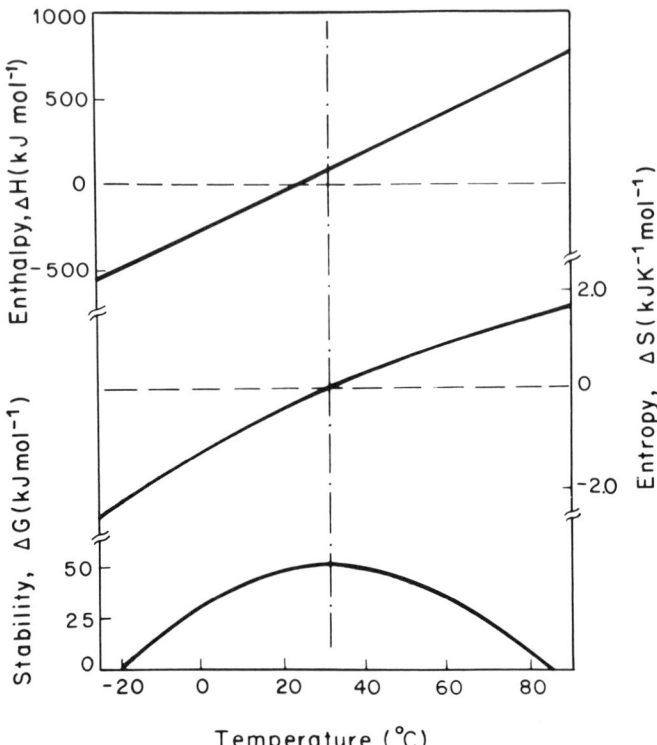

FIG. 12. The enthalpy, entropy, and Gibbs energy differences between native and denatured states of CN-metmyoglobin in 40 mM glycine solution, pH 10.7.

ity difference extrapolated to the transition midpoint between the native and denatured states. As already shown, the transition temperature T_t is usually close to the temperature of the maximal heat absorption T_{max} [see Eq. (52)].

Figure 12 shows the enthalpy, entropy, and Gibbs energy differences of the native and denatured states of a typical small compact globular protein, myoglobin, whose denaturation is very close to the two-state transition. As can be seen, the enthalpy and entropy differences are both increasing functions of temperature, while the Gibbs energy difference is a function with an extremum. The maximal value of the Gibbs energy difference is achieved at the temperature at which

$$\frac{\partial \Delta_N^D G(T)}{\partial T} = - \Delta_N^D S(T) = 0 \tag{105}$$

Since the Gibbs energy difference between the native and denatured states determines actually the stability of the cooperative structure, since it is the work required for its disruption, it follows that the maximal stability of the cooperative unit structure is achieved at conditions at which the native and denatured states do not differ in their entropy values (for details see Refs. 27 and 28).

[27] P. L. Privalov and N. N. Khechinashvili, *J. Mol. Biol.* **86,** 665 (1974).
[28] W. Pfeil and P. L. Privalov, *Biophys. Chem.* **4,** 23 (1976).

[3] Kinetic Mechanisms of Protein Folding

By Hiroyasu Utiyama and Robert L. Baldwin

Introduction

The main purpose in working out the kinetic mechanism of folding of a protein is to take the first step in determining its pathway of folding. The second step is to find conditions in which structural folding intermediates are well populated. The third step is to characterize the structures of the intermediates and to place the intermediates in order on the folding pathway. This chapter considers how to determine the kinetic mechanism of folding; later chapters consider how to characterize the folding intermediates (see, for example, the chapters on circular dichroism[1] and on amide proton exchange[2]). Another chapter[3] considers the use of subzero temperatures as a means of populating folding intermediates for long times. If a folding reaction shows no populated intermediates, the only direct information that can be learned about the folding pathway is that the process of folding is highly cooperative. The nature of the rate-limiting step can, however, be studied by investigating the dependence of the folding rate on temperature, pH, and denaturant concentration.

The goal of determining the pathway of folding by characterizing the structures of intermediates has now been pursued by protein chemists for more than two decades. Attention has centered on small, single-domain proteins in an effort to find out how the simplest proteins fold up. The conclusion has been widely accepted for a decade that folding intermediates can be detected only with difficulty, if at all, in equilibrium studies of single-domain proteins in aqueous solvents. (It remains to be seen whether the use of high-resolution two-dimensional (2-D) NMR will change this conclusion.) Consequently, there has been wide interest in the kinetic intermediates found in the folding reactions of single-domain pro-

[1] A. M. Labhardt, this volume [7].
[2] P. S. Kim, this volume [8].
[3] A. L. Fink, this volume [10].

teins. These were discovered when fast-reaction methods (temperature-jump, stopped-flow) were used to study unfolding and refolding reactions.[4-10] Earlier, slow kinetic methods (manual mixing or producing a temperature jump by switching thermostats) had detected only the slow phase of folding or unfolding and had given the erroneous impression that there were no kinetic intermediates.

The study of kinetic intermediates took a new turn in 1973 with the finding[11] that unfolded RNase A contains both fast-folding (U_F) and slow-folding (U_S) forms. Subsequent studies showed that the unfolding pathway is $N \rightleftharpoons U_F \rightleftharpoons U_S$[12-14] and supported the hypothesis[12] that the slow cis–trans isomerization of proline residues after unfolding accounts for the $U_F \rightleftharpoons U_S$ reaction. Proline isomerization is generally regarded as a nuisance by theorists who propose pathways for folding. For experimentalists engaged in determining folding pathways, proline isomerization is a central problem. A later chapter[15] considers how to detect and analyze $U_F \rightleftharpoons U_S$ reactions in unfolded proteins, and another chapter[16] considers how to analyze the cis/trans ratios of individual proline residues. Because proteins often have several *trans*-proline residues and sometimes have one or two *cis*-proline residues, the major form of an unfolded protein is not uncommonly U_S. For a U_S species, the problem of analyzing proline isomerization during folding is inextricably tied up with the determination of the folding pathway.

This situation has two important consequences for the determination of the folding pathway. First, the kinetic intermediates seen in protein folding reactions might not represent structural intermediates at all, but might instead arise only from $U_F \rightleftharpoons U_S$ reactions.[12] Second, the fact that proline isomerization is a very slow reaction at 25° or below may aid in populating folding intermediates: if the protein can fold partially before proline isomerization occurs, then proline isomerization should act as a

[4] A. Ikai and C. Tanford, *Nature (London)* **230**, 100 (1971).
[5] H. F. Epstein, A. N. Schechter, R. F. Chen, and C. B. Anfinsen, *J. Mol. Biol.* **60**, 499 (1971).
[6] T. Y. Tsong, R. L. Baldwin, and E. L. Elson, *Proc. Natl. Acad. Sci. U.S.A.* **68**, 2712 (1971).
[7] L. L. Shen and J. Hermans, Jr., *Biochemistry* **11**, 1836 (1972).
[8] T. Y. Tsong and R. L. Baldwin, *J. Mol. Biol.* **69**, 145 (1972).
[9] A. Ikai, W. F. Fish, and C. Tanford, *J. Mol. Biol.* **73**, 167 (1973).
[10] C. Tanford, K. C. Aune, and A. Ikai, *J. Mol. Biol.* **73**, 185 (1973).
[11] J.-R. Garel and R. L. Baldwin, *Proc. Natl. Acad. Sci. U.S.A.* **70**, 3347 (1973).
[12] J. F. Brandts, H. R. Halvorson, and M. Brennan, *Biochemistry* **14**, 4953 (1975).
[13] J.-R. Garel, B. T. Nall, and R. L. Baldwin, *Proc. Natl. Acad. Sci. U.S.A.* **73**, 1853 (1976).
[14] P. J. Hagerman and R. L. Baldwin, *Biochemistry* **15**, 1462 (1976).
[15] F. X. Schmid, this volume [4].
[16] J. F. Brandts and L.-N. Lin, this volume [6].

kinetic trap for structural intermediates.[17-21] Much of the work on kinetic intermediates in the last decade has centered on these issues: finding out whether or not structural intermediates are populated during folding, and asking whether proline isomerization is the correct explanation for the $U_F \rightleftharpoons U_S$ reactions of unfolded proteins. The evidence for populated structural intermediates is considered in a recent review.[22]

Nature of Kinetic Mechanisms in Protein Folding Experiments

Our discussion is limited to reversible folding reactions: ones in which the native and unfolded forms are in equilibrium inside the folding transition zone and for which refolding experiments produce native protein in essentially 100% yield. Kinetic experiments measuring irreversible denaturation (usually unfolding followed by aggregation) are useful for some purposes but not for determining the kinetic mechanism of unfolding and refolding. Our discussion is limited also to monomeric proteins: special techniques are needed for oligomeric proteins and are discussed in another chapter.[23]

The term "kinetic mechanism" is used here in its narrow sense. The following four examples provide illustrations of simple kinetic mechanisms. The first rule in kinetic experiments is to find the simplest mechanism that fits the data. In protein folding experiments one has to watch out that the simplest mechanism may be wrong because it is too simple.

Two-State Mechanism: $U \rightleftharpoons N$. If only the two species N (native) and U (unfolded) are detectable at any time in unfolding and refolding, then experiments will show only one kinetic phase and the reaction follows a two-state mechanism. This does not mean that unfolding or refolding proceeds in a single step without passing through intermediates, but only that the intermediates are not detectable. For example, the folding process could take place in two steps via the intermediate. I*, which may be considered as the intermediate present at the rate-limiting step.

$$U \underset{k_{21}}{\overset{k_{12}}{\rightleftharpoons}} I^* \underset{k_{32}}{\overset{k_{23}}{\rightleftharpoons}} N \tag{1}$$

The rate constants in the 2-state mechanism are k_r (refolding) and k_u (unfolding).

[17] K. H. Cook, F. X. Schmid, and R. L. Baldwin, *Proc. Natl. Acad. Sci. U.S.A.* **76,** 6157 (1979).
[18] F. X. Schmid and H. Blaschek, *Eur. J. Biochem.* **114,** 111 (1981).
[19] T. E. Creighton, *J. Mol. Biol.* **137,** 61 (1980).
[20] Y. Goto and K. Hamaguchi, *J. Mol. Biol.* **156,** 891 (1982).
[21] P. McPhie, *Biochemistry* **21,** 5509 (1982).
[22] P. S. Kim and R. L. Baldwin, *Annu. Rev. Biochem.* **51,** 459 (1982).
[23] R. Jaenicke and R. Rudolph, this volume [12].

$$U \underset{k_u}{\overset{k_r}{\rightleftharpoons}} N \tag{1a}$$

The steady-state approximation $d(I^*)/dt = 0$ can be used to express k_r and k_u in terms of the four rate constants of Eq. (1) because, if $(I^*) \cong 0$, then $d(I^*)/dt \cong 0$.

$$k_r = k_{12} \left(\frac{k_{23}}{k_{21} + k_{23}} \right) \tag{2a}$$

$$k_u = k_{32} \left(\frac{k_{21}}{k_{21} + k_{23}} \right) \tag{2b}$$

By analyzing the dependences of k_r and k_u on variables such as temperature, it may be possible to show that folding is not a single-step process.[24] Nevertheless, as long as I* is not detectable either in unfolding or refolding, then only one kinetic phase can be observed and folding allows a 2-state mechanism; I* is a hypothetical intermediate.

An example can be given. Consider the situation at the middle of the folding transition, and let $(U) = (N) = 0.49975$, $(I^*) = 0.0005$ at equilibrium; the concentrations are normalized so that the sum over all species is 1. Let the $U \rightleftharpoons I^*$ step be fast compared to the $I^* \rightleftharpoons N$ step, with $k_{21} \gg k_{23}$. Then, from Eqs. (2a) and (2b), $k_r = (k_{12}/k_{21})k_{23}$ and $k_u = k_{32}$; moreover, $k_r = k_u$ at the midpoint of the transition. At equilibrium, $(I^*)/(U) = 0.0005/0.49975 = 0.001 = k_{12}/k_{21}$, and $(N)/(I^*) = 0.49975/0.0005 = 1000 = k_{23}/k_{32}$. In refolding, U and I* equilibrate rapidly so that, after a short time of refolding, $(I^*) = (k_{12}/k_{21})(U) = 0.001$. If 0.001 is considered to be too small to detect, then I* is not detectable at any time, either in unfolding or refolding.

On-Pathway Intermediate: $U \rightleftharpoons I \rightleftharpoons N$. If I is detectable at equilibrium, inside the folding transition zone where U and N coexist, then I is an equilibrium intermediate. If I is not detectable at equilibrium but I is populated during unfolding or refolding, then I is a kinetic or transient intermediate. Two kinetic phases will be observed in experiments in which all three species (U, I, N) are populated.

The following example shows how a kinetic intermediate can be populated transiently but not at equilibrium. Let the situation at equilibrium inside the folding transition zone be the same as for I* in the example

---

[24] For example, in a single-step process the refolding rate constant should show a simple temperature dependence with a positive activation enthalpy. By contrast, experimental data for two proteins show complex temperature dependences for the refolding rate constant of the $U_F \rightarrow N$ reaction and the rate constant goes through a maximum as the temperature increases. This has been observed both for RNase A[14] and hen lysozyme.[25]

[25] S. Kato, N. Shimamoto, and H. Utiyama, *Biochemistry* **21**, 38 (1982).

above, so that I is not populated significantly at the transition midpoint. Use the notation of Eq. (1) but replace I* by I. Let I increase in stability relative to U in refolding conditions outside the transition zone, either at low temperature or at a low GdmCl[26] concentration; for example, let k_{12} = k_{21} in refolding conditions, so that (I)/(U) = 1 when I and U equilibrate. Let the U $\rightleftharpoons$ I step be fast compared to I $\rightleftharpoons$ N (i.e., let $k_{21} \gg k_{23}$), so that U and I equilibrate rapidly after the start of refolding. Then, after refolding for a short time, (I) = (k_{12}/k_{21})(U) or (I) = (U) = 0.5. The result is that I is well populated transiently in refolding conditions provided that refolding takes place well outside the folding transition zone. Nevertheless, I is not populated at equilibrium, either inside or outside the transition zone, if (U)/(N) = (I)/(N) $\cong$ 0 at equilibrium in refolding outside the transition zone.

Two Unfolded Forms: $U_S \rightleftharpoons U_F \rightleftharpoons N$. As regards kinetic mechanism of folding, U_S and U_F are different unfolded forms, and the folding mechanism is $U_S \rightleftharpoons U_F \rightleftharpoons N$, if U_S and U_F have these properties. (1) U_S and U_F have different rates of refolding. (2) Both are populated when the protein is completely unfolded, as in 6 M GdmCl. (3) In unfolding U_F is formed first (N $\rightarrow$ U_F) and U_S is formed from U_F. For this mechanism two kinetic phases are observed both in unfolding and refolding, but a special technique ("double-jump" method) is needed to monitor the slower phase when unfolding takes place outside the transition zone.

The $U_S \rightleftharpoons U_F \rightleftharpoons N$ mechanism can be distinguished from the U $\rightleftharpoons$ I $\rightleftharpoons$ N mechanism by the kinetic ratio test[27] provided that I differs both from U and from N in its extent of folding. Two probes are used to monitor folding, and the kinetic progress curves monitored by the two probes are compared. One probe detects N but not I or U: it could be a spectral property that changes when a specific ligand binds to an enzyme. The other probe detects both I and N: it could be the fluorescence of one or a few tryptophan groups. The kinetic progress curve of folding is given by $y(t)$ versus t, where $y(t)$ is defined so as to give the fraction of N present when folding follows a 2-state mechanism.

$$y(t) = \frac{B(t) - B(U)}{B(N) - B(U)} \tag{3}$$

B is the property used to follow the reaction: it should be a property such as absorbance or fluorescence that is proportional to concentration. For

---

[26] Abbreviations: GdmCl, guanidinium chloride; RNase A, bovine pancreatic ribonuclease A.

[27] A. M. Labhardt and R. L. Baldwin, *J. Mol. Biol.* **135**, 231 (1979).

the intermediate I, let $y(I)$ be defined as

$$y(I) = \frac{B(I) - B(U)}{B(N) - B(U)} \tag{3a}$$

For a probe that detects only N, $y(I) = 0$. For a probe that detects both I and N, $y(I)$ will have a nonzero value, often between 0 and 1.

The kinetic progress curves monitored by two such probes will not be superimposable for the $U \rightleftharpoons I \rightleftharpoons N$ mechanism but will be superimposable for the $U_S \rightleftharpoons U_F \rightleftharpoons N$ mechanism, if $y(U_S) = y(U_F) = 0$ for both probes. The reader should note that the comparison of kinetic progress curves monitored by two probes, one of which detects only native or native-like species and the other of which detects both native and also partly folded species, provides a standard test for a structural intermediate that is populated during folding.

Abortive Intermediate: $I \rightleftharpoons U \rightleftharpoons N$. In this mechanism an intermediate (I) is formed from U during refolding, but the intermediate does not lead from U to N. Two kinetic phases are observed in experiments for which all three species (I, U, N) are populated. The $I \rightleftharpoons U \rightleftharpoons N$ mechanism can be distinguished from the $U \rightleftharpoons I \rightleftharpoons N$ mechanism by a test based on the comparison of amplitudes from an unfolding and from a refolding experiment made in the same final conditions.[4,28] Early studies of the folding of horse cytochrome c[9] and hen lysozyme[10] were interpreted by this test and indicated that abortive intermediates were populated during refolding. But these experiments were made before the existence of U_F and U_S forms was known, and this test is not applicable if the unfolded protein contains both U_F and U_S forms.[14,25,29] Later work has shown that both horse cytochrome c[30] and hen lysozyme[25] do contain both U_F and U_S forms, as suggested earlier,[29] and at present there is no evidence for abortive intermediates in the refolding of these proteins.

This illustrates a basic problem in the study of kinetic mechanisms: conclusions based on a simple mechanism may be invalid if the actual mechanism is more complex.

Description of the Kinetic Mechanism

A kinetic experiment yields directly the number of kinetic phases ($i = 1, \ldots, n$) and the amplitude (A_i) and time constant (τ_i) of each phase. The time constant is the reciprocal of the apparent rate constant λ of a phase: $\tau_i = 1/\lambda_i$. If the reaction is monomolecular, the curve of $A(t)$ versus t can be represented by a sum of n exponential terms in which each term describes one kinetic phase.

[28] A. Ikai and C. Tanford, *J. Mol. Biol.* **73**, 145 (1973).
[29] P. J. Hagerman, *Biopolymers* **16**, 731 (1977).
[30] J. A. Ridge, R. L. Baldwin, and A. M. Labhardt, *Biochemistry* **20**, 1622 (1980).

$$A(t) - A(\infty) = \sum_{i=1}^{n} A_i \exp(-t/\tau_i) \tag{4}$$

$A(\infty)$ is the signal registered by the probe at the end of the reaction and A_i is the change in A that occurs in phase i. Description of the kinetic mechanism begins by giving the number of phases and the values of A_i and τ_i for each phase; this is done for each of several unfolding and refolding experiments. Usually the amplitudes are given as normalized or fractional amplitudes, α_i:

$$\alpha_i = A_i/[A(0) - A(\infty)] \tag{4a}$$

and the kinetic progress curve can be written

$$\alpha(t) = 1 - \sum_{i=1}^{n} \alpha_i \exp(-t/\tau_i) \tag{4b}$$

where $\alpha(t)$ changes from 0 at $t = 0$ to 1 at $t = \infty$. The value of α_i depends on the probe that is used, but τ_i is independent of the probe. The time constant[31] of a kinetic phase depends only on the final conditions, because it is a function only of the rate constants. In general, the normalized amplitude of a phase depends both on the initial and final conditions, although in special cases it depends only on the final conditions.

These kinetic data, taken in different unfolding and refolding conditions and with different probes, provide the raw material for determining the kinetic mechanism. Once the correct mechanism has been found, the kinetic data can be used to calculate the microscopic rate constants in the mechanism if it is sufficiently simple. Usually there are twice as many microscopic rate constants as kinetic phases, and it may not be possible to determine the rate constants if the mechanism is complex (for more detail, see a textbook on kinetics[33]). When the microscopic rate constants have been determined, they may be used to calculate the way in which the concentration of each species changes with time in a stopped-flow experiment.

A complete description of the mechanism consists of giving first the kinetic data for several sets of unfolding and refolding conditions and for different probes, and giving next the chosen mechanism, the tests on

[31] The time constant τ is sometimes termed the relaxation time. This is the proper usage for a relaxation experiment[32] in which only a small perturbation of the previous equilibrium is made, but protein folding experiments involve large perturbations.

[32] M. Eigen and L. DeMaeyer, in "Technique of Organic Chemistry" (S. L. Friess, E. S. Lewis, and A. Weissberger, eds.), Vol. 8, Part 2, p. 895. Wiley (Interscience), New York, 1963.

[33] J. W. Moore and R. G. Pearson, "Kinetics and Mechanism," 3rd Ed. Wiley, New York, 1981.

which it is based, and the alternative mechanisms that have been ruled out. Finally, the microscopic rate constants are given if it is possible to calculate them.

A kinetic phase will not be detected by a probe unless it monitors the species whose concentrations change in that phase. For example, when unfolding takes place outside the folding transition zone and follows the $N \rightleftharpoons U_F \rightleftharpoons U_S$ mechanism, N disappears completely in the fast phase of unfolding ($N \rightarrow U_F$). Spectral probes that monitor N but do not distinguish between U_F and U_S will record only the fast phase of unfolding and will miss the slow $U_F \rightleftharpoons U_S$ reaction (see below).

Procedure

Outline

1. Choose the kinetic method, the denaturant, and the probes.

2. Measure the equilibrium transition curve for unfolding and check reversibility.

3. Test and calibrate the stopped-flow apparatus.

4. Test procedures for resolving the kinetics into phases and determining their time constants and amplitudes.

5. Perform unfolding and refolding experiments on both sides of the transition zone and throughout the transition zone.

6. Test for fast-folding and slow-folding forms of the unfolded protein.

7. Use the kinetic data to test possible mechanisms. After choosing the best mechanism, calculate the microscopic rate constants if possible.

8. Design further experiments to test the mechanism and set limits on the conditions in which it is valid.

Description

Choice of Kinetic Method, Unfolding Agent, and Probes. Most studies of the kinetic mechanism of folding use the stopped-flow method because both refolding and unfolding can be studied, and because unfolding or refolding jumps across the entire transition zone can be made. The observable reaction times range from a few milliseconds to 100 sec. The shortest time is determined by the time required for complete mixing of reactants and the longest time by the disturbance of the mixed solution in the cell by diffusion of unreacted liquids. The temperature-jump method allows faster reactions to be studied (time range about 10^{-5} to 1 sec) but is limited to studies of unfolding.

Unfolding is usually achieved by adding a denaturant (GdmCl or urea). The denaturant concentration is decreased quickly (in a refolding experiment) or increased (in unfolding) by rapid mixing. The progress of protein refolding or unfolding after mixing is monitored with an appropriate probe. The temperature is kept constant during the observation.

In order to observe the complete refolding process, we must first completely unfold the protein. A thorough study of this problem has been made[34] and the conclusion was reached that 6 M GdmCl completely unfolds many globular proteins (but not membrane proteins). The current criterion for complete unfolding is that all secondary structure should be destroyed. Unfolding unavoidably exposes the hydrophobic residues of a protein to the solvent. This often results in aggregation, and a good denaturing agent must also prevent this both in the unfolded state and during stopped-flow dilution to the refolding conditions. Spontaneous aggregation sometimes occurs after prolonged standing of an unfolded protein, and this possibility must be checked. In early work it was tacitly assumed that incomplete unfolding could be achieved and that the residual structure could affect the refolding kinetics. Such an effect of residual structure in speeding up the refolding kinetics has not yet been observed for any single-domain protein. The special property of GdmCl that makes it widely used in studies of protein unfolding is that it both disrupts completely the native conformation and also protects many unfolded proteins from aggregation during stopped-flow dilution.

Other reagents for unfolding proteins besides GdmCl and urea are of limited usefulness, because intermolecular aggregation occurs frequently. Typical denaturants are acid or base and organic solvents such as dimethyl sulfoxide, dioxane, 2-chloroethanol, aliphatic acids, and aliphatic alcohols at acidic pH. Detergents cannot be used because they bind strongly to proteins to form a complex of specific conformation and they do not dissociate quickly after simple dilution. Both GdmCl and urea solutions should be prepared freshly and not allowed to stand; specially purified materials are needed and are commercially available. The urea must be free of cyanate ion, which is formed by hydrolysis of urea and which reacts at neutral pH with free amino groups of the protein. High purity of the denaturant is needed because of the disparity between the concentrations of the denaturant (6 M) and the protein (10^{-5} M); an impurity in the denaturant can easily be present at the same concentration as the protein.

Protein samples to be used for studies of folding kinetics may be purified by a standard method and stored in a lyophilized form. A single band in the electrophoretic pattern of a protein analyzed by the SDS–polyacrylamide gel method may be a good enough criterion for purity, but

[34] C. Tanford, *Adv. Protein Chem.* **23**, 121 (1968).

the amount of protein added to one slot in the gel should be varied by at least a factor of 10. In preparing protein solutions it is convenient first to dissolve the lyophilized material as a concentrated protein solution (10–30 mg/ml). It is often observed that lyophilized protein does not redissolve completely. Aggregated material is removed by centrifugation (4000 g, 10 min) followed by filtration through a glass filter of appropriate pore size to avoid contamination with large sedimented particles. Alternatively one may filter the solution through a Millipore filter (0.43 μm pore size). The protein concentration is determined spectrophotometrically by using the appropriate extinction coefficient. Denatured protein solutions are prepared by mixing the concentrated protein with denaturant in the same buffer solution. It is wise to repeat the step of centrifugation–filtration, or of filtration through a Millipore filter, just before a folding experiment. Aggregated particles in the stopped-flow cell after mixing may be detected by measuring absorbance at 340 nm.

For both kinetic and equilibrium studies, UV absorption and fluorescence bands are frequently used as probes of the folding reactions. The principle is that the absorption and fluorescence spectra of a chromophore vary with a change in its environment and thus change during folding. Proteins which contain many aromatic residues such as tryptophan, tyrosine, and phenylalanine are advantageous in this respect. Tryptophan has the largest extinction coefficient and is often buried inside the native molecule, becoming exposed when unfolded, and so it is a very useful probe of folding. If a protein contains several tryptophan residues, the working concentration can be reduced accordingly, which helps in preventing aggregation.

Circular dichroism provides very valuable information on protein conformation[1] but there are special problems in using circular dichroism as a kinetic probe for fast reactions because of optical artifacts caused by mixing: an efficient mixer for a cell whose path length is small is not available. Another problem is that the spectrum of the backbone amide bands (190–260 nm) may not be measurable at high concentrations of GdmCl because of light absorption by impurities present in even the most highly purified samples. Vibrational spectra of the backbone amide bands can now be obtained by the laser Raman method.[35] The use of the method has been demonstrated, and applications to protein folding experiments will probably be made as instruments become available. There are also site-specific probes of folding for certain proteins, such as the Soret absorbance band for heme proteins.

[35] R. W. Williams, *J. Mol. Biol.* **166**, 581 (1983).

Measurement and Analysis of the Equilibrium Transition Curve. In studies of kinetic mechanism, there are several reasons to measure the equilibrium transition curve, besides the usual reason which is to measure the stability of the protein. These are (1) to check reversibility of unfolding, (2) to test for equilibrium intermediates, (3) to find out if the total kinetic amplitude equals the static value, or whether some kinetic phase is too fast or too slow to be detected in stopped-flow experiments, (4) to relate the folding kinetics to position in the transition curve, and (5) to compute the apparent equilibrium constant $K_N = (N)/(U)$ for use in testing kinetic mechanisms. Experimental details of measuring transition curves and processing the data are discussed in a later chapter.[36]

In general, to interpret kinetic changes in protein conformation, it is necessary to know as much as possible about the conformation in the initial and final conditions. On the one hand, this means that the equilibrium transition curve for the conformational change must be obtained in the same conditions and with the same probe as in the kinetic experiment. On the other hand, it means that any conformational changes that take place within the stopped-flow mixing time must also be detected and characterized by comparing the total kinetic amplitude with the static amplitude.[37] To do this, it is necessary to make a baseline correction: thus, for an unfolding experiment, the absorbance of the native protein must be extrapolated to the value that it would have in the unfolding conditions.

To use the equilibrium transition curve to test for reversibility of unfolding, one simply checks that the same transition curve is obtained starting from both directions, using either native or unfolded protein to start with. This simple test is usually sufficient. Sensitive kinetic tests of reversibility are described below.

The equilibrium transition curve is also used to check for the presence of equilibrium intermediates. A standard qualitative test for intermediates is to find out whether or not the transition curves measured by different probes are superimposable. To do this, it is necessary to extrapolate the specific values of the property being measured for the native and unfolded forms into the transition zone. This must be done with care: the apparent transition curve can be shifted away from its correct position by an incorrect choice of either baseline. It is often advantageous to choose a wavelength where the specific property of the native form is independent of temperature or of GdmCl concentration.[38]

[36] C. N. Pace, this volume [14].
[37] S. Kato, M. Okamura, N. Shimamoto, and H. Utiyama, *Biochemistry* **20,** 1080 (1981).
[38] N. N. Khechinashvili, P. L. Privalov, and E. I. Tiktopulo, *FEBS Lett.* **30,** 57 (1973).

A standard quantitative test for equilibrium intermediates is to measure the transition curve by calorimetry[39] and then to compare the actual ΔH for unfolding with the apparent value calculated from the van't Hoff relation for a 2-state ($N \rightleftharpoons U$) reaction. When intermediates are populated, and have enthalpies different from both N and U, then the van't Hoff ΔH (ΔH_{vH}) will be less than the calorimetrically measured value (ΔH_{cal}). The ratio $\Delta H_{cal}/\Delta H_{vH}$ gives information about the intermediate(s).

Tests of the Stopped-Flow Apparatus. The two most commonly used commercial stopped-flow spectrophotometers are the Model RA-401 (Union Giken Company, Osaka) and the Gibson–Durrum stopped-flow instrument (Dionex Company, Palo Alto, CA). In the RA-401 the two reactant solutions, thermostatted separately in two reservoirs, are driven pneumatically at about 5 kg/cm² pressure of nitrogen gas. The pressure, and hence the liquid flow, is regulated with an air valve which is operated electromagnetically. In the Gibson–Durrum instrument, the two reactant solutions are mechanically injected into the mixer from two syringes. The pneumatic drive has the following advantages. (1) The mixing ratio of the two reactant solutions can be easily adjusted by inserting a plug (short polyethylene tubing with a fine diameter) in the flow channel between the sample reservoir and the mixer. (2) Since the solutions are under 5 kg/cm² pressure, cavitation is much less serious, and prolonged degassing of a sample solution is not necessary. The mixing ratio of the Gibson–Durrum instrument is more accurate. With the pneumatic drive, the mixing ratio is determined by comparing the protein concentrations before and after mixing. To reduce the amount of sample solution required for one mixing, similar air valves are used at the outlet of each reservoir, and the volume in the flow channel between the valve and the mixer is minimized. The longest observable time is limited by the disturbance in the observation cell that results from the diffusion of reactant solutions through the mixer. The horizontal orientation of the flow channel in the Gibson–Durrum instrument is advantageous in this respect. The efficiency of mixing becomes worse as the mixing ratio deviates from unity. One can improve the efficiency by using either a double mixer or a longer flow path between the outlet from the mixer to the cell.

For good temperature control, the temperatures of the observation chamber and the reservoir should be regulated separately by circulating liquids thermostatted at different temperatures with the use of two circulating baths in order to have the same final temperature in the observation chamber as in the reservoir. If measurements are made at either high or

---

[39] P. L. Privalov and S. A. Potekhin, this volume [2].

low temperatures, serious optical artifacts are likely to be observed[14] unless this precaution is taken. Temperature stability can be determined by measuring the absorbance of a 0.01% phenolphthalein solution in 50 mM glycine–NaOH buffer, pH 8.8, at 552 nm. The sensitivity of the measurement is 0.05 A/°C. One may measure the temperature of the observation cell by directly inserting a thermocouple.

The dead time of the stopped-flow apparatus can be measured by making use either of the reduction of sodium 2,6-dichlorophenolindophenol with L-ascorbic acid[40] which is a pseudo first-order reaction (described here) or by a two-step disulfide exchange reaction.[41] The rate constant, k, of the ascorbate reduction reaction depends strongly on pH. The observed amplitudes, X_1 and X_2, at two pH values, are given by

$$X_1 = X \exp(-k_1 t_d) \tag{5a}$$
$$X_2 = X \exp(-k_2 t_d) \tag{5b}$$

and the dead time t_d is found from these expressions as

$$t_d = (k_2 - k_1)^{-1} \ln(X_1/X_2) \tag{5c}$$

The absorption spectrum of sodium 2,6-dichlorophenolindophenol depends on pH, but measurements can be made at the isosbestic point, 524 nm. Since the dead time depends on the viscosity of the reactant solution, it is more convenient to determine the dead volume V_d, which is related to the flow rate, f, by

$$V_d = f t_d \tag{5d}$$

The flow rate can be measured from the volume of the reaction mixture drained during a certain period of flow duration.

It is important to test the reliability of the stopped-flow apparatus by control experiments that test the stability of the output signal when (1) two identical buffer solutions are mixed, (2) native protein is mixed with dilute buffer, (3) concentrated denaturant is mixed with a buffer solution, (4) denatured protein is mixed with a concentrated denaturant, and so on.

It is necessary to characterize the optical quality of the spectrophotometer used in the stopped-flow apparatus. Since the emphasis is placed on an intense incident beam and high speed of recording, the observed spectrum does not necessarily coincide with that obtained in a conventional spectrophotometer, and the difference is caused by the lower optical quality of the stopped-flow system. For an accurate comparison of the

[40] B. Tonomura, H. Nakatani, O. Nasatke, J. Yamaguchi-Ito, and K. Hiromi, *Anal. Biochem.* **84**, 370 (1978).
[41] C. Paul, K. Kirschner, and G. Haenisch, *Anal. Biochem.* **101**, 442 (1980).

kinetic amplitude with the static one, it is best to measure the static signal in the stopped-flow apparatus by alternately pushing through the flow cell each of the two solutions from their respective reservoirs.

Acquisition of the Kinetic Data. It is now routine to use a microcomputer for storage, averaging, and further processing the data from a stopped-flow spectrophotometer, via an appropriate interface. Since protein refolding reactions usually exhibit two or at most three kinetic phases, signals are recorded with double time bases with a few hundred points per time base. Acquisition of data in the long time range is started at a preset time after the last measurements in the short time range have been made. For small absorbance changes, the signal can be recorded in the transmittance mode rather than in the absorbance mode.

For two kinetic phases, the data are analyzed by fitting them to a sum of two exponential terms:

$$A(t) - A(\infty) = A_1 \exp(-t/\tau_1) + A_2 \exp(-t/\tau_2)$$

The equilibrium value $A(\infty)$ is first evaluated from data taken in the long time range as an average of the signal at times longer than $6\tau_1$, where τ_1 is the time constant of the slow phase. If the refolding reaction is not complete at the longest time allowed by the limits of the stopped-flow apparatus, then one has to compare the results obtained with manual mixing in a conventional spectrophotometer. A very small linear change in signal at long times usually results from an artifact such as aggregation. Thus it is difficult to determine a kinetic phase whose time constant is long and whose amplitude is small.[37]

The amplitude A_1 and the time constant τ_1 for the slower phase are determined from the linear region in the plot of $\log[A(t) - A(\infty)]$ vs time. The contribution to $A(t)$ of the slower phase evaluated in this way is subtracted to determine next A_2 and τ_2 in the same way for the faster phase. In using least-squares to fit the absorbance data to a first-order plot, we use a weighting factor of $(A - A_\infty)^2$, where A_∞ denotes the absorbance at the end of the phase being fitted.

The most serious problem in this type of analysis is the difficulty of resolving two exponential terms whose time constants are close to each other. To resolve two exponential terms of equal amplitude, the time constants should differ by at least 5-fold.

Perform Stopped-Flow Experiments in Different Unfolding and Refolding Conditions, and with Different Probes. The reasons for making these different types of stopped-flow experiments are given below. Basically three types of experiments are needed: (1) complete refolding, after a jump across the entire transition zone, (2) complete unfolding, after a jump in the opposite direction, and (3) partial unfolding and refolding studied in the same final conditions, after jumps into the transition zone. If a test for equilibrium intermediates is to be made, the opposite type of partial jump is also made, starting from inside the transition zone where the suspected intermediate may be populated.[9,10]

An important test of the reversibility of the protein folding process is to check that the same values of τ_1 and τ_2 are obtained both from partial unfolding and from partial refolding experiments made in the same final conditions. This is a stringent kinetic test for reversibility. Another check of reversibility is to vary the initial unfolding conditions when measuring complete refolding. Aggregation that occurs in only some initial conditions is likely to show up as an extra, very slow, phase of refolding.[42]

A useful test of the consistency of the kinetic data is to plot the τ_i and α_i values against the unfolding agent, temperature or GdmCl concentration, and to look carefully for discontinuities. A discontinuity may reflect mislabeling of the kinetic phases beyond that point. This can be checked against the predictions of the kinetic mechanism under investigation.

If a slow phase is incomplete at the end of the permissible time in a stopped-flow experiment, or if the existence of an even slower phase is suspected, manual mixing experiments should be made in a conventional spectrophotometer and carried out to long times. If all kinetic phases are accounted for, the total kinetic amplitude should agree with the static amplitude.

The usual wavelength used to monitor folding corresponds to the maximum or minimum in the static difference curve between native and unfolded protein. It may also be interesting to examine a wavelength at which the static difference is 0. If the protein binds a specific ligand, it is advantageous to monitor formation of the ligand binding site during folding. The ligand should bind rapidly (<1 msec) as well as specifically, and a wavelength should be used where there are no changes in the protein spectrum upon folding. Specific ligands have been used to monitor the folding of RNase A,[11,18] hen lysozyme,[25] and carbonic anhydrase.[43]

Look for U_F and U_S Forms of the Unfolded Protein. Procedures for doing this are discussed in later chapters.[15,44] Enough examples have been found of proteins that contain both U_F and U_S forms after unfolding[11,19,20,25,30,45] that it is obligatory to look for this possibility when investigating a new protein. Even if the protein shows only a U_S form at equilib-

[42] B. T. Nall, J.-R. Garel, and R. L. Baldwin, *J. Mol. Biol.* **118,** 317 (1978).

[43] B. P. N. Ko, A. Yazgan, P. L. Yeagle, S. C. Lottich, and R. W. Henkens, *Biochemistry* **16,** 1720 (1977).

[44] T. E. Creighton, this volume [9].

[45] M. Jullien and R. L. Baldwin, *J. Mol. Biol.* **145,** 265 (1981).

rium, it may give a U_F form immediately after unfolding at low temperatures, which then is converted slowly to a U_S form (see below).

Analysis of the Kinetic Mechanism and Calculation of the Microscopic Rate Constants. For illustration, we use the $U_S \rightleftharpoons U_F \rightleftharpoons N$ mechanism to show what calculations are made in testing the kinetic data to see if they fit a given mechanism. For RNase A,[14] hen lysozyme,[25] and the constant fragment of the immunoglobulin light chain,[20] this mechanism works well for studies close to or inside the folding transition zone, although it may be necessary to include unfolding[14] or refolding[17,18,20,46] intermediates in studies outside the transition zone. The properties of the $U_S \rightleftharpoons U_F \rightleftharpoons N$ mechanism have been contrasted with those of other simple mechanisms.[29] The four microscopic rate constants of this mechanism are:

$$U_S \underset{k_{21}}{\overset{k_{12}}{\rightleftharpoons}} U_F \underset{k_{32}}{\overset{k_{23}}{\rightleftharpoons}} N \tag{6}$$

The first test is, of course, that two kinetic phases should be observed in unfolding and refolding experiments. Provided that the two phases are well separated in time ($\tau_1 \gg \tau_2$), the expressions for τ_1 and τ_2 simplify to[14]

$$(1/\tau_2) = k_{23} + k_{32} \qquad (\tau_1 \gg \tau_2) \tag{7a}$$
$$(1/\tau_1) = k_{12} + k_{21}/(1 + K_{23}) \qquad (\tau_1 \gg \tau_2) \tag{7b}$$

We assume here that $\tau_1 \gg \tau_2$ unless otherwise specified. The two equilibrium constants are:

$$K_{23} = k_{23}/k_{32} = (N)/(U_F) \tag{8a}$$
$$K_{12} = k_{12}/k_{21} = (U_F)/(U_S) \tag{8b}$$

To begin with, one assumes provisionally that the $U_S \rightleftharpoons U_F \rightleftharpoons N$ mechanism is valid and makes the calculations described below.

Complete refolding. Beginning with completely unfolded protein, a jump is made across the transition zone. (1) The same value for the relative amplitude of the fast phase, $\alpha_2 = A_2/(A_1 + A_2)$, is given by all probes that do not distinguish between U_F and U_S. Because $A_2/A_1 = K_{12}$, one can use α_2 to calculate the ratio of the concentrations of the two unfolded species [Eq. (8b)]. For RNase A unfolded by GdmCl or urea,[13] K_{12} is independent within experimental error of variables such as pH, temperature, and GdmCl concentration. (2) In complete refolding experiments, it follows from $(N)/(U_F) \gg 1$ that $k_{23} \gg k_{32}$ and $(1/\tau_2) = k_{23}$ [Eq. (7a)]. Also $(1/\tau_1) = k_{12}$ [Eq. (7b)]. Thus, in conditions of complete refolding, the time constants of the two kinetic phases yield directly k_{12} and k_{23}, two of the four microscopic rate constants [Eq. (6)]. Since K_{12} is also

[46] F. X. Schmid, *Biochemistry* **22**, 4690 (1983).

known from A_2/A_1, three of the four microscopic rate constants are determined directly by a single stopped-flow experiment. As one would expect, $(1/\tau_2)$ for the fast phase yields the rate constant for folding and $(1/\tau_1)$ for the slow phase yields a rate constant for isomerization of the unfolded protein.

Complete unfolding. A jump across the transition zone is made in the opposite direction, beginning with N and ending with an equilibrium mixture of (U_F, U_S). (1) For spectral probes of folding that do not distinguish between U_F and U_S, the entire unfolding reaction occurs in the fast phase: $\alpha_2 = 1$ and $\alpha_1 = 0$. The slow phase is not detected outside the folding transition zone by such probes. (2) The slow phase can, however, be monitored by taking aliquots and measuring their refolding kinetics to determine the $(U_F)/(U_S)$ ratio at different times.[15] (3) In complete unfolding experiments, since $(N)/(U_F) \cong 0$, $k_{32} \gg k_{23}$ and it follows that $(1/\tau_2) = k_{32}$ [Eq. (7a)] and $(1/\tau_1) = k_{12} + k_{21}$ [Eq. (7b)]. Consequently, the most direct way to study the unfolding rate constant k_{32} as a function of temperature or of GdmCl concentration is to make measurements of τ_2 at different temperatures or GdmCl concentrations above the folding transition zone. Likewise, the dependence of the refolding rate constant k_{23} on temperature or GdmCl concentration can be found most directly by measuring τ_2 below the folding transition zone.

Partial unfolding compared to partial refolding. Jumps are made into the transition zone so that the unfolding and refolding kinetics can be compared in the same final conditions. A key quantity is α_2, the fraction of the total amplitude change that occurs in the fast phase of unfolding or refolding. (1) The same value of α_2 should be found in an unfolding experiment as in the refolding experiment made at the same point in the transition zone, provided that in refolding the $(U_F)/(U_S)$ ratio does not change between the initial and final conditions. (2) The experimental value of α_2 is compared with the value predicted from τ_1, τ_2, K_{12}, and k_{12}.[14] (3) The two different mechanisms $U_S \rightleftharpoons U_F \rightleftharpoons N$ and $U_S \rightleftharpoons N \rightleftharpoons U_F$ can be distinguished from each other if τ_1/τ_2 is sufficiently small (<20) because different values of α_2 are predicted for these two mechanisms.[14] Both RNase A and hen lysozyme have been shown in this way to follow the $U_S \rightleftharpoons U_F \rightleftharpoons N$ mechanism.[14,25] (4) The equilibrium transition curve can be predicted from the kinetic data and compared with the measured transition curve.[14,20] (5) It is also informative to compare the equilibrium transition curve with the curves described by extrapolating each kinetic phase to zero time.[20,47]

[47] C. Tanford, *Adv. Protein Chem.* **24,** 1 (1970).

These calculations provide a reasonably thorough test of the $U_S \rightleftharpoons U_F \rightleftharpoons N$ mechanism and they also show how the four microscopic rate constants of the mechanism can be determined.

When only one kinetic phase is observed, so that the reaction appears to follow a two-state ($U \rightleftharpoons N$) mechanism, and when the time constant for refolding at $25°$ is 10 sec or longer, it is important to test whether the unfolded species is actually a U_S species and the mechanism is actually a $U_S \rightleftharpoons U_F \rightleftharpoons N$ mechanism in which $(U_F)/(U_S) \cong 0$ at equilibrium. This test can be made by using a denaturant to produce rapid unfolding at $0°$, and then using refolding assays to test for a $U_F \rightarrow U_S$ reaction that occurs slowly after unfolding.[15]

When three kinetic phases are observed, so that the reaction is more complex than the $U_S \rightleftharpoons U_F \rightleftharpoons N$ system, one can see if the data fit either of the following 4-species mechanisms. Inside the folding transition zone, the kinetics of unfolding and refolding of the immunoglobulin light chain[20] fit the $U_S \rightleftharpoons U_F \rightleftharpoons N$ mechanism. At GdmCl concentrations below the transition zone, however, an intermediate kinetic phase appears and the relative amplitude of the slow phase decreases. The results can be explained at least in part[20] by the mechanism:

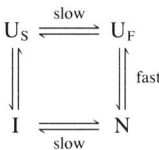

in which the intermediate I is stable only at low GdmCl concentrations outside the transition zone. In thermal unfolding at pH 3.0, RNase A rapidly forms an unfolding intermediate at the upper end of the transition zone and at higher temperatures.[14] An analysis of the kinetics has been given for the mechanism[14]:

$$N \overset{\text{fast}}{\rightleftharpoons} I \rightleftharpoons U_F \rightleftharpoons U_S$$

Other small proteins have also been shown to follow refolding mechanisms characterized by well-populated folding intermediates: α-lactalbumin,[48,49] penicillinase,[50] carbonate dehydratase,[43,51] pepsinogen,[21] and the α-subunit of tryptophan synthase.[52]

[48] K. Kuwajima, *J. Mol. Biol.* **114**, 241 (1977).
[49] D. A. Dolgikh, R. I. Gilmanshin, E. V. Brazhnikov, V. E. Bychkova, G. V. Semisotnov, S. Y. Venyaminov, and O. B. Ptitsyn, *FEBS Lett.* **136**, 311 (1981).

Microscopic rate constants are meaningless if calculated for the wrong kinetic mechanism. This point can be illustrated by reviewing what happened in early work on protein folding kinetics before stopped-flow and temperature-jump methods were used. The data were interpreted by the two-state mechanism

$$U \underset{k_u}{\overset{k_r}{\rightleftharpoons}} N$$

The microscopic rate constants k_r and k_u were calculated from τ_1 (only the slow kinetic phase was observed) and from the apparent equilibrium constant K_N by these relations.

$$K_N = (N)/(U) = k_r/k_u \tag{9a}$$
$$(1/\tau_1) = k_r + k_u \tag{9b}$$
$$(1/\tau_1) = k_u(1 + K_N) \tag{9c}$$

Today, stopped-flow experiments on these same systems typically would show at least one additional fast kinetic phase. If the $U_S \rightleftharpoons U_F \rightleftharpoons N$ mechanism is applicable, then the older results can be reinterpreted by expressing k_u and k_r [as defined by Eqs. (9a)–(9c)] in terms of the rate constants of the $U_S \rightleftharpoons U_F \rightleftharpoons N$ mechanism. The results are as follows. First, note that

$$K_N = (N)/[(U_F) + (U_S)] \tag{10a}$$
$$K_N = K_{12}K_{23}/(1 + K_{12}) \tag{10b}$$

Then, provided that $\tau_1 \gg \tau_2$, it follows that [see Eq. (7b)]

$$k_u = k_{12}/[K_N + K_{12}/(1 + K_{12})] \tag{11a}$$
$$k_r = k_{12}K_N/[K_N + K_{12}/(1 + K_{12})] \tag{11b}$$

In a complete unfolding experiment, for which $K_N \sim 0$

$$k_u = k_{12} + k_{21} \qquad (K_N \ll K_{12}) \tag{12a}$$

and in a complete refolding experiment, for which K_N is large

$$k_r = k_{12} \qquad (K_N \gg K_{12}) \tag{12b}$$

[50] B. Adams, R. J. Burgess, E. A. Carrey, I. R. Mackintosh, C. Mitchinson, R. M. Thomas, and R. H. Pain, in "Protein Folding" (R. Jaenicke, ed.), p. 447. Elsevier, Amsterdam, 1980.

[51] L. F. McCoy, E. S. Rowe, and K. P. Wong, *Biochemistry* **19**, 4738 (1980).

[52] C. R. Matthews, M. M. Crisanti, J. T. Manz, and G. L. Gepner, *Biochemistry* **22**, 1445 (1983).

Consequently, k_u and k_r are closely related to the isomerization rate constants of the $U_S \rightleftharpoons U_F$ reaction rather than being related to the folding and unfolding rate constants of the $U_F \rightleftharpoons N$ reaction. In older work k_u and k_r were interpreted as microscopic rate constants for unfolding and refolding, and dubious conclusions about the nature of the rate-limiting step in folding were drawn from the temperature dependences of k_u and k_r.

Further Tests of the Mechanism. Once a mechanism has been found that fits the kinetic data, the next step is to characterize the species that appear in the mechanism. If there are two unfolded species (U_F, U_S), it is important to find out if the $U_F \rightleftharpoons U_S$ reaction can be correlated with the cis–trans isomerization of proline residues in the unfolded protein.[16] If there is a folding intermediate I, it is important to find conditions that maximize the concentration of I and to characterize I by techniques such as circular dichroism[1] or amide proton exchange.[2] It is also important to set limits on the range of unfolding and refolding conditions in which the mechanism is valid. Any simple mechanism of folding is likely to be valid only in limited conditions; additional intermediates in unfolding or refolding are likely to be found as the range of conditions is extended.

Acknowledgments

We thank Drs. P. S. Kim, F. X. Schmid, and N. Shimamoto for their discussion. This work was supported in part by Research Grants to H.U. (538028) from the Ministry of Education, Science and Culture of Japan and to R.L.B. (GM 19988-23) from the U.S. National Institutes of Health.

[4] Fast-Folding and Slow-Folding Forms of Unfolded Proteins

By Franz X. Schmid

In some unfolded proteins fast-folding (U_F) and slow-folding (U_S) forms have been shown to coexist in slow equilibrium. These different unfolded species are discriminated by their relative rates of refolding: the fast-folding molecules regain the native state (N) in the time range of milliseconds, the slow-folding molecules refold in the time range of seconds. The rates of both reactions depend on the folding conditions. The fast- and slow-folding species, U_F and U_S, are also referred to as U_2 and U_1, respectively.

The first evidence for multiple unfolded forms was presented by Garel and Baldwin in 1973 for ribonuclease A.[1] They proposed a

$$U_S \underset{\text{slow}}{\rightleftharpoons} U_F \underset{\text{fast}}{\rightleftharpoons} N \qquad (1)$$

model to explain the kinetic properties of the folding transition of this protein. A plausible explanation for the occurrence of U_F and U_S species in unfolded proteins was given by the proline model (first proposed by Brandts and co-workers)[2-4]: U_F and U_S differ in the cis–trans configuration of X-proline peptide bonds; U_F has all essential prolines in the same configuration as in the native state, U_S molecules contain at least one incorrect proline isomer. Depending on the number and location of the essential prolines, different U_S species may exist. The role of proline isomerization for protein folding was discussed in a recent review.[5]

The existence of fast- and slow-folding species can be inferred from a kinetic analysis of the folding and unfolding reactions[1,6,7] (see below). The formation of slow-folding species U_S is detected easily in unfolding experiments. After the rapid $N \rightarrow U_F$ unfolding step [cf. Eq. (1)] which yields almost pure U_F species, U_S is formed from U_F in a slow equilibration reaction, which can be monitored by slow refolding assays.

Fast- and slow-folding species have recently been detected in the unfolded immunoglobulin light chain,[8] in lysozyme,[9] cytochromes c from horse[10] and yeast,[11] carp parvalbumins,[12] pepsinogen,[13] the α-subunit of tryptophan synthase,[14] and a modified trypsin inhibitor.[15]

[1] J.-R. Garel and R. L. Baldwin, *Proc. Natl. Acad. Sci. U.S.A.* **70**, 3347 (1973).

[2] J. F. Brandts, H. R. Halvorson, and M. Brennan, *Biochemistry* **14**, 4953 (1975).

[3] F. X. Schmid and R. L. Baldwin, *Proc. Natl. Acad. Sci. U.S.A.* **75**, 4764 (1978).

[4] K. H. Cook, F. X. Schmid, and R. L. Baldwin, *Proc. Natl. Acad. Sci. U.S.A.* **76**, 6157 (1979).

[5] P. S. Kim and R. L. Baldwin, *Annu. Rev. Biochem.* **51**, 459 (1982).

[6] P. J. Hagerman and R. L. Baldwin, *Biochemistry* **15**, 1462 (1976).

[7] P. J. Hagerman, F. X. Schmid, and R. L. Baldwin, *Biochemistry* **18**, 293 (1979).

[8] Y. Goto and K. Hamaguchi, *J. Mol. Biol.* **156**, 891 (1982).

[9] S. Kato, N. Shimamoto, and H. Utiyama, *Biochemistry* **21**, 38 (1982).

[10] J. A. Ridge, R. L. Baldwin, and A. M. Labhardt, *Biochemistry* **20**, 1622 (1981).

[11] B. T. Nall and T. A. Landers, *Biochemistry* **20**, 5403 (1981).

[12] L. N. Lin and J. F. Brandts, *Biochemistry* **17**, 4102 (1978).

[13] P. McPhie, *J. Biol. Chem.* **257**, 689 (1982).

[14] M. M. Crisanti and C. R. Matthews, *Biochemistry* **20**, 2700 (1981).

[15] M. Jullien and R. L. Baldwin, *J. Mol. Biol.* **145**, 265 (1981).

Detection of Fast- and Slow-Folding Forms

Analysis of Refolding Kinetics

The refolding of a mixture of U_F and U_S species results in complex folding kinetics which are composed of parallel fast and slow kinetic phases [Eq. (2)].

$$U_F \rightarrow N$$
$$U_S^X \rightarrow N \tag{2}$$
$$U_S^Y \rightarrow N$$

Sequential refolding reactions via transient intermediates [Eq. (3)] which are partly folded, however, can lead to similar complex kinetic patterns.

$$U \rightarrow I_X \rightarrow I_Y \rightarrow N \tag{3}$$

Furthermore any of the parallel reaction pathways sketched in Eq. (2) may actually involve sequential steps as in Eq. (3). In the following, criteria are listed which are useful to discriminate kinetic phases which originate from parallel refolding reactions [Eq. (2)] from phases which stem from a sequential pathway via transient intermediates [Eq. (3)].

1. The relative amplitudes of the fast- and slow-folding phases have to be independent of the probe that is used to follow the folding kinetics, provided that the different unfolded forms are identical with respect to these properties.

2. The relative amplitudes of the fast and slow folding reactions have to be independent of the *final folding conditions*. They depend only on the equilibrium distribution of U_F and U_S species in the unfolded protein. Cis ⇌ trans equilibria about X-proline peptide bonds are usually independent of pH (outside the range of acid catalysis[3]), temperature, or denaturant concentration,[3,16–18] therefore the relative amplitudes of folding reactions which originate from species with different proline isomers should in general also be independent of the *initial conditions of unfolding*. Exceptions from this rule may occur, e.g., if proline is next to charged residues in sequence. Proline isomerization in model peptides is known to be dependent on adjacent charged groups.[2,19]

3. Parallel fast and slow refolding reactions yield native protein. This is the most useful criterion to distinguish parallel folding reactions of U_F and U_S species from sequential reactions involving transient intermediates. The formation of native or native-like molecules during refolding can

[16] L. N. Lin and J. F. Brandts, *Biochemistry* **22**, 564 (1983).
[17] F. X. Schmid and R. L. Baldwin, *J. Mol. Biol.* **133**, 285 (1979).
[18] F. X. Schmid, M. H. Buonocore, and R. L. Baldwin, *Biochemistry* **23**, 3389 (1984).
[19] C. Grathwohl and K. Wüthrich, *Biopolymers* **15**, 2025 (1976).

be probed by measuring the regain of enzymatic activity or the binding of inhibitors that are specific for the active site of the protein. Three prerequisites have to be fulfilled for such experiments. (1) The refolding kinetics should not be altered by the presence of substrates or inhibitors, (2) enzymatic catalysis or binding of the inhibitor has to be fast in comparison to the rates of the fast and slow refolding reactions, and (3) intermediates with partial activity should not accumulate during folding. A rigorous test of whether native protein is formed in parallel refolding reactions is provided by kinetic experiments in the presence of varying concentrations of inhibitor. When the concentration of inhibitor is small ($[I] < [U_F]$), all the inhibitor is bound by the product of the fast refolding reaction, no slow phase should be detected by inhibitor binding. With increasing concentration of inhibitor the amplitude of slow refolding increases, until in the presence of excess inhibitor the relative amplitudes reflect the initial distribution of unfolded species (see preceding section). Such a test has been carried out for ribonuclease A.[20]

These criteria refer to refolding experiments which end up in the native baseline region. A discussion of the folding kinetics within the transition region with respect to the presence of fast- and slow-folding species was given by Hagerman and Baldwin.[6]

Slow Refolding Assays ("Double Jumps")

After the $N \rightarrow U_F$ unfolding step, U_S is formed slowly in the course of the

$$U_F \underset{k_{SF}}{\overset{k_{FS}}{\rightleftharpoons}} U_S \qquad (4)$$

chain equilibration reaction. Both U_F and U_S are equally unfolded species and show similar physical properties. Therefore the kinetics of the equilibration reaction usually cannot be measured directly. The amount of U_S formed after different times of unfolding can be determined by slow refolding assays. Samples are withdrawn from the unfolding solution at different time intervals after the initiation of unfolding; these samples are transferred to standard refolding conditions, and the amplitude of the slow refolding reaction, $U_S \rightarrow N$, is determined. Refolding of the U_F species is complete within the time of manual mixing. The amplitude of the slow reaction is proportional to the concentration of U_S which was present at the time when the sample was withdrawn for the assay. The dependence of these slow refolding amplitudes on the time of unfolding yields the kinetics of the $U_F \rightleftharpoons U_S$ reaction. The measured rate constant is equal to the sum of the rate constants in both directions, $\tau_{app}^{-1} = k_{FS} + k_{SF}$

[20] F. X. Schmid and H. Blaschek, *Eur. J. Biochem.* **114,** 111 (1981).

[cf. Eq. (2)]. When the equilibrium constant $K = [U_S]/[U_F]$ is known, the individual rate constants k_{FS} and k_{SF} can be derived.

Formation of U_S Species after Unfolding of Ribonuclease A. Measurement of the $U_F \rightleftharpoons U_S$ reaction of ribonuclease A after unfolding is described here as an example for the application of the double-jump technique. In these experiments the refolding assays are carried out under conditions where slow refolding is a single, monophasic reaction. Therefore, the assays yield the kinetics of overall formation of U_S species after unfolding. They do not discriminate between the different U_S species of ribonuclease A.[16,18,20]

Unfolding. To initiate unfolding, at time zero, 50 μl of native ribonuclease A (1.5 mM in H_2O, 0°) is mixed with 250 μl of 6 M GuHCl solution in 0.1 M glycine, pH 2 at 0°. The resulting unfolding conditions are 0.25 mM ribonuclease A in 5 M GuHCl, pH 2, 0°. At acid pH the unfolding step $N \rightarrow U_F$ is very rapid and complete before the onset of the $U_F \rightleftharpoons U_S$ equilibration; at low temperature the $U_F \rightleftharpoons U_S$ reaction is strongly decelerated, because of its high activation enthalpy.[2-4] Consequently, these are unfolding conditions where both reactions are completely decoupled and the $U_F \rightleftharpoons U_S$ equilibration can be measured in a convenient time range without interference from the preceding $N \rightarrow U_F$ unfolding step.

Refolding assays. After different time intervals 70 μl samples are withdrawn from the unfolding solution and quickly diluted into 830 μl of the refolding buffer (1.2 M GuHCl in 0.1 M cacodylate, pH 6.2) in the spectrophotometer cell, which is thermostatted at 25°. The slow refolding reaction of U_S is monitored at 287 nm. At the end of each refolding assay the absorbance of the refolded sample is recorded at 277 nm to determine the actual concentration of ribonuclease in each assay. Representative refolding assays, performed 2–60 min after the $N \rightarrow U_F$ unfolding, are shown in Fig. 1. The particular assay conditions were chosen, because at pH 6, 1.5 M GuHCl and 25°, slow refolding occurs in a single exponential phase in a time range, which is convenient for manual sampling techniques.

Data treatment. The refolding kinetic traces (Fig. 1) are linearized. The time constant for refolding, $\tau = 100 \pm 10$ sec, is independent of the time of sampling, as the final folding conditions were the same in all assays. The amplitudes of refolding (ΔA_{287}) increase with the time of unfolding. They are divided by the concentration of ribonuclease A present in the individual assays, to obtain the molar changes in absorbance, $\Delta\varepsilon_{287}[M^{-1}$ cm$^{-1}]$. The increase in $\Delta\varepsilon_{287}$ with the time of sampling yields the kinetics of the formation of slow-folding species U_S (Fig. 2). These kinetics can be fitted to a single exponential curve, yielding a time constant of 900 sec and an amplitude of $\Delta\varepsilon_{287} = 2050$ M^{-1} cm^{-1} (cf. the

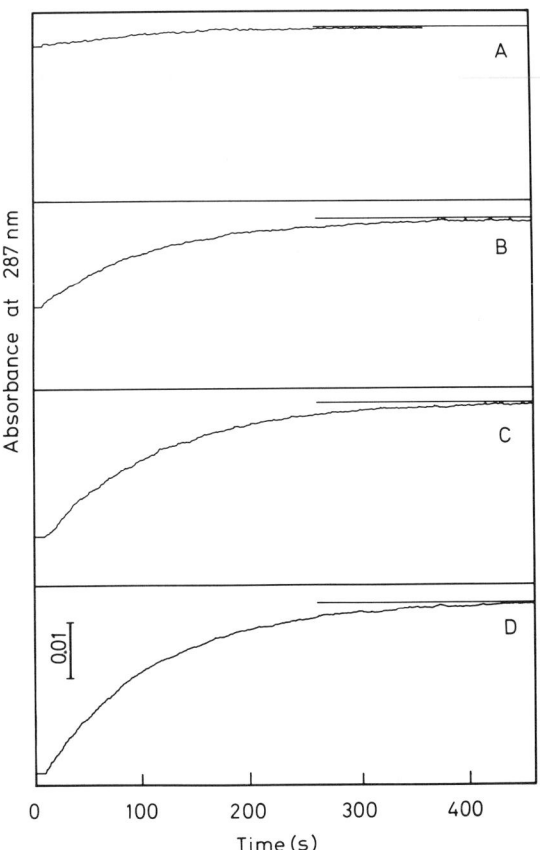

FIG. 1. Formation of slow-folding species U_S after unfolding of ribonuclease A (0.25 mM) in 5 M GuHCl and 0.1 M glycine, pH 2 at 0°. Representative slow refolding assays after (A) 120 sec, (B) 600 sec, (C) 1320 sec, and (D) 3600 sec of incubation in the unfolding solution are shown. The slow refolding assays were carried out by a 13-fold dilution to 1.5 M GuHCl and 0.1 M cacodylate, pH 6.2 at 25° (concentration of ribonuclease A: 19 μM). The slow refolding kinetics, measured at 287 nm, yield single exponential curves with a time constant of $\tau = 100 \pm 5$ sec.

continuous line in Fig. 2). The observed rate constant, τ^{-1}, is equal to the sum of the intrinsic rate constants [Eq. (4)].

$$\tau^{-1} = k_{FS} + k_{SF} = 11.1 \times 10^{-4} \text{ sec}^{-1} \tag{5}$$

With the known equilibrium constant,

$$K_{eq} = [U_S]/[U_F] = k_{FS}/k_{SF} = 4 \tag{6}$$

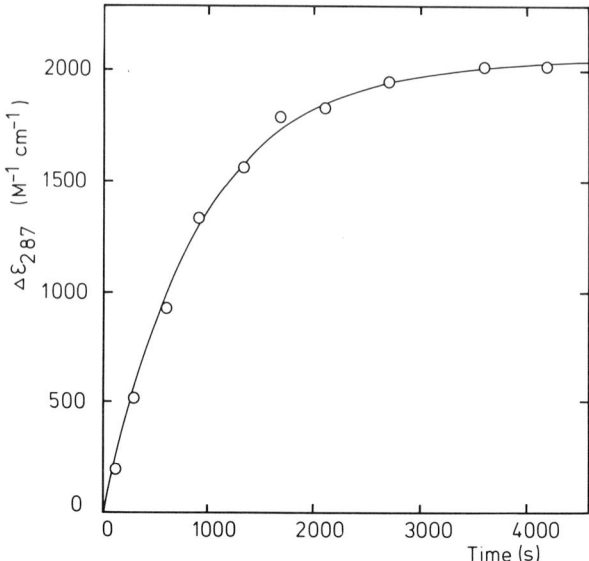

FIG. 2. Kinetics of formation of slow-folding species U_S after unfolding of ribonuclease A. The increase of the amplitudes of the slow refolding assays with the time of unfolding is shown. Slow refolding assays were carried out as shown in Fig. 1. Unfolding conditions: 5 M GuHCl and 0.1 M glycine, pH 2 at 0°. Refolding assays were performed at 1.5 M GuHCl and 0.1 M cacodylate, pH 6.2, 25°. The actual concentration of ribonuclease A was determined after each assay by absorbance at 277 nm by using $\varepsilon_{277} = 9700\ M^{-1}\ cm^{-1}$ (H. J. Sage and S. J. Singer, *Biochemistry* **1**, 305, 1962). The amplitudes of the refolding assays are expressed as molar changes in absorbance, $\Delta\varepsilon_{287}$. The line represents a single exponential curve, according to $\Delta\varepsilon_{287} = 2050(1 - e^{-t/900})$.

the individual rate constants are calculated to be $k_{FS} = 8.9 \times 10^{-4}\ sec^{-1}$ and $k_{SF} = 2.2 \times 10^{-4}\ sec^{-1}$.

Guidelines for the Design of Double-Jump Experiments. Unfolding conditions. Any unfolding conditions can be employed; however, for practical reasons it is of advantage to select conditions (1) where the N $\rightarrow$ U_F unfolding step is rapid and (2) where the $U_F \rightleftharpoons U_S$ equilibration is slow. Under such conditions, $U_F \rightleftharpoons U_S$ can be studied without interference from the preceding unfolding step and manual sampling techniques can be used to perform the slow refolding assays. Because of the high activation enthalpy of proline isomerization, the $U_F \rightleftharpoons U_S$ reaction is strongly decelerated by decreasing the temperature. Consequently, experiments at low temperature are usually of advantage.

Refolding conditions. The slow refolding assays are best carried out under conditions where (1) $U_F \rightarrow N$ refolding is complete within the time of manual mixing, and (2) $U_S \rightarrow N$ occurs in a single phase in a convenient time range. This facilitates the accurate determination of the refolding amplitudes. After each refolding assay the actual protein concentration should be determined to correct for variations in concentration, which may be caused by the various pipetting steps involved in the double-jump procedure.

In cases where more than one U_S species is formed after unfolding, the individual rates of formation of these species can be measured, when the refolding assays are carried out under conditions where the refolding of the various U_S species can be separated kinetically.[16]

Sequential Nature of Unfolding and Isomerization

The double jump technique can be used to probe the sequential nature of the $N \rightarrow U_F$ unfolding step and the subsequent $U_F \rightleftharpoons U_S$ reaction, according to the extended two-state model,

$$N \xrightarrow{k_{NU}} U_F \underset{k_{SF}}{\overset{k_{FS}}{\rightleftharpoons}} U_S \tag{7}$$

If conditions can be found, where both reactions, the $N \rightarrow U_F$ unfolding step and the $U_F \rightleftharpoons U_S$ equilibration are similar in rate, then the sequential model [i.e., first unfolding, then isomerization, Eq. (7)] predicts that formation of U_S should show an initial lag phase. The magnitude of the lag depends on the ratio of the rate constants. Equation (8) describes the kinetics of the sequential formation of U_S.

$$[U_S] = [U_S]_{eq} \left[1 + \frac{1}{k_{NU} - k'} (k' e^{-k_{NU}t} - k_{NU} e^{-k't}) \right] \tag{8}$$

$[U_S]_{eq}$ is the final concentration of U_S present at equilibrium; k' is the apparent rate constant of the $U_F \rightleftharpoons U_S$ reaction, which is equal to the sum of the individual rate constants, $k' = k_{FS} + k_{SF}$.

The $N \rightarrow U_F$ unfolding reaction of ribonuclease A is rapid at acid pH[6] and slow at neutral pH. Unfolding rates increase with the concentration of the denaturant GuHCl (see the table). The kinetics of formation of U_S species at pH 2, 5 M GuHCl, and at pH 6, 4.2–6 M GuHCl are compared in Fig. 3. At low pH, $N \rightarrow U_F$ unfolding is rapid; there is no coupling between folding and isomerization, and the double-jump kinetics can be fitted by a single exponential. At neutral pH, coupling between unfolding and the $U_F \rightleftharpoons U_S$ reaction leads to sigmoidal kinetics, which can be analyzed in terms of Eq. (9).

$$\Delta\varepsilon_{287} = \Delta\varepsilon_{287}^{\infty} \left[1 + \frac{1}{k_{NU} - k'} (k' e^{-k_{NU}t} - k_{NU} e^{-k't}) \right] \tag{9}$$

pH	GuHCl (M)	Direct unfolding[b] $1/k_{NU}$ (sec)	Slow refolding assays[c] $1/k_{NU}$ (sec)	Slow refolding assays[c] $1/k'$ (sec)
6	4.2	400	320	180
6	5.0	140	140	200
6	6.0	85	85	200
2	5.0	Fast	—	200

[a] Comparison of the N $\rightarrow$ U$_F$ unfolding kinetics with the U$_F \rightleftharpoons$ U$_S$ isomerization. At 10°, 0.1 M cacodylate, pH 6 or 0.1 M glycine, pH 2.

[b] Time constant of N $\rightarrow$ U$_F$ unfolding, measured directly by the decrease in absorbance at 287 nm.

[c] The time constants are derived from a fit of the results, shown in Fig. 3 to Eq. (9) according to the sequential model for unfolding [Eq. (7)].

Equation (9) is derived from Eq. (8) by substituting the amplitudes of the slow refolding assays, $\Delta\varepsilon_{287}$, for U$_S$. The values for $1/k_{NU}$ obtained from a fit of the experimental data according to Eq. (9) agree well with the directly determined $1/k_{NU}$ values (see the table) and the values for $1/k'$ are identical to $1/k'$ measured at pH 2, where N $\rightarrow$ U$_F$ unfolding is fast. These results support the sequential model for unfolding [Eq. (7)] of ribonuclease A and suggest that formation of U$_S$ (i.e., proline isomerization) can take place only after the N $\rightarrow$ U$_F$ unfolding step (which is monitored by the change in tyrosine absorbance).

In some conditions unfolding intermediates (I$_U$) may accumulate transiently prior to the formation of U$_F$.[6,7] The comparison of sigmoidal "double-jump" kinetics with unfolding kinetics measured directly (e.g., by spectroscopic techniques) provides a means to determine whether tertiary structural probes, such as tyrosine absorbance, really monitor the formation of U$_F$ or of an earlier unfolding intermediate I$_U$, where the proline residues are not yet free to isomerize.[21]

[21] A. Rehage and F. X. Schmid, *Biochemistry* **21**, 1499 (1982).

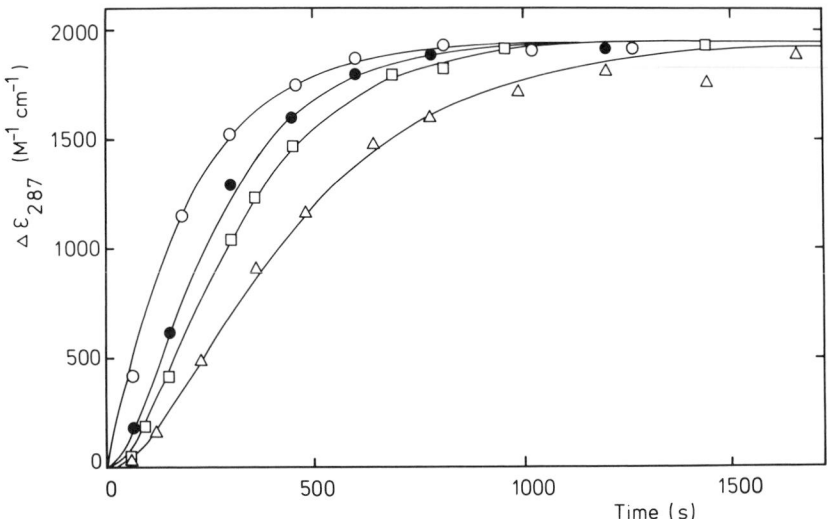

FIG. 3. Sequential formation of slow-folding species U_S after unfolding of ribonuclease A at 10°, 0.1 M cacodylate, pH 6 in the presence of (△) 4.2 M, (□) 5.0 M, (●) 6.0 M GuHCl. The increase of the amplitudes of the slow refolding assays with the time of unfolding is shown. Slow refolding assays were carried out as shown in Fig. 1. Refolding assay conditions: 1.5 M GuHCl and 0.1 M cacodylate, pH 6.0, 25°. The continuous lines represent best fits of the data to a sequential model [Eq. (7)] by using Eq. (9) and the parameters given in the table. The formation of U_S at pH 6 is compared to data, obtained at 5.0 M GuHCl and pH 2, 10° (○). These are conditions where formation of U_S is described by a single exponential curve with a time constant of 200 sec.

Assay for Fast-Folding Molecules

The amount of fast-folding molecules, U_F, both at equilibrium or in the time course of unfolding, can be determined by a kinetic assay.[22] This assay for U_F species makes use of the wide separation in rate between the $U_F \rightarrow N$ and the $U_S \rightarrow N$ refolding reactions. The basic outline of the assay procedure is as follows. Step (1), the protein is equilibrated (or incubated for a defined time) in the desired unfolding conditions. Step (2), a short refolding pulse is applied to refold U_F selectively. Step (3), the amount of native enzyme (N) formed during step (2) is determined by an unfolding assay in standard conditions. The amount of N detected by this procedure is equal to the amount of U_F present initially.

The unfolding assay for U_F was developed for ribonuclease A. The procedure and the results obtained for this enzyme are described here. The assays can be applied to any protein, provided that some require-

[22] F. X. Schmid, *Eur. J. Biochem.* **128,** 77 (1982).

ments are met, which pertain to the folding properties of the protein under investigation.

Requirements for the Unfolding Assays for U_F. The change in conditions from the initial unfolding (step 1) to the refolding pulse (step 2), as well as the refolding pulse itself, have to be short compared to the rate of the $U_F \rightleftharpoons U_S$ equilibration, in order not to shift the U_F/U_S distribution.

Refolding of U_F has to be complete within the refolding pulse (step 2), and refolding of U_S should almost be zero during this pulse.

The amplitudes of the unfolding assays in step (3) can only be correlated quantitatively to the amount of native protein formed in step (2) if the unfolding of the native protein occurs in a single first-order process, which accounts for the entire signal change upon unfolding.

Experimental Details of the Assay for U_F Molecules of Ribonuclease A. The assay for fast-folding U_F molecules consists of three steps.

1. *Equilibration of U_F and U_S.* Ribonuclease A (3 mM) is incubated in the particular unfolding solution (e.g., 6 M GuHCl, pH 6.0) at the desired temperature for sufficient time to allow for complete equilibration of U_F and U_S. The rate of equilibration is strongly dependent on temperature; therefore, 1–2 hr of incubation is required to attain equilibrium at 0–10°.

2. *Folding pulse to refold U_F selectively.* A 20-μl aliquot of the unfolded protein is withdrawn and mixed with 180 μl of 1.3 M GuHCl, pH 6, in a glass tube immersed in an ice/water mixture to assure a rapid drop in temperature to 0°. The protein is kept under the resulting conditions of 1.8 M GuHCl, pH 6, 0° for 20 sec. During this 20 sec pulse U_F refolds entirely to N, whereas U_S remains essentially unfolded, as the $U_S \rightarrow N$ reaction is extremely slow under the given folding conditions ($\tau = 2700$ sec). The rate of the $U_F \rightleftharpoons U_S$ reaction is strongly decreased by lowering the temperature ($\Delta H^{\ddagger} = 20$ kcal/mol); at 0°, it shows a time constant of 900 sec (cf. Fig. 2). Therefore, reequilibration between U_S and U_F during the 20 sec folding pulse is negligible.

3. *Unfolding assay for native molecules, formed in step (2).* After the 20 sec folding pulse (step 2), the sample is diluted tenfold into a final solution of 5 M GuHCl, 0.1 M cacodylate, pH 6, at 10°. The resulting unfolding reaction of N is monitored by the decrease in tyrosine absorbance at 287 nm. Under these conditions unfolding occurs in a single first-order process. The amplitude of the unfolding assay depends on the amount of native protein formed in step (2) and consequently is a quantitative measure for the amount of U_F present initially in step (1).

In reference experiments refolding is not stopped after 20 sec, but allowed to go to completion in step (2). The subsequent unfolding assays [step (3)] yield the total unfolding amplitude, giving a measure for the sum

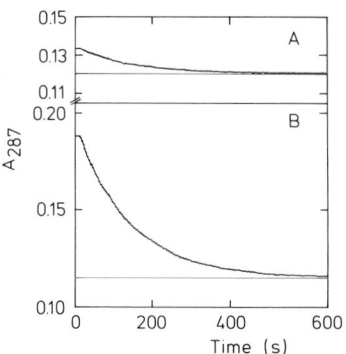

FIG. 4. Determination of the amount of U_F. Representative unfolding assays. Ribonuclease A (3 mM) was unfolded at equilibrium at 0°, 6.0 M GuHCl, 0.1 M cacodylate, pH 6. Sample (A) was exposed to a 20 sec refolding pulse at 1.8 M GuHCl, pH 6, 0.3 mM ribonuclease A, 10°; the changes in absorbance during the subsequent assay at pH 6, 5.0 M GuHCl, 29 μM ribonuclease A, 10°, are shown. Sample (B) was allowed to refold completely (for 30 min at 1.8 M GuHCl, pH 6); the corresponding unfolding assay [conditions as in (A)] yields the total unfolding amplitude. (Reproduced from Schmid.[22])

of fast-folding and slow-folding molecules, $U_F + U_S$, present at the beginning (step 1).

A representative assay for U_F is shown in Fig. 4. The assay consists of two measurements. The first sample was withdrawn after a 20 sec folding pulse ($U_F \rightarrow N$ is complete) and the amount of native protein formed at that time was determined by the unfolding assay shown in Fig. 4A. Analysis of the kinetics yields the time constant for unfolding, $\tau = 143$ sec, and the amplitude, $\Delta\varepsilon_{287} = 510$ M^{-1} cm^{-1}. The remainder of the protein was allowed to refold completely (for 30 min at 20°, under otherwise identical conditions; except that the temperature was raised to accelerate folding). The unfolding assay performed with this entirely refolded sample (Fig. 4B) yields $\tau = 141$ sec and $\Delta\varepsilon_{287} = 2850$ M^{-1} cm^{-1}, which reflects the total amount of unfolded ribonuclease A present initially ($U_F + U_S$). The ratio of the amplitudes of both unfolding assays, 0.18, is equal to the fraction of U_F molecules present in the initial unfolding conditions.

Kinetic Changes of U_F during Unfolding. The assay for U_F is useful not only to measure the amount of U_F in samples unfolded at equilibrium, but also to follow the kinetics of the decrease of U_F during unfolding, which is described by Eq. (7). After the fast unfolding reaction $N \rightarrow U_F$, the slow $U_F \rightleftharpoons U_S$ equilibration reaction proceeds until an equilibrium mixture of $U_F : U_S$ 20 : 80 is reached.[1] The increase in U_S during this reaction can be followed by slow refolding assays ("double-jumps" see above). The assay for U_F molecules allows the detection of the concomi-

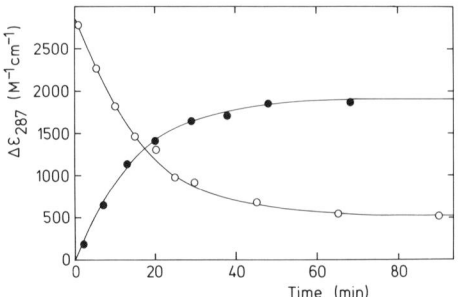

FIG. 5. Decrease of U_F after fast unfolding of ribonuclease A. The amounts of U_F still present after varying times of unfolding at 3.0 M GuHCl, 0.2 M glycine, pH 1.9, 0° were determined by unfolding assays (○) as described in the text. In a control experiment the formation of slow-folding species U_S was determined under identical conditions (●) by slow refolding assays as in Fig. 2. The sum of both amplitudes decreases slightly with time, as the total change in absorbance upon folding is smaller at 1.5 M GuHCl (slow refolding assays for U_S) than at 5.0 M GuHCl (assays for U_F). (Reproduced from Schmid.[22])

tant decrease of the amount of U_F during the $U_F \rightleftharpoons U_S$ equilibration. Figure 5 presents a comparison of the increase of U_S (measured by the slow refolding assays) with the decrease of U_F (measured by the assay for U_F) after the fast $N \rightarrow U_F$ unfolding step at 0°, pH 1.9, 3 M GuHCl. The amplitudes of the assays for U_F drop from $\Delta\varepsilon_{287} = 2900$ M^{-1} cm^{-1} at $t = 0$ to $\Delta\varepsilon_{287} = 520$ M^{-1} cm^{-1} at $t = \infty$ (i.e., from 100 to 18%). The rate constants of the decrease of U_F and of the increase of U_S are identical as expected from the model [Eq. (1)]. This shows that only U_F and U_S species participate in the slow interconversion reaction after the fast $N \rightarrow U_F$ unfolding step.

The assay for fast-folding molecules can be employed to determine the $U_F/(U_F + U_S)$ ratio under a wide variety of unfolding conditions (with regard to temperature, pH, and denaturant concentration). The relative amounts of fast-folding and slow-folding species can also be determined from an amplitude analysis of stopped-flow refolding kinetics. However, the range of unfolding conditions (e.g., temperature) accessible to stopped-flow experiments is usually restricted by mixing problems or temperature gradients. On the other hand, the assay for U_F can only be employed if conditions can be found, where fast and slow refolding differ widely in rate (for step 2) and conditions where unfolding occurs in a single slow reaction (for step 3).

Other Methods to Detect Fast- and Slow-Folding Forms. Fast-folding and slow-folding forms of unfolded proteins can also be detected and characterized by urea gradient gel electrophoresis (this volume [9]) and by isomer-specific proteolysis (this volume [6]).

[5] Disulfide Bonds as Probes of Protein Folding Pathways

By Thomas E. Creighton

Partially folded conformations of the type that would define the folding pathway of a small single-domain protein are inherently unstable thermodynamically, relative to the fully folded or unfolded states.[1] Consequently they are not populated significantly at equilibrium. They might accumulate transiently as kinetic intermediates, but only if they precede the rate-limiting step on the pathway and if, under the particular conditions, their free energy is lower than or very close to that of the starting protein. All other intermediates will not accumulate substantially, even kinetically. Those that do are likely to be very transient and elusive; moreover, their kinetic roles can be very difficult to ascertain. Due to their inherent instability, they are likely to be detectable at only extremes of conditions (e.g., temperature, pH, denaturant concentration) favoring either the folded or unfolded states, and different intermediates will accumulate under those two extremes of conditions. A rigorous kinetic analysis requires that the values of individual rate constants be known under all folding conditions, but, since each can be measured only within a limited range, they must be extrapolated to other conditions; how they vary with the folding conditions is generally not known. Consequently, the kinetic roles of any intermediates detected can be very uncertain.

These problems could be alleviated if it were possible to manipulate the rates of forming different interactions within the protein, to trap the intermediate conformations in a stable form, and to determine which interactions were present. It is easy to imagine how a protein folding pathway could be determined if intramolecular hydrogen bond formation, for example, could be manipulated in this way. If the rate of forming each hydrogen bond could be made to be rate-limiting and constant, molecules with 1, 2, 3, 4, ... hydrogen bonds would accumulate transiently to substantial levels at various times during refolding. If protein hydrogen bond formation, breakage, and rearrangement could be quenched rapidly at these times, the intermediates would be trapped. They would then only need to be identified and their kinetic roles determined, which would be possible if the rates of hydrogen bond formation and breakage could be manipulated.

No such method is available for hydrogen bonds, but this approach can be used with another type of interaction present in some proteins, that

[1] P. L. Privalov, *Adv. Protein Chem.* **33**, 167 (1979).

of disulfide bond formation between Cys residues.[2] The redox nature of this interaction means that its formation or breakage requires an appropriate electron donor or acceptor in the environment, which is under experimental control. Consequently, the rates of protein disulfide bond formation and breakage may be manipulated, all disulfides present at any time may be trapped in a stable form, and the disulfides may be identified chemically. The presence of a disulfide bond between two Cys residues requires that they be within a few angstroms of each other, so disulfides serve as probes of conformational transitions in proteins.

The approach is ideally suited to proteins that rely on the presence of disulfides for stability of the folded state, so that the reduced protein is unfolded, even in the absence of any denaturant. In this case folding accompanies disulfide bond formation, no denaturants are required, the folding conditions can be specified only by the redox potential of the environment, and how these conditions affect the rates of disulfide bond formation and breakage are known. Therefore, the kinetics are known throughout the folding transition region.

Proteins that adopt folded conformations in the absence of disulfides are not suitable for these studies, although much information can be gained in this way about the stabilities of the folded state with and without the disulfides.[3]

Disulfide Bond Formation, Breakage, and Rearrangement

The basic approach is to place a protein, either folded or unfolded, with or without disulfides, under various conditions and to follow as a function of time disulfide bond formation, breakage, and rearrangement. The prime requirement is that the pH be neutral to alkaline, because virtually all reactions involving thiols are dependent upon the ionized thiolate anion. A further consequence is the desirability of a moderate ionic strength to minimize electrostatic effects on the ionization of different thiols. Otherwise, a wide variety of conditions may be used.[4] The standard conditions adopted in this laboratory are a buffer containing 0.1 M Tris–HCl, 0.2 M KCl, and 1 mM EDTA at 25° and pH 8.7, where most thiols are substantially ionized. The actual methods used will depend on the protein and its properties in both the folded and reduced forms, particularly its solubility and net charge. Therefore, detailed methods will not

[2] T. E. Creighton, *Prog. Biophys. Mol. Biol.* **33**, 231 (1978).
[3] Y. Goto and K. Hamaguchi, *J. Mol. Biol.* **146**, 321 (1981).
[4] T. E. Creighton, this series, Vol. 107, p. 305.

be given here, but the basis of the approach and the interpretation of the results will be described in detail.

A large number of different ways of forming and breaking protein disulfides are known, but those found most suitable for studies of protein folding use thiol–disulfide exchange between the protein, P, and a reagent, R, in either the thiol or disulfide form.[2,4,5]

$$
P^{SH}_{SH} \xrightleftharpoons[R_{SSR}]{R_{SH}} P^{SSR}_{SH} \xrightleftharpoons{R_{SH}} P^{S}_{S}
\tag{1}
$$

Reduced protein   Mixed disulfide   Protein disulfide

All of the reactions described here actually involve only the thiolate anion. Therefore, the reaction rates are pH dependent, as are the equilibria if the thiols of the reactants and products have different pK_a values. These factors can be considered by writing the equations as involving only the thiolate anions, but having them in equilibrium with the protonated thiols; an example is illustrated in Fig. 1. It would be impractical to do this for all the reactions considered here, so this complication will be ignored; all rate constants will be those observed considering the total thiol concentrations, and protonated thiols will be indicated in the equations.

The intrinsic thiol–disulfide exchange reaction is rapid, reversible, extremely specific, free of side-reactions, and predictable in rate.[6–8] With knowledge of the intrinsic rate, the observed rate within a protein may be interpreted in terms of the effect of the protein conformation (see below).

Air oxidation of thiol groups is to be avoided, because it is difficult to control, catalyzed by trace metal ions, complicated in mechanism, impossible to interpret in terms of protein conformational transitions, and accompanied by the formation of hydrogen peroxide and radicals, such as superoxide anion, that can react with other groups on the protein.[9,10]

Disulfide rearrangements can occur solely intramolecularly in proteins with both thiols and disulfides, so their rates should be independent of the redox conditions.

[5] V. P. Saxena and D. B. Wetlaufer, *Biochemistry* **9**, 5015 (1970).

[6] T. E. Creighton, *J. Mol. Biol.* **96**, 767 (1975).

[7] Z. Shaked, R. P. Szajewski, and G. M. Whitesides, *Biochemistry* **19**, 4156 (1980).

[8] R. P. Szajewski and G. M. Whitesides, *J. Am. Chem. Soc.* **102**, 2011 (1980).

[9] H. P. Misra, *J. Biol. Chem.* **249**, 2151 (1974).

[10] M. Costa, L. Pecci, B. Pensa, and C. Cannella, *Biochem. Biophys. Res. Commun.* **78**, 596 (1977).

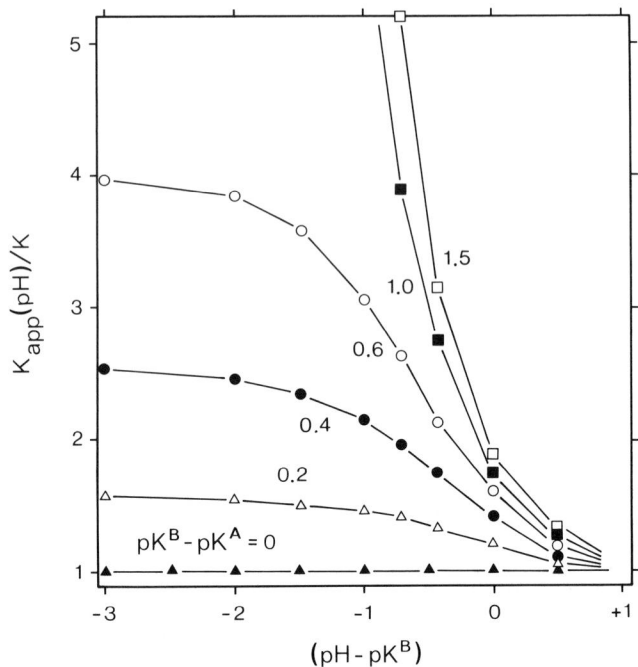

FIG. 1. Expected pH dependence of the thiol–disulfide exchange equilibrium between two different thiols with different pK_a values. Thiol–disulfide exchange occurs only with the ionized thiol, so the reaction can be written as follows:

$$As^- + BssB \xrightleftharpoons{K} AssB + Bs^-$$

$$K^A \uparrow\!\!\downarrow H^+ \qquad\qquad H^+ \uparrow\!\!\downarrow K^B$$

$$A_{SH} \qquad\qquad\qquad B_{SH}$$

K^A and K^B are the ionization constants of the two thiols, the negative logarithms of which give the pK_a values, pK^A and pK^B, respectively. K is the pH-independent equilibrium constant involving the ionized thiols. The apparent equilibrium constant, K_{app} (pH), measured by the total thiol concentrations, will vary at pH values near or below the pK_a values if they are different for the two:

$$K_{app} (pH) = K \left[\left(1 + \frac{[H^+]}{K^B} \right) \Big/ \left(1 + \frac{[H^+]}{K^A} \right) \right]$$

The limiting increase in K_{app} (pH) at very low pH will be K^A/K^B.

$$(2)$$

$$(3)$$

In proteins with no free thiols, disulfide rearrangements will not occur unless catalyzed by thiols or other nucleophiles of the solution capable of transiently breaking at least one of the disulfides.

Trapping Disulfides

Whatever disulfides are present at any instant of time may be trapped in a stable form by simply blocking, rapidly and irreversibly, all thiol groups in the solution.[2,4] This is accomplished readily and efficiently by adding concentrated solutions of iodoacetate or iodoacetamide to a final concentration of 0.1 to 0.2 M. Normal accessible protein thiol groups react at pH 8.7 and 25°C with rate constants of approximately 10 sec^{-1} M^{-1}, so the half-time for reaction should be 0.3 sec with 0.2 M iodoacetate.[11]

The quenching reaction should be as rapid as possible to ensure that it faithfully traps the species present at the time of addition of the quenching reagent.[4] Adding a slight excess of reagent[12,13] is unlikely to be satisfactory. Adding simultaneously a denaturant such as urea[12,13] is almost certain to generate spurious species by producing unfolding and disulfide rearrangements prior to trapping. Acidification[14] is more rapid, but is not irreversible and only slows thiol–disulfide exchange by lowering the concentration of thiolate anion. Intramolecular rearrangements can occur on the minute time scale even at pH 2.[4]

Iodoacetate, iodoacetamide, and virtually every other thiol-blocking reagent have the disadvantage of reacting slowing with protein groups

[11] T. E. Creighton, *J. Mol. Biol.* **96**, 777 (1975).
[12] A. S. Acharya and H. Taniuchi, *J. Biol. Chem.* **251**, 6934 (1976).
[13] W. L. Anderson and D. B. Wetlaufer, *J. Biol. Chem.* **251**, 3147 (1976).
[14] Y. Konishi, T. Ooi, and H. A. Scheraga, *Biochemistry* **21**, 4734 (1982).

other than just thiols. However, the proteins need be exposed to them for only a minute or so and they may then be removed by gel filtration. With trapped aliquots to be analyzed by electrophoresis without delay, placing the trapped mixtures on ice after 2 min is generally satisfactory. Reaction with amino groups at high pH can be minimized by dropping the pH to about 8 on adding the iodoacetate.

The major difficulty with this trapping procedure is that buried thiol groups do not react at a sufficient rate.[3] For this reason, one two-disulfide species of bovine pancreatic trypsin inhibitor (BPTI) was overlooked initially, because it adopted a stable native-like conformation with two thiols buried and unreactive.[15] They could be blocked by adding denaturants to the trapping solution, but considerable rearrangements of the two disulfides occurred before the trapping reaction was complete. This is an inherent problem with the disulfide approach to protein folding.

Nevertheless, once all free thiol groups are blocked irreversibly, the trapped species are stable indefinitely, in the absence of any reagents able to catalyze disulfide interchange. Consequently, they may be separated and characterized in detail.

Analyzing the Trapped Species

During the time course of any complex reaction, with multiple steps of similar rates, a large number of different species will be present simultaneously. Therefore, the mixtures trapped at various times need to be resolved into the individual species. This may be accomplished with disulfide-trapped species using the normal electrophoretic and chromatographic techniques of protein purification, especially taking advantage of the heterogeneity introduced by the blocking groups on the Cys thiols. Electrophoretic separations are sensitive to both net charge and protein conformation[16]; for example, reduced BPTI trapped by acid or by neutral iodoacetamide has essentially the same net charge as native BPTI, but only two-thirds of the electrophoretic mobility, due to being unfolded. When trapped with the acidic reagent iodoacetate, its mobility is further decreased by half due to the six acidic carboxymethyl (CM) groups. A comparison of iodoacetate- and iodoacetamide-trapped samples, otherwise identical, can determine how many disulfides each species contains. The difference in mobilities is assumed to be due to only the difference in charge introduced by the appropriate number of acidic CM groups on the free Cys residues, in proportion to the effect of six such groups on the

[15] D. J. States, C. M. Dobson, M. Karplus, and T. E. Creighton, *J. Mol. Biol.* **174,** 411 (1984).

[16] D. P. Goldenberg and T. E. Creighton, *Anal. Biochem.* **138,** 1 (1984).

fully reduced protein. Several proteins have been found to exhibit this behavior,[2,17,18] although it should not be expected that all will, since the blocking groups could also change the conformation. Electrophoretic analysis is ideally suited to kinetic studies, where the relative levels of all the species need to be determined in a large number of samples trapped at different times under a variety of folding conditions.

Chromatographic analysis can produce greater resolution of the trapped species,[2,4] plus preparative quantities for their characterization. Ion-exchange chromatography can utilize the charge differences produced by acidic CM groups on the free Cys residues, and a single separation was able to resolve most of the trapped BPTI species in essentially pure form.[19,20] The same procedure was nearly as effective with two related proteins,[18] but too many intermediates accumulated during refolding of reduced ribonuclease A to be resolved into species that could be characterized.[21] Most other types of chromatography should also be useful.

Identifying the Trapped Species

Which pairs of Cys residues are involved in disulfide bonds, and the chemical nature of the other Cys residues, whether reacted with the trapping reagent or in a mixed disulfide with the disulfide reagent, can be determined in a number of different ways. Radioactive disulfide reagents and quenching reagents can be very useful. However, diagonal electrophoresis[22] at pH 3.5 gives all this information on all the Cys residues of the protein and is currently the method of choice. Its efficacy has been demonstrated with BPTI[19,20] and several modifications,[23] with two homologous proteins,[18] and with ribonuclease.[24] It has been described in detail in another volume of this series.[4]

Reconstructing the Folding Pathway

Determining the kinetic roles of intermediates on a complex pathway can be difficult, but it is relatively straightforward with the disulfide interaction due to the ability to manipulate experimentally the rates of disulfide bond formation and breakage. The pathway deduced for BPTI is shown in Fig. 2.

[17] T. E. Creighton, *J. Mol. Biol.* **113,** 329 (1977).
[18] M. Hollecker and T. E. Creighton, *J. Mol. Biol.* **168,** 409 (1983).
[19] T. E. Creighton, *J. Mol. Biol.* **87,** 603 (1974).
[20] T. E. Creighton, *J. Mol. Biol.* **95,** 167 (1975).
[21] T. E. Creighton, *J. Mol. Biol.* **129,** 411 (1979).
[22] J. R. Brown and B. S. Hartley, *Biochem. J.* **101,** 214 (1966).
[23] T. E. Creighton and D. F. Dyckes, *J. Mol. Biol.* **146,** 375 (1981).
[24] T. E. Creighton, *FEBS Lett.* **118,** 283 (1980).

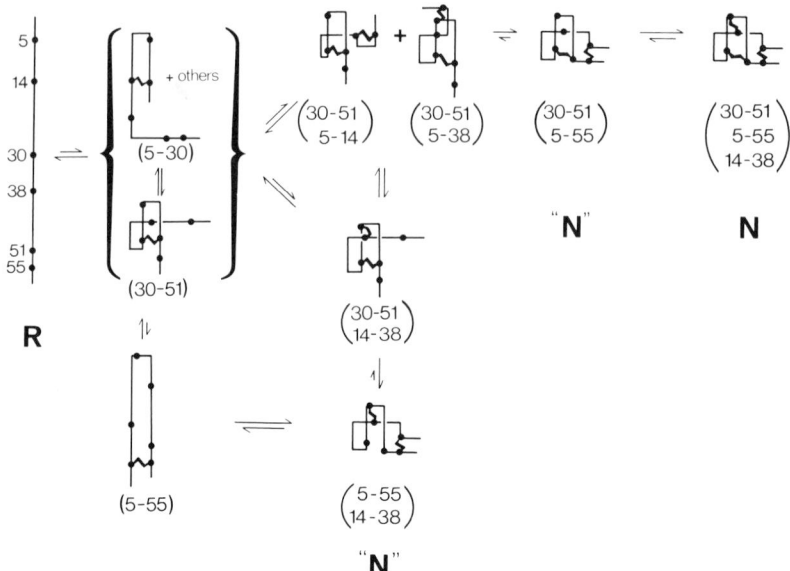

FIG. 2. The pathway for unfolding and refolding of BPTI accompanying disulfide bond breakage and formation, respectively. The polypeptide backbone is designated as a solid line, with the positions of the six Cys residues indicated. R is the fully reduced protein, N the fully folded protein with three disulfides. The numbers of the Cys residues paired in disulfides are indicated under each; intermediates that adopt native-like conformations are designated "N." The brackets enclose the major one-disulfide intermediates that are in equilibrium by rapid intramolecular disulfide interchange. Intermediate (5–55) is not included in the brackets because it is not a major species and because it is not equilibrated so rapidly. The evidence for this role for this intermediate is somewhat circumstantial, and the actual species may be (5–51); this disulfide would have to be rapidly interchanged to 5–55 in forming intermediates (5–55, 14–38). The "+" sign between (30–51, 5–14) and (30–51, 5–38) signifies that they both play the indicated kinetic role. Adapted from Ref. 31.

Steps involving disulfide bond formation generally occur with a rate proportional to the concentration of disulfide reagent, whereas the rates of steps involving disulfide breakage are usually proportional to the concentration of reagent in the thiol form; intramolecular steps, including disulfide rearrangements, should be independent of both. By varying the concentrations of both disulfide and thiol forms of the reagent during both unfolding and refolding, it is usually clear whether the rate-limiting steps in formation and disappearance of a species involve protein disulfide formation or breakage, or intramolecular steps. Where there are different alternative paths to a species, the dependence may not be clear cut, but with multiple, parallel paths, there are thermodynamic restrictions on the possible values of rate constants due to the necessity that there be no net free energy change around a cyclic pathway.

The kinetic behavior of an intermediate on a complex pathway can be affected by the rates of steps preceding its formation or after its disappearance, so it is necessary to demonstrate that the kinetic behavior of *all* the species under *all* the folding conditions, during *both* unfolding and refolding, can be adequately simulated simultaneously using a single set of rate constants for all of the steps. Each rate constant for disulfide bond formation or breakage should include the dependence of the rate on the concentration of the disulfide or thiol forms of the reagent. The values of these multiple-order rate constants may differ systematically with different reagents, but the values of all intramolecular steps should be independent of the reagent. By systematically varying the concentrations of thiol and disulfide forms of a reagent, values for the rate constants for all of the steps should be obtained by fitting the simulated kinetics to the observed behavior of all the species. We routinely carry out the kinetic simulations using simple numerical integration of the rate equations for the proposed pathway; more general computer programs are available.[25]

Knowing the disulfide bonds in the various species helps to infer a plausible pathway, for the pathway should provide for a step-wise process of disulfide bond formation, breakage, and rearrangement. However, the presence of two species differing in only a single disulfide bond should not be taken as indicating that they are interconverted directly. Also, crucial intermediates need not accumulate to detectable levels if they are interconverted to other species by rapid rearrangements.

To a first approximation, at least, all molecules with the same disulfide bonds are assumed to be equivalent and to represent a homogeneous species. However, this should not be assumed to be the case, for most species with incomplete disulfide bonds are likely to have flexible conformations, so molecules with the same disulfide bonds are likely to have very different conformations at any instant of time. They will still behave kinetically as a homogeneous species if the conformational interconversions are rapid on the time scale of disulfide formation; this is often the case, but there are exceptions. For example, cis–trans isomerization of peptide bonds preceding Pro residues is both intrinsically slow[26] and a significant factor in protein folding, producing conformational heterogeneity in unfolded proteins.[27] Such heterogeneity is apparent in the refolding of BPTI if the rate of disulfide formation is faster than the rate of cis–trans isomerization, when about 20% of the molecules form incorrect

[25] B. A. Barshop, R. F. Wrenn, and C. Frieden, *Anal. Biochem.* **130,** 134 (1983).
[26] J. F. Brandts, H. R. Halvorson, and M. Brennan, *Biochemistry* **14,** 4953 (1975).
[27] L.-N. Lin and J. F. Brandts, *Biochemistry* **22,** 559 (1983).

disulfide bonds[28]; with slower rates of disulfide formation, such heterogeneity is not as apparent. Also, there is always the possibility of covalent heterogeneity of the protein, other than that involving Cys residues. Studies involving disulfide bond formation and breakage in ribonuclease have been hampered by an unknown, but apparently covalent, modification of the protein.[28] In spite of these documented examples of conformational and covalent heterogeneity, plus many more possibilities, the kinetics of refolding of reduced ribonuclease have been erroneously analyzed with the assumption that all molecules with the same *number* of disulfides, irrespective even of their identities, have identical kinetic properties.[29] The not surprising observation that the kinetics of appearance of enzymatically active ribonuclease could not be fit with a single rate constant from the three-disulfide intermediates led Konishi *et al.*[29] to conclude that rate-limiting steps must also involve one- and two-disulfide intermediates, even though this has been shown directly not to be the case.[17]

The pathway of disulfide formation and breakage can be further dissected using molecules in which one or more of the Cys residues are blocked irreversibly, so as to prevent their involvement in disulfide bonds. For example, the rates of disulfide formation and breakage were dramatically decreased in BPTI with both Cys-14 and Cys-38 blocked, illustrating the importance of these two Cys residues in the disulfide rearrangements that are energetically the most favorable means of unfolding and refolding.[30] Molecules with either Cys residue blocked were observed to refold at only slightly diminished rates, indicating that both pathways of rearrangement, using either Cys-14 or Cys-38, were nearly equally utilized. On the other hand, molecules with Cys-5 and Cys-55 blocked were able to form both the 30–51 and 14–38 disulfides at nearly normal rates, indicating that their formation did not require Cys-5 and Cys-55 (Fig. 2).

The trapped intermediates are useful sources of such altered proteins. The kinetics of breakage and of re-formation of their disulfide bonds can be determined in isolation with the purified trapped species and compared with the kinetics of the unmodified protein. Some caution must be exercised in interpreting these results, as the blocking groups on the Cys residues could affect the folding pathway. A specific procedure for converting Cys residues to Ser would be very useful for these studies, but is not currently available.

In some cases, intermediates may be isolated without modifying their free Cys residues. The BPTI intermediate (5–55, 14–38) acquires a native-

[28] T. E. Creighton, *J. Mol. Biol.* **113,** 295 (1977).
[29] Y. Konishi, T. Ooi, and H. A. Scheraga, *Biochemistry* **21,** 4734 (1982).
[30] T. E. Creighton, *J. Mol. Biol.* **113,** 275 (1977).

like conformation[15] and buries the Cys-30 and Cys-51 thiols, making them unreactive.[31] Consequently, it is isolated unmodified along with the fully refolded BPTI, so its kinetic properties could be determined directly, with refolded BPTI as an useful internal control.[31]

Once a pathway consistent with all the experimental observations has been found, it should be remembered that kinetic analysis cannot prove the validity of any mechanism, but merely disprove others.

Interpretation of Observed Rates

The observed rate of forming or breaking a protein disulfide bond is determined both by the protein folding transitions that occur and by the chemical reactions that are involved. Disulfide bonds are useful for protein folding only if they reflect the folding transitions, not the chemistry involved. Accordingly, it is important to understand the chemical reactions so as to be able to infer the folding transitions. The following interpretation is based both upon first principles and upon chemical studies with small molecules and has been amply verified in studies with BPTI, several homologous and modified proteins, and ribonuclease. The rate constants that apply to the BPTI pathway of Fig. 2 are given in the table.

Two fundamental types of thiol and disulfide reagents are useful. In one the disulfide bond links two molecules that are separated in the thiol form and is considered here to be intermolecular, such as in glutathione disulfide (GSSG) and oxidized mercaptoethanol. In the other, the disulfide is intramolecular between two thiols on the same molecule, such as oxidized dithiothreitol (DTT_S^S). The process of forming or breaking a protein disulfide is exactly analogous with both types of reagents, but there are important differences that result in them providing complementary kinetic information. The following analysis assumes that they differ only in being intra- or intermolecular; this assumption is verified by the results obtained with BPTI.

It will be assumed that there are no specific interactions between the protein and the thiol/disulfide reagent, other than that involving the sulfur atoms. Exceptions should be apparent from unexpected values of the thiol–disulfide exchange reactions and from inconsistencies using different reagents. The kinetics and intermediates of the BPTI pathway have been shown to be the same with glutathione and mercaptoethanol as reagents, and to be remarkably consistent with the kinetics determined with dithiothreitol. Therefore, no specific interactions occur in this instance, but this should not be assumed in others.

[31] T. E. Creighton and D. P. Goldenberg, *J. Mol. Biol.* **179**, 497 (1984).

RATE CONSTANTS FOR FORMING AND BREAKING THE DISULFIDE BONDS IN THE FOLDING TRANSITION OF BPTI

| | Rate constants | | | | | |
| | Glutathione | | | Dithiothreitol | | |
Step[a]	Formation (sec⁻¹ M⁻¹)	Breakage (sec⁻¹ M⁻²)	Intramolecular[b] (sec⁻¹)	Formation (sec⁻¹ M⁻¹)	Breakage (sec⁻¹ M⁻¹)	Intramolecular[c] (sec⁻¹)
R ⇌ I	75	60	3.3	0.022	20	8.1
I ⇌ II	90	600	4.2	0.03	250	9.0
I ⇌ (5–55, 14–38)	3	0.6	200	0.048	12	14.4
(30–51, 5–55) ⇌ N	28	0.15	2.0×10^3	5.7	30	1.7×10^3

[a] R is fully reduced BPTI, N is fully folded with three disulfides; I is all the one-disulfide intermediates; II is intermediates (30–51, 14–38), (30–51, 5–14), and (30–51, 5–38). Taken from Ref. 31.

[b] Calculated from the observed rate constant for breakage, using Eq. 7 of the text, $k_{ex} = 20$ sec⁻¹ M^{-1}, and the rate constant for reduction by dithiothreitol as the value for k_r.

[c] Calculated from the observed rate constant for disulfide formation, using Eq. 17 of the text and 600 M^{-1} as the value of K_{DTT}.

Intermolecular Disulfide Reagents

Protein disulfides are formed and broken in two sequential thiol–disulfide exchanges with a reagent. With an intermolecular reagent RSSR, the two steps may be represented in the ideal case as follows:

$$\text{(protein)} \overset{SH}{\underset{HS}{}} + \text{RSSR} \underset{\frac{1}{2}k_{ex}}{\overset{2k_{ex}}{\rightleftharpoons}} \overset{SSR}{\underset{HS}{}} + \text{RSH} \qquad (4)$$

$$\overset{SSR}{\underset{SH}{}} \underset{k_r}{\overset{k_{intra}}{\rightleftharpoons}} \overset{S}{\underset{S}{}} + \text{RSH} \qquad (5)$$

Three of the four steps are simply bimolecular thiol–disulfide exchange between the protein and the reagent, and their rates depend upon the concentration of either the thiol or disulfide forms of the reagent. This reaction occurs with a characteristic rate constant, k_{ex}: its value depends somewhat upon the electrostatic properties of both the thiol and disulfide molecules,[6–8] but with typical Cys residues of proteins and of glutathione, the second-order rate constant has the value 20 $\sec^{-1} M^{-1}$ at pH 8.7 and 25°. It can be measured readily by the reduction of appropriate model disulfides by dithiothreitol (DTT_{SH}^{SH}), following spectrophotometrically the formation of its disulfide form (DTT_S^S).[6,32] The factors of 2 and $\frac{1}{2}$ are statistical factors that reflect the number of thiol groups and disulfide sulfur atoms.

The value of k_r should be the same as k_{ex} for a normal disulfide, but may be different if strained or inaccessible. This rate constant can be measured directly from the rate of reduction by dithiothreitol (see below).

The intramolecular step in which a second Cys thiol displaces the mixed-disulfide is the primary factor that distinguishes one protein disulfide from another. Its rate constant, k_{intra}, is the one that is most pertinent for protein folding, for it is the only one that involves the protein conformational transitions that bring two Cys residues into proximity for forming a disulfide bond.

The kinetics of disulfide bond formation and breakage depend upon the relative rates of the intra- and intermolecular steps. If k_{intra} is very fast, as is often the case in the absence of conformational restrictions, the rate of forming a protein disulfide bond with a reagent RSSR is determined by the initial bimolecular reaction between a protein thiol and the disulfide reagent; the mixed disulfide does not accumulate significantly. The rate therefore reflects only the thiol–disulfide exchange reaction and the num-

[32] K. S. Iyer and W. A. Klee, *J. Biol. Chem.* **248**, 707 (1973).

ber of Cys thiols, n, and is similar for forming each disulfide in a protein. This rate is not of interest for folding, but its constant value has the advantage of ensuring significant accumulation of intermediates with all possible numbers of disulfide bonds. The rate of forming a specific disulfide bond will be lower than expected from the total population of precursors if only a fraction of the precursors are able to form that disulfide. This can be useful for determining that only one species of, say, one-disulfide intermediates is the precursor for a specific two-disulfide intermediate.[31]

In this case, the varying stabilities of different disulfides are reflected only in their rates of breakage by the thiol reagent. Initial breakage of the protein disulfide by RSH to generate the transient mixed disulfide is rapidly reversed, due to the large value of k_{intra}. The complete breakage of the disulfide requires that the competing reaction of the mixed disulfide with a second molecule of RSH be sufficiently fast. Consequently, the observed rate depends upon the square of the thiol reagent concentration

$$\text{Rate} = k_{obs} \, [\text{GSH}]^2 \, [\text{protein disulfide}] \qquad (6)$$

The observed rate constant can be used to calculate the apparent value of k_{intra}:

$$k_{intra}^{app} = \frac{k_{ex} k_r}{2 k_{obs}} \qquad (7)$$

using the standard value for k_{ex} and the value of k_r measured by the rate of reduction by dithiothreitol (see below). The more stable the disulfide, the slower the rate of breakage by monothiols.

Where a number of different disulfide species are in rapid equilibrium, relative to the rate of disulfide reduction, either by intramolecular interchange of protein disulfides [Eq. (3)] or mixed disulfides [Eq. (2)], only a single rate constant will be measurable. The apparent value of k_{intra} determined with Eq. (7) will be the sum of the individual values divided by $n/2$, where n is the number of Cys residues that participate as the mixed disulfides.

The other extreme, where the value of k_{intra} is very small, occurs only if there are conformational restrictions preventing Cys residues from forming disulfide bonds, or if the Cys residues are very far apart on the polypeptide chain. In this case, the mixed disulfide is formed at the expected rate, k_{ex} times the number of Cys thiols, but accumulates. If the mixed disulfide accumulates to detectable levels, the value of k_{intra} may often be estimated directly from its rate of disappearance. Accumulation of the mixed disulfide is positive evidence that nonformation of a disulfide is due to conformational restrictions, not simply unreactivity of the thiols.[33]

If other free Cys residues also cannot form disulfides readily, they also will form mixed disulfides; this protein species may be trapped with stable mixed disulfides on all free Cys residues if no thiol reagent is present to reverse the process.

$$\text{(structure, SSR, SH)} + \text{RSSR} \rightleftharpoons \text{(structure, SSR, RSS)} + \text{RSH} \qquad (8)$$

The reduction by RSH of a protein disulfide with a small value of k_{intra} occurs sequentially in two steps, each with a rate proportional to the monothiol concentration.

When interpreting the kinetics of a complex process like disulfide formation in a protein with several Cys residues, it is important to ensure that any mixed disulfide species detected is actually the precursor of a protein disulfide bond. For example, Konishi et al.[29] attempted to measure the levels of accumulation of the various ribonuclease species with different numbers of protein disulfides and mixed disulfides, without distinguishing between which Cys residues were involved. They simulated the kinetic behavior assuming a single step interconverting each of the presumed homogeneous species with the same numbers of mixed and protein disulfides and analyzed the energetics of the process.[34] However, virtually all of their presumed single species are undoubtedly heterogeneous, with different Cys residues involved in the disulfides.[21] Some disulfides are formed with large values of k_{intra}, so these species will accumulate in the disulfide form. Others, with low values of k_{intra}, will be present as the mixed disulfide.[21] The value of any rate or equilibrium constant derived on the assumption that the latter are precursors of the former will be meaningless.

With BPTI, the identities of all the protein and mixed disulfides that accumulate have been determined. The six thiols of reduced BPTI are equivalent and as reactive as model thiols,[11] so presumably a reagent like RSSR reacts equally likely with all of them. The observed rate is nearly that expected (see the table). Yet no mixed disulfide accumulates, so each Cys residue must be able to make readily a disulfide with at least one other Cys residue of the reduced protein, with a value of k_{intra}^{app} of at least 0.03 sec^{-1}; the rate of reduction by GSH indicates $k_{intra}^{app} = 3.3$ sec^{-1}. That

[33] T. E. Creighton, J. Mol. Biol. 151, 211 (1981).
[34] Y. Konishi, T. Ooi, and H. A. Scheraga, Biochemistry 21, 4741 (1982).

all Cys residues participate undoubtedly reflects the disordered nature of the fully reduced protein, with no restrictions on which Cys residues can come into proximity. Within the predominant one-disulfide intermediate, (30–51), only Cys-55 was found to accumulate as a mixed disulfide. This indicated that it cannot form readily a disulfide with any other free Cys residue, which was confirmed using modified proteins with Cys residues blocked. In contrast, the three other free Cys residues, 5, 14, and 38, must be able to form disulfides readily; the resulting disulfides 5–14, 5–38, and 14–38 were found in the major two-disulfide intermediates, and they are formed collectively with a value of k_{intra}^{app} of 4.2 sec^{-1}; since they are in rapid equilibrium and three Cys residues are involved, the sum of the individual values of k_{intra} should be 6.3 sec^{-1}.

A fourth two-disulfide intermediate (5–55, 14–38) is formed at only 3% of the rate of the others, suggesting that it is formed from only a minor one-disulfide intermediate.[31] It could be (5–55), (14–38), or (5–51), which has been detected at those levels; if the last, the 5–51 disulfide would have to be rapidly interchanged. If only two Cys thiols are involved in making the second disulfide, the precursor could be present as 5% of the one-disulfide intermediates and k_{intra} would be 200 sec^{-1}. If all four Cys thiols participate, the respective numbers would be 2.5% and 400 sec^{-1}.

The three two-disulfide intermediates (30–51, 14–38), (30–51, 5–14), and (30–51, 5–38) tend to accumulate with mixed disulfides on their two free Cys residues. With very high concentrations of RSSR reagents, both Cys residues will react rapidly and the double mixed disulfide species are effectively trapped. The value of k_{intra} for forming a third disulfide bond in each of these species is negligible. Their inability to form third disulfides is not due to inaccessibility of the Cys thiol groups, but due to conformational restrictions on the transitions required to form the disulfides.[33]

These species rearrange their second disulfide bonds in an intramolecular step, independent of the reagent, to the native-like (30–51, 5–55). Its two Cys residues are on the surface of the protein and in reasonable proximity for forming a second disulfide; no conformational changes in the protein are required, so the 14–38 disulfide is formed rapidly and no mixed disulfide accumulates.

Reduction of native BPTI by monothiols such as GSH is very unfavorable. The high value for k_{intra} for forming the 14–38 disulfide bond, approximately 2×10^3 sec^{-1}, results in it being virtually impervious to reduction. Significant breakage of the 14–38 disulfide requires a GSH concentration of at least 0.1 M.[31,35]

[35] T. E. Creighton, D. A. Hillson, and R. B. Freedman, *J. Mol. Biol.* **142**, 43 (1980).

Intramolecular Disulfides and Effective Concentrations

The 14–38 disulfide of BPTI is more stable than that of GSSG, and the other two BPTI disulfides even more so, because protein disulfides are intramolecular, whereas that of the reagent RSSR is intermolecular. The equilibrium constant for their interconversion [K in Eqs. (9) and (10)] compares the relative free energies of their disulfides; it can be measured directly in some instances or indirectly from the rate constants in the two directions. As with any comparison of otherwise identical inter- and intra-molecular reactions,[36] this gives the effective concentration of the pair of Cys thiols in the protein without the disulfide. This parameter reflects the tendency of the protein to keep the Cys thiols in proximity even in the absence of the disulfide bond. Within the reduced BPTI, this value averages about 0.1 M, slightly higher than is observed with other disordered polymers.[37] However, in the folded conformations, the values are 2.3×10^2, 1.7×10^3, and $4.6 \times 10^5 M$, for the 14–38, 30–51, and 5–55 disulfides, respectively.[31] Such high values are not unexpected,[36-38] and they explain why proteins such as BPTI can form stable disulfide bonds within the seemingly reducing environment of most cells.[39] These considerations are also applicable to other interactions within proteins, such as hydrogen bonds and salt bridges, and can account for protein stability and its temperature-dependence.[40]

$$\underset{\substack{P \\ SH}}{\text{\Large$\sim$}}\kern-1em\text{SH} + \text{RSSR} \underset{}{\overset{K}{\rightleftharpoons}} \underset{P \underset{S}{\overset{S}{|}}}{\text{\Large$\sim$}}\kern-1em S{-}S + 2\text{RSH} \tag{9}$$

$$K = \frac{\left[P\,{\overset{S}{\underset{S}{|}}}\right]\left[\text{RSH}\right]^2}{\left[P\,{\overset{SH}{\underset{SH}{}}}\right]\left[\text{RSSR}\right]} \tag{10}$$

[36] A. J. Kirby, *Adv. Phys. Org. Chem.* **17**, 183 (1980).
[37] G. Illuminati and L. Mandolini, *Acc. Chem. Res.* **14**, 95 (1981).
[38] M. I. Page and W. P. Jencks, *Proc. Natl. Acad. Sci. U.S.A.* **58**, 1678 (1971).
[39] T. E. Creighton, *in* "Functions of Glutathione: Biochemical, Physiological, Toxicological, and Clinical Aspects" (A. Larsson, S. Orrenius, A. Holmgren, and B. Mannervik, eds.), p. 203. Raven, New York, 1983.
[40] T. E. Creighton, *Biopolymers* **22**, 49 (1983).

Intramolecular Disulfide Reagents

Dithiothreitol (DTT_{SH}^{SH}) forms a stable intramolecular disulfide bond because of the high effective concentration of its two thiol groups,[41] even though its disulfide may be slightly strained.[42] The equilibrium constant for thiol–disulfide exchange with glutathione

$$DTT_{SH}^{SH} + GSSG \rightleftharpoons DTT_{S}^{S} + 2GSH \qquad (11)$$

has been measured to be $1.3 \times 10^4 M$ (ref. 41) or $8.8 \times 10^3 M$ (ref. 8) at pH 7. We have indirectly measured at pH 8.7 very consistent values of between 0.98×10^3 and $1.25 \times 10^3 M$, with a mean of $1.16 \times 10^3 M$, using the relative equilibrium constants for forming four different disulfides in the BPTI pathway with both reagents.[31] The discrepancy with the values at pH 7 may be due to the pH difference if the two thiol reagents have different pK_a values (Fig. 1). Konishi et al.[34] and Scheraga et al.[43] prefer the value of 6 mol fraction units because it gives better agreement with theoretical calculation of the effects of cross-linking on the conformational entropy of the polypeptide chain. However, the basis for these calculations is not sufficiently sound to warrant dismissing experimental measurements. Also, this value was obtained erroneously by equating the free energies of the disulfide in GSSG with that between Cys-14 and Cys-38 in folded BPTI,[34] completely ignoring the entropy difference in forming these inter- and intramolecular disulfides and the difference in units of the two equilibrium constants. That the 14–38 disulfide of BPTI is not equivalent to that of GSSG is shown by the equilibrium constant of $2.4 \times 10^2 M$ for thiol–disulfide exchange between them.[31]

The effective concentration of the two thiol groups in DTT_{SH}^{SH}, $C_{eff,DTT}$, is one-half the equilibrium constant for Eq. (11), since GSSG has two equivalent sulfur atoms; dithiothreitol is assumed to act kinetically like a monothiol, since only one thiol is likely to be ionized at normal pH values. The high value of $C_{eff,DTT}$ results in mixed disulfides involving it to be intrinsically unstable, for its cyclic disulfide is formed intramolecularly and rapidly

$$RSH + DTT_{S}^{S} \underset{k_{DTT}}{\overset{k_{ex}}{\rightleftharpoons}} DTT_{SH}^{SSR} \qquad (12)$$

$$k_{DTT} = k_{ex}C_{eff,DTT} \qquad (13)$$

$$K_{DTT} = k_{ex}/k_{DTT} = 1/C_{eff,DTT} \qquad (14)$$

The reverse rate constant, k_{DTT}, is expected at pH 8.7 and 25° to be between 1.3×10^5 and $1 \times 10^4 \sec^{-1}$, i.e., half-times of only 5.4 and 70

[41] W. W. Cleland, Biochemistry **3**, 480 (1964).
[42] S. Capasso and Z. Zagari, Acta Crystallogr. **B37**, 1437 (1981).
[43] H. A. Scheraga, Y. Konishi, and T. Ooi, Adv. Biophys. **18**, 21 (1984).

μsec, depending upon the value used for $C_{\text{eff,DTT}}$ and the pK_a value of its thiol. We use the value of 600 M for $C_{\text{eff,DTT}}$ that is measured directly under our standard conditions using Eq. (11).

Because the mixed disulfide involving dithiothreitol is so unstable, it never accumulates substantially, and protein disulfide formation using DTT_S^S as disulfide reagent is proportional to its concentration and is slow, for the intramolecular step must be sufficiently rapid to compete with the reversal of the mixed disulfide.

$$\text{(structure)} + DTT_S^S \underset{k_{\text{DTT}}}{\overset{2k_{\text{ex}}}{\rightleftharpoons}} \left[\text{(structure)} \right] \underset{k_r}{\overset{k_{\text{intra}}}{\rightleftharpoons}} \text{(structure)} + DTT_{SH}^{SH} \tag{15}$$

$$k_{\text{obs}} = 2k_{\text{intra}}K_{\text{DTT}} \tag{16}$$

As a result, the value of k_{intra} may be estimated from the observed second order rate constant and the equilibrium constant for mixed-disulfide formation from DTT_S^S

$$k_{\text{intra}} = k_{\text{obs}}/2K_{\text{DTT}} \tag{17}$$

Therefore, it provides a method for measuring large values of k_{intra}, up to approximately 10^5 sec^{-1}; the limit is set by the requirement that k_{intra} for the protein disulfide be no greater than that for closing the ring of DTT_S^S.

When more than one disulfide is being formed in a step, as in forming the first disulfide bond in reduced BPTI or ribonuclease, the apparent value of k_{intra} is the sum of the individual values for all the disulfides contributing. For example, the value of $k_{\text{intra}}^{\text{app}}$ for forming all the initial disulfides in reduced BPTI obtained in this way is 8.1 sec^{-1}, very similar to the value 9.9 sec^{-1} obtained by GSH breakage after correcting for the involvement of all six Cys residues.

With DTT_S^S as disulfide reagent, there is no requirement for protein Cys residues to form disulfides, either intramolecular or with the reagent. Only favorable intramolecular disulfides are formed at significant rates. This property was used to show that there was no energetically favorable pathway for forming the four disulfides of RNase in which the rate-limiting step occurred within the one- and two-disulfide intermediates.[17] In the presence of high concentrations of purified DTT_S^S, reduced RNase formed rapidly the first and second disulfide bonds, but no third disulfides. Instead, an apparent equilibrium was attained between the reduced protein and the one- and two-disulfide species. Formation of third and fourth disulfides is energetically unfavorable, yet there is no more favorable pathway into and out of the native conformation. Any more favorable pathway, involving rapid formation of first, second, or third disulfides, should have been apparent from the kinetics with DTT_S^S.

Because the mixed disulfide involving dithiothreitol is so unstable, and the disulfide of DTT_S^S formed so fast, the rate of reducing a protein disulfide by DTT_{SH}^{SH} is nearly always determined by the initial exchange, i.e., k_r of Eqs. (5) and (15). This should be the case for all protein disulfides less stable than that of DTT_S^S; more stable protein disulfides should be reduced more slowly, since the reversal of the initial cleavage, k_{intra}, will compete with closure of the disulfide of DTT_S^S. In general, the rate constant for reduction by DTT_{SH}^{SH} will be given by

$$k_{obs} = k_r \Big/ \left(1 + \frac{k_{intra}}{k_{DTT}}\right) \tag{18}$$

Because $k_{intra} < k_{DTT}$ in each case, the rate constants for reduction of all disulfides in the BPTI folding pathway are very similar and close to the value of the intrinsic exchange reaction, k_{ex}. Two exceptions are the 30–51 and 5–55 disulfides within the native-like conformation. They are reduced much more slowly, if at all, due to their inaccessibility.

Strain in a disulfide causes it to be reduced more rapidly by DTT_{SH}^{SH} than expected from k_{ex}. For example, the disulfide of the five-membered ring of lipoic acid is reduced 160-fold more rapidly than otherwise expected.[6] The initial two-disulfide intermediates in the BPTI folding pathway are reduced 10-fold more rapidly than expected (see the table), suggesting the presence of conformational strain. This should also affect the initial rate of breakage of the disulfide by a monothiol, so the measured rate constant for reduction by DTT_{SH}^{SH} should be used in Eq. (7) as the value for k_r when estimating the value of k_{intra} in that way.

The rate of reduction of a protein disulfide by DTT_{SH}^{SH} gives little or no information about the disulfide's stability and rate of formation, but does depend on the conformational properties of the protein with the disulfide.

Microscopic and Apparent Rate Constants

Two methods have been described here for determining the intramolecular rate constant for forming a protein disulfide, k_{intra}: from the rate of its reduction with a monothiol and by the rate of its formation with DTT_S^S. In the first case, the mixed disulfide is in unfavorable equilibrium with the protein containing the disulfide; in the second, with the protein lacking the disulfide. The values determined in these two ways should be the same for a single step. This is the case with steps involving single disulfide bonds in the BPTI folding transition.[31] However, the two values need not be the same with a complex pathway with multiple parallel steps where the precursor and product of forming a specific disulfide bond are only a fraction of the total molecules. The value determined by reduction will generally be that of the microscopic rate constant, while that from disulfide formation will be the apparent macroscopic value.

The difference between the two can be illustrated with formation of the second disulfides in the folding pathway of BPTI.[31] In which precursors the second disulfides are formed is impossible to determine kinetically, because the one-disulfide intermediates are in rapid equilibrium. Consequently, they have identical kinetic properties and are classed together for kinetic simulations. Intermediate (30–51) represents about 60% of these intermediates, (5–30) about 30%, and several minor species the remaining 10%. Three different second disulfides are probably formed directly in (30–51), to form three two-disulfide intermediates classed together as II, while a minor intermediate, either (5–55) or (5–51), forms the 14–38 disulfide to generate species (5–55, 14–38), perhaps via (5–51, 14–38):

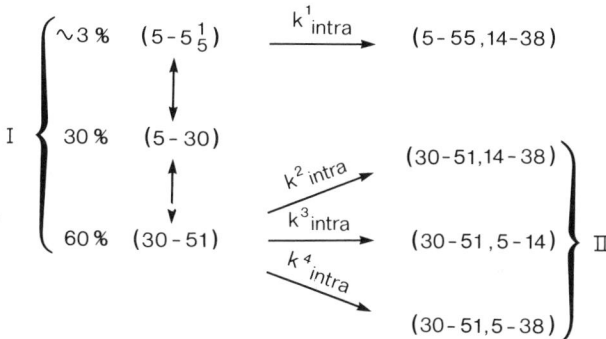

The apparent intramolecular rate constant for forming (5–55, 14–38) from I using DTT_S^S is only 14.4 sec^{-1}, while that determined by reduction of (5–55, 14–38) with GSH is 200 sec^{-1}. The discrepancy is due to the former rate constant being measured as if all the one-disulfide intermediates were contributing to its formation, whereas the rate of formation with GSSG indicates that only a small fraction of one-disulfide intermediates are forming (5–55, 14–38). Correcting for this yields a value of 288 sec^{-1}, in reasonable agreement with the value of 200 sec^{-1} determined with GSH.

Although formed more rapidly than (5–55, 14–38) with GSSG, the other three two-disulfide intermediates, designated as II, are formed with DTT_S^S at a slower rate, indicating a lower value of k_{intra}. If formed only in (30–51), this macroscopic value, 9.0 sec^{-1}, would be expected to be 60% of the sum of the individual values, $k_{intra}^2 + k_{intra}^3 + k_{intra}^4$; i.e., 15 sec^{-1}. The single value of 4.2 sec^{-1} obtained from the reduction by GSH must be corrected for the involvement of three Cys residues, to give 6.3 sec^{-1}. The agreement is reasonable, but would be better if most of the one-disulfide intermediates and all four Cys residues were involved. Without these independent kinetic rate constants, there would have been no evidence as to which one-disulfide intermediate was the precursor for the various two-disulfide intermediates.

Conformational Forces Directing Disulfide Formation

A very nonrandom pathway of disulfide formation under the conditions described here implies that it is being directed by the conformational properties of the protein, but this needs to be demonstrated and the nature of the conformational forces determined. The primary structure of a protein plus the environment determine its conformational properties, so the folding pathway should be affected by changes in either. Not all amino acid residues are expected to be crucial for folding, but modification of Met-52 in BPTI has been shown to affect significantly the spectrum of one-disulfide intermediates in a manner suggestive that the hydrophobic nature of this side chain is a factor contributing to the stability of the predominant intermediate (30–51).[23] Homologous proteins with the same backbone conformation but only one-third of their residues identical to BPTI have significant differences in the rates and energetics of various steps of the pathway, but the major productive pathway has been conserved.[18] Covalent cross-links between groups distant in the primary structure have substantial effects on the flexibility of the unfolded protein, and their effects on the folding pathway and on its energetics can elucidate the normal tendency of these groups to come into proximity during the various stages of folding.[44]

Changing the environment affects BPTI folding in a manner consistent with the presence of varying amounts of normal conformational interactions throughout the pathway. Denaturants such as 8 M urea and 6 M guanidinium chloride randomize disulfide formation. The energetic consequences vary, indicating under normal conditions a gradual acquisition of nonrandom conformation as disulfide formation progresses.[45] Lower temperatures favor nonrandom conformation: the effects of salts according to the Hofmeister series and the apparent heat capacity changes indicate a role for hydrophobic interactions.[46] However, the more negative enthalpy of the nonrandom conformation implies that it also has greater van der Waals interactions and more stable hydrogen bonds than the unfolded

[44] D. P. Goldenberg and T. E. Creighton, *J. Mol. Biol.* **179,** 527 (1984).
[45] T. E. Creighton, *J. Mol. Biol.* **113,** 313 (1977).
[46] T. E. Creighton, *J. Mol. Biol.* **144,** 521 (1980).

state.[40] In spite of the normal presence of nonrandom stabilizing conformation in the BPTI intermediates, they always have higher free energies than either the fully folded or fully unfolded, reduced forms, and the folding transition is cooperative,[28] in the same way as those of other small proteins not involving disulfide bonds.[1] In this case, the disulfide bonds have permitted us to trap and to characterize the inherently unstable, otherwise transient intermediates.

Although only the disulfide bonds have been trapped, so in effect have the conformational forces that favored these disulfides. It is virtually a thermodynamic requirement that whatever conformation favored certain disulfides be likewise favored to the same extent by the presence of the disulfide, as in the trapped intermediates, for the two are linked functions (Fig. 3). The only way that this relationship could not be true would be if there were slow conformational changes after formation of the disulfides, but none has been detected with BPTI. The trapped intermediates have demonstrated the presence of nonrandom conformation by their electrophoretic mobilities,[2] immunochemical analysis,[47] absorbance,[48] circular dichroism,[49–51] and nuclear magnetic resonance.[52] However, there are not yet available methods for determining detailed conformations of flexible polypeptides in solution. In contrast, the trapped ribonuclease intermediates, except for one with three correct disulfide bonds and a nativelike conformation,[24] have demonstrated no significant nonrandom conformation, even less than thermally denatured ribonuclease,[53] which probably reflects the few conformational restraints on disulfide formation in this protein.[21]

One complication is the presence of the blocking groups on the Cys residues of the trapped protein, which might affect its conformational properties. Thus far, there are remarkably few indications that the CM groups on the BPTI intermediates (5–55, 14–38) and (30–51, 14–38) disrupt substantially their tendencies to adopt the native-like conformation, even though the blocked Cys residues are normally within the close-packed interior; nevertheless, they must have some effect. Unfortunately, blocking groups are necessary to prevent disulfide rearrangements, especially intramolecularly; even at pH 2, such rearrangements in BPTI intermediates probably take place on the minute time scale. In this

[47] T. E. Creighton, E. Kalef, and R. Arnon, *J. Mol. Biol.* **123**, 129 (1978).
[48] P. A. Kosen, T. E. Creighton, and E. R. Blout, *Biochemistry* **19**, 4936 (1980).
[49] P. A. Kosen, T. E. Creighton, and E. R. Blout, *Biochemistry* **20**, 5744 (1981).
[50] P. A. Kosen, T. E. Creighton, and E. R. Blout, *Biochemistry* **22**, 2433 (1983).
[51] M. Hollecker, T. E. Creighton, and M. Gabriel, *Biochemie* **63**, 835 (1981).
[52] D. J. States, Ph.D. thesis, Harvard University (1983).
[53] A. Galat, T. E. Creighton, R. C. Lord, and E. R. Blout, *Biochemistry* **20**, 594 (1981).

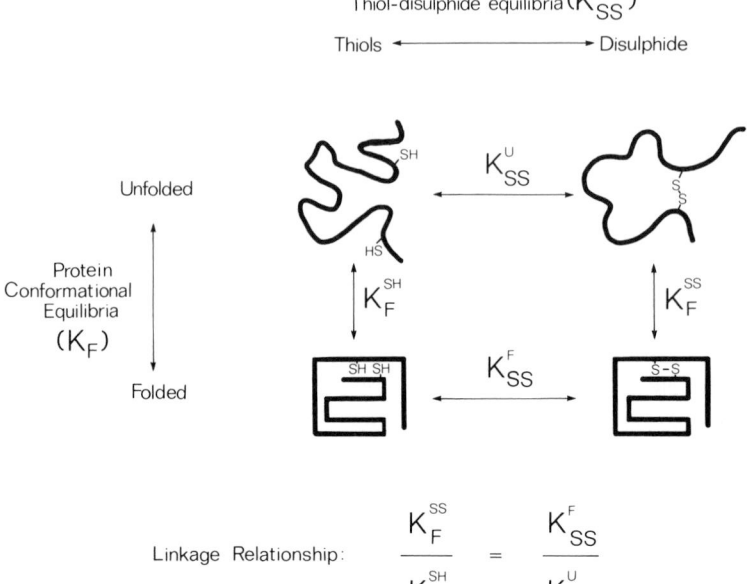

FIG. 3. Linkage between disulfide formation and conformational equilibria. Whatever nonrandom conformation that favors formation of a specific disulfide bond must be favored to the same extent by the presence of that disulfide, as in the trapped intermediates.

case, binding an unblocked intermediate to an antibody molecule[54] is likely to perturb the conformation and to induce disulfide interchange; disulfide interchange might also take place with the antibody molecule or other components of the mixture. This may account for the inconsistencies reported on unblocked ribonuclease intermediates using antibodies[54,55] and spectral measurements.[56,57]

A method is needed of removing the blocking group and the sulfur atom, to leave an Ala or Ser residue, to permit full exploitation of the disulfide-trapped intermediates.

[54] L. G. Chavez and H. A. Scheraga, *Biochemistry* **16**, 1849 (1973).
[55] L. G. Chavez and H. A. Scheraga, *Biochemistry* **19**, 996 (1980).
[56] Y. Konishi and H. A. Scheraga, *Biochemistry* **19**, 1308 (1980).
[57] Y. Konishi and H. A. Scheraga, *Biochemistry* **19**, 1316 (1980).

[6] Proline Isomerization Studied with Proteolytic Enzymes

By JOHN F. BRANDTS and LUNG-NAN LIN

Introduction

It is generally accepted that the peptide bond formed by amino acid residues other than proline is thermodynamically preferred in the planar trans configuration. The planarity is due to the double bond character of the peptide linkage, and the preferability of the trans isomer is due to more favorable interactions and a larger conformational entropy relative to the cis form. Except for three bonds in native carboxypeptidase A[1] and one in native concanavalin A[2] that have been seen from X-ray analysis of the crystalline forms, cis peptide bonds for nonproline residues have never been experimentally detected in solution or in solid forms of peptides or proteins, within the sensitivity of NMR and X-ray methods. However, it has been argued that the presence of a small amount of cis form of nonproline peptide bonds (<1%) in polypeptides in solution cannot yet be ruled out.[3]

In contrast to nonproline peptide bonds, the peptide bond formed between a proline residue and the immediately preceding amino acid residue is more likely to exist as a mixture of cis and trans forms in solution because of the added steric restraints associated with linkage of the proline side chain to the peptide nitrogen atom. Although in most cases the trans isomer is predominant, the cis form is usually present to an extent of 10% or more[3] and, in certain cases, equilibrium concentrations as high as 90% have been seen.[4] The process of proline isomerization, which is characterized by slow kinetics and a high energy barrier, has been well documented in the literature only for very small peptides.[3,5,6]

The possible role which proline isomerization may play in protein structure and enzyme regulation had received very little attention until 1975 when Brandts *et al.*[3] suggested from indirect evidence that proline isomerization is the rate-limiting step in protein folding from the dena-

[1] D. C. Rees, M. Lewis, R. B. Honzatko, W. N. Lipscomb, and K. D. Hardman, *Proc. Natl. Acad. Sci. U.S.A.* **78,** 3408 (1981).

[2] G. N. Reeke, Jr., J. W. Becker, and G. M. Edelman, *Proc. Natl. Acad. Sci. U.S.A.* **75,** 2286 (1978).

[3] J. F. Brandts, H. R. Halvorson, and M. Brennan, *Biochemistry* **14,** 4953 (1975).

[4] L. N. Lin and J. F. Brandts, *Biochemistry* **23,** 5713 (1984).

[5] H. N. Cheng and F. A. Bovey, *Biopolymers* **16,** 1465 (1977).

[6] C. Grathwohl and K. Wüthrich, *Biopolymers* **20,** 2623 (1981).

tured state. More recently, indirect evidence has suggested that proline isomerization may also play an important role in conferring kinetic regulation on subtle conformational processes in native proteins. For example, it has been shown[7] that concanavalin A has two similar "native states" with quite different saccharide binding activity and the slow equilibrium between the two forms was originally suggested to be mediated by proline isomerization. A kinetic study of the fluorescence quenching of bovine prothrombin fragment I[8] also showed that this protein can exist in solution in two native-like forms with different metal ion affinity, and interconversion between these two forms may be rate limited by the isomerization of a proline residue. Thus, cis–trans isomerization of proline bonds not only may dictate the structure of peptides and proteins, but may also regulate their function in some cases.

In spite of its potential importance in protein structure and function, the isomerization process is very difficult to "see" experimentally, so that little definitive information is now available. Among traditional spectroscopic methods employed to investigate the isomerization process, only the NMR technique has proven to be a potentially powerful tool for distinguishing the cis and trans isomers of prolyl residues in solution. However, its effectiveness so far has been limited to small peptides where overlapping resonances are not a serious problem. For large polypeptides and proteins, the role of cis–trans isomerization in most cases has only been indirectly inferred by comparing various thermodynamic and kinetic parameters with those of model peptides. Unfortunately, the properties of proline isomerization are now known to be greatly influenced by various parameters such as amino acid sequence, state of ionization, chain length and chain dynamics, and disulfide bond linkage. Because of this, the isomerization of proline residues in proteins can be very much different from those in model peptides, especially from those in X-Pro dipeptides, which have most frequently been used as the comparative model peptides (see the table). The inability to obtain direct information on proline isomerization has already led to controversy in the interpretation of protein folding data,[9-12] where it has proven difficult to determine which kinetic phases involve isomerization, which involve formation of structural intermediates, and which may involve both. Clearly, a new experimental

[7] R. D. Brown, III, C. F. Brewer, and S. H. Koenig, *Biochemistry* **16**, 3883 (1977).

[8] H. C. Marsh, M. E. Scott, R. G. Hiskey, and K. A. Koehler, *Biochem. J.* **183**, 513 (1979).

[9] B. T. Nall, J. R. Garel, and R. L. Baldwin, *J. Mol. Biol.* **118**, 317 (1978).

[10] K. H. Cook, F. X. Schmid, and R. L. Baldwin, *Proc. Natl. Acad. Sci. U.S.A.* **76**, 6157 (1979).

[11] F. X. Schmid and H. Blaschek, *Eur. J. Biochem.* **114**, 111 (1981).

[12] L. N. Lin and J. F. Brandts, *Biochemistry* **22**, 564, 573 (1983).

method which is directly sensitive to cis and trans isomers of prolyl residues in proteins is needed for understanding the role of isomerization in protein structure and regulation. During the past few years, our laboratory has developed a chemical method, which we call isomeric specific proteolysis (ISP), that is based on the finding that many different proteases have an absolute specificity for peptide bonds in the trans configuration. This new technique is the subject of this article.

Principles of the ISP Method

It was established[13-15] in 1979 that prolidase,[16] a dipeptidase which hydrolyzes only X-Pro dipeptides (X = any amino acid except proline) and aminopeptidase P (APP),[17] a proline-specific exopeptidase which can hydrolyze an N-terminal X-Pro bond (X = any amino acid including proline) regardless of the size of the peptide chain, exhibit absolute specificity toward the trans form of the X-Pro peptide bond, while the cis form is not a true substrate and must isomerize to the trans form before it can be hydrolyzed. It was shown later that endopeptidases such as trypsin[18] and proline-specific endopeptidase from *Flavobacterium*[19,20] exhibit isomeric specificity toward a neighboring X-Pro bond, i.e., trypsin can only cleave an active Lys-X bond in a substrate with Lys-X-Pro sequence when the following X-Pro bond is in the trans form, while the proline-specific endopeptidase can only cleave an active Pro-Y bond in an X-Pro-Y sequence when the preceding X-Pro bond is in the trans configuration. Taking advantage of the isomeric specificity of enzyme activity, thermodynamic and kinetic properties of isomerization for particular proline residues in peptides or proteins may be obtained simultaneously when hydrolysis as a function of time is carried out at high enzyme activity under conditions where the rate of isomerization is slow relative to the rate of hydrolysis. Two kinetic phases will be seen if the X-Pro bond of the substrate exists as a mixture of cis and trans forms. A fast phase, of which the rate of hydrolysis strongly depends on enzyme concentration, corresponds to direct hydrolysis of the trans form. A slow phase, of which the relaxation time is independent of enzyme concentration, corresponds to the cis-to-

[13] L. N. Lin and J. F. Brandts, *Biochemistry* **18**, 43 (1979).
[14] L. N. Lin and J. F. Brandts, *Biochemistry* **18**, 5037 (1979).
[15] L. N. Lin and J. F. Brandts, *Biochemistry* **19**, 3055 (1980).
[16] N. C. Davis and E. L. Smith, *J. Biol. Chem.* **224**, 261 (1957).
[17] A. Yaron and D. Mlynar, *Biochem. Biophys. Res. Commun.* **32**, 658 (1968); A. Yaron and A. Berger, this series, Vol. 19, p. 521.
[18] L. N. Lin and J. F. Brandts, *Biochemistry* **22**, 553 (1983).
[19] L. N. Lin and J. F. Brandts, *Biochemistry* **22**, 4480 (1983).
[20] T. Yoshimoto, R. Walter, and D. Tsuru, *J. Biol. Chem.* **255**, 4786 (1980).

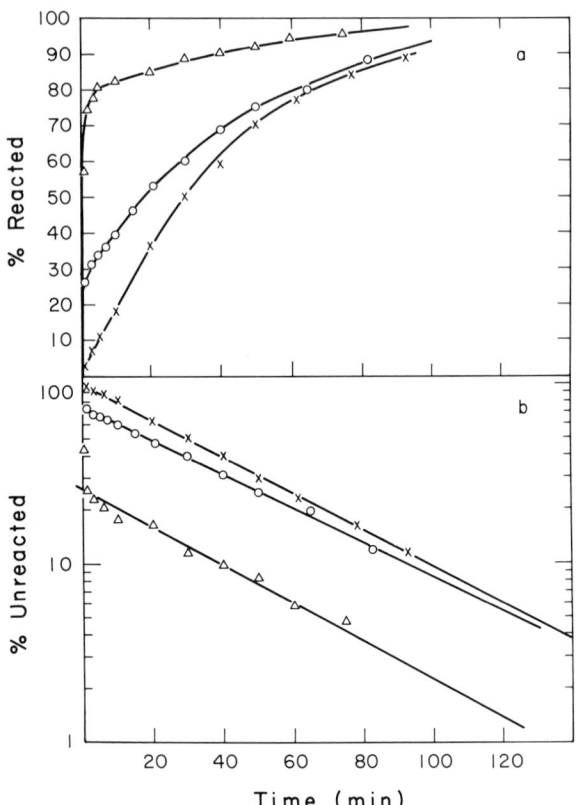

FIG. 1. Rate of hydrolysis of L-phenylalanyl-L-proline catalyzed by aminopeptidase P at 23°. (a) Degree of hydrolysis vs time; (b) semilog plot of substrate concentration vs time. The ratio of aminopeptidase P activity to substrate concentration was 15 units/μmol. The concentration of L-phenylalanyl-L-proline was 2.1 mM in 1.23 ml of 0.05 M veronal buffer containing 2.5×10^{-4} M DTT, 8.1×10^{-3} M manganous chloride, and 3.3×10^{-2} M sodium citrate at pH 8.6. Triangles, circles, and crosses represent the cationic, zwitterionic, and solid forms of substrate, respectively, which were present when hydrolysis was initiated. Reprinted with permission from Lin and Brandts.[14] Copyright 1979 American Chemical Society.

trans isomerization and subsequent hydrolysis. A semilog plot of substrate concentration vs time should show that the fast and slow phases can be treated as first-order processes with respect to time (see Fig. 1). The amplitudes and the relaxation times of the two phases can be determined by using the peel-off method. The relative amplitude of the fast to slow phase is equal to the trans/cis ratio existing for the peptide in the initial state just before the enzyme solution is added. The relaxation time of the slow phase is equal to the inverse rate constant for the cis-to-trans isomerization in solution since the backward reaction (i.e., trans-to-cis) will not occur in the presence of high enzyme activity. The relaxation time of the two-way reaction cis $\rightleftarrows$ trans can then be calculated from the trans/cis ratio and the one-way cis-to-trans relaxation time.

General Strategy for Protein Substrates

The application of the ISP method to a specific proline residue in a large protein will usually require the use of several proteolytic enzymes in tandem, in contrast to the single enzyme approach which is frequently adequate for small peptides. There are several reasons for the more complicated approach that is necessary for investigating proteins. First, most native and reversibly unfolded proteins are not good substrates for proteases such as trypsin or the proline-specific endopeptidase in the buffers which are necessary to maintain high protease activity. One way to circumvent this problem is to pretreat the native or reversibly unfolded protein with a short (~20 sec) but intensive pepsin pulse in a buffer (pH 2.0, 4.5 M urea) which is denaturing to the protein substrate but not to pepsin.[21] This renders the protein irreversibly unfolded, whereupon it becomes a good substrate for other proteases such as trypsin in the usual buffers used for carrying out stereospecific hydrolysis. (Alternatively, we have recently found for RNase that the pepsin pulse can be replaced by a treatment which causes fast reduction of disulfide bonds prior to ISP proteolysis.[4] This method is discussed in the section on methods.)

Second, prior cleavage may be necessary when the proline-specific endopeptidase is to be used in the isomer-specific step. Although an endopeptidase, this enzyme apparently will not cleave a Pro-Y bond if that bond is very far removed from the carboxyl terminus of the chain. Thus, in the investigation of proline-114 in RNase[4] (i.e., -Gly[112]-Asn[113]-Pro[114]-Tyr[115]-Val[116]-) it was necessary to cleave the Tyr-Val bond for all RNase molecules with chymotrypsin before adding the proline endopeptidase for stereospecific cleavage of the Pro-Tyr bond; the latter cleavage being specific for the trans isomer of the preceding Asn-Pro bond.

Third, it is sometimes desirable to use an additional protease after the stereospecific cleavage has been carried out in order to facilitate the assay for cleavage. In the study of proline-93 in RNase[18,21] (i.e., -Lys[91]-Tyr[92]-Pro[93]-), for example, the Lys-Tyr bond was stereospecifically cleaved by trypsin for those molecules having the Tyr-Pro bond in the trans form. Without further treatment the cleavage assay would now require the identification and quantitation of the long peptides that are produced by the

[21] L. N. Lin and J. F. Brandts, *Biochemistry* **22**, 559 (1983).

trypsin cleavage at position 91–92. The difficult problems associated with peptide assays (see the methods section) were circumvented by incubating for a long time with APP subsequent to trypsin hydrolysis, which causes the release of tyrosine-92, as the free amino acid, from all chains that had previously been cleaved by trypsin at the 91–92 position. Since APP is absolutely specific for an N-terminal X-Pro sequence, there was no tyrosine other than tyrosine-92 that was released. By precolumn labeling of the final hydrolyzate with OPA (o-phthalaldehyde and mercaptoethanol) and using fluorescence detection, the assay for cleavage can be carried out by measuring free tyrosine released, using as little as 10^{-11} mol of RNase for each HPLC run.

Thus, the possible steps that might be required to investigate a single proline isomerization in a native or reversibly unfolded protein are (1) pepsin cleavage (or disulfide reduction) to irreversibly unfold the substrate, (2) chain-trimming cleavage, (3) time-dependent stereospecific cleavage, and (4) assay-facilitating cleavage. The first three steps must, of course, be carried out in times which are short relative to the time required for the proline to isomerize.

Methods

The stereospecific hydrolysis experiments are initiated by mixing a substrate solution with an enzyme solution at a controlled temperature. At suitable time intervals, aliquots of the reacting solution are quickly pipetted into an inhibitor solution or into an organic solvent to stop enzyme activity. As mentioned above, in the case of protein substrates in which the active bonds are not initially accessible to rapid proteolysis, they must first be quickly and irreversibly unfolded before the hydrolysis is initiated. Hydrolysis products may be assayed directly, or alternatively the resulting peptide fragment(s) can be further hydrolyzed to release a free amino acid which may be more conveniently assayed. Factors which are important in carrying out ISP hydrolysis reactions and in the assay of hydrolytic products are discussed below. Readers interested in more detailed procedures of a particular experiment should consult the original publications.[4,13–15,18,19,21]

pH. Although the initial pH of substrate solutions may be varied, hydrolysis should be carried out near the optimum pH for the working enzyme. If several enzymes are used in tandem in a hydrolysis sequence, the pH may be adjusted accordingly for each step. For example, in studying proline-93 isomerization for RNase,[21] the protein was subjected to a pepsin pulse for 20 sec at pH 2.0, whereupon the pH of the solution was taken to about 8 when trypsin was added, and then to 8.5 when APP was added.

Temperature. While enzymes usually have higher activity as the temperature is increased, the activation energy for enzyme-catalyzed proteolysis usually is only about half of that for proline isomerization ($\sim$20 kcal/mol). Therefore, proteolysis is speeded up relative to isomerization as the temperature is lowered, so that the best separation of fast and slow phases in ISP studies is achieved at low temperatures of $\sim$0–10°. In addition to the importance of the relative rates of proteolysis and isomerization, the absolute rate of isomerization can also be critical since the hand-mixing procedures are slow and preclude the measurement of isomerization reactions with a relaxation time faster than $\sim$30 sec no matter how fast proteolysis occurs. This additional factor also favors the use of low temperature.

Buffer Solutions and Denaturants. Other solution variables besides pH and temperature may also affect enzyme activity. It was found[4] that APP has higher activity in veronal buffer and proline-specific endopeptidase has higher activity in phosphate buffer, while trypsin and chymotrypsin were found to have about equal activity in the two buffers. Since prolidase and APP require the presence of Mn^{2+} for high activity, phosphate buffer should not be used for these two enzymes. Although it is well-known that Ca^{2+} can increase trypsin activity, it was avoided in studies where trypsin and APP were used in tandem because of its inhibition of APP activity. The choice of buffer solution also depends on the assay method. For example, Tris buffer can be used for the acid ninhydrin assay, while it should not be used whenever OPA reagent is employed.

Pepsin is an exceptional protease since it still has nearly full activity in moderately high concentrations of denaturants such as GuHCl and urea. Our experiments show that the activities of enzymes such as trypsin, chymotrypsin, and APP toward peptide substrates are greatly decreased even in 1 M urea solution. Therefore, ISP hydrolysis should be carried out in low denaturant concentration (0.2 M urea or below) whenever these enzymes are employed. GuHCl should not be used at all for trypsin studies because the guanidinium ion is a strong competitive inhibitor of trypsin.

Enzyme Solution. Since the ISP method is based on the requirement that the rate of hydrolysis must be much faster than that of the isomerization to be studied, very high enzyme concentration is usually employed with molar ratios of enzyme/substrate sometimes in excess of 50. Actual enzyme concentrations used depend on the intrinsic enzyme activity toward the substrate, number of competing cleavage sites, the assay method, and the enzyme purity since contaminating molecules and contaminating activities can be critical. For example, the intrinsic activity of chymotrypsin is much lower for most substrates than that of trypsin for its substrates. Higher chymotrypsin concentration is therefore necessary to achieve equally fast cleavage. However, the relaxation time for cleavage of a particular bond can only be accelerated to a certain point by increasing the concentration of a particular enzyme. Additional enzyme concentration above a certain point may not accelerate the cleavage very much and may cause some problems in assaying for hydrolysis, especially for those experiments using several enzymes in tandem and protein as substrate. It is well known that commercial proteases such as trypsin, chymotrypsin, and pepsin are contaminated with products of low molecular weight. We found, for example, that some of the contaminants become fluorescent when derivatized with OPA reagent and elute in the time range of OPA-labeled amino acids on HPLC columns. They may therefore interfere with the assay of certain hydrolysis products.

It has been found[4,22] that the degree of contamination of proteases varies considerably for different preparations of the same protease, and even among different lots from the same preparation. This was found to be a particular problem for α-chymotrypsin since removal of contaminants by gel filtration is not very effective, and since very high concentrations of enzyme, and therefore of contaminants, are necessary to achieve rapid cleavage. Thus, it is advantageous to select and use only the cleanest preparations of chymotrypsin. Although pepsin preparations are also strongly contaminated, gel filtration was found to be more effective in removing contaminants. Porcine trypsin was found to be much cleaner than bovine trypsin, and was therefore used exclusively in our studies.[21] Because of the higher intrinsic activities of APP and the proline endopeptidase, these can be used at much lower enzyme/substrate ratios and contaminants are not usually a problem.

In principle then, there will frequently exist an optimum concentration range for the working enzyme, which represents a compromise between the desire to achieve very rapid cleavage of substrate and the desire to minimize the interference of contaminants in the assay. In particular instances, it was found[4] that the effect of contaminants can be minimized by increasing substrate concentration at constant enzyme concentration. At low substrate concentration (i.e., below the effective K_m) where many of these studies are carried out, the relaxation time or half-time for cleavage is nearly independent of substrate concentration, as expected. At any rate, various control experiments without substrate should be run in par-

[22] A. Yapel, M. Han, R. Lumry, A. Rosenberg, and D. F. Shiao, *J. Am. Chem. Soc.* **88,** 2573 (1966).

allel with sample runs so that contributions from intrinsic contaminants, and from other interfering species that are produced during proteolysis, can be subtracted out of the assay data. These control runs should be carried out over the entire time interval for proteolysis, since the concentration of interfering species will frequently depend on the time of exposure to proteolysis.

Irreversible Substrate Modifications. Model peptides and chemically modified denatured proteins such as oxidized RNase, in which peptide bonds are accessible to enzyme cleavage, can be directly used as substrates in contrast to native and reversibly unfolded proteins which must first be chemically modified so they will be good substrates for isomer-specific proteases in the native-state buffers which must be used to maintain high enzyme activity. It was shown[21] that pepsin, an acid endopeptidase which has high activity at acid pH and in the presence of moderately high denaturant concentrations (up to 4.5 M urea) can render all RNase (>97%) irreversibly unfolded after just a short pulse of 20 sec at 10°. Recently, it was also shown[4] that native or reversibly unfolded RNase can be completely and irreversibly unfolded by incubating with 0.2 M mercaptoethanol (in 8 M urea, pH 12) for 20 sec immediately prior to incubation with a protease in a native-state buffer.

Inhibition and Precipitation of Protease. Proteases such as trypsin and chymotrypsin rapidly inactivate APP, prolidase, or the proline-specific endopeptidase. Before these enzymes can be added into solutions containing trypsin or chymotrypsin, the activity of these latter two enzymes must be inhibited. Trypsin activity can be completely and rapidly inhibited by soybean trypsin inhibitor at a 2 to 1 ratio, by weight, while chymotrypsin activity can be effectively inhibited by lima bean trypsin inhibitor at a 3 to 1 ratio. Addition of methanol or ethanol, followed by heating to 85–90°, has also been used at various stages of experiments to stop enzyme activity and (or) to precipitate proteins and high-molecular-weight peptides, which also facilitates the HPLC assay of free amino acids. Ethanol should not be used with phosphate buffer solutions (0.05 M or above) because of the low solubility of phosphate ion. The resulting phosphate gel can trap low-molecular-weight hydrolysis products, and thereby lead to an erroneous assay.[4]

Assay for Hydrolysis Products. Depending on the substrates, a variety of assay methods have been developed to follow the rate of hydrolysis. For hydrolysis of X-Pro dipeptides and polyproline, it is convenient to assay the free proline which is released by using the acid ninhydrin method.[13,14,23] For Gly-Pro-Ala hydrolyzed by APP, chromotropic acid

[23] W. Troll and T. Lindsley, *J. Biol. Chem.* **215**, 655 (1955).

was used to assay the released glycine,[14,24] although HPLC would be a more sensitive and convenient method. In assaying for hydrolysis of Gly-Gly-Lys-Phe-Pro by trypsin,[18] the resulting Phe-Pro was further cleaved by prolidase after trypsin activity was inhibited, and the free proline which is released was assayed by the acid ninhydrin method. For the substrate Cbz-Gly-Pro-Leu-Gly, hydrolyzed by the proline-specific endopeptidase, the dipeptide Leu-Gly was directly derivatized with OPA reagent and quantitated by HPLC.

In contrast to these short peptides, protein substrates have many cleavage points when subjected to proteases such as trypsin and pepsin, so the final assay mixture may be extremely heterogeneous. Rather than attempting to resolve, identify, and quantitate the peptide(s) of interest, it may in some cases be easier to add another protease which results in release of a smaller product that is easier to deal with. As previously mentioned, this was done in the study of proline-93 of RNase,[18,21] where the assay was based on release of free tyrosine-92 (by subsequent APP hydrolysis) from the N-terminus of the peptide of interest, rather than attempting to identify the peptide itself in the heterogeneous hydrolysate.

In general, an assay based on the release of a free amino acid is more reliable and convenient than one based on a large peptide, for the following reasons. (1) Identification is a trivial problem, because direct comparison can be made to free amino acid standards. (2) Separation from interfering species in the HPLC chromatogram is expected to be better, because most hydrolysis products are peptides and are not free amino acids. (3) Sensitivity of detection is much higher for free amino acids, since precolumn-labeled OPA derivatives of amino acids give a very intense and highly reproducible fluorescent signal. Peptide-OPA derivatives (except those containing lysine) not only have a much lower quantum yield, but give highly erratic results so that, particularly for reasonably large peptides, absorbance detection is preferred. Even at 210 nm, the sensitivity of detection is normally lower by a factor of 10 or more, depending on the MW of the peptide. This means that the substrate concentration in the hydrolysis solutions must be increased, which can in turn produce longer relaxation times for proteolytic cleavage at equivalent protease activity. Nevertheless, it will not always be possible to assay chain cleavage by release of free amino acids, so that methods will have to be developed in the future to facilitate assays based on peptide release.

---

[24] P. Dehm and A. Nordwig, *Eur. J. Biochem.* **17**, 364 (1970).

Isomerization of Proline Residues in Model Peptides

Although proline isomerization for most short peptides can readily be studied by NMR, the ISP method does offer several potential advantages. (1) Only a minimum amount of peptide is needed for the ISP method. Depending on the type of assay, the quantity of peptide needed for an experiment is from 10^{-6} to 10^{-11} mol. Thus, the ISP method is capable of studying in aqueous solution the isomerization of peptides which have low solubility in water or those which are difficult to obtain in quantity. (2) The thermodynamics and kinetics of isomerization can be simultaneously determined with the ISP method, while the NMR technique has difficulty in obtaining kinetic data with high accuracy.[6] (3) Sample impurity may cause serious problems for NMR studies, while the specificity inherent in enzymatic hydrolysis means that high concentrations of most impurities can be tolerated in the ISP method.

A brief discussion of proline isomerization for several model peptides investigated by the ISP method is presented below.

Proline-Specific Exopeptidases (Prolidase and APP). Prolidase is a dipeptidase, so its effectiveness is limited to the study of X-Pro dipeptides, while APP can be used to examine an amino terminal X-Pro bond for a peptide of any length. These two enzymes have proven useful both in determining the isomeric state in small peptides[13-15] and also in assaying hydrolysis products for proteins and peptides.[18,21] Thus far prolidase has been used to study isomerization of both Gly-Pro and Phe-Pro.[13] APP has been used to investigate Phe-Pro, Gly-Pro-Ala, and poly(L-proline).[14,15] Several other peptides (unpublished data) have also been examined by these two enzymes.

A typical APP hydrolysis study is shown in Fig. 1 for Phe-Pro which was preincubated under three different sets of conditions, but hydrolyzed in the same buffer in each case. As expected, the slopes of semilog plots for the slow phase, corresponding to the rate constant of cis-to-trans isomerization in the hydrolysis buffer, are the same for all three forms of Phe-Pro. The intercepts, corresponding to the percentage cis isomer when hydrolysis is initiated, are different. It is typical of dipeptides that the zwitterionic form has a larger cis content than the cationic form[25] and this is seen to be the case for Phe-Pro. Apparently, the crystalline form of Phe-Pro is cis, since it shows no fast phase of hydrolysis. Thermodynamic and kinetic data for Phe-Pro and other peptides obtained from the ISP method are listed in the table, along with corresponding data obtained from NMR. The agreement between the two methods is generally quite good. It should be noted that the relaxations of X-Pro dipeptides are generally

[25] C. Grathwohl and K. Wüthrich, *Biopolymers* **15**, 2025 (1976).

KINETIC AND THERMODYNAMIC DATA OF PROLINE ISOMERISM FOR MODEL PEPTIDES AND PROTEINS

Compound	Solvent/temp. (°C)	cis (%) Cation	cis (%) Zwitterion	Relaxation time (sec) cis → trans (final pH or pD)	ΔE_a (kcal/mol)	Method	Reference
Gly-L-Pro	D$_2$O/25	15	37	454 (7.5)	18.5	NMR	6
	D$_2$O/23	—	—	70	20.0	NMR	c
	aq./23.5	18	39	321 (8.1)	21.5	ISP	13
	H$_2$O/22.5	—	—	380 (4.5)[a]	—	pH jump	3
L-Phe-L-Pro	D$_2$O/25	25	77	5263 (7.0)	19.6	NMR	6
	aq./23.5	30	73	2640 (8.5)	21.5	ISP	13
	H$_2$O/22.5	—	—	2780 (8.4)[b]	—	pH jump	13
L-Tyr-L-Pro	D$_2$O/25	23	78	2857 (7.5)	19.4	NMR	6
Gly-L-Pro-Ala	D$_2$O/25	16	15	—	—	NMR	25
L-Phe-L-Pro-L-Ala	aq./5	—	18	300 (8.6)	—	ISP	14
Gly-Gly-L-Lys-L-Phe-L-Pro	D$_2$O/25	—	34	222 (7.0)	18	NMR	6
	aq./32	45	58	440 (8.0)	18	ISP	18
Gly-Gly-L-Pro-L-Ala	D$_2$O/25	—	20	33 (7.2)	16.7	NMR	6
Cbz-Gly-L-Pro-L-Leu-Gly	aq./5	—	19	300 (8.5)	—	ISP	19
L-Thr-L-Lys-L-Pro-L-Arg	D$_2$O/25	—	0	—	—	NMR	d

Ac-L-Tyr-L-Pro-L-Asn	D₂O/25	—	29	—	NMR	29
Ac-L-Asn-L-Pro-L-Tyr-NHMe	D₂O/25	—	13	—	NMR	29
Pro-114 in oxidized RNase fragment 105–124	D₂O/25	—	12	—	NMR	29
Pro-24 in corticotropin fragment 1–32	D₂O/20	25 (pH ≤ 7)	43 (pH ≤ 11)	—	NMR	28
Proline-93 in oxidized RNase fragment 92–98	aq./10	—	37	660 (8.5)	ISP	18
Proline-93 in oxidized RNase	aq./10	—	35	—	ISP	18
Proline-93 in the denatured RNase (8.5 M urea, pH 4.0)	aq./10	—	70	—	ISP	21

[a] Calculated from the two-way relaxation time by assuming cis to trans ratio of 37:63.
[b] Calculated from the two-way relaxation time by assuming cis to trans ratio of 77:33.
[c] H. P. Bächiger, P. Bruckner, R. Timpl, and J. Engel, *Eur. J. Biochem.* **90**, 605 (1978).
[d] M. Blumenstein, P. P. Layne, and V. A. Najjar, *Biochemistry* **18**, 5247 (1979).

much slower than for prolines located inside peptide chains, and this effect is clearly seen for Gly-Pro-Ala, which isomerizes about 15 times faster than does Gly-Pro.[14]

Non-Proline-Specific Endopeptidases. So far, only trypsin has been shown to exhibit isomeric specificity toward a following X-Pro bond.[18] However, it is expected that most endopeptidases, if not all, will display such stereospecificity. Only peptides which have an active bond immediately preceding the X-Pro bond of interest can be studied by these enzymes, so their application in ISP studies of proteins is strongly dependent upon amino acid sequence. The isomerization of a pentapeptide, Gly-Gly-Lys-Phe-Pro, has been investigated with trypsin, and the results are listed in the table. The cis/trans ratio of this peptide was found to be slightly dependent on temperatures.[18]

Proline-Specific Endopeptidases. No endopeptidase which will cleave a peptide bond on the N-terminal side of a prolyl residue is known. However, several proline-specific endopeptidases (also called postproline cleaving enzymes), which will cleave a peptide bond on the carboxyl side of prolyl residues, have been isolated from bacteria[20] and mammalian tissues.[26,27] The proline-specific endopeptidase isolated from *Flavobacterium*[20] has been used in our studies. Besides cleaving prolyl peptide bonds at the carboxyl side, this enzyme can also hydrolyze an Ala-Ala bond in a peptide chain, but at a much slower rate. The enzyme also will not directly hydrolyze prolyl bonds in denatured proteins such as lysozyme, or cytochrome *c* where such bonds are far removed from the carboxyl terminus. Despite these shortcomings, proline-specific endopeptidases may ultimately become the most important enzyme for the ISP method, because of their potential applicability to any proline residue, no matter what the amino acid sequence. If protein substrates are first "trimmed" into suitably short peptide segments with other proteases, then proline-specific endopeptidases may be able to examine the isomerization of the X-Pro bonds in large proteins, where sequence restrictions may not allow the study of such bonds using only non-proline-specific endopeptidases. Furthermore, isomerization of several X-Pro bonds in the same protein may be simultaneously studied using these enzymes, which can greatly increase the efficiency of the ISP method.

Isomerization of two peptides, N-Cbz-Gly-Pro-MCA (MCA = 4-methylcoumarinyl-7-amine) and N-Cbz-Gly-Pro-Leu-Gly, have been investigated by proline-specific endopeptidase.[19] The results are listed in the table. For assaying the hydrolysis of N-Cbz-Gly-Pro-MCA, the in-

---

[26] M. Koida and R. Walter, *J. Biol. Chem.* **251**, 7593 (1976).

[27] P. C. Andrews, C. M. Hines, and J. E. Dixon, *Biochemistry* **19**, 5494 (1980).

crease in fluorescence intensity resulting from the release of MCA was followed continuously until completion. For N-Cbz-Gly-Pro-Leu-Gly, the resulting Leu-Gly dipeptide was derivatized with OPA reagent and quantitatively determined by HPLC.

Isomerization of Proline Residues in Proteins

As mentioned, the NMR technique is still unable to directly observe isomerization of proline residues in proteins because of overlapping resonances. In searching the literature, it was found that corticotropin fragment 1–32 is the largest protein fragment in which isomerization of a proline residue has been observed by NMR.[28] The aromatic resonances of tyrosine-23 and the methyl resonances of valine-22 were found to be sensitive to the geometry of proline-24, and were used to obtain the ratio of the cis–trans isomers (25% cis at pH $\leq$ 7, 43% cis at pH $\geq$ 11; see the table). More recently, isomerization of proline-114 for the RNase fragment 105–124 was also observed from NMR, along with its short peptide analogs[29] (see the table), although the resolution of peaks for the fragment was very marginal. No kinetic data were determined for either of these fragments, and it seems unlikely that this can be done effectively by NMR even for such relatively small peptides.

Recently, the ISP method has been successfully applied to study proline-93 isomerization in oxidized RNase and its fragment 92–98,[18] RNase,[21] and reversibly denatured RNase.[21] The kinetics of isomerization of proline-93 during RNase refolding and unfolding were also obtained.[21] A brief review of these results is presented below.

Isomerization of Proline-93 in Oxidized RNase, Native RNase, and Reversibly Denatured RNase. While proline-93 isomerization for oxidized RNase[18] can be readily studied without first treating with a short pulse of pepsin activity, this proved not to be possible for native RNase and reversibly denatured RNase. For comparison purposes, hydrolyses of these three forms of RNase were carried out using the same experimental protocol: i.e., 20 sec pepsin pulse, various times of exposure to high trypsin activity for stereospecific cleavage, and finally assaying for hydrolysis by releasing Tyr[92] from the cleaved fragment with APP. Figure 2 shows the semilog plots of the percentage of free Tyr[92] released as a function of the incubation time with trypsin. Since trypsin can only cleave the Lys[91]-Tyr[92] bond when the Tyr[92]-Pro[93] bond is trans, the slow kinetic

[28] F. Toma, S. Fermandjian, M. Low, and L. Kisfaludy, *Biochim. Biophys. Acta* **534**, 112 (1978).

[29] E. R. Stimson, G. T. Montelione, Y. C. Meinwald, R. K. E. Rudolph, and H. A. Scheraga, *Biochemistry* **21**, 5252 (1982).

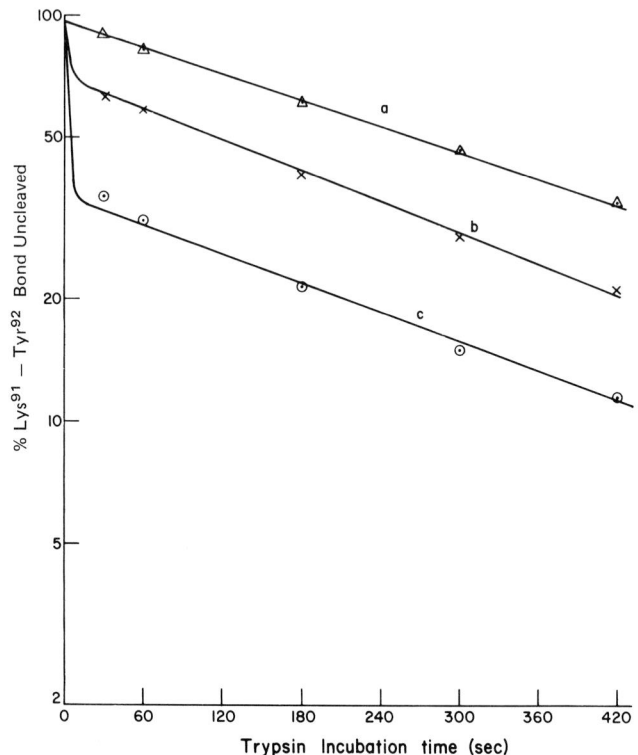

FIG. 2. Rates of hydrolysis of the Lys[91]-Tyr[92] bond of RNase catalyzed by trypsin at 10°. Initial states of RNase: (a) native form; (b) reversibly denatured form; (c) oxidized form. All samples were first treated with pepsin for 20 sec immediately before the addition of trypsin at zero time. Reprinted with permission from Lin and Brandts.[21] Copyright 1983 American Chemical Society.

phases seen in Fig. 2 can be attributed to the cis-to-trans isomerization of the Tyr[92]-Pro[93] bond and subsequent hydrolysis of the Lys[91]-Tyr[92] bond. The fast hydrolysis, implicit in the data of Fig. 2, corresponds to the rapid cleavage of the trans form present at the time the incubation began. The total amplitude of the slow phase, obtained by extrapolation to zero time, is then the percentage of cis form present in the original, intact RNase molecule. The estimates of cis form are 97% for the native, 70% for the reversibly denatured with disulfide bonds intact, and 36% for oxidized RNase. The relaxation time for the slow phase is nearly identical for all three species, as expected, with a value of 6.0 ± 0.5 min. It should be noted that these relaxation times are not for the whole RNase polypeptide chain, but for the pepsin-produced fragment containing the Tyr[92]-Pro[93] bond.

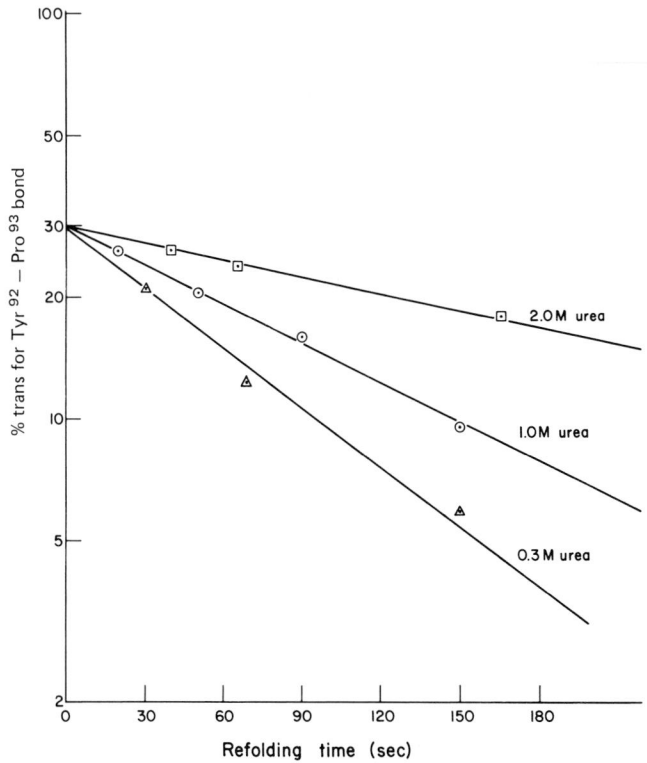

FIG. 3. The semilog plot of trans-to-cis isomerization of Tyr^{92}-Pro^{93} bond during RNase refolding at 10°. RNase was initially unfolded in 8.5 M urea solution, pH 4.0, and then refolded in 0.3 M (triangles), 1 M (circles), and 2 M (squares) urea solution by dilution. The content of the trans form of the Tyr^{92}-Pro^{93} bond at various refolding times was assayed by the method described for Fig. 2. Reprinted with permission from Lin and Brandts.[21] Copyright 1983 American Chemical Society.

Isomerization of Proline-93 during RNase Refolding and Unfolding. The isomerization of proline-93 during RNase refolding and unfolding was investigated by using the same experimental approach. The only difference was that RNase was refolded (or unfolded) for a period of time before the pepsin-trypsin treatment was initiated. The amount of cis proline-93 at each refolding or unfolding time was obtained by extrapolating the slow phase to zero trypsin incubation time, in the same manner as shown in Fig. 2.

The results of refolding experiments are shown in Fig. 3. In each case, RNase was equilibrated in 8.5 M urea at 10°, pH 4, to achieve denaturation. At zero time, the denatured RNase was transferred to a native-state buffer at 10°, pH 5, which contained low urea concentrations of either 0.3, 1.0, or 2.0 M. The abscissa represents the time allowed for refolding before the pepsin was added. These semilog plots are linear and show that the isomerization of proline-93 during refolding is a first-order process. It is seen that the relaxation time for disappearance of the trans isomer during refolding varies significantly with urea concentration, ranging from 90 sec at 0.3 M to 130 sec at 1 M to 310 sec at 2 M urea. This effect of urea on the rate of isomerization of proline-93 arises indirectly, as the result of an effect of urea on another process which becomes rate limiting at high urea concentrations.[12] The extrapolated amplitudes are identical, about 30%, which is consistent with previous finding from data in Fig. 2 that reversibly unfolded RNase has a 30% trans content.

Experiments similar to those described above for refolding of RNase have also been carried out in the unfolding direction and are shown in Fig. 4. For final denaturing conditions of 8.5 M urea, 10° and pH 2, relaxation time of 140 sec is obtained from the first order plot of Fig. 4.

In summary, the above data show that reversibly unfolded RNase in 8.5 M urea contains 70% of the Tyr^{92}-Pro^{93} bond in the cis form, compared to 35% for oxidized RNase; local interactions or steric hindrance associated with the disulfide bond 40–95 could be responsible. These results emphasize the fact that isomerization can be strongly influenced by residues other than those directly involved in the peptide bond. The conventional assumption that denatured proteins have prolines which are only ~10–20% cis seems to be unjustified, at least for those prolines known to be cis in the native form. Another indication of a long-range effect on isomerization is evident from comparing two-way relaxation times for isomerization of proline-93 in various polypeptide chains. These values range from ~7 min in fragment 92–98 to 3.2 min in oxidized RNase to 1.8 min for the pepsin fragment with disulfides intact to 1.1 min for reversibly unfolded RNase.

Future Prospects

The ISP method is a very new technique, having thus far been used in only one laboratory and having been applied in detail to only a single proline in a protein, i.e., proline-93 in ribonuclease. Studies are now completed on both proline-114 and -117 in ribonuclease,[4] where the method has also worked successfully.

The use of a non-proline-specific endopeptidase, such as trypsin, for the stereospecific cleavage step greatly limits the applicability of the method, and necessitates development of a different protocol for each proline to be examined. For the future, it seems possible to develop a

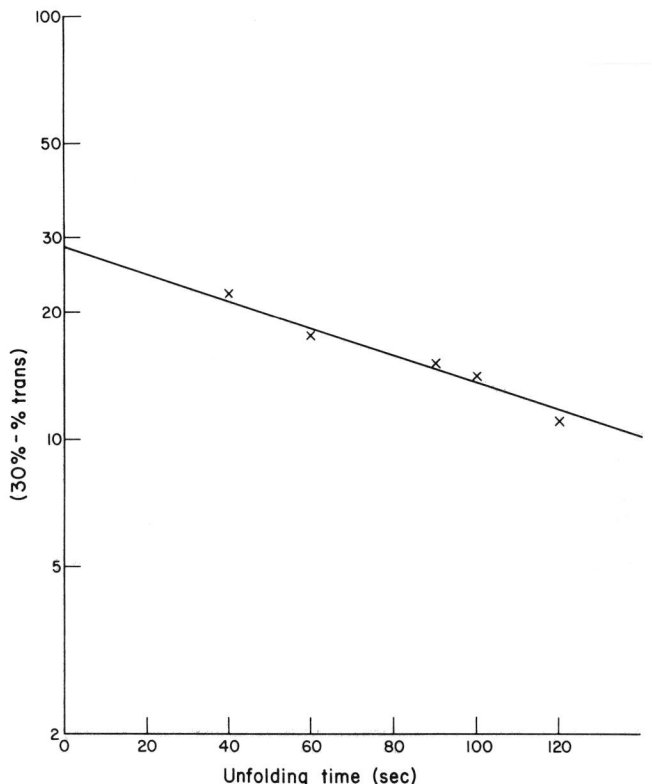

FIG. 4. The semilog plot of cis-to-trans isomerization of Tyr[92]-Pro[93] bond during RNase unfolding at 10°. Native Rnase in H_2O was unfolded in 8.5 M, pH 2, urea solution. At various times, the amounts of cis and trans forms of Tyr[92]-Pro[93] bond were determined by the method described for Fig. 2. Reprinted with permission from Lin and Brandts.[21] Copyright 1983 American Chemical Society.

general method based on the use of the proline-specific endopeptidase in the stereospecific cleavage steps. This would mean that many or all prolines in a single protein can be studied simultaneously and that essentially the same protocol can be used for all proteins except perhaps in the chain-trimming step. On the negative side, it seems likely that such a general method will require assays based on the identification and quantitation of peptide cleavage products, rather than free amino acids.

Although the ISP method was developed for the immediate purpose of understanding the slow phase seen in major unfolding and folding reactions of proteins, it may have a more important use over the long term. Many proteins and polypeptide hormones are known to exist in two or more slowly equilibrating, native-like states, having vastly different activity but only slightly different structure, i.e.,

$$\text{Active} \overset{\text{slow}}{\rightleftharpoons} \text{inactive}$$

In at least some cases where such subtle equilibria are known to be important in regulating biological activity [i.e., for concanavalin A,[7] prothrombin fragment I,[8] and superoxide dismutase[30] among proteins, and for the peptide hormone(s) angiotensin II and its analogs[31]] there is already indirect evidence to suggest that peptide bond isomerization may be the rate-limiting process. If nature has generally taken advantage of the high activation barrier of isomerization for providing kinetic control of activity-regulating equilibria, then the ISP method should prove to be quite valuable in the study of such phenomena.

[30] J. S. Valentine and M. W. Pantoliano, *Met. Ions Biol.* **3**, 291 (1981).
[31] H. E. Bleich, R. J. Freer, S. S. Stafford, and R. E. Galardy, *Proc. Natl. Acad. Sci. U.S.A.* **75**, 3630 (1980).

[7] Folding Intermediates Studied by Circular Dichroism

By A. M. LABHARDT

In protein folding investigations two kinds of information are of special interest: (1) the character of stable intermediate states of partial folding, and (2) the kinetics of the folding pathway.

(1) Local folding is presumed to seed the folding pathway.[1-4] The seeding structure has been termed the kernel if it is stable by itself and the nucleus if it is not.[5] Various suggestions have been made regarding the local structures (hydrophobic cluster,[6-8] α-helix,[9-11] β-hair-pins,[12-14] or

[1] S. Tanaka and H. A. Scheraga, *Proc. Natl. Acad. Sci. U.S.A.* **72**, 3802 (1975).
[2] M. Karplus and D. L. Weaver, *Nature (London)* **260**, 404 (1976).
[3] M. I. Kanehisa and T. Y. Tsong, *J. Mol. Biol.* **124**, 177 (1978).
[4] N. Go, *Annu. Rev. Biophys. Bioeng.* **12**, 183 (1983).
[5] P. S. Kim and R. L. Baldwin, *Annu. Rev. Biochem.* **51**, 459 (1982).
[6] W. Kauzmann, *Adv. Protein Chem.* **14**, 1 (1959).
[7] R. R. Matheson, Jr. and H. A. Scheraga, *Macromolecules* **11**, 819 (1978).

reverse turns[7]) involved in kernels or nuclei of folding. Kernels have been searched for with circular dichroism (CD) by seeking particularly stable elements within the overall structure in unfolding conditions,[15–17] in the reduced protein in conditions that favor folding[18–23] and in proteolytic fragments or synthetic peptides.[24–26]

Folding intermediates have been characterized by CD at enthalpic minima in nonnative conditions[17,27,28] or preceding a high activation barrier (either an intrinsic barrier of the folding reaction[29,30] or an artificial barrier produced by blocking the disulfide bond formation during reoxidation of reduced protein[23,31–34]; for a review see, e.g., Ref. 35) or using fragments of the native protein.[36–45]

[8] H. A. Scheraga, in "Protein Folding" (R. Jaenicke, ed.), p. 261. Elsevier, Amsterdam, 1980.

[9] D. Kotelchuck and H. A. Scheraga, Proc. Natl. Acad. Sci. U.S.A. 62, 14 (1969).

[10] V. I. Lim, FEBS Lett. 89, 10 (1978).

[11] V. I. Lim, in "Protein Folding" (R. Jaenicke, ed.), p. 149. Elsevier, Amsterdam, 1980.

[12] O. B. Ptitsyn and A. V. Finkelstein, in "Protein Folding" (R. Jaenicke, ed.), p. 101. Elsevier, Amsterdam, 1980.

[13] O. B. Ptitsyn and A. V. Finkelstein, Q. Rev. Biophys. 13, 339 (1980).

[14] O. B. Ptitsyn, A. V. Finkelstein, and P. Falk, FEBS Lett. 101, 1 (1979).

[15] K. Kuwajima, K. Nitta, M. Yoneyama, and S. Sugai, J. Mol. Biol. 106, 359 (1976).

[16] K. Kuwajima, Y. Ogawa, and S. Sugai, J. Biochem. 89, 759 (1981).

[17] R. W. Henkens, B. B. Kitchell, S. C. Lottich, P. J. Stein, and T. J. Williams, Biochemistry 21, 5918 (1982).

[18] C. B. Anfinsen, E. Haber, M. Sela, and F. H. White, Proc. Natl. Acad. Sci. U.S.A. 47, 1309 (1961).

[19] A. Fontana, C. Vita, and D. Dalzoppo, Proc. Int. Symp. Biomol. Struct. Interactions, Suppl. J. Biosci. 8, 57 (1985).

[20] A. M. Tamburro, E. Boccu, and L. Celotti, Int. J. Pept. Protein Res. 2, 157 (1970).

[21] S. W. Schaffer, Int. J. Pept. Protein Res. 7, 179 (1975).

[22] S. Takahashi, T. Kontani, M. Yoneda, and T. Ooi, J. Biochem. 82, 1127 (1977).

[23] M. Hollecker, T. E. Creighton, and M. Gabriel, Biochimie 63, 835 (1981).

[24] J. E. Brown and W. A. Klee, Biochemistry 10, 470 (1971).

[25] A. Bierzynski, P. S. Kim, and R. L. Baldwin, Proc. Natl. Acad. Sci. U.S.A. 79, 2470 (1982).

[26] P. S. Kim, A. Bierzynski, and R. L. Baldwin, J. Mol. Biol. 162, 187 (1982).

[27] M. Desmadril and J. M. Yon, Biochemistry 23, 11 (1984).

[28] S. Era, H. Ashida, S. Nagaoka, H. Inouye, and M. Sogami, Int. J. Pept. Protein Res. 22, 333 (1983).

[29] L. F. McCoy, E. S. Rowe, and K.-P. Wong, Biochemistry 19, 4738 (1980).

[30] S. Gerard, D. Puett, and W. M. Mitchell, Biochemistry 20, 1857 (1981).

[31] A. Galat, T. E. Creighton, R. C. Lord, and E. R. Blout, Biochemistry 20, 594 (1981).

[32] A. Ménez, F. Bouet, W. Guschlbauer, and P. Fromageot, Biochemistry 19, 4166 (1980).

[33] P. A. Kosen, T. E. Creighton, and E. R. Blout, Biochemistry 20, 5744 (1981).

[34] P. A. Kosen, T. E. Creighton, and E. R. Blout, Biochemistry 22, 2433 (1983).

[35] T. E. Creighton, Prog. Biophys. Mol. Biol. 33, 231 (1978).

[36] R. G. Reed, R. C. Feldhoff, O. L. Clute, and T. Peters, Jr., Biochemistry 14, 4578 (1975).

(2) Analyzing the kinetics of folding by CD gives information on the structure built up during the rate-limiting step(s). Two possibilities exist: Formation of the structure itself is rate limiting or else a transient intermediate impedes folding. For example, the folding of S-peptide of ribonuclease S when added to folded S-protein is a rapid reaction[44]; however, during folding of ribonuclease S the α-helix 3–13 of S-peptide is formed in a slow reaction[45] although combination between S-peptide and refolding S-protein occurs rapidly.[46] Evidently a transient intermediate is detected in this way.

Alternatively the absence of CD detected kinetics has been used to make inferences on secondary structure formation preceding the rate-limiting step(s).[29,41,45,47]

Depending on the perspective, equilibrium or kinetic (stopped flow) CD has been used, in connection with various chemical and biochemical modifications of the proteins, to search for intermediates. The CD of equilibrium folding intermediates has been treated in some cases[17,22,23,28,43,48] by methods established for folded proteins[49–55] of extracting the secondary structure composition from CD spectra. Implicit assumptions and potential pitfalls are discussed in the section below. The use of near-UV CD is not reviewed here.

For fast kinetic measurements, home built or modified CD spectrome-

[37] E. A. Carrey and R. H. Pain, *Biochem Soc. Trans.* **5,** 689 (1977).

[38] A. Hogberg-Raibaud and M. E. Goldberg, *Biochemistry* **16,** 4014 (1977).

[39] A. M. Labhardt, *in* "Protein Folding" (R. Jaenicke, ed.), p. 401. Elsevier, Amsterdam, 1980.

[40] B. Adams, R. J. Burgess, E. A. Carrey, I. R. Mackintosh, C. Mitchinson, R. M. Thomas, and R. H. Pain, *in* "Protein Folding" (R. Jaenicke, ed.), p. 447. Elsevier, Amsterdam, 1980.

[41] K. O. Johanson, D. B. Wetlaufer, R. G. Reed, and Th. Peters, Jr., *J. Biol. Chem.* **256,** 445 (1981).

[42] A. M. Labhardt, *Biopolymers* **20,** 1459 (1981).

[43] A. M. Labhardt, *J. Mol. Biol.* **157,** 331 (1982).

[44] A. M. Labhardt, *J. Mol. Biol.* **157,** 357 (1982).

[45] A. M. Labhardt, *Proc. Natl. Acad. Sci. U.S.A.* **81,** 7674 (1984).

[46] A. M. Labhardt, J. A. Ridge, R. N. Lindquist, and R. L. Baldwin, *Biochemistry* **22,** 321 (1983).

[47] M. Erard, E. Burggraf, and J. Pouyet, *FEBS Lett.* **149,** 55 (1982).

[48] A. J. Hillquist Damon and G. C. Kresheck, *Biopolymers* **21,** 895 (1982).

[49] N. Greenfield and G. D. Fasman, *Biochemistry* **8,** 4108 (1969).

[50] V. P. Saxena and D. B. Wetlaufer, *Proc. Natl. Acad. Sci. U.S.A.* **68,** 969 (1971).

[51] Y. H. Chen, J. T. Yang, and H. M. Martinez, *Biochemistry* **11,** 4120 (1972).

ters with mechanical stopped-flow devices[47,56,57] have been used. For a recent review of fast CD instrumentation, see Bayley.[58]

Characterization of Folding Kernels

By definition, only a folding kernel can be observed and characterized by equilibrium spectroscopy. A nucleus will decay or rapidly induce further folding and hence will not accumulate. Such presumed kernels have been investigated with CD by unfolding whole proteins by extremes of pH,[28,44,45,59] by the action of heat[39,42,43] or denaturants such as GuHCl or urea,[15-17,27,29,60] by the reduction of disulfide bonds,[20-23,32-34,61,62] or by characterizing the conformation of protein fragments.[24-26,37,41]

The interpretation of the spectral properties of the unfolded state is difficult. It has been suggested on the basis of CD and ORD data that pH and heat unfolded proteins[15,16,39,42,43,62-65] as well as reduced proteins[20-23] retain structure capable of being disrupted in concentrated solutions of GuHCl. CD and optical rotatory dispersion data taken with solutions of denaturants must be interpreted with caution since anomalous optical effects have been proposed to arise from the interaction of proteins with denaturants, such as GuHCl, urea, and sodium dodecyl sulfate (SDS).[64,66-68] For arguments for the contrary, see Ref. 48 and references cited therein. Enhancements of aromatic bands upon denaturation of pro-

52 Y. H. Chen, J. T. Yang, and K. H. Chau, *Biochemistry* **13**, 3350 (1974).
53 C. T. Chang, C.-S. C. Wu, and J. T. Yang, *Anal. Biochem.* **91**, 13 (1978).
54 S. Brahms and J. Brahms, *J. Mol. Biol.* **138**, 149 (1980).
55 J. P. Hennessey and W. C. Johnson, *Biochemistry* **20**, 1085 (1981).
56 J. Luchins and S. Beychok, *Science* **199**, 425 (1978).
57 H.-P. Bächinger, H.-P. Eggenberger, and G. Hänisch, *Rev. Sci. Instrum.* **50**, 1367 (1979).
58 P. M. Bayley, *Prog. Biophys. Mol. Biol.* **37**, 149 (1981).
59 K. Nitta, T. Segawa, K. Kuwajima, and S. Sugai, *Biopolymers* **16**, 703 (1977).
60 F. X. Schmid, *FEBS Lett.* **139**, 190 (1982).
61 N. Okabe, E. Fujita, and K.-I. Tomita, *Biochim. Biophys. Acta* **700**, 165 (1982).
62 Ch. Tanford and K. C. Aune, *Biochemistry* **9**, 206 (1970).
63 K. C. Aune, A. Salahuddin, M. H. Zarlengo, and Ch. Tanford, *J. Biol. Chem.* **242**, 4486 (1967).
64 Ch. Tanford, *Adv. Protein Chem.* **23**, 121 (1968).
65 Ch. Tanford, *Adv. Protein Chem.* **24**, 1 (1970).
66 B. Jirgensons and S. Capetillo, *Biochim. Biophys. Acta* **214**, 1 (1970).
67 M. L. Tiffany and S. Krimm, *Biopolymers* **12**, 575 (1973).
68 D. Balasubramanian, *Biopolymers* **13**, 407 (1974).

teins with alkali have been observed.[69-73] To assess the contributions of aromatic residues or disulfide bonds to the weak amide region of unfolded proteins the disulfide bonds can be reduced[20] or the tyrosine residues can be titrated.[15,16,71] Nonrandom CD spectra in the unfolded state when generated by residual structure may be protease sensitive.[24,44]

Characterization of Equilibrium Folding Intermediates by CD

CD is becoming increasingly popular as a spectral probe to monitor equilibrium transitions. In single wave length experiments, the strong α-helix band is usually selected, generally at 222 nm.[27,29,37,48,74]

Recently attempts have been made to analyze and quantitate the secondary structure of folding intermediates by amide CD measured at multiple wavelengths.[17,19,22,23,28,39,43,48] Spectral decompositions have been carried out using the methods of Provencher and Glöckner[75] and Chen et al.[51,52] or modifications thereof.[39,43] These procedures use the CD spectra of proteins of known conformation as a reference base. All procedures are reported to be successful in estimating the correct α-helix and β-pleated sheet contents. In general, varying success is reported when such additional information is extracted as the average length of helical segments,[52] the fraction of β-turns,[53,54,76] or differentiation between parallel and antiparallel β-sheets.[77] For this reason it seems inevitable to limit the CD analysis of folding intermediates to monitoring the changes of α-helix and β-sheet contents.

The following problems and potential pitfalls associated with spectral decomposition must be clearly realized.

1. Strong aromatic contributions to the amide region generally cannot be excluded. For example, the CD of native BPTI is considered abnormal[33,34,78] and an incorrect analysis for secondary structure is obtained from CD data.[23] A comparative study of the homologous proteins toxins I and K from black mamba has led to the suggestion that the unusual CD of

[69] B. Jirgensons, *Biochim. Biophys. Acta* **317**, 131 (1973).

[70] Y.-Y. T. Su and B. Jirgensons, *Macromol. Chem.* **180**, 367 (1979).

[71] B. Jirgensons, *Biochim. Biophys. Acta* **625**, 193 (1980).

[72] E. H. Strickland, *CRC Crit. Rev. Biochem.* **2**, 113 (1974).

[73] Y. Tamura and B. Jirgensons, *Arch. Biochem. Biophys.* **199**, 413 (1980).

[74] J. L. Barbero, L. Franco, F. Montero, and F. Moran, *Biochemistry* **19**, 4080 (1980).

[75] S. W. Provencher and J. Glöckner, *Biochemistry* **20**, 33 (1981).

[76] S. Brahms and J. G. Brahms, *J. Chim. Phys.* **76**, 841 (1979).

[77] I. A. Bolotina, V. O. Chekhov, and V. Y. Lugauskas, *Int. J. Quantum Chem.* **16**, 819 (1979).

[78] H. Rosenkranz, *in* "Protein Inhibitors" (H. Fritz, H. Tschesche, L. J. Greene, and E. Truscheit, eds.), p. 458. Springer-Verlag, Berlin, 1974.

BPTI and toxin K results from Tyr-35.[23] In BPTI this residue is immobilized.[79] Toxin I but not K lacks Tyr-35. A strong and broad positive band around 226 nm is thought to be generated by tyrosine residues interacting with peptide bonds.[72,73,80,81] Likewise the far-UV CD of RNase A may be influenced by broad positive contributions from tyrosines[82] which overlap with the more narrow amide bands. As a result, the region of cancellation around 235 nm may be particularly sensitive to changes in CD which remain undetected at shorter wavelengths.[83] For bovine carbonate dehydratase, Henkens et al.[17] report slight but systematic deviations due to aromatic contributions when fitting the amide CD by the method of Chen et al.,[51] and the estimates for helix and sheet fractions remain lower than the crystal structure values.[84]

In particular, single wavelength measurements made in the amide region, but outside the strong α-helix bands, or with proteins that lack significant fractions of helix do not necessarily report secondary structure changes.

2. When using reference spectra determined from native proteins to analyze folding intermediates, the following additional problems exist. (1) During unfolding new conformations may arise and dominate the CD spectrum. Tests must be made for the presence of CD active conformations in the unfolded protein (compare section on kernels). (2) By using "native" basis spectra for α-helix and β-sheet, without correction for size, exposure to solvent, or involvement in supersecondary structure, one implicitly assumes that these factors are unimportant. Evidence to the contrary has been given. Chen et al.[52] have shown that the α-helical CD spectrum depends on helix length. Manavalan and Johnson[85] summarize evidence for the dependence of the CD spectra on the supersecondary structure and on the conformational class (all-α, $\alpha + \beta$, α/β, all β).[86] The CD spectra of some all-β proteins resemble the spectra of models for the random coil.[85] Analysis of the CD spectrum of reduced toxin I predicts 50% β-sheet.[23] As pointed out by the authors, this is probably incorrect. Similarly, the calculated β-pleated sheet content of reduced and

[79] G. H. Snyder, R. Rowan, M. Karplus, and B. D. Sykers, Biochemistry 14, 3765 (1975).
[80] M. Baba, K. Hamaguchi, and T. Ikenaka, J. Biochem. 65, 113 (1969).
[81] R. W. Woody, Biopolymers 17, 1451 (1978).
[82] C. R. Cantor and S. N. Timasheff, "The Proteins" (H. Neurath and R. L. Hill, eds.), Vol. 5, 3rd Ed., p. 145 (1982).
[83] M. N. Pflumm and S. Beychok, J. Biol. Chem. 244, 3973 (1969).
[84] K. K. Kaunan, A. Liljas, I. Waara, P.-C. Bergsten, S. Lovgren, B. Standberg, U. Bengtsson, U. Carbom, K. Fidborg, L. Jarup, and M. Petef, Cold Spring Harbor Symp. Quant. Biol. 136, 221 (1972).
[85] P. Manavalan and W. C. Johnson, Jr., Nature (London) 305, 831 (1983).
[86] M. Levitt and C. Chothia, Nature (London) 261, 552 (1976).

carboxymethylated RNase A, has a value between 0 and 25% depending on the reference spectra used.[22] (3) The CD of the nonperiodic structure in the native protein (referred to as irregular structure, excluding the α-helix and β-sheet regions) may be significantly different from that of the unfolded "random coil" protein. Fitting the CD spectra of RNase A or S either in the thermally denatured state (pH 1.7, 60°), or in the presence of 5 M GuHCl, or after tryptic digestion by the method of Chen *et al.*,[51] would incorrectly indicate that one-third to one-half of the native β-sheet content is retained.[87] This result has been traced to the inappropriate use of the irregular structure reference spectrum for the unfolded random coil state. Irregular structure reference spectra vary significantly,[51–53,77] depending on the types and number of reference proteins. They tend to deviate more or less systematically from CD spectra obtained from polypeptides that are considered to be models for the random coil state. It has been suggested[39,43] that the reference spectrum for the irregular structure in the native state should be taken from the measured random coil spectrum of the same protein, since inadequacies in the presence of the strong helix and sheet CD spectra become relatively smaller.

3. Alternatively one can depart completely from the use of pure-state reference spectra and instead decompose the spectral changes measured on unfolding into fractional changes based on sufficiently orthogonal spectra obtained from known mixed conformations. Using native RNase S and folded and thermally denatured S-protein as references, the spectra in the unfolding transitions of RNase S and A were decomposed in this way[43] and given a structural interpretation. As noted by Era *et al.*,[28] extension of this approach to a basis set of five reference proteins yields a decreased signal-to-noise ratio and the conceptually equivalent regulator method of Provencher and Glöckner[75] should be used.

Characterization of Transient Folding Intermediates by Stopped Flow CD

CD as a kinetic technique has several practical difficulties. Before deciding to start such investigations, these problems should be carefully considered.

1. The sensitivity in the amide region is low when short response times are needed.

2. As a consequence, high protein concentrations (above 1 mg/ml) and multiple signal averaging are necessary to obtain reliable results.

---

[87] A. M. Labhardt, unpublished results.

3. The instrumentation needed for *fast* kinetics is currently still beyond the reach of most laboratories.[56–58]

The following points have been emphasized as advantages of the CD stopped-flow technique for investigations of folding. It is a multiparameter approach. Differences in the kinetics can be monitored by aromatic absorbance, aromatic CD, and amide CD. This allows one to order, respectively, the burial of the aromatic groups in the interior of the protein, the fixation of aromatic groups in a folded environment, and the generation of the backbone conformation. Intrinsic fluorescence from tyrosine residues has also been used to monitor hydrogen bonding of these residues during folding.

The kinetics of heme-induced refolding of human α-globin has been investigated[88] using the far-UV instrument described by Luchins and Beychok.[56] The kinetics were observed by CD at 222 nm and compared to the kinetics measured by absorbance in the Soret band and by quenching of tryptophan fluorescence. The instrument was calibrated at 290 nm against *d*-10-camphorsulfonic acid and at 222 nm against a solution of methemoglobin. The signal-to-noise ratio at 222 nm was optimized by testing samples of various concentrations with absorbance in the range 0.15 to 0.5 OD. Data acquisition used a Nicolet Explorer 1090A digital oscilloscope. Based on the measured rates, the following interpretation of events was given. (1) First the heme enters the pocket with a half-time of 10 msec at 2.4 μM concentration, 4°. (2) Subsequently the pocket assumes its final conformation (halftime 40 sec). (3) The induced growth of the α-helices is slow and biphasic with the major component having a half-time of about 160 sec.

Slow helix formation has also been observed for the S-peptide helix 3–13 of RNase S[45]: In pH-jumps from low to neutral pH this helix forms in the terminal phase of folding with the same time constant measured for the burial of the tyrosines in the S-protein moiety (42 ± 6 sec). The regain of affinity for 2'-CMP occurs in the same terminal phase of folding of the S-peptide : S-protein complex. In this work the far-UV instrument of Bächinger *et al.*[57] was used. Up to 30 single shots, each one consuming 0.4 ml of solution, were averaged to improve the signal-to-noise ratio in the millisecond time range. Data collection used a Datalab transient recorder model DL 905 interfaced to an LSI 11/23 computer. A significant improvement of the signal-to-noise performance was achieved by executing a sliding average (integration) over 5 to 10 subsequent points of the digital transients. In order to avoid distortion of the progress curves, care had to be taken that the time interval corresponding to the length of the

[88] Y. Leutzinger and S. Beychok, *Proc. Natl. Acad. Sci. U.S.A.* **78**, 780 (1981).

sliding average did not exceed $0.1t_{min}$, where t_{min} is the fastest relaxation time. An alternative smoothing procedure[89] has been used by Hasumi in kinetic CD studies of the alkaline isomerization of ferricytochrome c[90] and spinach ferredoxin.[91]

Terminal secondary structure formation may be the exception rather than the rule. In refolding of bovine carbonate dehydratase B from 6 M GuHCl,[29] secondary structure forms (as monitored by CD_{222}) before the burial of the chromophores (as monitored by absorbance at 291.5 nm) and clustering of the aromatic residues (as measured by CD_{270}). The kinetic CD measurements were made by rapid manual mixing, with a dead time of 20 sec. Data were collected either at single wavelength settings, or—because of the very slow kinetics—by a 8.8 min scan of the wavelength region 240 to 205 nm on a Jasco J 20 spectropolarimeter.

Goto and Hamaguchi[92] report the refolding kinetics of the constant fragment of the immunoglobin light chain. Of the backbone CD 70% returns in the dead time of the experiment (1 min). Refolding was monitored by manually diluting the 7 mg/ml protein solution from 4 M GuHCl to 0.1 M GuHCl. The CD signal was recorded with a Jasco J-20 photometer with CD attachment at 17 wavelengths between 235 and 208 nm.

The refolding of swine pepsinogen was investigated in a similar fashion.[93] The protein was unfolded at pH 8 or 11.5 by 6 M urea. Dilution to 0.29 M urea produces within the manual mixing dead time a folding intermediate. Its UV-CD spectrum was recorded at 13 wavelengths between 240 and 210 nm. It closely resembles the CD spectrum of native pepsinogen. Spectral decomposition was attempted as described by White.[94] Refolding monitored by other probes appears slow.

Kuwajima *et al.*[95] have introduced a rapid dilution device consisting of a conventional cuvette and a magnetic stirrer. The dead-time is decreased to 3 sec. They have investigated the refolding of α-lactalbumin and lysozyme by a 20-fold dilution of the protein solutions containing 6 M GuHCl. In the aromatic region almost the full changes expected from equilibrium studies are detected kinetically with time constants larger than 10 and 15 sec, respectively. In contrast most of the ellipticity changes in the peptide region occur in the dead-time. They extrapolate the CD kinetics to the dilution time point to get the CD spectra of the early folding intermedi-

[89] A. Savitzki and M. J. E. Golay, *Anal. Chem.* **36,** 1627 (1964).
[90] H. Hasumi, *Biochim. Biophys. Acta* **626,** 265 (1980).
[91] H. Hasumi, *J. Biochem.* **92,** 1049 (1982).
[92] Y. Goto and K. Hamaguchi, *J. Mol. Biol.* **156,** 891 (1982).
[93] P. McPhie, *Biochemistry* **21,** 5509 (1982).
[94] F. H. White, *Biochemistry* **15,** 2906 (1976).
[95] K. Kuwajima, Y. Hiraoka, M. Ikeguchi, and S. Sugai, *Biochemistry* **24,** 874 (1985).

ates. In the case of α-lactalbumin this spectrum is compatible with the spectrum of the A state. The latter is an equilibrium state of intermediate folding observed in 2 M GuHCl below pH 5[15] and above pH 8.5.[16] Its spectral characteristics are the absence of aromatic CD and the presence of most of the peptide CD.

It had been observed quite early by CD that secondary structure formation precedes the regain of activity during reoxidation of reduced RNase A.[18,21,22] This is consistent with the common opinion that local structural preferences drive the selection of disulfide bond formation (e.g., Ref. 35). Note that the time range for these experiments is hours instead of seconds for refolding with the disulfide bonds intact.

The unfolding of apomyoglobin[96] and myoglobin[97,98] after a pH jump from neutral to low pH has been reported by Kihara et $al.$ using stopped flow measurements of CD at 222 nm. A Union Giken CD-1002 stopped flow spectrophotometer with 2 msec dead-time and an electronic response time of 1/13 msec was used. In such experiments, the helical structure of apomyoglobin breaks down too rapidly to be followed kinetically. In the case of myoglobin, however, part of the helical CD disappears in the second time range.

The unfolding of the RNase S-peptide helix 3–13 when lowering the pH from 6.6 to 1.7 could be monitored at 13° by CD at 225 nm.[45,57] This reaction has a time constant of 513 ± 150 msec and represents the first conformational change detected in the unfolding of RNase S.

At pH 3 the tetrameric lectin concanavalin A (Con A) dissociates into dimers with no gross conformational change of the subunits.[99] The kinetics of unfolding of dimeric Con A by 4 to 8 M urea was investigated.[100] The CD signal was measured after manual mixing (30 sec) at 218, 225, and 283 nm with a Jasco 500 dichrometer. The traces were digitized and smoothed by the Savitzky–Golay procedure.[89]

The denaturation of TMV coat protein by urea or by GuHCl has been reported using stopped-flow measurements of CD at 222 nm.[101] An entirely commercial setup with a dead time of 20 msec and averaging capabilities of up to 64 transients was used.

[96] H. Kihara and E. Takahashi, $Biochem.$ $Biophys.$ $Res.$ $Commun.$ **95,** 1687 (1980).

[97] H. Kihara, E. Takahashi, K. Yamamura, and I. Tabushi, $Biochim.$ $Biophys.$ $Acta$ **702,** 249 (1982).

[98] H. Kihara, E. Takahashi-Ushijima, and S. Saigo, $Jichi$ $Med.$ $J.$ **6,** 143 (1983).

[99] H. E. Auer and T. Schilz, $Int.$ $J.$ $Pept.$ $Protein$ $Res.$ **24,** 462 (1984).

[100] H. E. Auer and T. Schilz, $Int.$ $J.$ $Pept.$ $Protein$ $Res.$ **24,** 569 (1984).

[101] Y. Sano and H. Inoue, $Chem.$ $Lett.$ 1087 (1979).

[8] Amide Proton Exchange as a Probe of Protein Folding Pathways[1]

By PETER S. KIM

Introduction

The practical problems in determining a pathway of protein folding are quite formidable: a pathway has not been determined for any protein in structural terms. The methods that one would like to use (X-ray crystallography, NMR) to study structural aspects of protein folding are intrinsically slow, so that equilibrium folding intermediates are desired. In general, however, equilibrium intermediates are not significantly populated for single domain proteins, most likely due to marginal stability of intermediates inside the unfolding transition zone.

Kinetic studies of folding have the advantage that they can be carried out in conditions where the folded state is strongly favored and intermediates are more likely to be stable.[2] Although well-populated intermediates have been detected in kinetic folding experiments,[2a] the intermediates are transient and steps in folding are often fast (msec–min). Strategies for dealing with these problems include slowing down folding (e.g., at subzero temperatures in cryosolvents[3]) or adapting spectroscopic methods so that rapid measurements can be made (e.g., stopped-flow CD studies[4]). A third approach is summarized here: the protection from labeling that is provided by structure is measured at different stages during folding. The location and extent of labeling are determined after folding has gone to completion. Thus, analysis of the partially labeled protein is used to infer the structural state of the protein when it was labeled.

The choice of a labeling reagent is crucial. It is important that folding is not affected by the labeling reaction or conditions. Steps in folding are often fast, so the labeling reaction must be fast, and the ability to quench labeling is highly desirable. Since the labels will be located after folding has gone to completion, it is important that the native structure of the protein is not perturbed by labeling. An ideal labeling reaction, with regard to the above requirements, is amide proton exchange.

[1] Dedicated to the memory of Michael P. Graf.
[2] R. L. Baldwin, *Annu. Rev. Biochem.* **44**, 453 (1975).
[2a] Reviewed by P. S. Kim and R. L. Baldwin, *Annu. Rev. Biochem.* **51**, 459 (1982).
[3] A. L. Fink, this volume [10]
[4] A. M. Labhardt, this volume [7].

METHODS IN ENZYMOLOGY, VOL. 131

Methods for studying protein folding using amide proton exchange are described here. The feasibility of using amide proton exchange to study folding has been demonstrated using the radioactive isotope, ^3H. Recent work has begun to take advantage of the fact that the extent of labeling for individual amide protons can be determined using high resolution ^1H NMR and deuterium labeling. A strategy to characterize structural intermediates in folding using ^1H NMR and D_2O as a labeling reagent has been proposed.[5] It seems likely that these methods will provide a detailed and relatively high resolution picture of the folding process for small monomeric proteins.

Amide Proton Exchange

The probes for folding used here are the amide protons (peptide NH) of the polypeptide backbone. These nitrogen-bound protons exchange with solvent protons[6]; the exchange reaction is acid and base catalyzed, with a minimum exchange rate occurring near pH 2–3 (pH$_{min}$). Amide protons can thus be labeled with deuterium or tritium, using isotopic water (D_2O or ^3H_2O).

The hydrogen-bond (H-bond) donors in both α-helices and β-sheets are the amide protons. When amide protons are H-bonded, their exchange rates drop dramatically, since the H-bond must break before exchange can occur; exchange involves the addition or removal of a proton by standard proton transfer mechanisms.[7,8] Solvent exclusion also retards exchange, since some solvent must be accessible to the peptide NH group in order for exchange to occur. H-bonding and solvent exclusion are thought to be chiefly responsible for the enormous reduction of exchange rates (up to 10^{10} times slower than the corresponding rate in a freely exposed amide) for some protons in native proteins. Amide proton exchange provides a very sensitive probe of structure formation.

The pH dependence of amide proton exchange rates offers a large dynamic range. In model compounds, exchange rates increase 10-fold for each pH unit increase in the base-catalyzed pH region (and vice versa in the acid-catalyzed region) of exchange. The half-time for exchange can be varied from greater than 1 hr (pH 3, 0°) to less than 1 msec (pH 10, 0°) by changing pH.[6] This makes it feasible to label a transient folding intermedi-

[5] K. Kuwajima, P. S. Kim, and R. L. Baldwin, *Biopolymers* **22**, 59 (1983).
[6] Reviewed by S. W. Englander, N. W. Downer, and H. Teitelbaum, *Annu. Rev. Biochem.* **41**, 903 (1972).
[7] M. Eigen, W. Kruse, G. Maass and L. DeMaeyer, *in* "Progress in Reaction Kinetics" (G. Porter, ed.), Vol. 2, p. 286. Pergamon, Oxford, 1964.
[8] M. Eigen, *Angew. Chem Int. Ed.* **3**, 1 (1964).

ate. For many small proteins it is also possible to quench exchange, if the protein can refold at low pH.

The replacement of ^1H with ^2H or ^3H is a minor perturbation as compared to most chemical labeling procedures. Where it has been checked, the structures of proteins are the same before and after the amide protons are replaced by deuterium atoms, as judged by X-ray crystallography.[9–11] Nevertheless, isotope effects and solvent perturbations (i.e., H_2O vs D_2O) must be considered in any amide proton exchange experiment (see Isotope and Solvent Effects).

Since deuterium atoms are not observed in an ^1H NMR spectrum, it is possible to use ^1H NMR to quantitate the extent of labeling at individual sites in a protein. Recent developments of ^1H NMR techniques to study proteins (in particular, the application of two-dimensional NMR techniques[12] by Wüthrich and co-workers[13,14] should make it feasible to "track" many individual amide protons in the folding processes of small, globular proteins.[5]

Exchange from the Native State

The mechanism of amide proton exchange from proteins, in conditions where the native conformation is stable, is the subject of much investigation and discussion.[15–20] For now, we simply note that studies of amide proton exchange in numerous proteins have shown that there is often a group of protons with substantially reduced rates of exchange (slowly exchanging protons). For example, approximately 50 amide protons in RNase A are resistant to exchange-out for at least 6 hr at pH 6, 10°. These protons have exchange rates that are reduced by at least a factor of $\sim 10^4$, since the average half-time for exchange in a freely exposed amide proton is a few seconds at pH 6, 10°.

The slowly exchanging protons are particularly well suited as probes for the folding process. (1) They are sensitive probes of structure formation; the exchange rates of these protons decrease by factors of 10^4 to 10^{10}

[9] A. A. Kossiakoff, *Nature (London)* **296**, 713 (1982).

[10] A. Wlodawer and L. Sjölin, *Biochemistry* **22**, 2720 (1983).

[11] B. P. Schoenborn, *Cold Spring Harbor Symp. Quant. Biol.* **36**, 569 (1971).

[12] W. P. Aue, J. Karhan, and R. R. Ernst, *J. Chem. Phys.* **64**, 4226 (1976).

[13] K. Wüthrich, G. Wider, G. Wagner, and W. Braun, *J. Mol. Biol.* **155**, 311 (1982).

[14] K. Wüthrich, *Biopolymers* **22**, 131 (1983).

[15] F. M. Richards, *Carlsberg Res. Commun.* **44**, 47 (1979).

[16] G. Wagner and K. Wüthrich, *J. Mol. Biol.* **160**, 343 (1982).

[17] C. Woodward, I. Simon, and E. Tüchsen, *Mol. Cell. Biochem.* **48**, 135 (1982).

[18] G. Wagner, *Q. Rev. Biophys.* **16**, 1 (1983).

[19] R. B. Gregory, L. Crabo, A. J. Percy, and A. Rosenberg, *Biochemistry* **22**, 910 (1983).

[20] S. W. Englander and N. R. Kallenbach, *Q. Rev. Biophys.* **16**, 521 (1983).

as the protein folds. (2) The slowly exchanging protons often represent a substantial fraction of the backbone amide protons in a protein ($\sim$50/119 in RNase A and $\sim$30/53 in BPTI). (3) Since they are resistant to exchange in the native state, analysis of these protons can be carried out using methods that are intrinsically slow (e.g., NMR). (4) Most of the highly protected amide protons are H-bonded in secondary structure (α-helices, β-sheets) of the native protein in known cases (trypsin,[9] BPTI,[16,18] RNase A[10]). Therefore, exchange measurements of these protons in folding intermediates is likely to provide a readily interpretable picture, in terms of secondary structure units.

Overview of the Experimental Strategies

The Competition Method

Schmid and Baldwin[21] were the first to study a kinetic folding process with amide proton exchange. The experimental design was to set up a competition between refolding and exchange-out of the amide protons. Amide protons of the slow-folding species of RNase A were labeled with ^{3}H. The labeled, unfolded protein (in GuHCl and at the pH_{min}) was then rapidly diluted with nonradioactive buffer. This initiates the competition: the protein refolds since [GuHCl] is low, and exchange-out from the amide protons occurs at a rate that depends on the pH of the solution. In the absence of folding intermediates, the competition experiment can be represented as:

$$U^* \xrightarrow{k_{folding}} N^*$$
$$k_{HX} \downarrow$$
$$U$$

$$(1)$$

where the asterisk indicates retention of label and $k_{folding}$ and k_{HX} refer to the rate constants for folding and amide proton exchange, respectively. Label trapped in N^* remains, since only the stable amide protons are studied.

In Eq. (1), a single reactant (U^*) directly produces two products (U and N^*). Thus, the relative concentrations of the two products gives the relative rate constants. Since those molecules that lose label ($U^* \rightarrow U$) will refold to give nonradioactive protein (N), we have

$$N^*/N = k_{folding}/k_{HX} \qquad (2)$$

This equation assumes that there are no structural intermediates between U and N so that trapping of amide protons follows the same kinet-

---

[21] F. X. Schmid and R. L. Baldwin, *J. Mol. Biol.* **135**, 199 (1979).

ics as other probes of folding (e.g., tyrosine absorbance changes). The results obtained by Schmid and Baldwin[21] showed that substantially more protons were trapped than predicted by Eq. (2). For example, at pH 6, 10°, fewer than 3 protons are predicted to be trapped[21a] in the competition experiment using Eq. (2). The experimental results show that 20 protons are trapped under these conditions. A plot of the number of protons remaining ($^3H_{rem}$) vs pH showed that the experimental points were shifted to higher pH values by ~1.3 pH units from the curve predicted with Eq. (2). These results indicate that the effective rate of amide proton trapping during refolding was ~20-fold faster than $k_{folding}$ determined by tyrosine absorbance.[22] In other words, folding under these conditions is not an "all or none" process, but involves the rapid formation of at least one folding intermediate which protects protons from exchange.

The competition method is well suited for studying early folding intermediates. The drawbacks of the competition method are that it cannot be easily adapted to the study of later intermediates in folding, and it requires knowledge of exchange rates for different classes of amide protons in the unfolded protein (the empirical rules of Molday et al.[23] give reasonable estimates of these rates).

The Pulse-Labeling Method

A complementary method, pulse-labeling, has also been used to study the folding of RNase A.[24] Whereas the competition method measures trapping of protons by folding intermediates, the pulse-labeling method measures exclusion of label from amide groups by formation of folded structure.

The experimental design is as follows. Refolding is allowed to occur for a variable period of time before a labeling "pulse" is applied. Those protons whose exchange rates are retarded by structure will be protected from labeling. At the end of the pulse, exchange is quenched by lowering the pH, so that further exchange is slow compared to refolding. The extent of labeling is assayed after folding has gone to completion.

[21a] The term "trapped protons" is used throughout this chapter, and refers to the retention of label (i.e., during the competition) by amide protons that are slow to exchange in the native state. When reference is made to the number of trapped protons, it should be noted that if tritium methods are used, this number could correspond to partial labeling of many amides.

[22] This result is consistent with several mechanisms. The effective rate of amide proton trapping depends on both the rates of formation of the intermediates and the stability of the intermediates (see Ref. 21).

[23] R. S. Molday, S. W. Englander, and R. G. Kallen, *Biochemistry* **11,** 150 (1972).

[24] P. S. Kim and R. L. Baldwin, *Biochemistry* **19,** 6124 (1980).

When folding of the slow-folding species of RNase A is studied, the kinetics of protection from amide proton exchange measured with pulse-labeling are faster than the kinetics of tyrosine absorbance change.[24] This comparison (kinetic ratio test[25]) demonstrates that structure is formed, which protects amide protons from exchange, before burial of the tyrosine residues. The pulse-labeling results showed that ~20 protons were protected from label within 4 sec after folding was initiated,[24] confirming the existence of an early folding intermediate previously detected with the competition method.[21] This confirmation does not rely on knowing the exchange rates for amide proton exchange in the unfolded protein, since pulse-labeling conditions were chosen so that any freely exposed amide proton would be completely labeled by the pulse.

It may not always be possible to quench exchange at the pH_{min} (e.g., if the protein is not stable at low pH). McPhie[26] has used amide proton exchange without a quench to study the folding of pepsinogen (pepsinogen activates itself to pepsin at low pH). Here the "pulse" was 15 min (long enough to permit complete refolding of the protein), and labeling was stopped by gel filtration. It may also be possible to use a partial quench (e.g., decreasing the pH as much as possible, followed by gel filtration). Optimal quench conditions are readily evaluated (see Experimental Details).

The advantages of the pulse-labeling method are that it can be used to study intermediates at any time in folding and does not require accurate knowledge of exchange rates in the unfolded protein.

Medium Resolution Studies Using ^3H and HPLC

Amide proton exchange measurements have been classified as low, medium, or high resolution.[27] High-resolution techniques are those that monitor individual amide protons (e.g., ^1H NMR, neutron diffraction). Low-resolution methods monitor exchange from the entire protein, usually with ^3H as a label. Medium-resolution methods[28,29] use HPLC to separate labeled fragments of a protein, which can be obtained with proteases.

The folding of RNase S has been studied with medium-resolution

[25] A. M. Labhardt and R. L. Baldwin, *J. Mol. Biol.* **135**, 231 (1979).
[26] P. McPhie, *Biochemistry* **21**, 5509 (1982).
[27] N. M. Allewell, *J. Biochem. Biophys. Methods* **7**, 345 (1983).
[28] J. J. Rosa and F. M. Richards, *J. Mol. Biol.* **133**, 399 (1979).
[29] S. W. Englander, D. B. Calhoun, J. J. Englander, N. R. Kallenbach, R. K. H. Liem, E. L. Malin, C. Mandel, and J. R. Rogero, *Biophys. J.* **32**, 577 (1980).

methods.[30] RNase S[31] is an enzymatically active derivative of RNase A cleaved by subtilisn at the peptide bond between residues 20 and 21. The cleavage products can be separated and are referred to as S-peptide (residues 1–20) and S-protein (residues 21–124). S-peptide and S-protein recombine to form the noncovalent complex, RNase S.

In the medium-resolution study of RNase S folding by Brems and Baldwin,[30] it was not necessary to use a protease to obtain fragments. After pulse-labeling, S-peptide and S-protein were separated rapidly by HPLC, under conditions where exchange is slow (pH 2.7, 0°). Previous NMR studies[32] had shown that the stable amide protons in the S-peptide moiety of RNase S correspond to residues 7–14, and therefore ³H in the HPLC peak of S-peptide corresponds predominantly to H-bond donors of the 3–13 α-helix (i.e., amide protons of residues 7–13). It was thus possible to investigate the rate at which the 3–13 α-helix is stabilized in the folding of RNase S. The results (refolding conditions: pH 6, 10°) show that whereas S-protein (residues 21–124) protects amide protons from exchange early in the kinetic folding process, the S-peptide α-helix is not protected from exchange until a late stage in folding.

If proteases are used to produce fragments of labeled proteins, then an acid protease such as pepsin is desirable since digestion can be done near the pH$_{min}$ of exchange. Medium-resolution techniques (proteolysis and HPLC separation of fragments; introduced by Rosa and Richards[28]) are reviewed in detail by Rogero et al.[33]

Matthews and co-workers[34] have used medium-resolution ³H methods to characterize domains in kinetic folding intermediates of the α-subunit of tryptophan synthase. Limited trypsin digestion (pH 5.5, 0°) cleaves the α-subunit into three fragments. Two of these fragments result from the NH$_2$-terminal domain, and the third fragment is the COOH-terminal domain of the α-subunit. All three fragments are structurally stable enough to permit quantitative recovery of ³H label following HPLC separation (125 protons are stable to exchange-out for 6 hr at pH 5.5, 0°; all of these protons can be recovered following trypsin digestion and HPLC separation[34]). This demonstrates the feasibility of using medium-resolution techniques in proteins that are not stable at acid pH (e.g., the α-subunit), provided that stable fragments can be obtained. The results of Beasty and Matthews[34] show that the NH$_2$-terminal domain is significantly more resistant to exchange than the COOH-terminal domain in early folding inter-

[30] D. N. Brems and R. L. Baldwin, J. Mol. Biol. **180,** 1141 (1984).
[31] F. M. Richards and P. J. Vithayathil, J. Biol. Chem. **234,** 1459 (1959).
[32] K. Kuwajima and R. L. Baldwin, J. Mol. Biol. **169,** 281 (1983).
[33] J. R. Rogero, J. J. Englander, and S. W. Englander, this volume [22].
[34] A. M. Beasty and C. R. Matthews, Biochemistry **24,** 3547 (1985).

mediates of the intact α-subunit. This supports the model for α-subunit folding proposed earlier,[35] based on hydrodynamic and spectroscopic properties of the intermediates.

Medium-resolution methods promise to be useful tools for analysis of folding intermediates labeled with ^3H_2O. These methods are particularly useful when studying large multidomain proteins, as demonstrated by the characterization of domain folding in the α-subunit of tryptophan synthase. The pulse-labeling studies of RNase S folding demonstrate the feasibility of monitoring individual structural units during folding, even without NMR techniques.

NMR Studies

In order to use amide proton exchange to characterize structures of folding intermediates by NMR, it is necessary to change from H_2O to D_2O (or vice versa) during folding, and then allow folding to go to completion. The ^1H NMR spectrum of the protein is used to analyze the location and extent of amide proton labeling at individual sites; deuterium is not detected in a ^1H NMR spectrum. Note that the NMR spectrum is taken after folding is complete—resonance assignments for folding intermediates are not needed.

Analysis of the ^1H NMR spectrum requires that amide proton resonances are resolved (i.e., not overlap other resonances) and assigned to individual protons in the protein. A solution to both of these requirements is provided by the elegant two-dimensional (2D) NMR studies of BPTI by Wagner, Wüthrich and co-workers.[13,14,16,18,36] These 2D-NMR techniques have been used to obtain resonance assignments in a systematic manner for almost all protons in BPTI.[36] Then, by measuring the amide proton–C^α proton cross-peak intensities in 2D-homonuclear correlated ^1H NMR spectra (COSY), it was possible to make quantitative exchange measurements for 38 of the 53 backbone amide protons in BPTI.[16] Overlapping resonances in a one-dimensional NMR spectrum are usually resolved in crosspeaks of the two-dimensional spectrum, so that the extent of labeling for many more amide protons can be determined. Although 2D-NMR methods have been used primarily to study small proteins like BPTI (58 residues), work has begun in several laboratories to apply these techniques to larger proteins.

One-dimensional NMR techniques are capable of characterizing trapped amide protons, if the number of protons studied in a given spec-

---

[35] C. R. Matthews, M. M. Crisanti, J. T. Manz, and G. L. Gepner, *Biochemistry* **22**, 1445 (1983).

[36] G. Wagner and K. Wüthrich, *J. Mol. Biol.* **155**, 347 (1982).

trum is limited to a few amides ($\sim$10 at a time). This is demonstrated by Kuwajima and Baldwin's ^{1}H NMR studies of the eight amide protons in residues 7–14 of RNase S.[32,37] Samples were prepared in which only these eight amide protons were in the ^{1}H form; the remaining amide protons of RNase S were deuterated. By using differential exchange, it is sometimes possible to divide amide protons of a protein into classes that are small enough to be resolved using one-dimensional NMR.[38] Neutron diffraction studies of trypsin[9] and RNase A[10] give the locations of stably protected amide protons and present the possibility of assigning amide protons by complementary ^{1}H NMR and neutron diffraction techniques.

Roder's[39,39a] study of BPTI is the first direct demonstration of the feasibility of using NMR and amide proton exchange to study kinetic folding processes. The competition method was used, and trapping of ^{1}H label in eight individual amide groups was analyzed with one-dimensional NMR. A rapid mixing apparatus was used to study the fast-folding (U_F) species of BPTI, which is known to be the major unfolded species.[40] BPTI was unfolded in H_2O (40% v/v n-propanol, pH 2, 70°) and the refolding/ exchange-out competition was initiated by mixing with D_2O buffer (final conditions: 70°, 13% v/v n-propanol, pH 4 to 7.5). Samples prepared in this manner were lyophilized and redissolved in D_2O. ^{1}H NMR was used to determine amide proton resonance intensities, with nonexchangeable protons as an internal concentration standard.

The results of this competition experiment[39,39a] were analyzed using intrinsic exchange rates obtained for the same amide protons in thermally unfolded RCAM-BPTI.[39,41] The apparent rates of amide proton trapping during refolding [i.e., corresponding to $k_{folding}$ in Eq. (1)] were similar for seven of the eight amide protons studied ($k = 30$–70 sec^{-1}); these seven amides are in the β-sheet of BPTI. One of the amide protons studied (Met-52) which is in the C-terminal helix of BPTI, showed a reduced rate of trapping during refolding ($k = 15$ sec^{-1}).

Definitive conclusions about the nature of intermediates in the folding of BPTI cannot yet be made, since only one amide proton in the α-helix was studied and since the kinetics of folding have not been measured under these conditions by other probes (e.g., tyrosine absorbance). Nevertheless, these experiments[39,39a] represent an important first step in the

[37] K. Kuwajima and R. L. Baldwin, *J. Mol. Biol.* **169,** 299 (1983).
[38] A. Bierzynski, N. R. Kallenbach, P. S. Kim, and R. L. Baldwin, unpublished results with RNase A.
[39] H. Roder, Ph.D. thesis No. 6932, ETH Zürich (1981).
[39a] H. Roder and K. Wüthrich, personal communication.
[40] M. Jullien and R. L. Baldwin, *J. Mol. Biol.* **145,** 265 (1981).
[41] H. Roder, G. Wagner, and K. Wüthrich, *Biochemistry* **24,** 7407 (1985).

application of NMR and amide proton exchange to the study of folding pathways.

Determining the Degree of Protection for Structural Units in Intermediates

The methods described here can be used to determine the degree of protection from exchange for different amide protons in a folding intermediate (cf. Ref. 37). We define the degree of protection from exchange for a given amide proton, $[\theta]_P$, as the ratio of the intrinsic exchange rate, k_{int} (i.e., in the absence of structure) to the observed exchange rate for the amide, k_{obs}:

$$[\theta]_P = k_{int}/k_{obs} \tag{3}$$

In the simplest model of exchange from an intermediate, where exchange occurs only after unfolding of the intermediate, we have

$$I \underset{k_{21}}{\overset{k_{12}}{\rightleftharpoons}} U \xrightarrow{k_{int}} U^* \tag{4}$$

where k_{12} and k_{21} refer to the rates of unfolding and refolding, respectively, of the intermediate. If $k_{21} \gg k_{int}$, then the observed rate of exchange is given as[42]

$$k_{obs} = K_{12}k_{int} \tag{5}$$

where $K_{12} = k_{12}/k_{21}$. Thus, $[\theta]_P = 1/K_{12}$, and the degree of protection is equal to the stability constant of the intermediate ($= 1/K_{12}$), in this simple model for exchange.

If direct exchange from I is not negligible, however, or if exchange can occur from an intermediate between I and U in Eq. (4), then Eq. (5) is not valid. Since the mechanism for exchange from folding intermediates is unknown, and likely to be different for different intermediates, we cannot directly relate $[\theta]_P$ to the stability constant of the intermediate.

Nevertheless, $[\theta]_P$ changes dramatically during folding (for the stable amide protons, $[\theta]_P = 1$ in U, and $[\theta]_P = 10^4$ to 10^{10} in N). Determination of $[\theta]_P$ for different amide protons at successive stages in folding will give information about the apparent stabilities of structural units in intermediates.

---

[42] This case is analogous to the EX2 mechanism of amide proton exchange from proteins (see Ref. 43). In the limit, $k_{int} \gg k_{21}$ (analogous to the EX1 mechanism), biphasic exchange kinetics will be observed for individual amide protons, with rate constants k_{int} and k_{12}. The relative amplitudes of the two phases will depend on the [U] : [I] ratio.

[43] Reviewed by A. Hvidt and S. O. Nielsen, *Adv. Protein Chem.* **21**, 287 (1966).

The pulse-labeling method provides one way of obtaining the degree of protection, $[\theta]_P$. By varying the duration of the pulse (or by changing the pH of pulse labeling), it is possible to change the sensitivity of the pulse-labeling assay. For example, in the experimental conditions used to study RNase A[24] (pH 7.5, 10°) the average half-time for exchange in unfolded RNase A is ~0.1 sec. Therefore, a 10 sec pulse will label amide protons whose exchange rates are retarded by a factor of ~100 or less (i.e., $[\theta]_P <$ 100). In principle, it is possible to titrate the stability of an intermediate, and determine $[\theta]_P$ for different structural units in the intermediate, by changing the sensitivity of the pulse-labeling assay.

For small single domain proteins, determination of $[\theta]_P$ for structural units (e.g., α-helices, β-sheets) in folding intermediates is best done using NMR techniques to observe individual amide protons. This eliminates the problems associated with ^3H methods, involving heterogeneity of intrinsic exchange rates. $[\theta]_P$ for the S-peptide α-helix in the equilibrium unfolding transition of RNase S has been determined by ^1H NMR measurements of individual amide proton exchange rates,[37] using RNase S that was specifically labeled as described earlier. When measurements are made within the unfolding transition zone, $[\theta]_P$ is the same for all seven amide protons in the S-peptide helix, within a factor of two.[37]

These results demonstrate the feasibility of using NMR and amide proton exchange to determine apparent stabilities ($[\theta]_P$) of structural units in equilibrium unfolding intermediates. The extension to kinetic intermediates is straightforward, but includes the usual difficulties encountered with a short-lived species. With kinetic intermediates, the labeling time must be short compared to individual steps in refolding (e.g., a short pulse should be used).

As a first step to obtain $[\theta]_P$ in kinetic intermediates, the sensitivity of the labeling assay could be changed (e.g., by changing the pH or duration of the pulse), and the extent of labeling for individual amide protons determined by ^1H NMR. A series of measurements with different sensitivities would give the observed exchange rate (k_{obs}) for an amide proton, at a particular point in folding.[44] $[\theta]_P$ could then be determined using Eq. (3).

Characterization of Folding Intermediates

As mentioned previously, the major factors responsible for retardation of amide proton exchange rates in folded proteins are believed to be H-bonding and solvent exclusion. Proposed mechanisms for exchange include penetration of solvent to the site of exchange,[15,17] and local unfold-

---

[44] The observed rate of exchange may be complicated by the presence of multiple populated intermediates.

ing of structural units[20] in the otherwise folded protein. These and other mechanisms are used to explain how exchange occurs under conditions where the native conformation is very stable. Under conditions where the protein is marginally stable (e.g., high temperatures) there is more general agreement—most workers believe that some form of unfolding (local or total) is responsible for exchange.

The rate of amide proton exchange from *unfolded* proteins can be predicted quite well using data obtained with solvent-exposed model compounds, as demonstrated by Molday *et al.*[23] Small amounts of residual structure have been detected in thermally unfolded proteins as compared to their denaturant-unfolded counterparts.[45] Amide proton exchange from thermally unfolded RNase S, however, is independent of urea concentration (0–5 M urea), after correction for the effects of urea on intrinsic exchange rates.[46,47] The polypeptide chain of unfolded RNase in aqueous solutions is accessible to solvent, at least with regard to amide proton exchange.

The retardation of exchange rates observed in kinetic folding experiments with RNase A and RNase S are large (>100-fold), demonstrating that populated folding intermediates exist. It seems likely that the major determinant retarding exchange in folding intermediates is H-bonding, and that methods described here will lead to determination of the H-bonded structures in intermediates. Exclusion of solvent may also contribute to inhibition of amide proton exchange in folding intermediates, since there is a large decrease in solvent accessible surface area when an α-helix or β-sheet is formed.[48] In any case, these methods will show what parts of the molecule are involved in structure that retards exchange.

For RNase A, results obtained with both the competition[21] and pulse-labeling[24] methods demonstrate that exchange rates for many amide protons are retarded by structure formed early in the folding process, before tyrosine absorbance changes are complete. The aromatic absorbance changes observed during folding occur when solvent is excluded from the environment of aromatic rings.[49] These data suggest that a substantial part of the secondary structure framework of RNase A is formed before tertiary structure in folding. Most results obtained with other methods also support a framework model of folding for single domain proteins.[2a] Amide

[45] Reviewed by C. Tanford, *Adv. Protein Chem.* **23**, 121 (1968).
[46] D. J. Loftus, G. O. Gbenle, P. S. Kim, and R. L. Baldwin, *Biochemistry* **25**, 1428 (1986).
[47] These effects were evaluated in the model compounds poly(DL-alanine) and *N*-acetyl-lysine methyl ester.
[48] Reviewed by F. M. Richards, *Annu. Rev. Biophys. Bioeng.* **6**, 151 (1977).
[49] S. Yanari and F. A. Bovey, *J. Biol. Chem.* **235**, 2818 (1960).

proton exchange methods should permit characterization of these early H-bonded frameworks.

Concluding Remarks

The methods described here permit labeling of transiently populated kinetic intermediates in a relatively nonperturbing manner. The large dynamic range of amide proton exchange rates makes it feasible to study fast steps in folding. Although most studies of folding by amide proton exchange have centered on slow-folding species, some preliminary results have been obtained with the fast-folding species (U_F) of BPTI.

A major feature of using amide proton exchange to study folding is that the locations and amount of label can ultimately be obtained with high resolution and precision, using ^1H NMR techniques after folding has gone to completion. A fairly detailed picture of protein folding pathways should emerge from these studies, since individually assigned amide protons in different parts of the protein can be studied.

It may be possible to use amide proton exchange methods to evaluate the stability of structural units in intermediates, in addition to the rate at which they form during folding. Eventually, one would like to quantitate the degree of protection from exchange ($[\theta]_P$) for individual amide protons, at successive stages in folding.

Low and medium resolution information can be obtained without NMR, using ^3H_2O as a labeling reagent. This offers a probe of secondary structure formation in most proteins. When combined with methods based on proteolysis and HPLC, information about the folding of structural units and domains can be obtained even in proteins that are too large to be studied by traditional NMR methods. ^3H methods are particularly important in determining suitable conditions for high resolution studies.

Experimental Details

^3H Methods: General Aspects

As with all studies involving amide proton exchange, it is important that the pH and temperature of solutions are carefully controlled. An error of 0.1 pH unit can result in a 25% change in intrinsic exchange rate, and the activation energy for exchange in model compounds in ~20 kcal/mol.[6] Buffers and temperature baths should be used, and the pH of solutions should be checked in trial runs containing protein (^3H_2O can be eliminated in these trials). Since charge effects on amide proton exchange can be substantial,[50,51] a moderate ionic strength (>0.1 M) should be maintained. The required concentration of ^3H_2O depends on the number of protons that are labeled and on the amount of protein that is assayed. For RNase A, we have found that an ^3H_2O concentration of ~20 mCi/ml, and a sample size of ~1 mg (~0.1 μmol) works well. Since ^3H_2O is volatile, all manipulations should be done in a fume hood or enclosed glove box.

Typically the protein is unfolded with GuHCl or urea, under conditions where previous equilibrium measurements (e.g., of a spectroscopic probe) have demonstrated that the protein is completely unfolded. Thermal unfolding can be used, but should be avoided where possible, since the rates of refolding and amide proton exchange are both temperature dependent. In general, the protein should be kept unfolded long enough to allow complete equilibration of fast- and slow-folding species.[52]

Refolding is initiated by adding a small aliquot of unfolded protein to refolding buffer, so that the final denaturant concentration is low (a 1:10 to 1:20 dilution is usually used). In general, folding intermediates are stabilized by low temperatures (0–10°), low denaturant concentrations, and stabilizing salts [e.g., $(NH_4)_2SO_4$].

Labeled protein is separated from excess ^3H_2O on a short Sephadex column,[6] equilibrated at low temperature and low pH (if possible). Coarse or medium grade Sephadex will result in a faster separation, and up to 5 lb/in.[2] of nitrogen pressure can be applied to the top of the column to increase the flow rate. It is possible to separate RNase A from ^3H_2O in less than 5 min using this procedure.[21]

At the end of all manipulations (including an exchange-out procedure if only the stable amide protons are to be studied), the protein must again be separated from tritium that has exchanged-out. A second Sephadex column[6] or a convenient filter paper assay[53] (utilizing cation-exchange paper) can be used. The filter assay is very fast and works very well with RNase A; however, each protein should be checked since some proteins do not give consistent results with the filter assay.[54,55]

It is necessary to correct for differences in protein recovery. This can be done spectroscopically, using known molar extinction coefficients or by including trace amounts of protein labeled with another isotope. Radioactive protein can be made by chemical modification (e.g., we reduc-

[50] P. S. Kim and R. L. Baldwin, *Biochemistry* **21**, 1 (1982).
[51] J. B. Matthew and F. M. Richards, *J. Biol. Chem.* **258**, 3039 (1983).
[52] F. X. Schmid, this volume [4].
[53] A. A. Schreier, *Anal. Biochem.* **83**, 178 (1977).
[54] M. Lennick and N. M. Allewell, *Proc. Natl. Acad. Sci. U.S.A.* **78**, 6759 (1981).
[55] K. R. Shoemaker and R. L. Baldwin, unpublished results.

tively methylate[56] the lysines of RNase A with $H^{14}CHO$ and $NaCNBH_3$) or, if possible, by growth of microorganisms in the presence of radioactive amino acids.

The Competition Method[21]

This method uses labeled, unfolded protein that has been separated from excess ^3H_2O. For this reason, it is desirable that the protein can be unfolded with a denaturant at low pH (pH_{min}) and low temperature. It is also desirable that refolding rates are independent of pH in the pH range where the competition will take place (or at least that the pH dependence of refolding rates is known).

If the slow-folding species (U_S) are to be studied, it may be advantageous to use the selective labeling procedure of Schmid and Baldwin[21] which does not label the fast-folding species (otherwise $U_F^* \rightarrow N^*$ contributes to the protons trapped during folding). The procedure is as follows. (1) Unfold in the absence of ^3H_2O until the $U_F \rightleftharpoons U_S$ reaction has reached equilibrium. (2) Refold for a short time under conditions where $U_F \rightarrow N$ is fast but $U_S \rightarrow N$ is slow. (3) Add ^3H_2O under conditions where exchange into U_S is fast, but exchange into the stable protons of N (formed from U_F) is slow. (4) Adjust pH to the pH_{min} and add denaturant to keep the protein unfolded. (5) Rapidly separate the labeled protein from ^3H_2O on a Sephadex column at $0°$, equilibrated with denaturant at the pH_{min}. (6) This procedure gives U_F and U_S^*, which is used immediately in the competition experiment. Selective labeling of U_S can be obtained in this manner only if the $U_S \rightleftharpoons U_F$ equilibrium is slow compared to the column separation time.

The alternative to selective labeling of U_S is to label both U_S and U_F, and then correct for contribution of the $U_F^* \rightarrow N^*$ reaction to the amount of label trapped. This approach is simpler than selective labeling, but is less precise when the concentration of U_F is significant. Unfolded protein ($U_F \rightleftharpoons U_S$ equilibrium) is labeled with ^3H_2O and then separated from excess label on a Sephadex column equilibrated (pH_{min}, $0°$) with denaturant to keep the protein unfolded. The competition experiment is carried out with the labeled mixture (U_F^*, U_S^*). The contribution of the $U_F^* \rightarrow N^*$ reaction can be evaluated if refolding of U_F is fast compared to exchange in the pH range used, and if the relative concentration of U_F molecules in the equilibrium unfolded mixture is known. Schmid and Baldwin[21] also used completely labeled RNase A (U_F^*, U_S^*) in their competition experiments. After correction for label trapped by the $U_F^* \rightarrow N^*$ reaction, the amount of label trapped by U_S^* was the same, within experi-

[56] N. Jentoft and D. G. Dearborn, *J. Biol. Chem.* **254**, 4359 (1979).

mental error, as that obtained using selectively labeled slow-folding species (U_S^*).

If the fast-folding reaction is to be studied, then the completely labeled protein (U_S^*, U_F^*) can usually be used (cf. Refs. 39, 39a), since the competition conditions required to study the $U_F \rightarrow N$ reaction will typically result in complete loss of label from U_S^* species. If the concentration of U_F is low in the equilibrium-unfolded mixture, then a high concentration of U_F (and hence, U_F^*) can be obtained using the "double-jump" technique that has been used extensively in kinetic studies of refolding.[52] Briefly, this involves using unfolding conditions (high denaturant concentrations and low temperature) where $N \rightarrow U_F$ is fast and $U_F \rightleftharpoons U_S$ is slow, to populate U_F kinetically. The folding of U_F can then be studied without interference from U_S.

Competition between exchange-out and refolding is initiated by diluting the unfolded, labeled protein with refolding/exchange-out buffer. Typically, refolding is allowed to continue to completion, although exchange-out can be quenched after a shorter time of competition. When refolding is complete, the partially labeled protein is exchanged-out[56a] so that only the stably protected protons remain. Exchange-out can be carried out at the pH of competition. Alternatively, one set of exchange-out conditions can be used for all samples. In either case, the number of protons that remain after the competition is compared to that obtained with labeled native protein (N^*), exchanged-out in the same conditions.

It is necessary to obtain values for k_{HX} and $k_{folding}$ in order to predict trapping of label in the absence of populated intermediates [Eqs. (1) and (2)]. Reasonable estimates of k_{HX} can be obtained using the rules of Molday et al.[23] It is also possible to measure amide proton exchange directly from the unfolded protein. This can be done by measuring exchange from (1) thermally unfolded protein—significant errors may result from uncertainties in extrapolation to lower temperatures, (2) proteins unfolded due to chemical modification (e.g., reduction and blocking of S–S bonds), where errors (usually small) may result from residual structure, and (3) urea-unfolded proteins, where errors may result from uncertainties in the correction for urea effects on exchange rates (these effects have been calibrated in model compounds[46] and are small compared to temperature effects).

Values for $k_{folding}$ are obtained preferably with a probe for folding that monitors formation of tertiary structure. These include absorbance or fluorescence from chromophores (e.g., aromatic residues) in the protein,

---

[56a] This final exchange-out procedure should not be confused with exchange that occurs during the competition.

and probes of the active site (e.g., specific inhibitor binding, enzymatic activity). If there are multiple unfolded forms of the protein, then it is important to determine accurately the concentration as well as folding rates of each unfolded species. The pH dependence of relative amplitudes and rates of folding should be determined.

The amount of label trapped in the absence of folding intermediates is predicted using the "null equation."[21] The null equation is essentially Eq. (2), modified to include the following sources of complexity. (a) Exchange rates for amide protons in the unfolded protein are not all the same, so that k_{HX} in Eq. (2) is replaced by a summation over all amides, each with its own k_{HX}. (b) If there are multiple unfolded species, then $k_{folding}$ in Eq. (2) is also a summation, weighted according to the relative concentration of each unfolded species. (c) Only a fraction of the amide protons is stable to the exchange-out procedure used. This fraction is determined in control experiments with N*, as described previously. Unless there are data to suggest otherwise, one assumes that the stable protons are a representative sample of all amide protons, with regard to their exchange rates in the unfolded protein. When NMR methods are used to monitor trapping by individual amide protons, the only source of complexity comes from multiple unfolded species.

A good control is to repeat the competition experiment under conditions where intermediates are not populated during folding (i.e., to directly test the null equation). For this purpose, marginally native folding conditions have been used. The refolding/exchange-out competition is performed just outside the unfolding transition zone, where folding is slower but goes to completion. In these conditions, folding usually occurs without populated intermediates, which are destabilized near the transition zone. Moderate concentrations of denaturants can be included in refolding/exchange-out buffers to achieve marginally native folding conditions. Corrections for the effects of denaturants on k_{HX} can be made by using the data of Loftus et al.,[46] who calibrated GuHCl and urea effects on intrinsic exchange rates. When the competition experiment was carried out with RNase A in the presence of 2.5 M GuHCl (conditions where folding goes to completion), the amount of ^{3}H label retained coincided with that predicted by the null equation.[21]

The Pulse-Labeling Method[24]

From a practical point of view, the pulse-labeling method offers the advantage that unfolding can be done at any pH, and refolding can be studied at one pH, so that the pH dependence of refolding rates does not need to be investigated. The method does not require knowledge about differences in exchange rates of amide protons in the unfolded protein. Amplitudes and rate constants for individual kinetic phases in refolding do not need to be characterized extensively. This is because intermediates that protect amide protons are demonstrated by a direct comparison of kinetic progress curves (kinetic ratio test[25]) rather than by calculation of expected results in the absence of intermediates. In general, the pulse-labeling method is more time consuming than the competition method, since the samples taken at each time point must be individually manipulated through all steps that follow.

Unfolded protein ($U_S \rightleftharpoons U_F$ equilibrium) is refolded by rapid dilution into buffer. It is preferable, but not necessary, to refold at the same pH as the labeling pH to be used. Since the $U_F \rightarrow N$ reaction typically occurs in a fraction of a second, it does not usually interfere with measurements of the $U_S \rightarrow N$ reaction.

The protein is allowed to refold for a variable period of time before the labeling pulse is applied. For a given set of labeling conditions, the time of refolding before addition of label is the only variable. If the pH of refolding is the same as the labeling pH, then the pulse can be initiated by adding ^3H_2O directly to the buffered refolding solution. If the labeling pH is different from refolding pH, then ^3H_2O should be added as a buffered solution. Errors in pH and/or duration of the pulse are the largest sources of scatter in the data.

Some guidelines for initial pulse-labeling studies are as follows. The temperature of labeling and refolding should be the same. When possible, the pulse should be short compared to folding (if necessary, a rapid-mixing device can be used). The pH of labeling is chosen based on the average exchange rate of an exposed amide proton. We want to label completely any amide proton that is not protected from exchange (i.e., $[\theta]_P = 1$), without labeling protected protons in the native protein (i.e., $[\theta]_P = 10^4$ to 10^{10}). A good starting point is to use a pulse that on average labels protons with $[\theta]_P < 50$. The sensitivity (S) of the pulse-labeling assay is defined here as the average value of $[\theta]_P$ for amide protons that are half-labeled by the pulse:

$$S = t/(t_{1/2})_{HX} \qquad (6)$$

where $(t_{1/2})_{HX}$ is the average half-time for exchange in the absence of structure at the pH and temperature used in the pulse, and t is the duration of the (short) pulse. Values of $(t_{1/2})_{HX}$ are readily available from model compound studies of amide proton exchange.[6,23] Since the kinetic ratio test[25] will be used, it is not necessary to know the distribution of intrinsic exchange rates [i.e., different $(t_{1/2})_{HX}$ values]. However, it should be recognized that S will be different for different amide protons in the protein due to individual differences in $(t_{1/2})_{HX}$. Since amide proton exchange rates are similar[6] within a factor of 10 (with a few exceptions), values of S

below 50 are not generally recommended. Two control experiments should be mentioned here. It is important to demonstrate that the pulse does not significantly label stable protons in the folded protein (N), and that the pulse completely labels unfolded protein (U). The latter can be checked by pulse-labeling U in the presence of denaturants, since the effects of GuHCl and urea on intrinsic exchange rates are small.

At the end of the labeling pulse, exchange is quenched, if possible by lowering the pH to the pH_{min} (for RNase A, pH_{min} 2.7). A weak acid (e.g., formic) or buffered solution can be used. If the protein is not stable at the pH_{min}, then a partial quench can be achieved by lowering the pH as much as possible. Decreasing the temperature may also help. The best pH to use for a quench will usually be one where the ratio of folding rate to intrinsic amide proton exchange rate is largest.

The effectiveness of the quench can be evaluated by measuring the amount of label that is incorporated into protein that has been refolded at the quench conditions, in the presence of ^3H_2O. Exchange-in rather than exchange-out is measured since the former has a lower background and is therefore more sensitive. Checking the quench in this manner gives the maximum effect of incomplete quenching on pulse-labeling, since this test starts with unfolded protein whereas intermediates (with some protected protons) are quenched in pulse-labeling experiments. When RNase A ($U_S \rightleftharpoons U_F$) is refolded at pH 2.7, 3° (i.e., conditions corresponding to the quench), in the presence of ^3H_2O, approximately 3 protons are stably trapped.[24] The most likely sources of these protons are (1) nonspecific, low-level labeling of backbone amide protons and (2) specific labeling of some protons with significant exchange rates at pH 2.7, especially the side-chain primary amide protons of asparagine and glutamine residues.[57]

Exchange-in experiments during refolding can be used to define the optimum pH for quenching exchange relative to folding. This is particularly useful for proteins that are not stable at the pH_{min}.

The partially labeled protein is separated from excess ^3H_2O on a Sephadex column, equilibrated at the quench pH. The column is kept cold, and refolding continues during the separation. The separated, pulse-labeled protein solution is adjusted to exchange-out conditions that have been determined previously, so that only the stable amide protons re-

---

[57] The side-chain amide protons of Asn and Gln residues have minimum exchange rates near pH 4–5 (Ref. 23). At pH 2.7, 0°, exchange from these side-chain protons has a rate approximately 1000-fold faster than the average rate of exchange from backbone amide protons. In native proteins, exchange from side-chain amides can also be retarded by structure; for example, the side-chain amide of Asn-43 is one of the slowly exchanging protons in BPTI (Ref. 58).

[58] R. Richarz, P. Sehr, G. Wagner, and K. Wüthrich, *J. Mol. Biol.* **130**, 19 (1979).

main. The number of stable protons trapped in pulse labeling is determined as a function of refolding time before the pulse was applied.

The pulse-labeling results are analyzed with the kinetic ratio test[25] which states that, if normalized kinetic progress curves obtained with two probes are not superimposable (i.e., the kinetic ratio test is positive), then a populated folding intermediate is detected. The kinetic ratio test requires that (1) the reaction is monomolecular (e.g., folding of a polypeptide chain) and (2) different initial species (i.e., unfolded proteins) are indistinguishable by the probes used.

The kinetic ratio test can be extended to multimolecular processes only if the observed time-dependent changes arise from the participation of exactly one of the components (e.g., in the RNase S pulse-labeling study,[30] S-peptide contains no tyrosine, and the amounts of label in S-peptide and S-protein can be quantitated separately following HPLC separation). If there are multiple unfolded forms (e.g., different U_S species) which refold at different rates, the kinetic ratio test is still valid provided that different unfolded forms are indistinguishable by the probes used[24,25] (e.g., same number of exposed amide protons, same absorbance extinction coefficient, lack of enzymatic activity). For the derivation, and a more complete discussion of the kinetic ratio test, see Labhardt and Baldwin.[25]

The pulse-labeling data are normalized and compared to the normalized changes of another probe of folding. Since pulse-labeling is likely to be most sensitive to secondary structure formation, it is usually best to use a tertiary structure probe in the kinetic ratio test.

As a control, it is desirable to find conditions where the kinetic ratio test is negative. This is likely to occur in marginally native folding conditions. When the folding of RNase A (U_S) was studied in 2.5 M GuHCl (pH 7.5, 10°), the normalized kinetics of protection from exchange and of tyrosine absorbance changes during folding were superimposable.[24]

Isotope and Solvent Effects

The equilibrium isotope effect for amide proton exchange has been measured ($^3H/^1H = 1.21$; $^3H/^2H = 1.05$), and results should be corrected accordingly.[6] Kinetic isotope effects have also been measured and found to be small.[59,60] When changing solvent from H_2O to D_2O (or vice versa), it is probably best *not* to correct pH for the glass electrode effect—both exchange rates[60] and apparent ionization constants[61] in model compounds

[59] R. S. Molday and R. G. Kallen, *J. Am. Chem. Soc.* **94**, 6739 (1972).
[60] J. J. Englander, D. B. Calhoun, and S. W. Englander, *Anal. Biochem.* **92**, 517 (1979).
[61] A. Bundi and K. Wüthrich, *Biopolymers* **18**, 285 (1979).

are similar in H_2O and D_2O when pH meter readings without corrections are used.

It is possible that the stability of intermediates will be altered by changes in solvent (H_2O to D_2O) or deuteration of H-bond donors. ^3H_2O is used only in tracer amounts and is likely to be much less perturbing than D_2O. In general, D_2O has a stabilizing effect on native proteins as compared to H_2O.[43] The isotope-induced effects on stability are likely to be minor perturbations as compared to the large stability changes that occur in protein folding. Nevertheless, it will be important to check this directly in labeling experiments utilizing ^1H NMR. This can be done by using H_2O when D_2O is called for (and vice versa) in folding studies. If complementary experiments give the same results, this suggests that isotope-dependent effects do not significantly perturb intermediates in the folding pathways.

Acknowledgments

It is a pleasure to thank Dr. Robert L. Baldwin for his advice and discussions. I am also grateful to A. Bierzynski, D. N. Brems, J. Carey, K. H. Cook, N. R. Kallenbach, K. Kuwajima, D. Loftus, C. R. Matthews, C. Mitchinson, H. Roder, F. X. Schmid, and K. R. Shoemaker for discussions. This research was supported by grants from the National Institutes of Health (GM19988-22) and the National Science Foundation (PCM77-16834) to R. L. Baldwin, and by a Medical Scientist Training Program Grant from the National Institutes of Health (GM07365). Use of the Stanford Magnetic Resonance Facility (supported by NSF Grant GP23633 and NIH Grant RR00711) is acknowledged.

[9] Detection of Folding Intermediates Using Urea-Gradient Electrophoresis

By THOMAS E. CREIGHTON

The pathway of protein folding must be studied kinetically because the partially folded intermediates that define it are inherently unstable and not present in significant concentrations at equilibrium.[1] Unstable intermediates might accumulate transiently as kinetic intermediates in either unfolding or refolding, although there is no fundamental requirement that they do so. The only intermediates that will accumulate are those that both precede the rate-limiting step in the pathway and, under the final conditions, have free energies close to, or lower than, the initial form of the protein. These are severe restrictions, for the fully folded and unfolded states under most conditions differ only slightly in free energy.[1]

[1] P. L. Privalov, *Adv. Protein Chem.* **33,** 167 (1979).

Furthermore, unless the kinetics are extremely favorable, any such intermediates will be transient and populated by only a small fraction of molecules, coexisting at all times with the fully folded and unfolded states. Consequently, the identification and characterization of such intermediates must be done within a mixture that is always changing with time. Most of the physical methods that can provide measurements of conformation sufficiently rapidly, primarily spectral, provide only average properties of the total population of molecules, so the properties of individual transient species can be ascertained only by deconvoluting the time dependence of the measurements. Each measurement provides only a limited amount of information about each species, so as many parameters as possible need to be measured. The kinetic roles of such intermediates can be very difficult to discern from their kinetic properties, which need to be known throughout the transition region, often under conditions where the intermediates are not detectable.

These complications can be overcome if a specific conformational interaction can be manipulated experimentally to control its kinetics and energetics and to trap otherwise unstable intermediates. Disulfide bonds between Cys residues satisfy these criteria,[2] but not all proteins have disulfides, so the generality of observations made with them must be demonstrated, and other approaches must be used to overcome or to minimize the inherent difficulties.

Characterization of kinetic intermediates is easier if the rates of the reactions are slowed down so that the intermediates accumulate for longer periods of time. This can be accomplished in a number of ways, but all are liable to alter the pathway to at least some extent. For example, denaturants slow protein refolding, but by destabilizing the important intermediates that are to be identified and characterized. Lowering the temperature is the most general way to slow a reaction, but this also affects protein conformation. Fortunately, lower temperatures generally increase the stabilizing interactions and favor the nonrandom conformation both in native proteins[1] and in partially folded intermediates,[3] probably due to the increased stability of intramolecular hydrogen bonds[4] and van der Waals interactions,[5] so partially folded intermediates are more likely to be populated. However, addition of cryosolvents to permit temperatures lower than 0° undoubtedly affects hydrogen bonding in water and the hydrophobic effect, and thereby could have drastic effects on protein conformation.

[2] T. E. Creighton, this volume [5].
[3] T. E. Creighton, *J. Mol. Biol.* **144,** 521 (1980).
[4] T. E. Creighton, *Biopolymers* **22,** 49 (1983).
[5] J. Bello, *J. Theor. Biol.* **68,** 139 (1977).

The kinetics of protein refolding are complicated by the presence of kinetically distinguishable forms of the unfolded protein.[6] The inherent conformational heterogeneity of an unfolded protein could easily be imagined to produce immense kinetic complexities, if the very many conformations adopted at different times and by the many molecules of the population were interconverted more slowly than folding occurred. However, the only major factor described thus far is the inherently slow cis–trans isomerization of peptide bonds preceding Pro residues.[7,8] Virtually all proteins contain Pro residues, so this should be a factor in most protein folding transitions. However, the consequences of the phenomenon appear to be different with different peptide bonds and different proteins. Some wrong isomers may block refolding, others may have little or no effect if the folded conformation is compatible with an incorrect isomer. Nonrandom conformation in a protein can be imagined either to slow or to accelerate the isomerization. If any conformational interaction is more favorable with one isomer, this interaction necessarily must favor that isomer to the same extent (e.g., the interaction involving Tyr-92 in ribonuclease A[9] and its effect on cis–trans isomerization of the adjacent Pro peptide bond[8]).

Multiple kinetic states of the unfolded protein complicate all kinetic measurements of the average properties of the population, producing overlapping multiple phases that will obscure kinetic complexities arising from intermediates within any one phase. Multiple species are ideally suited for analysis by techniques that use separation, but isomers can only be resolved with separations that require less time than the isomers do to interconvert. Fortunately, the rate of Pro cis–trans isomerization is inherently very temperature dependent, and the half-time for isomerization of an individual bond should be of the order of 20 min at 0°. With larger numbers of Pro residues that are required for folding to be of the right isomer, the slow-refolding species should be a greater fraction of the total population of unfolded molecules, and folding should occur more slowly.[10]

Electrophoretic separations are possible on this time scale, and the mobility of a protein through a polyacrylamide gel is sensitive to its conformation. Unfolded proteins nearly always migrate more slowly than when folded, and intermediates have intermediate mobilities.[11] Consequently, slowly interconverted isomers with different conformational

[6] J. R. Garel and R. L. Baldwin, *Proc. Natl. Acad. Sci. U.S.A.* **70**, 3347 (1973).
[7] J. F. Brandts, H. R. Halvorson, and M. Brennan, *Biochemistry* **14**, 4953 (1975).
[8] L.-N. Lin and J. F. Brandts, *Biochemistry* **22**, 559 (1983).
[9] A. Rehage and F. X. Schmid, *Biochemistry* **21**, 1499 (1982).
[10] T. E. Creighton, *J. Mol. Biol.* **125**, 401 (1978).
[11] D. P. Goldenberg and T. E. Creighton, *Anal. Biochem.* **138**, 1 (1984).

properties might be resolved. Incorporation of a linear gradient of an agent that affects protein stability, such as urea,[12] across the gel provides continuously varying conditions under which to compare the electrophoretic properties of each species.[11]

Urea Gradient Electrophoresis: Theory

A slab gel of polyacrylamide is prepared, incorporating a linear gradient of urea concentrations across it; detailed directions are given below. A sample of protein, either folded or unfolded, is layered evenly and continuously across the top and subjected to electrophoresis through the gel.

The protein sample may consist of several species resolvable by electrophoresis, either differing covalently or being isomers that are only slowly interconverted. Each may produce an individual electrophoretic band. The following discussion of the principles of urea gradient electrophoresis pertains to a single electrophoretic species.

The protein migrates at different urea concentrations across the gel, which may cause it either to unfold or to refold. At sufficiently low urea concentrations, the folded state predominates at equilibrium, whereas it is the unfolded state at sufficiently high urea concentrations. If unfolding and refolding are rapid relative to the duration of the electrophoretic separation, all the molecules interconverted in this way have the same mobility at each urea concentration and will define a single band of protein. The mobility at each urea concentration will be the average of that of the molecules throughout the duration of the separation.

Rates of unfolding and refolding are generally dependent upon the urea concentration, so they will vary across the gel. If they are sufficiently rapid at all urea concentrations, a continuous band of protein across the gel will result. It will give a continuous graphic illustration of the change of mobility with urea concentration. Proteins that undergo no conformational change, that are fully folded even at high urea concentration or fully unfolded even in the absence of urea, show no change in mobility across the gel, other than that produced directly by the varying urea concentration, probably due to the variation in viscosity. The phenomenon may be compensated by including an inverse gradient of acrylamide concentration, e.g. from 15 to 11%.

Proteins that unfold in a single, cooperative, two-state transition exhibit one abrupt decrease in mobility, with a single inflection point at the midpoint of the transition. At this point, the molecules are fully folded

[12] T. E. Creighton, *J. Mol. Biol.* **129,** 235 (1979).

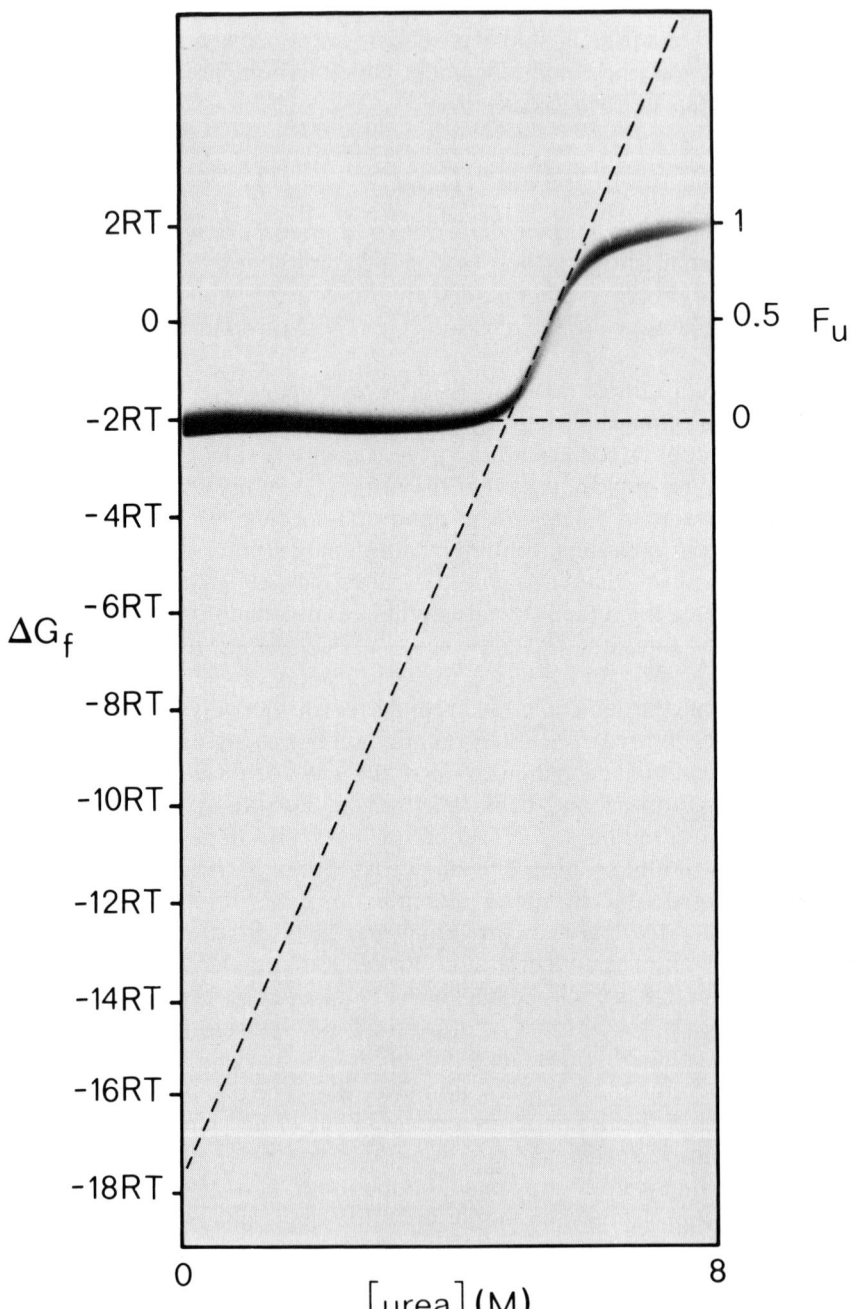

half the time, fully unfolded the other half, and the folded and unfolded states have equal free energies. Their relative free energies at other adjacent urea concentrations can be estimated from their relative proportions, as given by the relative electrophoretic mobility. The variation of their relative stabilities with urea concentration, defined by the shape of the band through the transition region, can be extrapolated to the other urea concentrations, including to the edge where there is no urea. Such extrapolations are most convenient, and at least as valid as any other, if it is assumed that the relative free energy of the folded and unfolded states varies linearly with the urea concentration. Then a straight line through the linear midpoint of the transition curve gives this value directly, with the zero value defined by the transition midpoint and the free-energy scale defined as $-2RT$ by the mobility of the native protein and $+2RT$ by that of the unfolded protein.[13] An example using horse ferricytochrome c is given in Fig. 1.

Multidomain proteins may exhibit complex equilibrium folding transitions if the individual domains unfold independently. "Intermediate" conformations may be present at equilibrium in which one or more domains are folded, others unfolded. Examples that have been detected by urea-gradient electrophoresis are the H state of penicillinase (Fig. 7), calmodulin,[11] tryptophan synthase α-subunit,[14] and transferrins.[15]

Molecules differing covalently are readily separated by this method, producing different bands across the gel, and so should molecules differing conformationally but interconverted only very slowly. If the interconversions are negligible during the separation, each population will give a discrete band (Fig. 2). With faster rates of interconversion, so that some

[13] M. Hollecker and T. E. Creighton, *Biochim. Biophys. Acta* **701**, 395 (1982).
[14] C. R. Matthews and M. M. Crisanti, *Biochemistry* **20**, 784 (1981).
[15] R. W. Evans and J. Williams, *Biochem. J.* **189**, 541 (1980).

FIG. 1. Estimation of the net stability of cytochrome c by extrapolation from the urea-induced unfolding transition. Horse ferricytochrome c was electrophoresed through a urea-gradient gel at pH 4.0 at 15° to produce a profile of the equilibrium unfolding transition. The mobility of the protein at each urea concentration indicates the fraction of protein unfolded, F_u, as indicated on the right hand scale. Near the midpoint of the transition, where $F_u = 0.5$ and the net stability of the folded state, ΔG_f, equals 0, F_u is approximately a linear function of urea concentration. The net stability in the absence of denaturant is estimated by extrapolating ΔG_f from the transition region to the position of zero urea concentration, assuming a linear relationship between ΔG_f and urea concentration. The plateau positions of the U and N states correspond to the positions on the left hand scale where $\Delta G_f = 2RT$ and $-2RT$ (R is the gas constant and T is the absolute temperature), as derived in Ref. 13. Taken from Goldenberg and Creighton.[11]

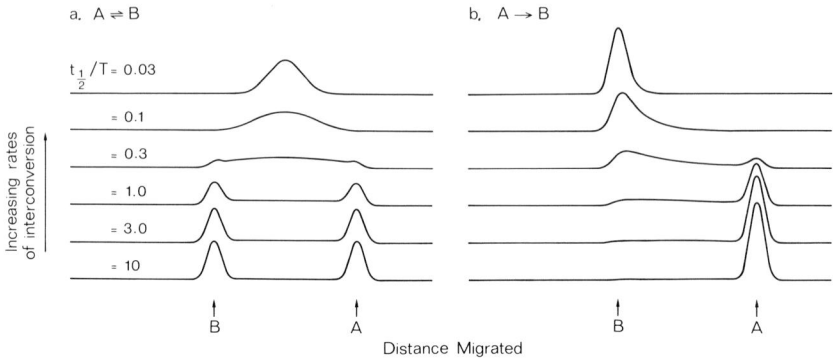

FIG. 2. Calculated electrophoretic profiles of interconverting molecules. The expected distributions of protein after electrophoresis through a gel were calculated using the probabilistic model of R. M. Mitchell, *Biopolymers* **15**, 1717 (1976). The profiles on the left (a) are those expected if an equilibrium mixture composed of equal amounts of the fast (A) and slow (B) migrating forms of the protein were applied to the gel, and reversible interconversion between the two forms took place during electrophoresis, at the indicated relative rates. The profiles on the right (b) are those expected if the fast migrating form (A) was applied to the gel and was irreversibly converted to the slow migrating form (B) during electrophoresis. Such a profile would be generated, for instance, if a native protein was electrophoresed into a gel containing urea, where irreversible unfolding took place. For all of the calculated profiles, the two forms of the protein, A and B, were taken to have electrophoretic mobilities (indicated by the arrows) of 0.2 and 0.1 mm/min, respectively, and both to have diffusion coefficients of 3.3×10^7 cm²/sec. The time of electrophoresis (T) in all cases was 200 min. The rates of interconversion are indicated as the half time ($t_{1/2}$) relative to the time of electrophoresis. Taken from Goldenberg and Creighton.[11]

molecules isomerize during the electrophoresis, there is a smearing of part of the band of starting protein toward the position expected for the other conformer. With half-times for interconversion comparable to the duration of the experiment, most of the molecules isomerize once or only a few times during the separation and consequently are smeared out between the two limiting positions. With faster transitions, with half-times of roughly 0.1 the duration of the separation, the molecules isomerize a number of times, have similar average mobilities, and define a somewhat diffuse band. With half-times lower by a factor of 10, the width of the band is generally determined only by the diffusion coefficient of the protein. Estimates of the rates of unfolding and refolding may be obtained by comparing the observed pattern at each urea concentration with those in Figs. 2 and 3 predicted for various rates relative to the duration of the separation.

The predicted effects on urea gradient electrophoresis of different rates of unfolding and refolding of a protein in a single two-state transition

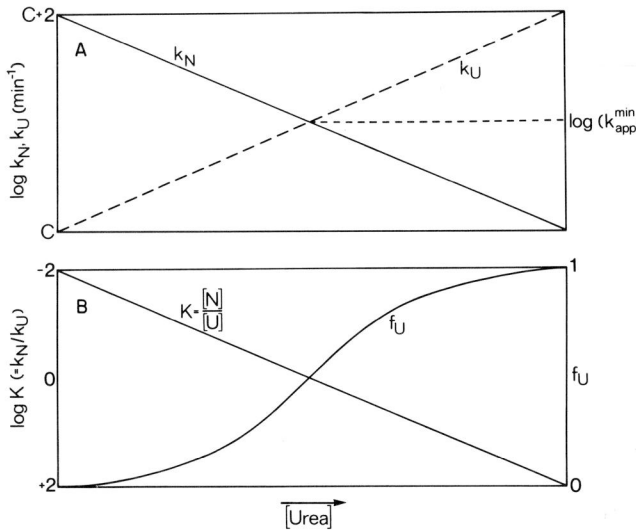

FIG. 3. Simulated urea-gradient gel electrophoresis patterns expected for two-state behavior for the range of rate constants of folding and unfolding in which the kinetics affect the
electrophoretic pattern. (A, B) Illustration of assumed dependence of the kinetics of folding
and unfolding as a function of the concentration of urea. In (A), the logarithms of the rate
constants for folding (k_N) and unfolding (k_U) are plotted as a function of the linear concentration of urea, showing the presumed dependence of both rate constants. Only shown is the
(unspecified) urea concentration range in which the values of both rate constants vary
100-fold around that concentration at which they are equal. The absolute values of the rate
constants were taken to vary, as indicated by the parameter C. The minimum value of the
apparent rate constant for the folding transition ($k_{app}^{min} = k_N + k_U$, where $k_N = k_U$) is indicated.
Whatever the value of C, the relative values of k_N and k_U in (A) define the 10^4-fold dependence of the equilibrium constant K shown in (B). This then defines the indicated fraction of
molecules that are unfolded at equilibrium, f_U. (C) The expected profiles at each concentration of urea were calculated according to R. M. Mitchell, *Biopolymers* **15,** 1717 (1976).
Native, folded protein (N) was present initially for the patterns on the left, fully unfolded
protein (U) on the right. The dependence of the rate constants for refolding (k_N) and unfolding (k_U) on the concentration of urea was that illustrated in (A). The absolute values of both
rate constants were varied to give the indicated minimum values of the apparent rate constant for the folding transition ($k_N + k_U$) at the midpoint of the transition. All of the electrophoretic patterns are those expected after electrophoresis for 200 min for a protein of which
the folded (N) and unfolded (U) form have electrophoretic mobilities of 0.2 and 0.1 mm/min,
respectively, and both have an effective diffusion coefficient of 3.3×10^{-7} cm^2/sec. Taken
from Creighton.[12]

are shown in Fig. 3. With slow folding transitions, different patterns are
expected starting with either folded or unfolded proteins. A smearing of
protein between the mobilities of the folded and unfolded states is predicted. Where a band of intermediate mobility is observed, it should fol-

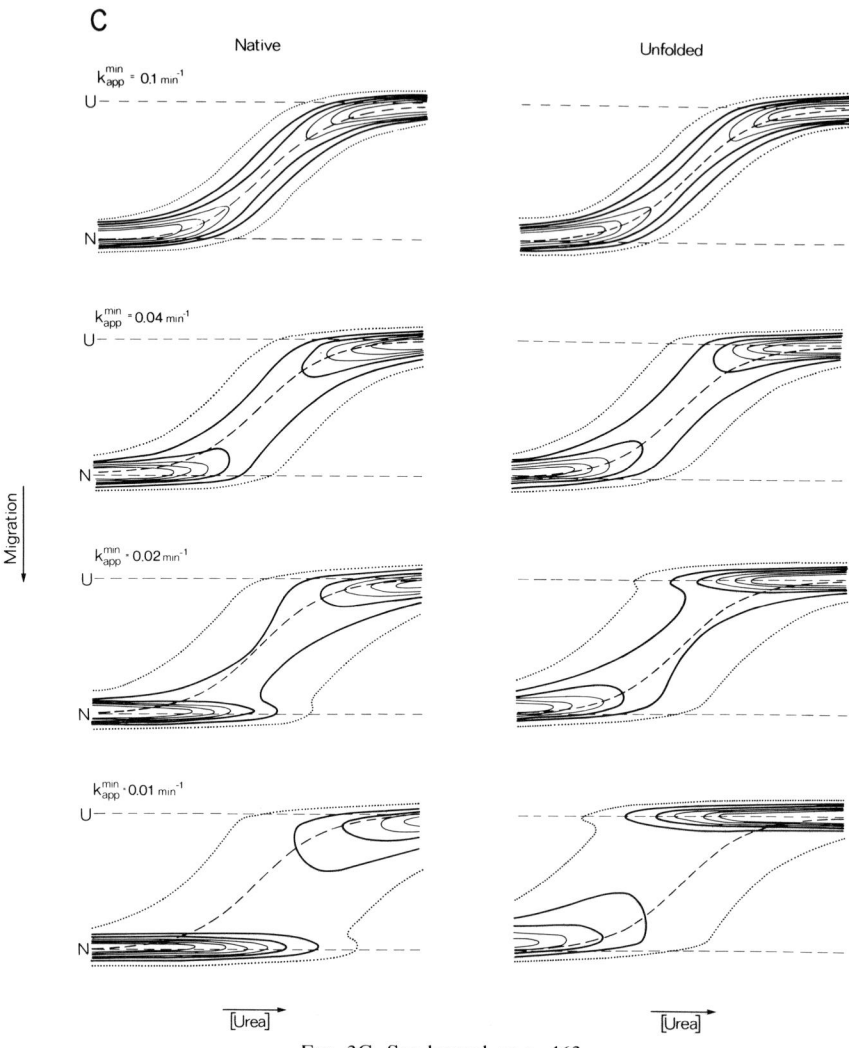

FIG. 3C. See legend on p. 163.

low closely the expected equilibrium average mobility. No new bands are predicted to result from slow folding transitions.

Schematics of the patterns expected with some of the different plausible folding transitions are illustrated in Fig. 4. The ability to distinguish between very fast and very slow transitions makes the technique complementary to other kinetic methods that follow the time dependence of some parameter, where all reactions must occur on the appropriate time scale in order to be observed; very fast reactions can be missed with such

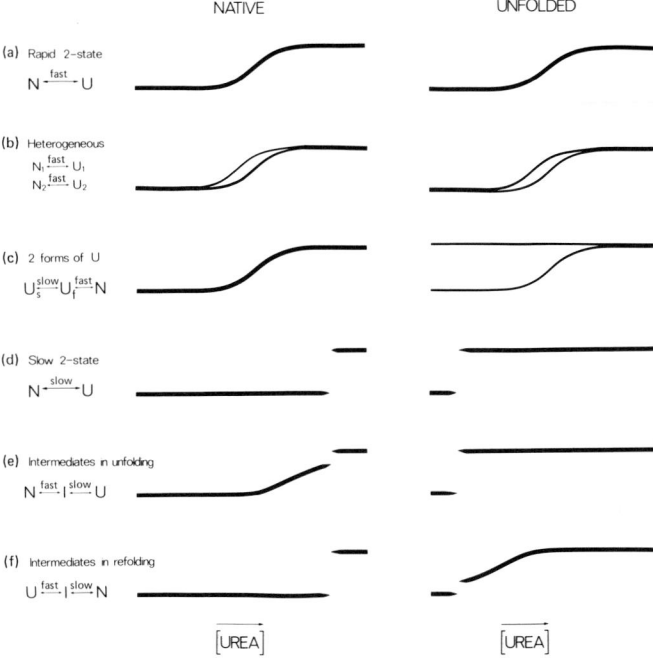

NATIVE　　　　　　　　　UNFOLDED

(a) Rapid 2-state

$N \xrightarrow{\text{fast}} U$

(b) Heterogeneous

$N_1 \xrightarrow{\text{fast}} U_1$
$N_2 \xrightarrow{\text{fast}} U_2$

(c) 2 forms of U

$U_s \xrightarrow{\text{slow}} U_f \xrightarrow{\text{fast}} N$

(d) Slow 2-state

$N \xrightarrow{\text{slow}} U$

(e) Intermediates in unfolding

$N \xrightarrow{\text{fast}} I \xrightarrow{\text{slow}} U$

(f) Intermediates in refolding

$U \xrightarrow{\text{fast}} I \xrightarrow{\text{slow}} N$

$\boxed{\text{UREA}}$　　　　　　$\boxed{\text{UREA}}$

FIG. 4. Urea gradient electrophoresis patterns expected for some folding transitions. The protein applied to the gel is either fully folded (left) or unfolded (right). The apparent rate constants for the transitions are described as ''fast'' or ''slow'' relative to the duration of the electrophoretic separation. The equilibrium midpoint of the transition is near the middle of the urea gradient in each case. (a) A rapid two-state transition produces a smooth continuous pattern, which is the same whether native or unfolded protein is applied to the gel. (b) If the protein population is heterogeneous, with the various species unfolding at different urea concentrations, but all of the molecules have the same native and unfolded mobilities, the band of protein will split in the transition region. (c) Two slowly interconverted populations of unfolded protein, one which refolds rapidly (U_f) and the other slowly (U_s), as expected if incorrect proline isomers block refolding, will generate two bands at low urea concentrations when electrophoresis is begun with the unfolded protein. The U_s form will remain unfolded during the time of electrophoresis and will produce a straight band across the gel, while the U_f form will be in rapid equilibrium with the native conformation and will generate a pattern like that in (a). If the electrophoresis begins with folded protein, only the rapid transition between N and U_f will be observed, since U_s will not be formed during the time of electrophoresis. (d) If the rates of unfolding and refolding are slow, except at the highest and lowest, respectively, urea concentrations, then a band of the original form of the protein will extend past the transition midpoint, starting with either the native or unfolded form. At the urea concentrations where the protein begins to either unfold or refold, but where the rates are not rapid, the protein will be smeared out, rather than produce a smooth band between the native and unfolded forms. (e) Rapid and reversible partial unfolding of the native conformation to less compact intermediate conformations, prior to a slow unfolding step, will result in the band of folded protein curving upwards and extending past the midpoint of the equilibrium transition. The intermediate will not be detected when the experiment is begun with unfolded protein. (f) More compact intermediates in rapid equilibrium with the unfolded state will cause bending downward of the band of unfolded protein, when the unfolded form is applied to the gel. Taken from Creighton.[17]

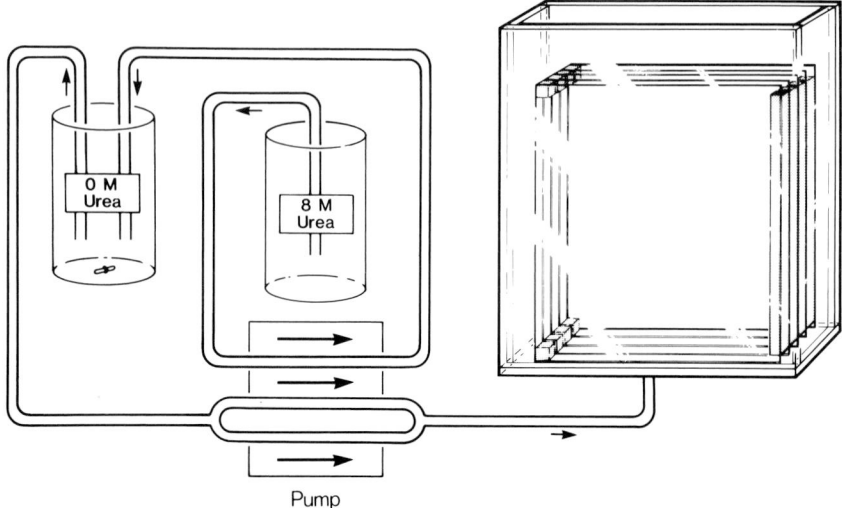

Fɪɢ. 5. Apparatus for preparing urea-gradient gels. See text for a description of the method. Taken from Goldenberg and Creighton.[11]

methods. Two or more steps are observed only if both occur within the appropriate time range.

Intermediates that are detectable by such orthodox kinetic methods, because they are formed and disappear with comparable rates, might be missed by separation methods such as urea-gradient electrophoresis. Simulations of the expected behavior have not been carried out, because the problem is nontrivial mathematically, but two sequential steps with similar rates, comparable in half-time to the separation time, would be expected to produce a very diffuse protein profile in which the intermediate would not be apparent as a discrete band. Consequently, urea-gradient electrophoresis cannot claim to be a substitute for orthodox kinetic procedures, but to complement them.

Preparation of Transverse Urea-Gradient Gels

Gradient gels are generally cast in a clear plastic box designed to hold a number of gels at a time, such as that manufactured by Pharmacia Fine Chemicals, so that several comparable gels are prepared at the same time. A vertical gradient is prepared between square plates (Fig. 5), and after polymerization the gels are turned 90° and fitted to the electrophoresis tank so that the gradient is perpendicular to the direction of electrophoresis. The glass plates are held apart by three plastic spacers (1–3 mm thick); a long spacer on the right, which when removed forms a trough on what will be the top of the gel, for the sample to be layered into, and two small spacers on the left corners. The plates and spacers are held together with tape on the side edges (not shown). After polymerization the long spacer is replaced by two small ones at each end, the tape is removed, and fresh tape is applied to the edges that will be on the sides during electrophoresis.

The gradient is prepared from two stock solutions, with compositions of the ends of the desired gradient. Many variations are possible but we usually use urea gradient gels in which the first solution contains 15% acrylamide and no urea and the second solution contains 11% acrylamide and 8 M urea; the other components are identical. The inverse gradient of acrylamide is an empirical correction for the tendency of urea to slow the migration of all components, even the tracking dye, probably due to the increased viscosity. A small amount of the first solution, sufficient to fill the top 1 cm of the gels, is first pumped into the apparatus. The gradient is then generated by pumping from the low concentration solution into the gel forms, using two channels of a peristaltic pump, while simultaneously mixing the high concentration solution, pumped through a single channel of the same pump, with the low concentration solution (Fig. 5). The volume of the gradient is chosen to fill all but 2 cm of the gel forms. After the gradient is formed, more of the high concentration solution is pumped in, so that the gel is finally composed of a 1 cm length of the low concentration solution, the gradient, and a 1 cm length of the high concentration solution. The acrylamide is overlayed with a 20% ethanol solution, placed in the apparatus before the gel solutions are pumped in.

Either direct sunlight or the light from five 8-W "cool white" fluorescent tubes is used to photopolymerize the gels; consequently, polymerization can be prevented until after the gradient is formed. The gels contain 5 mg/liter riboflavin and 0.12% (v/v) N,N,N',N'-tetramethylethylenediamine (TEMED) as polymerization catalysts. These catalysts alone will usually cause photopolymerization of gels of relatively high acrylamide concentration (>10%) and low pH (<5) in about 15 min. However, for gels of lower concentration or higher pH, it is often necessary to add a small amount of ammonium persulfate, typically about 0.005%. Because the rate of polymerization is sensitive to pH, temperature, acrylamide concentration, and urea concentration, the exact conditions should be tested with a small volume before preparing the gradient gels.

The final urea gradient can be checked by including a dye in one of the original acrylamide solutions and measuring its concentration by densitometry. The concentration gradient of urea would be expected to be diminished gradually by diffusion, so the gels are used as soon as possible. On occasions, they have been stored at 4° overnight before use, with no deterioration apparent in the final pattern, but no systematic study of their stability has been made. Urea solutions are well known to produce cyanate, which reacts with protein amino groups and then introduces electrophoretic charge heterogeneity.[16] The only precautions we take are to use the highest grade of solid urea commercially available and to keep all urea solutions for the minimum amount of time. Ionic impurities within the gels may be removed by preelectrophoresis, although this may permit diffusion of the urea gradient. It has not been generally required, and covalent modifications of proteins have not been a problem. The exception has been horse ferricytochrome c, where the unfolded protein often yielded a second, more cationic band, apparently a result of covalent modification. This phenomenon was irreproducible, but could be minimized by preelectrophoresis of the gel and by electrophoresis of the protein in the dark; the latter observation suggests involvement of the heme group.

We have used only continuous buffer systems in which the same buffer is present throughout the gel and the electrode compartments. However, there seems no reason why other, discontinuous buffer systems could not be used, although preelectrophoresis would not then be possible.

The protein sample is ideally in a solution of as low as ionic strength as possible to produce sharpening of the band upon the start of the electrophoresis. If necessary, glycerol is added to about 10% (v/v) to provide sufficient density for the sample to layer on top of the gel and under the buffer. An appropriate tracking dye (e.g., methyl green for cationic proteins, bromophenol blue for anionic) is included to monitor both application of the sample and the progress of the electrophoresis. A protein concentration of 1 to 2 mg/ml is nearly optimal, permitting application of an appropriate sample volume. The sample is generally applied by carefully dispensing it from a microsyringe along the top of the gel. The gel surface must be flat so that the sample is distributed uniformly over it by gravity.

Electrophoresis is carried out at voltages, currents, temperatures, and durations appropriate to the experiment and to the protein and buffer system used. Voltages up to 400 V have been used, currents of 10 to 50 mA per 8 cm square gel, and temperatures of from 0 to 35°.

Urea may interfere with the fixation and staining of some proteins, especially small ones. A staining procedure that has been universally successful is to immerse the gels in a fresh solution of 0.1% (w/v) Coomassie Blue in 10% (w/v) trichloroacetic acid and 10% (w/v) sulfosalicylic acid for at least 2 hr. Besides being very effective, it has the advantage that

---

[16] N. L. Anderson and B. J. Hickman, *Anal. Biochem.* **93**, 312 (1979).

protein bands are visible after just a few minutes, without destaining. Nevertheless, the gels are generally destained by first rinsing off precipitated dye from the surface of the gel with a solution of 45% (v/v) methanol and 9% (v/v) acetic acid, then by diffusion into 5% (v/v) methanol and 7.5% (v/v) acetic acid.

Detection of Intermediates

Kinetic studies of protein folding using urea-gradient electrophoresis require that several gels be run for different periods of time, starting with both the folded and unfolded proteins. The two patterns should be different if the unfolding and refolding are slow on the time scale of the electrophoresis. The appropriate types of intermediates that are populated significantly are expected to produce protein bands with mobilities intermediate between those of the fully folded and unfolded forms.

Some protein folding transitions are slow even at room temperature and ideally suited to electrophoretic analysis.[17] β-Lactoglobulin demonstrates the presence of two different forms of the unfolded protein, U_S and U_F (Fig. 6). Starting with native protein, a single rapid, apparently two-state transition, $N \leftrightarrow U_F$ is observed. However, starting with the unfolded protein, two major bands are observed, one identical to that obtained with the native protein; it would have arisen from U_F molecules, which comprised about half the population. The other half, designated U_S, did not refold during the electrophoresis, and its conversion to U_F must have been negligible. These U_S molecules probably do not differ in Pro isomers, even though there are eight Pro residues. They may have originated by disulfide interchange in the unfolded protein, since the U_S species was not observed if unfolding were at acid pH; β-lactoglobulin has two disulfides and one free thiol which could interchange at neutral or alkaline pH. Also, the electrophoretic mobilities of the U_F and U_S species were detectably different at high urea concentrations (Fig. 6). The U_S species was also heterogeneous, because it diverged into at least two bands at low urea concentrations. The most significant observation is that the mobilities of all the U_S species increased at low urea concentrations, indicating a rapid equilibration with more compact, but nonnative, conformations. The nature and kinetic significance of these conformations are not clear from these experiments, but the urea dependence of their stability suggests that they result from noncovalent interactions between different parts of the polypeptide chain like those present in folded proteins. Consequently, there are no true kinetic intermediates detectable with β-lactoglobulin

[17] T. E. Creighton, *J. Mol. Biol.* **137**, 61 (1980).

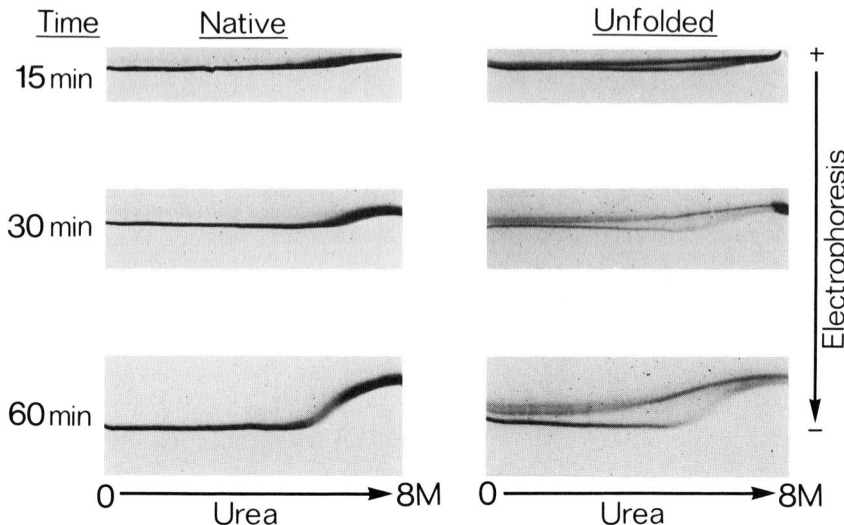

FIG. 6. Urea gradient electrophoresis of native and unfolded bovine β-lactoglobulin B. The initially native (left) and unfolded (right) proteins were subjected to electrophoresis at 400 V in 1-mm-thick urea-gradient gels at pH 4 (0.05 M Tris titrated to pH 4.0 with acetic acid) at 2° for the periods of time indicated on the left. Taken from Creighton.[17]

under these conditions, but the tendency of the U_S species to adopt, rapidly and reversibly, nonrandom conformations would suggest their existence.

A complex folding transition is exhibited by normal penicillinase from *Staphylococcus aureus,* which unfolds and refolds only very slowly.[18] Starting with native protein, N, unfolding occurred only at high urea concentrations, well above the transition midpoint of about 2 M urea. The half-times for unfolding were estimated to increase from about 2 min at 8 M urea to about 20 min at 4 M. The mobility of the band of residual folded protein was constant across the gel, indicating no tendency to unfold partially before complete unfolding, even at urea concentrations of greater than 6 M. Consequently, no intermediates in unfolding appear to be populated kinetically.

Very different behavior is exhibited by unfolded penicillinase, U. The protein remained unfolded, with constant electrophoretic mobility, at urea concentrations greater than 4 M. At lower urea concentrations the mobility increased nearly 2-fold, in a sharp, continuous transition, to reach a new plateau value at about 3 M urea; this indicates rapid equilibration of state U with a more compact conformational state H. State H had

[18] T. E. Creighton and R. H. Pain, *J. Mol. Biol.* **137,** 431 (1980).

been known previously, as it is the most stable state at intermediate denaturant concentrations; its nature is unknown, but is probably related to the multiple-domain structure of the folded protein. At a lower urea concentration of about 2 M, the mobility of the band increased abruptly to a plateau value of about 0.8 of that of N. Again, the transition was sharp and continuous, indicating rapid, urea-dependent equilibration between states U and H and a third, even more compact state, designated I. This previously undetected species was only metastable, for its band became smeared out by complete refolding to state N. The simultaneous existence of all four states U, H, I, and N could be demonstrated by electrophoresis of a mixture of unfolded and refolded protein samples.

Consequently, at least two intermediate states, H and I, can be demonstrated directly with normal penicillinase. The kinetic roles of these intermediates are impossible to infer directly from these observations, since they are in rapid equilibrium with each other and with U. However, the increasing compactness of the protein and the greater accumulation of these species at lower urea concentrations, where refolding is increasingly more rapid, would be consistent with roles as productive intermediates. In both unfolding and refolding, the rate-limiting step is the interconversion of these states with the fully folded state, N.

Only very small fractions of slow-refolding molecules were observed with unfolded penicillinase. They were visible as faint bands retaining the mobilities of states U and H at very low urea concentrations. The protein has nine Pro residues, so incorrect cis–trans peptide bond isomers do not appear to block refolding.

Other proteins, such as chymotrypsinogen and ribonuclease A, demonstrated both fast- and slow-refolding forms.[17] However, a number of others did not, even though they contained multiple Pro residues that might have been expected to slow or to block refolding at the low temperatures used, due to the presence of incorrect cis or trans peptide bond isomers. All slow unfolding transitions demonstrated an absence of partially folded intermediates in unfolding, but a universal tendency of the unfolded protein to equilibrate rapidly with compact, but nonnative, conformations at low urea concentrations below the transition midpoint. This suggests that the rate-limiting transition in both unfolding and refolding is very near to the folded state, when measured by compactness of the protein molecule.[19]

This survey of the kinetics of folding of a number of proteins provided some general information about the energetics of protein folding transitions and about the roles of cis–trans isomers of peptide bonds in refold-

[19] T. E. Creighton, *Adv. Biophys.* **18**, 1 (1984).

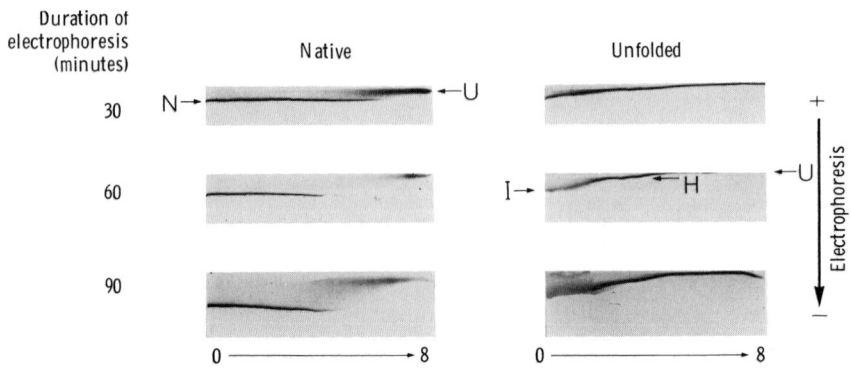

FIG. 7. Unfolding and refolding of penicillinase. The native and unfolded proteins were applied to 1-mm-thick urea-gradient gels and electrophoresed at 300 V at pH 7.2 (0.05 M bis-Tris and 1 mM EDTA adjusted to pH 7.2 with acetic acid) at 2° for the indicated times. The native protein, N, shows a slow unfolding transition, with no accumulation of intermediates. During refolding, the unfolded protein, U, rapidly equilibrates with two partially folded species, H and I, which are only slowly converted to the native conformation, so that only little native protein appears during the period of electrophoresis. Taken from Creighton and Pain.[18]

ing. It would have been an immense task using orthodox kinetic methods, requiring large numbers of kinetic measurements of both unfolding and refolding at many different urea concentrations. Moreover, the tendency of unfolded molecules to adopt compact conformations rapidly under refolding conditions appears not to have been detected by many of the spectral probes that have been used, for it had not been particularly noticed. The state I of penicillinase (see Fig. 7) also was undetected until observed directly by urea-gradient electrophoresis. Consequently, the method has the twin advantages of being rapid and of measuring a useful parameter of protein conformation, its compactness. It should be especially useful in surveys of the folding transitions of mutant or modified proteins.[13]

[10] Protein Folding in Cryosolvents and at Subzero Temperatures

By ANTHONY L. FINK

Introduction

Recent investigations of ribonuclease A folding have shown that aqueous–organic cryosolvents and subzero temperatures may provide unique opportunities to stabilize intermediate states in protein folding.[1-5] These studies have resulted in the first direct observation of partially folded intermediates in protein folding. The underlying basis of these observations is believed to be the existence of partially folded states with exposed nonpolar groups. In aqueous solution the unfavorable free energy of such species results in their rapid conversion to species (including the native state) in which the hydrophobic groups are predominantly buried. In the presence of the cosolvent these conformations would be more stable due to the more favorable interactions between the nonpolar protein groups and the more hydrophobic cryosolvent. In other words the driving force for completion of folding of partially folded species would be decreased in the presence of methanol. This expectation is borne out by the observation of intermediates revealed by NMR,[2,3] absorbance,[3,5] circular dichroism,[4] and the decrease in the cooperativity of the denaturation transition.[4]

If secondary structure is formed early in folding as suggested by several recent investigations[4,6] additional stabilization might be expected on the basis of the following reasoning. Low temperatures strengthen hydrogen bonds and weaken hydrophobic interactions. Therefore if the partially folded intermediate states are composed of predominantly secondary structure (stabilized by H-bonds) and have exposed hydrophobic regions, which coalesce through hydrophobic interactions in subsequent formation of the tertiary structure, the intermediates would be anticipated to be stabilized at low temperatures relative to both native and unfolded states.

The advantage of studying folding at subzero temperatures is that the rates of un- and refolding will be much slower than at ambient tempera-

[1] A. L. Fink and B. G. Grey, in "Biomolecular Structure and Function" (P. Agris, ed.), p. 471. Academic Press, New York, 1978.
[2] R. G. Biringer and A. L. Fink, J. Mol. Biol. **160**, 87 (1982).
[3] R. G. Biringer and A. L. Fink, Biochemistry **21**, 4748 (1982).
[4] A. L. Fink and B. Painter, submitted (1986).
[5] R. Biringer, C. Austin, and A. L. Fink, submitted (1986).
[6] P. S. Kim and R. L. Baldwin, Annu. Rev. Biochem. **51**, 459 (1982).

tures. Several types of media are available which are fluid at subzero temperatures and are compatible with proteins (see Ref. 7 for summary). These include aqueous solutions of high salt concentration, supercooled water in reversed micelles, and mixed aqueous–organic solvents. Based on their utility in studying enzymes at subzero temperatures[7,8] aqueous methanol mixtures, which have low viscosities and low freezing points, are the cryosolvents of choice. Investigations of protein folding at subzero temperatures have only been reported for aqueous–organic cryosolvents; the majority of these data concerns the ribonuclease A/methanol system, to which this chapter will be restricted. However, the techniques should be applicable to many other systems; the main limitation will likely be the limited solubility of the protein in the cryosolvent.

Effects of Cryosolvent on the Native Structure

High resolution proton NMR provides a detailed means of comparing the conformational state of the native protein in cryosolvents to that in aqueous solution. The temperature at which the protein is fully native can be estimated from thermal denaturation transition (melting) curves, obtained using absorbance, for example. A more direct method is to collect the NMR spectrum at successively lower temperatures until there is no evidence of any nonnative material.[2]

Ribonuclease A was purified by a modification of the method of Taborsky.[2,9] Solutions of deuterium-exchanged ribonuclease A in 25 and 35% methanol-d_4 were prepared at room temperature as follows. The cryosolvents were mixed on a v/v basis, i.e., 25:75 methanol-d_4:D_2O (99.9%) and adjusted to a predetermined ionic strength using KCl and then mixed with lyophilized enzyme. DSS was added to the solvent prior to mixing with RNase. The pH* was adjusted with DCl or NaOD as necessary. Solutions of the enzyme in 50% methanol-d_4 were prepared at 1° by mixing the cryosolvent with the dry enzyme. If the enzyme and methanol ($\geq$50%) cryosolvent were mixed at 25°, where the protein is well into its denaturation transition, a gel was formed. If the enzyme is prepared in 50% methanol at 1°, where no significant denaturation has yet occurred, gel formation is not observed at any subsequent temperature. In order to exclude moisture from the air, the samples can be prepared in a dry-box under an atmosphere of nitrogen. The pH* values reported are the observed pH-meter readings obtained at 25° using a glass combination

[7] A. L. Fink and S. J. Cartwright, *CRC Crit. Rev. Biochem.* **11**, 145 (1981).
[8] P. Douzou, "Cryobiochemistry: An Introduction." Academic Press, New York, 1977.
[9] G. Taborsky, *J. Biol. Chem.* **234**, 2652 (1959).

electrode and thus represent the apparent protonic activity in the deutero-cryosolvent.[10]

RNase A proton NMR spectra were obtained with a standard 1-pulse sequence at 360 MHz, with a pulse delay of $\geq$3 sec. The long pulse delay allowed complete equilibration of the His–C-2 resonances prior to the start of the next pulse sequence. Gated decoupling was used to eliminate solvent resonances. Sample concentrations were generally 20 mg/ml.

The similarities between the spectra of the native state in aqueous and methanolic solvents (Fig. 1) indicate that the presence of methanol does little to perturb the native structure.[2] A detailed analysis of the His, Tyr, and Phe contributions to the aromatic region of the spectrum using resolution enhancement (Biringer and Fink, unpublished data) indicates that there are no *significant* changes in the chemical shifts of these residues on going from aqueous to methanolic solvent systems. The apparent broadening of the resonances from Tyr and Phe, especially notable in spectra below 0°, is due to the close proximity of several resonances, such that a small increase in linewidth, due to decreased temperature, increased solvent viscosity, and decreased ring-flipping rates, results in considerable overlap of adjacent resonances.

The lack of adverse effects of ethanol and methanol on the native protein reflects the preferential exclusion of the cosolvent from the vicinity of the protein,[7,11] i.e., charged groups on the surface result in unfavorable interactions with the relatively more hydrophobic cosolvent (compared to water), leading to the exclusion of the cosolvent molecules from such areas, and hence preferential hydration.

Unfolding in Aqueous Methanol under Equilibrium Conditions

Spectrophotometric Measurements. The denaturation of RNase A in aqueous methanol solvents is fully reversible as judged by the complete reversal of changes in the near-UV spectrum and return of catalytic activity.[1,4] The unfolding has been monitored by absorbance, fluorescence, and circular dichroism.[4] Of particular note is the noncorrespondence of the unfolding transition when monitored by absorbance and fluorescence compared to that by circular dichroism.[4]

The standard protocol used in carrying out the unfolding is as follows: an aliquot of a concentrated stock solution of RNase A, in aqueous or cryosolvent solution, is added to an appropriate volume of the solvent at a temperature below the beginning of the unfolding transition, the final protein concentration being in the 0.05 to 0.5 mg/ml range. (The absence

[10] G. Hui Bon Hoa and P. Douzou, *J. Biol. Chem.* **248,** 4649 (1973).
[11] E. P. Pittz and S. N. Timasheff, *Biochemistry* **17,** 615 (1978).

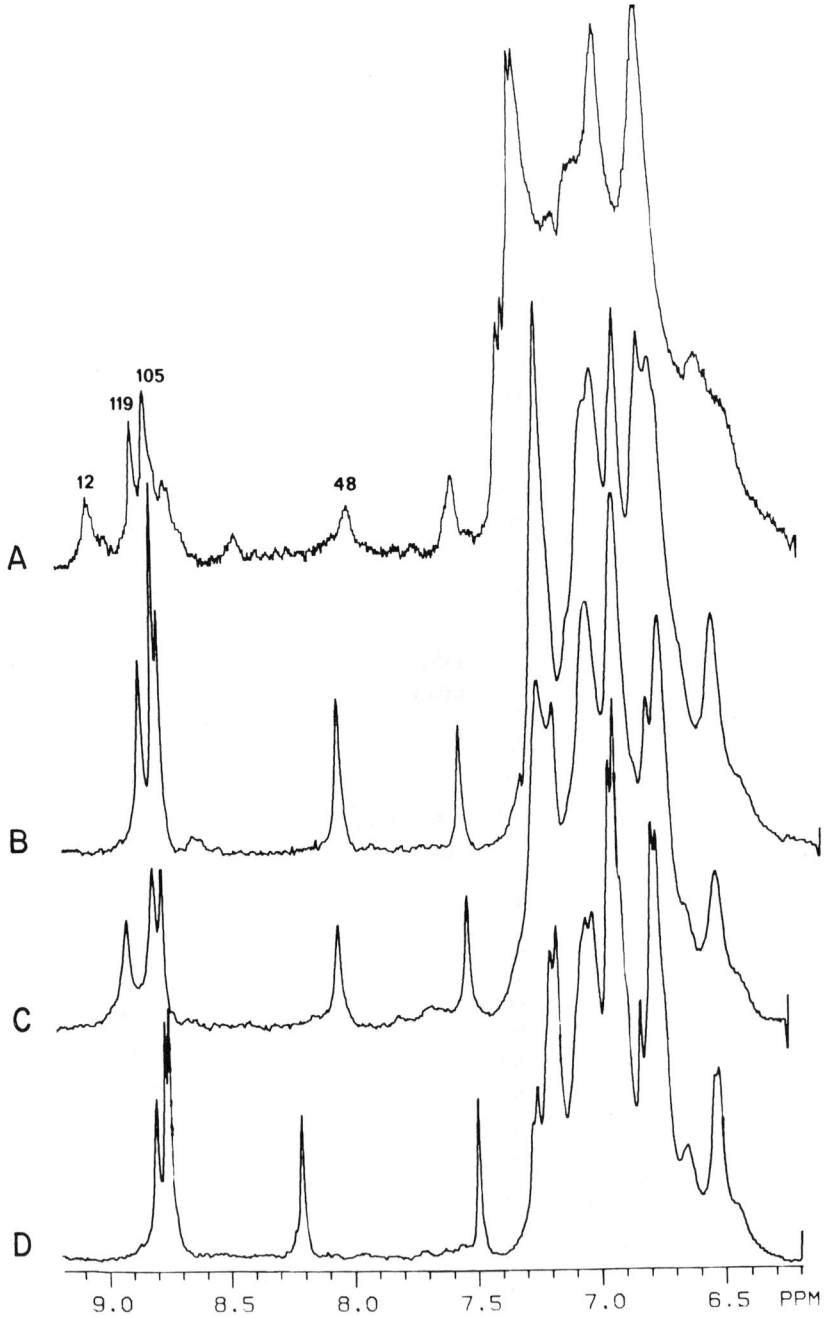

FIG. 1. The effect of methanol on the aromatic region of the 360 MHz ^{1}H NMR spectrum of ribonuclease A. Spectrum D, D_2O, 22°, pH 3.00, 0.1 M KCl; spectrum C, 25% methanol-d_4, 12°, pH* 2.80, 0.1 M KCl; spectrum B, 35% methanol-d_4, 12°, pH* 2.81, 0.1 M KCl; spectrum A, 50% methanol-d_4, −3°, pH* 3.00, 0.1 M KCl. From Biringer and Fink.[2]

of aggregation under the experimental conditions can be confirmed by light scattering experiments using fluorescence or ultracentrifugation.[2]) The temperature is raised in small increments and the change in spectral signal noted. After each new temperature equilibration has been reached a reading is taken only after the signal has been constant for 10 min. The sample can be stirred with a small magnetic stirring bar, or a programmed temperature gradient can be used. Temperature can be measured (and recorded if necessary) using a thermocouple and digital thermometer.

Changes in absorbance at 286 nm, and fluorescence with excitation at 280 nm and emission at 310 nm, were used to follow the exposure of buried tyrosine residues. The unfolding transition has also been followed with circular dichroism in the 250–190 nm region[4] (Fink, Moore, and Lustig, unpublished observations).

Control experiments are highly desirable to determine whether observed partially folded species are on the productive folding pathway, i.e., between N and U or arise from the folding of U into some state not on the pathway between N and U due to the influence of the cosolvent. It is known that alcohols have the potential to induce structure in polypeptides. In the case of ribonuclease A the reduced, alkylated protein is essentially devoid of secondary and tertiary structure in aqueous solution and provides an experimental model. At low pH* the properties of the methanol-induced structure of the reduced, alkylated material are very different from those of the partially-folded species observed with the native enzyme (Fink and Moore, unpublished observations).

Proton NMR Measurements. The thermal denaturation of RNase A has been investigated with 360 MHz proton NMR in 25, 35, and 50% methanol.[2] The NMR spectrum of a protein contains a large amount of potential structural information. If partially folded intermediates are present during refolding, new transient resonances may be observed (either in slow or fast exchange, or both), whose chemical shifts are different from those of the unfolded and native states.

The unfolding process was also found to be completely reversible by NMR; in fact the same sample could be completely unfolded, refolded, and partially unfolded again to give essentially identical spectra at the same temperature several times. The chemical shifts of the signals corresponding to the His imidazole C-2 protons in the *native* conformation showed temperature dependence, consistent with a fast-exchange process. The experimental setup for these experiments was similar to that described above for examination of the native structure in aqueous methanol.

New resonances, in slow exchange, appear in the spectra as the temperature is raised (Fig. 2). These are ascribed to the presence of partially

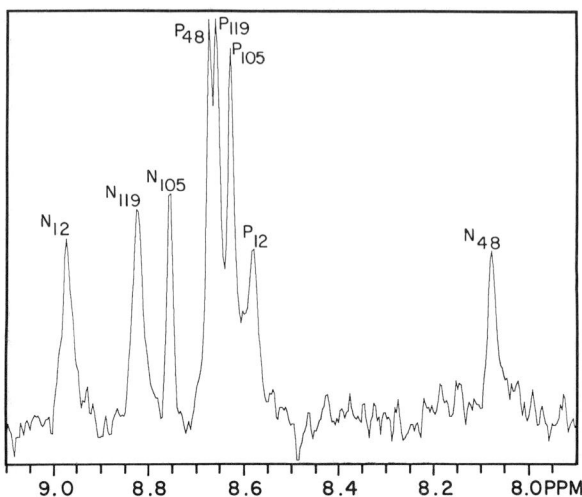

FIG. 2. The 360 MHz NMR spectrum of the His–C-2 region of ribonuclease A showing resonances from partially folded states. The experimental conditions were 50% methanol-d_4, pH* 3.0, 33.1°. The resonances labeled N correspond to native material, those labeled P refer to resonances from partially folded material. Reprinted with permission from Biringer and Fink.[3] Copyright 1982 American Chemical Society.

folded species and have been assigned to particular residues using selective deuteration, as outlined below. The temperature dependence of the chemical shifts of the partially folded species show complex, multiphasic behavior. Above some threshold temperature, all the resonances from the native state will have disappeared leaving only resonances from the partially folded states visible, for example in 50% methanol, pH* 3.0 this occurs at 40°.[2] The data in 35 and 50% methanol reveal at least three distinct partially folded substates. In fact, comparison of the results as the concentration of methanol is progressively increased reveals progressively more stabilization of the partially folded states.

Experiments to Detect Aggregation. A number of methods were used to assess the state of association of ribonuclease A at the high concentrations used in these NMR experiments. Laser light scattering was used to determine the diffusion coefficient with a light scattering photometer, equipped with a laser.[2,3] Ultracentrifugation (sedimentation equilibrium) experiments were carried out with a Spinco Model E instrument, equipped for absorbance monitoring. In each case samples of 20 mg/ml in aqueous and aqueous–methanol, pH* 3.0 were used. Columns (1 × 30 cm) of BioGel P60 were used for gel permeation chromatography, also to assess the state of protein association. The sample was 0.5 ml of 40 mg/ml protein. These experiments were performed at 25°, pH* 3.0 (0.05 M formate), 0.1 M KCl, in aqueous and aqueous–methanol solvents and revealed no aggregation under the experimental conditions.[2]

Selective Deuteration Studies. The identity of the His residue responsible for each of the slow-exchanging resonances (Fig. 2) was assigned by selective deuteration experiments. When incubated in D_2O at pH* 8, the imidazole C-2 protons of RNase A exchange with solvent at different rates, in the order His 105 > 119 > 12 > 48.[12] It was thus possible to prepare samples in which the C-2 protons of the imidazole side chains could be unambiguously identified on the basis of the relative peak areas.

Selective deuteration was accomplished as follows: RNase A was dissolved in D_2O (99.8%) (40 mg/ml) and the pH* adjusted to 8.00 with NaOD. The solution was then incubated at 40°. Aliquots were removed at 3, 6, and 16 day intervals. Each aliquot was allowed to cool to ambient temperature, and the pH* adjusted to 3.0 to quench the exchange reaction. A white precipitate forms during the process and was removed by centrifugation. The supernatant was further purified by ultrafiltration using an Amicon PM10 membrane, and washed with deionized, distilled water. The solution was then lyophilized, and the backbone protons exchanged by the usual procedure.

Kinetics of Refolding at Subzero Temperatures

Refolding of ribonuclease A has been monitored by absorbance (ΔA_{286}, which reflects the solvent exposure of Tyr residues), fluorescence (which also reflects Tyr solvent exposure), circular dichroism (far-UV, which reflects secondary structure), absorbance of nitrotyrosine derivatives, and by the return of inhibitor binding and catalytic activity. By using probes located in different regions of the molecule it is possible to ascertain whether different regions of the molecule refold at different rates. The results[5] indicate that this is indeed the case, in that different regions of the protein attain their native environment (signal) at different rates (Fig. 3).

Spectrophotometric Measurements. Cryosolvent solutions were prepared on a v/v basis as described previously.[2,13] By using buffer systems based on carboxylate the temperature effect on pH* can be minimized. Absorbance measurements of the folding were performed at 286 nm, fluorescence experiments were done with excitation at 279 nm and emission at 303 nm, circular dichroism kinetics were followed at 220 nm. An insulated, thermostatted brass-block holder was used to maintain constant temperature in the sample cuvette. The coolant used was ethanol from

[12] J. L. Markley, *Biochemistry* **14**, 3554 (1975).
[13] A. L. Fink and M. A. Geeves, this series, Vol. 63A, p. 336.

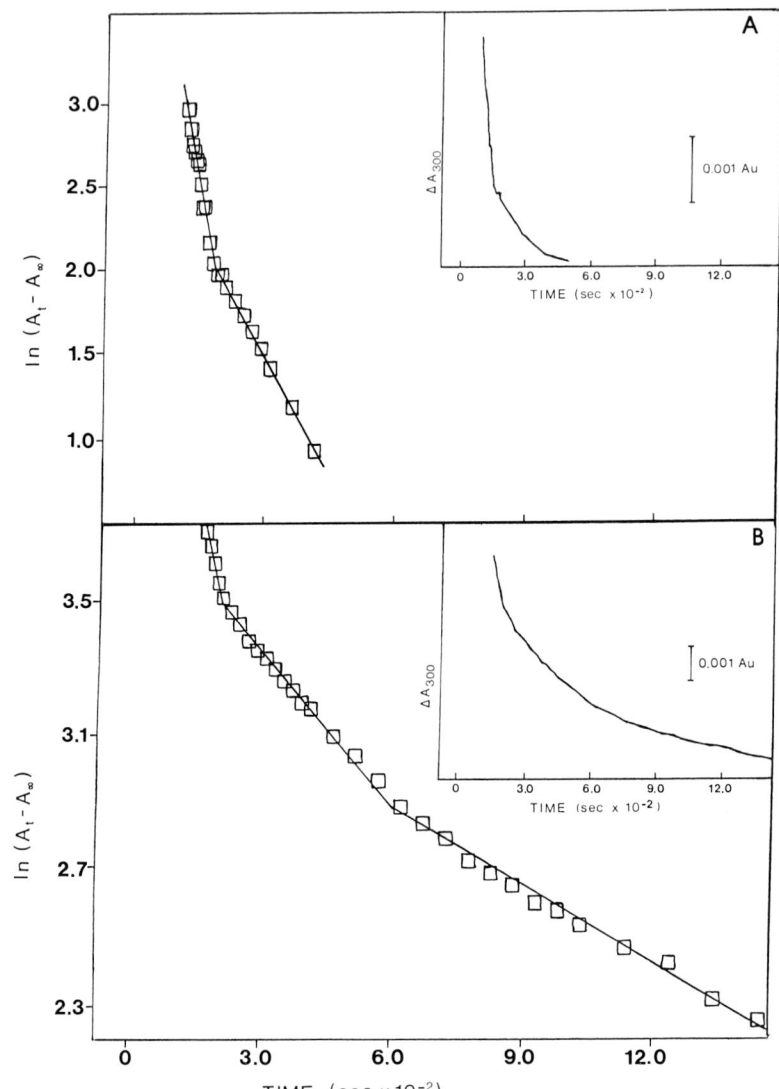

FIG. 3. Time-dependent changes in the refolding of nitrotyrosine derivatives of ribonuclease A, 35% methanol, pH* 3.0, −15°. (A) C$^{\varepsilon 115}$-nitrotyrosine ribonuclease A, biphasic kinetics are observed, the native state is environment in 900 sec. (B) C$^{\varepsilon 115,76}$-nitro-Tyr RNase A, triphasic kinetics are observed, the native state is environment in 6000 sec.

Neslab RTE-8 or ULT-80 baths. Dry gas is used to purge the sample compartment to prevent condensation on the optical faces. In a typical refolding experiment 20 μl of RNase A (0.75 mM, pH* 3.0, in 35 or 50% aqueous methanol) was taken up in a microsyringe and unfolded by incubation at 70° for 10 min. The syringe contents were then injected into the thermostatted cuvette containing 1.0 ml of 35 or 50% methanol cryosolvent at the desired temperature. After mixing the contents, the absorbance changes were followed. Typical concentrations of RNase A were 15 μM. Protein stock solutions were made up immediately before use. Mixing was done with a precooled plumper (Calbiochem) or vibrating stirrer (Helma) for 40 sec. this long time period was used to ensure adequate mixing of the hot unfolded material with the cold cryosolvent.

Refolding kinetics can be measured using the binding of the competitive inhibitor 2'-CMP to monitor formation of the substrate binding site. The protein was unfolded as above with the exception that the stock solution used was 7 mg/ml. The unfolded enzyme was injected into 0.8 ml of 35% methanol, 0.033 M in formate, pH* 3.0, and 100 μM in 2'-CMP. The reaction was monitored at 271.5 nm. The isosbestic point for *both* enzyme unfolding and 2'-CMP binding is 254 nm under these conditions ($-15.2°$). A minimal change in absorbance due to enzyme refolding is observed at 271.5 nm whereas a large change due to 2'-CMP binding can be seen. The small contributions of the absorbance changes at 271.5 nm due to protein refolding were eliminated from the 2'-CMP binding data by subtraction of the average absorbance change from several folding experiments performed in the absence of 2'-CMP. At pH* 6.0 the binding of 2'-CMP was followed at 254 nm, the isosbestic point for folding of the protein alone.

The rate of return of catalytic activity during refolding has been determined in the following manner. An aliquot (25 μl) of stock solution of RNase A consisting of 2 mg/ml protein in 35% methanol, pH* 3.0, was heated in a glass syringe for 10 min at 70°. The sample was then injected into 225 μl of cryosolvent, e.g., 35% methanol, pH* 3.0, precooled to the desired temperature in the spectrophotometer. The protein was allowed to refold for various time periods, then 1.0 ml of precooled substrate (5 mM 2',3'-CMP) in 35% methanol, pH* 6.0, 0.033 M acetate was added. The reaction was monitored at 296 nm and the slope of the initial linear portion taken as a measure of the enzymatic activity. The activity is expressed as a percentage of that expected for the native enzyme under the assay conditions. It was found necessary to carry out the assays at pH* 6.0 rather than at pH* 3.0 in order to obtain linear initial velocities.

Preparation of Nitrated Ribonuclease. Nitrated derivatives of RNase A (C^{ε}-NO$_2$^{115}-RNase A and C^{ε}-NO$_2$^{115,76}-RNase A) were prepared by reac-

tion with tetranitromethane. RNase A (50 mg) was dissolved in 10 ml of
0.13 M TRIS buffer and the pH adjusted to pH 8.0. A 30-fold excess of
tetranitromethane was added as a 10% (w/v) solution in 905 ethanol. The
reaction was allowed to proceed with constant stirring for 20 min. The
reaction was then quenched by passing the solution through a Sephadex
G-25 column (2.5 × 6.0 cm) with 0.1 M ammonium hydroxide solution.
The protein fraction was acidified to pH 7.0 and ultrafiltered 4–5 times
with a 50-fold excess of distilled water in an Amicon ultrafiltration apparatus with a PM10 membrane, then lyophilized. The ultrafiltration step is
necessary to remove a small amount of nitroformate which remains
tightly bound to the enzyme after the initial column purification. These
particular conditions maximize the yield of mono- and dinitrated protein
while minimizing the production of polynitrated and covalently cross-
linked enzyme.[5]

In order to remove covalently cross-linked protein the lyophilized
powder was dissolved in 1–2 ml of 0.13 M phosphate and purified on a
Sephadex SPC-25 column. Cross-linked protein eluted well ahead of the
monomeric nitrated derivative which elutes just prior to unmodified protein. Phosphate was removed by repetitive ultrafiltration.[2]

Fractionation of the various nitrated derivatives was accomplished
with preparative thin-layer isoelectric focusing. A glass plate (10 × 20 cm)
was coated with a 0.3 cm layer of 6.4% (w/v) Sephadex IEF suspension
containing 6.4% (v/v) of carrier ampholyte, pH 8–9.5 (LKB). Lyophilized protein from two successive preparations was combined and dissolved in a minimum of carrier ampholyte and then mixed with a small
amount of gel suspension. The resulting mixture was applied to a 1 × 8 cm
trough cut in the gel bed. Voltage was applied through platinum wires
inserted through filter paper tubes which had been previously soaked in
0.1% (w/v) carrier ampholyte, pH 6–8 at the anode, and pH 8–10 at the
cathode. Focusing was performed for 16–18 hr at 300 V, followed by an
additional 4 hr at 800–1000 V. Good separation is observed between three
yellow bands.

The major bands (C^ε-NO_2^{115}-RNase A and $C^{\varepsilon 115,76}$-RNase A) were removed and separated from the gel beads by filtration through a sintered
glass filter. The volume of the protein solution was reduced by ultrafiltration and then passed through a Sephadex G-25 column with 0.1 M ammonia to remove the major portion of the ampholyte. Ammonia and residual
ampholytes were removed by ultrafiltration, and the protein was then
lyophilyzed. Each protein band was further purified by a second focusing,
using the same procedure as outlined above, with the exception that a 5 ×
20 cm plate was used and the protein was applied as a concentrated
solution in 6.4% carrier ampholyte (pH 8–9.5).

^{1}H NMR Measurements. Results from investigation of the refolding of RNase A at subzero temperatures, as monitored by His–C-2 resonances, reveal the following main points of interest: (1) considerable native protein is formed very rapidly even at $-16°$; (2) there is a relatively compact intermediate (or intermediates) which can be observed by NMR during refolding in aqueous methanol in the -10 to $-20°$ region, thus the folding pathway must be intermediate-controlled; and (3) the multiphasic changes observed during refolding are due to both refolding of different configurational isomers at different rates and to multiple populated intermediate states during refolding.[3,5]

All NMR spectra were collected at 360 MHz using a Redfield 21412 pulse sequence in which the shortest pulse length was 71 μsec. The temperature was calibrated with an external methanol standard. The time required for temperature equilibration in refolding experiments was determined as follows. An aliquot of the sample solution was placed in an NMR tube and allowed to equilibrate at the desired subzero temperature. The dispersion signal used for locking the sample was centered on the oscilloscope. The spectrometer was then tuned to this sample, which was then replaced by the quenched refolding sample. The dispersion signal returned to the center of the oscilloscope within 3 min indicating that temperature equilibration of the refolding solution had been achieved by this time. Spectra were collected at 1 min intervals, with 60 scans per spectrum.

Several different procedures were investigated for rapidly changing the sample from denatured to native conditions. The two best methods, which gave similar results, were (1) heat denaturation of the protein in a syringe (70°, 10 min) and injection of the contents into the NMR tube positioned in liquid nitrogen; the frozen sample was then placed in the NMR probe at the desired temperature; or (2) heat denaturation of the sample in the NMR tube, followed by quenching of the tube in liquid nitrogen to a few degrees below the desired temperature, prior to placing it in the NMR probe. Control experiments indicated that the sample reached $-20°$ within 6 sec.

Quenching techniques involving dilution of concentrated protein samples (e.g., to allow unfolding by high guanidinium chloride concentrations) were found unsatisfactory due to limited solubility of the enzyme, and to insufficient final sample concentrations. Refolding experiments were carried out at several temperatures in the 0 to $-20°$ range, with both 35 and 50% methanol solvent systems.[3] In general the data are quite similar (although of course the rates are different at different temperatures). Similar results were obtained from either technique.

During refolding spectral acquisitions were collected in blocks of 60,

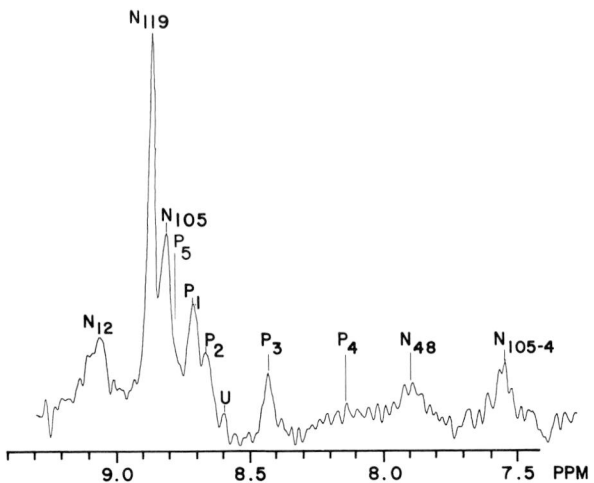

FIG. 4. Proton NMR spectrum (360 MHz) of the His–C-2 region of ribonuclease A during refolding in 35% methanol-d_4, pH* 2.8, −16°. This spectrum is the sum of 5 spectra collected between 5.1 and 10.1 min after initiation of folding. N refers to a resonance from the native state, P refers to A resonance from a partially folded state. Reprinted with permission from Biringer and Fink.[3] Copyright 1982 American Chemical Society.

giving spectra at 60 sec intervals. This was the minimum time interval in which a resolved spectrum could be accumulated under the experimental conditions. Some improvement in S/N could be obtained by adding 2 or 3 such spectra together, however the rapidly changing peak areas severely limits such signal averaging. Due to the short acquisition periods, and fast relaxation caused by reduced mobility at subzero temperatures, the S/N is rather poor.

Spectra from the earliest time periods in refolding showed new resonances not observed from the native or unfolded protein (Fig. 4). The chemical shift of the His–C-2 signal of unfolded RNase A is expected at 8.63 ppm. Even in the earliest spectra obtained (3 min) there is no evidence of any significant concentration of unfolded protein. The new resonances are attributed to His residues in a partially folded state. Examination of the data indicated that the chemical shifts of all the resonances were constant with time, whereas the peak areas showed time-dependent changes. Because of temperature equilibration the first useful spectra are obtained in 3–4 min.

The peak areas of the resonances assigned to partially folded species, which are in slow exchange with the native material, decrease as refolding progresses, whereas the areas of the native peaks increase. Ultimately the spectrum becomes essentially identical to that of the native protein under similar experimental conditions. In order to obtain an unambiguous assignment of the resonances from the partially folded material we used a selectively deuterated sample of RNase A. This deuteration results in the reduction of the His–C-2 protons to the following levels when compared to His-105–C-4: His-105 0.05 proton, His-12 0.12 proton, His-119 0.65 proton, and His-48 0.91 proton. Refolding of this sample in 35% methanol, pH* 2.8, −15° revealed two major resonances from partially folded species at 8.80 and 8.70 ppm. The resonances from the partially folded material were assigned on the basis of the relative peak areas.

Acknowledgments

Supported in part by grants from the National Science Foundation and the National Institutes of Health.

[11] Complementation in Folding and Fragment Exchange

By HIROSHI TANIUCHI, GARY R. PARR, and MARCEL A. JUILLERAT

General Consideration

Richards and colleagues have demonstrated that noncovalent interactions of the two fragments, S-protein and S-peptide, formed by cleavage between residues 20 and 21 of RNase A with subtilisin, generate the structure with the function, closely resembling the native enzyme.[1–4] This first fragment complementing system of RNase A and a second of staphylococcal nuclease discovered later[5–8] have served as excellent testing objects for the principles underlying the structure–function and dynamics of

[1] F. M. Richards, *Proc. Natl. Acad. Sci. U.S.A.* **44**, 162 (1958).

[2] F. M. Richards and R. J. Vithayathil, *J. Biol. Chem.* **241**, 4389 (1959).

[3] H. W. Wyckoff, K. D. Hardman, N. M. Allewell, T. Inagami, L. N. Johnson, and F. M. Richards, *J. Biol. Chem.* **242**, 3984 (1967).

[4] H. W. Wyckoff, D. Tsernoglou, A. W. Hanson, J. R. Knox, B. Lee, and F. M. Richards, *J. Biol. Chem.* **245**, 305 (1970).

[5] H. Taniuchi, C. B. Anfinsen, and A. Sodja, *Proc. Natl. Acad. Sci. U.S.A.* **58**, 1235 (1967).

[6] H. Taniuchi and C. B. Anfinsen, *J. Biol. Chem.* **243**, 4778 (1968).

[7] H. Taniuchi, D. R. Davies, and C. B. Anfinsen, *J. Biol. Chem.* **247**, 3362 (1972).

[8] H. Taniuchi and G. M. Cohen, *Abstr. Am. Crystallogr. Assoc. Summer Meet., Calgary, Alberta, 1980.*

proteins (see a review by Richards and Wyckoff[9] in the first system and those by Anfinsen et al.[10] and Taniuchi[11,12] in the second).

Use of the fragment complementing systems for the studies of the structure–function and dynamics of proteins is justified in a view that if the structure and functions of the fragment complexes resemble native proteins, interatomic interactions responsible for the formation of the fragment complexes are relevant for native proteins.[13] In fact, as the phenomena with native proteins are characteristically specific, there is the specificity of the mechanism in determining the fragment complexes as follows. First, the functional fragment complexes can be formed by a combination of the fragments only in such a way that the discontinuity of the polypeptide chain occurs in specific regions of the three-dimensional structure[12,14,15] (Fig. 1[16–20]). There are probably only a few of such permissible regions for cleavage of the polypeptide chains without disruption of the ordered structure, and these are usually located in surface loops of the three-dimensional structures.[4,8,12,14,19–27] Second, if there are two permissible regions for cleavage, a three-fragment complex having the discontinuity of the polypeptide chain at these two permissible regions can be formed. In this case all three of the fragments are required for formation

[9] F. M. Richards and H. W. Wyckoff, in "The Enzymes" (P. D. Boyer, ed.), Vol. 4, p. 647. Academic Press, New York, 1971.

[10] C. B. Anfinsen, P. Cuatrecasas, and H. Taniuchi, in "The Enzymes" (P. D. Boyer, ed.), Vol. 4, p. 177. Academic Press, New York, 1971.

[11] H. Taniuchi, PAABS Rev. 1, 419 (1972).

[12] H. Taniuchi, in "The Impact of Protein Chemistry on the Biomedical Sciences" (A. N. Schechter, A. Dean, and R. F. Goldberger, eds.), p. 67. Academic Press, New York, 1984.

[13] H. Taniuchi, J. Biol. Chem. 248, 5164 (1973).

[14] G. Andria, H. Taniuchi, and J. L. Cone, J. Biol. Chem. 246, 7421 (1971).

[15] D. Parker, A. Davis, and H. Taniuchi, J. Biol. Chem. 256, 4557 (1981).

[16] J. L. Cone, C. L. Cusumano, H. Taniuchi, and C. B. Anfinsen, J. Biol. Chem. 246, 3103 (1971).

[17] J. L. Bohnert and H. Taniuchi, J. Biol. Chem. 247, 4557 (1972).

[18] A. Davis, I. B. Moore, D. S. Parker, and H. Taniuchi, J. Biol. Chem. 252, 6544 (1977).

[19] A. Arnone, C. J. Bier, F. A. Cotton, V. W. Day, E. E. Hazen, Jr., D. C. Richardson, J. S. Richardson, and A. Yonath, J. Biol. Chem. 246, 2302 (1971).

[20] F. A. Cotton, E. E. Hazen, Jr., and M. J. Legg, Proc. Natl. Acad. Sci. U.S.A. 76, 2551 (1979).

[21] H. Taniuchi and C. B. Anfinsen, J. Biol. Chem. 246, 2291 (1971).

[22] H. Taniuchi, D. S. Parker, and J. L. Bohnert, J. Biol. Chem. 252, 125 (1977).

[23] A. Wlodawer, R. Bott, and L. Sjolin, J. Biol. Chem. 257, 1325 (1982).

[24] R. R. Hantgan and H. Taniuchi, J. Biol. Chem. 252, 1367 (1977).

[25] G. R. Parr, R. R. Hantgan, and H. Taniuchi, J. Biol. Chem. 253 (1978).

[26] M. Juillerat, G. R. Parr, and H. Taniuchi, J. Biol. Chem. 255, 845 (1980).

[27] T. Takano and R. E. Dickerson, J. Mol. Biol. 153, 95 (1981).

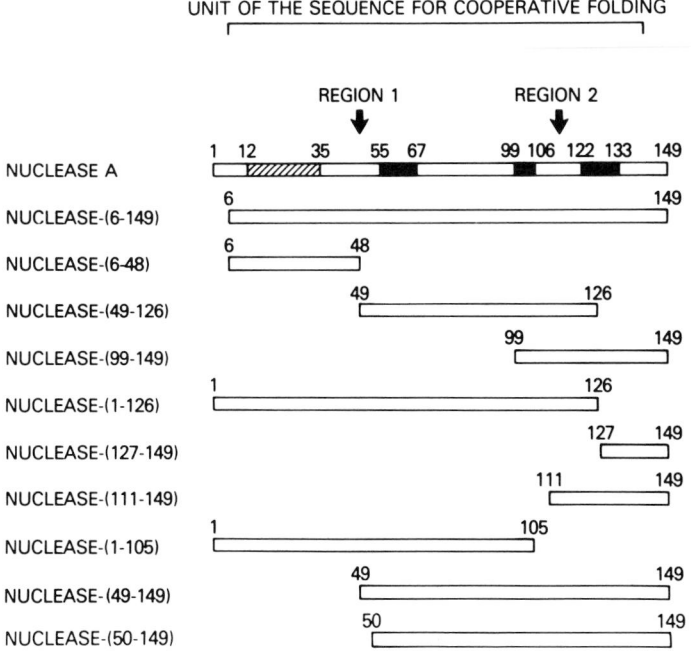

FIG. 1. The fragment system of staphylococcal nuclease A. Nuclease A contains 149 amino acid residues and is devoid of sulfhydryl groups and disulfide bonds.[16–18] Only combination of two or three fragments which allows the discontinuity of the polypeptide chain to occur in the permissible regions for cleavage is productive.[12] For example, while combination of nuclease-(1–105) and nuclease-(99–105) is nonproductive, combination of nuclease-(1–105) and nuclease-(50–149) or that of nuclease-(1–126) and nuclease-(99–149) is productive. Three helical segments and the segment of β-structure of nuclease A[19,20] are indicated by solid dark and hatched areas, respectively.

of the productive complex.[14,26] Third, if the two fragments overlap in such a way that the two permissible regions are contained in both fragments, two alternative complementing structures with a discontinuity in one of the permissible regions are simultaneously formed (Fig. 2). Fourth, in the case of cytochrome *c* it appears that such permissible regions have been conserved during evolution.[24–26,28]

Although even cleavage at such a permissible region is not without a degree of perturbation of interatomic interactions operative in the three-dimensional structure, the fragment complementing systems provide the information which would be difficult to obtain using native proteins. First, with the fragment complementing systems it is possible to study the folding and unfolding under the same physiological conditions (even at 4°). To

[28] G. R. Parr and H. Taniuchi, *J. Biol. Chem.* **258**, 3759 (1983).

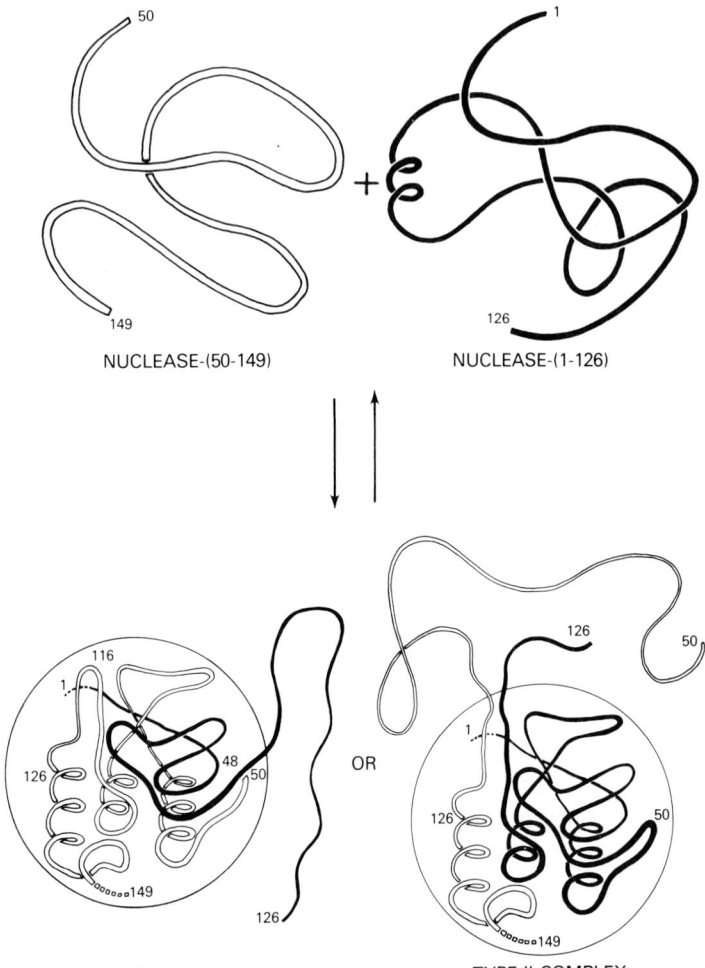

NUCLEASE-(50-149)   NUCLEASE-(1-126)

TYPE I COMPLEX   TYPE II COMPLEX

FIG. 2. Dynamic equilibrium through folding and unfolding of two types of complementing structures simultaneously formed from nuclease-(50–149) and nuclease-(1–126).[21,22] Two
possible types of complexes are simultaneously formed in such a way that the polypeptide
chain incorporated into the ordered structure changes from one fragment to the other with a
discontinuity in one of the two permissible regions for cleavage (see Fig. 1). In type I
structure the portion of residues 1–48 of fragment (1–26) interacts with the entire polypeptide chain of fragment (50–149). In type II, the portion of residues 1 to, say, 115 of fragment
(1–126) interacts with residues, say, 116 to 141 of fragment (50–149). In both complexes, the
redundant portions flexibly protrude from the ordered structure. The rate of folding of each
complex is independent of the stabilizing forces which come to existence only after completion of folding and also independent of temperature and the presence and absence of ligands,
pdTp and Ca^{2+}.[22] On the other hand, the rate of unfolding is strongly dependent on temperature and suppressed by binding with ligands. Note that the nuclease-(1–126) chain of type I

study folding and unfolding of native proteins[13,29] (with the exception of hydrogen exchange[30-36] and formation of disulfide bonds[37-42]) changes of the conditions from or to the extreme conditions (high temperature, extreme pH and denaturants) have to be applied.[43-47] Then, two processes, the conformational transition between the physiological and the nonphysiological conditions and the transition between the native and the nonnative conformations under physiological conditions, would overlap and could not be easily resolved.

Second, since the rate of unfolding of intact proteins can be measured only in the range of the transition temperature under a set of conditions, the temperature parameter which can be varied is limited (e.g., $\Delta T < 10°$). On the other hand the fragment complementing system permits the trans-

[29] A. Light, H. Taniuchi, and R. F. Chen, *J. Biol. Chem.* **249**, 2285 (1974).
[30] A. Hvidt and K. Linderstrom-Lang, *Biochim. Biophys. Acta* **14**, 574 (1954).
[31] A. Hvidt and S. O. Nielsen, *Adv. Protein Chem.* **21**, 287 (1966).
[32] S. W. Englander, N. W. Downer, and H. Teitelbaum, *Annu. Rev. Biochem.* **41**, 903 (1972).
[33] A. N. Schechter, L. Moravek, and C. B. Anfinsen, *J. Biol. Chem.* **244**, 4981 (1969).
[34] J. J. Rosa and F. M. Richards, *J. Mol. Biol.* **145**, 835 (1981).
[35] A. Wlodower and L. Sjolin, *Proc. Natl. Acad. Sci. U.S.A.* **79**, 1418 (1982).
[36] G. Wagner and K. Wüthrich, *J. Mol. Biol.* **160**, 343 (1982).
[37] C. B. Anfinsen, *Science* **181**, 223 (1973).
[38] H. Taniuchi, *J. Biol. Chem.* **245**, 5459 (1970).
[39] T. E. Creighton, *J. Mol. Biol.* **95**, 167 (1975).
[40] S. S. Ristow and D. B. Wetlaufer, *Biochem. Biophys. Res. Commun.* **50**, 544 (1973).
[41] A. S. Acharya and H. Taniuchi, *Proc. Natl. Acad. Sci. U.S.A.* **74**, 2362 (1977).
[42] G. Andria and H. Taniuchi, *J. Biol. Chem.* **253**, 2262 (1978).
[43] W. Kauzman, *Adv. Protein Chem.* **14**, 1 (1959).
[44] C. Tanford, *Adv. Protein Chem.* **23**, 121 (1968).
[45] F. M. Pohl, *Eur. J. Biochem.* **4**, 373; **7**, 146 (1968).
[46] J. F. Brandts, *in* "Structure and Stability of Biological Macromoles" (S. N. Timasheff and G. D. Fasman, eds.), p. 213. Dekker, New York, 1969.
[47] C. B. Anfinsen and H. A. Scheraga, *Adv. Protein Chem.* **29**, 205 (1975).

structure and both the nuclease-(1–126) and the nuclease-(50–149) chains of type II are partly folded and partly unfolded (flexible). Such partial folding of the polypeptide chains would be thermodynamically unfavorable unless it is coupled with a sufficiently strong interaction. Thus, this phenomenon suggests that the energy state of one region of the ordered structure is coupled with the states in distant regions throughout the structure. Furthermore, the fact that there are only two permissible regions for cleavage, to permit such fragment complexes to be formed, points to the specific nature of the mechanism of this hypothetical global coupling. The figure is taken from H. Taniuchi, A. S. Acharya, G. Andria, and D. S. Parker, *Adv. Exp. Med. Biol.* **86A**, 51 (1977), with permission.

formation to be observed in a substantially increased range of temperature (e.g., $\Delta T > 20°$) (below the transition temperature), resulting in an increase in accuracy of the thermodynamic or activation thermodynamic values to be measured.

Third, to study the effect of substitution of amino acids on the structure–function and dynamics, the fragment complementing system is particularly powerful in that the fragment can be chemically synthesized with reasonable yields.[48–55] In contrast, the total synthesis of native proteins would be a formidable task if not impractical. Substitution of amino acids would be possible using genetic manipulation.[56–58] However, it would still be difficult to substitute amino acids with nonnatural amino acids (e.g., norvaline[59]) in such a genetic approach.

Fourth, the fragment complexes exhibit a dynamic equilibrium of global folding and unfolding under physiological conditions, as demonstrated by the fragment exchange measurements.[13,55,60,61] It is important to unequivocally establish the reversibility of overall folding and unfolding under the same physiological conditions, for making a framework of protein folding in theoretical considerations.[12,13] Hydrogen exchange of proteins is also related to various modes of partial as well as overall unfolding.[30–36] Therefore, the hydrogen exchange method has a more resolving power in analyzing the concerted molecular motions than the fragment exchange technique. However, it is not always easy to resolve the overall unfolding from partial unfolding by a hydrogen exchange technique. Moreover, hydrogen exchange methods cannot differentiate whether overall unfolding occurs through sequential unfolding or whether there is

[48] F. M. Finn and K. Hofmann, in "The Proteins" (H. Neurath and R. L. Hill, eds.), Vol. 2, p. 105. Academic Press, New York, 1976.

[49] B. W. Erickson and R. B. Merrifield, in "The Proteins" (H. Neurath and R. L. Hill, eds.), Vol. 2, p. 257. Academic Press, New York, 1976.

[50] P. G. Katsoyannis and G. P. Schwartz, this series, Vol. 47, p. 501.

[51] M. S. Doscher, this series, Vol. 47, p. 578.

[52] A. R. Mitchell, B. W. Erickson, M. N. Ryabtsev, R. S. Hodges, and R. B. Merrifield, J. Am. Chem. Soc. **98**, 7357 (1976).

[53] D. A. Ontjes and C. B. Anfinsen, Proc. Natl. Acad. Sci. U.S.A. **64**, 428 (1969).

[54] M. Pandin, E. A. Padlan, C. DiBello, and I. M. Chaiken, Proc. Natl. Acad. Sci. U.S.A. **73**, 1844 (1976).

[55] M. Juillerat and H. Taniuchi, Proc. Natl. Acad. Sci. U.S.A. **79**, 1825 (1982).

[56] R. C. Mulligan, B. H. Howard, and P. Berg, Nature (London) **277**, 108 (1979).

[57] C. Kondor-Koch, H. Riedel, K. Soderberg, and H. Garoff, Proc. Natl. Acad. Sci. U.S.A. **79**, 4525 (1982).

[58] S. W. Liu and G. Milman, J. Biol. Chem. **258**, 7469 (1983).

[59] M. Juillerat and H. Taniuchi, Abstr. FEBS Meet., 15th, Brussels 214 (1983).

[60] H. Taniuchi and J. L. Bohnert, J. Biol. Chem. **250**, 2388 (1975).

[61] R. R. Hantgan and H. Taniuchi, J. Biol. Chem. **253**, 5373 (1978).

a concerted motion such that unfolding occurs through simultaneous disordering of many regions of the structure (nonsequential). On the other hand, application of the fragment exchange method to the complexes containing three or more fragments could detect nonsequential unfolding through concerted motion.[55] Thus, the hydrogen exchange and the fragment exchange method are complementary.

Fifth, the complementing fragments alone are generally disordered and flexible with the exception that the fragments containing disulfide bonds are apparently more restricted in conformation than those without disulfide bonds. However, these complementing fragments have the potential to fold to the native structures upon interactions with appropriate second fragments. In this sense, the conformational state of the fragments may be considered as mimicking an intermediate in which progression of folding would have been arrested due to the incomplete information. It is not easy to study an intermediate with a short life which would be formed during refolding of native proteins. Thus, characterization of large fragments would provide an insight into the intermediate state.[15,42]

The fragments of large proteins which represent domains have been reviewed by Wetlaufer.[62] An excellent review of the fragment system of RNase A is given by Richards and Wyckoff.[9] Kinetic aspects of folding and unfolding of the RNase A fragment complex is reviewed by Kim and Baldwin.[63] The mode of regeneration of disulfide bonds of the reduced RNase A fragment is reviewed by Acharya and Taniuchi.[64] In this chapter we describe the preparation and properties of the fragment complementing systems of staphylococcal nuclease and cytochrome *c* as examples of the methods of complementation in folding and fragment exchange and the interpretation of the data obtainable by the methods. There are many interesting fragment complementing systems of other proteins including thioredoxin,[65,66] tryptophan synthase,[67] cytochrome b_5 reductase,[68] human pituitary growth hormone,[69] and hemoglobin.[70] These papers of other fragment complementing systems should be consulted for fuller understanding of fragment complementation and its implications.

[62] D. B. Wetlaufer, *Adv. Protein Chem.* **34**, 61 (1981).
[63] P. S. Kim and R. L. Baldwin, *Annu. Rev. Biochem.* **51**, 459 (1982).
[64] A. S. Acharya and H. Taniuchi, *Mol. Cell. Biochem.* **44**, 129 (1982).
[65] I. Slaby and A. Holmgren, *Biochemistry* **18**, 5584 (1979).
[66] A. Holmgren and I. Slaby, *Biochemistry* **18**, 5591 (1979).
[67] D. A. Jackson and C. Yanofsky, *J. Biol. Chem.* **244**, 4539 (1969).
[68] P. Strittmatter, R. E. Beary, and D. Corcoran, *J. Biol. Chem.* **247**, 2768 (1972).
[69] C. H. Li and T. A. Bewley, *Proc. Natl. Acad. Sci. U.S.A.* **73**, 1476 (1976).
[70] C. S. Craik, S. R. Buchman, and S. Beychok, *Proc. Natl. Acad. Sci. U.S.A.* **77**, 1384 (1980).

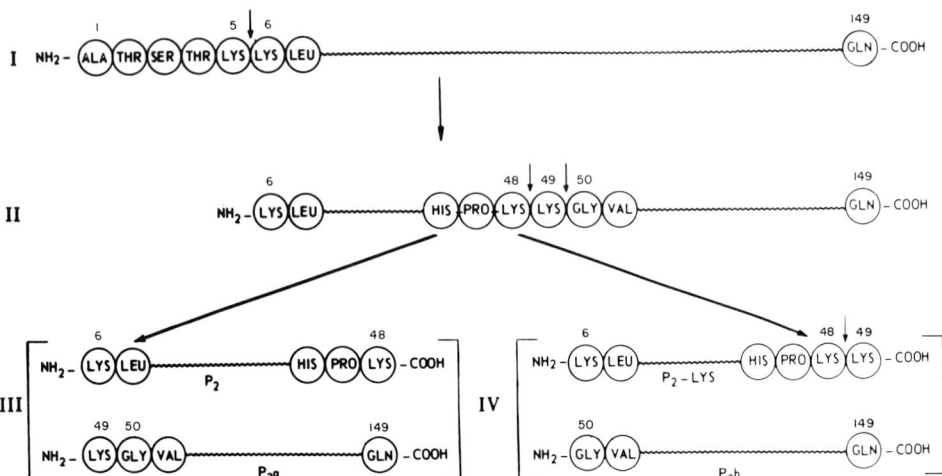

FIG. 3. Steps of formation of nuclease T from nuclease A by limited digestion with trypsin. Trypsin cleaves the peptide bonds of the specific regions of nuclease A bound with pdTp and Ca^{2+}, to form nuclease T. (**I**) Fast cleavage of the peptide bond between residues 5 and 6 yields nuclease-(6–149) exhibiting the enzymatic activity of the native protein. This step is followed by (**II**) cleavage of the peptide bond between residues 48 and 49 or between residues 49 and 50, resulting in a mixture of two similar complexes, (**III**) and (**IV**). Subsequently trypsin removes lysine 49 from complex IV. The final products nuclease T, complexes nuclease-(6–48) · nuclease-(49–149) and nuclease-(6–48) · nuclease-(50–149) exhibit 8% of the enzymatic activity of nuclease A. The data are taken from Taniuchi and Anfinsen.[6]

Preparation of the Fragment Complementing Systems

As described earlier only specific regions of proteins are permissible for cleavage without disruption of forces to stabilize the structure. Therefore, combinations of two or more fragments have to be chosen for the productive complexes in such a way that the discontinuity of the polypeptide chains occurs in these permissible regions (allowed complementing structures) (Figs. 1 and 2). There is no rule known at the present time for locating such a permissible region with the exception that the permissible regions are usually located in loop regions and may be conserved during evolution of individual proteins. One empirical procedure for finding the permissible regions under an ideal situation is as follows.

If one can remove amino acid residues one by one from the NH$_2$-terminus of a protein, the shortened protein would still exhibit the ordered structure up to a certain step of removal. Then sudden disordering of the structure would occur upon removal of the next one or two residues, yielding a large, disordered COOH-terminal fragment. Similarly, one can obtain a large, disordered NH$_2$-terminal fragment. Combination of these two overlapping fragments would simultaneously yield most, if not all, allowed complementing structures (Fig. 2). The redundant portions protruding from the ordered structure can be easily digested away with a proteolytic enzyme, if the ordered structure is protected against proteolysis. Separation and identification of the shortened fragments of the derived complexes would then establish a fragment complementing system.[21,24–26]

Staphylococcal Nuclease A

There are two permissible regions for cleavage in staphylococcal nuclease which are located between residues 48 and 50[6] and between residues 115 and 117[21,71] (Fig. 1). It is also known that removal of five NH_2-terminal or one COOH-terminal residues does not cause significant destabilization of the structure or the function.[6,72] Thus, combination of the fragment containing residues 6 to 48, nuclease-(6–48) with nuclease-(49–149) or nuclease-(50–149), yields a type I complex, called nuclease T, exhibiting 8% of the enzymatic activity[6] (Fig. 1). Combination of nuclease-(1–126) or nuclease-(1–123) with nuclease-(99–149), nuclease-(111–149) or nuclease-(114–149) yields a second type of the complementing structure with 10% of the enzymatic activity[21] (Fig. 1). Combination of nuclease-(1–126) and nuclease-(50–149) simultaneously forms these two types of the complementing structures[21,22] (Fig. 2). Combination of three fragments, nuclease-(6–48), nuclease-(49–126) [or nuclease-(50–126)] and nuclease-(99–149), yields a three-fragment complex with a low level of enzymatic activity.[14]

Preparation of Nuclease-(6–48), Nuclease-(49–149), and Nuclease-(50–149). Trypsin specifically cleaves the peptide bonds at one of the permissible regions, residues 48 and 50 of nuclease A,[5,6] when nuclease is stabilized by binding with deoxythymidine 3′,5′-diphosphate (pdTp) and Ca^{2+}.[73] Nuclease T is a mixture of approximately equal amounts of two complexes, nuclease-(6–48) · nuclease-(49–149) and nuclease-(6–48) · nuclease-(50–149) thus formed (Fig. 3). An example is as follows.

Purified nuclease A[18] (12 mg/ml) is incubated with trypsin (1% w/w) [Worthington, TRTPCK, treated with L-(1-tosylamido-2-phenyl)ethyl chloromethyl ketone] in the presence of pdTp (10% w/w), 0.01 M $CaCl_2$, and 0.05 M NH_4HCO_3 (pH 8.0) at 25°. The progress of cleavage of the peptide bonds is followed by assaying the enzymatic activity by the method of Cuatrecasas *et al.*[74] (a decrease to 8%). After incubation for

[71] I. Parikh, L. Corley, and C. B. Anfinsen, *J. Biol. Chem.* **246**, 7392 (1971).
[72] H. Taniuchi and C. B. Anfinsen, *J. Biol. Chem.* **241**, 4366 (1966).
[73] P. Cuatrecasas, H. Taniuchi, and C. B. Anfinsen, *Brookhaven Symp. Biol.* **21**, 172 (1968).
[74] P. Cuatrecasas, S. Fuchs, and C. B. Anfinsen, *J. Biol. Chem.* **242**, 1541 (1967).

2 hr, the trypsin action is quenched by addition of soybean trypsin inhibitor, STI. Then, gel filtration on a Sephadex G-75 (Pharmacia) column (2 × 200 cm), using 0.1 M ammonium acetate, pH 8, at 6–8°, separates nuclease T from other components including the trypsin–STI complex and nuclease-(1–5) (see Fig. 3) but excess of STI.

Purification of nuclease T is achieved by affinity chromatography on Sepharose-pdTp [Sepharose 4B (Pharmacia) coupled to 3′-(4′-aminophenylphosphoryl)deoxythymidine 5′-phosphate (Ash Stevens)] prepared according to Cuatrecasas et al.[75] The nuclease T sample obtained by lyophilization after gel filtration is dissolved (27 mg/ml) in a buffer of 0.1 M ammonium acetate/0.005 M $CaCl_2$, pH 8, and applied to a Sepharose-pdTp column (2 mg of nuclease T/ml of bed volume) equilibrated with the same buffer at 23°. After washing with the same buffer, adsorbed nuclease T is eluted with 0.5 M guanidine–HCl, pH 7.0,[18] and lyophilized. The yield is almost quantitative.

The sample of nuclease T thus obtained is subjected to gel filtration on a Sephadex G-50 (fine) (Pharmacia) column (diameter, 2 to 4 cm; height, 200 cm) using 30% acetic acid at 6–8°, to separate a mixture of nuclease-(49–149) and nuclease-(50–149) from nuclease-(6–48), the former mixture eluting earlier than the latter fragment. Separation of nuclease-(49–149) and nuclease-(50–149) is accomplished by ion-exchange chromatography on phosphocellulose (Whatman, Chromedia PI), using a gradient elution with 0.3 M, pH 6.0, and 1 M, pH 8, ammonium acetate at 6–8°. Nuclease-(6–48) may be purified by repeating gel filtration. Nuclease-(6–48), nuclease-(49–149), and nuclease-(50–149) are identified by amino acid analysis, two-dimensional peptide mapping after digestion with trypsin,[6] and measurements of generation of the enzymatic activity with their complementing fragments (dissociation constant, 10^{-6}–10^{-7} M at pH 8 at 25°).[13,29]

A complex of nuclease-(6–48) and nuclease-(49–149) can be crystallized.[7] X-Ray diffraction analysis of the crystals has shown that the three-dimensional structure of this complex closely resembles the native nuclease A.[7,8]

Preparation of Nuclease-(1–126).[15,76] Cleavage of trifluoroacetylated nuclease[77] by trypsin is faster at arginine-126 than other arginine residues, resulting in a degree of accumulation of fragment (1–126) in the early stage of digestion.[72] The following is an example.

[75] P. Cuatrecasas, M. Wilchek, and C. B. Anfinsen, *Proc. Natl. Acad. Sci. U.S.A.* **61**, 636 (1968).

[76] H. Taniuchi and C. B. Anfinsen, *J. Biol. Chem.* **244**, 3864 (1969).

[77] H. Taniuchi, C. B. Anfinsen, and A. Sodja, *J. Biol. Chem.* **242**, 4752 (1967).

Trifluoroacetylation of nuclease A is carried out according to the method of Goldberger and Anfinsen.[78] Briefly, an aqueous solution containing purified nuclease A[18] (22 mg/ml) is adjusted to pH 10 with 1 N KOH. A volume of redistilled ethylthiol trifluoroacetate (Pierce) (20 μl/ mg nuclease A) is at once added to the solution. The mixture is vigorously stirred and kept at pH 10 by addition of 1 N KOH at 25°. After consumption of KOH stops (60 min), the mixture is dialyzed against 6 liter of 0.1 N acetic acid at 6° overnight with six changes of 0.1 N acetic acid. Then, the dialyzed mixture (containing precipitates of trifluoroacetylated nuclease A) is lyophilized.

Trifluoroacetylated nuclease A is dissolved in 0.05 M NH$_4$HCO$_3$, pH 8 (13 mg nuclease A equiv./ml) and incubated with diisopropyl fluorophosphate treated trypsin[79] (or TRTPCK trypsin) (23 μg/ml) at 25°. After incubation for 10 min, the digestion is quenched by addition of STI. Deblocking of ε-NH$_2$ groups is attained by incubating with 1 M piperidine (Eastman, redistilled) at a concentration of 11 mg nuclease A equiv./ml for 2 hr in an ice bath. Then, piperidine is removed by lyophilization.

The deblocked mixture is chromatographed on phosphocellulose. The sample of nuclease-(1–126) [containing contaminating intact nuclease and nuclease-(1–105)], thus obtained is treated with a Sepharose-pdTp column in a manner similar to that described with nuclease T. Nuclease-(1–126) and contaminating nuclease-(1–105) are passed through the affinity column while nuclease is absorbed. Nuclease-(1–105) is removed by gel filtration on a Sephadex G-75 column (2 × 200 cm), using 30% acetic acid at 6–8°.[15] Nuclease-(1–126) thus purified is identified by amino acid analysis, two-dimensional peptide mapping after digestion with trypsin,[72] and measurements of generation of the enzymatic activity with nuclease-(99–149) (dissociation constant, 10^{-6}–10^{-7} M at pH 8 at 25°[21]) [nuclease-(1–105) does not generate enzymatic activity with nuclease-(99–149)[15]]. The overall yield of purified nuclease-(1–126) ranges from 12 to 20%.

Ion-exchange chromatography of deblocked nuclease-(1–126) sometimes results in two fractions comprising two partially resolved absorbance peaks (see Fig. 1 of Ref. 76). The samples purified by affinity chromatography and gel filtration (see above) from these two fractions cannot be distinguished from each other on the basis of amino acid analysis, two-dimensional peptide map of the tryptic digest, and the ability to generate enzymatic activity with nuclease-(99–149).[15] Since deblocked nuclease-(1–126) is homogeneous by polyacrylamide gel electrophoresis at pH 2.3,[76] such heterogeneity by ion-exchange chromatography could be due

[78] R. F. Goldberger and C. B. Anfinsen, *Biochemistry* **1**, 401 (1962).
[79] J. T. Potts, A. Berger, J. Cooke, and C. B. Anfinsen, *J. Biol. Chem.* **237**, 1851 (1962).

to partial deamidation of the asparagine or glutamine residues (which could occur in a random fashion during blocking or deblocking of ε-NH$_2$ groups and could not be detected by peptide mapping) rather than partial deblocking of ε-NH$_2$ groups. In fact, contaminating nuclease A (remaining undigested) is also heterogeneous by ion-exchange chromatography as two enzymatic activity peaks appear in the elution profile.

Preparation of Nuclease-(99–149) and Nuclease-(111–149).[21] Nuclease-(99–149) can be obtained by digestion of intact nuclease A or nuclease-(50–149) [or nuclease-(49–149)] with cyanogen bromide in 70% formic acid for 20 hr at 25° followed by gel filtration on Sephadex G-50 (fine) using 0.01 N acetic acid/0.1% ammonium acetate at 25°. Purification of the sample is attained by repeating gel filtration.[15]

To prepare nuclease-(111–149), a complex of nuclease-(1–126) and nuclease-(99–149) is digested with trypsin in the presence of pdTp and Ca^{2+} at 25° for 10 min and then the digestion is quenched by addition of STI. The derived complex containing nuclease-(1–126) and nuclease-(111–149) is separated by gel filtration on Sephadex G-50, using 0.01 N acetic acid/1% ammonium acetate at 25°, and then purified by ion-exchange chromatography on phosphocellulose at 6° (the ion-exchange chromatography at 25° results in dissociation of the derived complex[21]). The component fragments, nuclease-(1–126) and nuclease-(111–126), are separated by gel filtration on a Sephadex G-75 column (2 × 200 cm) using 30% acetic acid at 6°.[15,21]

Partial cleavage of the peptide bond of nuclease-(1–126) between residues 48 and 49 or between residues 49 and 50 or both would occur in addition to complete cleavage of the peptide bond between residues 110 and 111 of nuclease-(99–149) during this limited digestion, resulting in a fractional population of a three-fragment complex composed of nuclease-(1–48), nuclease-(49–126) (or 50–126), and nuclease-(111–149).[15,22] This contaminating three-fragment complex is apparently removed by ion-exchange chromatography.[21]

Nuclease-(99–149) and nuclease-(111–149) are identified on the basis of amino acid analysis, two-dimensional peptide map of the tryptic digest, and generation of enzymatic activity with nuclease-(1–126).[21]

Preparation of Nuclease-(50–126) [or Nuclease-(49–126)].[14] Nuclease-(50–126) [or (49–126)] can be prepared from trifluoroacetylated nuclease-(50–149) [or (49–149)] by limited digestion with trypsin followed by deblocking in a manner similar to preparation of nuclease-(1–126). Deblocked nuclease-(50–126) [for (49–126)] is purified by gel filtration on a Sephadex G-50 column (2.2 × 300 cm) using 0.01 N acetic acid at 25°. Removal of contaminating intact nuclease-(50–149) [or (49–149)] is achieved by affinity chromatography on Sepharose 4B (Pharmacia) coupled to nuclease-(6–48) by the established method.[80,81] Nuclease-(50–126) [or (49–126)] is identified on the basis of amino acid analysis, two-dimensional peptide mapping after digestion with trypsin, α-NH$_2$- and COOH-terminal groups analyses, and generation of enzymatic activity upon mixing with the two fragments, nuclease-(6–48) and nuclease-(99–149). Only a combination of all three of these fragments generates the active site or the ligand binding site.

Cytochrome c

So far, two permissible regions for cleavage of the polypeptide chain without disruption of characteristics of horse cytochrome c have been found between residues 23 and 25 and between residues 39 to 55.[24–26] Interestingly, the structural region corresponding to one of these permissible regions for cleavage, namely residues 38–57, is missing in *Pseudomonas aeruginosa* cytochrome c_{551}[82] and the two regions missing in *Chlorobium thiosulfatophilum* cytochrome c_{555}[83] exactly correspond to the two permissible regions. These truncated procaryotic cytochrome c species still fold to the three-dimensional structures comparable to eukaryotic cytochrome c (cytochrome c fold).[84] This fact, if not accidental, suggests that the information critical for the cytochrome c fold is conserved in the nonpermissible regions for cleavage. Being consistent with this idea, the two complementing fragments, heme fragment containing residues 1 to 25, (1–25)H and apofragment (23–104) are completely exchangeable between horse and tuna cytochrome c with respect to formation of hybrid complexes.[28]

In preparation of the fragment complementing system of cytochrome c, the principle described earlier is applied. That is, if two large overlapping fragments, one containing the NH$_2$-terminus and the other containing the COOH-terminus, are mixed, the sytem finds by itself the complementing structure(s) in which the discontinuity of the polypeptide chains folded to the ordered structure occurs in the permissible region(s). On the basis of the characteristics of native cytochrome c, the ferrous form of which is resistant to proteolysis,[85] digestion of the fragment complexes

[80] J. Porath, R. Axen, and S. Ernback, *Nature (London)* **215,** 1491 (1967).
[81] I. Kato and C. B. Anfinsen, *J. Biol. Chem.* **244,** 5849 (1969).
[82] R. J. Almassy and R. E. Dickerson, *Proc. Natl. Acad. Sci. U.S.A.* **75,** 2674 (1978).
[83] Z. R. Korszun and F. R. Salemme, *Proc. Natl. Acad. Sci. U.S.A.* **74,** 5244 (1977).
[84] R. E. Dickerson, T. Takano, D. Eisenberg, O. B. Kallai, L. Samson, A. Cooper, and E. Margoliash, *J. Biol. Chem.* **246,** 1511 (1971).
[85] M. Nozaki, H. Mizushima, H. Horio, and K. Okunuki, *J. Biochem. (Tokyo)* **45,** 815 (1958).

with trypsin after reduction with ascorbate should remove only the redundant portions.[24,25]

Complex (1–53)H · (54–104). The procedure of Hantgan and Taniuchi[24,61] is briefly described. Apocytochrome *c* (1–104) is prepared from purified horse heart cytochrome c[86,87] using treatment with $Ag_2SO_4/7\%$ acetic acid,[87,88] followed by reduction with dithiothreitol[87] and alkylation with *N*-eythylmaleimide.[89] Cyanogen bromide heme fragment (1–65)H is prepared according to Corradin and Harbury.[90] Equimolar amounts of heme fragment (1–65)H and apocytochrome *c* (1–104) are mixed in 0.05 *M* potassium phosphate, pH 7.8, to give a concentration of complex 2.6 × 10^{-4} *M* and then deaerated, equilibrated with argon (nitrogen may also be used), reduced with 8 × 10^{-4} *M* ascorbate, and incubated with trypsin (final concentration, 50 μg/ml) for 60 min at 25°. Then, the trypsin action is stopped with STI. The derived complexes are separated by gel filtration on Sephadex G-75 using 0.05 *M* ammonium acetate, pH 7.0, at 6–8° and lyophilized. To separate the component fragments the derived complexes are dissolved in 0.02 *M* potassium phosphate, pH 7.2/8 *M* urea and subjected to ion-exchange chromatography on a SP-Sephadex C-25 column (1 × 50 cm for 13 μmol of the original parent complex), using a gradient elution with 0.02 and 0.2 *M* potassium phosphate, pH 7.2/8 *M* urea (200 ml each) at 6–8°. The fragments thus separated are desalted on Sephadex G-10 using 0.05 *M* ammonium acetate, pH 6.9, at 6–8°. In separation of the fragments, the urea serves two functions: first it disrupts interactions between the fragments; second, apofragments (non-heme-containing fragments) appear to be irreversibly bound to the gel matrix in the absence of a high concentration of urea.

The derived complexes formed under the conditions of the limited digestion consist of two major populations, complexes (1–53)H · (56–104) and (1–53)H · (54–104) and two minor populations, complexes (1–53)H · (40–104) and (1–53)H · (39–104) as analyzed by ion-exchange chromatography on SP-Sephadex C-25 in the absence of urea. Thus, the yields of separated component fragments (1–53)H, (56–104), (53–104), (40–104), and (39–104) were 37, 18, 15, 11, and 3%, respectively, in a preparation carried out using 30 μmol of the parent complex.[24] These separated fragments can be identified using amino acid analysis, three steps of Edman degradation, and peptide mapping after digestion with trypsin, and COOH-terminal analysis.

[86] T. Flatmark, *Acta Chem. Scand.* **18,** 1517 (1964).
[87] W. R. Fisher, H. Taniuchi, and C. B. Anfinsen, *J. Biol. Chem.* **248,** 3188 (1973).
[88] K. G. Paul, *Acta Chem. Scand.* **4,** 239 (1950).
[89] E. Roberts and G. Rouser, *Anal. Chem.* **30,** 1292 (1958).
[90] G. Corradin and H. Harbury, *Biochim. Biophys. Acta* **221,** 489 (1970).

Horse heart cytochrome c contains a single tryptophan residue at position 59. In the native protein, the fluorescence of this residue is nearly fully quenched due to its proximity to the heme group.[87,91] Therefore, the decrease in fluorescence intensity on adding aliquots of heme fragment to a solution of the apofragment can be used to measure the stoichiometric binding and the dissociation constant. For complex ferri(1–53)H · (54–104), a dissociation constant of at least $3 \times 10^{-7} M$, and possibly as low as $3 \times 10^{-8} M$ is found at pH 7.8 at 25°.[24] The biological activity of these complexes can be determined with a yeast lactate dehydrogenase (cytochrome b_2) assay system. Values of approximately 25% of that of the native protein are found at pH 7.8,[24] but this figure increases to 62% at pH 6.8.[61] The difference in these results is probably due to the pH dependence of the ligation of methionine 80 to the heme iron,[92] a necessary condition for biological activity.[93,94] Finally, the UV-CD spectrum of the complex (1–53)H · (54–104) is very similar to that of the native protein.[61]

These results clearly demonstrate that a complex closely resembling the native protein, in both structure and function, is formed by complementing heme fragment ferri(1–53)H with either apofragment (54–104) or (56–104). We have used the term "productive" when referring to such complexes. Thus, the region of the amino acid sequence between residues 54 and 55 represents a permissible region for cleavage. In advent of discovery of a second permissible region for cleavage between residues 23 and 25 as described later in this chapter, it is puzzling that only one type of complementing structure corresponding to the derived complex (1–53)H · (54–106) or (1–53)H · (56–104) is found in combination of (1–65)H · (1–104). However, it cannot be ruled out that a second complex corresponding to complex (1–25)H · (23–104) is formed in a small population in this system. Indeed, recent reexamination has shown that approximately 6% of the complex formed from heme fragment (1–65)H and apocytochrome c not at 25 but at 10° yields the derived complex (1–25)H · (23–104) upon digestion with trypsin in the reduced form.[95] It is unknown whether this small population reflects equilibrium between two types of complexes corresponding to the derived complexes (1–25)H · (23–104) and (1–53)H · (56–104) or the relative rates of formation of the two complexes.

[91] T. Y. Tsong, *Biochemistry* **14**, 1542 (1975).

[92] G. R. Parr and H. Taniuchi, manuscript in preparation (1986).

[93] E. Shechter and P. Saludjian, *Biopolymer* **5**, 788 (1967).

[94] N. Osherhoff, D. Borden, W. H. Koppenol, and E. Margoliash, *J. Biol. Chem.* **255**, 1689 (1980).

[95] M. Juillerat and H. Taniuchi, manuscript in preparation (1986).

Complexes (1–25)H · (23–104) and (1–38)H · (56–104). Harris and Offord have reported that N^ε-acetimidyl heme fragment (1–38)H binds with
N^ε-acetimidyl apofragment (39–104) to form a productive complex.[96]
Wilgus *et al.* also have shown that heme fragment (1–38)H and apocytochrome *c* (1–104) form a productive complex.[97] The latter combination
of two overlapping fragments has been used to find a second permissible
region. It turns out that this system simultaneously yields two alternative
complementing structures, represented by the derived complexes (1–
25)H · (23–104) and (1–38)H · (56–104)[25] as briefly described in the following.

Preparation.[25,26] Horse heart cytochrome *c* is reacted with citraconyl
anhydride in the presence of 6 *M* guanidine-HCl.[26,98] Citraconylated cytochrome *c* is digested with trypsin to yield citraconylated heme fragment
(1–38)H and citraconylated apofragment (39–104) which are separated by
gel filtration on Sephadex G-50 (fine) using 0.05 *M* ammonium bicarbonate, pH 8.3, at 6–8°.[26] Then, citraconylated (1–38)H are deblocked in 30%
acetic acid. The deblocked fragment is purified by gel filtration on a
Sephadex G-50 (fine) using 10% formic acid at 6–8° followed by ion-
exchange chromatography on SP-Sephadex C-25 using a gradient elution
with 0.02 and 0.3 *M* ammonium bicarbonate at 6–8°.

Apofragment (39–104) can be prepared by deblocking of the citraconylated (39–104) and then purifying it by gel filtration using 10% formic acid
and ion-exchange chromatography in the presence of 8 *M* urea, followed
by desalting on Sephadex G-10. Alternatively, apocytochrome *c* is digested with clostripain[99] which selectively cleaves the peptide bond between arginine-38 and lysine-39.[25] The resulting apofragment (39–104) is
purified by gel filtration on Sephadex G-50 (fine) using 10% formic acid at
6–8°.

When the combination of fragments (1–38)H and apocytochrome *c* is
subjected to the complementation–reduction–digestion procedure described above, two derived complexes (1–25)H · (23–104) and (1–
38)H · (56–104) are formed.[25] In this case, the size differences of the component fragments permit an initial separation on Sephadex G-50 in 2.0 *M*
acetic acid at 6–8° (10% formic acid may also be used with somewhat
better results). Additional separation and purification are then achieved
with ion-exchange chromatography on SP-Sephadex C-25 in the presence

[96] D. F. Harris and R. E. Offord, *Biochem. J.* **161,** 21 (1977).
[97] H. Wilgus, J. S. Ranweiler, G. S. Wilson, and E. Stellwagen, *J. Biol. Chem.* **253,** 3265 (1978).
[98] M. Z. Atassi and A. F. S. A. Habeeb, this series, Vol. 25, p. 546.
[99] W. M. Mitchell and W. F. Harrington, *J. Biol. Chem.* **243,** 4683 (1968).

of 8 M urea followed by desalting. Fragments are identified as described earlier.

Properties of the Complexes.[25] Complexes ferri(1–25)H · (23–104) and ferri(1–38)H · (56–104), each reconstituted from the component fragments, exhibit visible absorption spectra, UV-CD spectra, and fluorescence properties similar to the native protein. The absorption band at 695 nm, indicative of ligation of methionine 80 to the heme iron,[93,94] is equivalent in strength to that of the native protein for complex ferri(1–25)H · (23–104), but is relatively weak for complex ferri(1–38)H · (56–104). Dissociation constants of $<3 \times 10^{-8}$ M are measured for both complexes at pH 7.0 at 25°. Furthermore, both complexes could be reduced by sodium ascorbate, indicating an oxidation–reduction potential in the vicinity of that of the native protein. Relative activities obtained with the yeast lactate dehydrogenase assay systems are approximately 60 and 46%, respectively. Thus, a new allowed cleavage site between residues 23 and 25 is defined, and the entire region from residue 39 to 55 could be eliminated while retaining substantial structure and function. As noted earlier, the evolutionarily deleted residues of *Pseudomonas* cytochrome c_{551}[83] correspond exactly to the latter region.

The initial ratio of two types of complexes, A and B, corresponding to derived complexes (1–38)H · (56–104) and (1–25)H · (23–104), respectively, simultaneously formed from heme fragment (1–38)H and apocytochrome c (1–104) is unknown. However, equilibrium between complexes A and B is dependent on temperature and appears to be attained without going through dissociation of the two fragments in a short time.[95] The ratio of type B to type A complexes in the equilibrium state is greater (approximately 2 : 1) at 6 than at 25° (approximately 1 : 1).[25,95,100]

A productive complex can also be formed from fragments (1–38)H and (39–104),[25] as would be expected from the above result. Such a complex has been shown to be biologically active in a depleted mitochondrial system[96] and with cytochrome oxidase and succinate-cytochrome c reductase systems.[101] It has also been shown that residues 56–58 may also be deleted without destroying the ability to form a productive complex.[102]

We consider that the complexes (1–38)H · (39–104) and (1–38)H · (56–104) are essentially the same species while complex (1–53)H · (54–104) [or (56–104)] is a distinct species. Digestion of complex (1–38)H · (39–104) in the ferrous form with trypsin yields the derived complex consisting of

[100] G. R. Parr and H. Taniuchi, *J. Biol. Chem.* **256,** 125 (1981).
[101] L. W. Westerhuis, G. I. Tesser, and R. J. F. Nivard, *Rec. Trav. Chim. Pays Bas* **98,** 109 (1978).
[102] L. W. Westerhuis, G. I. Tesser, and R. J. F. Nivard, *Int. J. Pept. Protein Res.* **19,** 290 (1982).

heme fragment (1–38)H and apofragment (56–104).[25] This suggests that the region of (39–55) is flexible in complex (1–38)H · (39–104). On the other hand, no significant amount of (1–38)H is detected when ferrous complex (1–53)H · (54–104) or (56–104) is subjected to extensive trypsin digestion.[61] Apparently, in this complex, the region of (39–55) is not flexible. Harris has reported that heme fragment, N^{ε}-acetimidyl-(1–37)H and apofragment N^{ε}-acetimidyl-(38–104) forms a complex which restores the succinate oxidation in cytochrome c-depleted mitochondria close to the level of native cytochrome c and higher than complex of heme fragment, N^{ε}-acetimidyl-(1–38)H and apofragment N^{ε}-acetimidyl-(39–104).[103] Thus, it is possible that in this complex residues 39 to 55 are not flexible. At present, the factors involved in stabilization or destabilization of the segment of residues 39 to 55 of these complexes are not clear. It should be noted that these residues are reported to be involved in a concerted change of the atomic coordinates upon oxidation and reduction of heme iron.[27]

Three-Fragment Complex (1–25)H · (28–38) · (39–104). Since there are two permissible regions for cleavage in the amino acid sequence of cytochrome c, it is possible to construct a three-fragment complex. It turns out that the three-fragment complex of cytochrome c^{26} is more stable (presumably due to the interatomic interactions involving a heme moiety), and, therefore, more suitable, as a model, for investigation of the thermodynamic properties and the molecular motion than the three-fragment complex of nuclease A[14] or RNase A.[104,105]

Heme fragment (1–25)H and apofragment (39–104) form a complex with a dissociation constant of $7.5 \times 10^{-7} M$ at pH 7.0 at 25° which is more than an order of magnitude greater than that for the productive complexes described above. This complex exhibits a level of UV-CD approaching the productive complexes.[25] However, this complex is considered nonproductive in that the Soret absorbance maximum is shifted to lower wavelengths than those of the productive complexes by 3 nm, no 695 nm absorbance is observed, the complex is not reducible by ascorbate, and there is no biological activity.[25] These results, in conjunction with kinetic experiments described later in this chapter, suggest that while the gross tertiary structure of this nonproductive complex may approximate that of the native protein, the detailed structure of the heme environment necessary for biological activity is not established. As described later, binding of a third fragment containing residues 28 to 38 with this nonproductive

[103] D. F. Harris, *in* ''Peptide-Structure and Biological Function'' (E. Gross and J. Meienhofer, eds.), p. 613. Pierce Chemical Co., Rockford, Illinois, 1979.
[104] M. C. Lin, B. Gutte, S. Moore, and B. Merrifield, *J. Biol. Chem.* **245**, 5169 (1970).
[105] H. Taniuchi, A. Davis, and G. R. Parr, unpublished results.

complex restores the properties of the productive complexes.[26] Thus, the native state of ligation of methionine 80 to the heme iron atom appears to be coupled with the positioning of residues 28 to 38 in the opposite side of the heme disc.

Preparation.[26] Apofragment (1–65) is obtained after cyanogen bromide cleavage of apocytochrome *c* and purified by gel filtration on a Sephadex G-50 column (1 × 200 cm), using 10% formic acid at 6–8°. Heme fragment (1–25)H and apofragment (39–104) are prepared as described earlier. An equimolar mixture of heme fragment (1–25)H and apofragments (1–65) and (39–104) yields a three-fragment complex which, upon gel filtration on a column of Sephadex G-75 using 0.1 *M* ammonium acetate, pH 7.0, at 6–8°, elutes as a single peak in 90% yield. A UV-CD spectrum indicates a helical content of the complex close to cytochrome *c*. The ferric form of the complex exhibits an absorption maximum at 695 nm. Binding of constituting fragments is measured in titration experiments, monitoring either tryptophan fluorescence quenching or change in absorbance of the Soret band. Apparent dissociation constants, 7.6×10^{-7} and 10^{-5} *M* at pH 7.0 at 25° corresponding to the reaction ferri(1–25)H + (39–104) $\rightleftharpoons$ ferri(1–25)H · (39–104) and ferri(1–25)H · (39–104) + (1–65) $\rightleftharpoons$ (1–25)H · (1–65) · (39–104), respectively, have been determined. Since binding of (1–65) with (1–25)H or (39–104) is weak, complexes (1–25)H · (1–65) and (39–104) · (1–65) are assumed not to be present in experimentally measurable quantities.

To remove the redundant portions, this three-fragment complex (3 μmol) is digested with trypsin in the presence of ascorbate under anaerobic conditions for 70 min at 25°. The digestion is quenched by addition of STI. The derived complex thus formed is purified by gel filtration on a column of Sephadex G-75 (1.8 × 200 cm) using 0.05 *M* ammonium acetate, pH 7.0, at 6–8°, followed by ion-exchange chromatography on a column of SP-Sephadex C-25 (1.0 × 90 cm), using a gradient elution with 0.03 and 0.3 *M* ammonium acetate, pH 7.0 (300 ml each). Separation of the constituting fragments is achieved by gel filtration on a column of Sephadex G-50 (1 × 200 cm) usig 10% formic acid at 6–8°. Three fragments, (1–25)H, (56–104), and (28–38), thus separated, are each identified by amino acid analysis.

To reconstitute the nonoverlapping three-fragment complex, heme fragment (1–25)H and apofragment (56–104) are obtained as described above. Apofragment (28–38) is obtained upon trypsin digestion of heme fragment (1–38)H followed by ion-exchange chromatography on an SP-Sephadex C-25 column (1 × 100 cm) at 25°, using a gradient elution with 400 ml each of 0.01 *M*, pH 4, and 0.1 *M*, pH 7.0, ammonium acetate. The fractions containing fragment (28–38) are located by Sakaguchi and Pauli reactions. Fragment (28–38) thus separated is purified by gel filtration on Sephadex G-25 (fine) using 10% formic acid.

Heme fragment (1–25)H and two apofragments, (28–38) and (56–104), are mixed in equimolar ratio. The resulting three-fragment complex is isolated as a single complex after ion-exchange chromatography followed by gel filtration as described earlier. Again the reconstituted complex exhibits an absorption maximum at 695 nm in the oxidized form and is reducible by ascorbate and lactate dehydrogenase. It also shows a helical content, as indicated by a negative ellipticity at 222 nm, close to that of native cytochrome c.

Properties.[26] Affinity of fragment (28–38) for heme fragment (1–25)H or apofragment (56–104) is low, as measured on the basis of change in absorbance at 398 nm and by gel filtration, respectively. In contrast, the apparent equilibrium constant of complex ferri(1–25)H · (56–104) is 7.5 × 10^{-7} M at pH 7.0 at 25° as described earlier. An overall equilibrium constant of complex ferri(1–25)H · (28–38) · (56–104) is found to be $<10^{-12}$ M^2 at pH 7.0 at 25° on the basis of change in absorbance at 398 nm.

Ferrous heme fragment (1–25)H (reduced with sodium dithionite) and apofragment (56–104) form a tightly bound complex with an apparent dissociation constant $<10^{-8}$ M. $K_{dissociation}$ for the reaction ferro(1–25)H · (28–38) · (56–104) $\rightleftharpoons$ ferro(1–25)H · (56–104) + (28–38) is found to be 1.3 × 10^{-7} M at pH 7.0 at 25° by equilibrium dialysis.

Folding Kinetics and Intermediates of Fragment Complexes

The intermediate states through the entire folding process of the fragment complexes cannot be the same as those of intact proteins, since folding of the fragment complexes involves the interaction between two or more polypeptide chains while folding of intact proteins is the phenomenon involving uncleaved polypeptide chains. Nonetheless, it is remarkable that the ordered structures of the fragment complexes closely resemble the native proteins.[3,4,7,8] This fact implies that (1) the pathway for folding is intrinsically flexible or manifold and (2) although the events in the earlier phase of folding may be different, the events in the later phase would be similar between the fragment complexes and native proteins.

It may also be considered that the early intermediates would be the states in which there are portions of the polypeptide chain which would not interact with the rest of the polypeptide chain. In this sense, a flexible, large fragment, lacking a portion of COOH- or NH_2-terminal amino acid sequence but having a potential to fold with a complementing fragment, may mimic such an intermediate in which the COOH- or NH_2-terminal portion does not interact with the rest of the polypeptide chain. One of the component fragments of the three-fragment complexes which lacks both NH_2- and COOH-terminal sequence of the native proteins may also simulate an intermediate state in a similar sense. Thus, using the fragment complementing systems it is possible to investigate the significant events in the early as well as the late phase of the folding process as described below.

Folding of the Fragment Complexes of Staphylococcal Nuclease A[29]

Folding of each of the complexes, nuclease-(6–48) · nuclease-(50–149), nuclease-(6–48) · nuclease-(49–149), nuclease-(1–126) · nuclease-(99–149), or nuclease-(1–126) · nuclease-(50–149), upon mixing the two fragments of concentrations from 1×10^{-5} to 7×10^{-5} M at pH 8, can be followed by measuring increases in fluorescence intensity due to formation of the hydrophobic environment for a single tryptophan at position 140, using a stopped flow system (0.1 sec to 1 min). A prominent feature is that the kinetic profile exhibits a first-order rate process (half-life, 11 to 27 sec), suggesting that the step of combination of the two fragments is fast and that this step is followed by the first-order process involving burying the tryptophan residue in the interior of the three-dimensional structure.[19,20] This first-order process is not accelerated in the presence of ligands, pdTp and Ca^{2+}, which stabilize the native structure. The rate of this process is also independent of temperature as measured with complex nuclease-(6–48) · nuclease-(50–149) [or nuclease-(6–48) · nuclease-(49–149)].[29] Apparently, the energy barrier of this first-order rate process is entropic. Folding of intact nuclease is also reported to involve a temperature-independent kinetic phase.[106] This entropic energy barrier is assumed to be related to ordering of the polypeptide chain in the activated state.[12]

Folding of Two Fragment Complexes of Cytochrome c[28,100,107–110]

Complementation of ferric heme fragments and apofragments of horse heart cytochrome c as investigated using stopped flow techniques occurs in the two consecutive processes: a second-order reaction followed by first-order kinetic processes. The former process represents formation of largely folded, but rather motile intermediate complexes and the latter represents ligation of methionine 80 to the heme iron atom to establish the

[106] H. F. Epstein, A. N. Schechter, R. F. Chen, and C. B. Anfinsen, J. Mol. Biol. 60, 499 (1971).
[107] G. R. Parr and H. Taniuchi, J. Biol. Chem. 254, 4836 (1979).
[108] G. R. Parr and H. Taniuchi, J. Biol. Chem. 255, 2616 (1980).
[109] G. R. Parr and H. Taniuchi, J. Biol. Chem. 255, 8914 (1980).
[110] G. R. Parr and H. Taniuchi, J. Biol. Chem. 257, 10103 (1982).

native-like heme environment with an increase in redox potential, a decrease in motility, and a decrease in enthalpy.

The Second-Order Rate Process.[107,108] When changes in fluorescence are monitored at pH 7.0 at 25°, a single phase exhibiting second-order kinetics is observed for all cases of productive complexes. The rate constants are independent of the presence of overlapping amino acid sequences on the apofragments, but are strongly dependent on the amino acid sequences of the heme fragments. Thus for fragment ferri(1–38)H, a rate constant of $2.1 \times 10^5 \, M^{-1} \, sec^{-1}$ is observed with apofragments (1–104), (23–104), and (56–104).

The First-Order Rate Processes.[109,110] By using high concentrations of the fragments as well as a large excess of one fragment, the second-order step is forced to completion in very short times and thus effectively decoupled from the ensuing first-order kinetic events. First-order kinetic processes are thus observed by following changes in Soret absorbance. Furthermore, when the complementation reaction is monitored by following absorbance changes at 695 nm (which measures ligation of methionine 80), the entire change is found to occur in first-order rate processes. Since absorbance changes at 695 nm are relatively weak, high concentrations of fragments are necessary to observe the reactions.

Hybrid Complexes.[28] A noncovalently bound, ordered complex consisting of heme fragment ferri(1–25)H and apofragment (14–103) derived from tuna cytochrome c can be prepared similarly to that of horse cytochrome c. The equilibrium properties of this complex are indistinguishable from those of the corresponding complex, ferri(1–25)H · (23–104), of horse cytochrome c. The tuna species possesses a second tryptophan residue at position 33 whose fluorescence is also essentially fully quenched in the native form and in the complex. The fluorescence of these two tryptophans is fully quenched during the second-order kinetic phase of the reaction mechanism, being consistent with the compactly folded intermediate complex described earlier.

Although the horse and tuna proteins exhibit sequence differences at nineteen positions (17% of the total sequence), qualitatively, the folding mechanism of the tuna complex is very similar to that of the horse complex. Quantitatively, there are several differences. The second-order rate constant measured for the tuna complex is 2-fold lower than that of the horse complex. In hybrid complexes, i.e., tuna ferri(1–25)H plus horse (23–104) and horse ferri(1–25)H plus tuna (14–103), the rate is dependent on the species of the heme fragment. Moreover, for the tuna complex the profile of the first-order processes is different in terms of both rate constants and relative amplitudes. Again by using hybrid complexes, it is possible to show that at least one of these reactions is dependent only on the species of the apofragment. These results serve to highlight the subtle effects exerted on the folding process by a relatively small number of changes in the amino acid sequence. However, systematic studies of fragment combinations with only a single amino acid substitution would be required to pinpoint the role of individual residues.

The Effect of Reduction of Heme Iron.[100,110] The studies of equilibrium and kinetic properties of ferrous complexes are complicated by the fact that the sodium dithionite used to reduce the heme fragments irreversibly decomposes the heme group at a slow rate at 25°. Furthermore, the reduced heme fragments appear to aggregate in solution. This aggregation can be investigated using stopped flow relaxation techniques. Data for heme fragment ferro(1–38)H and ferro(1–53)H can be fit to a dimerization model. Data for ferro(1–25)H suggest higher aggregation processes. No change in binding strength of productive complexes can be detected due to limitations of the techniques employed. However, the binding strength of the nonproductive ferrous complex is found to increase by approximately an order of magnitude, approaching that of the productive ferric complexes. The kinetics of formation of the various complexes utilizing ferrous heme fragment is, in the presence of 50 mM imidazole, remarkably similar to the kinetics observed with ferric heme fragments in the presence of imidazole (no aggregation of ferrous heme fragment is observed in the presence of 50 mM imidazole).

Fragment Exchange

Nuclease T, a complex of nuclease-(6–48) · nuclease-(50–149), was the first system to be quantitatively analyzed for the fragment exchange.[13,60] This analysis has revealed the following. (1) The fragment exchange occurs through dissociation of the complex and involves the overall folding and unfolding. (2) The energy barrier of this unfolding involves a major disruption of stabilizing forces of the three-dimensional structure without a large change in conformation. (3) This unfolding occurs under physiological conditions, pH 7.0, even at 6°, well below the transition temperature. (4) The rate constant of this unfolding is highly dependent on temperature, and the activation enthalpy is virtually independent of temperature as measured at 6 to 20°. That is, there is a high degree of compensation of activation enthalpy, favoring folding and activation entropy, favoring unfolding as is the case with unfolding of proteins observed in the range of the transition temperature.[111] (5) The interatomic interactions with the ligand at the ligand binding site are connected to the

[111] A. Stearn and H. Eyring, *J. Chem. Phys.* **5**, 113 (1937).

interatomic interactions at the distant region of the three-dimensional structure of the complex, namely those between the two fragments, nuclease-(6–48) and (50–149) to strengthen each other.

Mathematical Tools

McKay's first order law of isotopic exchange[112] can be used to analyze the data of fragment exchange

$$A^* \cdot B + A \rightleftharpoons A \cdot B + A^*$$

where $A^* \cdot B$ and $A \cdot B$ denote complex of fragments A^* and B and that of fragments A and B, respectively, and A^* is fragment A labeled with isotope. Let the concentration of A^* and A^*B at time t be x and y, respectively, and let the concentration of the sum of A^* and A and that of the sum of A^*B and AB be a and b, respectively. The rate of the exchange reaction, R, is a function of the concentrations a and b but a constant in any given experiment as

$$R = Kf(a)g(b) \tag{1}$$

where K is a constant and f and g can be any functions depending on the mechanism of isotopic exchange.

Then, an increase in A^* must be equal to a decrease in $A^* \cdot B$ as

$$\frac{dx}{dt} = -\frac{dy}{dt} = -\frac{x}{a}R + \frac{y}{b}R \tag{2}$$

Assuming $x = 0$ at $t = 0$ and letting $x = x_\infty$ at $t = \infty$, we have

$$x + y = b$$

and

$$x_\infty = \frac{ab}{a + b} \tag{3}$$

Then, Eq. (2) is solved as

$$-\ln(1 - \gamma) = R\left(\frac{a + b}{ab}\right)t \tag{4}$$

where $\gamma = x/x_\infty$. Equations (1) and (2) ignore isotopic effects. The mechanism of isotopic exchange can be examined by determining R as a function of a and b. For example, in the case of exchange by dissociation

$$A \cdot B \xrightarrow{k_1} A + B \tag{5}$$

$$A + B \xrightarrow{k_2} A \cdot B \tag{6}$$

---

[112] H. A. C. McKay, *Nature (London)* **142**, 997 (1938).

$$A^* + B \xrightarrow{k_3} A^*B \tag{7}$$

$$A^*B \xrightarrow{k_4} A^* + B \tag{8}$$

where k_1, k_2, k_3, and k_4 are the rate constants of the corresponding reactions and $k_1 = k_4$ and $k_2 = k_3$ (no isotopic effect),

$$R = k_1 a^0 b \tag{9}$$

as shown later. That is, R is linear with respect to b and independent of a.

Isotope exchange through dissociation [Eqs. (5) to (8)] is treated by Harris.[113] Let the changing concentrations of A^*B, A^*, $A \cdot B$, and A be $b - x$, x, x, and $a - x$, respectively at time t. In the case of no isotopic effect on the basis of Eqs. (5) to (8) it is easy to derive

$$\frac{dx}{dt} = k_1 b \left[1 - \left(\frac{a + b}{ab} \right) x \right] \tag{10}$$

and $x_\infty = ab/(a + b)$ as in Eq. (3). Then, the solution of Eq. (10) is

$$-\ln(1 - \gamma) = k_1 b \left(\frac{a + b}{ab} \right) t \tag{11}$$

where $\gamma = x/x_\infty$ as in Eq. (4). Comparison of Eqs. (4) and (11) gives Eq. (9). Thus, if a and b are known, a plot of $\ln(1 - \gamma)$ versus time yields a slope from which the value of k_1 can be calculated [Eqs. (4) and (9) or Eq. (11)]. The value for γ at a given time can be obtained by measurements of the extent of exchange, x, at a given time.

Experimental Procedures

To measure the extent of fragment exchange, x, as a function of time, two procedures have to be designed on the basis of the properties of individual fragment complementing systems, one, preparation of the labeled complex and the other, separation of the complex and the free fragment without permitting a significant exchange to occur during separation.

Ideally, labeling of the fragment should not introduce a perturbation to the conformation (i.e., interatomic interactions) of the complex. This can be achieved by substituting one of the amino acids or a group of the fragment with the same amino acid or group containing isotope (e.g., ^{3}H, ^{14}C, or ^{13}C). If this type of substitution is not feasible, attachment of a radioactive group to the α-NH_2 group of the fragment can be used with a minimal degree of perturbation of the conformational properties.

[113] G. M. Harris, *Trans. Faraday Soc.* **47,** 716 (1951).

In the latter type of labeling, the reaction conditions for attachment of labeled groups sometimes causes aggregation of the fragment, resulting in a poor yield. This problem may be overcome by using the complex instead of the free fragment for labeling and then separating the labeled fragment under the conditions used for dissociation of the complex. One of the labeled fragments thus obtained can be combined with the complementing fragment (nonlabeled) to form the labeled complex.

It should be tested whether isotopic effects or the effect of the labeled, extra group are significant with respect to the binding force of a complex. This test can be done by measuring the specific radioactivity or isotopic content of the complex and the free fragment after the equilibrium of the label between the complex and the free fragment having been attained through the fragment exchange.

In fragment exchange experiments, the labeled complex is mixed with the nonlabeled free fragment (or alternatively the nonlabeled complex can be mixed with the labeled free fragment), redistribution of the labeled fragment between the complex and the free fragment (the extent of exchange) is measured as a function of time. For this purpose aliquots are taken from the incubation mixture and subjected to treatment to separate the complex and the free fragment under conditions which suppress or minimize the fragment exchange. Separation of a complex and a free fragment has been achieved by gel filtration, ion-exchange chromatography, or affinity chromatography. High-pressure liquid chromatography is expected to be of use in the future studies. Use of NMR permits the extent of the fragment exchange to be determined without separation of the complex and the fragment.[114]

Treatment of Exchange Data[61]

To treat the exchange data it is convenient to define $E(t)$, the normalized extent of exchange at time t as

$$E(t) = A(t) - A(0)$$

where $A(t)$ = [the isotopic quantity (e.g., radioactivity) of the free fragment]/(the sum of the isotopic quantity of the complex and the free fragment) at time t and $A(0) = A(t)$ at time 0. If $A(0) = 0$,

$$E(t) = \frac{C(0) - C(t)}{C(0)}$$

[114] C.-H. Niu, H. Shindo, S. Matsunra, and J. S. Cohen, *J. Biol. Chem.* **225**, 2036 (1980).

where $C(t)$ = the specific radioactivity (or isotopic content) of the complex at time t and $C(0) = C(t)$ at time 0. Then, by definition $A(0) = 0$

$$x = bE(t) \tag{12}$$

and

$$x_\infty = bE(\infty) \tag{13}$$

From Eqs. (3) and (13), we obtain

$$E(\infty) = \frac{a}{a + b} \tag{14}$$

Also, by definition and from Eqs. (12) and (13)

$$\gamma \equiv \frac{E(t)}{E(\infty)} \equiv \frac{x}{x_\infty} \tag{15}$$

By substitution of Eq. (4) using Eq. (15) we obtain

$$-\ln \left(\frac{E(\infty) - E(t)}{E(\infty)}\right) = R \left(\frac{a + b}{ab}\right) t$$

Thus, plotting $\ln\{[E(\infty) - E(t)]/E(\infty)\}$ versus t gives a line with a slope of $R[(a + b)/ab]$. Since a and b are known quantities and $E(\infty)$ is obtained from Eq. (14) [$E(\infty)$ can also be experimentally determined], R can be calculated from the slope.

In fragment exchange between a fragment complex and a free fragment, it has been shown that R is first- and zero-order with respect to the concentrations of the complex, b, and the fragment, a, respectively,[13] being consistent with Eq. (9). That is,

$$R = k_1 b$$

Thus, from the values for R and b the value for k_1, the rate constant of dissociation (unfolding), can be calculated.

It is recommended that the order of dependency of R with respect to a and b be experimentally determined with each fragment complementing system for unambiguous interpretation. An example is given in the following.

The Fragment Complexes of Nuclease A[13,60]

Preferential attachment of [1-^{14}C]acetyl groups to the NH_2-termini of nuclease T is achieved by incubation of 0.1 ml (1.1 mmol) of [1-^{14}C]acetic anhydride (7.7×10^8 cpm) with 12 μmol of nuclease T [complex nuclease-(6–48)·nuclease-(50–149)] in a total volume of 5.1 ml in the presence of saturated solution of sodium acetate, pH 5.8, in an ice bath for 1 hr on the basis of the method of Fraenkel-Conrat.[115] Then, the two labeled fragments are separated by gel filtration using 50% acetic acid at 6°. One of these fragments, [1-^{14}C]acetyl-(50–149) is combined with unlabeled nuclease-(6–48) to form labeled nuclease T, which is purified by affinity chromatography using Sepharose-pdTp as described earlier. Actually, fragment (50–149) thus labeled contains three [1-^{14}C]acetyl groups per molecule, one at the NH$_2$-terminus, the second at the ε-NH$_2$ group of residue 127, and the third at the ε-NH$_2$ group of another lysine residue.

In the fragment exchange experiment, 1.6×10^{-4} M the labeled nuclease T (1.1×10^4 cpm, 1.4 nmol) is incubated with 1.3×10^{-4} M the free unlabeled fragment nuclease-(50–149) in a total volume of 90 μl in the presence of 0.08 M ammonium acetate, pH 7.8, at a given temperature. After incubation for a given period of time the mixture is quickly cooled in an ice bath and 1 μl of 1 M CaCl$_2$ is added. This mixture is applied to a Sepharose-pdTp column (1 × 7 cm) equilibrated with a buffer, 0.1 M ammonium acetate/0.005 M CaCl$_2$, pH 7.8, at 6°. Up to 100 nmol of the labeled nuclease T can be adsorbed to this column. The loaded column is initially eluted with 24 to 30 ml of the buffer to obtain the free fragment which passes through the column without adsorption. Then, the column is eluted with 0.1 N NH$_4$OH to obtain the adsorbed nuclease T. The radioactivities of nuclease T and the free fragment thus separated are measured to determine the extent of exchange. The labeled nuclease T may contain a small amount (6%) of the radioactivity which is not adsorbed to the affinity column. This radioactivity is subtracted from the radioactivity of the free fragment. Guanidine-HCl (0.5 M), instead of 0.1 N NH$_4$OH, may also be used to elute nuclease T.[60]

When pdTp is present in the exchange reaction mixture to test the effect of ligands, a two-zone column (1 × 9 cm) containing Sepharose-nuclease and Sepharose-pdTp in the upper (4 cm height) and lower (5 cm height) zones, respectively, is used. PdTp contained in the incubation mixture is removed by adsorption to the Sepharose-nuclease zone. This removal of pdTp permits nuclease T in the mixture to be adsorbed to the Sepharose-pdTp zone. Sepharose-nuclease is prepared by cyanogen bromide activation[80,81] and has a capacity to bind 57 nmol of pdTp per 1 ml bed volume in the presence of 0.005 M CaCl$_2$ at pH 8 at 6°.

This system of the fragment exchange has revealed the following in addition to the findings (1, 2, 3, 4, and 5) described earlier. (6) Nuclease T bound with one or both of ligands pdTp and Ca^{2+} can unfold without going through dissociation of ligands (the activated state still binds ligands with

---

[115] H. Fraenkel-Conrat, this series, Vol. 4, p. 247.

weakened binding forces). These findings (1 through 6) taken together with the concept of the permissible regions for cleavage (e.g., see Fig. 2) have led to a hypothesis that the energy state of one region of the three-dimensional structure would be coupled to the states of other regions throughout the structure after folding of almost the entire amino acid sequence of nuclease A providing extra force to bind the polypeptide chain in the ordered state[13,60] (coupling of interatomic interactions throughout the structure or global cooperative interactions).

Two-Fragment Complex of Horse Cytochrome c

The complex of heme fragment (1–53)H and apofragment (54–104) was the second fragment exchange system studied. The following is the experimental procedure according to Hantgan and Taniuchi.[61]

To only the α-NH$_2$ group (residue 54) of complex (1–53)H · (54–104) (native horse cytochrome c contains an acetyl group at the NH$_2$-terminus),[116] a [^{14}C]carbamoyl group (NH$_2$-^{14}CO-) is attached by incubation of 3.6 mM the complex (1.8 μmol) with 0.22 M labeled potassium cyanate (K^{14}CNO, 6.7 Ci/mol) in the presence of 0.2 M potassium acetate, pH 5.8, at 25° for 1 hr using the method described by Stark.[117] The reaction is quenched by addition of glycylglycine (0.5 M). Then, after desalting by gel filtration, the [^{14}C]monocarbamoylated complex is purified by ion-exchange chromatography on a SP-Sephadex C-25 column (1 × 50 cm) using a gradient elution with 0.02 M, pH 7.0, and 0.2 M, pH 6.85, potassium phosphate (200 ml each) at 6–8°.

To overcome insolubility problems, a stock solution of unlabeled free apofragment (54–104) (33.7 nmol) is prepared by first dissolving the fragment in 0.05 ml of 0.05 M acetic acid and then adjusting the pH of the solution to 7.0 with addition of 1.0 ml of 0.06 M potassium phosphate, pH 7.0, and 50 μl of 0.05 M KOH. The radioactive complex (30.2 nmol, 3.6 × 10^{12} cpm/mol) is dissolved in 0.2 ml of 0.06 M potassium phosphate, pH 7.0. The fragment exchange mixture is made to contain 2.5 × 10^{-5} M of the radioactive complex (ferric form) and 2.5 to 5.0 × 10^{-5} M of free, nonlabeled fragment (54–104) in a total volume of 1.2 ml. After incubation for a desired time at a given temperature, the mixture is cooled in an ice bath and applied to a SP-Sephadex C-25 column (0.7 × 10 cm) equilibrated with 0.06 M potassium phosphate, pH 7.0, at 6–8°. Elution of the free fragment is effected with the same buffer. Then, the complex is eluted with 0.2 M potassium phosphate, pH 7.0. A small amount of the radioactivity (6%) elutes with the first buffer when the radioactive com-

---

[116] E. Margoliash, E. L. Smith, G. Kreil, and H. Tuppy, *Nature (London)* **192**, 1125 (1961).
[117] G. R. Stark, this series, Vol. 11, p. 590.

plex alone is applied to the column. This radioactivity is corrected for in the treatment of data.

For fragment exchange with the complex in ferrous form, the radioactive complex is reduced with 50-fold molar excess ascorbate under argon and then incubated with unlabeled apofragment (54–104) under constant flushing with argon. After incubation, the mixture is cooled on ice and the complex is reoxidized with addition of a 3-fold molar excess of potassium ferricyanide over ascorbate before application to the ion-exchange column for separation. Absorbance of the fragment exchange mixture is measured at 550 nm at the beginning and the end of incubation in order to confirm that the complex remains reduced.

Unfolding of the ferrous complex (1–53)H · (54–104) is strongly suppressed as compared with the ferric complex.[61] This is also consistent with the fact that stability of native cytochrome c increases with reduction of heme iron without a large change in conformation (cf. Ref. 27).

Three-Fragment Complex of Horse Cytochrome c^{55}

In order to attach a radioactive label to apofragment (56–104) or (39–104) of three-fragment complex (1–25)H · (28–38) · (39–104) or (1–25)H · (28–38) · (56–104) the methyl groups of methionine are substituted with tritiated methyl groups according to Jones et al.[118] [14C]-labeled apofragment (28–38) containing [14C]phenylalanine at position 36 can also be chemically synthesized using Merrifield's solid phase synthesis and purified by complex formation and gel filtration.

Motility of Fragment [14C](28–38) in Complex Ferro-(1–25)H · [14C](28–38) · (56–104). To 20 μl of 0.05 M potassium phosphate, pH 7.0, containing 23.8 nmol of labeled complex (1–25)H · [14C](28–38) · (56–104) (970 dpm/nmol) and 5 mg of sodium ascorbate which has been equilibrated at 3° for 20 min, is added 7.5 μl of 3.2 mM (24 nmol) of unlabeled fragment (28–38). This fragment exchange mixture is incubated for a desired time at 3°. Then the complex and free (28–38) is separated by gel filtration on a column (0.5 × 18 cm) of Sephadex G-50 (superfine), using 50 mM potassium phosphate, pH 7.0, and 3° for measurement of the radioactivities. The measured dissociation–association reaction of fragment (28–38) is fast. Exchange of fragment [14C](28–38) reaches an equilibrium state within the time (19 min) limited by this method. Thus the rate constant for dissociation of (28–38) has to be greater than 10^{-3} sec^{-1}.

Unfolding of the Complex Ferro-(1–25)H · (28–38) · [3H](56–104) Trapping Method. Since direct measurement of exchange by the proce-

---

[118] C. W. Jones, T. M. Rothgeb, and F. R. N. Gurd, *J. Biol. Chem.* **251**, 7452 (1976).

dures described earlier is hampered by aggregation of free fragment (56–104), a trapping method has been developed.

Its main principle is to in the first step trap free [^{3}H](56–104) generated during unfolding of the three-fragment complex, as a stable complex. This procedure is based on a difference in rate of unfolding of the three-fragment complex and the stable two-fragment complexes. For example, complex ferro(1–25)H · (28–38) · [^{3}H](39–104) dissociates much faster than ferro(1–38)H · [^{3}H](39–104).[95] Furthermore, association is much faster than dissociation[107] (these second-order rate constants with ferri two-fragment complexes are applicable to systems where ascorbate is the reducing agent, as reduction with ascorbate occurs only after association of the fragments[100,110]). Since "trapping fragment" is in large excess, the reversal, namely reassociation of the three-fragment complex, is considered to be negligible. Thus, dissociated fragment [^{3}H](56–104) is trapped with heme fragment ferri(1–38)H to form complex (1–38)H · [^{3}H](56–104) which is then reduced wih ascorbate present in the system. The second step is quenching the trapping reaction, at a given time, with addition of excess of apocytochrome c (1–104) which binds with heme fragments (1–38)H as well as (1–25)H. In the third step the temperature of the system is raised to 37°. Under these conditions, the remaining ferro three-fragment complex fully dissociates and heme fragment (1–25)H thus released is bound with apocytochrome c (1–104). As a result, labeled fragment [^{3}H](56–104) of the remaining three-fragment complex is converted to the free state.

The best method to separate complexes and free fragment is gel filtration. In order to scale down the size of column and decrease separation time, another difference in behavior of ferrous complexes versus free fragments can be utilized. That is, the former complexes are completely resistant to trypsin digestion at 37° for 1 hr, while the latter fragments are fully digested within 10 min.

In a typical run, 31 nmol of labeled complex (1–25)H · (28–38) · [^{3}H](56–104) (1150 dpm/nmol), dissolved in 1 ml of 0.1 M potassium phosphate (pH 7.0) and placed in a 15 ml conical tube, is mixed with 250 μl of 1 M sodium ascorbate/0.1 M potassium phosphate, pH 7.0. After equilibration for 15 min under N$_2$ at a given temperature, 600 μl of 1 mM of heme fragment (1–38)H, preequilibrated at the same temperature, is added. This mixture of the labeled ferrous three-fragment complex and trapping heme fragment is incubated for a desired time at the same temperature under N$_2$ (step 1). Then an aliquot (250 μl) is withdrawn and mixed with 45 μl of 5 mM apoprotein (1–104)/0.1 M ascorbate (step 2). The mixture is incubated at 37° for 20 min (step 3). Then, trypsin (50 μg) is added. After further incubation for 20 min at 37°, the digestion is stopped by addition of STI (200 μg). The digested fragments and the complex is separated by gel filtration on a column (0.5 × 50 cm) of Sephadex G-50 (superfine), using 0.03 M potassium phosphate (pH 7.0) at 6°, to determine their radioactivities.

Direct Measurement. Fragment exchange of the three-fragment complex can also be directly measured using complex ferro(1–25)H · (28–38) · [³H](39–104).[55] This is possible because fragment (39–104) exhibits a higher solubility than fragment (56–104) and the complex containing either (39–104) or (56–104) is expected to show similar properties. That is, the portion of residues (39–55) of ferro(1–25)H · (28–38) · (39–104) is presumably flexible and protruding[25,26] as described earlier.

Solution A is prepared by mixing 15 μl of 1 M potassium phosphate, pH 7.0, and 155 μl of 0.1 M sodium dithionite, pH 7.0, with 450 μl of a solution containing 15.2 nmol of labeled complex (1–25)H · (28–38) · [³H](39–104) (23,800 dpm/nmol) and 0–440 nmol of fragment (28–38). Solution B is 256 μM of unlabeled (39–104) (aqueous). Then a 200 μl of solution B is mixed with the entire solution A at 15°. This fragment exchange mixture, consisting of 18.5 μM of radioactive ferrous three-fragment complex and 62.4 μM of free fragment (39–104) is incubated for a desired time. Then an aliquot (130 μl) of the mixture is withdrawn, mixed with 50 μl of 0.02 M sodium dithionite (pH 7.0), and incubated with trypsin (5 μl of 0.3% solution) for 15 min at 15°. After addition of STI, the complex and digested fragments are separated by gel filtration to determine their radioactivities. Control experiments without addition of the unlabeled fragment are used to correct for a partial digestion of the complex.

Dynamic Properties of the Three-Fragment Complex. As revealed by the fragment exchange experiments, three-fragment complex, ferro(1–25)H · (28–38) · (39–104) exhibits a smaller and a larger cooperative motion, i.e., a faster dissociation–association reaction of fragment (28–38) and a slower overall unfolding–folding through virtually simultaneous dissociation of all three of the fragments at pH 7.0. These two concerted motions are modulated by temperature. For example, in the presence of excess of free fragment (28–38) and below 30°, unfolding occurs mainly without going through the two-fragment complex (1–25)H · (39–104). This mode of unfolding is not a reversal of the major folding pathway through an intermediate, two-fragment complex (1–25)H · (39–104). Since the binding interaction of fragment (28–38) with complex (1–25)H · (39–104) appears to be linked with the interatomic interactions between heme fragment (1–25)H and apofragment (39–104) in the three-fragment complex, it may be thought that this coupling of interatomic interactions would come to existence after completion of folding. Once this coupling would be established, the unfolding of the complex would no longer be restricted to the pathway via reversal of the past folding pathway. Thus, the molecule could unfold through reversal of any folding pathway depending on the experimental conditions.

Arrhenius plots for this virtually simultaneous dissociation of all three of the fragments are linear from 10 to 30°. On the basis of the reasoning described elsewhere,[55,61] this linearity indicates that the conformation of the activated state would be similar to the ground state. Thus, using the three-fragment complex system, it is possible to detect the concerted global motion such as nonsequential unfolding following activation without a large change in conformation.

As described earlier, nuclease T bound with ligands can unfold without going through dissociation of ligands.[60] These two phenomena of unfolding through apparent nonreversal of the folding pathway, one with the three-fragment complex of cytochrome c and the other with liganded nuclease T, may be considered as two different aspects of disruption of the coupling of interatomic interactions throughout the structure.

Remarks

As being tested with the three-fragment complex ferro(1–25)H · (28–38) · (39–104) of cytochrome c, substitution of evolutionary invariant leucine-32 with isoleucine has a profound effect in destabilizing the structure while substitution of variant leucine-35 with isoleucine has only a small effect.[59] Both leucine-32 and -35 are buried in the interior of the structure.[27,82] Since the fragment exchange and the hydrogen exchange phenomenon measures different aspects of the concerted motion of the molecule and both of them are sensitive to the change of the stability of the structure,[60,61,95,119] a systematic study of the fragment exchange and the hydrogen exchange as a function of such substitution of an amino acid using the same fragment complementing systems would help greater understanding of the concerted molecular motions. This understanding would, in turn, give an insight into the mechanism of the hypothetical global cooperative interactions (the hypothetical ordered structure mediated interaction between distant residues[120]), the elucidation of which is a central problem of protein folding.

Acknowledgment

We thank Mrs. Dorothy Stewart for her help in preparation of the manuscript.

[119] N. Roy, C. DiBello, and H. Taniuchi, *Int. J. Pept. Protein Res.* **27,** 165 (1986).
[120] E. Poerio, G. R. Parr, and H. Taniuchi, *J. Biol. Chem.,* in press (1986).

[12] Refolding and Association of Oligomeric Proteins[1]

By Rainer Jaenicke and Rainer Rudolph

Definitions

Linderstrøm-Lang and Bernal were the first to recognize a hierarchy of structural levels of organization within proteins. Present knowledge allows us to add two more levels to their original four so that six levels can be distinguished. The *primary structure* refers to the specific amino acid sequence along the covalent polypeptide chain; the *secondary* and *supersecondary structure* defines regular arrangements of the polypeptide backbone and its energetically and topologically preferred aggregates in terms of clusters of helices and β-structures; the *tertiary structure* refers to the three-dimensional chain fold, with structural or functional *"domains"* as spatially separated entities (sometimes with specific function, such as ligand binding); the *quaternary structure* describes the stoichiometry and geometry of assemblies built up from subunits characterized by the foregoing levels of structural organization.

The cotranslational or posttranslational acquisition of the functional state of the nascent polypeptide chain is based solely on the amino acid sequence and the solvent environment at the site of translation or posttranslational processing. We call the transition from the one-dimensional to the three-dimensional structure "folding."

In going from single-chain proteins to dimers, tetramers, or (more general) oligomers and multimers, subunit association has to be considered in addition to folding. Both processes must be properly coordinated because the formation of the native quaternary structure requires the surfaces of the structured monomers to be preformed in the correct way such that specific recognition is achieved.

Aims of Reconstitution Studies on Oligomeric Proteins

Folding Code

Considering the vast amount of crystallographic data obtained for well over 100 distinct structures of globular proteins, the stage seems to be set for determining the pathway of folding that should eventually reveal the mechanism of folding and help in elucidating the "code" by which

[1] Dedicated to Professor Helmut Zahn on the occasion of his seventieth birthday.

METHODS IN ENZYMOLOGY, VOL. 131

the amino acid sequence of a protein specifies its three-dimensional structure.[1a-3]

This folding code will not only aid in engineered changes in protein structure using recombinant DNA techniques, but also provide a tool to specify hypothetical functions for given amino acid sequences deduced from DNA sequence analysis. At present, the code of protein folding is totally unknown, and there is no hint how it might be deciphered, even for small single-chain proteins. In the case of typical oligomeric proteins, domains and subunit interactions have to be considered; evidently they complicate the problem due to their specific distribution of hydrophobic residues in clusters in the interior and in patches in the intersubunit interfaces.

The experimental methods we are going to deal with in the following sections make use of the reversibility of protein denaturation focusing on refolding and reassociation rather than folding and association. The reason is that so far no general approach is available to study the acquisition of the native structure of the nascent polypeptide chain directly, either *in vivo* or *in vitro*.[4,5]

Refolding refers to the regain of the native structure of a given protein after denaturation. Therefore, the *in vitro* reconstitution may differ from the vectorial *in vivo* folding process, because refolding starts from the denatured protein while folding of the nascent chain is expected to occur as a cotranslational event.[5] The question of whether refolding indeed reflects the *in vivo* process cannot be answered unambiguously. However, certain evidence strongly suggests that the vectorial nature of translation does not represent a necessary requirement for the formation of the native structure, because (1) peptide synthesis in the reverse direction (C → N, according to Merrifield), as well as coupling of fragments yield at least closely similar, if not identical products compared with the native protein; (2) under suitable experimental conditions the kinetics of reconstitution are in the same time range as the folding of nascent polypeptide chains *in vivo;* and (3) the final product of reconstitution after denaturation–renaturation is found to be indistinguishable from the initial native state. Both, the acquisition of the native structure during translation and

[1a] R. L. Baldwin and T. E. Creighton, *in* "Protein Folding" (R. Jaenicke, ed.), p. 217. Elsevier, Amsterdam, 1980.

[2] P. S. Kim and R. L. Baldwin, *Annu. Rev. Biochem.* **51,** 459 (1982).

[3] C. Chothia, *Annu. Rev. Biochem.* **53,** 537 (1984).

[4] C. Ghélis and J. Yon, "Protein Folding," p. 562. Academic Press, New York, 1982; T. E. Creighton, "Proteins-Structures and Molecular Properties," p. 515. Freeman, San Francisco, 1984.

[5] D. B. Wetlaufer and S. Ristow, *Annu. Rev. Biochem.* **42,** 135 (1973).

the *in vitro* reconstitution, occur spontaneously without the need for "morphopoietic factors" or any additional information beyond that contained in a given primary structure on one hand, and the aqueous or nonpolar environment (for cytoplasmic or membrane proteins, respectively), on the other.[6,7]

Specificity of Subunit Recognition

The spatial arrangement and the subunit interactions defining the native quaternary structure are highly specific.[8–10]

With regard to the complementary geometry of subunits, tight packing is observed due to the energetic tendency to exclude water from hydrophobic intermolecular surfaces.[11] As a consequence, subunit association may be considered as entropy driven.

In order to achieve the unique complementarity of the subunit interfaces in oligomeric proteins, the pathway of quaternary structure formation must include well-populated folded intermediates, capable of "subunit recognition."[12] This is clearly indicated by the fact that neither *in vivo* nor *in vitro* major structural heterogeneity of subunit assembly occurs. Similarly, no multifunctional "chimeric" enzymes are observed when mixtures of enzymes are applied in reconstitution experiments.[13,14]

As in monomeric systems, where structural intermediates have been used to analyze the folding pathway,[2] intermediates of association may allow the identification of single events in the multistep folding and association process leading to oligomeric enzymes. Obviously, the observed folded intermediates are well defined with respect to their quaternary structure, i.e., there must be a unique pathway of association following the pathway of folding.[15] Both may be described by a sequence of unimolecular and bimolecular (folding and association) reactions which may be analyzed by spectroscopic and electrophoretic techniques (see below).

[6] D. B. Wetlaufer, ed., "The Protein Folding Problem," p. 203. AAAS Selected Symp. 89, Westview Press, Bolder, Colorado, 1984.

[7] R. Jaenicke, *Biophys. Struct. Mech.* **8**, 231 (1982).

[8] R. Jaenicke, *Angew. Chem.* **96**, 385 (1984); *Angew Chem. Int. Ed. (Engl.)* **23**, 395 (1984).

[9] C. B. Anfinsen and H. A. Scheraga, *Adv. Protein Chem.* **29**, 205 (1975)

[10] K. A. Thomas and A. N. Schechter, *Biol. Regul. Dev.* **2**, 44 (1980).

[11] F. M. Richards. *Annu. Rev. Biophys. Bioeng.* **6**, 151 (1977).

[12] R. Jaenicke, *in* "Mobility and Recognition in Cell Biology" (H. Sund and C. Veeger, eds.), p. 67. De Gruyter, Berlin, 1983.

[13] R. A. Cook and D. E. Koshland, Jr., *Proc. Natl. Acad. Sci. U.S.A.* **64**, 247 (1969).

[14] R. Jaenicke, R. Rudolph, and I. Heider, *Biochem. Int.* **2**, 23 (1981); M. Gerl, R. Rudolph, and R. Jaenicke, *Biol. Chem. Hoppe-Seyler* **365**, 447 (1985).

[15] R. Jaenicke and R. Rudolph, Mosbach Colloq. **34**, 62 (1983).

The kinetic analysis provides us with an elegant approach to discriminate the merging of structural domains from subunit association. In principle there is no difference between both processes[16]; however, they obey different reaction orders so that discrimination is rendered possible.

There are cases where the distinction is blurred by the fact that assembly structures are synthesized and folded as single polypeptide chains which are subsequently cleaved by a protease. Thus, a multidomain monomer may generate independent globular entities which may then assemble, e.g., forming the symmetrical coat of a virus.[17-19]

As a result of reconstitution studies, structural relationships may be established in order to characterize the specificity of subunit recognition.

Intermediates of Association

Considering the overall folding/association reaction of oligomeric proteins, early steps on the reconstitution pathway are presumably similar to those involved in the refolding of monomeric systems. As a consequence, intermediates on the association pathway are expected to represent true folding intermediates comparable to those investigated in detail, e.g., for ribonuclease.[2]

In order to minimize or avoid formation of "wrong aggregates,"[20] independent refolding of the subunits of oligomeric systems must occur before assembly takes place. The "structured monomers" as intermediate products of this folding process are expected to be different from the structure in the oligomeric state, simply due to the changed environment in the assembly compared to the separate polypeptide chain in free solution.

Apart from the question regarding the evolutionary advantage of quaternary structure formation (which is closely related to the question of whether or not isolated subunits show full biological function), reconstitution studies are concerned with the identification of the conformational refinements induced by subunit association. Polymorphic systems, like tobacco mosaic virus protein, as well as allosterically regulated oligomers, clearly prove that there may be considerable rearrangements within a given quaternary structure. In this connection, the inherent flexibility of

[16] G. E. Schulz and R. H. Schirmer, "Principles of Protein Structure," p. 314. Springer, New York, 1979.

[17] M. F. Jacobsen and D. Baltimore, *Proc. Natl. Acad. Sci. U.S.A.* **61,** 77 (1968).

[18] J. C. S. Clegg and S. I. T. Kennedy, *J. Mol. Biol.* **97,** 401 (1975).

[19] J. Bergsma, W. G. J. Hol, J. N. Jansonius, K. H. Kalk, J. H. Ploegman, and J. D. G. Smit, *J. Mol. Biol.* **98,** 637 (1975).

[20] G. Zettlmeissl, R. Rudolph, and R. Jaenicke, *Biochemistry* **18,** 5567 (1979).

the folded protein takes care of the mobility required to switch from one structural state to another.[21]

Since subunit association is a second-order process, intermediates of association are accessible to structural investigation upon reconstitution at low protein concentration. As indicated by circular dichroism, fluorescence spectroscopy, and enzymatic activity, folding intermediates on the assembly pathway differ from the native state with respect to all spectral and functional properties. Changed solvent conditions, e.g., addition of "structure making ions," may shift the properties of structured intermediates toward the native state. As indicated, intrinsic differences are clearly predictable from the increase in accessible surface area in the dissociated state. The corresponding energetic differences are reflected by changes in structure and stability, parallelled by altered ligand binding capacity, etc.

In summary, reconstitution studies (applying equilibrium and kinetic approaches) give access to intermediates of association which are at the same time intermediates of folding, undergoing "reshuffling" upon regain of their native quaternary structure. Attempts to characterize the two types of intermediates may provide us with criteria describing the structural requirements for the catalytic and regulatory capacity of a given oligomeric enzyme.

Oligomeric Proteins as Models for Complex Assemblies

Within the hierarchy of levels of protein structure, oligomeric proteins may be considered as model systems for multimeric assemblies of higher complexity. The specificity achieved by complementary surface shapes, hydrophobic "patches," hydrogen bond donors and acceptors, and ion pairs is expected to hold for both oligomeric and multimeric systems. However, in most cases a detailed structural analysis of the latter is not amenable to high resolution X-ray crystallography. Oligomeric systems may then provide a key to some clearer understanding of the energetics and kinetics of association, as well as the structural principles at atomic resolution. The respective thermodynamic quantities follow from direct calorimetric measurements or from van't Hoff enthalpies (calculated from equilibrium constants). The kinetics of the various association steps (monomer $\rightarrow$ dimer $\rightarrow$ tetramer $\rightarrow$ ... $\rightarrow$ multimer) may

---

[21] R. Huber and W. S. Bennett, Jr., in "Mobility and Recognition in Cell Biology" (H. Sund and C. Veeger, eds.), p. 21. De Gruyter, Berlin, 1983; W. S. Bennett and R. Huber, CRC Crit. Rev. Biochem. 15, 291 (1984).

involve a spectrum of rate constants requiring conventional mixing, as well as stopped-flow techniques for their determination. In these experiments, solvent effects and effects of temperature may be most significant for a number of reasons. (1) Denaturation conditions strongly influence the extent of unfolding; they may equally well affect the yield and rate of reconstitution. In certain cases, effects of residual denaturant on the kinetics of reconstitution have to be carefully considered.[22] (2) Since the activation energy of reconstitution may be exceedingly high, temperature control is crucial in obtaining reproducible results. (3) Denaturation may involve irreversible chemical modification, e.g., by changes in the redox state of thiol or disulfide groups, or irreversible loss of essential ligands. (4) Reconstitution experiments may require long incubation periods, especially at low protein concentrations; correspondingly, the experimental conditions have to take care of optimum long-term stability of the protein in its native and intermediate states. (5) Apart from stabilizing agents, optimum solvent conditions may imply specific components like cofactors or substrates; they may increase the rate and yield of reconstitution, either by "nucleation" or by shifting equilibria in the overall reconstitution reaction toward the native end product.[15,23–25]

The highest level of structural organization of proteins occurs in the interaction of multiple, nonidentical, polypeptide chains, such as in ribosomes or certain viruses. Because of their size and complexity, little direct evidence is available as to the details of such structures, as well as the pathway of their formation *in vivo* or *in vitro*. As shown by the successful reassembly and reactivation of complex heterogeneous multisubunit systems, like pyruvate dehydrogenase from *Bacillus stearothermophilus* (M_r 10^7)[26] or tobacco mosaic virus (M_4 4×10^7),[27] it is the complexity of the assembly structure rather than particle size that compete with reconstitution. There is no reason to doubt that the interpolypeptide contacts in these particles utilize the same types of interactions as those observed in the interior or in subunit contacts of less complex

[22] G. Zettlmeissl, R. Rudolph, and R. Jaenicke, *Eur. J. Biochem.* **100,** 593 (1979).

[23] J. Gerschitz, R. Rudolph, and R. Jaenicke, *Eur. J. Biochem.* **87,** 591 (1978).

[24] R. Rudolph, J. Gerschitz, and R. Jaenicke, *Eur. J. Biochem.* **87,** 601 (1978).

[25] R. Jaenicke, H. Krebs, R. Rudolph, and C. Woenckhaus, *Proc. Natl. Acad. Sci. U.S.A.* **77,** 1966 (1980).

[26] R. Jaenicke and R. N. Perham, *Biochemistry* **21,** 3378 (1982); M. Koike and K. Koike, *Adv. Biophys.* **9,** 187 (1976).

[27] P. J. G. Butler and A. C. H. Durham, *Adv. Protein Chem.* **31,** 188 (1977); P. J. G. Butler, *J. Gen. Virol.* **65,** 253 (1984); K. C. Holmes, *in* "Structural Molecular Biology" (D. B. Davies, W. Saenger, and S. S. Danyluk, eds.), p. 475. Plenum, New York, 1982.

oligomeric structures. However, there is clear evidence that complex multimers show ordered assembly mechanisms; in certain cases, morphopoietic factors have been found to be indispensable for successful reconstitution.[28–30]

Technological Aspects

The yield and rate of reconstitution depend on the mode of denaturation, as well as the solvent conditions and enzyme concentration during renaturation. The most prominent side reaction competing with reconstitution is the formation of "wrong aggregates" generated by weak intermolecular interactions of groups which in the native molecule tend to be involved in intramolecular or interdomain contacts. The effect becomes increasingly important with increasing size of the protein subunits, because "folding-by-parts"[4,5,31] then dominates, giving rise to wrong interactions.

The determination of optimum reconstitution conditions is of importance for several reasons. (1) *In vivo* folding and association produce the native protein without significant side reactions. In many cases, the reconstitution leads to a product indistinguishable from the native starting material. Therefore, reconstitution after previous denaturation (and separation of byproducts) may be used in the purification of proteins. (2) In cases where proteins (enzymes) are prone to proteolytic degradation during purification, separation from proteases may be achieved under strongly denaturing conditions (e.g., in the presence of 6 M guanidine-HCl or urea), using reconstitution as the final step of purification.[32] It is obvious that in order to apply this method, the proteases and the subunits of the enzymes under consideration must differ sufficiently in size or charge to achieve separation. (3) In connection with overproducing strains of microorganisms, reconstitution may be the method of choice to get access to proteins deposed as aggregates ("refractile bodies," "inclusion bodies") within the derepressed cells. Dissolution of aggregates by strong denaturants and subsequent reconstitution by dilution (or dialysis) has been successfully applied to gain access to "irreversibly denatured" material.

[28] K. N. Nierhaus, *Curr. Top. Microbiol. Immunol.* **97**, 81 (1982).
[29] E. Kellenberger, *CIBA Found. Symp.* p. 192 (1966).
[30] M. Nomura and W. A. Held, *in* "Ribosome" (M. Nomura, A. Tissières, and P. Lengyel, eds.), p. 193. Cold Spring Harbor Laboratory, Cold Spring Harbor, New York, 1974.
[31] D. B. Wetlaufer, *Adv. Protein Chem.* **34**, 61 (1981); H. Neurath, *in* "Molecular Evolution of Life" (H. Jörnvall, ed.). Cambridge Univ. Press, Cambridge, 1986.
[32] P. Bartholmes, H. Böker, and R. Jaenicke, *Eur. J. Biochem.* **102**, 167 (1979).

Denaturation/Dissociation and Renaturation/Reassociation of
  Oligomeric Proteins

Equilibrium Studies

Transition of Dissociation/Reassociation. Intermediates of folding of
monomeric proteins, populated at equilibrium, may be identified by analyzing the unfolding transition.[2,33] Upon dissociation and denaturation of
oligomeric proteins, intermediates of folding and association may be
present in a defined concentration range of the dissociating agent. Intermediates of association can be easily identified by investigating the molecular mass as a function of the denaturant concentration. The structural or
functional properties of the intermediates may be obtained by the comparison of the transitions of dissociation, deactivation, and denaturation. If
dissociation, as an example, precedes deactivation, the intermediates of
association must be catalytically active. However, this approach can only
be applied if the intrasubunit interactions are affected by the dissociating
agent to a far lesser extent than the intersubunit interactions.

In principle, protein secondary and tertiary, as well as quaternary
structures are stabilized by virtually the same side-chain interactions.
Therefore dissociation of oligomeric proteins by denaturants will usually
cause at least partial unfolding of the subunits (Fig. 1).[34,35] As a consequence, it is difficult to find out whether dissociation per se is accompanied by the loss of structural or functional properties.[36] Structurally intact
subunits may be obtained by addition or removal of specific ligands that
affect only the quaternary structure. The ATP-dependent dissociation of
glyceraldehyde-3-phosphate dehydrogenase may serve as an example for
such behavior.[37,38]

Dissociation by "Chemical Denaturants." Dissociation of oligomers
may be achieved by the same denaturants used for reversible unfolding of
monomeric proteins.[39] Since dissociation of large oligomers is frequently
superimposed by side reactions such as aggregation (see below), the different parameters of dissociation and denaturation should be determined
under strictly identical conditions. Batchwise incubation at increasing

[33] R. L. Baldwin, *Annu. Rev. Biochem.* **44,** 453 (1975).
[34] M. Engelhard, R. Rudolph, and R. Jaenicke, *Eur. J. Biochem.* **67,** 447 (1976).
[35] R. Rudolph, A. Haselbeck, F. Knorr, and R. Jaenicke, *Hoppe-Seyler's Z. Physiol. Chem.* **359,** 867 (1978).
[36] R. Jaenicke and R. Rudolph, *Int. Symp. Pyridine Nucleotide Dependent Dehydro-genases, 2nd* p. 351 (1977).
[37] P. Bartholmes and R. Jaenicke, *Biochem. Biophys. Res. Commun.* **64,** 485 (1975).
[38] P. Bartholmes and R. Jaenicke, *Eur. J. Biochem.* **87,** 563 (1978).
[39] C. Tanford, *Adv. Protein Chem.* **23,** 121 (1968).

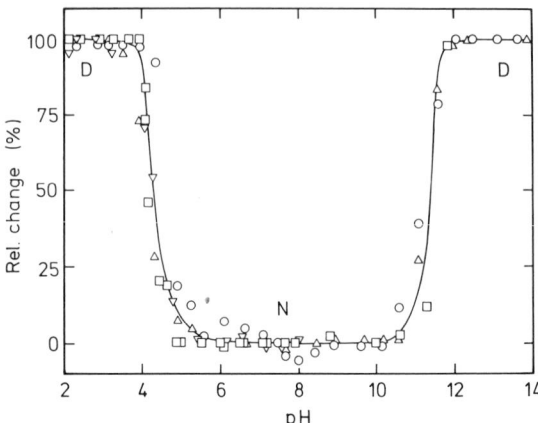

FIG. 1. pH-dependent dissociation, deactivation, and denaturation of rabbit muscle aldolase.[34,35] Incubation for up to 24 hr at the given pH in 0.1 M phosphate, containing 1 mM EDTA and 1 mM dithioerythritol, at a protein concentration of 0.02–0.5 mg/ml, 20°. (□) Sedimentation coefficient; (○) enzymatic activity; (△) maximum of fluorescence emission, λ_{max}, upon excitation at 275 nm; (▽) circular dichroism at 281 nm. 0 and 100% refer to the native (N) and the dissociated (D) states, respectively.

denaturant concentrations is to be preferred over the titration method, which should only be applied to oligomers characterized by completely reversible dissociation transitions. The various properties (e.g., M_r, UV absorbance, fluorescence emission, circular dichroism, activity) must be determined at one and the same protein concentration, since dissociation equilibria depend per definition on the concentration of the dissociating particles. In addition the given parameters should be determined after various periods of time, since equilibration may be extremely slow.

Most extensive unfolding (and dissociation) may be achieved by incubation with concentrated guanidine-HCl where proteins retain little or no elements of residual secondary structure.[40] Oligomeric proteins frequently undergo dissociation at relatively low guanidine concentrations ($\cong 1$ M); the transition curves are sometimes characterized by discrete steps of denaturation and dissociation.[41–44]

[40] C. Tanford, K. Kawahara, and S. Lapanje, *J. Am. Chem. Soc.* **89,** 729 (1967); C. M. Dobson, P. A. Evans, and K. L. Williamson, *FEBS Lett.* **168,** 331 (1984); C. M. Dobson, P. A. Evans, and R. O. Fox, in "Structure and Motion: Nucleic Acids and Proteins" (E. Clementi, G. Corongiu, M. H. Sarma, and R. H. Sarma, eds.), p. 265. Adenine Press, Albany, New York, 1985.

[41] E. Appella and C. L. Markert, *Biochem. Biophys. Res. Commun.* **6,** 171 (1961).

[42] G. R. Parr and G. G. Hammes, *Biochemistry* **14,** 1600 (1975); **15,** 857 (1976).

[43] G. Zettlmeissl, W. Teschner, R. Rudolph, R. Jaenicke, and G. Gäde, *Eur. J. Biochem.* **143,** 401 (1984).

[44] R. Jaenicke, W. Vogel, and R. Rudolph, *Eur. J. Biochem.* **114,** 525 (1981).

If kinetic experiments are aimed at the folding of subunits, apart from their assembly, the monomers should be maximally unfolded, e.g., by ≥ 4 M guanidine-HCl. This way perturbation of the folding mechanism by elements of residual backbone structure will be reduced to a minimum. Some examples for the application of guanidine-HCl in reversible dissociation experiments are given in the following references.[23,24,45-54]

Urea may be used as an alternative to guanidine-HCl for subunit dissociation and unfolding.[55-58] However, it has the general disadvantage that isocyanate, which is in equilibrium with aqueous urea, can cause covalent modifications of amino or thiol groups.[59]

Oligomeric proteins may also be dissociated by pH variation. In most cases, pH-dependent dissociation experiments have been performed at acid pH.[34,60-74] The reason is that upon prolonged incubation at alkaline

[45] F. J. Castellino and R. Barker, *Biochemistry* **7**, 2207 (1968).
[46] J. W. Teipel and D. E. Koshland, Jr., *Biochemistry* **10**, 792 (1971); **10**, 798 (1971).
[47] S. Lapanje, "Physicochemical Aspects of Protein Denaturation," p. 331. Wiley, New York, 1978.
[48] R. Rudolph, E. Westhof, and R. Jaenicke, *FEBS Lett.* **73**, 204 (1977).
[49] R. Rudolph, I. Heider, and R. Jaenicke, *Eur. J. Biochem.* **81**, 563 (1977).
[50] S. Zabori, R. Rudolph, and R. Jaenicke, *Z. Naturforsch.* **35c**, 999 (1980).
[51] G. Zettlmeissl, R. Rudolph, and R. Jaenicke, *Biochemistry* **21**, 3946 (1982).
[52] G. Zettlmeissl, R. Rudolph, and R. Jaenicke, *Eur. J. Biochem.* **125**, 605 (1982).
[53] R. Hermann, R. Rudolph, R. Jaenicke, N. C. Price, and A. Scobbie, *J. Biol. Chem.* **258**, 11014 (1983); R. Hermann, R. Jaenicke, and N. C. Price, *Biochemistry* **24**, 1817 (1985).
[54] N. C. Price and E. Stevens, *Biochem. J.* **209**, 763 (1983).
[55] O. P. Chilson, G. B. Kitto, and N. O. Kaplan, *Proc. Natl. Accad. Sci. U.S.A.* **53**, 1006 (1965).
[56] O. P. Chilson, G. B. Kitto, J. Pudles, and N. O. Kaplan, *J. Biol. Chem.* **241**, 2431 (1966).
[57] A. Heil and W. Zillig, *FEBS Lett.* **11**, 165 (1970).
[58] D. L. Burns and H. K. Schachman, *J. Biol. Chem.* **257**, 8648 (1982).
[59] G. R. Stark, W. H. Stein, and S. Moore, *J. Biol. Chem.* **235**, 3177 (1960).
[60] M. L. Anson, *Adv. Protein Chem.* **2**, 361 (1945).
[61] S. Anderson and G. Weber, *Arch. Biochem. Biophys.* **116**, 207 (1966).
[62] R. Jaenicke, *Eur. J. Biochem.* **46**, 149 (1974).
[63] A. Levitzki and H. Tenenbaum, *Isr. J. Chem.* **12**, 327 (1974).
[64] R. B. Vallee and R. C. Williams, Jr., *Biochemistry* **14**, 2574 (1975).
[65] H. Tenenbaum-Bayer and A. Levitzki, *Biochim. Biophys. Acta* **445**, 261 (1976).
[66] K. C. Ingham, B. D. Weintraub, and H. Edelhoch, *Biochemistry* **15**, 1720 (1976).
[67] R. Rudolph, M. Engelhard, and R. Jaenicke, *Eur. J. Biochem.* **67**, 455 (1976).
[68] R. Rudolph and R. Jaenicke, *Eur. J. Biochem.* **63**, 409 (1976).
[69] R. Rudolph, I. Heider, E. Westhof, and R. Jaenicke, *Biochemistry* **16**, 3384 (1977).
[70] R. Jaenicke, R. Rudolph, and I. Heider, *Biochemistry* **18**, 1217 (1979).
[71] G. Bernhardt, R. Rudolph, and R. Jaenicke, *Z. Naturforsch.* **36c**, 772 (1981).
[72] G. Zettlmeissl, R. Rudolph, and R. Jaenicke, *Eur. J. Biochem.* **121**, 169 (1981).
[73] R. Hermann, R. Jaenicke, and R. Rudolph, *Biochemistry* **20**, 5195 (1981).
[74] R. Hermann, R. Rudolph, and R. Jaenicke, *Hoppe-Seyler's Z. Physiol. Chem.* **363**, 1259 (1982).

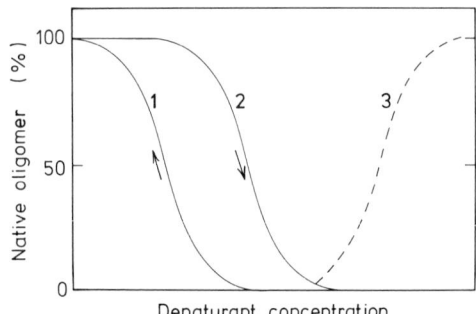

FIG. 2. Dissociation/reassociation of an oligomeric protein superimposed by a side-reaction (aggregation) in the transition range. (1) Reassociation after complete dissociation; (2) dissociation of native oligomer after incubation for a certain time t; (3) reassociation of the oligomer in the absence of denaturant, after dissociation at the given denaturant concentration. The "trough" defined by curves (1) and (3) characterizes the range where the side reaction (aggregation) occurs.

pH, the reversibility of dissociation may be decreased by chemical modification of amino acid side-chains or proteolysis of alkali-labile peptide bonds.[35,75–77] Glucose dehydrogenase is an example for reversible dissociation in weak alkaline solution (0.05 M glycine-Tris, pH 9.0, plus 0.05 M NaCl)[78]; similarly the reversible transition from double disks to A-protein in the case of tobacco mosaic virus protein should be mentioned here.[79]

Upon alkali or acid dissociation, as well as after dissociation by other denaturants, the dissociation transition of oligomeric proteins is frequently superimposed by side reactions such as aggregation.[5,7,23,35,46,49,69,80–84] If slow subunit dissociation causes the aggregation of native oligomers in the transition region to be slow, differences be-

[75] J. P. Greenstein and M. Winitz, "Chemistry of Amino Acids," p. 2872. Wiley, New York, 1961.

[76] R. E. Rosenfield, Jr., R. Parthasathy, and J. D. Dunitz, *J. Am. Chem. Soc.* **99,** 4860 (1977).

[77] H. E. Sine and L. F. Hass, *J. Biol. Chem.* **244,** 430 (1969).

[78] H. E. Pauly and G. Pfleiderer, *Biochemistry* **16,** 4599 (1977).

[79] A. C. H. Durham, J. T. Finch, and A. Klug, *Nature (London) New Biol.* **229,** 37 (1971).

[80] P. Elödi and G. Jécsai, *Acta Physiol. Hung.* **17,** 175 (1960).

[81] J. London, C. Skrzynia, and M. E. Goldberg, *Eur. J. Biochem.* **47,** 409 (1974).

[82] Y. Yano and M. Irie, *J. Biochem.* **78,** 1001 (1975).

[83] J. Gerschitz, R. Rudolph, and R. Jaenicke, *Biophys. Struct. Mech.* **3,** 291 (1977).

[84] R. Jaenicke, *Naturwissenschaften* **65,** 569 (1978); R. Jaenicke, *FEBS Symp.* **52,** 187 (1979).

tween the transition profiles of dissociation and reassociation may occur (Fig. 2).[84,85] The apparent differences in the transition curves caused by aggregation should not be confused with hysteresis effects caused by metastable states, e.g., in the salt-induced transition between active dimers and inactive monomers of halophilic malate dehydrogenase.[86]

Dissociation by charge repulsion of ionizable groups may leave certain elements of backbone structure intact. After short-term acid dissociation of, e.g., porcine muscle lactate dehydrogenase the monomers are still in a "correct" structure. In this case, no rate-limiting folding steps are involved in the reassociation reaction, so that high yields of reconstitution are obtained.[15,68] On the other hand, long-term acid dissociation has been found to result in slow structural rearrangements from "correct" to "incorrect" monomer structures.[15] These rearrangements are reflected by a decrease of the far-UV circular dichroism, and by alterations of the proteolytic susceptibility of certain interdomain peptide bonds.[72,87] The given structural changes are responsible for a decrease of both the rate and yield of reconstitution.

Independent of the time of incubation at low pH, the "correct" structure of the monomers can be preserved by adding stabilizing salt (e.g., 1 M Na_2SO_4) to the dissociation medium.[71,73] At neutral pH, dissociation into correctly structured monomers may be achieved by the addition of moderately denaturing salts, such as NaSCN or $NaClO_4$ at 1.0–1.8 M concentration,[88] or high concentrations of KI or LiCl.[26,56,89]

Detergents such as SDS have rarely been applied to reversibly dissociate soluble globular proteins, mainly because of the poor reversibility, and the difficulties connected with the removal of the denaturant. In some cases, reassociation of SDS-dissociated oligomers has been reported, in particular after chain separation by electrophoresis.[90–94] The solubilization of membrane proteins by detergents and their subsequent recon-

[85] R. Rudolph, I. Fuchs, and R. Jaenicke, *Biochemistry* **25,** 1662 (1986).

[86] M. Mevarech and E. Neumann, *Biochemistry* **16,** 3786 (1977).

[87] G. Zettlmeissl, R. Rudolph, and R. Jaenicke, *Arch. Biochem. Biophys.* **224,** 161 (1983).

[88] D. L. Burns and H. K. Schachman, *J. Biol. Chem.* **257,** 8638 (1982).

[89] A. S. Levi and N. O. Kaplan, *J. Biol. Chem.* **246,** 6409 (1971).

[90] L. F. Lykins, C. W. Akey, E. G. Christian, G. E. Duval, and R. W. Topham, *Biochemistry* **16,** 693 (1977).

[91] A. Rigo, F. Marmocchi, D. Cocco, P. Viglino, and G. Rotilio, *Biochemistry* **17,** 534 (1978).

[92] D. A. Hager and R. R. Burgess, *Anal. Biochem.* **109,** 76 (1980).

[93] S. A. Lacks and S. S. Springhorn, *J. Biol. Chem.* **255,** 7467 (1980).

[94] R. E. Manrow and R. P. Dottin, *Proc. Natl. Acad. Sci. U.S.A.* **77,** 730 (1980).

stitution can only be mentioned here.[95-98] It requires a more detailed treatment.

Dissociation by Changes in Physical Parameters. Contrary to dissociation experiments involving chemical denaturants, clear-cut thermodynamic and kinetic data describing the equilibrium of dissociation/reassociation may be obtained from experiments using either temperature or hydrostatic pressure as variables affecting subunit dissociation.

Incubation of oligomeric proteins at elevated temperatures is in most cases superimposed by "irreversible aggregation"[99,99a] Much more frequently, low temperature (including freezing) causes reversible dissociation of oligomeric proteins which is often mediated by the stabilizing or destabilizing effect of specific ions or ligands.[101] The subtle interplay between temperature and ligand effects is illustrated by the reversible cold dissociation of tetrameric glyceraldehyde-3-phosphate dehydrogenase from yeast in the presence of ATP.[37,38,102]

The requirement for specific "effectors" prevents low temperature from being applicable as a general means to achieve subunit dissociation. The "endothermic polymerization" of tubulin, tobacco mosaic virus protein, bacterial surface layer proteins, and other multimeric assemblies may serve to illustrate the importance of temperature in both the preparation and thermodynamic analysis of the "phase diagrams" of the respective proteins.[27,103-106]

High hydrostatic pressure has been occasionally used in studies on the

[95] Y. Hatefi and W. G. Hanstein, this series, Vol. 31, p. 770.

[96] E. Racker, this series, Vol. 55, p. 699.

[97] Y. Kagawa, C. Ide, T. Hamamoto, M. Rögner, and N. Sone, *Cell Surf. Rev.* **8**, 137 (1982).

[98] E. Racker, *Fed. Proc., Fed. Am. Soc. Exp. Biol.* **42**, 2899 (1983).

[99] R. Jaenicke, *J. Polymer Sci. Part C* **16**, 2143 (1967).

[99a] "Irreversible aggregation" is a misleading term for the temperature-dependent precipitation, since aggregates of otherwise unmodified polypeptide chains may be redissolved (e.g., by 6 *M* guanidine-HCl) and reactivated.[100]

[100] R. Rudolph, G. Zettlmeissl, and R. Jaenicke, *Biochemistry* **18**, 5572 (1979).

[101] R. Jaenicke, *Annu. Rev. Biophys. Bioeng.* **10**, 1 (1981).

[102] G. M. Stancel and W. C. Deal, Jr., *Biochemistry* **8**, 4006 (1969).

[103] M. A. Lauffer, "Entropy-Driven Processes in Biology," p. 264. Springer-Verlag, Berlin and New York, 1975.

[104] R. C. Weisenberg, *Science* **117**, 1104 (1972); M. L. Shelanski, F. Gaskin, and C. R. Cantor, *Proc. Natl. Acad. Sci. U.S.A.* **70**, 765 (1973).

[105] R. Jaenicke, R. Welsch, M. Sára, and U. B. Sleytr, *Biol. Chem. Hoppe-Seyler's* **366**, 663 (1985); K. Hecht, F. Wieland, and R. Jaenicke, *Biol. Chem. Hoppe-Seyler's* **367**, 33 (1986).

[106] R. B. Scheele and G. G. Borisy, *in* "Microtubules" (K. Roberts and J. S. Hyams, eds.), p. 175. Academic Press, London, 1979.

reversible dissociation of oligomeric proteins.[101,107,108] Since pressure is a common parameter in the biosphere, these studies may be of direct biological significance.

High hydrostatic pressure affects proteins in a complex way. In different pressure ranges, the effects tend to be antagonistic: at low pressure stabilization may occur, while extreme pressure causes irreversible denaturation.[101] In principle, the effects can be described in terms of structural changes involving exposure of groups, changed affinity for the solvent, or changes in electrostriction accompanying the ionization of amino acid side chains and buffer components.[108,109] In the case of oligomeric or multimeric proteins, high hydrostatic pressure may cause subunit dissociation.[108,110-120] Details of the equipment used for high-pressure studies (autoclaves, quench cells, transmission and fluorescence cuvettes) are given by Hawley,[121] Heremans,[122] Jaenicke et al.,[108,114,116,119] and Weber and Drickamer.[123]

The reversible high-pressure dissociation of tetrameric porcine heart and skeletal muscle lactate dehydrogenase has been most thoroughly studied.[114-116] The dissociation transition of the two isoenzymes has been determined by quenching experiments and direct fluorescence measurements. Incubation of the samples in an autoclave for defined periods of time causes dissociation and deactivation, as determined after fast pressure release. Deactivation strictly parallels the tetramer–monomer transition, which is found to be completely reversible over the whole transition

[107] R. Jaenicke and R. Rudolph, in "Protein Folding" (R. Jaenicke, ed.), p. 525. Elsevier, Amsterdam, 1980.
[108] R. Jaenicke, Naturwissenschaften 70, 332 (1983).
[109] F. H. Johnson, M. Eyring, and B. J. Stover, "The Theory of Rate Processes in Biology and Medicine," p. 273. Wiley, New York, 1974.
[110] R. Jaenicke and R. Koberstein, FEBS Lett. 17, 351 (1971).
[111] J. T. Penniston, Arch. Biochem. Biophys. 142, 322 (1971).
[112] E. Schulz, H.-D. Lüdemann, and R. Jaenicke, FEBS Lett. 64, 40 (1976).
[113] E. Schulz, R. Jaenicke, and W. Knoche, Biophys. Chem. 11, 253 (1976).
[114] B. C. Schade, R. Rudolph, H.-D. Lüdemann, and R. Jaenicke, Biochemistry 19, 1121 (1980); Biophys. Chem. 11, 257 (1980).
[115] K. Müller, H.-D. Lüdemann, and R. Jaenicke, Biochemistry 20, 5411 (1981).
[116] K. Müller, H.-D. Lüdemann, and R. Jaenicke, Biophys. Chem. 14, 101 (1981); 16, 1 (1982).
[117] R. Jaenicke, H.-D. Lüdemann, and B. C. Schade, Biophys. Struct. Mech. 7, 195 (1981).
[118] J. Kornblatt, J. Kornblatt, and G. Hui Bon Hoa, Eur. J. Biochem. 128, 577 (1982).
[119] K. Müller, T. Seifert, and R. Jaenicke, Eur. Biophys. J. 11, 87 (1984).
[120] T. Seifert, P. Bartholmes, and R. Jaenicke, Biophys. Chem. 15, 1 (1982); FEBS Lett. 173, 381 (1984); Biochemistry 24, 339 (1985).
[121] S. A. Hawley, this series, Vol. 49, p. 15.
[122] K. A. H. Heremans, Annu. Rev. Biophys. Bioeng. 11, 1 (1981).
[123] G. Weber and H. G. Drickamer, Q. Rev. Biophys. 16, 1 (1983).

range. Assuming a two-state equilibrium for the reversible dissociation allows one to compute the thermodynamic quantities of the pressure-induced dissociation.[116] Comparing homologous subunit structures, small oligomers (dimers) in general exhibit higher stability toward pressure-induced dissociation than larger polymers (tetramers or multimeric systems).[124] As a consequence, the dimer–monomer transition (if it occurs) is accompanied by irreversible pressure denaturation. Thermophilic enzymes seem to exhibit increased pressure stability.[119] Further insight into the mechanism of high-pressure dissociation may be obtained from binding studies with anilinonaphthalenesulfonic acid (ANS).[116] The label is known to bind to hydrophobic surface areas.[61] Upon binding, the fluorescence is drastically enhanced so that exposure of hydrophobic groups may be quantitatively determined.

Dissociation by Ligand Effects. Dissociation of oligomeric enzymes without gross perturbation of the subunit structure can be achieved by the addition (subtraction) of specific destabilizing (stabilizing) ligands, such as coenzymes, substrates, allosteric effectors, or other prosthetic groups. The displacement of dissociation equilibria by metabolites may be one of the mechanisms of enzyme regulation *in vivo.*[125]

As an example, adenine nucleotides have been shown to affect the reversible polymerization of glutamate dehydrogenase or phosphofructokinase.[126-130] Another example of a highly specific ligand effect is the interconversion of glycogen phosphorylase which is brought about by serine phosphorylation.[131]

The dissociation transition can also be affected by (specific) binding of buffer ions. The stabilization of the tetrameric state of porcine muscle lactate dehydrogenase by phosphate during acid or high-pressure dissociation may serve as an example.[68,114] Because of the inherent specificity, no general method of protein dissociation by ligand effects can be given.

Dissociation by Dilution. According to Ostwald's dilution law dissociation should occur upon decrease of the protein concentration. Evidently, dissociation by dilution is applicable only if the dissociation constant of the given oligomer allows dissociation in a concentration range, where accurate measurements are still feasible. Examples for concentration-

[124] K. Müller, H.-D. Lüdemann, and R. Jaenicke, *Naturwissenschaften* **68**, 524 (1981).
[125] C. Frieden, *Annu. Rev. Biochem.* **40**, 653 (1971).
[126] C. Frieden, *J. Biol. Chem.* **234**, 809 (1959).
[127] R. P. Aaronson and C. Frieden, *J. Biol. Chem.* **247**, 7502 (1972).
[128] R. J. Cohen and G. B. Benedek, *J. Mol. Biol.* **108**, 151 (1976).
[129] D. Thusius, *J. Mol. Biol.* **115**, 243 (1977).
[130] L. K. Hesterberg and J. C. Lee, *Biochemistry* **21**, 216 (1982).
[131] E. H. Fischer, P. Cohen, M. Fosset, L. W. Muir, and J. C. Saari, *in* "Metabolic Interconversion of Enzymes" (O. Wieland, E. Helmreich, and H. Holzer, eds.), p. 11. Springer-Verlag, Berlin and New York, 1972.

dependent dissociation are the depolymerization of multimeric tobacco mosaic virus protein[132] and phosphofructokinase[133,134] into monomers and dimers, respectively, and of glutamate dehydrogenase into hexamers.[135,136] Reports on the concentration-dependent dissociation of certain dimeric and tetrameric enzymes had to be refuted after careful reevaluation.[70,137,138] For analyses of the molecular mass at extreme dilution see Teller.[139-141]

As will be shown, the most versatile method to produce intermediates is the kinetic approach using reassociation instead of dissociation. Data obtained from such "kinetic intermediates" are more reliable than those determined for "equilibrium intermediates" produced by carefully selected, but ambiguous solvent conditions.

Kinetic Studies

Refolding of monomeric proteins is determined by sequential or parallel first-order folding reactions as rate-limiting steps. In the process of reconstitution of oligomers, second-order association steps may be rate-determining besides folding. Folding steps must, on the one hand, precede association in order to provide the surface areas required for correct intersubunit recognition; on the other hand, they must succeed association, as a consequence of the exclusion of water from the subunit interfaces that serve to stabilize the native quaternary structure. Apart from the given sequence of first- and second-order processes, dissociation–association equilibria between intermediates of association may participate in the mechanism of reconstitution.

The first step in analyzing the kinetic mechanism of subunit association is the determination of the order of the rate-limiting reconstitution reaction(s). For this purpose it is indispensable to analyze reconstitution at various subunit concentrations, and to calculate the apparent reaction order from the slope of a double logarithmic plot of the initial velocity of reconstitution vs subunit concentration.[142]

[132] A. T. Ansevin and M. A. Lauffer, *Nature (London)* **183**, 1601 (1959).
[133] D. P. Bloxham and H. A. Lardy, *in* "The Enzymes" (P. D. Boyer, ed.), Vol. 8, p. 239. Academic Press, New York, 1973.
[134] N. Kono and K. Uyeda, *J. Biol. Chem.* **248**, 8592 (1973).
[135] H. Sund, K. Markau, and R. Koberstein *in* "Subunits in Biological Systems" (S. N. Timasheff and G. D. Fasman, eds.), Part C, p. 225. Dekker, New York, 1975.
[136] H. Eisenberg, R. Josephs, and E. Reisler, *Adv. Protein Chem.* **30**, 101 (1976).
[137] P. Bartholmes, H. Durchschlag, and R. Jaenicke, *Eur. J. Biochem.* **39**, 101 (1973).
[138] N. C. Price and R. Jaenicke, *FEBS Lett.* **143**, 283 (1982).
[139] D. C. Teller, this series, Vol. 27, p. 346.
[140] L. M. Gilbert and G. A. Gilbert, this series, Vol. 27, p. 273.
[141] R. Valdes, Jr. and G. K. Ackers, this series, Vol. 61, p. 125.
[142] K. J. Laidler, "Chemical Kinetics" (2nd Ed.). p. 566. McGraw-Hill, New York, 1965.

The reconstitution kinetics of oligomeric proteins may exhibit the following characteristics.

1. First-order folding reactions are rate determining for the recovery of the structural and functional integrity of the native oligomer. The level of association where the observed folding step occurs can only be determined by comparison with the state of association which has to be investigated by an independent method (see below).

2. Second-order association steps are rate determining. In this case the question of whether certain structural and functional properties of the protein appear before or after the rate-limiting association reaction can be answered without knowing the exact mechanism of association. As an example, both reactivation and regain of native fluorescence of porcine muscle lactate dehydrogenase after short-term acid dissociation obey second-order kinetics.[68] This proves (1) that the monomers must be essentially inactive under the standard test conditions applied, and (2) that at the monomer level the tryptophan residues are not yet in their native environment which is only formed after subunit association.

3. Consecutive first- and/or second-order reactions are rate limiting for the formation of the native oligomer. In this case the structural or functional properties may be recovered by different kinetics, depending on whether or not the respective property is regained on the level of the intermediate state. The comparison of various kinetic parameters allows the determination of the properties of the intermediate.

The reactivation kinetics of porcine muscle lactate dehydrogenase after denaturation in 6 M guanidine-HCl may serve as an example.[51] Reassociation, as determined by cross-linking experiments (see below), is characterized by slow monomer folding, followed by fast dimer formation. The final step of reconstitution consists of slow dimer–tetramer association. The dimers, which therefore accumulate as intermediates of association, are enzymatically inactive under standard test conditions. This is clearly indicated by the fact that reactivation strictly parallels tetramer formation (Fig. 3). Adding 1.5 M $(NH_4)_2SO_4$ to the test mixture stabilizes the backbone structure and induces 50% of the native activity already at the level of the dimeric intermediate of association.[143,144] As a consequence, the kinetics of reactivation change their shape from sigmoidal to nonsigmoidal profiles (Fig. 3).

[143] R. Girg, R. Jaenicke, and R. Rudolph, *Biochem. Int.* **7**, 433 (1983).
[144] R. Girg, R. Rudolph, and R. Jaenicke, *FEBS Lett.* **163**, 132 (1983).

The given example illustrates how kinetic analysis of reconstitution data may be used to determine the structural and functional properties of intermediates of association.

Mechanism of Association

Cross-Linking with Bifunctional Reagents. "Snap-shot" fixation of reassociating oligomers, and subsequent analysis by SDS–polyacrylamide electrophoresis has become a powerful tool for the analysis of assembly processes.[51,53,73,88,145] The following scheme illustrates the rationale of the procedure:

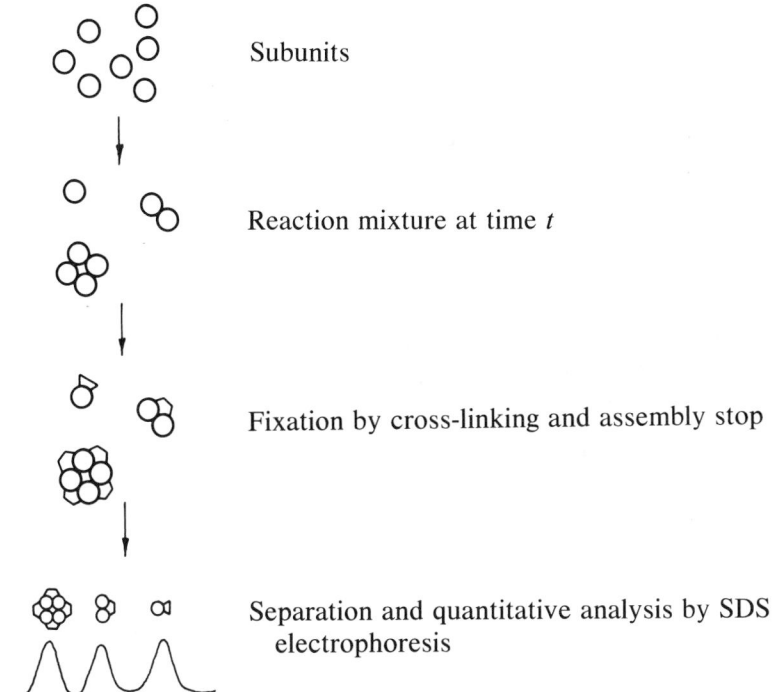

Subunits

Reaction mixture at time *t*

Fixation by cross-linking and assembly stop

Separation and quantitative analysis by SDS electrophoresis

Cross-linking with glutaraldehyde as bifunctional reagent allows the detection and quantitative evaluation of intermediates of association even at very low enzyme concentrations.[146] The experimental conditions have to be specifically optimized for a given oligomer (e.g., with respect to cross-linking time or glutaraldehyde concentration) to satisfy the following requirements for a quantitative analysis: (1) completeness of the intramolecular cross-linking of associated particles, (2) negligible interparticle

[145] R. Hermann, R. Rudolph, and R. Jaenicke, *Nature (London)* **277**, 243 (1979).
[146] R. Hermann, R. Jaenicke, and G. Kretsch, *Naturwissenschaften* **70**, 517 (1983).

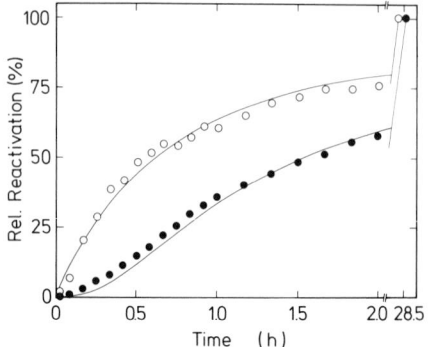

FIG. 3. Reactivation of porcine muscle lactate dehydrogenase after denaturation in 6 M guanidine-HCl (5 min, 0.1 M sodium phosphate, pH 6.2, 1 mM dithioerythritol, 1 mM EDTA, 20°). Reactivation at 20° in 0.1 M sodium phosphate, pH 7.6, containing 1 mM dithioerythritol and 12 mM residual guanidine-HCl. Enzyme concentration, 1.7 μg/ml; yield 40%. (●) Reactivation analyzed by standard assay. Curve calculated for tetramer formation according to a consecutive first-order/second-order mechanism with $k_1 = 8 \times 10^{-4}$ sec^{-1} and $k_2 = 3 \times 10^4$ M^{-1} sec^{-1} for the monomer folding and dimer → tetramer association, respectively.[51] (○) Activity determined in the presence of 1.5 M (NH$_4$)$_2$SO$_4$. Curve calculated using the same constants k_1 and k_2, but assuming the dimers to possess 50% activity.

cross-linking, and (3) sufficiently high rate of intramolecular cross-linking compared to the rate of association.

In a first series of experiments the glutaraldehyde concentration required for the quantitative fixation of the native enzyme has to be determined (Fig. 4).

Quantitative cross-linking has not been possible for certain oligomeric proteins, most probably due to the unfavorable distribution of reactive lysine side chains on the surface or inside the protein.[53,54,88,147] Despite the incomplete fixation, reassociation kinetics could be determined by cross-linking, provided that the data are corrected accordingly, i.e., on the basis of the particle distribution obtained for the native oligomer.[88]

Quantitative fixation of the native oligomer does not necessarily imply that intermediates of association are quantitatively cross-linked as well. On one hand, different orientations of lysine residues may cause different fixation patterns; on the other hand, changes in polarity and dielectric constant of the buffer upon addition of glutaraldehyde could induce dissociation of some labile intermediates before cross-linking can occur. The latter artifact can be excluded by simultaneous cross-linking of multiple samples of the reassociating oligomer with increasing amounts of the bifunctional reagent. The relative amounts of monomers, association in-

[147] S. Pillai and B. K. Bachhawat, *J. Mol. Biol.* **131**, 877 (1979).

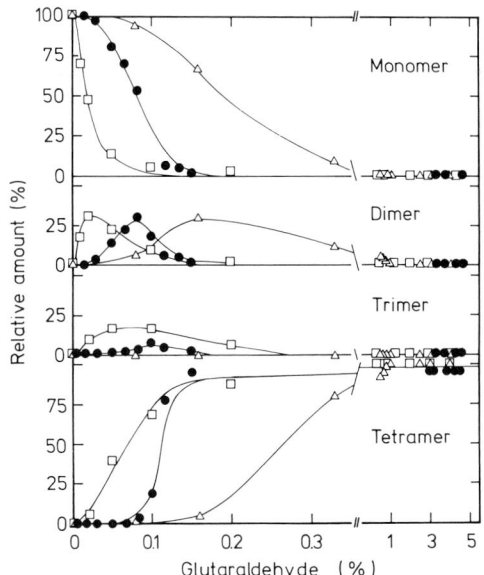

FIG. 4. Influence of glutaraldehyde concentration on the covalent fixation of tetrameric enzymes.[53,71,73] (●) Porcine heart muscle lactate dehydrogenase, $c = 4.7$ μg/ml; (△) porcine skeletal muscle lactate dehydrogenase, $c = 10$ μg/ml; (□) yeast phosphoglycerate mutase, $c = 20$ μg/ml. Cross-linking at 20° by 2-min incubation in 0.1 M sodium phosphate, pH 7.6, containing 1 mM EDTA, in the presence of varying amounts of glutaraldehyde. The relative amounts of monomers, dimers, trimers, and tetramers were determined by densitometry after SDS–polyacrylamide gel electrophoresis. At higher glutaraldehyde concentrations (>0.5%, w/v), octamers (<5%) could in some cases be detected.

termediates, and native oligomers should reach a plateau above a critical concentration of the cross-linking reagent, as illustrated for phosphoglycerate mutase from yeast (Fig. 5).

The kinetics of reassociation of some tetrameric enzymes were analyzed by the following procedure[51,53,73,146]: At various times, aliquots of the reassociating oligomers, containing 20–50 μg of protein, are rapidly mixed with glutaraldehyde (25%, w/v) to give a final concentration of 1%. After 2 min (20°), the cross-linking reaction is quenched by adding a $\cong$10-fold molar excess of NaBH$_4$ (solid or dissolved in 0.1 M NaOH). This way, the cross-linking products are stabilized by reduction of hydrolyzable imines to the corresponding amines; at the same time, unreacted glutaraldehyde is quenched by reduction. After 20 min incubation, excess NaBH$_4$ is destroyed by acidification, e.g., by adding 2 M NaH$_2$PO$_4$ to a final concentration of 7.5% (v/v). Noncovalently linked associates are dissociated by SDS at 1% (w/v) final concentration. To remove buffer salts and reactants, the reaction mixture is dialyzed for 2 days against 0.2% SDS (several changes) at 40–60°. After lyophilization, the material is

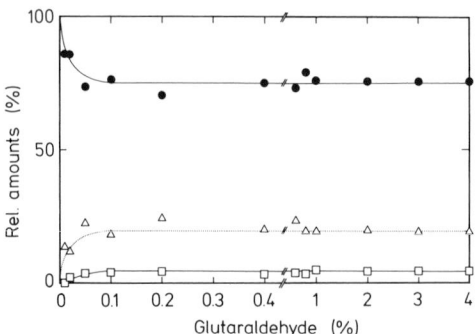

FIG. 5. Influence of glutaraldehyde concentration on the covalent fixation of reassociating tetrameric phosphoglycerate mutase from yeast.[53] Reconstitution at 20° in 0.05 M sodium phosphate, pH 7.5, containing 1 mM dithioerythritol at c = 20 μg/ml, after 15 min denaturation by 4 M guanidine-HCl. After 3.5 min of reassociation, samples were simultaneously cross-linked by increasing amounts of glutaraldehyde, as described in Fig. 4. (●) Monomers; (△) dimers; (□) tetramers.

redissolved in 20–50 μl water, containing 50 mM dithioerythritol, followed by 5 min incubation at 100° (water bath).

An alternative, more convenient way of removing salts and concentrating the protein is deoxycholate precipitation.[148] After NaBH$_4$ reduction, deoxycholate is added to a final concentration of $\cong$0.1% (w/v). The pH is lowered by adding trichloroacetic acid (78%, w/v) or concentrated H$_3$PO$_4$, until deoxycholate and protein precipitate. At the same time excess NaBH$_4$ is destroyed. After centrifugation, the pellet is washed with ice-cold acetone, and again centrifuged at 4000 rpm for 20 min. After redissolving in 20–50 μl 0.1 M Tris–HCl, pH 8.6, plus 1% (w/v) SDS and 50 mM dithioerythritol at 100°, the samples may be immediately applied to conventional homogeneous or gradient gel electrophoresis.

The cross-linking procedure has been applied both after acid and guanidine-induced dissociation. Residual guanidine (at concentrations up to 0.1 M) does not interfere with the fixation reaction. The band pattern obtained by covalent fixation of reassociating porcine muscle lactate dehydrogenase is illustrated in Fig. 6.

Observed deviations of the migration behavior of the cross-linked proteins from that of conventional molecular weight markers may be ascribed to the increased compactness of the proteins after extensive cross-linking. The lower Stokes radius would enhance the migration velocity, while decreased SDS binding as well as the molecular weight increment caused by polymerized glutaraldehyde would cause retardation.

[148] A. Bensadoun and D. Weinstein, *Anal. Biochem.* **70**, 241 (1976).

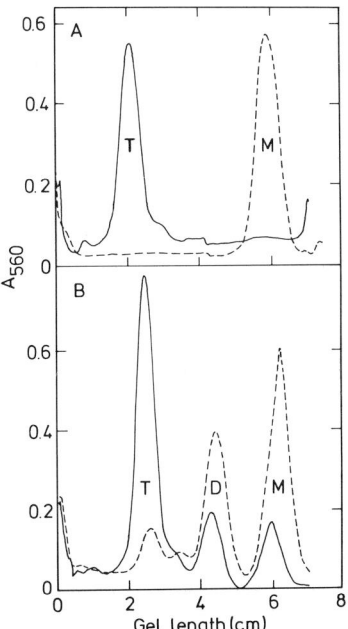

FIG. 6. SDS–polyacrylamide gel electrophoresis of cross-linked porcine muscle lactate dehydrogenase.[145] Cross-linking at 20° by 2 min incubation in the presence of 3–4% (w/v) glutaraldehyde. Correlation of the bands to the tetrameric (T), dimeric (D), and monomeric (M) fractions was achieved by calibration with proteins of defined molecular mass. (A) Cross-linking of native tetramers (solid line), and SDS–denatured monomers (dashed line); protein concentration, 2.8 μg/ml. (B) Cross-linking of reassociating enzyme after acid dissociation; reactivation at 20°, pH 7.6; protein concentration, 12 μg/ml. Cross-linking after 210 sec (dashed line) and 1 hr (solid line) of reassociation.

The application of the cross-linking technique to the analysis of the reassociation kinetics of two tetrameric enzymes, lactate dehydrogenase from pig muscle and phosphoglycerate mutase from yeast, is illustrated in Fig. 7.

Reassociation of both enzymes can be quantitatively described by the following kinetic model, involving monomers (M), dimeric intermediates (D), and native tetramers (T):

$$4M \underset{k_{-1}}{\overset{k_1}{\rightleftharpoons}} 2D \overset{k_2}{\rightarrow} T \tag{1}$$

A modified version of the cross-linking procedure has been successfully applied to the determination of the assembly mechanism of the catalytic trimers of aspartate carbamoyltransferase.[88] Experiments with dimethyl suberimidate instead of glutaraldehyde as cross-linking reagent gave only low yields of covalent fixation, even after long reaction times.[149]

Hybridization Experiments with Isoenzymes or with Chemically Modified Subunits. The rationale of the hybridization procedure is to "stop" reassociation of a given oligomer by a "chase" with an excess of electrophoretically different subunits. These may consist either of isoenzymes, or of chemically modified chains of one and the same oligomer. The following scheme illustrates the underlying procedure:

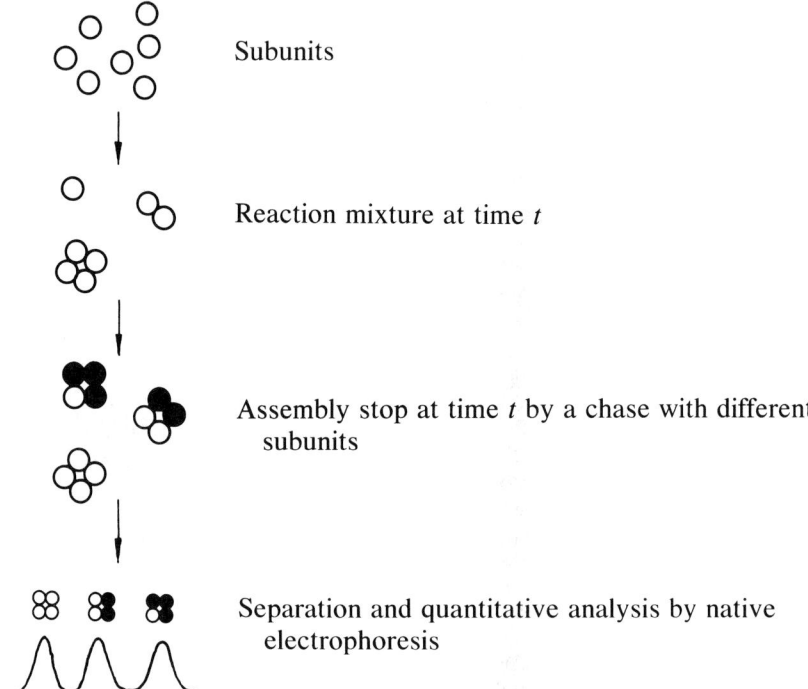

Subunits

Reaction mixture at time *t*

Assembly stop at time *t* by a chase with different subunits

Separation and quantitative analysis by native electrophoresis

The electrophoretically distinct subunits have to be added at various reassociation times in a sufficiently high excess (subunit ratio 1 : 5 to 1 : 30). This way subunits, not yet fully associated at the time of mixing, will associate preferentially with the "quencher." If the intermediates of association are in a fast dissociation/association equilibrium of the type proposed in Eq. (1), the hybridization data will not precisely reflect the actual particle distribution at the given time due to redistribution. Similarly, erroneous results may be obtained if the quencher molecules fold slowly during the "chase."

[149] G. F. Bickerstaff, C. Paterson, and N. C. Price, *Biochim. Biophys. Acta* **621,** 305 (1980).

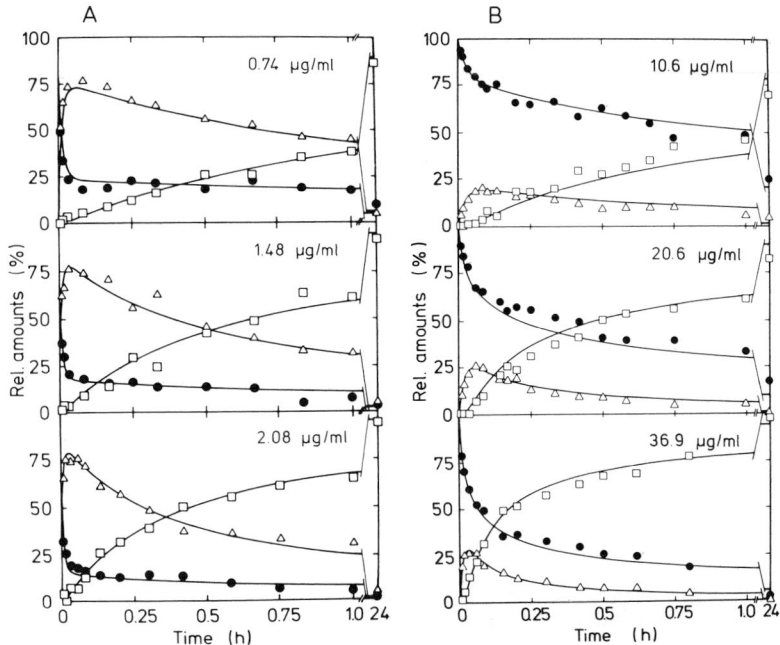

FIG. 7. Kinetics of reassociation of porcine muscle lactate dehydrogenase (A) and yeast phosphoglycerate mutase (B), as determined by cross-linking with glutaraldehyde.[53,146] Dissociation by 15 min incubation at 20° in 0.1 M H_3PO_4, containing 1 M Na_2SO_4 (A), or in 4 M guanidine-HCl (B). Reassociation at 20° determined by 2 min cross-linking in 1% (w/v) glutaraldehyde and subsequent SDS–polyacrylamide gel electrophoresis. (●) Monomers; (△) dimers; (□) tetramers. The full lines are calculated according to Eq. (1), with the following rate constants: for lactate dehydrogenase, $k_1 = 1.4 \times 10^6$ M^{-1} sec^{-1}, $k_{-1} = 2.15 \times 10^{-3}$ sec^{-1}, $k_2 = 3.15 \times 10^4$ M^{-1} sec^{-1}; for phosphoglycerate mutase, $k_1 = 6.25 \times 10^3$ M^{-1} sec^{-1}, $k_{-1} = 6.0 \times 10^{-3}$ sec^{-1}, $k_2 = 2.75 \times 10^4$ M^{-1} sec^{-1}.

Qualitative data on the association of porcine lactate dehydrogenase and bovine pyruvate kinase have been obtained by the given techniques using electrophoretically distinct isoenzymes.[63,65,150] A quantitative analysis of the reassociation of the catalytic trimers of aspartate transcarbamoylase made use of succinylated subunits of the enzyme.[58,88] Reassociation of the β_2-subunit of tryptophan synthase after high-pressure dissociation was studied by quenching the reaction with β-subunits which had been modified by reducing the internal aldimine between pyridoxal 5'-phosphate and lysine 86.[120]

Apart from the previously mentioned quenching technique, reassociation may alternatively be determined by mixing simultaneously, but separately reassociating isoenzymes after defined times of reconstitu-

[150] J. M. Cardenas, D. R. Hubbard, and S. Anderson, *Biochemistry* **16**, 191 (1977).

tion.[150–152] Using this approach, Hermann et al.[74] were able to confirm the reassociation mechanism of lactate dehydrogenase which has been previously determined by chemical cross-linking.

The kinetic interpretation of the hybridization data is complicated by the fact that the probability of the formation of the various hybrid species is strongly dependent on the experimental conditions. Due to the difference in both rate and activation energy of the rate-limiting folding and association steps of the various hybrid species, it is obvious that a binomial hybrid distribution is the exception rather than the rule. For a quantitative analysis it is therefore indispensable to determine empirically the dependence of hybrid formation on the experimental conditions (subunit ratio, concentration, and temperature).[74]

If binomial recombination is warranted, as in the case of the closely related allosteric L-lactate dehydrogenases of *Lactobacillus curvatus* and *Lactobacillus casei,* hybridization data may be directly interpreted by simple mathematical models.[152]

Hybridization as a tool for the analysis of association is by far less reliable and versatile than the cross-linking techniques described above.

Limited Proteolysis. Limited proteolysis has been successfully applied in a quantitative analysis of the reassociation of tetrameric porcine muscle lactate dehydrogenase.[153] In this case, the reassociation of the dimeric intermediate was quenched by the addition of thermolysin (10°). The optimum thermolysin : lactate dehydrogenase ratio was found to be 1 : 10. After 3 min of proteolysis, thermolysin was inactivated by adding solid EDTA to a final concentration of 10 mM. Isolation of the products of proteolysis under native conditions may be achieved by gel chromatography on Sephadex G-100 superfine (4°), or by affinity chromatography at 20°, using Procion green H-E 4 BD coupled to Sepharose 4B.[143,153] Further association of the dimers to form the native tetramer is prevented by the removal of an N-terminal extension of 10 amino acid residues. According to Rossmann[154] the N-terminal "arm" is responsible for the dimer–dimer interactions stabilizing the native tetramer.

The approach to analyze intermediates of association by proteolysis can hardly be applied in general. However, the reconstitution of the native quaternary structure may be generally determined by proteolysis,

[151] A. Levitzki, *FEBS Lett.* **24,** 301 (1972).

[152] U. Mayr, R. Hensel, H. Pruscha, and O. Kandler, *Eur. J. Biochem.* **115,** 303 (1981).

[153] R. Girg, R. Rudolph, and R. Jaenicke, *Eur. J. Biochem.* **119,** 301 (1981); R. Rudolph, *Biochem. Soc. Trans.* **13,** 308 (1985); U. Opitz, R. Rudolph, R. Jaenicke, L. Ericsson, and H. Neurath, submitted for publication (1986).

[154] J. J. Holbrook, A. Liljas, S. J. Steindel, and M. G. Rossmann, *in* "The Enzymes" (P. D. Boyer, ed.), Vol. 11, p. 191. Academic Press, New York, 1975.

provided that the following requirements are met: (1) the native oligomer must be considerably more stable toward proteolytic attack than the intermediates of association; (2) proteolytic degradation of the intermediates of association must be fast compared to reassociation; and (3) there must be a rapid method for inactivating the protease in order to prevent further degradation. The amount of fully reconstituted (proteolytically inert) oligomeric species present at the time of proteolysis may then be easily determined by conventional separation techniques such as affinity chromatography or gel filtration.

Further Methods to Analyze Dissociation/Association. A number of experimental techniques have been applied to follow the kinetics of dissociation/association, making use of either spectroscopic or scattering methods.

In certain cases, UV difference spectra as well as fluorescence emission and circular dichroism may reflect the state of association, thus providing practical approaches to monitor the kinetics of dissociation/reassociation.

In the case of multimeric systems showing a high level of scattering, inelastic and quasielastic light scattering have been widely used to monitor association.

The easy way to make use of particle scattering is to measure the turbidity spectrophotometrically. In quantifying such data, corrections for absorbance and interparticle interference (Rayleigh and Mie scattering) have to be carefully considered.

In connection with oligomeric systems, "wrong aggregation" causes severe problems in monitoring the kinetics of association. In most cases, the low protein concentrations required to reduce aggregate formation (cf. Fig. 9) do not allow scattering measurements of sufficient accuracy to guarantee unambiguous results. High energy sources, like synchroton radiation, promise to overcome this problem. Measurements involving, e.g., the assembly of microtubules, muscle proteins, or chromatin may serve as examples to illustrate the versatility and the potential of this method.[155,156]

---

[155] E.-M. Mandelkow, A. Harmsen, E. Mandelkow, and J. Bordas, *Nature (London)* **287,** 595 (1980); E. Mandelkow, E.-M. Mandelkow, and J. Bordas, *J. Mol. Biol.* **167,** 179 (1983).

[156] H. E. Huxley, A. R. Faruqi, M. Kress, J. Bordas, and M. H. J. Koch, *J. Mol. Biol.* **158,** 637 (1982); H. E. Huxley, A. R. Simmons, A. R. Faruqi, M. Kress, J. Bordas, and M. H. J. Koch, *J. Mol. Biol.* **169,** 469 (1983); J. Lowy and F. R. Poulsen, *Nature (London)* **299,** 308 (1982); F. R. Poulsen and J. Lowy, *Nature (London)* **303,** 146 (1983); J. Bordas, L. Perez-Grau, M. H. J. Koch, M. C. Vega, and C. Nave, *Eur. Biophys. J.* **13,** 157, 175 (1986).

In the given context, electron microscopy may offer a valuable additional approach, because it requires intrinsically high dilution of the samples, and may allow to distinguish correct assemblies from wrong aggregates in a clear-cut and direct way.[26] Of course, the time resolution in this case is low.

A further direct measure for the particle distribution, besides chemical cross-linking or hybridization, has been recently proposed: separation of association intermediates by high-performance gel permeation chromatography.[157] Using, e.g., a TSK SW-3000 column (7.5 × 600 mm), separation may be performed in less than 1 hr. Provided that reassociation can be slowed down sufficiently during HPLC analysis, e.g., by a fast decrease in temperature,[158] the state of association of samples taken at various times may be directly determined.

Determination of the Kinetics of Reactivation and Refolding of Oligomeric Proteins

Comparison of the reassociation data with the kinetics of reactivation or refolding allows us to determine the extent to which structure and function are regained at the different levels of association.

Reactivation. Reactivation of oligomeric enzymes may be analyzed by two procedures. First, a small aliquot of the denatured subunits may be directly added to the test mixture. If the activity, as an example, is recorded by the change of absorbance of NAD, the regain of activity is directly reflected by the change in E_{340} with time.[62,159] The half-time of reactivation is defined by the time at which the slope of the enzyme assay is half the final slope. Using this approach one has to make sure that the enzyme is always working under saturating conditions, and that the final value of reactivation is determined correctly. In addition, one has to consider that the rate or reactivation may be influenced by the substrates or coenzymes present in the test solution.

Far more reliable data on the reactivation of oligomers can be obtained by the second procedure. In this case, reactivation is followed by subjecting aliquots of the reassociating enzyme to the standard test. This way, reactivation can be studied under controlled conditions, e.g., in the absence of coenzymes or substrates, and the final value of reactivation can be determined after prolonged reassociation time. The reactivation kinetics of dimeric porcine mitochondrial malate dehydrogenase (Fig. 8) may be taken to illustrate the analysis of the reactivation kinetics at various enzyme concentrations. Using the sampling technique, one has to make

[157] M. A. Recny and L. P. Hager, *J. Biol. Chem.* **257,** 12878 (1982).
[158] R. Rudolph, I. Heider, and R. Jaenicke, *Biochemistry* **16,** 5527 (1977).
[159] S. G. Waley, *Biochem. J.* **135,** 165 (1973).

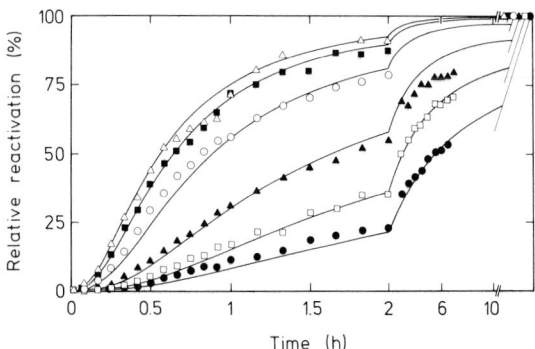

FIG. 8. Kinetics of reactivation of porcine mitochondrial malate dehydrogenase[70] after 5 min deactivation in 1 M glycine/H_3PO_4, pH 2.3, 1 mM EDTA, 1 mM dithioerythritol at 20°, and dilution with 0.2 M phosphate buffer, pH 7.6, 10 mM EDTA, 10 mM dithioerythritol at varying enzyme concentrations (μg/ml): ($\triangle$) 5.0; ($\blacksquare$) 3.1; ($\bigcirc$) 1.2; ($\blacktriangle$) 0.35; ($\square$) 0.14; ($\bullet$) 0.07. Reactivation was calculated relative to final values, determined after a reactivation time of up to 290 hr. Solid lines are calculated according to a consecutive first-order/second-order mechanism with $k_1 = 6.5 \times 10^{-4}$ sec^{-1} and $k_2 = 3 \times 10^4$ M^{-1} sec^{-1}.

sure that no further reactivation occurs during the assay. If reactivation is determined by rate-limiting association reactions, simple dilution of the samples in the assay may decelerate further association sufficiently. If reactivation in the enzymatic assay is indicated by curved profiles, this reaction may be suppressed by adding denaturant or a protease to the test mixture. Addition of either urea or trypsin to the assay mixture at a concentration where native enzyme is still active (e.g., 2.3 M urea, or 20 μg/ml trypsin, respectively) has been shown to prevent further reactivation completely.[160,161] The respective amounts of denaturant or protease which are required to quench reactivation without affecting the fully reassociated oligomer have to be determined for each system individually.

In the case of rabbit muscle aldolase, addition of 2.3 M urea has been shown to inactivate partially active intermediates of association (monomers), apart from preventing reactivation during the test.[48,161] Similarly partially active intermediates of association of yeast phosphoglycerate mutase are eliminated by adding 20 μg/ml trypsin to the assay mixture.[53] Under certain conditions, reconstitution of the active center of an enzyme may be followed by substrate or coenzyme binding. This approach can be applied only if substrate or coenzyme binding is accompanied by a significant spectral effect, such as the "Racker band"[162] in the case of glyc-

[160] W. C. Deal, W. J. Rutter, V. Massey, and K. E. van Holde, *Biochem. Biophys. Res. Commun.* **10**, 49 (1963).
[161] W. W. -C. Chan, J. S. Mort, D. K. K. Chong, and P. D. M. MacDonald, *J. Biol. Chem.* **248**, 2778 (1973).
[162] E. Racker and I. Krimsky, *J. Biol. Chem.* **198**, 731 (1952).

eraldehyde-3-phosphate dehydrogenase, or changes in fluorescence emission upon NAD binding to dehydrogenases.[163]

Refolding. The common spectroscopic techniques used to monitor protein folding in the case of single chain proteins may also be applied to analyze the regain of the native backbone structure in the assembly of oligomers. As examples, UV absorption or fluorescence may be used to follow the folding reactions accompanying subunit association. More detailed information regarding the structural integrity of individual intermediates of association may be obtained from the kinetic analysis of the recovery of the far-UV circular dichroism. Since, for a number of proteins, a major part of the native backbone structure is found to be restored on the monomer level in a rapid reaction (half-time <1 sec), devices for fast kinetics (stopped flow) are desirable for such measurements. Subsequent slow folding and association reactions that are rate determining for reassociation and reactivation can be investigated after manual mixing (cf. A. Labhardt, this volume [7]).

Comparison of the Kinetics of Reassociation and Reactivation/Refolding

The properties of the intermediates of association can only be derived from the given kinetic data by proper mathematical analysis. First, the association data, as determined by cross-linking or other techniques, have to be described by a kinetic model.

In the most simple case this is an irreversible first-order folding or second-order association mechanism. Consecutive unimolecular–bimolecular or bimolecular–bimolecular reaction sequences are frequently encountered upon reassociation of oligomeric proteins.[107] A complete analytical solution of these reaction types has been put forward by Chien.[164] In more complex cases, where the complete analytical solution of the rate equation is not accessible, the kinetic data of association must be analyzed by numerical approximatation.[53,66,146]

Once the kinetic mechanism of reassociation has been established for a given oligomer, the kinetics of refolding and reactivation can be correlated with the rate-limiting association steps. In those cases, where a functional property such as enzymatic activity is restored parallel to the formation of the native oligomer, the correlation is trivial: all the intermediates must be devoid of the given property. If, on the other hand, the intermediates participate in the regain of the respective properties, the exact quantity of recuperation at the level of any intermediate can be computed. Figure 3 illustrates the determination of the activity of the

[163] A. D. Winer, G. W. Schwert, and D. B. S. Millar, *J. Biol. Chem.* **234,** 1149 (1959).
[164] J.-Y. Chien, *J. Am. Chem. Soc.* **10,** 2256 (1948).

dimeric intermediates of lactate dehydrogenase, observed in the presence of stabilizing salt. Similarly, all spectroscopic properties of the intermediates can be derived by kinetic modeling.

Side Reactions during Reconstitution

The structural or functional properties of a given oligomeric protein, i.e., its immunological, hydrodynamic, spectral, and enzymatic characteristics, are hardly ever fully restored after dissociation and subsequent reconstitution.

In most cases the reconstituted material is heterogeneous, consisting of structurally modified (inactive) and fully native fractions in varying proportions.[15] After separation of the inactive part (in case of inactive aggregates simply by gel filtration) the reactivated oligomers have been shown to be indistinguishable from the initial native protein.[36] Although it has been proven for many oligomeric enzymes that native and reactivated material is one and the same, this fundamental prerequisite for reassociation studies should be verified for each individual case.

The yield of correct association may be limited by the following mechanisms: (1) kinetic competition of correct subunit folding and aggregation; (2) aggregation at intermediate denaturant concentrations; and (3) chemical modification during dissociation or reassociation.

Kinetic Competition of Correct Subunit Folding and Aggregation. "Wrong aggregation" as the major side reaction seems to be a general feature of *in vitro* reconstitution of oligomeric proteins. For the reconstitution of tetrameric lactate dehydrogenase it could be shown that aggregation to inactive material (which is characterized by a reaction order ≥ 2) is in kinetic competition with a first-order folding step on the correct pathway.[20] This kinetic mechanism is clearly documented by the inverse proportion of reactivation, i.e., tetramer formation, and "wrong aggregation" with increasing enzyme concentration (Fig. 9). For the given example, the refolding reaction of the monomers competing with aggregation is not determined by cis–trans isomerization around X-Pro peptide bonds (cf. F.X. Schmid, this volume [4]). This was shown by double-jump experiments including reconstitution after very short times of guanidine denaturation, where aggregation occurs despite that all X-Pro peptide bonds are expected still to be in their native isomeric form.[52]

The inactive aggregates are partially structured as shown by circular dichroism analysis of the "wrong aggregates."[20] Apparently, the encounter of incompletely structured monomers stabilizes incorrect interparticle interactions, instead of allowing the correct hydrophobic core of the globular protein to be formed. Since formation of "wrong aggregates" is increased at high subunit concentrations, reconstitution by dilution

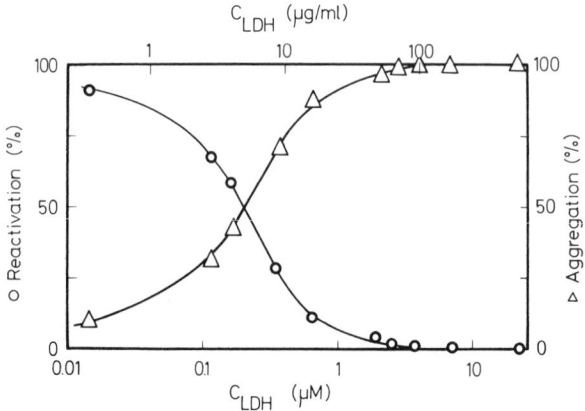

FIG. 9. Effect of subunit concentration on the extent of reactivation (○) and aggregation (△) of porcine muscle lactate dehydrogenase after 5 min dissociation at 20° in 0.1 M H_3PO_4, pH 2.5, containing 1 mM EDTA and 1 mM dithioerythritol.[20] Reactivation in 0.1 M phosphate buffer, pH 7.0, containing 1 mM EDTA and 1 mM dithioerythritol, 20°, was determined after up to 192 hr.

should be performed quickly in order to prevent high local subunit concentrations.

Similar to the effect of protein concentration, the yield of reconstitution can be influenced by temperature variation if the competing folding and aggregation reactions differ in their respective energies of activation.[52]

Another possibility to increase the yield is to prevent competing side reactions entirely by starting reconstitution from structured monomers. Correctly structured subunits can be produced by short-term dissociation under moderately denaturing conditions.[42,44,88,108] For example, the reconstitution of lactate dehydrogenase after acid dissociation (pH 1.0–2.3) in the presence of stabilizing salt (1 M Na_2SO_4) is not perturbed by side reactions even after extended acid incubation.[71,73] Upon prolonged dissociation in the absence of stabilizing additives, slight changes of the residual monomer structure can result in lower yields. In this case, the subunits become trapped in "wrong" structures from which refolding is decelerated so that aggregation predominates.[42,72,87]

Finally, the kinetic competition of reconstitution on one hand, and side reactions to inactive material on the other can be influenced by specific ligand effects.[23,25,83]

In certain cases, side reactions during reconstitution *in vitro* may predominate despite all conceivable experimental variations.[165]

[165] K. Müller and R. Jaenicke, Z. *Naturforsch.* **35c,** 222 (1980).

Aggregation at Intermediate Denaturant Concentrations. In case of oligomeric proteins where the dissociation transition is superimposed by side reactions such as aggregation (cf. Fig. 2), maximum yields of correct reassembly may be obtained by avoiding long-term incubation at the critical denaturant concentrations where dissociation and aggregation parallel each other. Dissociation/reassociation should then be performed under essentially irreversible conditions, i.e., far from the range of transition by quickly diluting out the denaturant. In certain cases, removal of the denaturant by dialysis (involving prolonged incubation at moderate denaturant concentrations) may result in high reactivation yields by allowing reequilibration of the aggregates, as described by curve 1 in Fig. 2.

Chemical Modification during Dissociation or Reassociation. Modification of the constituent polypeptide chains may prevent reassociation of an oligomeric protein. Compared with the native assemblies, unfolded proteins during denaturation/renaturation are far more susceptible toward, e.g., proteolysis by proteases or thiol modification.[87,166] Minute amounts of a protease (which may not affect the native oligomer) may very well resist denaturation, or they may refold at a higher rate than the protein under consideration. The protein preparations should therefore be essentially devoid of proteases. Thiol modification of cysteine-containing proteins should be prevented by adding 1–10 mM amounts of EDTA and sulfhydryl reagent (e.g., dithioerythritol). Furthermore, the buffer solutions should be degassed and saturated with nitrogen. The decrease of reversibility of dissociation caused by chemical modification or by surface denaturation can be avoided by adding, e.g., serum albumin, or by prerinsing the glassware with reassociating protein.[167]

Concluding Remarks

The acquisition of the three-dimensional structure of oligomeric proteins has been shown to be a sequential process involving well-defined folding and association steps. Folding studies on small single-chain molecules clearly prove that folding intermediates are well populated. This is expected to hold equally for the early steps on the pathway of reconstitution of multidomain proteins and subunit assemblies. Kinetic methods have been described to elucidate the consecutive mechanism of folding and association. They provide a means to determine the particle distribution in the process of reconstitution, and to characterize intermediates of

[166] R. Rudolph and I. Fuchs, *Hoppe-Seyler's Z. Physiol. Chem.* **364,** 813 (1983).
[167] P. Bernfeld, B. J. Berkeley, and R. E. Bieber, *Arch. Biochem. Biophys.* **111,** 31 (1965).

association with respect to their structural and functional properties. Evidently, the conformational refinements and the functional achievements accompanying quaternary structure formation are closely related to the central problem of the evolutionary role of molecular assocation.

[13] Mutational Analysis of Protein Folding Pathways: The P22 Tailspike Endorhamnosidase

By JONATHAN KING and MYEONG-HEE YU

Introduction

Two major unsolved questions in protein biochemistry are (1) the actual pathways polypeptide chains follow in achieving their native conformation, and (2) the nature of the amino acid sequence code that directs the folding process.

Since the initial work of Beadle and Tatum[1] genetic approaches have provided a powerful means of analyzing biochemical pathways, primarily through the characterization of intermediates accumulating before the step blocked by mutation. The analysis of pathways of polypeptide chain folding should be accessible to genetic analysis. Practically, this requires systems in which folding intermediates can be stabilized and trapped, or where the off-pathway forms can be characterized.

The deciphering of the genetic code, specifying the correspondence between the sequence of nucleotides in DNA and the linear sequence of amino acids in polypeptide chains, was complete by 1965. In contrast, the protein folding code through which the amino acid sequence of the polypeptide chain directs its three-dimensional conformation remains obscure. Certain general features, for example, high redundancy and context dependence, emerge from inspection of the amino acid sequences of related families of proteins, such as the globins. In this article we treat the code as specifying kinetic steps in the pathway of protein folding, as well as interactions stabilizing the native conformation.

The general nature of the genetic code was elucidated in part by examining mutants in the T4 rII gene which altered the reading frame and readout. Features of the amino acid code for protein folding should also be accessible by analyzing mutants blocked in the folding of the mutant gene product.

Until recently, progress has been very limited in the use of mutations to determine folding pathways or the folding code. We identify below

[1] G. W. Beadle and E. L. Tatum, *Proc. Natl. Acad. Sci. U.S.A.* **27**, 499 (1941).

some of the barriers to this analysis. Then we consider aspects of the isolation and use of mutants which need to be taken into consideration regardless of the particular system under study. We describe the temperature-sensitive folding mutants of the P22 tailspike protein, which permit direct probing of the protein folding code, and lastly, we describe critical features of other experimental systems which are amenable to mutational analysis of protein folding or protein stability.

Note that most of the discussion refers to *in vivo* folding pathways since mutant strains are generally identified with respect to the behavior of the mutant protein within the organism.

Few Proteins Have Been Subjects of Both Crystal Structure Determination and Fine Structure Genetic Analysis

The proteins that were initially studied successfully by crystallographers were generally eukaryotic proteins available in large amounts, such as the hemoglobins, cytochrome *c*, bovine pancreatic ribonuclease, and hen egg white lysozyme. None of these proteins comes from organisms in which it is possible to carry out fine structure genetics. The proteins of the fruit fly *Drosophila* have until recently been avoided by protein biochemists due to difficulties in obtaining sufficient quantities.

On the other hand, the proteins whose genes were subjects of fine structure genetic analysis have been almost exclusively from microorganisms. These include β-galactosidase of *E. coli*, tryptophan synthase of *E. coli*, the RII protein of phage T4, and the lac and lambda repressors. For many years, despite extensive efforts none of these was obtained as large crystals. The first prokaryotic proteins whose atomic structures were solved by crystallographers were staphylococcal nuclease and aspartate transcarbamoylase. Ironically, neither of these proteins was the subject of fine structure genetic analysis, the former because of its host, *Staphyloccus,* and the latter because of the difficulty, until recently, of obtaining mutants. It is also unfortunate that the structure of the lac repressor, far and away the genetically best characterized of all known genes, with over 2000 mapped and characterized mutants, has not been solved.[2]

Difficulties in Characterizing Unfolded or Partially Folded Chains

Polypeptide chains blocked prior to folding into their native state are not likely to maintain the distinctive properties of their native state. This has often made it difficult to characterize or even detect such chains in cells or cell lysates. Thus in the characterization of mutants of *E. coli,*

[2] J. H. Miller and U. Schmeissner, *J. Mol. Biol.* **131,** 223 (1979).

Salmonella, and *Neurospora,* which lacked enzymatic activity, a common assay was to test serologically for CRM (cross-reacting material).[3] Those mutants which failed to demonstrate CRM, that is which showed no activity with respect to wild type, were generally not studied further. In many cases the lack of an assay made it very difficult to determine if the polypeptide chain had been synthesized. Only with the development of SDS–polyacrylamide gel electrophoresis was it routinely possible to ascertain the existence of a complete polypeptide chain, in the absence of any activity.[4] In fact one of the very first applications of high resolution SDS–gel electrophoresis resulted in the identification of a gene of T4, gene 57, whose product was needed for folding or maturation of three other T4 fibrous structural proteins.[5]

Mutational Analysis

Biases Due to Selection

The conditions under which a mutant strain is initially isolated are very important constraints on the properties of the mutant. For example, the general method of identifying hemoglobin variants in the human population is to screen blood samples by gel electrophoresis for bands with altered electrophoretic mobility. This technique requires that the hemoglobin chains be organized into native or native-like molecules. Molecules with defects which prevented the formation of native like molecules, for example blocked in folding, are not routinely detected. Similarly, selections which rely on immunological identification of a gene product may not detect mutations which prevent acquisition of native structure.

General Missense Mutations

The most general class of mutation affecting protein structure is missense mutation, in which the polypeptide chain is synthesized but has an alteration in amino acid sequence. This can be a single substitution, or a more complex change such as an addition or deletion of a number of residues. Large collections of missense mutations have been isolated in microbial genes coding for metabolic enzymes whose product can be provided from the medium. The loss of enzymatic activity gives evidence of the mutation. However the strain can be propagated by providing the product, for example histidine, from the medium.

---

[3] D. Perrin, *Ann. N.Y. Acad. Sci.* **103,** 1058 (1963).
[4] U. K. Laemmli, *Nature (London)* **227,** 680 (1970).
[5] J. King and U.K. Laemmli, *J. Mol. Biol.* **62,** 465 (1971).

If the gene product is essential for the growth of the organism, for example, a ribosomal structural protein, or RNA polymerase, missense mutations preventing function will be lethal. In haploid organisms such as bacteria and viruses this class of mutations will not be routinely recovered.

Conditional Lethal Mutations

Conditional lethal mutations were developed by Epstein and Edgar and their co-workers[6] as a means of studying the essential genes of haploid organisms, such as bacteria and viruses. They require two conditions of growth: one in which the altered protein is functional, allowing the organism to propagate; and a second, in which the mutant protein is defective, generally preventing growth. The most common classes are temperature-sensitive mutations (both heat sensitive and cold sensitive) and nonsense mutations, although other classes of conditional lethal mutations have been described, such as osmotic remedial and pH dependent.

Temperature-Sensitive Mutations. These are a subclass of missense mutations in which the mutant phenotype is a property of the temperature of growth of the mutant organism. These were developed initially by Horowitz and Leupold and later by Edgar and Lielausis.[7] The most common class has a functional protein formed at low permissive temperature, whereas active protein does not accumulate at a higher restrictive temperature. The reverse cold-sensitive class, defective at low temperature, is less frequently isolated, but has been found for many proteins.[8,9]

The *ts* mutations were early divided into two classes by Sadler and Novick[10]: TL (for thermolabile) and TSS (for temperature-sensitive synthesis). With TL mutants, the active protein produced at permissive temperature is inactivated on exposure to the higher restrictive temperature. However, it is quite common for *ts* mutant proteins to maintain their activity when exposed to restrictive temperature. In these TSS mutants, the defect is expressed only if the mutant protein is synthesized or matured at restrictive temperature. Smith and King[11] showed that at least some of these mutations are defective in the polypeptide folding pathway

[6] R. H. Epstein, A. Bolle, C. M. Steinberg, E. Kellenberger, E. Boy de la Tour, R. Chevalley, R. S. Edgar, M. Susman, G. Denhardt, and A. Lielausis, *Cold Spring Harbor Symp. Quant. Biol.* **28,** 375 (1963).

[7] R. S. Edgar and A. Lielausis, *Genetics* **49,** 649 (1964).

[8] C. Guthrie, H. Nashimoto, and M. Nomura, *Proc. Natl. Acad. Sci. U.S.A.* **63,** 384 (1969).

[9] J. H. Cox and H. B. Strack, *Genetics* **67,** 5 (1971).

[10] J. R. Sadler and A. Novick, *J. Mol. Biol.* **12,** 305 (1965).

[11] D. H. Smith and J. King, *J. Mol. Biol.* **145,** 653 (1981).

at higher temperatures and these are referred to as TSF mutations, for temperature-sensitive folding.[12]

Nonsense (Amber) Mutations. These create an additional stop codon which prematurely terminates the polypeptide chain in the wild-type host. This is the *restrictive* condition. In order for the mutant strain to be recovered and propagated there must be a permissive host in which the nonsense codon can be circumvented. These *permissive* hosts have an alteration in their protein synthetic apparatus, generally in one of the tRNAs. The mutant tRNA inserts its amino acid at the stop codon generating a complete polypeptide chain. The requirement for growth on two different hosts means that nonsense mutants are most commonly isolated in viruses.

Though many sets of amber fragments have been characterized for the presence or absence of activities of the mature protein, much less has been done in terms of characterizing the inactive fragments. Sets of amber fragments of increasing length are a potentially valuable resource for studying protein folding.

As already noted, isolation of strains carrying amber mutations requires the existence of a permissive host whose protein translation apparatus inserts an amino acid at the premature stop codon. The well-characterized *E. coli* suppressors—all mutant tRNAs—insert serine, glutamine, tyrosine, or leucine.[13] This generates a protein with a single amino acid substitution at the mutant site. To recover the amber mutant this suppressed protein must be biologically active, to allow propagation of the mutant. In fact the suppressor tRNA may insert the wild-type amino acid sequence. More generally, the suppressor inserts another amino acid, generating an altered protein. Thus the propagation of nonsense mutations in the permissive strains generates a protein that is the formal equivalent of a nonlethal missense mutation.

It is important to bear in mind that the isolation of nonsense mutations generally requires that the suppressed polypeptide chain be functional. Thus the constraints are that the codon can mutate to stop, and that the polypeptide chain is relatively insensitive to the residue at that site. Though such mutations may occasionally identify critical sites in polypeptide chains, the nature of the isolation process mitigates against identifying such sites with nonsense mutations.

The efficiency of suppression varies for different hosts, so that quantitative yields are almost always depressed with respect to wild type. Furthermore, the suppressing strain may be partially defective due to the

[12] M.-H. Yu and J. King, *Proc. Natl. Acad. Sci. U.S.A.* **81,** 6584 (1984).
[13] L. Gorini, *Annu. Rev. Genet.* **4,** 107 (1970).

incorrect reading of true stop signals, generating extra long polypeptide chains.

If the protein is not essential for growth, the requirement that the suppressed protein be active does not hold. This factor was, in part, an aspect that allowed Miller and co-workers[14] to isolate a very large collection, over 300 nonsense mutants of the lac repressor.

Temperature-Sensitive Folding Mutants

The existence of TSS mutants in the genes of diverse organisms suggested that they might represent folding mutants. To unambiguously establish this required an experimental system in which the native protein was quite heat stable, so that one could demonstrate that the mutation affected the pathway and not the native protein. The initial isolation of *ts* mutants requires that the mutant polypeptide chain forms the native structure at permissive temperature. The critical experiment is then to compare the mutant protein in its native state, matured at permissive temperature, with the mutant polypeptide chain matured at restrictive temperature.

The identification of temperature-sensitive mutants which interfere with protein folding has been carried out in an experimental system which is quite different from the traditional systems for the study of protein folding. The target protein, the P22 tailspike, is a large structural protein; the experimental characterization is of the *in vivo* folding and subunit assembly pathway; and the mutations, which do affect the pathway, do not affect the native protein once matured.

Properties of the Tailspike Proteins of Salmonella Phage P22

The thermostable tailspikes of phage P22 provide the apparatus for cell attachment. The tailspike protein is composed of three identical polypeptide chains of molecular weight 71,600, the products of gene 9 of P22.[15,16] The native spike cannot be dissociated into chains unless it is fully denatured, so that there is no species corresponding to a "native" monomer. One end of the native tailspike binds irreversibly but noncovalently to the head of P22. Iwashita and Kanegasaki[17] identified an enzymatic activity with the tailspikes which cleaves the O-antigen of *Salmo-*

---

[14] J. H. Miller, C. Coulondre, M. Hofer, U. Schmeissner, H. Sommer, and A. Schmitz, *J. Mol. Biol.* **131,** 191 (1979).

[15] D. Botstein, C. H. Waddell, and J. King, *J. Mol. Biol.* **80,** 669 (1973).

[16] P. B. Berget and A. R. Poteete, *J. Virol.* **34,** 234 (1980).

[17] S. Iwashita and S. Kanegasaki, *Biochem. Biophys. Res. Commun.* **55,** 403 (1973).

nella outer membrane at α-rhamnosyl 1-3-galactose linkage. Both the free spike and the particle bound form have the endorhamnosidase activity which is believed to represent the mechanism of cell attachment and host range specificity.

Soluble tailspike protein is obtained from *Salmonella* cells infected with mutants defective in phage head assembly. When P22 capsid assembly is blocked by mutation, the level of tailspike gene expression increases and the tailspike protein becomes a major protein synthesized inside the infected cells. The free protein is quite soluble and solutions of up to 50 mg/ml can be obtained without difficulty.

The physical properties of the tailspikes are noteworthy. First, native trimeric protein is resistant to dissociation by SDS. At room temperature the protein remains native in concentrations of SDS which denature most *Salmonella* and phage proteins.[18] However, if the protein is heated to boiling in SDS, complete dissociation is obtained. This property of the protein has been exploited in distinguishing mature native tailspikes from partially folded intermediates or incorrectly folded byproducts found intracellularly. The native spike is also very resistant to protease digestion, while the denatured chain is susceptible.

The tailspike protein is thermostable, with a half time of irreversible inactivation of 5 min at 90°. Though not intuitively obvious, this thermostability was critical in the isolation of the *ts* folding mutants. Single amino acid substitution is unlikely to yield a protein whose melting temperature had dropped 40° to the restrictive temperature for growth of the mutant phage. Thus the heat stability of the protein creates a situation where screening for *ts* mutants selects for those affecting the folding pathway, rather than the native structure.

No proteolytic or other covalent modifications have been identified in the tailspike. The intracellular maturation pathway represents only chain folding and chain association steps.

Determination of in Vivo Folding and Subunit Assembly Pathway

Under the conditions of SDS–gel electrophoresis most nonnative states of the gene 9 polypeptide chain form a normal SDS–polypeptide complex, while the mature spike migrates more slowly, probably due to decreased SDS binding. Figure 1 shows the distribution of nonnative and native forms of the polypeptide chain as a function of time after synthesis. Quantitation of the two bands showed that the sum of both remains the same, indicating that the SDS–polypeptide chain complexes represent species which within the cell were precursors of the native spike. The

[18] D. P. Goldenberg, P. B. Berget, and J. King, *J. Biol. Chem.* **257**, 7864 (1982).

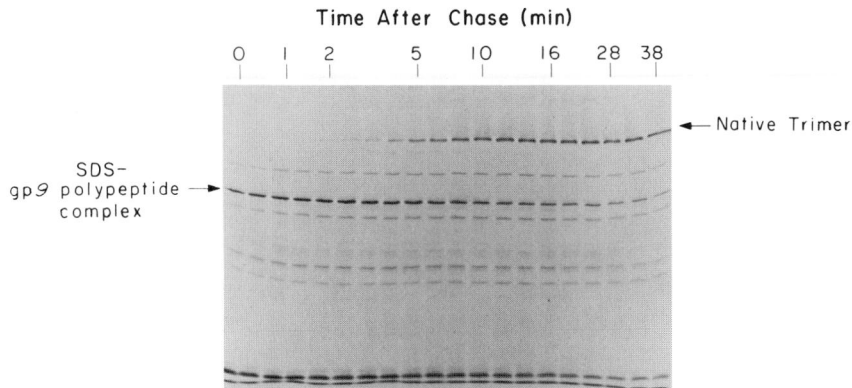

FIG. 1. Kinetics of *in vivo* formation of the SDS-resistant native tailspike.[19]

time course of the overall *in vivo* maturation process for spikes is slow compared to those of smaller proteins such as ribonuclease. The process has a half-time of 5 min at 30°. This strongly indicates the existence of long-lived intermediates in the tailspike folding and subunit assembly pathway.

One putative intermediate could be trapped and stabilized by chilling the infected cells. The trapped intermediate was separable from the native form on polyacrylamide gel electrophoresis in the absence of detergent. This was not surprising, considering that folding intermediates are likely to have a different conformation from the final native structure and the property of sieving through polyacrylamide matrix should be different.

Pulse-chase labeled cells were frozen in liquid nitrogen and lysed at 4°, and then immediately fractionated by native gel electrophoresis at 4°. Figure 2 shows a strong band which increases in intensity with time after the chase and represents the native tailspike. Migrating more slowly than mature tailspike is another species, the amount of which increases and then decreases. Both species are missing in the control culture, infected with a nonsense mutant of gene 9. On second-dimensional analysis on SDS polyacrylamide gel the precursor species migrates with the tailspike polypeptides. The formation and disappearance of the species had a precursor–product relationship with the native spike.[19]

Further analysis of the precursor revealed that it represented a partially folded intermediate in which the three chains are already associated. This species, the protrimer, converts *in vitro* to the native spike on incubation at physiological temperatures. This represents a subunit folding and subunit assembly pathway different from that generally found with

[19] D. Goldenberg and J. King, *Proc. Natl. Acad. Sci. U.S.A.* **79**, 3403 (1982).

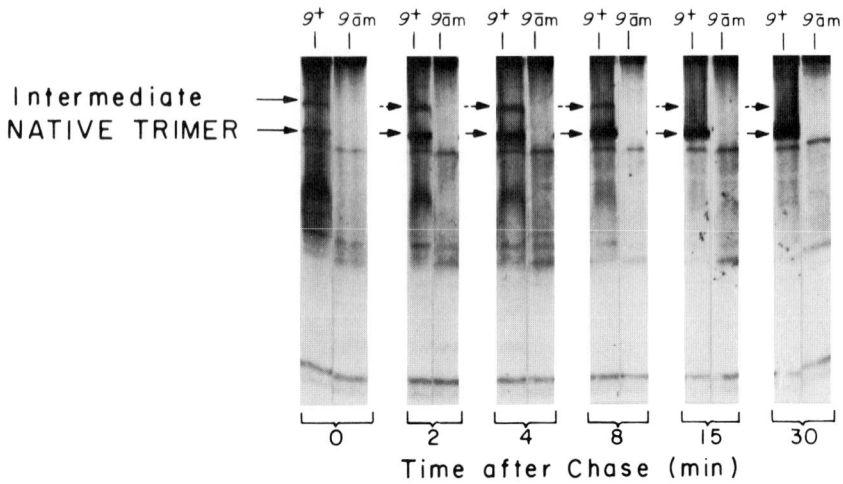

FIG. 2. Analysis of kinetic intermediates by native gel electrophoresis.[19]

soluble enzymes, where the chain folds into a subunit and then subunits associate in a postfolding process. The complete folding of the tailspike chain is coupled to chain association, somewhat reminiscent of the collagen maturation pathway.

Though the native protein is quite thermostable, the formation of the protein is thermolabile. The conversion of newly synthesized polypeptide chains to native spikes falls off with increasing temperatures, and is less than 25% at 40°. This property indicates that either an intermediate in the wild-type pathway is thermolabile, or some off-pathway step is speeded up at elevated temperature.

Isolation of ts Folding Mutants

The temperature-sensitive folding mutants in gene 9 were not the product of a special selection. They were isolated as general *ts* mutants, capable of propagating at a permissive temperature of 30° but not at the restrictive temperature of 39–40°. However, given the high stability of wild-type tailspike protein it was thought likely that these would be of the TSS category.

Surprisingly these mutants mapped throughout gene 9. Smith *et al.*[20] described 38 *ts* mutations which mapped at 33 sites in the gene. At present there are more than 100 *ts* mutations which have been mapped to a minimum of 55 sites.

[20] D. H. Smith, P. B. Berget, and J. King, *Genetics* **96,** 331 (1980).

Purification of Wild-Type and Mutant Tailspike Proteins

The procedures described below, modified from Berget and Poteete,[16] are for the purification of mutant tailspike proteins formed at *permissive* temperature, and yield the native form of the mutant protein. We do not yet have reliable procedures for the purification of nonaggregated states of the unfolded mutant polypeptide chains formed at restrictive temperature. The phage and bacterial strains described are available from the authors.

Preparation of Infected Cells. The normal hosts for P22 infections are derivatives of *Salmonella* LT2. The strain used at M.I.T. for preparation of tailspikes is DB7000, a suppressor⁻ strain. A saturated "overnight" culture is prepared by innoculating 50 ml of broth with cells scraped from a colony on a petri dish or slant using a sterilized platinum loop. This is incubated with aeration or shaking for 8–16 hr until saturated. Such a culture can be used as starter culture for about a week, if kept cold.

For preparation of 5–10 mg of protein, inoculate 1 liter of superbroth with 10 ml of overnight culture and incubate with aeration at 27–28°, until the cell density reaches 4×10^8 cells/ml.

The phage strains used for preparation of protein carry an amber mutation in gene 13, which results in delayed lysis in the su^- host. This increases the protein yield and also the recovery of infected cells. The phage strains also carry an additional amber mutation in gene 5 specifying the coat protein. In the absence of head assembly, the tailspike protein remains as a major soluble product inside the infected cells. In addition, the tailspike is overproduced in such cells, in which DNA is not packaged.

Infect the host cells with an appropriate phage strain at an m.o.i. (multiplicity of infection) of about 7 to ensure that all the cells are infected. Incubate at 27–28° with vigorous aeration. After 3 hr chill the infected cells with continued aeration and then collect them by centrifugation at 4000 g. Cell pellets can be frozen in an ethanol–dry ice bath and stored at −20° until use.

Preparation of Cell Lysates. All the buffers used contained 1 mM sodium azide and 1 mM 2-mercaptoethanol, unless otherwise specified. Thaw the frozen cells in 20 ml of 50 mM Tris–HCl, 2 mM EDTA, 1% (v/v) glycerol, pH 8.0 (buffer M). Saturate the resuspended pellets with chloroform about 1% (v/v) to promote the lysis of infected cells. The sample will be very viscous due to the unpackaged phage DNA. Shake the suspension at room temperature for 20 min to allow lysis to continue. Add magnesium sulfate to 20 mM and DNase and RNase to 20 μg/ml. Shake the lysates further for 30 min and then chill.

Purification. Remove cell debris and chloroform by centrifugation at 4°, at 27,000 *g* for 20 min. The supernatant is then further cleared by ultracentrifugation in an angle head at 4°, at 159,000 *g* for 2 hr.

At this point the tailspike protein should be a major protein in the supernatant. Bring the supernatant to 30% saturation by adding solid ammonium sulfate and store it at 4°. Protein precipitates are collected by centrifugation at 4°, at 27,000 *g* for 15 min, resuspended in 5 ml of M25 (buffer M containing 25 m*M* NaCl), and dialyzed against the same buffer.

The dialysate is applied on a column of DEAE-cellulose (1.6 × 10 cm) equilibrated with M25 at 4°. The column is washed with 20 ml of M25, and then developed with 250 ml of linear gradient of M25-M200 at 4°. The column fractions are monitored by measuring absorbance at 280 nm and analyzed by SDS–gel electrophoresis.

Pool the tailspike-containing peak fractions and add solid ammonium sulfate to 35% saturation at 4°. Collect precipitates, resuspend in M25, pH 7.6, and dialyze against the same buffer. Apply the dialysate to a 10 ml hydroxylapatite column equilibrated with M25, pH 7.6. In this buffer, the tailspike protein does not absorb to the hydroxylapatite and elutes free of contaminants.

This protocol yields only native spikes; partially folded or misfolded molecules aggregate and are fractionated out. In general, the native mutant proteins behave similarly to the wild-type protein during the purification scheme. The yields expected are in the range of 7–15 mg from 1 liter of super broth infection.

Characterization of ts Defects

The most distinctive feature of the *ts* folding mutants is the absence of detectable defects in the protein formed at permissive temperature.[20a] Over 20 mutants have been studied in detail. Though all of these carry the mutant amino acid substitution, they were indistinguishable from the wild-type protein with respect to all its measurable activities: head binding, absorption to the host, and, most dramatically, heat stability. That is, the spikes carrying the *ts* substitutions, once matured at low temperature, are as heat stable as wild type. All of the mutant proteins retained endorhamnosidase activity.

At the restrictive temperature, the situation is just the contrary. Though the mutant polypeptide chains are synthesized at the normal rate, and are not degraded within the cell, they exhibit none of the properties or activities of the same polypeptide chains matured at low temperature. Particularly striking is their inability to react with antibody against the native protein.

[20a] D. P. Goldenberg and J. King, *J. Mol. Biol.* **145**, 633 (1981).

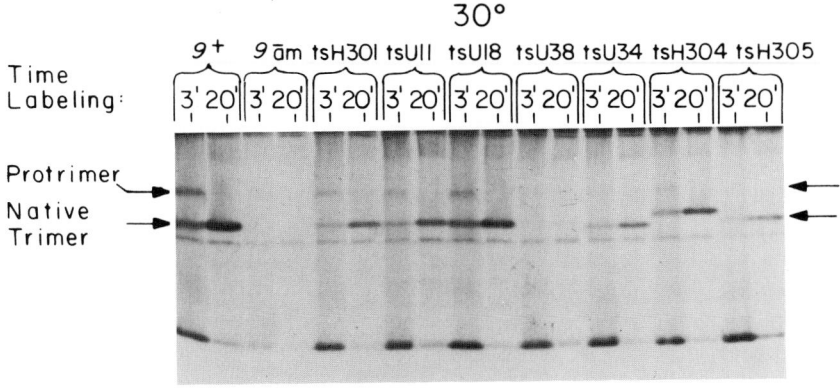

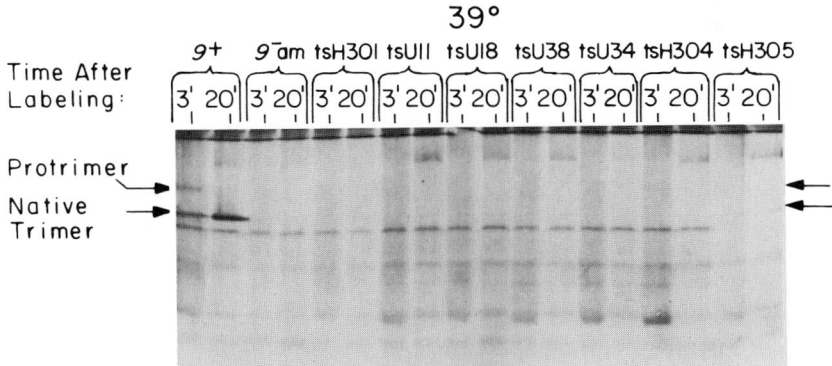

FIG. 3. Native gel analyses with the *ts* mutant-infected cells. They do not accumulate protrimers at restrictive temperatures. In one mutant (H304) the amino acid substitution also caused an electrophoresis mobility change in the mutant protein made at permissive temperature.[21]

Since the protrimer intermediate can be resolved at 39°, we first looked for protrimer formation by these mutants. The native gel analyses with lysates of ^{14}C-labeled amino acid infected cells are shown in Fig. 3. At 30°, all of the mutations characterized directed the formation of protrimer, although the amount varied among different mutants. However, none of the mutants made protrimers at the restrictive temperature. Most of the tailspike gene products made at 39° did not penetrate into polyacrylamide gel and stayed on top of the gel as large aggregates. The result suggests that the mutations block the folding pathway prior to or at protrimer formation.

Reversibility of the Folding Defects

Incorrectly or incompletely folded polypeptide chains could be related to intermediates in the folding pathway, or could be off-pathway dead ends. In the case of the TSF mutants, these can be distinguished by temperature shift-down experiments, in which the fate of chains synthesized at restrictive temperature is followed after a shift to permissive temperature. For many of the gene 9 *ts* mutants, the inactive polypeptide chains can be recovered as active tailspikes after shift-down.[11] These chains must be either intermediates in the folding pathway, or reversibly related to them.

In such shift-down experiments the protrimer appears after the shift to permissive temperature.[21] Thus the chains which accumulate at restrictive temperature represent an earlier intermediate in the pathway. Figure 4 schematically shows the *in vivo* folding pathway of the tailspike and the possible locations of *ts* defects in mutants.

The ts Defects Affect the Thermolabile Step in the Wild-Type Pathway

In order to understand if the rate of tailspike folding was affected by the mutations, the kinetics of tailspike formation have been examined at various temperatures. The folding kinetics of the *ts* mutant tailspike protein can be expressed as the increase in the percentage of mature tailspikes among the total population of gene 9 polypeptide chains. The formation of wild-type tailspikes is itself temperature sensitive, as noted above. This reflects not a decrease in rate but a decrease in the final yield of tailspikes at restrictive temperatures. This indicates the existence of thermosensitive intermediates or steps in the wild-type folding pathway. For four different *ts* mutations examined, at higher temperatures the final yield of the native form of the mutant tailspike decreased compared to the wild-type infection. However, the rates were not affected by the mutations. The result is consistent with the idea that mutant amino acid substitutions further destabilize a folding intermediate at higher temperatures, which already limits the wild-type pathway. Alternatively, there is a competing off-pathway reaction whose rate increases with temperature and by mutant amino acid substitutions.

Interpretation of Sequence Data

The gene[9] sequence was determined by Sauer *et al.*[22] The TSF amino acid substitutions have been identified by sequencing fragments of the

[21] D. P. Goldenberg, D. H. Smith, and J. King, *Proc. Natl. Acad. Sci. U.S.A.* **80**, 7060 (1983).

[22] R. T. Sauer, W. Krovatin, A. R. Poteete, and P. B. Berget, *Biochemistry* **21**, 5811 (1982).

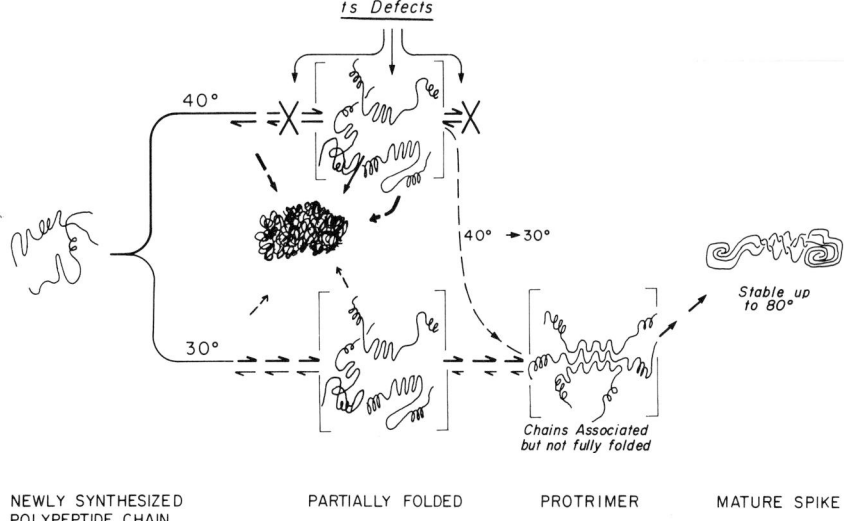

FIG. 4. Tailspike folding and subunit assembly pathway and the effects of *ts* mutations.

mutant genes containing the mutation.[12] Two distinctive classes of sites have threonines and glycines as the wild-type residue at the TSF sites. Three substitutions of threonine were by residues which cannot form a hydrogen bond, two isoleucines and one alanine. This suggests a role for the threonine side chain hydrogen bonds in maintaining the folding pathway at restrictive temperature. Perhaps they stabilize the thermolabile intermediates in the pathway.

Four glycines were replaced by charged bulkier residues, two arginines, one glutamic acid, and one aspartic acid. These substitutions for glycine also cause a change in isoelectric points in the mutant proteins made at permissive temperature. These substitutions probably end up on the surface of the native spike, explaining the ability of the chain to accommodate bulky substitutions for glycines and the mobility change. However, at restrictive temperature the bulky or polar substitutions appear to interfere with a kinetic step in the pathway.

In interpreting the *ts* substitution data, we assumed that the information encoded in the amino acid sequence determines the kinetic steps of the folding pathway, as well as interactions in the final structure. In small proteins the sequence information determining the folding pathway may be nested within residues participating in the activity of the folded protein and it may be difficult to separate out these roles. However in the larger, more complex proteins like tailspikes it is likely that there is some partitioning of these roles along the sequence. We define a local set of residues which control a kinetic step in the folding pathway as a ''foldon.'' One such candidate was identified in the present set of mutant sequences. The local sequences at the two Thr → Ile substitutions are highly homologous. The same substitution in the two sequences causes similar ts folding defects in different regions of the polypeptide chain. The sequences are Try(Trp)-Gln-Pro(Gly)-Thr-Val. Further analysis of ts folding mutants will help identify more candidates for foldons.

Other Systems for the Genetic Analysis of Protein Folding and Stability

A related subject, protein complementation, is covered in a separate chapter in this volume. Another relevant subject, mutations in signal peptides of proteins which enter membranes, has been extensively reviewed by many authors.[23]

Four proteins, in addition to the tailspike protein of P22, have been the subject of a serious attempt to use mutants to study folding processes; these are the α-subunit of tryptophan synthase of *E. coli,* the lysozyme of phage T4, the repressor of phage lambda, and staphylococcal nuclease.

Tryptophan Synthase

The α-subunit of tryptophan synthase is a monomeric protein of molecular weight 28,700 that contains no prosthetic groups. The activity of the subunit can be assayed independently of the whole enzyme molecule. The protein can be reversibly unfolded and refolded by varying the solution conditions. One, and probably two, metastable intermediates have been demonstrated in the transition zone, and both folding and unfolding kinetics have been well characterized.[24,25] Matthews *et al.* have taken the classic set of tryptophan synthase mutants originally isolated by Yanofsky and co-workers[26] and studied their effects on the *in vitro* unfolding and refolding pathway. The original criteria for the isolation of these mutations were the loss of enzymatic activity, and the presence of some native-like properties. However as Matthews *et al.*[27] showed, the replacement of glycine by arginine or glutamic acid at position 211 has substantial effect on the thermal unfolding transition although it has little effect on the stability. Further study showed that the glutamic acid replacement causes

[23] T. J. Silhavy, S. A. Benson, and S. D. Emr, *Microbiol. Rev.* **47,** 313 (1983).

[24] C. R. Matthews and M. M. Crisanti, *Biochemistry* **20,** 784 (1981).

[25] M. M. Crisanti and C. R. Matthews, *Biochemistry* **20,** 2700 (1981).

[26] C. Yanofsky, G. R. Drapeau, J. R. Guest, and B. C. Carlton, *Proc. Natl. Acad. Sci. U.S.A.* **57,** 296 (1967).

[27] C. R. Matthews, M. M. Crisanti, G. L. Gepner, G. Velicelebi, and J. M. Sturtevant, *Biochemistry* **19,** 1290 (1980).

changes in unfolding and refolding rates.[28] The residue 211 appears to affect the folding pathway somewhat independently of the activity, since mutant proteins are not that destabilized. Since the *in vitro* unfolding and refolding pathways are well defined, it will be possible to distinguish between, for example, stability of intermediates and of end products.

T4 Lysozyme

The crystal structure of T4 lysozyme has been determined by Matthews and co-workers.[29] They also characterized the crystal structure of *ts* mutant lysozymes in an effort to detect the structural difference which might cause the *ts* phenotype.[30] However, the structural differences between mutant and wild-type enzyme are very small. In all cases the mutant lysozymes were isolated at permissive temperature under which conditions they are both native and functional. Crystals were obtained at pH 6.7. The mutant lysozymes are more thermolabile than wild type at acidic pH.[31] At pH 7.0, the protein aggregates during denaturation and the careful measurement of melting temperature under these physiological conditions are difficult. Thus it is not clear whether *in vivo* these mutations act by reducing the stability of the native proteins, or they in fact interfere with folding at restrictive high temperature.

Recently Alber and Wozniak,[32] using an assay that detects lysozyme activity on petri dishes, developed a method to screen for thermostable mutant lysozymes. These mutant lysozymes were active at a temperature (55°) where wild-type protein was inactivated. Analysis of such mutants with respect to amino acid substitution and structure will identify interactions that stabilize proteins.

Lambda Repressor

Mutants defective in lambda repressor activity have been characterized in some detail by Sauer and co-workers.[33] Using an efficient scheme of mutagenizing a gene cloned into a plasmid, they isolated repressor mutants which lost the activity of binding lambda operator DNA. Of the

[28] C. R. Matthews, M. M. Crisanti, J. T. Manz, and G. L. Gepner, *Biochemistry* **22,** 1445 (1983).
[29] S. J. Remington, W. Anderson, J. Owen, L. Ten Eyk, C. Grainger, and B. Matthews, *J. Mol. Biol.* **118,** 81 (1978).
[30] M. G. Grutter, L. H. Weaver, T. M. Gray, and B. W. Matthews, *in* "Bacteriophage T4" (C. K. Matthews, E. M. Kutter, G. Mosig, and P. B. Berget, eds.), p. 365. American Society for Microbiology, Washington, D.C., 1983.
[31] R. Hawkes, M. G. Grutter, and J. Schellman, *J. Mol. Biol.* **175,** 195 (1984).
[32] T. Alber and J. A. Wozniak, *Proc. Natl. Acad. Sci. U.S.A.* **82,** 747 (1985).
[33] M. H. Hecht, H. C. M. Nelson, and R. T. Sauer, *Proc. Natl. Acad. Sci. U.S.A.* **80,** 2676 (1983).

mutants characterized 30 or so represent missense mutations in N-terminal domains. Half of the amino acid substitutions are located at the surface of the protein and appear to interfere with DNA binding directly. The other set of mutations is located partially or completely inside of the protein. Originally, they expected that in the second set of mutations mutant polypeptide chains might not fold into their proper three-dimensional structure, resulting in the loss of activity. However, to their surprise, these buried mutants formed native-like structure at the temperature at which they were isolated. Additional studies showed that these mutant proteins have decreased thermostability.[34]

Staphylococcal Nuclease

The gene encoding staphylococcal nuclease has been isolated and cloned by Shortle.[35] The protein was particularly suitable for a genetic analysis because of its known crystal structure and extensive previous folding studies. By applying *in vitro* mutagenesis he was able to obtain a large collection of structure gene mutants. The majority of substitutions occur in regions of the protein away from the active site. They appear to disrupt activity by reducing protein stability.[36]

[34] M. H. Hecht, J. M. Sturtevant, and R. T. Sauer, *Proc. Natl. Acad. Sci. U.S.A.* **81,** 5685 (1984).
[35] D. Shortle, *Gene* **22,** 181 (1983).
[36] D. Shortle and B. Lin, *Genetics* 539 (1985).

[14] Determination and Analysis of Urea and Guanidine Hydrochloride Denaturation Curves

By C. N. Pace

Introduction

The purpose of this chapter is to provide practical information for determining and analyzing urea and guanidine hydrochloride (GdnHCl) denaturation curves of proteins. What may we learn about a protein from such studies? First, it is generally possible to obtain an estimate of the conformational stability, $\Delta G_D^{H_2O}$, of the protein, i.e., how much more stable the globular, native conformation of the protein is than unfolded, denatured conformations. Denaturation curves are especially good for measuring the differences in conformational stability among proteins differing slightly in chemical structure because of differences in amino acid

METHODS IN ENZYMOLOGY, VOL. 131

sequence or alterations resulting from chemical modification. Second, even though equilibrium measurements are made, it is possible to draw deductions about the mechanism of folding of the protein, e.g., it can be shown that unfolding does not follow a two-state mechanism. Third, inferences can be drawn about the structure of the protein, e.g., when the denaturation curve has more than one step, it generally indicates that the protein has more than one domain and it is possible to estimate the relative stabilities of the domains. Finally, unexpected results may provide valuable information about the protein. For example, when the reduction of S–S bonds and unfolding of insulin were found to be irreversible, it suggested that insulin was synthesized in a different form, a finding that was soon confirmed.

Selecting a Technique to Monitor Unfolding

Most of the physical properties change substantially when a protein unfolds. Consequently, many techniques can be used to follow unfolding; those used most often are ultraviolet difference spectroscopy,[1,2] circular dichroism,[3] optical rotation,[3] fluorescence,[4] and NMR.[5] Figure 1 shows the fluorescence emission spectra of native and denatured ribonuclease T_1. The native protein has a much greater intrinsic fluorescence than the denatured protein. In general, the greater the change in the physical property when the protein unfolds, the better the technique is for following unfolding. In this case, a wavelength of 320 nm was selected to follow unfolding because the change in fluorescence intensity is greater (7.3-fold) than it is at 332 nm (5.2-fold) or 354 nm (2.4-fold) which are the wavelength maxima for the native and denatured states, respectively. Based on these results, it appeared that fluorescence might also be useful for following the denaturation of egg white cystatin which like ribonuclease T_1 has a single tryptophan residue and several tyrosine residues. However, for cystatin there is an unusably small increase (1.1-fold) in the fluorescence intensity when the protein unfolds. This suggests that the Trp in cystatin is located on the surface of the molecule exposed to solvent, whereas the Trp in ribonuclease T_1 is known to be buried in the interior of the molecule.[6] This illustrates that the first step in deciding on a technique and optimal experimental conditions for monitoring unfolding should always be to examine the spectra of the native and denatured states.

[1] J. W. Donovan, this series, Vol. 27, p. 497.
[2] T. T. Herskovits, this series, Vol. 11, p. 748.
[3] A. J. Adler, N. J. Greenfield, and G. D. Fasman, this series, Vol. 27, p. 675.
[4] L. Brand and B. Witholt, this series, Vol. 11, p. 776.
[5] K. Wütrich, G. Wagner, R. Richart, and W. Braun, *Biophys. J.* **32**, 549 (1980).
[6] U. Heinemann and W. Saenger, *Nature (London)* **299**, 27 (1982).

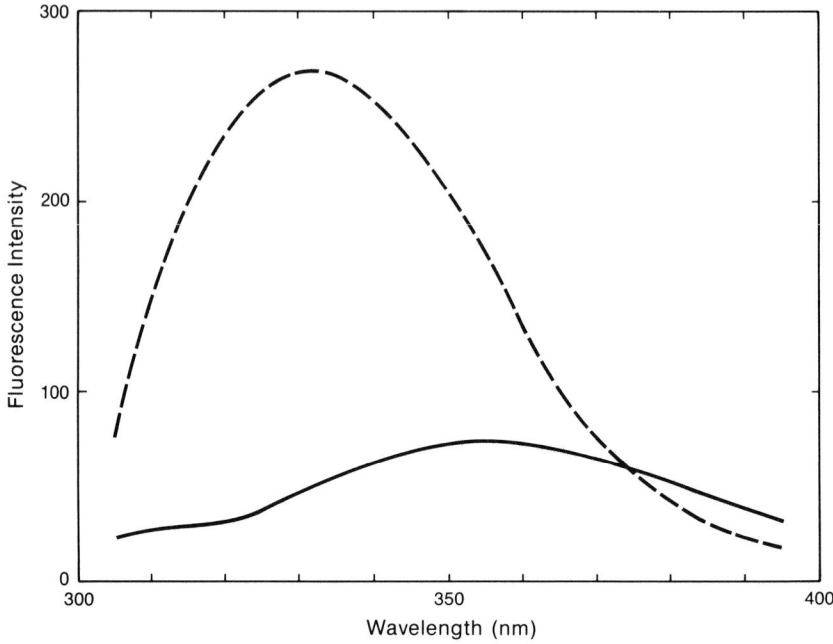

Fig. 1. Fluorescence emission spectra for native (dashed line) and denatured (6.4 M GdnHCl) (solid line) ribonuclease T_1 (excitation wavelength = 278 nm).

The sensitivity of the technique selected becomes an important consideration when limited amounts of protein are available. The urea denaturation curve of ribonuclease T_1 using fluorescence for following unfolding is shown in Fig. 2. Less than 1 mg of protein was used for this curve despite the fact that each point was determined on a separate solution. Unfolding can also be conveniently followed by measuring the optical rotation near 300 nm where there is a 6-fold decrease in the specific rotation on unfolding. However, it would have required about 100 mg of ribonuclease T_1 to determine a denaturation curve similar to that shown in Fig. 2 because the technique is much less sensitive. Optical rotation becomes much more sensitive at lower wavelengths, but the difference in rotation between the native and denatured states is generally diminished. In general, fluorescence is the best technique to minimize the amount of protein used. Optical rotation, circular dichroism, and difference spectroscopy generally require considerably more but comparable amounts of protein. NMR requires even greater amounts of protein, but the technique can provide much more information about the details of unfolding, especially in cases where unfolding is more complex than a simple two-state mechanism.

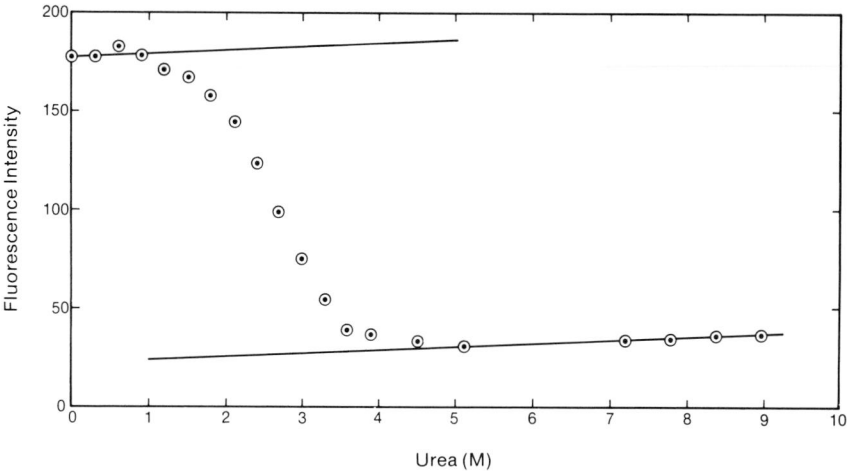

FIG. 2. Urea denaturation curve for ribonuclease T_1 in 0.1 M Tris (pH 8.05), 30°. The fluorescence intensity was measured at 320 nm with excitation at 278 nm.

Urea and Guanidine Hydrochloride Stock Solutions

Both urea and GdnHCl can be purchased commercially in high purity forms. Nevertheless, some batches of GdnHCl contain fluorescent impurities and some batches of urea contain significant amounts of metallic impurities, cyanates, etc.[7] Nozaki[8] has described methods for checking the purity of GdnHCl samples and for recrystallization when it is necessary. GdnHCl stock solutions are stable for months at room temperature. Prakash et al.[7] have described a procedure for purifying urea. The slow formation of cyanate and ammonium ions from the decomposition of urea is a problem which must be considered because cyanate can chemically modify the amino groups of a protein.[9] The equilibrium constant for this reaction is 3.65×10^{-5} M at 25°, pH 7.0, and $I = 0.25$.[10] Consequently, an 8 M urea solution would contain 17 mM cyanate at equilibrium. Taking into account both the rate of formation and decomposition of cyanate in an 8 M urea solution, calculations show that the cyanate concentration will reach 4 mM after 1 week and 7.5 mM after 2 weeks at 25°, pH 7.0, and $I = 0.25$.[10] Stark[9] has investigated the kinetics of the reaction of cyanate with amino groups.

[7] V. Prakash, C. Loucheux, S. Scheufele, M. J. Gorbunoff, and S. N. Timasheff, *Arch. Biochem. Biophys.* **210,** 455 (1981).

[8] Y. Nozaki, this series, Vol. 26, p. 43.

[9] G. R. Stark, *Biochemistry* **4,** 1030 (1965).

[10] P. Hagel, J. J. T. Gerding, W. Fieggen, and H. Bloemendal, *Biochim. Biophys. Acta* **243,** 366 (1971).

Table I summarizes useful information for preparing urea and GdnHCl stock solutions. Urea stock solutions are easily prepared by weight, but GdnHCl is hygroscopic and it is easier to determine the concentration from refractive index measurements after the solution has been prepared.

Experimental Aspects

A typical urea denaturation curve is shown in Fig. 2. Each measurement is generally made on a separate solution, but this is not necessary if the amount of protein available is limited.[11] If care is taken, the experimental solutions can be prepared volumetrically from denaturant and protein stock solutions with sufficient accuracy so that the overall accuracy is limited by the measurement of the physical property used to follow unfolding. Careful measurements in the transition region are crucial, but it is also important that the linear portions of the curve at low and high denaturant concentration be accurately determined since they must be extrapolated into the transition region in order to analyze the denaturation curve (see below). It follows that a minimum of 10 to 15 data points will be required to adequately define the denaturation curve.

It is important to be certain that the reaction is at equilibrium before measurements are made. This may require patience. The time required to reach equilibrium can range from milliseconds to days depending on the protein. Also, the time required may be markedly temperature dependent. With ribonuclease T_1, for example, the time needed to reach equilibrium increases from minutes to hours when the temperature is decreased from 30 to 20°.

It is also important to establish whether unfolding is reversible. This is done by allowing the reaction to reach equilibrium at a denaturant concentration sufficient to promote complete unfolding and then diluting the sample to a lower denaturant concentration where the protein is folded. Care must be taken with any protein containing free sulfhydryl groups because of the possibility of forming disulfide bonds and especially with proteins which contain both sulfhydryl groups and disulfide bonds because disulfide interchange is possible. The rates of both of these reactions decrease as the pH is lowered.[12] With β-lactoglobulin at pH 3, denaturation is found to be almost completely reversible if the sample is diluted immediately after unfolding is complete but the extent of reversibility decreases the longer the protein is exposed to unfolding conditions.[13] The irreversibility has been shown to result from chemical

[11] Y. Saito and A. Wada, *Biopolymers* **22**, 2123 (1983).
[12] T. E. Creighton, *Prog. Biophys. Mol. Biol.* **33**, 231 (1978).
[13] C. N. Pace and C. Tanford, *Biochemistry* **7**, 198 (1968).

TABLE I

UREA AND GUANIDINE HYDROCHLORIDE SOLUTIONS

Property	Urea	GdnHCl
Molecular weight	60.056	95.533
Solubility (25.0°)	10.49 M	8.54 M
$d/d_0{}^a$	$1 + 0.2658W + 0.0330W^2$	$1 + 0.2710W + 0.0330W^2$
Molarity^b	$117.66(\Delta N) + 29.753(\Delta N)^2 + 185.56(\Delta N)^3$	$57.147(\Delta N) + 38.68(\Delta N)^2 - 91.60(\Delta N)^3$
Activity^c	$0.9815(M) - 0.02978(M)^2 + 0.00308(M)^3$	$0.6761(M) - 0.1468(M)^2 + 0.02475(M)^3 - 0.00132(M)^4$
Grams of denaturant per gram of water to prepare		
6 M	0.495	1.009
8 M	0.755	1.816
10 M	1.103	—

^{a} W is the weight fraction denaturant in the solution, d is the density of the solution, and d_0 is the density of water. From K. Kawahara and C. Tanford, *J. Biol. Chem.* **241**, 3228 (1966).

^{b} ΔN is the difference between the refractive index of the denaturant solution and water at the sodium D line. The equation for urea solutions is based on the data of J. R. Warren and J. A. Gordon, *J. Phys. Chem.* **70**, 297 (1966); and the equation for GdnHCl solution is from Y. Nozaki, this series, Vol. 26, p. 43.

^{c} The equation for urea is based on the data of V. E. Bower and R. A. Robinson, *J. Phys. Chem.* **67**, 1524 (1963). The equation for GdnHCl gives the mean ion activity and fits the unpublished data of E. P. K. Hade above 0.5 M with an average deviation of 0.007 and the data of O. D. Bonner, *J. Chem. Thermodyn.* **8**, 1167 (1976) at 0.5 M and below with an average deviation of 0.006.

changes involving the free sulfhydryl group and the two disulfide bonds.[14] Another frequent cause of irreversibility is insolubility or aggregation of the unfolded protein. This is more of a problem with other types of denaturation than with urea or GdnHCl denaturation. For example, McCoy and Wong[15] have shown that the SDS-, acid-, or thermally-induced unfolding of carbonate dehydratase B cannot be reversed directly, but can be reversed by first transfering the protein to a 6 M GdnHCl solution.

Denaturation Curve Analysis

Figure 2 is a typical solvent denaturation curve. In this case fluorescence intensity was used for following unfolding, but any observable parameter y can be used. Many globular proteins have been found to closely approach a two-state mechanism

$$N \rightleftharpoons D \tag{1}$$

in which only the native state, N, and the denatured state, D, are present at significant concentrations in the transition region. Values of y characteristic of the native state, y_N, and of the denatured state, y_D, can be obtained in the transition region by extrapolation from the linear portions of the denaturation curve at low and high denaturant concentrations. These linear changes generally result from solvent effects on the properties of the native and denatured states. Sometimes they can be quite large, especially when difference spectroscopy is used to follow unfolding. (In this case, the linear portions become solvent perturbation studies of the native and denatured states and can be used to quantitate the number of exposed Trp and Tyr residues.[1]) For a two-state mechanism, $f_N + f_D = 1$, and $y = y_N f_N + y_D f_D$ where f_N and f_D represent the fraction of the protein present in the native and denatured states, respectively. Combining these equations, $f_D = (y - y_N)/(y_D - y_N)$ and $f_N = (y_D - y)/(y_D - y_N)$. Thus, an equilibrium constant, K_D, and a free energy of unfolding, ΔG_D, can be calculated using

$$K_D = e^{-\Delta G_D/RT} = f_D/f_N = (y - y_N)/(y_D - y) \tag{2}$$

The data from Fig. 2 have been analyzed in this way and the results are shown in Fig. 3.

For many proteins the denaturation curve shows a single step (Fig. 1) and unfolding has been found to closely approach a two-state mechanism. In other cases, the mechanism has been shown to be more complex than a two-state mechanism even though a single step denaturation curve is

[14] H. A. McKenzie, G. B. Ralston, and D. C. Shaw, *Biochemistry* **11**, 4539 (1972).
[15] L. F. McCoy and K. P. Wong, *Biochemistry* **20**, 3062 (1981).

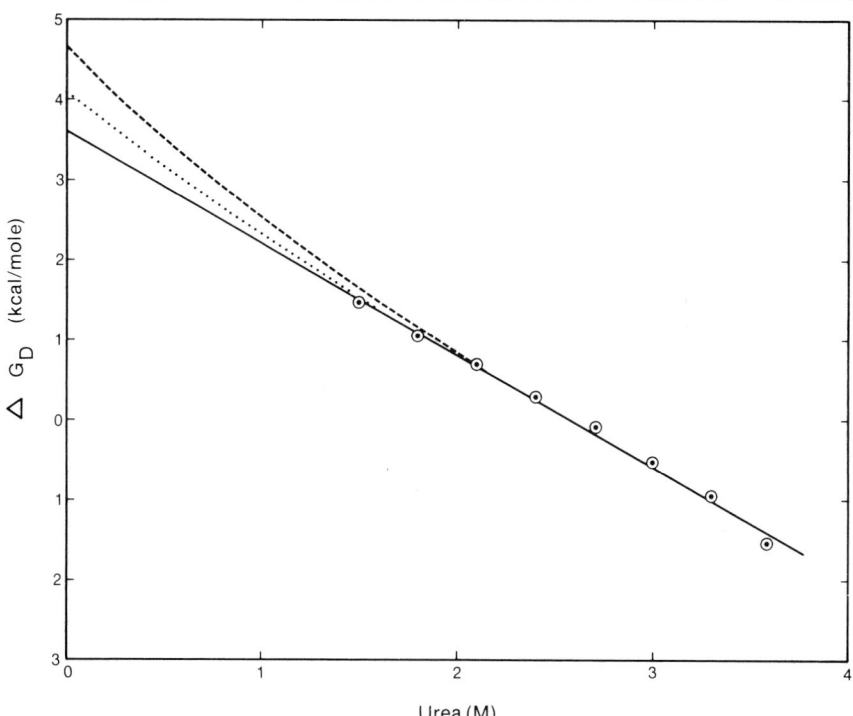

FIG. 3. ΔG_D as a function of urea molarity. ΔG_D was calculated from the data in Fig. 2 using Eq. (2). The solid line represents Eq. (7) with $m = 1400$ cal/mol and $\Delta G_D^{H_2O} = 3.6$ kcal/mol. The dotted line represents Eq. (6) with $k = 0.1$, $\Delta n = 32$, and $\Delta G_D^{H_2O} = 4.1$ kcal/mol. The dashed line represents Eq. (5) with $\bar\alpha = 0.34$ and $\Delta G_D^{H_2O} = 4.7$ kcal/mol.

observed, e.g., carbonate dehydratase B.[16] In still other cases, more than a single step is observed in the denaturation curve and the mechanism must be more complex than a two-state mechanism, e.g., tryptophan synthase.[17,18] With immunoglobulin light chains, both one- and two-step denaturation curves have been observed depending on the relative stabilities of the constant and variable domains.[19,20] Saito and Wada[11] have followed the GdnHCl denaturation of 17 globular proteins using four different spectroscopic techniques to follow unfolding and use the results to classify each of the proteins into one of these three categories.

[16] K. P. Wong and C. Tanford, *J. Biol. Chem.* **248**, 8518 (1973).
[17] K. Yutani, K. Ogasahara, M. Suzuki, and Y. Sugino, *J. Biochem.* **85**, 915 (1979).
[18] C. R. Matthews and M. M. Crisanti, *Biochemistry* **20**, 784 (1981).
[19] T. Azuma, K. Hamaguchi, and S. Migita, *J. Biochem.* **72**, 1457 (1972).
[20] E. S. Rowe and C. Tanford, *Biochemistry* **12**, 4822 (1973).

Even for the more complex mechanisms, it is generally possible to obtain quantitative estimates of the conformational stability from an analysis of the solvent denaturation curve. When stable intermediate states, X_i, each characterized by the property y_i and concentration f_i are present, the observed extent of unfolding, $f_{obs} = (y - y_N)/(y_D - y_N)$, becomes $f_{obs} = f_D + \Sigma_i f_i z_i$ where $z_i = (y_i - y_N)/(y_D - y_N)$. Thus, f_{obs} will differ from f_D by an amount that depends on the concentration of the intermediates weighted by their z_i values. The z_i value for an intermediate is the fractional change in y in going from N to X_i and is likely to be between 0 and 1, because y_i will generally fall between y_N and y_D. For some techniques, such as optical rotation measurements above 300 nm or viscosity, the z_i value will be roughly proportional to the extent to which the intermediate is unfolded. For other techniques, such as enzyme activity measurements or ultraviolet difference spectroscopy, which are dependent on only a small number of residues, there may be little correlation between z_i and the extent of unfolding.

The relationship between K_{app} [$= f_{obs}/(1 - f_{obs})$] and K_D [$= f_D/(1 - f_D)$] is

$$K_{app} = K_D \frac{1 + \sum_i z_i K_i/K_D}{1 + \sum_i (1 + z_i)K_i} \tag{3}$$

where $K_i = f_i/f_N$. For reasonable values of z_i ($z_i = 0.1$ to 0.9), K_D and K_{app} will be equal somewhere near the midpoint of the transition ($K_{app} = 1$), but K_D will be less than K_{app} below the midpoint and greater than K_{app} above the midpoint of the transition. Thus, the presence of intermediates will cause the slope of plots such as that in Fig. 3 to be less than the slope would be for a plot of ΔG_D versus denaturant concentration. This means that estimates of $\Delta G_D^{H_2O}$ derived from a two-state analysis of a system with stable intermediates will be smaller than the true value of $\Delta G_D^{H_2O}$. As an example,[18] a two-state analysis of the unfolding of the α-subunit of tryptophan synthase led to $\Delta G_D^{H_2O} = 3.6$ kcal/mol, but when the intermediate is taken into account, $\Delta G_D^{H_2O} = 11.0$ kcal/mol. In most cases where intermediate states are present, it has been possible to estimate or determine y_i so that a complete analysis of the system was possible.[16-19]

Estimating the Conformational Stability, $\Delta G_D^{H_2O}$

Three methods are currently used to analyze data such as those shown in Fig. 3 in order to estimate ΔG_D at zero concentration of denaturant, $\Delta G_D^{H_2O}$. These are described briefly below and in more detail elsewhere.[21]

---

[21] C. N. Pace, *Crit. Rev. Biochem.* **3**, 1 (1975).

Tanford's Model

The aqueous solubility of almost all of the constituent parts of a protein increases with increasing GdnHCl or urea concentration. This explains the ability of these compounds to unfold proteins even though it is not clear how GdnHCl and urea exert their solubilizing effect. Tanford[22] has shown that the GdnHCl and urea denaturation of proteins can be described quantitatively using free energies of transfer derived from these solubility studies on amino acids and peptides. His approach provides a means of estimating $\Delta G_D^{H_2O}$ and has also proven useful for other purposes.[21] The basic relationship is

$$\Delta G_D = \Delta G_D^{H_2O} + \sum_i \alpha_i n_i \delta g_{tr,i} \qquad (4)$$

where $\delta g_{tr,i}$ is the free energy for transfer of a group of type i from H_2O to denaturant, n_i is the total number of groups of type i present in the protein, and α_i is the average fractional change in the degree of exposure of groups of type i when the protein unfolds.[22] The δg_{tr} values for a peptide group and the 14 amino acid side chains which will generally make the largest contribution to ΔG_D are given in Table II. The n_i values are readily available for most proteins. Assigning an α_i value for a given type of residue is more difficult. If the three-dimensional structure of a protein is known, methods have been developed to estimate which groups are accessible to solvent in the folded conformation, and, with less certainty, when the protein is unfolded.[23] However, this approach is rarely used in analyzing denaturation curves. Instead, a single average value of α_i, $\bar{\alpha}$, is used for all of the groups in Table II,[21] or a subset of these groups,[22] and Eq. (4) becomes

$$\Delta G_D = \Delta G_D^{H_2O} + \bar{\alpha} \sum_i n_i \delta g_{tr,i} \qquad (5)$$

The value of $\bar{\alpha}$ is chosen so that the dependence of ΔG_D on denaturant concentration calculated with Eq. (5) equals the experimentally determined value, i.e., the m value of Eq. (7) below. At the midpoint of the denaturation curve, $\Delta G_D = 0$ and Eq. (5) becomes $\Delta G_D^{H_2O} = \sum_i \bar{\alpha} n_i \delta g_{tr,i}$. Thus, evaluating this equation at the denaturant concentration at the midpoint of the transition yields an estimate of $\Delta G_D^{H_2O}$.

The values of $\bar{\alpha}$ resulting from an analysis of the GdnHCl and urea denaturation curves of the same protein are in good agreement.[24] This is

[22] C. Tanford, *Adv. Protein Chem.* **24**, 1 (1970).

[23] F. M. Richards, *Annu. Rev. Biophys. Bioeng.* **6**, 151 (1977).

[24] R. F. Greene and C. N. Pace, *J. Biol. Chem.* **249**, 5388 (1974).

TABLE II

FREE ENERGIES OF TRANSFER FROM WATER TO AQUEOUS UREA OR GdnHCl SOLUTIONS
FOR A PEPTIDE GROUP AND SOME AMINO ACID SIDE CHAINS

| | $-\delta g_{tr}$ (cal/mol) | | | | | | | |
| | (Urea)[a] | | | | (GdnHCl)[b] | | | |
Group	2 M	4 M	6 M	8 M	1 M	2 M	4 M	6 M
Peptide[c,d]	49	86	118	130	83	134	207	245
Ala	0	−15	−10	−10	10	20	30	45
Val[e]	60	85	125	160	85	115	195	265
Leu	110	155	225	295	150	210	355	480
Ile[e]	100	140	205	265	135	190	320	430
Met	115	225	325	415	150	245	400	535
½-Cys[e]	115	225	325	415	150	245	400	535
Phe	180	330	470	600	215	355	580	775
Tyr	225	395	580	735	235	385	605	770
Trp	270	505	730	920	400	630	980	1235
Pro[e]	75	105	155	200	100	140	240	320
Thr	40	60	90	115	65	90	120	125
His	100	160	205	255	180	285	385	420
Asn	135	225	330	430	200	320	490	645
Gln	80	130	190	230	135	215	315	360

[a] Y. Nozaki and C. Tanford, *J. Biol. Chem.* **238,** 4074 (1963).
[b] Y. Nozaki and C. Tanford, *J. Biol. Chem.* **245,** 1648 (1970).
[c] Based on *N*-acetyltetraglycine ethyl ester and ethyl acetate. D. R. Robinson and W. P. Jencks, *J. Am. Chem. Soc.* **87,** 2462 (1965).
[d] More recent measurements which take into account the activity coefficients for the model compounds lead to δg_{tr} = 42 and 71 cal/mol for transfer of a peptide group to 2 and 4 M urea, respectively. H. Schönert and L. Stroth, *Biopolymers* **20,** 817 (1981).
[e] The $\delta g_{tr,i}$ values for these side chains are estimates based on results for the other side chains and on the results at a single denaturant concentration by D. B. Wetlaufer, S. K. Malik, L. Stoller, and R. L. Coffin, *J. Am. Chem. Soc.* **86,** 508 (1964).

expected since GdnHCl and urea are thought to produce complete unfolding of the polypeptide chain.[25] In contrast, the values of $\Delta G_D^{H_2O}$ derived from GdnHCl data are consistently 20 to 50% higher than the values from urea.[21] This disagreement is puzzling and deserves further study.

Denaturant Binding Model

It is possible to account for many of the negative δg_{tr} values in Table II in terms of a denaturant binding mechanism, i.e., by assuming that GdnHCl and urea molecules bind to the peptide group or an amino acid

---

[25] C. Tanford, *Adv. Protein Chem.* **23,** 121 (1968).

TABLE III
ANALYSIS OF FREE ENERGY OF TRANSFER DATA IN TERMS OF DENATURANT BINDING

	k for binding of GdnHCl or urea[a]							
Solution (M)	Trp[b]	Tyr	Phe	Two peptide groups[c]	Met	Leu	His	Asn
GdnHCl								
1	0.76	0.92	0.83	0.61	0.54	0.54	0.67	0.76
2	0.75	0.97	0.87	0.61	0.54	0.45	0.66	0.76
4	0.80	1.11	1.04	0.63	0.60	0.51	0.57	0.80
6	0.78	1.10	1.11	0.53	0.61	0.52	0.43	0.81
Average	0.78	1.03	0.96	0.60	0.57	0.51	0.58	0.78
Urea								
2	0.14	0.25	0.19	0.10	0.12	0.11	0.09	0.13
4	0.15	0.26	0.20	0.09	0.13	0.08	0.08	0.13
6	0.16	0.30	0.22	0.09	0.13	0.08	0.08	0.14
8	0.16	0.33	0.23	0.07	0.13	0.09	0.07	0.14
Average	0.15	0.29	0.21	0.09	0.13	0.09	0.08	0.13

[a] Calculated from the data in Table II using $\delta g_{tr} = RT \ln(1 + ka)$.
[b] Calculated assuming two binding sites per Trp side chain.
[c] D. R. Robinson and W. P. Jencks [*J. Am. Chem. Soc.* **87**, 2462 (1965)] suggest that each GdnHCl or urea molecule binds to two peptide groups.

side chain.[22,26] This is illustrated in Table III where binding constants, k, have been calculated from some of the δg_{tr} values in Table II using $\delta g_{tr} = RT \ln(1 + ka)$, where a is the activity of the denaturant. Timasheff's group investigated the interaction of GdnHCl[27] and urea[7] with a number of proteins and found that proteins interact preferentially with denaturants relative to water. For both denaturants there is a good correlation between the number of denaturant molecules bound and a summation of one-half the number of peptide bonds plus the number of aromatic amino acid residues.[7,27] The interaction between urea or GdnHCl molecules, and peptide bonds has been investigated by Jencks and co-workers.[28] These results suggest that it might be reasonable to treat protein unfolding in terms of a denaturant binding model.

If it is assumed that unfolding results because there are a greater number of identical, noninteracting binding sites for denaturant molecules

[26] C. N. Pace and K. E. Vanderburg, *Biochemistry* **18**, 288 (1979).
[27] J. C. Lee and S. N. Timasheff, *Biochemistry* **13**, 257 (1974).
[28] D. R. Robinson and W. P. Jencks, *J. Am. Chem. Soc.* **87**, 2462 (1965); M. Roseman and W. P. Jencks, *J. Am. Chem. Soc.* **97**, 631 (1975).

on the unfolded conformation than there are on the globular conformation, then it follows that

$$\Delta G_D = \Delta G_D^{H_2O} - \Delta nRT \ln(1 + ka) \qquad (6)$$

where Δn is the difference in the number of denaturant binding sites, k is the equilibrium constant for binding at each site, and a is the activity of the denaturant.[22] (Equations to calculate the molarity-based activities for urea and GdnHCl from their molar concentrations are given in Table I.) Estimates of $\Delta G_D^{H_2O}$, Δn, and k are obtained by fitting data such as those in Fig. 3 to Eq. (6). This method was first used[29] to analyze the GdnHCl denaturation of lysozyme and led to a value of $k = 1.2$. This value was subsequently used to analyze GdnHCl denaturation curves in many papers. The k values in Table III range from 0.5 to 1.0 for GdnHCl and from 0.08 to 0.3 for urea. This and several other lines of evidence indicate that a k value smaller than 1.2 should be used in Eq. (6) to analyze GdnHCl denaturation curves.[26] When a protein unfolds, many more peptide groups will be freshly exposed to solvent than any single type of side chain. Consequently, the k values for binding of GdnHCl ($k = 0.6$) or urea ($k = 0.1$) to a dipeptide unit probably provide a reasonable average value to use in Eq. (6) for analyzing denaturation curves. The fit of Eq. (6) to data such as those in Fig. 3 does not depend significantly on the value of k over the range of values given in Table III.[18,21,30] For GdnHCl or urea denaturation, the estimates of $\Delta G_D^{H_2O}$ from the denaturant binding model are generally in good agreement with estimates based on Tanford's model.[21,26] However, as with Tanford's model, the estimates based on GdnHCl studies are substantially greater than those based on urea.[21]

Linear Extrapolation

The simplest method of estimating $\Delta G_D^{H_2O}$ is to assume that the linear dependence of ΔG_D on denaturant concentration observed in the transition region continues to zero concentration and fit the data to an equation of the form

$$\Delta G_D = \Delta G_D^{H_2O} - m \text{ (denaturant)} \qquad (7)$$

This method was proposed originally[24] because the estimates of $\Delta G_D^{H_2O}$ from urea and GdnHCl denaturation were in reasonable agreement[24,31] and it was not clear that either of the more complicated methods described above gave a more reliable estimate of $\Delta G_D^{H_2O}$. Schellman[32] has

[29] K. Aune and C. Tanford, *Biochemistry* **8**, 4586 (1969).
[30] F. Ahmad and P. McPhie, *Biochemistry* **17**, 241 (1978).
[31] F. Ahmad and C. C. Bigelow, *J. Biol. Chem.* **257**, 12935 (1982).
[32] J. A. Schellman, *Biopolymers* **17**, 1305 (1978).

since suggested that a linear dependence of ΔG_D on denaturant concentration may be expected on theoretical grounds. This approach always gives the lowest estimate of $\Delta G_D^{H_2O}$.

There is frequently an interest in determining the difference in conformational stability between two proteins which differ only slightly in chemical structure. These may be proteins which differ in amino acid sequence[33] or by minor changes achieved through chemical modification.[34] In these cases, the denaturation curves are generally quite similar but the midpoints, $D_{1/2}$, are shifted to higher or lower denaturant concentrations. The difference in conformational stability can be estimated by multiplying the difference between the $D_{1/2}$ values by the m value from Eq. (7). This gives the difference in stability in the presence of denaturant, but avoids the uncertainties inherent in estimating $\Delta G_D^{H_2O}$ values.[33,34]

It is not clear at present which of the three methods described above leads to the best estimate of $\Delta G_D^{H_2O}$. It can be seen in Fig. 3 that for urea denaturation the estimates of $\Delta G_D^{H_2O}$ for ribonuclease T_1 range from 3.6 kcal/mol (Eq. 7) to 4.1 kcal/mol [Eq. (6) with $k = 0.1$] to 4.7 kcal/mol [Eq. (5)]. This is typical of what is found when the urea denaturation curves of other proteins are analyzed by these three methods: the agreement is reasonably good but linear extrapolation always leads to the lowest estimate of $\Delta G_D^{H_2O}$. For GdnHCl, the difference is more pronounced. Using the denaturant binding model with $k = 0.6$ or Tanford's model leads to similar $\Delta G_D^{H_2O}$ values which are 35 to 75% higher than the values obtained by linear extrapolation.[21] A comparison of the estimates of $\Delta G_D^{H_2O}$ from GdnHCl and urea studies with those from calorimetric studies of the thermal denaturation does not resolve this problem. For α-chymotrypsin and lysozyme, GdnHCl results analyzed by the denaturant binding model or Tanford's model give the best agreement.[35,36] For cytochrome c[35] and phage lysozyme,[32] linear extrapolation gives the best agreement.

The values of $\Delta G_D^{H_2O}$ obtained from an analysis of the urea and GdnHCl denaturation curves for the same protein should generally be similar. However, only when linear extrapolation is used are the results in good agreement.[21,31] For this and other reasons, I suggest, for the present, that linear extrapolation be used to estimate $\Delta G_D^{H_2O}$ from urea and GdnHCl denaturation curves. However, it is essential to keep in mind that these estimates may well prove to be too low.

[33] J. A. Knapp and C. N. Pace, *Biochemistry* **13**, 1289 (1974).
[34] J. F. Cupo and C. N. Pace, *Biochemistry* **22**, 2654 (1983).
[35] P. L. Privalov, *Adv. Protein Chem.* **33**, 167 (1979).
[36] W. Pfeil, *Mol. Cell. Biochem.* **40**, 3 (1981).

Concluding Remarks

Urea and GdnHCl are used for a variety of purposes by protein chemists. They offer several advantages over other means of unfolding a protein such as acid, heat, or detergents. First, the product is better defined because the degree of unfolding is maximized. Proteins in 8 M urea or 6 M GdnHCl with their disulfide bonds broken approach a randomly coiled conformation.[26] Second, unfolding is more likely to approach a two-state mechanism. And, third, denaturation is more likely to be completely reversible. These features are especially important in attempting to estimate the conformational stability of a protein.

GdnHCl is generally 1.5 to 2.5 times more effective as a protein denaturant than urea, with the difference being greater for more polar proteins.[24] Several more potent denaturants with structures related to GdnHCl and urea are known.[37,38] Some of these compounds will probably prove more useful than GdnHCl and urea in certain applications. Guanidine thiocyanate which is 2.5 to 3.5 times more effective than GdnHCl as a protein denaturant is a good example.[31]

[37] C. N. Pace and H. F. Marshall, *Arch. Biochem. Biophys.* **199,** 270 (1980).
[38] C. Mitchinson, R. H. Pain, J. R. Vinson, and T. Walker, *Biochim. Biophys. Acta* **743,** 31 (1983).

Section II

Structural Dynamics and Mobility of Proteins

[15] Internal Dynamics of Proteins

By MARTIN KARPLUS

Introduction

Globular proteins have a wide variety of internal motions. These can be classified, for convenience, in terms of their amplitude, energy, and time scale or by their structural type. The table lists the ranges involved for these quantities. One expects that an increase in one quantity (e.g., the amplitude of the fluctuation) will correspond to an increase in the others (e.g., a larger energy and a longer time scale). This is true in many cases, but not always. There are motions that are slow because they are complex, involving the correlated displacements of many atoms. An example might be partial to total unfolding transitions. In such a case, the correlation of amplitude, energy, and time scale is expected to hold. However, there are much more localized events, often involving small displacements of a few atoms, in which the motion is slow because there is a high activation barrier; an example is provided by the aromatic ring flips in certain proteins.[1,2] It is important to note that in this case the macroscopic rate can be very slow ($k \sim 1 \ sec^{-1}$ at 300 K) not because an individual event is slow (a ring flip occurs in $\sim 10^{-12}$ sec), but because the probability is very small ($p \sim 10^{-12}$) that a ring has sufficient energy to get over an activation barrier that is on the order of 16 kcal.

At any given time, a typical protein exhibits a wide variety of motions; the motions range from irregular elastic deformations of the whole protein driven by collisions with solvent molecules to chaotic librations of interior groups driven by random collisions with neighboring atoms in the protein. Considering only typical motions at physiological temperatures, the smallest effective dynamic units in proteins are those that behave nearly as rigid bodies due to their internal covalent bonding. Examples include the phenyl group in the side chain of phenylalanine, the isopropyl group in the side chains of valine or leucine, and the amide groups of the protein backbone. Except for the methyl rotations in the isopropyl group, these units display only relatively small internal motions due to the high energy cost associated with deformations of bond lengths, bond angles, or dihedral angles about multiple bonds. The important motions in proteins involve relative displacements of such groups associated with torsional

---

[1] G. H. Snyder, R. Rowan, S. Karplus, and B. D. Sykes, *Biochemistry* **14**, 3765 (1975).
[2] B. R. Gelin, and M. Karplus, *Proc. Natl. Acad. Sci. U.S.A.* **72**, 2002 (1975).

METHODS IN ENZYMOLOGY, VOL. 131

CLASSIFICATION OF INTERNAL MOTIONS OF GLOBULAR PROTEINS

Scales of motions (300 K)	
Amplitude	0.01 to 100 Å
Energy	0.1 to 100 kcal
Time	10^{-15} to 10^3 sec
Types of motions	
Local	Atom fluctuations, side-chain oscillations, loop and "arm" displacements
Rigid body	Helices, domains, subunits
Large-scale	Opening fluctuation, folding and unfolding
Collective	Elastic-body modes, coupled atom fluctuations, nonlinear motional contributions

motions about the rotationally permissive single bonds that link the groups together. High frequency vibrations occur within the local group, but these are not of primary importance in the relative displacements.

Most groups in a protein are tightly encaged by atoms of the protein or of the surrounding solvent. At very short times ($\leq 10^{-13}$ sec), such a group may display a rattling motion in its cage, but such motions are of relatively small amplitude (≤ 0.2 Å). More substantial displacements of the group occur over longer time intervals; these displacements involve concomitant displacements of the cage atoms. Broadly speaking, such "collective" motions may have a local or rigid-body character. The former involves changes of the cage structure and relative displacements of neighboring groups, while the latter involves relative displacements of different regions of the protein but only small changes on a local scale.

The presence of such motional freedom implies that a native protein at room temperature samples a range of conformations. Most of these are in the general neighborhood of the average structure, but at any given moment an individual protein molecule is likely to differ significantly from the average structure. This in no way implies that the X-ray structure, which corresponds to the average in the crystal, is not important. Rather, it suggests that fluctuations about that average are likely to play a role in protein function. In a protein (as in any polymeric system in which rigidity is not supplied by cross-linking due to covalent bonds) significant fluctuations cannot be avoided; they are likely, therefore, to have been taken into account in evolutionary development.

Although the existence of the fluctuations is now well established, our understanding of their biological role in specific areas is rather limited. Both conformational and energy fluctuations with local to global character are expected to be important. In a protein, as in other nonrigid condensed systems, structural changes arise from correlated fluctuations.

Perturbations, such as ligand binding, that produce tertiary or quaternary alterations do so by introducing forces that bias the fluctuations in such a way that the protein makes a transition from one structure to another. Alternatively, the fluctuations can be regarded as searching out the path or paths along which the transition takes place.

Methodology

To study theoretically the dynamics of a macromolecular system, it is essential to have a knowledge of the potential energy surface, the energy of the system as a function of the atomic coordinates. The potential energy can be used directly to determine the relative stabilities of the different possible structures of the system; the relative populations of such structures under conditions of thermal equilibrium are given in terms of the potential energy by the familiar Boltzmann distribution law.[3] The mechanical forces acting on the atoms of the systems are simply related to the first derivatives of the potential with respect to the atom positions. These forces can be used to calculate dynamic properties of the system, e.g., by solving Newton's equations of motion to determine how the atomic positions change with time.[3,4] From the second derivatives of the potential surface, the force constants for small displacements can be evaluated and these can be used to find the normal modes[5]; this serves as the basis for an alternative approach to the dynamics in the harmonic limit.[5-7]

Although quantum mechanical calculations can provide potential surfaces for small molecules, empirical energy functions of the molecular mechanics type are the only possible source of such information for proteins and their solvent surroundings. Since most of the motions that occur at ordinary temperatures leave the bond lengths and bond angles of the polypeptide chains near their equilibrium values, which appear not to vary significantly throughout the protein (e.g., the standard dimensions of the peptide group first proposed by Pauling in 1951),[8] the energy function representation of the bonding can be hoped to have an accuracy on the order of that achieved in the vibrational analysis of small molecules. Where globular proteins differ from small molecules is that the contacts among nonbonded atoms play an essential role in the potential energy of

[3] D. A. McQuarrie, "Statistical Mechanics." Harper, New York, 1976.
[4] J. P. Hansen and I. R. McDonald, "Theory of Simple Liquids." Academic Press, New York, 1976.
[5] R. M. Levy and M. Karplus, *Biopolymers* **18**, 2465 (1979).
[6] N. Go, T. Noguti and T. Nishikawa, *Proc. Natl. Acad. Sci. U.S.A.* **80**, 3696 (1983).
[7] B. Brooks and M. Karplus, *Proc. Natl. Acad. Sci. U.S.A.* **80**, 6571 (1983).
[8] L. Pauling, R. R. Corey, and H. C. Branson, *Proc. Natl. Acad. Sci. U.S.A.* **37**, 205 (1951).

the folded or native structure. From the success of the pioneering conformational studies of Ramachandran and co-workers in 1963[9] that made use of hardsphere nonbonded radii, it is likely that relatively simple functions (Lennard–Jones potentials supplemented by special hydrogen-bonding terms and electrostatic interactions) can adequately describe the interactions involved.

The energy function used for proteins are generally composed of terms representing bonds, bond angles, torsional angles, van der Waals interactions, electrostatic interactions, and hydrogen bonds. The resulting expression has the form[10]

$$E(\mathbf{R}) = \frac{1}{2} \sum_{\text{bonds}} K_b(b - b_0)^2 + \frac{1}{2} \sum_{\substack{\text{bond} \\ \text{angles}}} K_\theta(\theta - \theta_0)^2$$

$$+ \frac{1}{2} \sum_{\text{torsional}} K_\phi[1 + \cos(n\phi - \delta)]$$

$$+ \sum_{\substack{\text{nb pairs} \\ r < 8 \text{ Å}}} \left(\frac{A}{r^{12}} - \frac{C}{r^6} + \frac{q_1 q_2}{Dr} \right) + \sum_{\substack{\text{H} \\ \text{bonds}}} \left(\frac{A'}{r^{12}} - \frac{C'}{r^{10}} \right) \qquad (1)$$

The energy is a function of the Cartesian coordinate set, $\mathbf{R}$, specifying the positions of all the atoms involved, but the calculation is carried out by first evaluating the internal coordinates for bonds (b), bond angles (θ), dihedral angles (ϕ), and interparticle distances (r) for any given geometry, $\mathbf{R}$, and using them to evaluate the contributions to Eq. (1), which depends on the force constants K_b, K_θ, K_ϕ, Lennard–Jones parameters A and C, atomic charges q_i, dielectric constant D, hydrogen-bond parameters A' and C', and geometrical reference values b_0, θ_0, n, and δ. For most protein atoms an extended atom representation has been used; i.e., one extended atom replaces a nonhydrogen atom and any hydrogens bonded to it. However, although the earliest studies employed the extended atom representation for all hydrogens, present calculations treat hydrogen-bonding hydrogens explicitly and generally use a more accurate function to represent hydrogen bonding interactions (e.g., angular terms are included) than that given in Eq. (1).[10]

Given a potential energy function, one may take any of a variety of approaches to study protein dynamics. The most exact and detailed information is provided by molecular dynamics simulations, in which one uses a computer to solve the Newtonian equations of motion for the atoms of

[9] G. N. Ramachandran, C. Ramakrishnan, and V. Sasisekharan, *J. Mol. Biol.* **7**, 95 (1963).

[10] B. Brooks, R. E. Bruccoleri, B. D. Olafson, D. J. States, S. Swaminathan, and M. Karplus, *J. Comp. Chem.* **4**, 187 (1983).

the protein and any surrounding solvent.[11-13] With currently available computers, it is possible to simulate the dynamics of small proteins for periods of up to a few hundred picoseconds. Such periods are long enough to characterize completely the librations of small groups in the protein and to determine the dominant contributions to the atomic fluctuations. To study slower and more complex processes in proteins, it is generally necessary to use other than the straightforward molecular dynamics simulation method. A variety of dynamic approaches, such as stochastic dynamics,[14] harmonic dynamics,[5-7] and activated dynamics,[15] can be introduced to study particular problems.

Molecular Dynamics

Since the molecular dynamics simulation has so far provided the most detailed and interesting results on protein motions, we describe the methodology in some detail. To begin a dynamic simulation, one must have an initial set of atomic coordinates and velocities. These are obtained from the X-ray coordinates of the protein by a preliminary calculation that serves to equilibrate the system.[10] The X-ray structure is first refined using an energy minimization algorithm to relieve local stresses due to nonbonded atomic overlaps, bond length distortions, etc. The protein atoms are then assigned velocities at random from a Maxwellian distribution corresponding to a low temperature, and a dynamic simulation is performed for a period of a few picoseconds. The equilibration is continued by alternating new velocity assignments, chosen from Maxwellian distributions corresponding to successively increased temperatures, with similar intervals of dynamical relaxation. The temperature, T, for this microcanonical ensemble is measured in terms of the mean kinetic energy for the system composed of N atom as

$$\frac{1}{2} \sum_{i=1}^{N} m_i \langle v_i^2 \rangle = \frac{3}{2} N k_B T \tag{2}$$

where m_i and $\langle v_i^2 \rangle$ are the mass and average of the velocity squared of the ith atom, and k_B is the Boltzmann constant. Any residual overall translational and rotational motion can be removed to simplify analysis of the subsequent conformational fluctuations. The equilibration period is con-

[11] J. A. McCammon, B. R. Gelin, and M. Karplus, *Nature (London)* **267**, 585 (1977).

[12] J. A. McCammon, P. G. Wolynes, and M. Karplus, *Biochemistry* **18**, 927 (1979).

[13] W. F. van Gunsteren and M. Karplus, *Macromolecules* **15**, 1528 (1982).

[14] S. Chandrasekhar, *Rev. Mod. Phys.* **15**, 1 (1943).

[15] S. H. Northrup, M. R. Pear, C. Y. Lee, J. A. McCammon, and M. Karplus, *Proc. Natl. Acad. Sci. U.S.A.* **79**, 4035 (1982).

sidered finished when no systematic changes in the temperature are evident over a time of about 10 psec (slow fluctuations could be confused with continued relaxation over shorter intervals); it is necessary also to check that the atomic momenta obey a Maxwellian distribution and that different regions of the protein have the same average temperature. The actual dynamic simulation results (coordinates and velocities for all the atoms as a function of time) for determining the equilibrium properties of the protein are then obtained by continuing to integrate the equations of motion for the desired length of time. The available simulations for proteins range from 25 to 300 psec. Several different algorithms for integrating the equations of motion in Cartesian coordinates are being used in protein molecular dynamics calculations. Most common are the Gear predictor–corrector algorithm, familiar from small molecule trajectory calculations[12] and the Verlet algorithm, widely used in statistical mechanical simulations.[16]

Stochastic Dynamics

In certain cases it is advantageous to simplify the dynamic treatment by separating the system under study into two parts. One part is that whose dynamics are to be examined and the other serves as a heat bath for the first; this could be a protein in a solvent or one portion of a protein with the surrounding protein serving as the heat bath. In such an analysis (e.g., of a tyrosine side chain in a protein) the displacement of the part whose dynamics is to be studied relative to its neighbors is presumed to be analogous to molecular diffusion in a liquid or solid. The allowed range of motion can be characterized by an effective potential energy function termed the ''potential of mean force''[3,17]; this potential corresponds to the free energy of displacement of the elements being studied in the average field due to surrounding bath atoms. The motion of the group under study is determined largely by the time variation of its nonbonded interactions with the neighboring atoms. These interactions produce randomly varying forces that act to speed or slow the motion of the group in a given direction. In favorable cases, these dynamic effects can be represented by a set of Langevin equations of motion.[3,17,18] For a particle in one dimension, we can write

$$m \frac{d^2x}{dt^2} = F(x) - f \frac{dx}{dt} + R(t) \qquad (3)$$

[16] W. F. van Gunsteren and H. J. C. Berendsen, *Mol. Phys.* **34**, 1411 (1977).
[17] R. M. Levy, J. A. McCammon, and M. Karplus, *Chem. Phys. Lett.* **65**, 4 (1979).
[18] G. Lamm and K. Schulten, *J. Chem. Phys.* **75**, 365 (1981).

Here, m and x are the mass and position of the particle, respectively, and t is the time; thus, the term on the left is simply the acceleration of the particle. The term $F(x)$ represents the systematic force on the particle derived from the potential of mean force. The terms $-f dx/dt$ and $R(t)$ represent the effects of the varying forces caused by the bath acting on the particle; the first term is the average frictional force caused by the motion of the particle relative to its surroundings (f is the friction coefficient), and $R(t)$ represents the randomly fluctuating force. The Langevin equation and its generalized forms are phenomenological in character but they are consistent with more detailed models for the atomic dynamics.

The Langevin equation also provides a useful focal point in the discussion of large-scale motions.[19,20] For displacements of whole sections of a polypeptide chain away from protein surface (local denaturation), the terms corresponding to the one on the left of Eq. (3) are typically negligible in comparison with the others.[21] The motion then has no inertial character and the chain displacements have the particularly erratic character of Brownian motion. For elastic deformations of the overall protein shape, such as those involved in interdomain or hinge-bending motions, the potential of mean force may have a simple Hooke's law or springlike character.[19] The larger scale structural changes involved in protein folding (e.g., the coming together of two helices connected by a coil region to form part of the native structure) are also likely to have Brownian character.[22,23]

Harmonic Dynamics

An alternative approach to the dynamics of a protein or one of its constituent elements (e.g., an α-helix) is to assume that the harmonic approximation is valid. Early attempts to examine dynamic properties of proteins or their fragments used the harmonic approximation. They were motivated by vibrational spectroscopic studies,[24] where the calculation of normal mode frequencies from empirical potential functions has long been a standard step in the assignment of infrared spectra.[25] In calculating the

[19] J. A. McCammon, B. R. Gelin, M. Karplus, and P. G. Wolynes, *Nature (London)* **262**, 325 (1976).

[20] J. A. McCammon and P. G. Wolynes, *J. Chem. Phys.* **66**, 1452 (1977).

[21] J. A. McCammon, S. H. Northrup, M. Karplus, and R. M. Levy, *Biopolymers* **19**, 2033 (1980).

[22] M. Karplus and D. L. Weaver, *Nature (London)* **260**, 404 (1976).

[23] M. Karplus and D. L. Weaver, *Biopolymers* **18**, 1421 (1979).

[24] T. Miyazawa, in "Poly-α-Amino Acids" (G. D. Fasman, ed.). Dekker, New York, 1967.

[25] E. B. Wilson, J. C. Decius, and P. C. Cross, "Molecular Vibrations." McGraw-Hill, New York, 1955.

normal vibrational modes of a molecule, one assumes that the vibrational displacements of the atoms from their equilibrium positions are small enough that the potential energy can be approximated as a sum of terms which are quadratic in the displacements. The coefficients of these quadratic terms form a matrix of force constants which, together with the atomic masses, can be used to set up a matrix equation for the vibrational modes of the molecule.[25] Solution of this equation generally requires the diagonalization of a $3N$ dimensional matrix, where N is the number of atoms in the molecule. The result is obtained as a set of $3N$ eigenvalues (vibrational frequencies) and $3N$ eigenvectors (normal modes). Six of these eigenvectors are associated with eigenvalues of zero, corresponding to overall translation and rotation of the molecule. The remaining $3N-6$ eigenvalues are the internal vibrational frequencies of the molecule; the associated eigenvectors give the directions and relative amplitudes of the atomic displacements in each normal mode.

Although the harmonic model may not provide a complete description of the motional properties of a protein because of the contribution of anharmonic terms to the potential energy [Eq. (1)], it is nevertheless of considerable importance because it does serve as a first approximation for which the theory is highly developed. Further, the harmonic model is essential for some quantum mechanical treatments of vibrational contributions to the heat capacity and free energy[26,27] and for certain treatments of unimolecular reactions.[28]

Activated Dynamics

Enzyme-catalyzed reactions generally involve some processes in which the rate is limited by an energy barrier. In many cases the phenomenological time scale of such activated events is a microsecond or longer. Such processes that are intrinsically fast but occur rarely (i.e., with an average frequency much less than 10^{11} sec^{-1}) are not observed often enough for adequate characterization in an ordinary molecular dynamics simulation. To study such processes, alternative dynamic methods can be employed.

It is often possible to identify the particular character of the structural change involved (e.g., the reaction path) and then to approximate the associated energy changes. In the adiabatic-mapping approach, one calculates the minimized energy of the protein consistent with a given struc-

[26] R. M. Levy, D. Perahia, and M. Karplus, *Proc. Natl. Acad. Sci. U.S.A.* **79**, 1346 (1982).
[27] N. Go and H. Scheraga, *Macromolecules* **9**, 535 (1976).
[28] D. L. Bunker and F.-M. Wang, *J. Am. Chem. Soc.* **99**, 7457 (1977).

tural perturbation.[29-31] Minimization allows the remainder of the protein to relax in response to the assumed structural change and the resulting energy provides a rough approximation to the potential of mean force. Accurate potentials of mean force can be calculated by means of specialized molecular dynamics calculations,[15] but the computational requirements are greater. To analyze the time dependence of the process, the potential of mean force is incorporated into a model for the dynamics such as the familiar transition state theory.[29,32] A more detailed understanding of the process can be obtained by analyzing trajectories chosen to sample the barrier region.[15,33] The trajectory analysis displays the space and time correlations of the atomic motions involved and provides experimentally accessible quantities such as rate constants and activation energies.

Simplified Model Dynamics

To simulate processes that are intrinsically complicated (i.e., that involve the sampling of many configurations) it is sometimes possible to use simplified models for the structure and energetics of the protein. In one model of this kind, each residue in the protein is represented by a single interaction center and these centers are linked by virtual bonds.[34,35] The energy function for this model is obtained by averaging interresidue interactions over all the local atomic configurations within each residue.[21,35-37] Thus, the model incorporates the assumption of separated time scales for local and overall chain motions. The reduced number of degrees of freedom allows rapid calculation of the energy and forces, so that significantly longer dynamic simulations are possible than with a more detailed model. Such an approach is particularly useful for studying unfolding or folding of proteins[22,23] and their secondary structural elements.[21]

Examples of Protein Motions

In what follows we illustrate the theoretical techniques for studying the internal motions of proteins by specific examples.

[29] B. R. Gelin and M. Karplus, *Proc. Natl. Acad. Sci. U.S.A.* **72,** 2002 (1975).
[30] A. Warshel and M. Karplus, *J. Am. Chem. Soc.* **96,** 5677 (1974).
[31] D. A. Case and M. Karplus, *J. Mol. Biol.* **132,** 343 (1979).
[32] P. Pechukas, *in* "Dynamics of Molecular Collisions" (W. H. Miller, ed.), Part B. Plenum, New York, 1976.
[33] J. A. McCammon and M. Karplus, *Biopolymers* **19,** 1375 (1980).
[34] P. J. Flory, "Statistical Mechanics of Chain Molecules." Wiley, New York, 1969.
[35] M. Levitt, *J. Mol. Biol.* **104,** 59 (1976).
[36] M. Levitt, "Models for Protein Dynamics." CECAM Workshop Report, Univ. Paris XI, Orsay, 1976.
[37] M. R. Pear, S. H. Northrup, J. A. McCammon, M. Karplus, and R. M. Levy, *Biopolymers* **20,** 629 (1981).

Atomic Fluctuations

A quantitative measure of the atomic motions can be obtained from the mean square fluctuations of the atoms relative to their average positions. These can be related to the atomic temperature or Debye–Waller factors, B, determined in an X-ray diffraction study of a protein crystal.[38-40] The mean-square positional fluctuation, $\langle \Delta r^2 \rangle_{dyn}$, with the assumption of isotropic and harmonic motion can be written

$$\langle \Delta r^2 \rangle_{dyn} = \frac{3B}{8\pi^2} - \langle \Delta r^2 \rangle_{dis} \tag{4}$$

where $\langle \Delta r^2 \rangle_{dis}$ is the contribution to B from lattice disorder and other effects that are difficult to evaluate experimentally. For a number of proteins,[38-40] the measured value of $(3B/8\pi^2)$ average over all of the nonsurface atoms of the protein is in the range 0.48–0.58 Å^2. Comparison of this result with the mean value of $\langle \Delta r^2 \rangle_{dyn}$ from protein simulations (0.28–0.36 Å^2),[38,41,42] suggests that the nonmotional contribution to the B factor, $\langle \Delta r^2 \rangle_{dis}$, is in the range 0.20–0.25 Å^2. The only experimental estimate of $\langle \Delta r^2 \rangle_{dis}$ is from Mössbauer data for the heme iron in myoglobin[39]; for that one atom a somewhat smaller value (0.14 Å^2) was obtained. Thus, in the cases examined, approximately half of the experimental B factor is associated with thermal fluctuations in the atomic positions and half with other sources. However, some protein crystals, particularly those with a high percentage of water, appear to have a larger disorder contribution.

For most proteins studied, there is an increase in the magnitude of the experimental and theoretical fluctuations with distance from the center of the molecule. The magnitudes of the rms fluctuations range from ~0.4 Å for backbone atoms to ~1.5 Å for the ends of long sidechains. The hydrogen-bonded secondary structural elements (α-helices, β-sheets) tend to have smaller fluctuations than the random coil parts of the protein. More generally, the magnitude of the fluctuations varies widely throughout the protein interior, suggesting that the system is inhomogeneous and that some regions are considerably more flexible than others.

To examine the effect of bond length and bond angle fluctuations on the magnitude of the root mean square atomic displacements in the protein interior, simulations were performed on the bovine pancreatic trypsin

[38] S. H. Northrup, M. R. Pear, J. A. McCammon, M. Karplus, and T. Takano, *Nature (London)* **287**, 659 (1980).

[39] H. Frauenfelder, G. A. Petsko, and D. Tsernoglou, *Nature (London)* **280**, 558 (1979).

[40] P. J. Artymiuk, C. C. F. Blake, D. E. P. Grace, S. J. Oatley, D. C. Phillips, and M. J. E. Sternberg, *Nature (London)* **280**, 563 (1979).

[41] T. Ichiye, B. Olafson, S. Swaminathan, and M. Karplus, *Biopolymers* (in press).

[42] W. F. van Gunsteren and M. Karplus, *Nature (London)* **293**, 677 (1981).

inhibitor (PTI) in which the bond lengths or both the bond lengths and the bond angles were fixed at their average values.[13] It was found that use of fixed bond lengths (normal fluctuations ± 0.03 Å) does not significantly alter the dynamic properties on a time scale longer than 0.05 psec, but that constraint of the bond angles (normal fluctuations $\pm 5°$) reduces the mean amplitude of the atomic motions by a factor of two. This result demonstrates that in a closely packed system, such as a protein in its native configuration, the excluded volume effects due to the repulsive van der Waals interactions introduce a strong coupling between the dihedral angle and bond angle degrees of freedom.

Comparison of the calculated and X-ray experimental rms fluctuations on a residue by residue basis has been done for a number of molecules (e.g., cytochrome c, Ref. 38). There is generally good agreement for the relative values of the fluctuations, although the absolute values are uncertain because of disorder corrections. This correlation serves to confirm the reliability of both the simulation results and the temperature factors as detailed measures of the internal mobility of proteins. However, there are also significant differences in that the largest fluctuations may be shifted by a residue or two in the calculations relative to experiment.

Since most of the molecular dynamics simulations have been done for a protein in vacuum, it is expected that, particularly for the exterior residues, the results will be in error owing to the absence of solvent and, with regard to X-ray temperature factors, the absence of the crystal environment. For cytochrome c, the most prominent differences between theoretical and experimental mean displacements involve the residues calculated to have very large fluctuations; these are all charged side chains (particularly lysines) that protrude from the protein and so are not correctly treated in the vacuum simulation. This result is confirmed by molecular dynamics simulations of PTI in a Lennard–Jones solvent and in a crystal environment. The simulations show that the motion of the outside residues is significantly perturbed by the surrounding medium[42,43]; in particular, the interaction between charged side chains of a given protein and its crystal neighbors can produce a reduction in the rms values.[43,44] Such results for the external residues contrast with those for the protein interior, where the environmental effects on the amplitude of the fluctuations are found to be small. The dominant medium effect on the equilibrium properties of the PTI molecule is that the average structure in the solvent or crystal field is significantly closer to the X-ray structure than is the vacuum result (e.g., for the C^α atoms, the vacuum simulation has an

[43] W. F. van Gunsteren and M. Karplus, *Biochemistry* **21**, 2259 (1982).
[44] B. Gelin and M. Karplus, *Biochemistry* **18**, 1256 (1979).

rms deviation from the X-ray structure equal to 2.2 Å, while those for the solvent and crystal simulations are 1.35 and 1.52 Å, respectively). Recently, a crystal simulation of PTI, including the water molecules, was completed.[45] Although the simulation is too short (12 psec) for definitive conclusions, the magnitude of the fluctuations corresponded to those found in the earlier simulations of PTI, while the dynamic average structure was somewhat closer to the X-ray result (C^a atom rms deviation is 1.06 Å).

Of interest also are the results from the dynamic simulation concerning deviations of the atomic motions from the isotropic, harmonic behavior assumed in most X-ray analyses of proteins. The motions of many of the atoms were found in the simulations to be highly anisotropic and somewhat anharmonic. The rms fluctuation of an atom in its direction of largest displacement is typically twice that in its direction of smallest displacement; larger ratios are not uncommon.[41,43,46] It is sometimes possible to rationalize these directional preferences in terms of local bonding, e.g., torsional oscillation of a small group around a single bond.[46] In most cases, however, the directional preferences appear to be determined by larger scale collective motions involving the atom and its neighbors.[46–48] The atom fluctuations are generally somewhat anharmonic; that is, the potentials of mean force for the atom displacements deviate from the simple parabolic forms that would obtain at sufficiently low temperature.[41,43,49] The most markedly anharmonic atoms are those having multiple minima in their potentials of mean force. The shape of the PTI potential surface in the region of the native structure indicates that the anharmonicity is primarily associated with the softest collective modes of displacement in the protein.[50]

Time-Dependence; Local and Collective Effects

Analyses of the time development of the atomic fluctuations have been made for PTI[43,47] and cytochrome c.[48] It has been shown that the atomic

[45] W. F. van Gunsteren, H. J. C. Berendsen, J. Hermans, W. G. J. Hol, and J. P. M. Postma, *Proc. Natl. Acad. Sci. U.S.A.* **80**, 4315 (1983).

[46] S. H. Northrup, M. R. Pear, J. D. Morgan, J. A. McCammon, and M. Karplus, *J. Mol. Biol.* **153**, 1087 (1981).

[47] S. Swaminathan, T. Ichiye, W. F. van Gunsteren, and M. Karplus, *Biochemistry* **21**, 5230 (1982).

[48] J. D. Morgan, J. A. McCammon, and S. H. Northrup, *Biopolymers* **22**, 1579 (1983).

[49] B. Mao, M. R. Pear, J. A. McCammon, and S. H. Northrup, *Biopolymers* **21**, 1979 (1982).

[50] T. Noguti and N. Gō, *Nature (London)* **296**, 776 (1982).

fluctuations which contribute to the temperature factor (thermal ellipsoid) can be separated into local oscillations superposed on motions with a more collective character. The former have a subpicosecond time scale; the latter, which can involve only a few neighboring atoms, a residue, or groups of many atoms in a given region of the protein, have time scales ranging from 1 to 10 psec or longer ($\bar{\nu} \cong 3$ to 30 cm^{-1}). By following the time development of the atomic fluctuations of PTI from 0.2 to 25 psec, it has been shown that the high-frequency oscillations, which contribute about 40% of the average rms fluctuations of main chain atoms, tend to be uniform over the structure. It is the longer time scale, more collective motions which introduce the variations in the fluctuation magnitudes that characterize different parts of the protein.[47]

Biological Function

Although many of the individual atom fluctuations observed in the simulations or obtained from temperature factors may in themselves not be important for protein function, they contain information that is of considerable significance. The calculated fluctuations are such that the conformational space available to a protein at room temperature includes the range of local structural changes observed on substrate (inhibitor) binding for many enzymes. Also, the calculations show that there are local correlations in the fluctuations and that some regions of the protein are particularly flexible. This is found to be true not only in comparing the inside and outside of a protein but one interior region with another. Further, the small amplitude fluctuations are essential to all other motions in proteins; they serve as the "lubricant" which make possible larger scale displacements, such as domain motions (see the table), on a physiological time scale. In addition, it may be possible to extrapolate from the short time fluctuations to larger scale protein motions.

The collective modes are likely to be of particular significance in the biological function of the protein; e.g., they may be involved in the displacements of side chains, loops, or other structural units that are required for the transition from an inactive to the active configuration of a globular protein and in the correlated fluctuations that could play a role in enzyme catalysis. Further, the extended nature of these motions makes them more sensible to the environment (e.g., differences in the simulation results between vacuum and solution for PTI).[43,47] Because they involve sizable portions of the protein surface, the collective motions may be involved in transmitting external solvent effects to the protein interior. They might also be expected to be quenched at low temperature by freezing of the solvent. Their important contribution to the mean square fluctuations could explain the transition observed near 200 K in the temperature dependence of the fluctuations in proteins like in myoglobin.[51]

Side Chain Motions

The motions of aromatic side chains serve as a convenient probe of protein dynamics. The side chain motions span a time range from picoseconds, during which local oscillations occur, to milliseconds or longer required for 180° rotations. To cover this range of motions requires use of a variety of approaches that complement each other in the analysis of protein dynamics. Further, the results obtained are typical of a class of motional phenomena that do play a biological role (see below).

Tyrosines and Tryptophans

The torsional librations of buried tyrosines in PTI[12] and of tryptophans in lysozyme have been studied in some detail. Comparison of the potential energy contour maps for the side-chain dihedral angles χ_1 and χ_2 of tyrosines in PTI in a dipeptide fragment and in the protein[2] showed that the minimum energy conformations are very similar in the two cases; this appears to be true for most interior residues of proteins. Where the results differ is that the side chain is much more rigidly fixed in position by its nonbonded neighbors in the protein than it is by interactions with the backbone of the dipeptide. Comparison of the torsional motion of the ring in the protein shows that it is less regular when it is surrounded by the protein matrix than in the separated fragment. In PTI, the rms fluctuation of the aromatic ring orientation of a particular tyrosine (Tyr-21) is 12°, while that for the tyrosine fragment is 15°. This relatively small difference in amplitudes as compared with the forms of the rigid rotation potentials indicates that protein relaxation involving correlated fluctuations plays an important role in the ring oscillations. Further, characteristics of the ring motion in the protein are consistent with a torsional Langevin equation that contains a harmonic restoring force [Eq. (3)].[12]

The frictional random force terms are similar to those expected for ring rotation in an organic solvent; this is in accord with the hydrophobic environments of the rings in the protein. The time correlation functions for the torsional fluctuations decay to small values in a short time (~0.2 psec). Although such rapid decay is observed in the correlation function for the ring angle, the quantities involved in the relaxation times measured

[51] F. Parak, E. N. Frolov, R. L. Mössbauer, and V. I. Goldanskii, *J. Mol. Biol.* **145**, 825 (1981).

in fluorescence depolarization[43,52,53] (trigonometric functions of the angles) decay much more slowly. For the tyrosine rings in PTI there is a rapid partial decay in less than a picosecond to a plateau value equal to about 75% of the initial value; this behavior has recently been confirmed by fluorescence depolarization measurements.[54] Corresponding calculations[52] for the fluorescence depolarization of the tryptophan residues in lysozyme based on a molecular dynamics simulation indicate a wide range of variation in the depolarization behavior. Since there are six tryptophans in a variety of environments, the behavior is expected to correspond to that which occurs generally in proteins. Certain interior tryptophans have almost no decay over the time scale of the simulation while one in the active site (Trp-62) has its anisotropy reduced to 0.6 after 5 psec.

Torsional fluctuations of tyrosine and phenylalanine rings in which the ring rotates by 180° are known to occur in proteins.[1,55,56] Such rotational isomerizations or ring "flips" occur very infrequently because of the large energy barrier due to steric hindrance.[2,15] The long time intervals separating flips preclude systematic study by conventional molecular dynamics methods. A modified molecular dynamics method has been developed to handle such local activated processes.[15] This method is similar to adiabatic mapping in that one starts with an assumed "reaction coordinate" that defines the fundamental structural changes involved. It differs from the adiabatic method in that it involves consideration of all thermally accessible configurations and not just the minimum energy one for each value of the reaction coordinate, and also in that it provides a detailed description of the structural and dynamic features of the process. In this method, one calculates separately the factors in the following expression[15,57] for the rate constant of the process:

$$k = \frac{1}{2} \kappa \langle |\dot{\xi}| \rangle [\rho(\xi^{\ddagger})/\int_i \rho(\xi)d\varepsilon] \tag{5}$$

Here, ξ is the reaction coordinate, $\dot{\xi} = d\xi/dt$, and $\xi^{\ddagger}$ is the value of ξ in the transition state region for the process. The factor in square brackets is the probability that the system will be in the transition state region, relative to the probability that it is in the initial stable state. This factor corresponds roughly to the $\exp(-\Delta G^{\ddagger}/RT)$ factor in more familiar expressions for rate

[52] T. Ichiye and M. Karplus, *Biochemistry* **22**, 2884 (1983).
[53] R. M. Levy and A. Szabo, *J. Am. Chem. Soc.* **104**, 2073 (1982).
[54] A. Kasprzak and G. Weber, *Biochemistry* **21**, 5924 (1982).
[55] I. D. Campbell, C. M. Dobson, and R. J. P. Williams, *Adv. Chem. Phys.* **39**, 55 (1978).
[56] F. R. N. Gurd and T. M. Rothberg, *Adv. Protein Chem.* **33**, 73 (1979).
[57] D. Chandler, *J. Chem. Phys.* **68**, 2959 (1978).

constants; it can be calculated by carrying out a sequence of simulations in which the system is constrained to stay near particular values of ξ. The remaining factors can be evaluated by analysis of trajectories initiated in the transition state region.[15,33] The transmission coefficient κ is equal to one in ideal transition state theory (equilibrium is maintained in the stable states and there are only uninterrupted crossings of the transition state region); for real systems κ is less than one.

Application of this modified molecular dynamics method to the flipping of a tyrosine ring in PTI shows that the rotations themselves require only 0.5 to 1.0 psec.[33] At the microscopic level, the processes responsible for flipping are the same as those responsible for the smaller amplitude librations. The ring is driven over the barrier not as the result of a particularly energetic collision with some cage atom, but as the result of a transient decrease in frequency and intensity of collisions that would drive the ring away from the barrier. These alterations of the collision frequency are due to small, transient packing defects. The packing defects help to initiate ring rotation, but they are much too small to allow free rotation of the ring by a simple vacancy or free-volume mechanism. The ring in fact tends to be tightly encaged even in the transition state orientation. Collisions with cage atoms in the transition state produce frictional forces similar to those that occur in the stable state librations; these frictional effects reduce the transition rate to about 20% of the ideal transition state theory value. As to the free energy of activation, the calculations suggest that the activation enthalpy contribution is similar to that found by adiabatic mapping techniques[2] and that the activation entropy is small.

Biological Function

Although no enzyme has as yet been studied by the techniques used for the tyrosine ring flips, it is clear that the methodology is applicable to the activated processes central to most enzymatic reactions. Further, many of the qualitative features found for the tyrosines (e.g., lowering of the potential of mean force by cage relaxation, slowing of the rate by frictional effects) should be present in general.

A biological problem where side chain fluctuations are important concerns the manner in which ligands like carbon monoxide and oxygen are able to get from the solution through the protein matrix to the heme group in myoglobin and hemoglobin and then out again. The high-resolution X-ray structure of myoglobin[58-61] does not reveal any path by which ligands

[58] M. F. Perutz and F. S. Mathews, *J. Mol. Biol.* **21**, 199 (1966).
[59] H. C. Watson, *Prog. Stereochem.* **4**, 299 (1969).
[60] T. Takano, *J. Mol. Biol.* **110**, 569 (1977).
[61] S. E. Phillips, *Nature (London)* **273**, 247 (1978).

such as O_2 or CO can move between the heme-binding site and the outside of the protein. Since this holds true both for the unliganded and liganded protein, i.e., myoglobin[60] and oxymyoglobin,[61] structural fluctuation must be involved in the entrance and exit of the ligands. Empirical energy function calculations[31] showed that the rigid protein would have barriers on the order of 100 kcal/mol; such high barriers would make the transitions infinitely long on a biological time scale. To analyze pathways available in the thermally fluctuating protein, ligand trajectories were calculated with a test molecule of reduced effective diameter to compensate for the use of the rigid protein structure. A trajectory was determined by releasing the test molecule with substantial kinetic energy (15 kcal/mol) in the heme pocket and following its classical motion for a suitable length of time. A total of 80 such trajectories were computed; a given trajectory was terminated after 3.75 psec if the test molecule had not escaped from the protein. Slightly more than half the test molecules failed to escape from the protein in the allowed time; 25 molecules remained trapped near the heme-binding site, while another 21 were trapped in two cavities accessible from the heme pocket. Most of the molecules that escaped did so between the distal histidine (E7) and the side chains of Thr-E10 and Val-E11. A secondary pathway was also found; this involves a more complicated motion along an extension of the heme pocket into a space between Leu-B10, Leu-E4, and Phe-B14, followed by squeezing out between Leu-E4 and Phe-B14.

In the rigid X-ray structure, the two major pathways have very high barriers for a thermalized ligand of normal size. Thus, it was necessary to study the energetics of barrier relaxation to determine whether either of the pathways had acceptable activation enthalpies. Local dihedral rotations of key side chains, analogous to the tyrosine side chain oscillations described above, were investigated; it was found that the bottleneck on the primary pathway could be relieved at the expense of modest strain in the protein by rigid rotations of the side chains of His-E7, Val-E11, and Thr-E10. In this manner a direct path to the exterior was created with a barrier of ~5 kcal/mol at an energy cost to the protein of ~8.5 kcal/mol, as compared with the X-ray structure value of nearly 100 kcal/mol. On the secondary path, however, no simple torsional motions reduced the barrier due to Leu-E4 and Phe-B14, since the necessary rotations led to larger strain energies. A test sphere was fixed at each of the bottlenecks and the protein was allowed to relax by energy minimization (adiabatic limit) in the presence of the ligand. Approximate values for the relaxed barrier heights were 13 and 6 kcal/mol for the two primary path positions and 18 kcal/mol for the secondary path position. These barriers are on the order of those estimated in the photolysis, rebinding studies for CO myoglobin by Frauenfelder *et al.*[62-64] Further, a path suggested by the energy calculations was found to correspond to a high mobility region in the protein as determined by X-ray temperature factors.[39] Also, two recent crystal structures with an imidazole[65] or a phenyl[66] group bound to the iron suggest that the primary pathway determined by the dynamics plays an important role.

The type of ligand motion expected for such a several-barrier problem can be determined from the trajectory studies mentioned earlier. What happens is that the ligand spends a long time in a given well, moving around in and undergoing collisions with the protein walls of the well. When a protein fluctuation sufficient to significantly lower the barrier occurs, or the ligand gains sufficient excess energy from collisions with the protein, or more likely both at the same time, the ligand moves rapidly over the barrier and into the next well where the process is repeated. In a completely realistic trajectory involving a fluctuating protein and ligand–protein energy exchange, the time spent in the wells would be much longer than that found in the diabatic model calculations described above. From the complexity and the range of pathways in the protein interior, it is likely that the motion of the ligand has a diffusive character.

The analysis of myoglobin suggests that the native structure of a protein is often such that the small molecules that interact with the protein cannot enter or leave if the atoms are constrained to their average positions. Consequently, side chain and other fluctuations may be required for ligand binding by many proteins and for the entrance of substrates and exit of products from some enzymes.

Rigid Body Motions

A type of motion that plays an important role in proteins is referred to as a rigid body motion (see the table). It involves the displacement of one part of a protein relative to another in such a manner that a valid approximate description treats each moving portion as a rigid body. As has already been stated and will be made clear with specific examples, it is essential that smaller fluctuations accompany the "rigid body" motions to reduce the required energy and permit them to proceed at a sufficiently rapid rate.

[62] P. G. Debrunner and H. Frauenfelder, *Annu. Rev. Phys. Chem.* **33**, 283 (1982).

[63] R. H. Austin, K. W. Beeson, L. Eisenstein, H. Frauenfelder, and I. C. Gunsalus, *Biochemistry* **14**, 5355 (1975).

[64] D. Beece, L. Eisenstein, H. Frauenfelder, D. Good, M. C. Marden, L. Reinisch, A. H. Reynolds, L. B. Sorensen, and K. T. Yue, *Biochemistry* **19**, 5147 (1980).

[65] J. A. Bell, Z. R. Korszun, and K. Moffat, *J. Mol. Biol.* **147**, 325 (1981).

[66] D. Ringe, G. A. Petsko, D. E. Kerr, and P. R. Ortiz, *Biochemistry* **23**, 2 (1984).

Hinge-Bending

A large number of enzymes[67,68] and other protein molecules (e.g., immunoglobulins) consist of two or more distinct domains connected by a few strands of polypeptide chain which may be viewed as "hinges." In lysozyme, which will serve as an example, it was noted in the X-ray structure[69] that when an active site inhibitor is bound, the cleft closes down somewhat as a result of relative displacements of the two globular domains that surround the cleft. Other classes of proteins (kinases, dehydrogenases, citrate synthase) have considerably larger displacements of the two lobes on substrate binding than does lysozyme.

In the theoretical analysis of lysozyme,[70] which is the first protein whose hinge bending mode was studied, the stiffness of the hinge was evaluated by the use of an empirical potential energy function.[10] An angle bending potential was obtained by rigidly rotating one of the globular domains relative to a bending axis which passes through the hinge and calculating the changes in the protein conformational energy. This procedure is expected to overestimate the bending potential, since no allowance is made for the relaxation of unfavorable contacts between atoms that have been generated by the rotation. To take account of the relaxation, an adiabatic potential was calculated by holding the bending angle fixed at various values and permitting the positions of atoms in the hinge and adjacent regions of the two globular domains to adjust themselves so as to minimize the total potential energy. As in a previous adiabatic ring rotation calculation (see above),[2] only small (<0.3 Å) atomic displacements occurred in the relaxation process; the differences between the rigid and adiabatic bending potentials are largely due to small shifts in the relative positions of a few atoms which have been forced too close together by the rigid rotation model. The relief of these contacts can be effected by localized motions (e.g., bond angle and local dihedral angle deformations). The frequencies associated with these deformations (>100 cm^{-1}) are expected to be greater than the hinge bending frequency (≈ 5 cm^{-1}), so that the use of the adiabatic bending potential appears to be appropriate.

The rigid and adiabatic bending potentials were found to be approximately parabolic, with the restoring force constant for the adiabatic potential about an order of magnitude smaller than that for the rigid poten-

[67] C. M. Anderson, F. H. Zucker, and T. A. Steitz, *Science* **204**, 375 (1979).

[68] J. Janin and S. J. Wodak, *Prog. Biophys. Mol. Biol.* **42**, 21 (1983).

[69] I. Imoto, J. N. Johnson, A. C. T. North, D. C. Phillips, and J. A. Rupley, *in* "The Enzymes" (P. D. Boyer, ed.), Vol. 7, p. 665. Academic Press, New York, 1972.

[70] J. A. McCammon, B. R. Gelin, M. Karplus, and P. G. Wolynes, *Nature (London)* **262**, 325 (1976).

tial. However, even in the adiabatic case, the effective force constant is about 20 times as large as the bond-angle bending force constant of an α-carbon (i.e., N—C^α—C); the dominant contributions to the force constant come from repulsive nonbonded interactions involving on the order of 50 contacts. If the adiabatic potential is used and the relative motion is treated as an angular harmonic oscillator composed of two rigid spheres, a vibrational frequency of about 5 cm^{-1} is obtained. This is a consequence of the fact that, although the force constant is large, the moments of inertia of the two lobes are also large.

In considering the hinge bending motion it is essential to take account of the fact that lysozyme is normally in solution. Although fluctuations in the interior of the protein, such as those considered in myoglobin, may be insensitive to the solvent (because the protein matrix acts as its own solvent), the domain motion in lysozymes involves two lobes that are surrounded by the solvent. To take account of the solvent effect in the simplest possible way, the Langevin equation [Eq. (3)] for a damped harmonic oscillator was used. The friction coefficient for the solvent damping term was evaluated by modeling the two globular domains as spheres and calculating the viscous frictional drag accompanying the relative motion of these spheres by use of a modified Stokes law[71]; the internal friction of the protein was considered to be negligible compared to the hydrodynamic friction. From the adiabatic estimate of the hinge potential and the magnitude of the solvent damping, it was found that the relative motion of the two globular domains in lysozyme is overdamped; i.e., in the absence of driving forces the domains would relax to their equilibrium positions without oscillating. The decay time for this relaxation was estimated to be about 2×10^{-11} sec. Actually, the lysozyme molecule will experience a randomly fluctuating driving force due to collisions with the solvent molecules, so that the distance between the globular domains will fluctuate in a Brownian manner over a range limited by the bending potential; a typical fluctuation opens the binding cleft by 1 Å and lasts for 20 psec.

Biological Function

Since the hinge bending motion in lysozyme and in other enzymes involves the active site cleft, it is likely to play a role in the enzymatic activity of these systems. In addition to the possible difference in the binding equilibrium and solvent environment in the open and closed state, the motion itself could result in a coupling between the entrance and exit of the substrate and the opening and closing of the cleft. The interdomain

[71] P. G. Wolynes and J. A. McCammon, *Macromolecules* **10**, 86 (1977).

mobility in immunoglobulins has been suggested to be important in adapting the structure to make possible the binding of different macromolecular antigens and, more generally, it may play a role in the cross-linking and other interactions required for antibody function. In the coat protein of tomato bushy stunt virus, a two-domain structure with a hinge peptide has been identified from the X-ray structure[72] and rotations about the hinge have been shown to be involved in establishing different subunit interactions for copies of the same protein involved in the assembly of the complete viral protein shell.

Harmonic Dynamics

Although many aspects of protein motions involve anharmonic contributions to the potential,[41,43,49] main chain fluctuations appear to be approximately harmonic in character. A harmonic dynamic analysis of PTI using the empirical potential function of Eq. (1) has recently been performed in the full conformational space of the molecule[7]; that is, all bond lengths and angles, as well as dihedral angles, were included for the 580 atom system consisting of all heavy atoms and polar hydrogens.

The molecule was found to have an essentially continuous, though not completely uniform, distribution of vibrational frequencies between 3.1 and 1200 cm^{-1}. Between 1200 and 1800 cm^{-1}, the frequencies tend to come in groups, many of which are dominated by bond stretching vibrations. There are 20 modes between 3.1 and 13 cm^{-1} and there is a peak in the frequency distribution near 50 cm^{-1}. Because the structure used was not an absolute minimum, seven negative modes were found; energy searches along these modes, which are all local in character, indicated that their correct frequencies are in the range 20 to 40 cm^{-1}.

The root mean square (rms) atom fluctuations were calculated from the normal modes by evaluation of the classical expression[5]

$$\langle \Delta r_k^2 \rangle = k_\text{B} T \sum_i \frac{|\mathbf{a}_{ik}|^2}{\omega_i^2} \tag{6}$$

where $\mathbf{a}_{ik}$ is the vector of the projections of the ith normal mode with frequency ω_i on the Cartesian components of the displacement vector for the kth atom, k_B is the Boltzmann constant, and T is the absolute temperature; quantum corrections are negligible above 50 K.[26] The normal mode rms fluctuations were calculated at 300 K and compared with the results of a molecular dynamics simulation of PTI in a van der Waals solvent[43]; this simulation was used because its average structure is closest to that

[72] S. C. Harrison, *Biophys. J.* **32**, 139 (1980).

employed for the normal mode analysis. For the main chain fluctuations, the molecular dynamics and normal mode values are very similar; for the side chains, there is also a correspondence though the differences are more pronounced. The main chain values show that the carboxy-terminal end has large fluctuations, as does the loop region at the bottom of the molecule (residues 25–29) and the binding site in the neighborhood of residues 14 and 38 at the top of the molecule. By contrast the β-sheet residues (18–24, 29–35) show smaller fluctuations; the α-helices (3–7, 47–56) are intermediate.

The origin of the differences between the molecular dynamics and normal mode results is likely to have contributions from anharmonic and solvent effects and from the difference between the average dynamics structure and that used for the normal mode analysis. The main chain atoms apparently experience a potential of mean force that is closer to the harmonic potential than do the side chains; since the dynamics simulation was done in a van der Waals solvent, the exterior side chains are expected to be most perturbed. To investigate the nature of the potential in the neighborhood of the structure used for the harmonic model, an energy search was done along the low-frequency normal-mode displacements. The energy dependence showed significant anharmonic contributions. Fitting the resulting energies to a parabola generally led to an increase in the effective frequencies which somewhat reduces the rms fluctuations; the relative values of the fluctuations are essentially unchanged.

Analysis of the time scale as well as the magnitudes of the fluctuations has been made for the molecular dynamics results (see above). From the calculated time series and correlation functions,[47] it has been found that the atomic motions contributing to the rms displacements generally have a small local high frequency component ($\sim$0.2 psec or $\sim$150 cm^{-1}) on which are superposed motions of a more collective character with time scales ranging from 1 to 10 psec or 30 to 3 cm^{-1}. The present normal mode study essentially confirms the dynamic results. In most cases, the dominant contributions come from low frequency modes in the range 3 to 50 cm^{-1}, although nonnegligible contributions come from higher frequencies up to 130 cm^{-1}. For certain atoms (e.g., Tyr-21, $C^{\varepsilon 2}$), only a very small number of modes are important, while for other atoms (e.g., Ala-16 C^{β}, Asp-50 C^{β}) a range of frequencies is involved; for Leu-29 $C^{\delta 1}$, a mode at 44.5 cm^{-1} makes a very large contribution.

It is of considerable interest to examine the form of the normal modes themselves. This is of particular importance for the evaluation of the correlation between the motions of different atoms and different groups of atoms. Analysis of the dynamics results have indicated that the larger scale motions have a collective character that may involve a few neighboring atoms, a residue, or groups of many atoms in a given region of a protein. Most of the 300 lowest modes are delocalized; many are distributed over the entire molecule. The lowest real mode (at 3.1 cm^{-1}) mirrors the overall rms fluctuation. Other lower frequency modes, although they are also delocalized, are distributed somewhat differently over the various portions of the molecule. In considering the character of the individual modes, it must be recognized that because of the close spacing, relatively small effects, such as solvent damping or external perturbations (e.g., ligand binding), can lead to significant mode mixing. This may be of biological interest. It also suggests that rather than individual mode properties, those that involve averages over a range of modes with similar frequencies are likely to be most significant and least sensitive to anharmonic corrections.

The present results suggest that for certain motional properties, especially those involving averages over many modes, the harmonic model is a useful first approximation; this is certainly correct at low temperatures and in BPTI it appears true even at room temperature. Adjustment of the force constants to account for some anharmonic contributions can extend the range of the normal mode treatment ("quasi-harmonic model").[26,73] Phenomena such as ring flips or larger rearrangements involving barriers are clearly outside the harmonic realm, though even here the normal mode model may provide some insights. Whether it will be fruitful to approximate protein motions as involving fluctuations in multidimensional wells on a short time scale with transitions to other wells on a longer time scale awaits the results of future studies.

Simplified Dynamics

As an approach to the helix–coil transition in α-helices, a simplified model for the polypeptide chain was introduced to permit a dynamic simulation on the submicrosecond time scale appropriate for this phenomenon[21]: each residue is represented by a single interaction center ("atom") located at the centroid of the corresponding side chain and the residues are linked by virtual bonds,[34] as described earlier in the section on simplified model dynamics. The diffusional motion of the chain "atoms" expected in water was simulated by using a stochastic dynamics algorithm based on the Langevin equation with a generalized force term, Eq. (3). Starting from an all-helical conformation, the dynamics of the

[73] M. Karplus and J. N. Kushick, *Macromolecules* **14**, 325 (1981).

residues at the end of a 15-residue chain were monitored in several independent 12.5 nsec simulations at 298 K. The mobility of the terminal residue was quite large, with a rate constant $\approx 10^9$ sec^{-1} for the transitions between coil and helix states. This mobility decreased for residues further into the chain; unwinding of an interior residue required simultaneous displacements of residues in the coil, so that larger solvent frictional forces were involved. The coil region did not move as a rigid body, however; the torsional motions of the chain were correlated so as to minimize dissipative effects. Such concerted transitions are not consistent with the conventional idea that successive transitions occur independently. Analysis of the chain diffusion tensor showed that the frequent occurrence of the correlated transition results from the relatively small frictional forces associated with these motions.[37]

Conclusions

Theoretical approaches have been used in the last few years to explore dynamic phenomena in depth for a variety of proteins (protein inhibitors, transport and storage proteins, enzymes). The magnitudes and time scales of the motions have been delineated and related to a variety of experimental measurements, including NMR, X-ray diffraction, fluorescent depolarization, infrared spectroscopy, and Raman scattering. It has been shown how to extend dynamic methods from the subnanosecond time range accessible to standard molecular dynamics simulations to much longer time scales for certain processes by the use of activated, harmonic, and simplified model dynamics. Further, the effect of solvent has been introduced by stochastic dynamic techniques or accounted for in full dynamic simulations that may include the crystal environment. Concomitantly, a wealth of experimental information on the motions has been gathered. The resulting interplay between theory and experiment provides a basis for the present vitality of the field of protein dynamics.

On the subnanosecond time scale our basic knowledge of protein motions is essentially complete; that is, the types of motion that occur have been clearly presented, their characteristics evaluated, and the important factors determining their properties delineated. Simulation methods have shown that the structural fluctuations in proteins are sizable; particularly large fluctuations are found where steric constraints due to molecular packing are small (e.g., in the exposed side chains and external loops), but substantial mobility is also found in the protein interior. Local atomic displacements in the interior of the protein are correlated in a manner that tends to minimize disturbances of the global structure of the protein. This leads to fluctuations larger than would be permitted in a rigid polypeptide matrix.

For motions on a longer time scale, our understanding is more limited. When the motion of interest can be described in terms of a reaction path (e.g., hinge-bending, local-activated event), methods exist for determining the nature and rate of the process. However, for the motions that are slow as a result of their complexity and involve large-scale structural changes, extensions of the approaches described in this review are required. Harmonic and simplified model dynamics, as well as reaction-path calculations, can provide information on slower processes, such as opening fluctuations and helix–coil transitions, but a detailed treatment of protein folding is beyond the reach of present methods. It is to be hoped the required methodological developments and the experiments to test the results will not be too long in forthcoming.

[16] Observation of Internal Motility of Proteins by Nuclear Magnetic Resonance in Solution

By GERHARD WAGNER and KURT WÜTHRICH

Protein conformations are the result of a multitude of weak, nonbonding interactions between different atoms of the polypeptide chain and between the polypeptide and the surrounding medium. The latter may be, for example, an aqueous solvent, an ordered lipid lattice in biological membranes, or the crystal lattice in the single crystals used for X-ray studies. Specific interactions with substrates and effector molecules may also influence the protein conformation. Since the contribution of each individual nonbonding interaction to the free energy which stabilizes the conformation is typically of the same order of magnitude as the thermal energy at temperatures near 300 K, these "secondary bonds" are constantly being broken and reformed. Two consequences of the resulting dynamic nature of the spatial molecular structures are that protein conformations can readily adapt to changes of the environment and that the spatial structures of protein molecules in thermodynamic equilibrium fluctuate about a structure of minimum free energy.

Detailed descriptions of the conformational transitions that occur when the environment is varied have been presented for several proteins for which both the initial and final states could be studied by single-crystal

X-ray methods.[1] With the use of nuclear magnetic resonance (NMR) such studies can be further extended to proteins in noncrystalline environments, which may be more closely related to the physiological milieu.[2,3] NMR data will in general bear on both static and dynamic aspects of the protein conformations and the method offers a wide spectrum of possible applications. The following discussion, however, is limited to the use of NMR for investigations of intramolecular rate processes ("internal motility") in proteins which are in thermodynamic equilibrium.

Prominent NMR manifestations of internal protein motility are found in the spin relaxation,[4-10] in flipping motions of the aromatic side chains,[7,10,11-15] and in the exchange of labile protons with the solvent.[7,10,16-25] NMR studies of these processes at variable temperature and pressure result in a detailed, quantitative characterization, which includes activation enthalpies, activation entropies, and activation volumes.[14,23,25] For each of these different NMR measurements the information content is

[1] R. Huber, *Trends Biochem. Sci.* **4**, 227 (1979).

[2] K. Wüthrich, G. Wider, G. Wagner, and W. Braun, *J. Mol. Biol.* **155**, 311 (1982).

[3] W. Braun, G. Wider, K. H. Lee, and K. Wüthrich, *J. Mol. Biol.* **169**, 921 (1983).

[4] A. Allerhand, D. Doddrell, and R. Komoroski, *J. Chem. Phys.* **55**, 189 (1971).

[5] D. Doddrell, V. Glushko, and A. Allerhand, *J. Chem. Phys.* **56**, 3683 (1972).

[6] D. E. Woessner and B. S. Snowden, *Adv. Mol. Relaxation Processes* **3**, 181 (1972).

[7] K. Wüthrich, "NMR in Biological Research: Peptides and Proteins." Elsevier, Amsterdam, 1976.

[8] R. Richarz, K. Nagayama, and K. Wüthrich, *Biochemistry* **19**, 5189 (1980).

[9] A. A. Ribeiro, R. King, C. Restivo, and O. Jardetzky, *J. Am. Chem. Soc.* **102**, 4040 (1980).

[10] O. Jardetzky and G. C. K. Roberts, "NMR in Molecular Biology." Academic Press, New York, 1981.

[11] K. Wüthrich and G. Wagner, *FEBS Lett.* **50**, 265 (1975).

[12] I. D. Campbell, C. M. Dobson, and R. J. P. Williams, *Proc. R. Soc. London Ser. B* **189**, 503 (1975).

[13] I. D. Campbell, C. M. Dobson, G. R. Moore, S. J. Perkins, and R. J. P. Williams, *FEBS Lett.* **70**, 96 (1976).

[14] G. Wagner, A. DeMarco, and K. Wüthrich, *Biophys. Struct. Mech.* **2**, 139 (1976).

[15] K. Wüthrich and G. Wagner, *Trends Biochem. Sci.* **3**, 227 (1979).

[16] J. D. Glickson, C. C. McDonald, and W. D. Phillips, *Biochem. Biophys. Res. Commun.* **35**, 492 (1969).

[17] S. Karplus, G. H. Snyder, and B. D. Sykes, *Biochemistry* **14**, 3612 (1973).

[18] A. Masson and K. Wüthrich, *FEBS Lett.* **31**, 114 (1973).

[19] B. D. Hilton and C. K. Woodward, *Biochemistry* **18**, 5834 (1979).

[20] R. Richarz, P. Sehr, G. Wagner, and K. Wüthrich, *J. Mol. Biol.* **130**, 19 (1979).

[21] G. Wagner and K. Wüthrich, *J. Mol. Biol.* **130**, 31 (1979).

[22] G. Wagner and K. Wüthrich, *J. Mol. Biol.* **134**, 75 (1979).

[23] C. K. Woodward and B. D. Hilton, *Biophys. J.* **32**, 561 (1980).

[24] G. Wagner and K. Wüthrich, *J. Mol. Biol.* **160**, 343 (1982).

[25] G. Wagner, *Q. Rev. Biophys.* **16**, 1 (1983).

dramatically increased when sequence-specific resonance assignments have been established. With the presently available experimental techniques it is then possible to map the local motility across the entire protein structure and thus to investigate concerted motions which would involve larger areas of the molecular structure. This chapter describes primarily how such a result can be obtained.

NMR Time Scales

NMR experiments provide three separate "time windows" for observation of kinetic processes. Thus motional processes in proteins with widely different time constants, τ, can be observed: slow processes with $\tau \gtrsim 10$ min, events with medium time constants, $\tau \approx 1-10^{-5}$ sec, and fast processes with $\tau \lesssim 10^{-9}$ sec.

For *slow processes* where the state of the system does not change significantly during the time needed for recording a one-dimensional (1D) or two-dimensional (2D) NMR spectrum, the time course of the reaction can be monitored by comparison of different, consecutively recorded NMR spectra. This requires that spectral parameters change as a consequence of the kinetic process. One particular application is the measurement of hydrogen–deuterium exchange rates.

Processes on a *medium time scale* can be observed when a nucleus undergoes an exchange between two sites, A and B, with the resonance frequencies ν_A and ν_B, respectively. The rate of exchange, ν_e, can be determined by line shape analysis if it is in the range of the difference of the resonance frequencies, $\Delta\nu = |\nu_A - \nu_B|$. The expected line shapes are shown in Fig. 1. There are two kinetic limits. If the exchange rate is much slower than $\Delta\nu$, two separate resonances are observed at ν_A and ν_B. If the exchange rate ν_e is much faster than $\Delta\nu$, one averaged resonance is observed at $(\nu_A + \nu_B)/2$. Between these extreme situations the exchange broadens the resonances and then the exchange rate can be determined from line shape analysis. In the limit of slow exchange, the exchange rate ν_e is simply related to the line broadening, $\delta\nu$, of the resonances at ν_A and ν_B:

$$\nu_e = 2\pi\delta\nu \tag{1}$$

In the limit of fast exchange the exchange rate, ν_e, can be obtained from the line broadening, $\delta\nu$, of the averaged resonance at $(\nu_A + \nu_B)/2$:

$$\nu_e = \pi(\Delta\nu)^2/4\delta\nu \tag{2}$$

Between these limits, in the range of intermediate exchange, the exchange rates have to be determined by comparison of the line shapes with com-

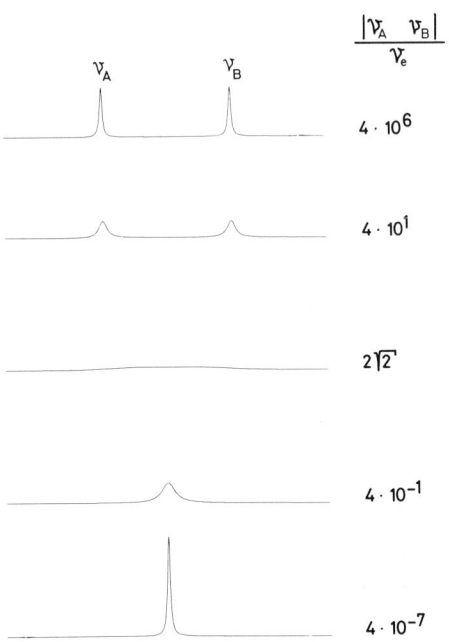

FIG. 1. Line shapes for a spin jumping between two equally populated states A (ν_A) and B (ν_B) for varying values of the quotient of the relative chemical shift in Hz $|\nu_A - \nu_B|$, and the exchange frequency, ν_e.

puter simulated spectra.[10,13,14] As long as the exchange is sufficiently slow so that two lines can be observed, saturation transfer techniques can also be used to determine two-site exchange rates.[26–28] With this method the resonance of one of the exchanging protons is saturated by selective radio frequency irradiation. This saturation is transferred to the other, corresponding resonance position by the exchange process. Measurement of the amount of saturation transfer allows determination of the exchange rates.[26–28]

Similar exchange effects are manifested in the averaging of spin–spin coupling constants. If the torsion angle between two coupled spins jumps between two orientations with the respective coupling constants J_A and J_B, an average coupling constant will be observed if the jump rate $\nu_e \gg |J_A - J_B|$. If $\nu_e \ll |J_A - J_B|$ both coupling constants will be manifested in the spectrum.[10,29]

[26] S. Forsén and R. A. Hoffman, *J. Chem. Phys.* **39**, 2892 (1963).
[27] I. D. Campbell, C. M. Dobson, R. G. Ratcliffe, and R. G. P. Williams, *J. Magn. Reson.* **29**, 397 (1978).
[28] J. J. Led and H. Gesmar, *J. Magn. Reson.* **49**, 444 (1982).
[29] K. Nagayama and K. Wüthrich, *Eur. J. Biochem.* **114**, 369 (1981).

Fast internal motions that occur on the time scale of the NMR frequencies can be studied with relaxation time measurements. In globular proteins in solution the relaxation times for spin 1/2 nuclei are mainly determined by the modulation of the dipole–dipole interactions between different nuclei by motional processes. The dominant contribution to these motional processes comes from the overall rotational diffusion of the globular protein (the correlation time for the overall rotation of a protein of ≈ 6000 Da is $\approx 3 \times 10^{-9}$ sec). If there are internal motions with similar or faster rates, these can be manifested in the relaxation times of the respective nuclei. The data can be obtained either from measurements of spin–lattice relaxation times, T_1, spin–spin relaxation times, T_2, or cross-relaxation rates in nuclear Overhauser enhancements.[4-10]

Exchange of Labile Protons

A characteristic feature of a globular protein structure is the network of intramolecular hydrogen bonds. The difference in Gibbs free energy of intramolecular hydrogen bonds relative to intermolecular hydrogen bonds with the solvent is of the order of only a few kcal/mol. Thus opening and reforming of these bonds have to be expected to be main features of internal motions in proteins. As a consequence of the opening of hydrogen bonds, contact between internal amide protons and the solvent may occur. If the protein is transferred from a protonated to a deuterated solvent, such internal amide protons will exchange against deuterium and their ^{1}H NMR signals will disappear.[30,31]

The amide protons of the peptide bond provide the most complete set of probes to study hydrogen exchange in proteins. Nevertheless, other labile protons have been used, such as the NH_2 groups of asparagine or the indole NH of tryptophan.[32,33]

In principle the NH_2 groups of glutamine and the labile protons of arginine or lysine side chains can also be studied when located in the interior of proteins.[31,32] All other labile protons, in particular those of hydroxyl or carboxyl groups, are expected to exchange too rapidly to be studied by NMR.[31,32]

In 1D NMR spectra the intensities of the absorption lines are directly proportional to the ^{1}H concentration at the respective peptide groups. This has commonly been used to measure exchange rates of well separated resonances. As an example, Fig. 2 shows the low field spectrum of a

[30] A. Hvidt and S. O. Nielsen, *Adv. Protein Chem.* **21,** 287 (1966).
[31] S. W. Englander, S. W. Downer, and H. Teitelbaum, *Annu. Rev. Biochem.* **41,** 903 (1972).
[32] K. Wüthrich and G. Wagner, *J. Mol. Biol.* **130,** 1 (1979).
[33] I. D. Campbell, C. M. Dobson, and R. J. P. Williams, *Proc. R. Soc. London Ser. B* **189,** 485 (1975).

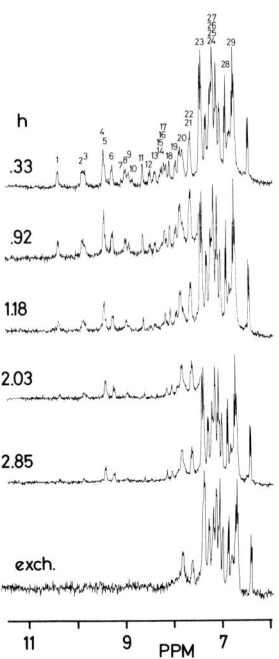

FIG. 2. Region from 6.0 to 11.0 ppm of the ^{1}H NMR spectra of a snail trypsin inhibitor at 36° and p^{2}H 5.0 recorded at different times after dissolving the protein in ^{2}H$_2$O. The resonances of the labile protons observed in the top spectrum are numbered in the order of the chemical shifts. The time which has elapsed between the sample preparation and the recording of the spectra is indicated on the left-hand side.[35]

snail trypsin inhibitor, recorded at different times after the protein was dissolved in ^{2}H$_2$O. The signals of labile protons which are numbered in the first spectrum disappear with time, and exchange rates can be calculated from a plot of the peak intensities vs time.[35] In two-dimensional correlated (COSY) spectra the relative intensity of each cross peak is determined by the degree of protonation of the peptide group. Therefore the time dependence of the intensity of each individual cross peak can be used for determining exchange rates. This has been done for the basic pancreatic trypsin inhibitor (BPTI),[24,25] as is demonstrated in Figs. 3 and 4. With this technique a complete survey of ^{1}H–^{2}H exchange rates has been obtained. These data are shown in Fig. 5, which is a plot of the logarithm of the exchange rates vs the amino acid sequence. Exchange rates faster than 10^{-1} min^{-1} cannot be measured by this method.[24] In Fig. 5 this is indicated

[34] R. E. Wedin, M. Delepierre, C. M. Dobson, and F. M. Poulsen, *Biochemistry* **21**, 1098 (1982).

[35] G. Wagner, K. Wüthrich, and H. Tschesche, *Eur. J. Biochem.* **89**, 367 (1978).

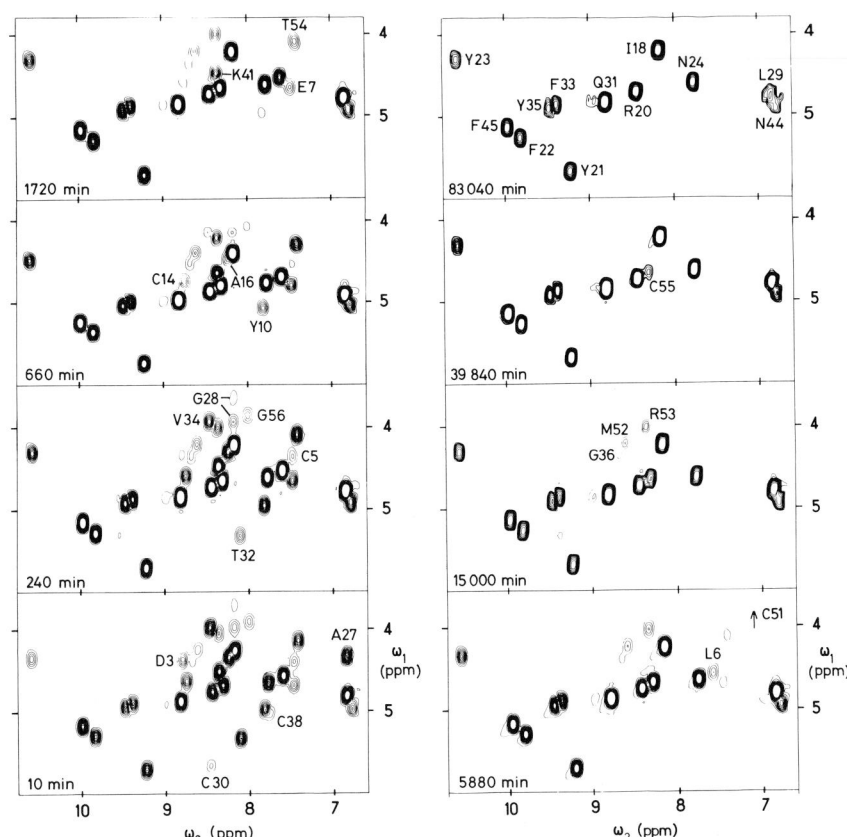

FIG. 3. Absolute value 500 MHz ^{1}H COSY spectra of 0.02 M solutions of BPTI in ^{2}H$_2$O recorded at different times after dissolving the protein. The solutions were freshly prepared at 24° and then kept at 36° to allow exchange of protein amide protons with ^{2}H of the solvent. At the times indicated in the figure, a particular solution was cooled to 24° and a spectrum was recorded in 12 hr. Only the region of the NH-C$^\alpha$H cross peaks is shown (ω_1 = 3.6–6.0 ppm, ω_2 = 6.6–10.8 ppm). The peaks that disappeared in the course of the experiment are identified in the last spectrum, where they can be observed readily. Thus this figure affords a qualitative survey of the exchange rates (for quantitative data, see Fig. 4). The peaks that did not disappear within 80,000 min are identified in the last spectrum.[24]

with an upward arrow. These rapidly exchanging amide protons corre-spond essentially to the protein surface, as can be seen from inspection of the solvent accessible surface area of the individual amide groups calcu-lated from the X-ray structure[24,36,37] (Fig. 5). All exchange rates slower

[36] B. Lee and F. M. Richards, *J. Mol. Biol.* **55**, 379 (1971).
[37] C. Chothia and J. Janin, *Nature (London)* **256**, 705 (1975).

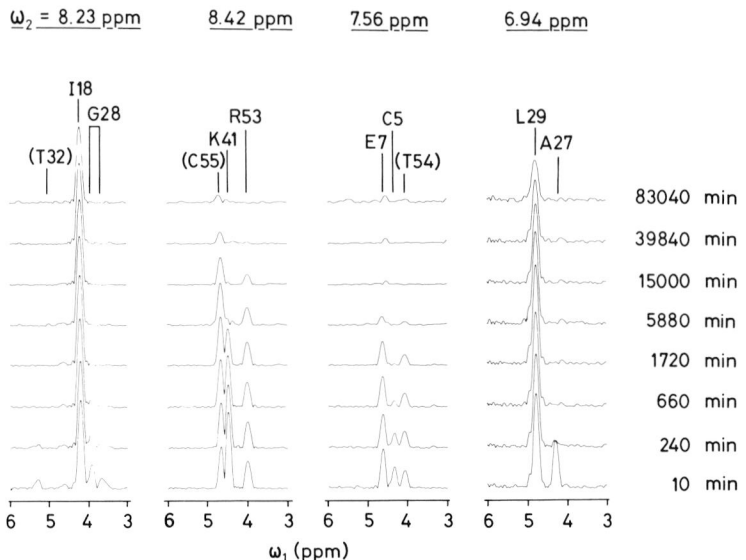

FIG. 4. Vertical cross-sections of the spectra shown in Fig. 3 taken at 4 different positions along the ω_2 axis. At the top of the figure, those residues are identified for which ω_2 coincides with ω_2 of the cross-section. The cross-peaks of the residues indicated in parentheses are located so close to the cross-sections that tails of the peaks are observed in these presentations.[24]

than 10^{-1} min^{-1} must correspond to amide groups that are shielded from solvent contact and therefore provide information on internal motions in the protein conformation. A meaningful interpretation of the exchange rates can be based on the following scheme of the exchange mechanisms[25,30,31]:

$$C(^1H) \underset{k_2}{\overset{k_1}{\rightleftharpoons}} O(^1H) \xrightarrow[^2H_2O]{k_3} O(^2H) \underset{k_1}{\overset{k_2}{\rightleftharpoons}} C(^2H) \qquad (3)$$

Closed states $C(^1H)$ are in equilibrium with open states $O(^1H)$. Exchange is possible only from the open states $O(^1H)$ and leads, in the presence of ^2H_2O, to deuteration of the peptide site. The experimentally observable overall exchange rate, k_m, is approximately[30,31]

$$k_m \cong \frac{k_1 k_3}{k_2 + k_3} \qquad (4)$$

There are two limiting kinetic situations. If $k_3 \gg k_2$ (EX_1 process) each opening of the "closed state," C, leads to an isotope exchange and we

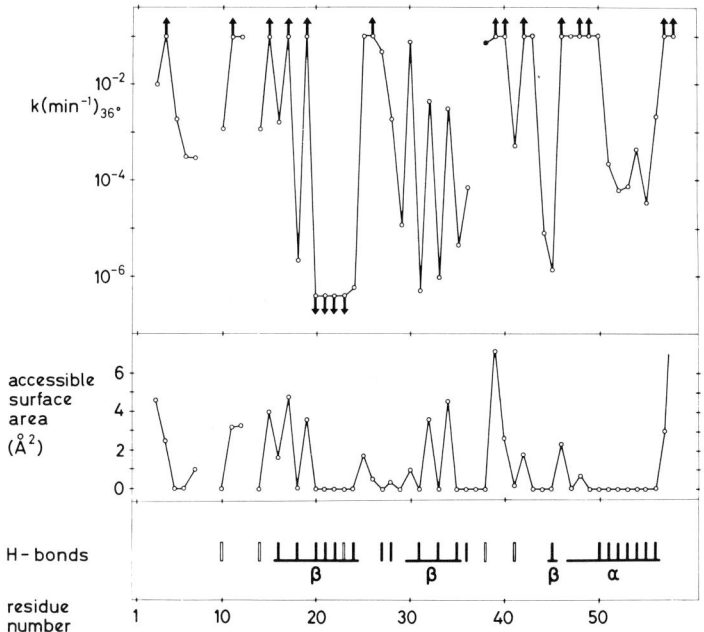

FIG. 5. Amide proton exchange data for BPTI at p^2H 3.5 and 36° and selected features of the crystal structure of BPTI. The horizontal scale represents the amino acid sequence of BPTI. The graph at the top is a logarithmic representation of the individual exchange rates, k_m, at 36°. Arrows pointing upward indicate that the exchange was too fast to be measured with the COSY experiments. Arrows pointing downward indicate that the exchange was too slow to be measured quantitatively within the maximum exchange time. In the lower part of the figure the static accessible surface area is plotted for the backbone peptide nitrogens, and the intramolecular hydrogen bonds in the crystal structure are indicated by filled vertical bars (intramolecular H-bonds with main chain carbonyls) or open bars (H-bonds to side chains or internal water molecules). Residues which are part of regular α or β secondary structure are joined by horizontal bars.[24,25]

have

$$k_m = k_1 \tag{5}$$

In this case the exchange rate gives directly the opening rate of the protein fluctuation. The pH and temperature dependences of the protein fluctuations are directly obtained from k_m. If two amide protons are adjacent in the three-dimensional structure of the protein and opening exposes both labile protons simultaneously, both protons should exchange in a correlated way.[38] If $k_3 \ll k_2$ (EX_2 process) only a small portion of all openings

[38] G. Wagner, *Biochem. Biophys. Res. Commun.* **97**, 614 (1980).

leads to exchange of internal protons and

$$k_m = (k_1/k_2)k_3 \qquad (6)$$

In this case no correlated exchange would be expected for neighboring protons,[39] and the parameters of the proton fluctuations are masked by the intrinsic exchange rates, k_3. If k_3 is known, then the equilibrium constant k_1/k_2 can be determined. The pH and temperature dependence of k_m contains contributions of k_1, k_2, and k_3. k_3 is directly proportional to the hydroxyl or hydronium ion concentration in the base or acid catalyzed regime, respectively. k_3 has been measured experimentally with small model peptides.[30,31,39]

A plot of $\log k_3$ vs p^2H is V shaped, it decreases with slope -1 below p^2H 3, and increases with slope $+1$ above p^2H 3. For EX_2 exchange in a protein a plot of $\log k_m$ vs p^2H should show the same V shape, which would, however, be shifted to slower rates by $\log (k_2/k_2)$.

Two criteria can be used for distinguishing experimentally between EX_1 and EX_2 processes, the p^2H dependence of the exchange rates[30,31] and measurements of nuclear Overhauser effects (NOE).[39] The appearance of a V-shaped curve for the pH dependence of the exchange rates indicates an EX_2 process. This criterion is, however, not a firm proof since k_1 may sometimes have a similar pH dependence as k_3. Another method for distinguishing between EX_1 and EX_2 processes analyzes the NOE between adjacent labile protons, i.e., the transfer of saturation from one labile proton to an adjacent one via dipole–dipole interaction.[39] This is demonstrated in Fig. 6. If two neighboring protons exchange in a correlated way (EX_1) the apparent NOE between the two protons should be the same in the fully protonated protein and in the case when the two peptide sites are half exchanged, since either both sites are protonated or both are deuterated in each individual molecule. In the uncorrelated case (EX_2) the NOE between the two labile protons should decrease in the partially deuterated protein, since mixed pairs of protonated and deuterated peptide sites appear for which no $^1H-^1H$ NOE can be obtained. With this method the exchange of labile protons has been analyzed in BPTI.[40] It was found that under most experimental conditions the exchange follows an EX_2 mechanism. However, correlated exchange (EX_1 mechanism) for some labile protons of the central β-sheet is observed over a small range of temperature and pH ($T > 50°$, pH $\sim7-10$), a fact that could not be derived from the pH dependence of the exchange rates alone.

[39] R. S. Molday, S. W. Englander, and R. G. Kallen, *Biochemistry* **11**, 150 (1972).
[40] H. Roder, Ph.D. thesis No. 6932, ETH Zürich, 1981.

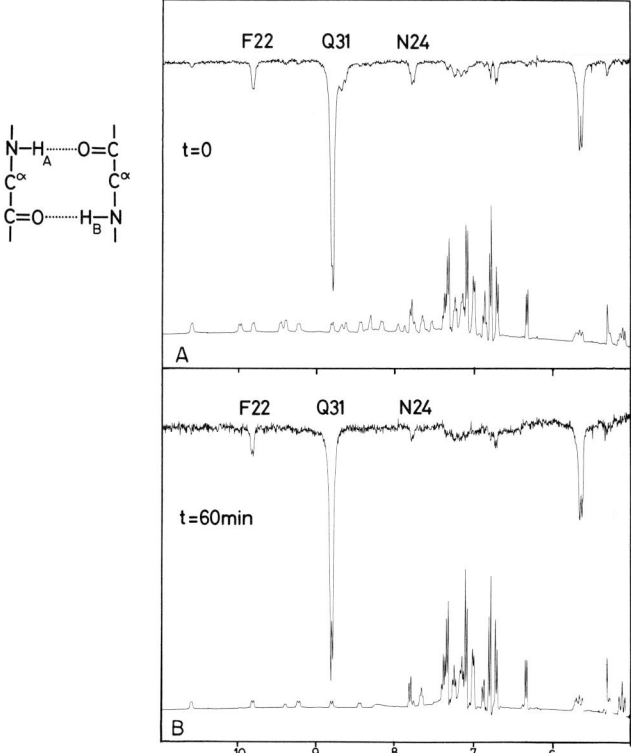

A

FIG. 6. (A) 360 MHz NOE differences spectrum of a 20 mM solution of BPTI in ^2H_2O at p^{2}H 4.6, 24°. Lower trace: reference spectrum; upper trace: NOE difference spectrum. (B) Same experiment but prior to the measurement the sample was kept at 60°, p^{2}H 8.0 for 1 hr to partially exchange the amide protons. Left-hand side: schematic representation of two adjacent hydrogen bonds connecting opposite strands of an antiparallel β-sheet. The two amide protons H_A and H_B are separated by approximately 2.6 Å. The relative magnetization transfer is the same in the fresh solution and in the partially exchanged sample (~12%). This indicates correlated exchange at these conditions.[38]

Aromatic Ring Flips and Other Rotational Motions of Amino Acid Side Chains

The rotational mobility of phenylalanine or tyrosine side chains in the anisotropic environment of the protein interior is manifested by the symmetry of the spin systems. A rotating tyrosine ring has a symmetric two-line spectrum of AA′BB′ symmetry, while a rigid side chain will generally show an unsymmetric four-line spectrum.[11-15] The transition from slow to fast rotation can be observed by variation of the temperature. This is demonstrated in Fig. 7 for the aromatic side chains of BPTI.[14,15] The

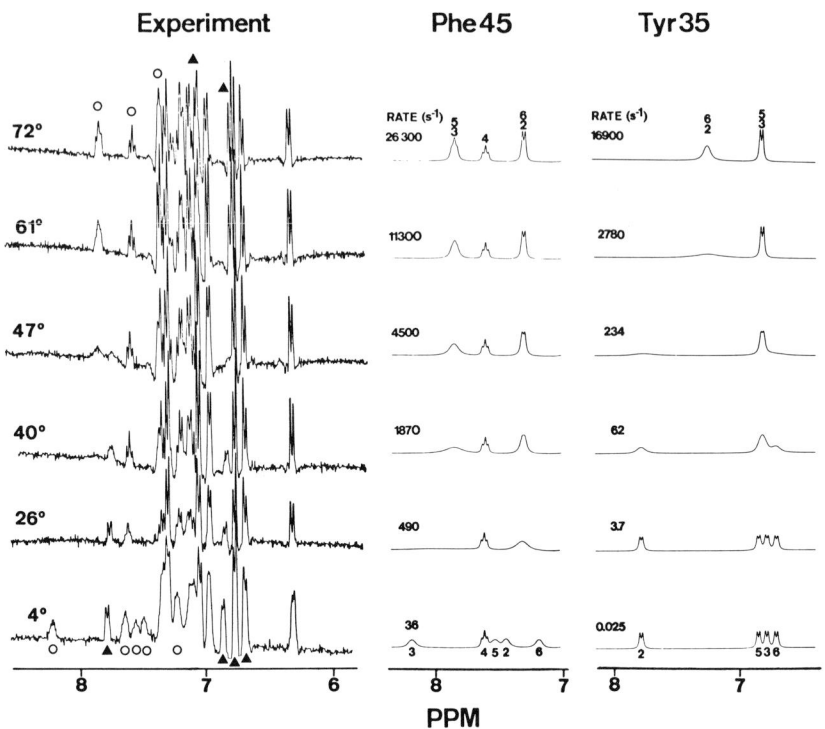

FIG. 7. Temperature dependence of the aromatic resonances in the 360 MHz ^{1}H NMR spectrum of BPTI. For Tyr-35 and Phe-45, the spectra are simulated individually and the flip rates at different temperatures obtained from the best fit with the experimental data are indicated. In the experimental spectrum at 4° the resonances of four protons of Phe-45 (○) and two protons of Tyr-35 (▲) are readily recognized, whereas the other lines are masked by resonances of the other aromatics in the protein. Most of the resonance lines of Phe-45 and Tyr-35 are also resolved in the spectra at higher temperatures and the transition from slow to rapid ring flipping is readily apparent.[15]

averaged resonances of the chemically equivalent ring protons are almost exactly in the middle of the resonance positions of the corresponding lines at low temperature. This is evidence that the averaging is only between two states, i.e., between two orientations of the ring corresponding to two indistinguishable energy minima. This means that the time spent in equilibrium orientation is long relative to the time used for the flip motion. Thus, ring flips are rare events compared with the lifetime of a particular ring orientation of minimum energy. The frequencies of the 180° flips (i.e., a two site exchange) can be determined quantitatively by line shape analysis.[10–15] Saturation transfer techniques[26–28] have also been used as an al-

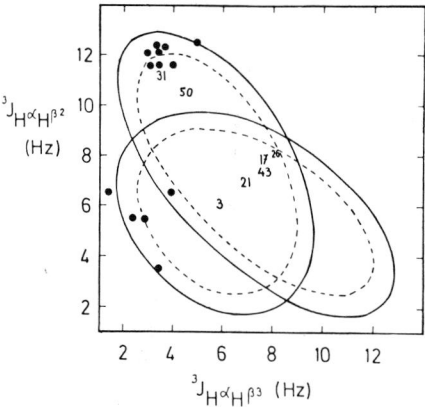

FIG. 8. Correlation diagram for the vicinal coupling constants $J_{H\alpha H\beta 2}$ and $J_{H\alpha H\beta 3}$ in amino acid residues. The solid curve represents the correlation between the two vicinal proton–proton coupling constants in the molecular fragment $C^\alpha H$-$C^\beta H_2$ for a rigid molecule. The broken line represents the correlation for a flexible molecule, where it was assumed that the time dependent variations of χ^1 extended over a range of $\chi_0^1 \pm 30°$ and that within this range each value of χ^1 was equally populated. (Obviously, for different types of fluctuations, e.g., for a harmonic fluctuation about χ_0^1, the dotted curve would correspond to a situation where considerably larger amplitudes than $\pm 30°$ might occur.) The diagram contains data for BPTI in ^2H_2O solution at p^2H 7.0 and 68°. Since the measured coupling constants had not been assigned stereospecifically, all the data were arbitrarily plotted in the upper left triangle of the correlation diagram. Data points located on the two curves or in the narrow band between the curves are indicated by the filled circles, the other ones by the position of the amino acid in the sequence.[29]

ternative method for measuring flip rates of some aromatic side chains of cytochrome c.[13]

In amino acids with a β-methylene group and hence two correlated vicinal coupling constants, $^3J_{H\alpha H\beta 2}$ and $^3J_{H\alpha H\beta 3}$, studies of the correlation between the two coupling constants (Fig. 8) provide a basis for qualitative distinction between different limiting dynamic situations.[29] In the correlation diagram of Fig. 8 data points located on the curves or in the narrow band between the solid and the dotted curve are compatible with a situation where the amino acid residues are locked in unique positions χ_0^1, with the internal mobility about χ^1 restricted to rapid fluctuations about this position. On the other hand, correlation points located within the area bounded by the peripheral branches of the dotted curve may indicate rapid averaging between two or several distinct values of χ^1 (e.g., the classical gauche, gauche and trans conformers about single bonds between tetrahedral carbon atoms). However, correlation points in the central area of Fig. 8 can also result for immobilized side chains if the two C^β methylene protons have identical chemical shifts. Therefore, in order to obtain unambiguous evidence for a mobile side chain, the data on the correlation of the spin–spin couplings must be complemented with additional experiments. Figure 8 shows data obtained for BPTI, where nearly all amino acid side chains in the interior of the protein were found to be locked into unique spatial orientations, with the mobility restricted to rapid rotational fluctuations about a unique value for the dihedral angle χ^1.

The most fruitful application of the correlation diagram of Fig. 8 appears to be for studies of surface residues in globular proteins, where this technique can provide an unambiguous identification of immobilized side chains. Measurements of the spin–spin coupling constants $^3J_{\alpha\beta}$ can be obtained with phase sensitive COSY experiments recorded with high digital resolution in ω_2,[41] or by 2D J-resolved spectroscopy.[29]

Spin Relaxation Measurements

For spin 1/2 nuclei the relaxation is usually dominated by internuclear interactions, so that the measured relaxation times must be correlated with both the internuclear distances and the effective correlation time for the modulation of the interactions. For studies of protein motility it is preferable to select, therefore, molecular fragments in which two or several nuclei are located at fixed relative distances by the covalent bonds. Carbon atoms which are covalently linked with hydrogen atoms are particularly suitable, since the ^{13}C relaxation is usually largely dominated by ^1H–^{13}C dipole–dipole coupling. ^{13}C relaxation parameters then vary with the correlation function and hence with the overall rotational mobility of the observed ^{13}C–^1H groups.[4,7] Figure 9 shows a measurement of the longitudinal relaxation times, T_1, for the methyl carbons in BPTI.[8] Even when T_1 measurements are obtained at different frequencies and complemented by studies of nuclear Overhauser enhancements and transverse relaxation times, T_2, the experimental relaxation parameters are usually not sufficient to characterize a unique type of motion for the observed group of atoms. A model which is compatible with the presently available relaxation data for the methyl carbons in BPTI is presented in Fig. 10.[8] In addition to the overall rotational tumbling of the protein and rotation of the methyl groups about the bond through which they are linked with the protein, the model invokes a "wobbling motion" of the methyl rotation axis. A sample of parameters obtained for the motility of the methyl groups is given in the legend to Fig. 10. The alanine methyls are of special interest, since their wobbling manifests flexibility of the polypeptide back-

[41] D. Marion and K. Wüthrich, *Biochem. Biophys. Res. Commun.* **113**, 967 (1983).

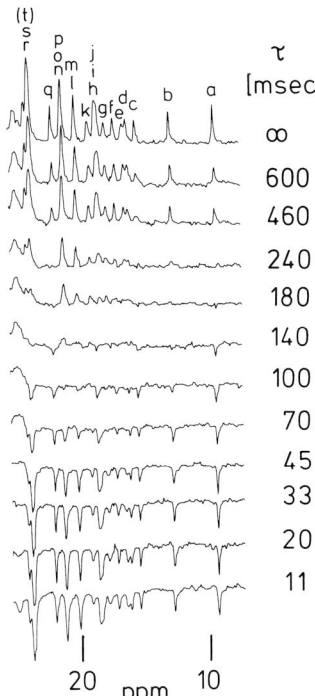

FIG. 9. Measurement of ^{13}C relaxation times T_1. High field region from 5 to 25 ppm of partially relaxed proton noise-decoupled ^{13}C NMR spectra at 25.1 MHz of a 0.025 M solution of BPTI in ^2H_2O, p^{2}H 4.2, $T = 40°$. The spectra were obtained with a $(180°–\tau–90°–$ acquire-t) pulse sequence, with $t = 1.6$ sec. The delay times τ are indicated on the right. The letters in the top trace indicate individual assignments of the methyl carbon lines. (a, b) Ile-18δ and Ile-19δ; (c) Met-52ε; (d) Ile-19γ; (e) Ala-48β; (f) Ile-18γ; (g) Ala-25β; (h, i) Ala-16β, Ala-40β; (j) Ala-58β; (k) Ala-27β; (m) Thr-11γ; (n) Thr-54γ; (o) Thr-32γ; (q) Leu-6δ; (l, p, r, s, t) Leu-6δ, Leu-29δ, Leu-29δ, Val-34γ, Val-34γ.[8]

bone. Other models have also been proposed which would be compatible with the presently available data on BPTI.[9,10]

In contrast to measurements of carbon longitudinal relaxation times, longitudinal ^1H relaxation times are less suitable for studies of dynamic phenomena since they are usually dominated by dipole–dipole interaction to more than one other nucleus. This is not the case when one measures individual cross-relaxation rates, σ_{ij}, between pairs of nuclei via NOE.[42,43] The cross-relaxation rate, σ_{ij}, between spins i and j with resonance frequencies ω_i and ω_j is determined by the intramolecular distance,

[42] G. Wagner and K. Wüthrich, *J. Magn. Reson.* **33**, 675 (1979).
[43] A. A. Bothner-By and J. N. Noggle, *J. Am. Chem. Soc.* **101**, 5152 (1979).

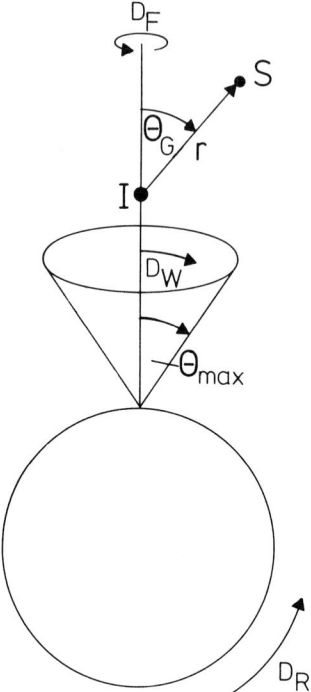

FIG. 10. "Wobbling in a cone" model used for the interpretation of ^{13}C relaxation data in BPTI. The relaxation of spin I ($\equiv ^{13}C$) by dipole–dipole coupling with spin S ($\equiv ^{1}H$) is considered. The two spins are located in a spherical particle which undergoes isotropic rotational motions with the diffusion constant D_R. The vector **r** which connects the two spins is attached at a fixed angle θ_G to an axis about which it rotates with a diffusion constant D_F. This rotation axis is further allowed to wobble with a diffusion constant D_w inside a cone defined by the half-angle θ_{max}. From an analysis of the methyl carbon relaxation parameters (Fig. 9) the following parameters for the molecular mobility of BPTI resulted from this analysis: for the overall rotational motions, $\tau_R = 4 \times 10^{-9}$ sec; for the librational "wobbling" of the backbone carbons of alanine in a cone with $\theta_{max} = 20°$, $\tau_w = 1 \times 10^{-9}$ sec; for the librational motions of individual aliphatic side chains in cones with θ_{max} varying between 30 and 60°, $\tau_w = 4 \times 10^{-10}$ to 3×10^{-9} sec; for methyl rotation about the C–C bond, $\tau_F \lesssim 1 \times 10^{-11}$ sec.[8]

r_{ij}, between the two spins and the rotational correlation time, τ_c, for rotational tumbling of the internuclear vector $\mathbf{r}_{ij}$:

$$\sigma_{ij} = \frac{\hbar^2 \gamma_i^2 \gamma_j^2}{10 r_{ij}^6} \left[\frac{6\tau_c}{1 + (\omega_i + \omega_j)^2 \tau_c^2} - \frac{\tau_c}{1 + (\omega_i - \omega_j)^2 \tau_c^2} \right] \tag{7}$$

The cross relaxation rate, σ_{ij}, can be determined experimentally from the initial build-up of the NOEs.[42,43] For nuclei at a well-defined distance, such as two methylene protons or two protons of an aromatic side chain, the value of σ_{ij} can be used to determine the rotational correlation time, τ_c.[44] If τ_c is considerably shorter than the correlation time for the overall rotation of the protein, an internal motion is indicated. This has been analyzed in some detail for the tryptophan side chains in lysozyme.[44,45]

Activation Enthalpies, Activation Entropies, and Activation Volumes from NMR Studies at Variable Temperature and Pressure

Whenever rates can be determined quantitatively, a further characterization of internal motions can be obtained by variation of temperature and pressure. According to Eyring's theory for rate processes[46] the rate k is given by

$$k = \frac{k_B T}{h} \exp - \left(\frac{\Delta H^{\ddagger}}{RT} - \frac{\Delta S^{\ddagger}}{R} \right) = \frac{k_B T}{h} \exp - \left(\frac{\Delta E^{\ddagger}}{RT} + \frac{P\Delta V^{\ddagger}}{RT} - \frac{\Delta S^{\ddagger}}{R} \right) \quad (8)$$

From a plot of $\ln k$ vs $1/T$ or vs P the activation enthalpy, $\Delta H^{\ddagger}$, the activation entropy, $\Delta S^{\ddagger}$, or the activation volume, $\Delta V^{\ddagger}$, respectively, can be determined. Eyring's theory allows a straightforward evaluation of the data. It has been pointed out, however, that it neglects frictional effects in kinetic processes and may thus lead to irrelevant values for the energy parameters of the activated state.[25,47-49] Kramers[50] has formulated an alternative theory for rate processes which would allow a better evaluation of kinetic data, provided that some information about local frictional coefficients or local viscosities, η, respectively, are available. In the latter theory we have

$$k = f(\eta) \exp - \left(\frac{\Delta E^{*}}{RT} + \frac{P\Delta V^{*}}{RT} - \frac{\Delta S^{*}}{R} \right) \quad (9)$$

where $f(\eta)$ is proportional to η or η^{-1} in the limits of low and high viscosity, respectively. Since the viscosity, η, may also vary with temperature

[44] E. T. Olejniczak, F. M. Poulsen, and C. M. Dobson, *J. Am. Chem. Soc.* **103**, 6574 (1981).
[45] C. M. Dobson, E. T. Olejniczak, F. M. Poulsen, and R. G. Ratcliffe, *J. Magn. Reson.* **48**, 97 (1982).
[46] H. Eyring, *J. Chem. Phys.* **3**, 107 (1935).
[47] D. Beece, L. Eisenstein, H. Frauenfelder, D. Good, M. C. Marden, L. Reinisch, A. H. Reynolds, L. B. Sorensen, and K. T. Yue, *Biochemistry* **19**, 5147 (1981).
[48] J. A. McCammon and M. Karplus, *Biopolymers* **19**, 1375 (1980).
[49] M. Karplus and J. A. McCammon, *FEBS Lett.* **131**, 34 (1981).
[50] H. A. Kramers, *Physica* **7**, 285 (1940).

or pressure, the values of $\Delta H^{\ddagger}$, $\Delta S^{\ddagger}$, or $\Delta V^{\ddagger}$ determined with Eyring's theory may be biased by the variation of η with temperature and pressure, respectively. Assuming that the internal viscosity in proteins behaves similarly to that of normal, liquid hydrocarbons[25,49] we have

$$\left(\frac{\partial \ln \eta}{\partial\ 1/T}\right)_{p=1\ \mathrm{bar}} \cong 800\ \mathrm{K} \tag{10}$$

and

$$\left(\frac{\partial \ln \eta}{\partial\ p}\right)_{T=25°} \cong 8 \times 10^{-4}\ \mathrm{bar}^{-1} \tag{11}$$

This corresponds to apparent activation enthalpies and activation volumes of 1.6 kcal/mol and 21 cm³/mol, respectively. Activation enthalpies that have been determined on the basis of Eyring's theory for flips of aromatic side chains[14] or for the exchange of labile protons are much larger than 1.6 kcal/mol. It appears therefore for these particular processes that Eyring's theory gives quite relevant energy parameters. Some activation volumes for internal motions in proteins that have been determined on the basis of Eyring's theory are not much larger than 21 cm³/mol and have therefore to be interpreted with caution.[25,51]

Sequence-Specific Resonance Assignments

NMR spectra contain a large number of resolved signals which can be used to study motional effects. The value of these studies depends crucially on the assignment of the resonances to individual atoms of the protein, since this allows a location of the various internal motions in the protein. In the last few years assignment techniques have become available where resonances are identified sequentially along the polypeptide backbone. These techniques are strongly facilitated by the use of two-dimensional NMR. At present nearly complete assignments of the proton NMR spectra are available for a number of small proteins. When assignments of the proton NMR spectrum are available, the resonances of protonated carbons can be assigned by heteronuclear spin decoupling in 1D experiments,[52] or by 2D heteronuclear correlated spectroscopy.[53]

[51] G. Wagner, *FEBS Lett.* **112**, 280 (1980).
[52] R. Richarz and K. Wüthrich, *Biochemistry* **17**, 2263 (1978).
[53] T.-M. Chau and J. L. Markley, *J. Am. Chem. Soc.* **104**, 4010 (1982).

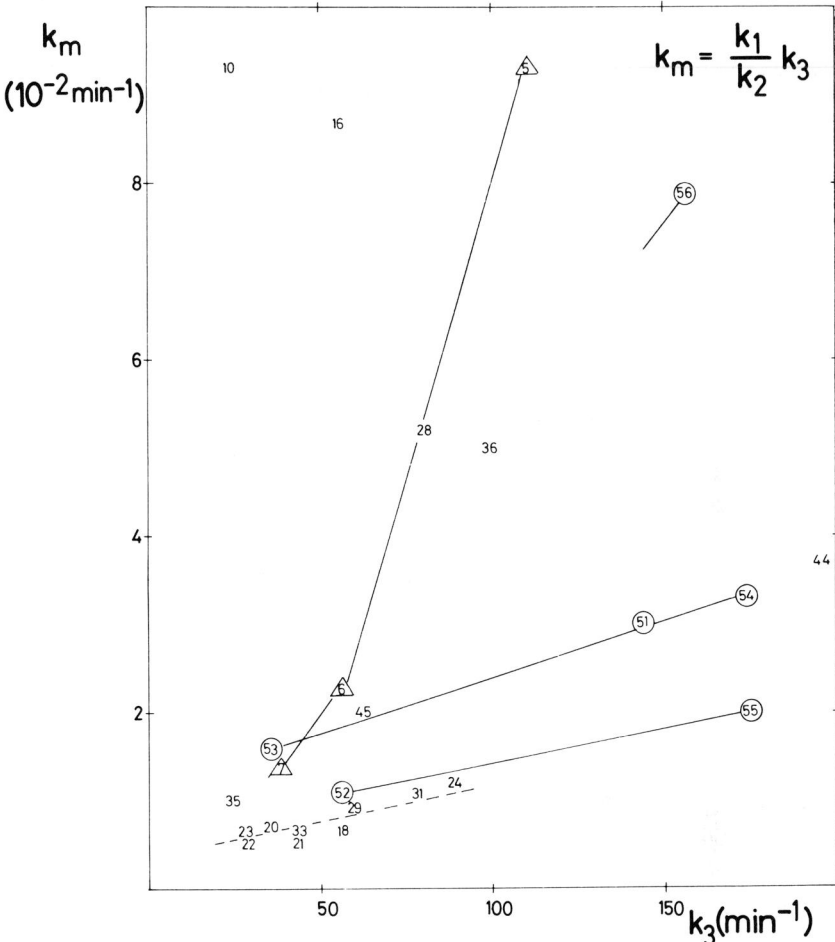

FIG. 11. Plot of the exchange rates, k_m, of the individual amide protons of BPTI vs their intrinsic exchange rates, k_3, at 68°, p^2H 3.5. The data points are identified with the number in the amino acid sequence. Location in the β-sheet, the C-terminal α-helix, and the N-terminal 3_{10}-helix is indicated by $\bigcirc$ and $\triangle$, respectively. In the domain of EX_2 exchange, amide protons that get solvent contact only by the same fluctuation should appear on a straight line. Groups of protons in the same secondary structure, for which such a correlation seems to be possible, are connected in the figure.[25]

Concerted Motions

The sequence-specific resonance assignments are particularly useful for amide exchange studies, since amide protons represent a large number of internal NMR probes for the mapping of internal motions.[25] In the regime of EX_2 exchange some information can be obtained about concerted motions by comparison of the exchange rates of individual amide protons. For this comparison we plot the exchange rates, k_m, vs the respective intrinsic exchange rates, k_3, for all internal amide protons (see Fig. 11 for BPTI). Since we have EX_2 exchange, the exchange rates of all amide protons that exchange due to the same internal motion should appear on a straight line, and the slope of this line would be the equilibrium constant, k_1/k_2, of the respective opening reaction [Eq. (6)]. Figure 11 shows that the exchange in the N-terminal 3_{10}-helix could be explained by a single fluctuation. In the C-terminal α-helix at least three different fluctuations have to be assumed. In the central β-sheet nine protons show a certain correlation with a correlation coefficient of 0.9 (Phe-22, Tyr-23, Arg-20, Tyr-21, Phe-33, Ile-18, Leu-29, Gln-31, and Asn-24). Additional fluctuations have to be assumed for the peripheral parts of the β-sheet (Tyr-35, Phe-45, Ala-16), and for the β-turn.[25]

Similar to the mapping of internal motility from amide proton exchange studies, different maps are obtained from evaluation of the flip frequencies for individually assigned aromatic rings[22] or the relaxation parameters of individually assigned carbon atoms.[8] Obviously, one then has the possibility of comparing the motility maps obtained from the different experiments. In the case of BPTI such comparisons have resulted in the proposal of a "hydrophobic domain architecture"[22,54] to explain the apparent occurrence of different types of concerted internal fluctuations.

Acknowledgments

Financial support by the Swiss National Science Foundation (project 3.284-0.82) is gratefully acknowledged.

---

[54] K. Wüthrich, G. Wagner, R. Richarz, and W. Braun, *Biophys. J.* **32**, 549 (1980).

[17] Protein Dynamics by Solid State Nuclear Magnetic Resonance

By STANLEY J. OPELLA

Introduction

The understanding of molecular structure and the understanding of molecular dynamics are inseparable. The descriptions of the structures of proteins seem to be highly refined[1]; however, this is due in part to the easily perceived representations that result from the analysis of X-ray diffraction data obtained from single crystals. In fact, descriptions of molecular properties based on such time-independent representations of structure are of limited value, since the functions of molecules such as enzymes occur in real time, which may vary from 10^{-3} to 10^{-12} sec, through fluctuations in structure.[2] Even if sufficient theoretical or experimental data were available, the time-dependent structures of molecules would be difficult to describe, because a language that is appropriate for this purpose is not fully developed as yet. This reflects the relatively early stage of research in the field.

The study of protein dynamics is now an extremely active area of research. Detailed descriptions of some of the motions present in native proteins are possible in favorable cases. This is primarily a consequence of the application of recently developed theoretical, spectroscopic, and scattering methods to proteins. Interestingly, the greatest advances have been made for proteins in the solid state. This is obvious in the case of highly refined diffraction studies of proteins that reveal small amplitude motions through temperature factors,[3] as well as for molecular dynamics calculations[4] that rely on the crystallographic coordinates to provide initial atomic positions. Solid state NMR studies of protein dynamics have also been successful in providing new information about the intramolecular motions in immobile samples. Here, the advantages of avoiding the influences of overall reorientation and analyzing the effects of motional averaging on solid state NMR spectral parameters greatly outweigh the technical difficulties inherent in the experiments. These solid state NMR studies of protein dynamics are the subject of this chapter.

[1] J. S. Richardson, Adv. Protein Chem. 34, 167 (1981).
[2] F. R. N. Gurd and T. M. Rothgeb, Adv. Protein Chem. 33, 73 (1979).
[3] G. A. Petsko and D. Ringe, Ann. Rev. Biophys. Bioeng. 13, 331 (1984).
[4] M. Karplus and J. A. McCammon, Ann. Rev. Biochem. 53, 263 (1983).

METHODS IN ENZYMOLOGY, VOL. 131

All of the diverse theoretical and experimental methods have detected intramolecular motions in native proteins. The motions of the polypeptide backbone and amino acid side chains range from very small amplitude angular dispersions of librational motions to large amplitude jumps of aromatic rings and aliphatic side chains and isotropic reorientation of some residues. Both large and small amplitude motions occur very rapidly, but vary enormously in frequency of occurrence reflecting the size of local energy barriers.

The view of protein dynamics that is emerging from these studies is that these large and highly structured molecules have many properties previously associated only with fluids. They are continuously fluctuating in the sense of multiple motions occurring frequently. Apparently these motions are in the form of discrete jumps among distinct stable conformations. Some of these processes, such as ring flips, result in the same structure, and some, such as asymmetric aliphatic side chains hopping around tetrahedral carbon centers, result in quite different local structures. Those portions of a protein undergoing rapid isotropic reorientation are, by definition, unstructured.

Solid State NMR of Biological Systems

Solid State NMR

The value of NMR spectroscopy in providing information about the motions of molecules has been appreciated from the very beginning.[5,6] The observables in NMR experiments reflect the nuclear spin interactions that are present. The spin interactions are strongly affected by motion. The motions of the spins, from whatever source, have two consequences; one is an averaging of the effects of spin interactions on the spectral line shapes and the other is the induction of spin relaxation. The dynamics of the molecules affect both the basic resolution and sensitivity of the experiments, as well as other aspects of the spectroscopy. If the molecular sites of interest do not reorient rapidly and isotropically, then their resonances will be naturally broad and weak. In liquids, drastic narrowing of resonances occurs because the large amplitude, rapid motions associated with the overall reorientation of the molecules completely average the nuclear spin interactions to their isotropic values. In high-resolution solid state NMR, radio frequency irradiations and mechanical sample spinning replace molecular motions as line narrowing mechanisms.

One of the goals of solid state NMR has always been to obtain well-

[5] N. Bloembergen, E. M. Purcell, and R. V. Pound, *Phys Rev.* **73,** 679 (1948).
[6] E. R. Andrew and R. G. Eades, *Proc. R. Soc. Ser. A* **218,** 537 (1953).

resolved spectra in situations not amenable to conventional high-resolution NMR experiments because of the absence of liquid-like motions. Thus, solid state NMR extends the range of the method to large biological structures that are immobile in solution, such as nucleoprotein or membrane–protein complexes, and to crystalline or amorphous solid samples of biological molecules.[7-13] In some solids, the molecules have so much motion that liquid-like signals are observed. By contrast, some molecular complexes tumble so slowly in solution because of their size and shape that essentially no averaging of the spin interactions occurs and the resonance properties are those of solids. Additional research goals are possible for solid state NMR studies because the lack of complete motional averaging of the nuclear spin interactions results in the angular dependence of the spin interactions being directly manifested in the observable spectral parameters. Fundamental information about molecular structure and dynamics can be obtained that is not present in the corresponding solution experiments.

This chapter is concerned with the design, execution, and interpretation of NMR experiments that give data on unaveraged or partially averaged nuclear spin interactions. The basic idea is that *all* manifestations of spin interactions are strongly influenced by motion. The precise nature of this influence depends on the spin interactions that are present at the site of interest, the rates or frequencies of motions, the amplitudes of motions, and experimental details such as the nuclear spin resonance frequencies. This chapter is not meant to encompass all NMR studies of dynamics in solid samples, but rather only those cases pertinent to current research on proteins. There is a general review on the subject of NMR studies of dynamics and it provides the necessary physical and mathematical detail for understanding the roles of motion in solid state NMR.[14]

The solid state NMR approaches that are most successful for studying molecular dynamics rely on the concept of dilute spins to provide a natural isolation of the nuclear spin interactions of interest. Proton NMR is almost always an abundant spin endeavor, where the mutual interactions among nearby protons are the dominant influence. Valuable, but limited information has been derived from direct studies of proton resonance

[7] D. A. Torchia and D. L. VanderHart, *Top. Carbon-13 NMR Spectrosc.* **3,** 3251 (1979).

[8] R. G. Griffin, this series, Vol. 72, p. 108.

[9] S. J. Opella, *Annu. Rev. Phys. Chem.* **33,** 533 (1982).

[10] D. A. Torchia, this series, Vol. 82, p. 174.

[11] E. Oldfield, R. A. Kinsey, and A. Kintanar, this series, Vol. 88, p. 310.

[12] C. A. Fyfe, "Solid State NMR for Chemists." CFC Press, Guelph, 1983.

[13] D. A. Torchia, *Ann. Rev. Biophys. Bioeng.* **13,** 125 (1984).

[14] H. W. Spiess, *NMR Basic Princ. Prog.* **15,** 55 (1982).

properties of proteins in the solid state.[15] The spin $S = 1/2$ nuclei of ^{13}C, ^{15}N, and ^{31}P are available in natural abundance or isotopically enriched sites in proteins. These dilute spins interact strongly with the abundant ^1H spins, but not among themselves. They usually have large, nonaxially symmetric chemical shielding tensors. Therefore, the dominant spin interactions are heteronuclear dipolar couplings and chemical shift anisotropy. The weak homonuclear dipolar couplings that are present can be detected in some cases. Deuterium with spin $S = 1$ has its NMR properties dominated by the effects of the nuclear quadrupole interaction. ^{14}N with spin $S = 1$ is the abundant form of nitrogen and even in those situations where it is not observed directly, its large quadrupole coupling has a substantial influence on spectral parameters of other nuclei.

Research in biological NMR spectroscopy is interdisciplinary. A variety of biological, chemical, and spectroscopic factors must be considered in the design of solid state NMR experiments. The biological factors are the most obvious, since they concern the choice of protein, the type of sample preparation, and what sites are crucial to the structure, function, or dynamics of the molecule. The source and preparation of the samples are also of importance in providing schemes for incorporation of stable isotopes. The chemical aspects are concerned with details of the sites of labels, the types of motions that are expected to be observed, and the properties of the labeled species. The spectroscopic considerations include the time scales of interest, the directions of the principal axes of the tensors, how the motions influence the spectral parameters, and how to carry out the experiments to observe the parameters of interest. The goals are to simplify the biophysical problems to manageable levels in order to obtain interpretable spectral data.

^2H NMR

Deuterium with spin $S = 1$ has a relatively large quadrupole coupling constant and a relatively small gyromagnetic ratio. Therefore, line shapes and relaxation properties of ^2H resonances are determined almost exclusively by the nuclear quadrupole interaction. ^2H NMR is a useful way of studying the dynamics of many biological systems.[8,16–18] Because C—^2H bonds can be arranged to replace C—^1H bonds by organic synthesis, site specific information can be obtained in molecules as complex as proteins.

[15] E. R. Andrew, D. N. Bone, D. J. Beyant, E. M. Casbell, R. Casper, and Q. A. Meng, *Pure Appl. Chem.* **54**, 585 (1982).

[16] J. Seelig, *Q. Rev. Biophys.* **10**, 353 (1977).

[17] H. R. Mantsch, H. Saito, and I. C. P. Smith, *Prog. NMR Spectrosc.* **11**, 211 (1977).

[18] R. E. Jacobus and E. Oldfield, *Prog. NMR Spectrosc.* **14**, 113 (1981).

Since the ^{2}H quadrupole interaction is so much larger than the chemical shift or dipolar interactions, ^{2}H NMR spectra reflect directly this quadrupole spin interaction without a great deal of spectroscopic manipulation. The quadrupole spin interaction arises from the electrostatic interaction of the nuclear quadrupole moment with the electric field gradient of the C—^{2}H bond. Since deuterium has nuclear spin quantum number $S = 1$, there are two allowed transitions resulting in two resonance frequencies symmetric about zero. The electric field gradient is characterized by a symmetric second rank tensor; as a consequence, the resonance frequencies depend on the orientation of the C—^{2}H bond with respect to the applied magnetic field.

In all cases of interest to the study of biological systems the C—^{2}H bond has favorable resonance properties. The quadrupole coupling constant ($\sim$180 kHz) is large enough to be the dominant spin interaction in a single resonance experiment. The frequency breadth of ^{2}H powder patterns and oriented doublets are sufficiently large to be measured accurately while being small enough to be measured with conventional spectrometers. Because of the frequency breadth of experimental spectra, they are recorded using the quadrupole echo pulse sequence.[19] The spectra result from Fourier transformation of the free induction decays. The quadrupole echo has the advantage of using the refocusing ability of the nuclear spins to avoid problems caused by the finite transient response time of the spectrometer. For a truly rigid lattice, excellent representations of the spectra can be obtained in this way. However, this pulse sequence has the disadvantage of allowing time to go by following the initial pulse before beginning to record the free induction decay. In mobile and strongly coupled systems the spins can undergo processes that affect the intensities and line shapes of the signals.[20]

^{13}C NMR

A much wider variety of NMR spectroscopy is performed on biological systems with carbon than with deuterium.[9,21] This reflects the opportunities presented by the various carbon sites, including carbonyl and quaternary carbons, and the presence of two spin interactions with ^{13}C chemical shift anisotropy and ^{13}C—^{1}H dipole–dipole couplings available for analysis. It also reflects the difficulties in dealing with a nucleus where approximately 1% of the sites in natural abundance are ^{13}C, since even in

[19] J. H. Davis, K. P. Jeffrey, M. Bloom, M. F. Valic, and T. P. Higgs, *Chem. Phys. Lett.* **42**, 390 (1976).
[20] H. W. Spiess and H. Sillescu, *J. Magn. Reson.* **42**, 381 (1981).
[21] S. J. Opella, J. G. Hexem, M. H. Frey, and T. A. Cross, *Philos. Trans. R. Soc. London Ser. A* **299**, 665 (1981).

labeled molecules the background from the natural abundance is a significant factor.

The chemical shift interaction is of primary interest in studying ^{13}C, ^{15}N, and ^{31}P sites. For ^{31}P it is essentially the sole interaction of interest, while most ^{13}C and ^{15}N sites have directly bonded protons resulting in the presence of the strong, competing heteronuclear dipolar coupling. The ^{13}C and ^{15}N spins interact strongly with the abundant ^{1}H spins, but not among themselves, and they typically have large, nonaxially symmetric chemical shielding tensors. Therefore, the dominant spin interactions at ^{13}C and ^{15}N sites are heteronuclear dipolar couplings and chemical shift anisotropy.

The chemical shift interaction reflects the local electronic environment. It provides a way of characterizing the bonding at a particular site. It also provides a means for distinguishing among chemically distinct sites and, importantly for protein NMR, among chemically identical residues in different locations in the protein. Isotropic chemical shift frequencies and multiplicities, and chemical shift anisotropy powder patterns are all sensitive to molecular dynamics.

The chemical shielding is described by a second rank tensor that is diagonal when oriented in the principal axis system of the molecule. The magnitudes of the principal values reflect the asymmetry of the electronic shielding. The isotropic chemical shift is the trace of the tensor and corresponds to the single resonance frequency observed in solution where complete motional averaging occurs.

Because dipolar couplings are generally larger than the chemical shift anisotropy, it is necessary to remove the broadening from the dipolar interactions to observe the line shape due to the chemical shift interaction.[22] The chemical shift anisotropy represents another broadening mechanism rather than a source of information when there is overlap among numerous sites or when the isotropic shifts are of interest. Magic angle sample spinning can replace isotropic molecular motion as a line-narrowing mechanism for chemical shift anisotropy and dipolar interactions. Proton decoupling and magic angle sample spinning can be combined in solid state NMR to effect complete line narrowing and give the isotropic chemical shift spectrum.[23] In general, the sensitivity of the experiment is greatly enhanced by the use of a cross-polarization procedure,[22] whereby proton magnetization is transferred to the dilute spins of interest. This procedure relies on dipole–dipole couplings to provide a

[22] A. Pines, M. G. Gibby, and J. S. Waugh, *J. Chem. Phys.* **59**, 569 (1973).
[23] J. Schaefer and E. O. Stejskal, *J. Am. Chem. Soc.* **98**, 1031 (1976).

transfer mechanism. Therefore, the rates and capabilities for transfer are sensitive to motion.

The measurement of spectra with chemical shielding as the main spectral feature requires many of the experimental resources of high-resolution solid state NMR. The couplings to ^1H need to be removed with strong irradiation at the proton resonance frequency. In some cases, magic angle sample spinning or selective acquisition procedures are needed to give sufficient resolution. In slow magic angle spinning, the sideband intensities are available for analysis.[24,25]

The resonance frequencies observed in single crystals depend on their orientation with respect to the applied magnetic field. Single crystal studies have been used to determine ^{13}C chemical shielding tensors in glycine,[26] alanine,[27] threonine,[28] and the dipeptide glycylglycine,[29] among other model systems. The ^{13}C chemical shielding tensors for most functional groups found in proteins have been determined and are described in standard references on solid state NMR.[30] The only unusual carbon sites in proteins are the carbonyl and α carbons of peptide bonds, and their chemical shift tensors are similar to those observed in model organic compounds.

It is generally difficult and not yet profitable to calculate chemical shift tensors. Therefore, in order to use the chemical shift interaction in studies of molecular dynamics, the static chemical shift tensor must be taken from a presumably rigid model compound, such as crystalline amino acids, and then the effects of motional averaging calculated. This differs from dipolar couplings, where everything can be done based on calculated parameters. The chemical shift tensors for some sites, especially aliphatic carbons, are small and difficult to utilize effectively in molecular studies. This is one of the reasons why most of the attention has been focused on aromatic and carbonyl carbon sites.

Both the ^{13}C—^1H and ^{15}N—^1H heteronuclear dipolar interactions are reliable sources of dynamic information. Heteronuclear dipolar coupling

[24] M. M. Maricq and J. S. Waugh, *J. Chem. Phys.* **70,** 3300 (1979).

[25] J. Herzfeld and A. E. Berger, *J. Chem. Phys.* **73,** 6021 (1980).

[26] R. A. Haberkorn, R. E. Stark, J. van Willigan, and R. G. Griffin, *J. Am. Chem. Soc.* **103,** 2534 (1981).

[27] A. Naito, S. Ganapathy, K. Akasaka, and C. A. McDowell, *J. Chem. Phys.* **74,** 3190 (1981).

[28] N. James, S. Ganapathy, and E. Oldfield, *J. Magn. Reson.* **54,** 111 (1983).

[29] R. E. Stark, L. W. Jelinski, D. J. Ruben, D. A. Torchia, and R. G. Griffin, *J. Magn. Reson.* **55** (1983).

[30] M. Mehring, "Principles of High Resolution NMR in Solids" (2nd Ed.). Springer-Verlag, Heidelberg, 1983.

is well understood, readily calculated, axially symmetric, and can be measured with two-dimensional separated local field experiments.[31] Because of complications from more than one bonded proton, the simple two spin heteronuclear dipolar couplings found in the α carbon of all of the amino acids except glycine and the aromatic rings have been examined most extensively.

Couplings between carbons and other nuclei also occur. In general, these are relatively small couplings because of the low gyromagnetic ratios and long bond distances involved. They have found relatively little use in studies of dynamics, except for very qualitative inferences about motion based on rapid ^{14}N relaxation decoupling ^{13}C—^{14}N interactions.

^{15}N NMR

^{15}N NMR spectroscopy is very attractive for biophysical studies because there are relatively few nitrogen atoms in proteins (compared to carbon or hydrogen) and they are often located in chemically interesting sites such as the functionally active residues of enzymes and dynamically interesting sites such as the amide groups of the polypeptide backbone. ^{15}N NMR has many features similar to those discussed for ^{13}C NMR. Most nitrogen sites have both chemical shift anisotropy and heteronuclear (^{15}N—^{1}H) dipolar interactions, enabling the full range of line shape, response, and high-resolution experiments to be carried out. The chemical shift tensor of ^{15}N in a peptide bond has been determined.[32]

The labeling opportunities vary between ^{13}C and ^{15}N sites. In both cases specific individual sites can be labeled through conventional synthetic or biosynthetic means. It is also possible to label uniformly all nitrogen sites in a biopolymer with ^{15}N. Very high enrichment with ^{15}N is advantageous from an NMR point of view because of the increased sensitivity due to the large number of spins with good resolution among the chemical shifts of various chemical types of nitrogens without the penalty associated with homonuclear couplings, as seen with uniform ^{13}C enrichment, since no nitrogens in biopolymers are bonded directly to other nitrogens. High-resolution spectra are obtained using proton decoupling and magic angle sample spinning.[33]

The availability of well resolved ^{15}N isotropic chemical shift spectra of biopolymers offers many possibilities for informative NMR experiments.

[31] J. S. Waugh, *Proc. Natl. Acad. Sci. U.S.A.* **73**, 1394 (1976).

[32] G. S. Harbison, L. W. Jelinski, R. E. Stark, P. A. Torchia, J. Herzfeld, and R. G. Griffin, *J. Magn. Reson.* **60**, 79 (1984).

[33] T. A. Cross, J. A. DiVerdi, and S. J. Opella, *J. Am. Chem. Soc.* **104**, 1759 (1982).

Dynamic information is available from the motional averaging of ^{15}N chemical shift and ^{15}N—^1H dipolar powder patterns, and the relaxation analyses of the isotropic ^{15}N resonances.

^{31}P NMR

There are few phosphorus atoms in proteins. However, in some enzymes, phosphorus atoms are on substrates or inhibitors. Studying these sites can give results valuable in the analysis of the structure and mechanism of the enzyme active site.

Solid state NMR studies of phosphorus are almost exclusively concerned with the large chemical shift anisotropy. Since there are generally no protons directly bonded to phosphorus atoms, the heteronuclear dipolar coupling is not useful.

Motional Averaging in Solid State NMR

Immobile Sites

The lack of motion in a molecular site must always be defined with respect to a frequency determined by the dominant spin interactions. Those sites that do not undergo large amplitude motions at a frequency fast compared to the "size" of the relevant spin interactions have unaveraged or static spin interactions as described by symmetric second-rank tensors.

When the molecules in a sample are disordered, as in polycrystalline or amorphous materials, the NMR spectra consist of overlapping powder patterns from the various nuclear spin interactions and sites. Static powder pattern line shapes are calculated from the principal values of the tensor by taking into account the probability density of all orientations of the tensor with respect to the applied magnetic field.[34] The frequency breadth of a powder pattern reflects the total range of resonance frequencies possible for the spin interactions at a particular site, since all possible angles are present in a powder. This breadth determines the lower limit for the frequencies of motions that can influence the powder pattern line shape.

Figure 1 presents calculated powder patterns for the three spin interactions at a "typical" carbon site with one bonded proton. The ^{13}C chemical shift powder pattern from an aromatic carbon is clearly nonaxially

[34] U. Haeberlen, "High Resolution NMR in Solids." Academic Press, New York, 1976.

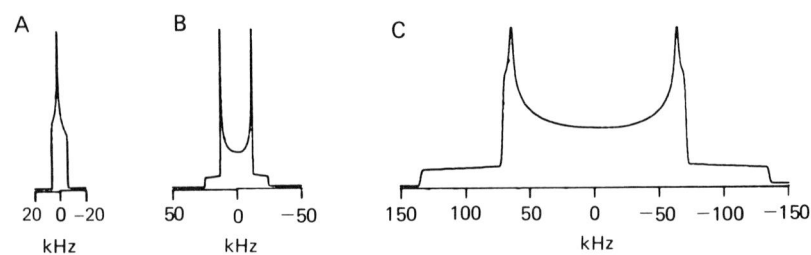

FIG. 1. Calculated static powder patterns for an aromatic carbon. (A) ^{13}C chemical shift anisotropy. (B) ^{13}C—^{1}H dipolar coupling. (C) ^{2}H quadrupolar coupling.

symmetric. The ^{13}C—^{1}H dipolar powder pattern is axially symmetric, while the ^{2}H quadrupolar powder pattern is very nearly axially symmetric. The spectra are displayed on a constant frequency scale. This clearly demonstrates the range of frequencies that the experiments are sensitive to, with chemical shift anisotropy affected by motions (10^{3} Hz) substantially slower than those that can affect the ^{2}H quadrupolar interaction (10^{6} Hz).

In addition to frequencies of the motions, the line shapes are affected by the amplitudes, directions, and types of the motions. In order to understand the influence of these factors specific models must be considered because of the intimate relationship among molecules, spin interactions, and the rates, directions, and amplitudes of the motions. For example, there are substantial differences between how slow, large amplitude motions and how rapid, small amplitude motions influence line shapes.[14] Examples of both situations are seen in proteins. A typical situation for otherwise static sites is the presence of small amplitude librational motions, such as in polypeptide backbone sites of native proteins. There are a variety of spin interactions and sites available for analysis. The ^{15}N chemical shift anisotropy of the amide group in the peptide bonds has proven useful in characterizing peptide backbone dynamics.[35,36] Figure 2 compares the effects of various degrees of motional averaging on the static chemical shift anisotropy powder pattern. Isotropic reorientation results in complete averaging of the powder pattern to a narrow single isotropic resonance line, as seen in liquid samples. Small amplitude motions, here modeled as librations about the ^{15}N chemical shift anisotropy principal axis system, that occur rapidly on the 10^{3} Hz time scale of this interaction alter the line shapes.

[35] T. A. Cross and S. J. Opella, J. Mol. Biol. **159**, 543 (1982).
[36] M. H. Frey, J. G. Hexem, G. C. Leo, P. Tsang, S. J. Opella, A. L. Rockwell, and L. M. Gierasch, in "Peptide Structure and Function: Proc. 8th. Am. Pept. Symp." (V. J. Hruby and D. H. Rich, eds.), p. 763. Pierce Chem. Co., Rockford, Illinois, 1983.

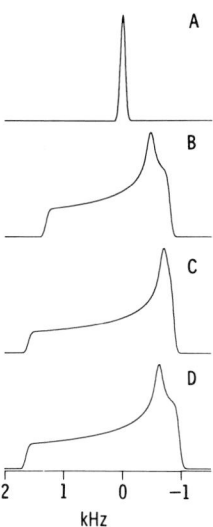

FiG. 2. Effects of motional averaging on the amide ^{15}N chemical shift anisotropy powder pattern. (A) Isotropic motion, (B) ±30° libration motion, (C) ±10° libration motion, (D) static. (From Frey *et al.*[36])

Aromatic Sites

There are four naturally occurring aromatic amino acids. They present two classes for motional analysis. The heterocyclic rings of histidine and tryptophan do not appear to have large amplitude motions in the examples studied to date. The possibilities for rapid librational motions are similar to those presented for static cases, such as the polypeptide backbone. By contrast, phenylalanine and tyrosine have six membered rings with C_2 symmetry and an obvious path for internal motion, reorientation about the C_β—C_γ bond axis. There are also possibilities for motion about the C_α—C_β bond and of the C_α backbone site itself.

There are two limiting cases for intramolecular reorientation about the C_β—C_γ bond axis for the phenylalanine and tyrosine rings: (1) jumps or 180° flips between two chemically indistinguishable sites, where the residence times are long compared to the transit times, and (2) continuous rotation, where the rings undergo unhindered diffusion around the C_β—C_γ bond. In samples where there is no overall reorientation on the NMR time scales, the intramolecular ring reorientations can be analyzed because of their effects on ^2H quadrupolar, ^{13}C chemical shift anisotropy, and ^{13}C—^1H dipole–dipole interactions. The presence of additional modes of motion, such as reorientation about the C_α—C_β bond axis or

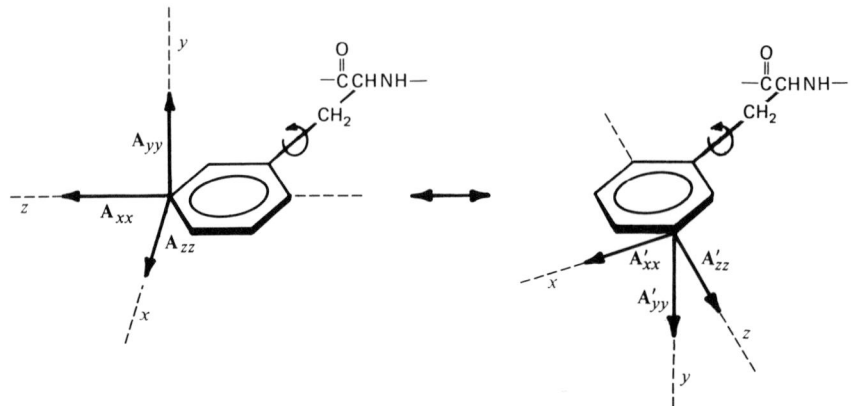

FIG. 3. The A tensor describing the interactions at the ε site of a phenyl ring, demonstrating the effect of 180° flips about the C_β—C_γ bond axis. (From Gall et al.[37])

backbone reorientation can be observed also in the spectra through additional averaging of the line shapes.

The basic procedure for analyzing the motions of the aromatic side chains also holds for the aliphatic side chains. ^{13}C- or ^{2}H-labeled sites are placed at the locations of interest through chemical or biosynthetic means. Powder patterns from the interactions are obtained. The interpretation of spectra is carried out by determining the principal values and their orientation in the molecular frame of the static tensor of the dominant spin interaction, calculating the influence of rotational or jump motions about a well-defined molecular axis on this tensor, and comparing the calculated powder patterns to those observed experimentally from the isotropically labeled sites. Since the final step is a line shape comparison, it lacks the properties of a complete proof of the motion. However, some of the line shapes are so distinctive, and the fits between experimental and calculated line shapes so close, that the motions can be described with great confidence.

The three interactions of interest for the C_δ and C_ε sites of the phenylalanine and tyrosine rings have tensors that are diagonal in the same principal axis system. This is shown in Fig. 3 with the A tensor.[37] The electric field gradient tensor for a C—^{2}H bond,[38] the chemical shielding tensor for an aromatic carbon,[39] and the heteronuclear ^{13}C—^{1}H dipolar interaction all have their largest (z) element along the C—H bond axis. The calcula-

[37] C. M. Gall, T. A. Cross, J. A. DiVerdi, and S. J. Opella, Proc. Natl. Acad. Sci. U.S.A. 79, 101 (1982).
[38] R. G. Barnes and J. W. Bloom, J. Chem. Phys. 57, 3082 (1972).
[39] S. Pausak, J. Tegenfeldt, and J. S. Waugh, J. Chem. Phys. 61, 1338 (1974).

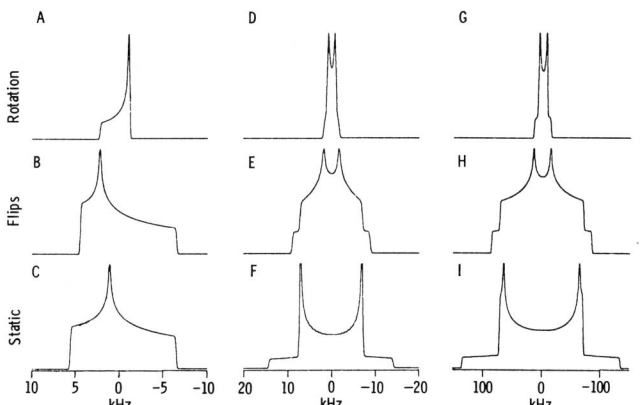

FIG. 4. Calculated powder patterns demonstrating the effects of motional averaging on the interactions of a δ or ε aromatic site. (A–C) ^{13}C chemical shift anisotropy, (D–F) $^{13}C-^{1}H$ dipole–dipole, (G–I) ^{2}H quadrupole. (From Frey et al.[42])

tions of the powder patterns can all be done in very similar ways starting with this A tensor. Motionally averaged powder patterns can be calculated from their static counterparts by taking into account the rates, amplitudes, and angle of the motions involved. When the rate of motion is much faster than the spectroscopic time scale, the problem is in the "fast-exchange" limit and only the angles are needed for line shape calculations.

The effect of fast diffusional rotation about an arbitrary axis on the tensor is calculated by transforming the tensor from its static molecular frame principal axis system to a rotating reference frame, generating a rotationally averaged tensor.[40] This new tensor, describing the spin interactions at the site, is always axially symmetric because the rotationally averaged x and y principal values are equal. Powder patterns are calculated for the rotationally averaged tensor just as for the static tensors. The effect of rapid jumps between discrete equivalent sites is also calculated using rotation matrices; however, flip averaging differs from rotational averaging in that the orientations of the residence sites rather than the actual path of reorientation are of concern. The "fast exchange" average of the coordinate systems for the two orientations is determined and the new tensor calculated.[41]

Figure 4 presents calculated powder patterns for the C_δ and C_ε sites of aromatic rings undergoing motion about the C_β—C_γ bond axis.[42] The

[40] M. Mehring, R. G. Griffin, and J. S. Waugh, J. Chem. Phys. 55, 746 (1971).
[41] T. Chiba, J. Chem. Phys. 36, 1122 (1962).
[42] M. H. Frey, J. A. DiVerdi, and S. J. Opella, J. Am. Chem. Soc. 107, 7311 (1985).

angles used in the calculations, and the resulting powder patterns, are the same for the C_δ and C_ε sites because of the symmetry of the rings. The ^{13}C chemical shift tensor is averaged in characteristic ways by motions about the C_β—C_γ bond axis. Both the static and flip-averaged powder patterns are nonaxially symmetric, while the rotationally averaged pattern is much reduced in frequency breadth and is axially symmetric. The ^2H quadrupolar and ^{13}C—^1H dipolar powder patterns for all three situations are quite similar in shape, although there are large differences in frequency breadth between them. It is clearly possible to find rings undergoing 180° flips fast enough to average the ^{13}C chemical shift and ^{13}C—^1H dipolar interactions but not the ^2H quadrupolar interaction at the same site. It is also possible to find "intermediate" exchange situations where unusual and rapidly relaxing line shapes are observed.

Aliphatic Sites

There are basically three motional regimes to be considered for aliphatic sites. They can be static, such as for C_α positions discussed above. Methyl groups undergo rapid reorientation about the C_3 symmetry axis. Methylene sites can undergo two site hops around the tetrahedral center.

Most studies of aliphatic side chains use ^2H NMR. Although all of the spin interactions are averaged by the motions, there are advantages to the use of ^2H NMR. The chemical shift tensors for aliphatic carbons are small. Therefore, it is experimentally difficult to measure their line shapes, especially with the narrowing that frequently accompanies motional averaging. The ^{13}C—^1H dipolar interaction is most simply observed and interpreted for carbon sites with only one attached proton, and many aliphatic sites have several bonded protons. Even with the use of ^{13}C labels, the natural abundance background from so many aliphatic sites with similar chemical shifts presents severe resolution problems for both the chemical shift and dipolar measurements. It is generally possible to place deuterons at specific sites by synthetic routes.

Figure 5 shows the possible ^2H NMR powder patterns for leucine methyl groups. These calculated spectra are representative of the motions that can occur at a tetrahedral side chain site.[43] The quadrupole coupling constants for all C—^2H bonds are similar, therefore the static powder pattern is as seen in Figs. 1 and 4. Rapid jumps among three or more equivalent sites results in the same averaging as seen for rotation.[40] As a result, the powder pattern for reorientation of the methyl group does not

[43] L. S. Batchelder, C. E. Sullivan, L. W. Jelinski, and D. A. Torchia, *Proc. Natl. Acad. Sci. U.S.A.* **79**, 386 (1982).

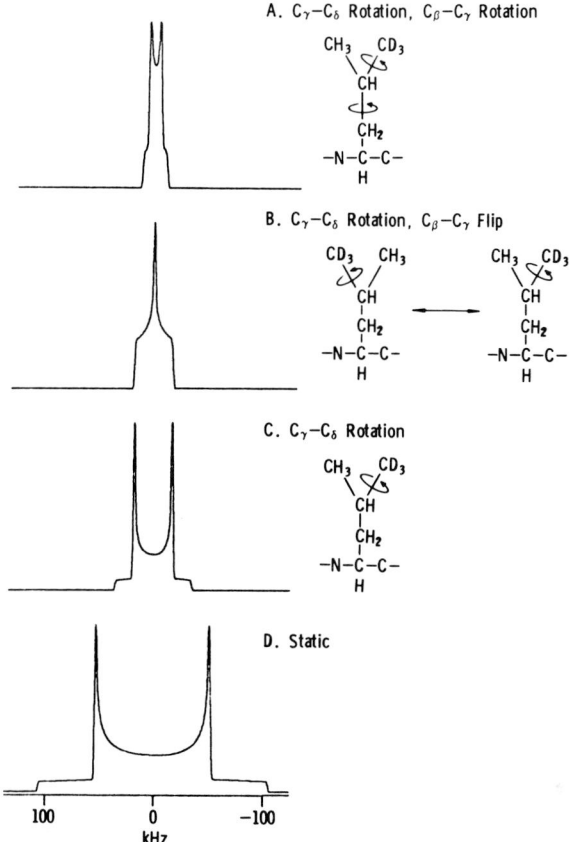

FIG. 5. Calculated powder patterns demonstrating the effects of motional averaging on the ^2H quadrupole interaction of a δ-CD_3 of leucine. (Based on Batchelder *et al.*[43])

differentiate between these types of motion. Leucine can have its side chain in two stable configurations and jump between them.[43] The drastic effect this has on the line shape is shown in Fig. 5B. It is also plausible for there to be two separate rotational motions in long aliphatic side chains and the combined effect is shown in Fig. 5A.

There are cases of interest where the jump angles for the motions of the aliphatic side chains deviate slightly from tetrahedral due intra- and interresidue constraints.[44] Somewhat different line shapes are calculated and observed in these cases. The plots in Fig. 6 relate the asymmetry parameter (η) and the breadth of the motionally averaged powder pattern with the jump angle θ for two-site jump motions where the rigid lattice

[44] L. A. Colnago, K. G. Valentine, and S. J. Opella, unpublished results.

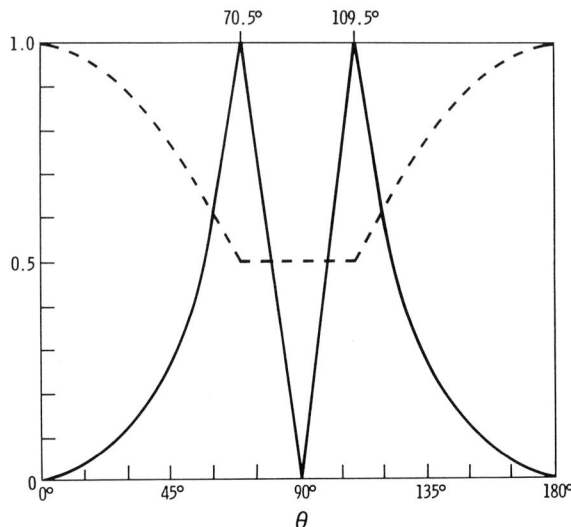

FIG. 6. Plots of the asymmetry parameter (η, solid line) and breadth of the motionally averaged powder pattern as a function of the angle of reorientation. The width (broken line) is relative to the rigid lattice powder pattern, or in the case of methyl groups, relative to the powder pattern resulting from threefold reorientation. (From Colnago et al.[44])

powder pattern has $\eta = 0$, as for —CD_3 groups. These plots clearly show that the tetrahedral jump motion ($\theta = 109.5$) is in a sensitive region of the plot, and small deviations from tetrahedral can have a large influence on the line shapes of the motionally averaged powder patterns. For example, the methyl group in methionine is attached to a sulfur atom with a linkage that differs slightly in geometry from tetrahedral. The motionally averaged powder pattern resulting from two-site jumps appropriate for the methionine methyl group ($\phi = 80°$, $\psi = 120°$, and $\theta = 117°$) is shown in Fig. 7B; it differs markedly from that for a tetrahedral two-site jump shown in Fig. 7C.

Typical Applications

Polypeptide Backbone

The polypeptide backbone of proteins provides structural constraints for the local and global folding of these large molecules. The dynamics of the backbone peptide linkages have been the focus of many experimental and theoretical studies. In native globular proteins, the peptide groups undergo rapid small amplitude motions that vary shortly among sites. The details of the processes at individual residues come from the temperature

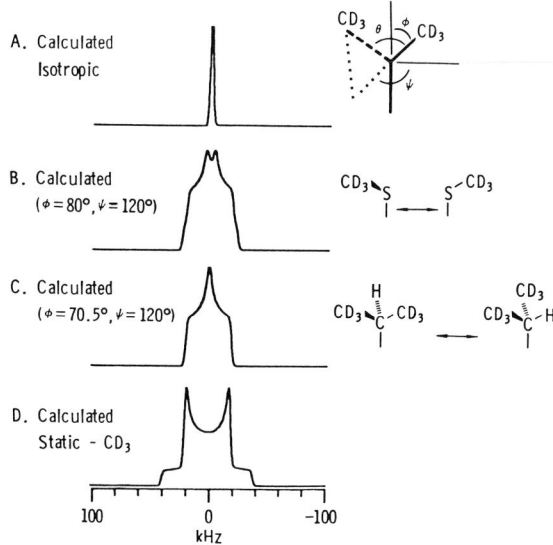

A. Calculated
   Isotropic

B. Calculated
   ($\phi = 80°, \psi = 120°$)

C. Calculated
   ($\phi = 70.5°, \psi = 120°$)

D. Calculated
   Static - CD_3

```
100        0        -100
         kHz
```

FIG. 7. Calculated ^{2}H NMR line shapes for CD_3-labeled methyl groups where the methyl groups are undergoing rapid threefold reorientation in addition to the motions shown. (From Colnago *et al.*[44])

factors of the scattering experiment and molecular dynamics calculations; only a few individual backbone sites have been resolved in NMR experiments and these have not really been subject to thorough analysis. Most NMR studies have treated the motions of the backbone as homogeneous throughout the sequence and used the resonance properties from groups of sites to describe the backbone motions. Since the backbone motions are generally of limited amplitude, the same backbone sites can be used to describe either segmented motions, such as for those portions of membrane proteins that are not immobilized by the lipids, or the overall reorientation of the protein molecule. The backbone dynamics are also of interest because of their influence on side chain dynamics.

The motions of the atoms that form peptide groups in amino acids and small peptides are taken as rigid lattice values with regard to large amplitude motions. However, there are small amplitude librational motions present. Some confidence can be ascribed to the rigid character of these groups because of the use of the temperature dependence and dipole–dipole couplings. The chemical shift anisotropy and quadrupole coupling are taken from model compounds and only at the very lowest temperatures available can these be assumed to be rigid. The backbone motions of several proteins have been analyzed in detail, and qualitative statements have been made about a number of others.

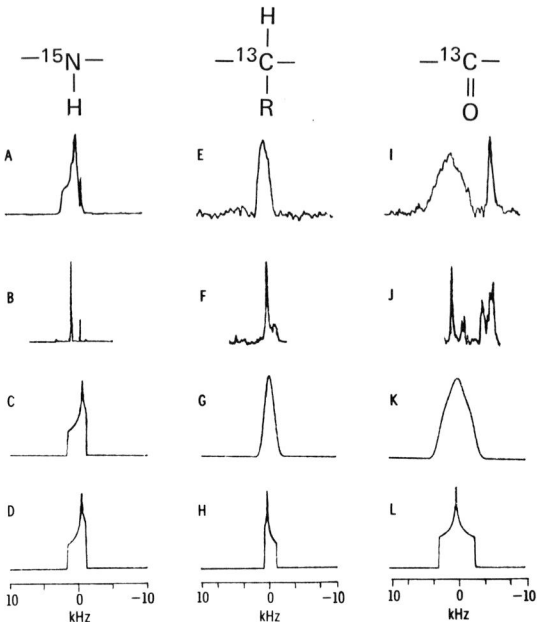

FIG. 8. Chemical shift anisotropy powder patterns for polypeptide backbone sites in isotopically labeled fd virus samples. (A and B) Uniformly ^{15}N; (E and F) $^{13}C_\alpha$ alanine; (I and J) ^{13}C=O valine. A, E, and I are stationary samples; B, F, and J are with magic sample spinning; (D, H, and L) calculated chemical shift powder patterns based on experimental spectra; (D) [^{15}N]acetylglycine; (H) $^{13}C_\alpha$ alanine; (L) ^{13}C=O tetraglycine. (C, G, and K) Broadened versions of D, H, and L. (From Cross and Opella.[45])

The dynamics of the coat protein of the filamentous bacteriophage fd have been extensively investigated. This is a tractable system because the protein is small with only 50 residues, and it can be readily labeled with stable isotopes because it is from a bacterial source. The coat protein dynamics are of biological interest because the protein undergoes a transition from a membrane-bound form to a structural form that interacts with DNA during the virus life cycle.

The experimental results on the backbone of the fd coat protein included measurements of the chemical shift anisotropy powder patterns for α and carbonyl carbon and the amide nitrogen sites, the ^{15}N—^{1}H dipole–dipole powder pattern, and both ^{13}C and ^{15}N relaxation parameters. Figure 8 shows the spectral comparisons for the chemical shift anisotropy of the backbone sites.[45] The ^{13}C and ^{15}N chemical shift anisotropy and ^{15}N—^{1}H dipolar powder patterns are reduced little or none in

[45] T. A. Cross and S. J. Opella, J. Mol. Biol. **159**, 543 (1982).

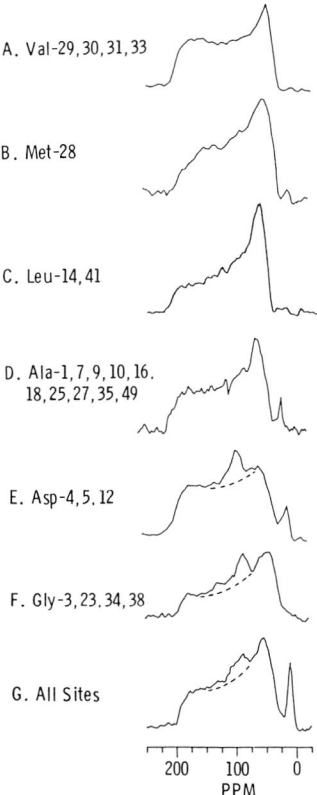

FIG. 9. Experimental ^{15}N NMR spectra of fd virus samples. The samples were labeled with ^{15}N at the sites indicated in the figure. (From Colnago *et al.*[44])

breadth compared to those of crystalline amino acids and peptides. This indicates an absence of backbone motions of substantial amplitude that are rapid on the 10^4 Hz time scale. The relaxation data for the α carbon and amide nitrogen sites were analyzed using a model for backbone motions where the C—H or N—H bond vectors wobble in a cone. The end picture is that otherwise stationary atoms of the backbone are undergoing relatively small amplitude librational motions.

The experimental ^{15}N NMR spectra in Figs. 9 and 10 were obtained from unoriented solutions of the virus where the coat proteins are labeled in various sites with ^{15}N.[44] These spectra can be interpreted directly by comparison with the calculated spectra in Fig. 2. Two distinct types of backbone sites are observed, narrow isotropic resonances and broad powder patterns in the spectra from various ^{15}N-labeled backbone sites of the coat protein in the virus. Most of the backbone sites are immobile and a

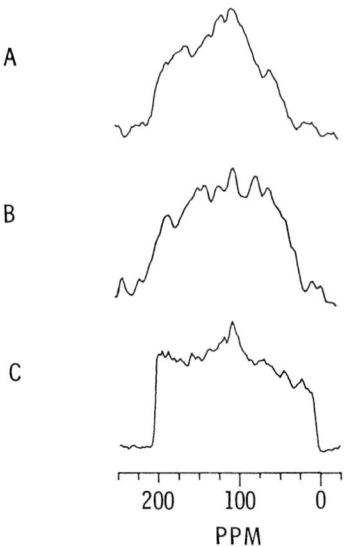

FIG. 10. Experimental ^{15}N NMR spectra of ^{15}N-Pro labeled samples. fd coat protein has a single Pro residue at position 6. (A) ^{15}N-Pro-6 labeled fd virus at 25°. (B) Same as in A except at −90°. (C) Ala-^{15}N-Pro polycrystalline model peptide. (From Colnago et al.[44])

few are isotropically reorienting on the 10^3 Hz time scale of the ^{15}N chemical shift interaction.

The uniformly ^{15}N-labeled virus sample gives the spectrum in Fig. 9G. This spectrum has resonance intensity near the isotropic position (100 ppm) superimposed on a powder pattern for the backbone ^{15}N amide sites. The resonance near 15 ppm in this sample is from the amino groups of the lysine side chains and the N-terminal alanine residue. Through incorporation of specifically ^{15}N-labeled amino acids, the dynamics of individual residues can be described. The spectra in Fig. 9E and F of Asp- and Gly-labeled coat proteins have an isotropic resonance from one mobile residue superimposed on the powder pattern from the other, immobile residues of each type (two of Asp and three of Gly). This estimate of intensities is reasonable since there are only three Asp residues and four Gly residues in the coat protein. The data in Figs. 9E–G clearly show the heterogeneous dynamics of coat protein backbone sites. These results show that residues 3 and 4 are mobile and that residue 5 is immobile.

The spectra in Figs. 9A–D show that residues 7–49 have immobile backbone sites. The N-terminal residue is Ala-1 and its amino group gives the narrow resonance near 15 ppm in the spectrum in Fig. 9D; since amino groups have very small chemical shift anisotropy the linewidth of this

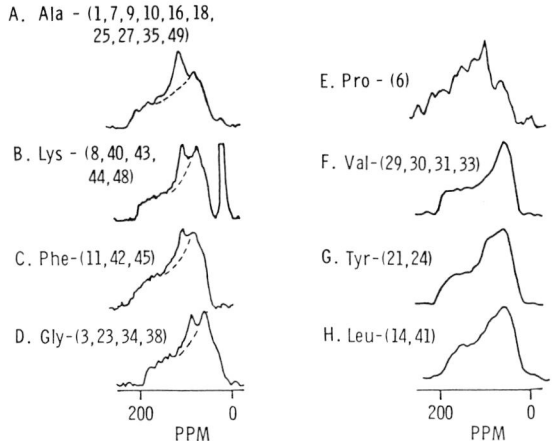

FIG. 11. ^{15}N NMR spectra of fd coat protein in DML bilayers. The samples were labeled with ^{15}N at the sites indicated in the figure. (From Leo et al.[46])

resonance does not discriminate between a rigid or mobile N-terminus. The amino acid proline has a different ^{15}N chemical shift tensor than the peptide linkages than the amino acids. This is apparent in the spectra in Fig. 10. Proline is residue 6 of fd coat protein and the spectrum in Fig. 10A shows that this site is immobile, since the breadth of the powder pattern is the same as for the rigid models in Figs. 10B and C.

Similar experiments have been performed on the membrane-bound form of fd coat protein. Important differences are observed in the backbone dynamics for the structural and membrane-bound forms of the coat protein.

The experimental data in Fig. 11A–D show that there are two distinct classes of amide resonances in the coat protein in bilayers, since a relatively narrow isotropic resonance is superimposed on a broad powder pattern. Immobile amide sites yield the broad static powder pattern. Motionally averaged amide sites yield the narrow resonance intensity near the amide isotropic chemical shift frequency (100 ppm).

The isotropic resonance intensity in the spectra in Fig. 11A–D arises from backbone sites with large amplitude motions that occur more frequently than about 10^3 Hz. The T_1 and NOE of the powder pattern compounds of the spectra in Fig. 11 indicate that small amplitude motions on the 10^9 Hz time scale are present in the immobile amide sites. The powder pattern lineshapes observed for most backbone sites of the coat protein in both the virus[45] and lipid bilayers[46] indicate the motional averaging is limited in amplitude. By comparison with the calculated lineshapes in Fig.

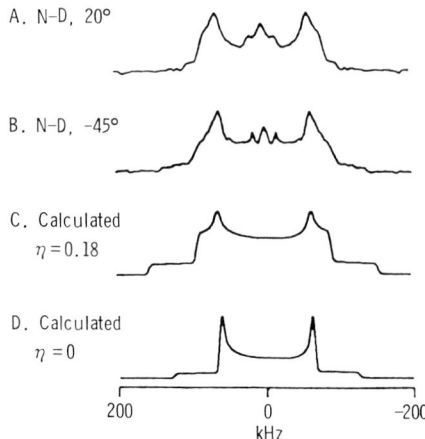

A. N–D, 20°

B. N–D, –45°

C. Calculated
$\eta = 0.18$

D. Calculated
$\eta = 0$

200   0   –200
kHz

FIG. 12. ^{2}H NMR spectra of amide N–D sites. (A and B) Experimental spectra obtained on samples of fd coat protein in DML bilayers that have been partially labeled by exchange at the amide N–D sites. (C and D) Calculated spectra. C is with parameters to match the experimental spectra. D is calculated with the same quadrupole coupling constant as in C, although $\eta = 0$. (From Leo et al.[46])

2, the experimental lineshapes are consistent with motional averaging over a 10–20° range. The relaxation parameters T_1 and NOE for these sites indicate that these small amplitude motions occur with frequencies around 10^9 Hz.

Figure 11 contains ^{15}N NMR spectra of specifically ^{15}N-labeled coat proteins in DML bilayers.[46] The four valine, two tyrosine, and two leucine residues all yield powder patterns characteristic of immobile amide sites in these spectra obtained by cross-polarization. The imide ^{15}N powder pattern for a sample labeled at Pro-6 is also characteristic for an immobile site. Immobile sites are readily observed because the powder pattern intensity can be easily obtained through cross-polarization.

The ^{15}N NMR spectra from the Ala, Lys, Gly, and Phe ^{15}N-labeled proteins are heterogeneous with isotropic amide resonances superimposed on powder patterns. The spectra in Fig. 11 clearly demonstrate that there are alanine, lysine, glycine, and phenylalanine residues in both the immobile and in the mobile regions of the coat protein.

Additional information about the backbone sites of the protein in phospholipid bilayers can be derived from ^{2}H NMR experiments. Figure 12 shows ^{2}H NMR spectra obtained from samples where the slowly exchanging amide sites on the coat protein in bilayers are labeled as N–D sites. The exchange with deuterium was performed in D_2O at high pH and elevated temperature for several hours. The temperature and pH were then lowered and the sample lyophilized. The sample was rehydrated with H_2O and lyophilized twice before final rehydration in deuterium-depleted water. This procedure removed the readily exchanged deuterons, which would also be the ones most likely to yield narrow isotropic resonances. The experimental spectra in Fig. 12A and B are representative powder patterns for the static N–D quadrupolar interactions. To determine the full magnitude of the N–D quadrupole interaction in peptide bonds, the spectrum was obtained from the same sample at −45° (Fig. 12B). There are only minor differences between the −45 and 20° spectra. The bottom two powder patterns in Fig. 12 are calculated and illustrate the differences between the powder patterns from axially symmetric and nonaxially symmetric quadrupolar interactions of similar magnitude. Figure 12C is a nonaxially symmetric quadrupolar powder pattern calculated using the magnitudes of the principal values obtained from Fig. 12B and from crystalline N-acetylglycine. Figure 12D was calculated using the same quadrupole coupling constant as for Figure 12C but with the asymmetry parameter (η) equal to zero. The comparisons between the experimental and calculated spectra indicate that the protein itself is not reorienting within the bilayer on the 10^6 Hz time scale of the N–D quadrupole interaction. In particular, the coat protein does not undergo rapid rotation about its long axis in DML bilayers, since the nonaxially symmetric line shape of the N–D quadrupole interaction is preserved in the hydrated protein–lipid complex at 25°.

There are clearly more mobile backbone sites in the membrane-bound form of the coat protein than in the structural form. By comparing results from specifically labeled sites and the use of proteolytic digestion experiments, it has been possible to identify the mobile and immobile domains in the membrane-bound form of the coat protein. Figure 13 directly compares the dynamics of the coat protein backbone in its two biological states. The major change in the coat protein upon going from the structural form in the virus to the membrane-bound form is that the C-terminal residues go from being rigidily constrained in the virus particle to highly mobile in the membrane-bound form. As shown in Fig. 13, the backbone of the protein is immobile from residues 5 or 6 through residues 42 or 43 in its membrane-bound form. This contrasts with the structural form of the coat protein in the virus where residues 5 through 50 have immobile backbone sites.

The molecular dynamics of the collagen backbone has been defined in

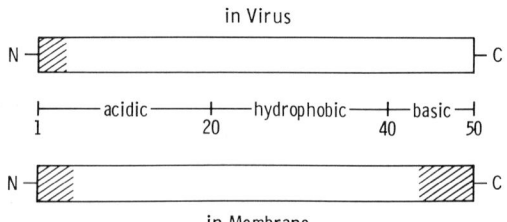

FIG. 13. Schematic drawing of distribution of mobile (hatched) and rigid (open) backbone sites for the structural (virus) and membrane-bound forms of fd coat protein.

a series of studies by Torchia and co-workers.[47–49] About one-third of the residues in collagen are glycines, therefore labeling with [1-^{13}C]glycine places many carbonyl chemical shift tensors into the protein backbone. The carbonyl line shapes in Fig. 14 are used to study the mobility as a function of cross-linking in soft tissue and of mineralization of collagen in bone. Axially asymmetric chemical shift powder patterns are observed for rat calvaria collagen, demineralized calvaria, and rat tail tendon collagen. While all of these samples have the same breadth to their powder patterns at −35°, at 22° the line width of the glycine carbonyl carbon resonance in rat calvaria collagen is significantly larger than the line width of demineralized rat calvaria, rat tail tendon, and reconstituted chick calvaria collagen. The differences in line shapes and widths among these samples are due to molecular motions. Calculations of line shapes based on varying angular dispersions in the azimuthal angle can be used to determine the amplitude of motions faster than the ~10^4 time scale imposed by the carbonyl chemical shift tensor. These results are consistent with earlier studies employing relaxation measurements.

Oldfield and co-workers[50] have extensively studied the backbone dynamics of the membrane protein, bacteriorhodopsin. They have observed ^{2}H NMR spectra from many labeled sites that show a narrow isotropic resonance superimposed on a broad powder pattern. These heterogeneous spectra are characteristic of some sites undergoing no or limited motional averaging and other sites undergoing large-amplitude rapid motions. A representative spectrum of glycine-labeled protein is shown in

[47] S. K. Sarkar, C. E. Sullivan, and D. A. Torchia, *J. Biol. Chem.* **258**, 9762 (1983).
[48] D. A. Torchia and D. L. VanderHart, *J. Mol. Biol.* **104**, 315 (1976).
[49] D. A. Torchia, Y. Hiyama, S. K. Sarkar, C. E. Sullivan, and P. E. Young, *Biopolymers* **24**, 65 (1985).
[50] M. A. Keniry, H. S. Gutowsky, and E. Oldfield, *Nature (London)* **307**, 383 (1984).

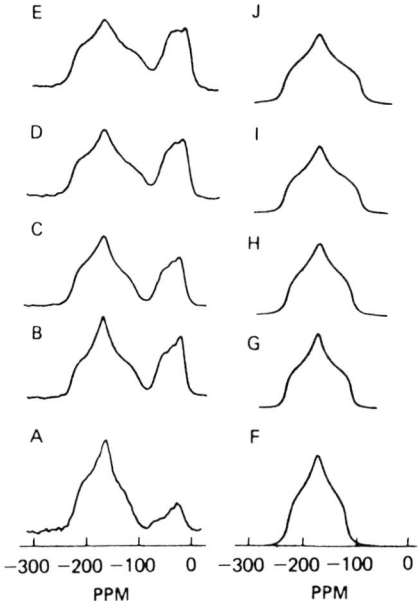

FIG. 14. Comparison of the experimental spectra (A–E) from [1-¹³C]glycine-labeled collagen with the line shapes calculated for the glycine carbonyl carbon (F–J). (A) Reconstituted collagen fibrils (22°); (B) rat tail tendon collagen (22°); (C) demineralized rat calvaria (22°); (D) mineralized rat calvaria (22°); (E) mineralized rat calvaria (−35°); (F) Δ = 108 ppm, γ_{rms} = 41°; (G) Δ = 124 ppm, γ_{rms} = 31°; (H) Δ = 120 ppm, γ_{rms} = 33°; (I) Δ = 140 ppm, γ_{rms} = 14°; (J) Δ = 145 ppm, γ_{rms} = 0° (Δ = frequency breadth of powder pattern; γ_{rms} = range of motion through azimuthal angle. (From Sarker *et al.*[47])

Fig. 15. These results have been interpreted in terms of structural models for bacteriorhodopsin by correlating the mobile sites with loop or end regions extending beyond the lipid bilayers and the rigid sites with the helices that span the bilayers.

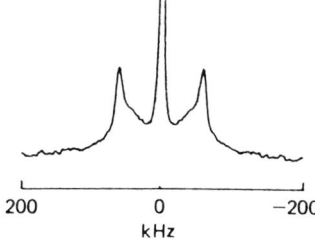

FIG. 15. ²H NMR spectrum of [α-²H₂]glycine-labeled bacteriorhodopsin. There is a narrow resonance superimposed on a static powder pattern. (From Keniry *et al.*[50])

Aromatic Sites

The motions of phenylalanine and tyrosine side chains have been studied extensively by solid state NMR.[21,37,42,51–59] Crystalline amino acids, crystalline peptide, peptides in bilayers, crystalline proteins, and proteins that are in large complexes have all been investigated. Almost all of the rings in these cases are either rigid or undergo 2-fold jumps about the C_β—C_γ bond axis. A few sites of peptides or proteins in membrane bilayers have additional modes of motion. Most of the studies have utilized ^2H NMR of specifically labeled rings. Figure 4 shows the characteristic motional averaging effects for the ^2H NMR powder patterns.

Polycrystalline tyrosine is rigid at room temperature. The ^2H NMR powder pattern from δ and ε labeled ring sites is a static powder pattern showing an absence of large amplitude motions on the 10^6 Hz time scale. Spectra of the ε ^{13}C NMR sites of tyrosine give a static ^{13}C chemical shift powder pattern. The high-resolution ^{13}C NMR spectra, where isotropic resonances are obtained in magic angle sample spinning experiments, have two peaks for the C_ε sites shown in Fig. 16.[21] The two separate resonances for the ε sites arise from the asymmetric influence of the hydroxyl group and the lack of motional averaging on the $\sim 10^2$ Hz time scale of the isotropic chemical shift differences. Tyrosine in crystalline enkephalin is rigid at room temperature, but undergoes ring flips on the 10^7 Hz time scale at 100°. These data, shown in Fig. 17, illustrate the spectral changes and distortions from the quadrupole echo experiment that accompany intermediate exchange rates for the ring.[51]

There are two tyrosines in the coat protein of fd bacteriophage. They are close in the sequence at locations 21 and 24. Many experiments show that they have similar properties. Figure 18 contains ^2H NMR spectra of

[51] D. M. Rice, R. J. Wittebort, R. G. Griffin, E. Meirovitch, E. R. Steinson, Y. C. Meinwald, J. H. Freed, and H. A. Scheraga, *J. Am. Chem. Soc.* **103**, 7707 (1981).

[52] L. W. Jelinski and D. A. Torchia, *J. Mol. Biol.* **133**, 45 (1979).

[53] C. M. Gall, J. A. DiVerdi, and S. J. Opella, *J. Am. Chem. Soc.* **103**, 5039 (1981).

[54] R. A. Kinsey, A. Kintanar, and E. Oldfield, *J. Biol. Chem.* **256**, 9028 (1981).

[55] D. M. Rice, A. Blume, J. Herzfeld, R. J. Witebort, T. H. Huang, S. K. Das Gupta, and R. G. Griffin, *Biomol. Stereodyn.* **II**, 255 (1981).

[56] S. Schramm, R. A. Kinsey, A. Kintanar, T. M. Rothgeb, and E. Oldfield, *Biomol. Stereodyn.* **II**, 271 (1983).

[57] L. M. Gierasch, M. H. Frey, J. G. Hexem, and S. J. Opella, *ACS Symp. Ser.* **191**, 233 (1982).

[58] L. M. Gierasch, S. J. Opella, and M. H. Frey, *in* "Peptides: Synthesis, Structure, and Function" (D. H. Rich and E. Gross, eds.), p. 267. Pierce Chem. Co. Rockford, Illinois, 1981.

[59] J. Schaefer, E. O. Stejskal, R. A. McKay, and W. T. Dixon, *J. Magn. Reson.* **57**, 85 (1984).

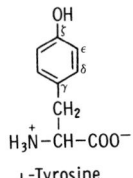

L-Tyrosine

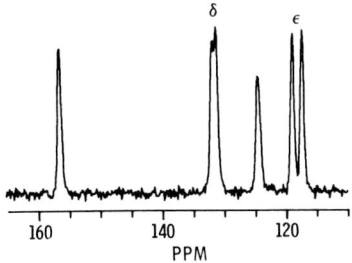

FIG. 16. Aromatic region of the high-resolution ^{13}C NMR spectrum of polycrystalline tyrosine. The split C_ε resonance shows the absence of motional averaging. (From Opella et al.[21])

these tyrosines labeled with deuterium in the assembled virus and in the protein–membrane complex. The comparison of these line shapes to that calculated for the flip averaged deuterium quadrupole tensor (Fig. 4) demonstrates that in both samples both tyrosines undergo rapid 180° flips and no other motions.

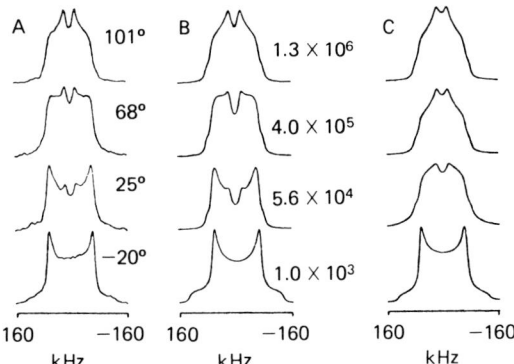

FIG. 17. (A) Experimental ^2H echo spectra of polycrystalline $[3,5\text{-}^2H_2]$tyrosine-labeled enkephalin as a function of temperature. (B) Simulations of the spectra using a 180° jump model including corrections for power roll off and for intensity destinations produced by the quadrupole echo. The jump rate at each temperature is indicated. (C) Line shapes at the same jump rate but without the echo distortion correction. (From Rice et al.[51])

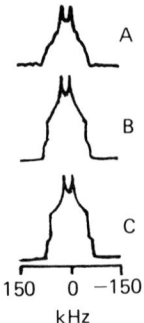

150   0 −150
kHz

FIG. 18. ^{2}H NMR spectra of Tyr-d_γ-labeled fd coat protein. (A) Protein in DML bilayers. (From Leo et al.[46]) (B) Protein in the fd virus. (From Gall et al.[37]) (C) Calculated flip averaged spectrum.

Bacteriorhodopsin has 11 tyrosines. The ^{2}H NMR spectrum of these labeled tyrosines is shown in Fig. 19. At low temperatures ($<-25°$) all of the tyrosine rings of this protein in the membrane are immobile. At moderate and elevated temperatures ($>37°$) all or nearly all of the tyrosine rings undergo rapid 180° flips.[56]

The dynamics of the phenylalanine side chains in the crystalline amino acid samples are complex. The ring motions vary with the type of crystal whether the zwitterion, the hydrochloride, or the form crystallized from ethanol and water. Phenylalanine ring dynamics have been studied using ^{2}H NMR by several groups[37,42,53,54,59] and a number of ^{13}C NMR experiments have been performed using both the ^{13}C chemical shift and ^{13}C—^{1}H dipolar interactions.[42,59]

There are two categories of deuterons on Phe-d_5. Those on the δ and ε

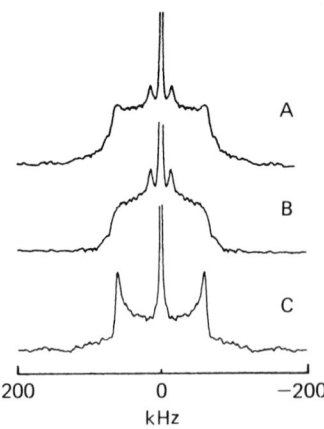

200   0   −200
kHz

FIG. 19. ^{2}H NMR spectra of bacteriorhodopsin labeled with deuterium. (A) Phe-d_5. (B) Tyr-d_2. (C) Trp-d_3. (From Schramm et al.[56])

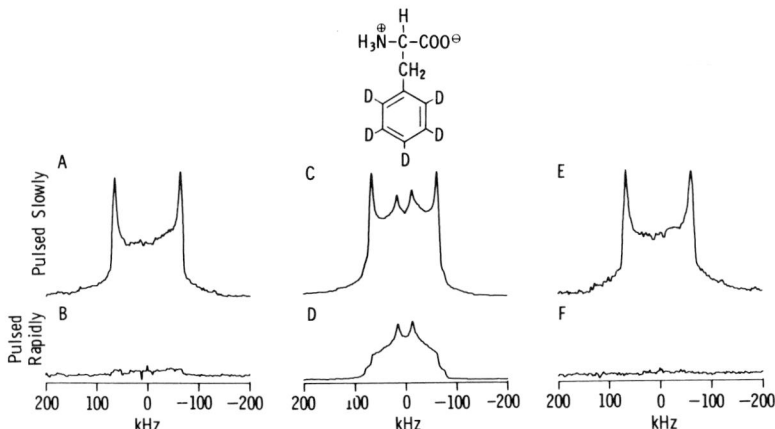

FIG. 20. ^{2}H NMR spectra of polycrystalline Phe-d_5. (A, B) Crystallized from ethanol/water. (C, D) Zwitterion crystallized from water. (E, F) Hydrochloride salt. (From Frey et al.[42])

sites are affected the same by motions about the C_β—C_γ bond axis. The single ξ deuteron is essentially unaffected by motions about this axis. Therefore, ^{2}H NMR spectra showing the effects of motion are heterogeneous, since there are superpositions of two types of spectra. In most cases this does not present a problem. However, in those situations where clear distinctions must be made, specifically d_4- and d_1-labeled phenylalanine can be made synthetically. Mobile and immobile sites in the same and different molecules can be differentiated on the basis of line shape and on the basis of relaxation rates. Figure 20 illustrates this for Phe-d_5 crystals. The top spectra were obtained with long recycle delays so that all deuterons give resonances while the bottom spectra were obtained with short recycle delays so that only deuterons with short T_1s give signals. Only the zwitterion species has mobile rings because it has deuterons with both short T_1s and motionally averaged powder patterns. All samples have immobile rings because there are static powder patterns with long T_{1s}. Both the phenylalanine hydrochloride and the sample recrystallized from ethanol/water have rigid rings. The situation for the zwitterion form is more complex. Spectrum C is a superposition of static and flip averaged powder patterns. However, the intensities are not in the 4:1 ratio if all the rings were undergoing 180° flips. Rather the intensities are better accounted for by half the rings being rigid and half undergoing flips, with the static powder pattern intensity from all five deuterons on the rigid rings and the one ξ deuteron on the flipping rings. These same considerations hold for peptides and proteins that have more than one phenylalanine residue, since dynamic heterogeneity is the rule rather than

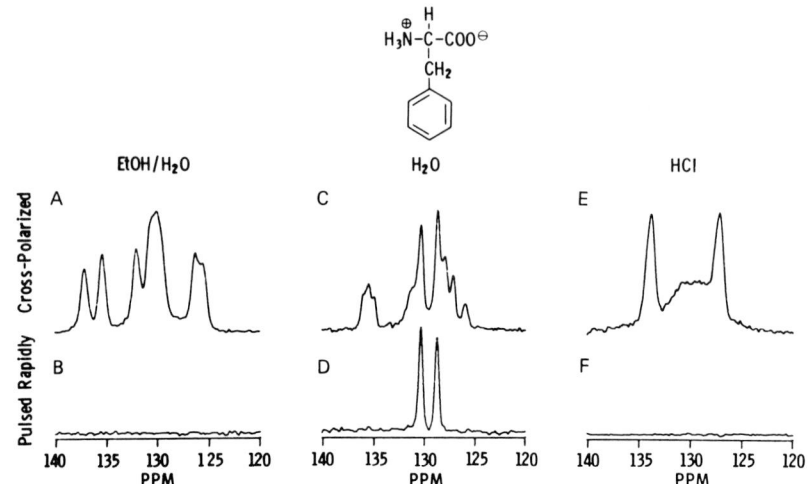

FIG. 21. ^{13}C NMR spectra of polycrystalline phenylalanine. (A, B) Crystallized from ethanol/water. (C, D) Zwitterion crystallized from water. (E, F) Hydrochloride salt. (From Frey et al.[42])

the exception. Spectral complexities are also observed in high-resolution ^{13}C NMR spectra of the amino acid and peptides containing phenylalanine. This is illustrated in Fig. 21 with the ^{13}C NMR spectra of the three types of phenylalanine crystals. The ethanol/water and zwitterion samples give complex spectra in the aromatic region showing the heterogeneities in the samples. Phenylalanine hydrochloride has a very unusual spectrum with narrow resonances from the γ and ξ carbons unaffected by motion about the C_β—C_γ bond axis and a broad resonance from the δ and ε sites. The hydrochloride has ring motions with frequencies near the 10^2 Hz time scale determined by chemical shift differences; the corresponding ^{2}H NMR spectra in Fig. 20 showed these same rings to be immobile on the 10^7 Hz time scale.

Similar phenomena are seen for the phenylalanine rings of cyclic pentapeptides when ^{13}C and ^{2}H NMR spectra are compared (Fig. 22).[57] Cyclo (D-Phe-Pro-Gly-D-Ala-Pro) has a phenylalanine side chain that is rigid on both the isotropic chemical shift and deuterium quadrupole line shape time scales. By contrast the similar peptide cyclo(D-Phe-Gly-Ala-Gly-Pro) has a ^{13}C NMR spectrum in the aromatic region very similar to that observed for phenylalanine hydrochloride, and like phenylalanine hydrochloride has a static ^{2}H NMR spectrum.

The phenylalanine ring dynamics have been studied in the same proteins as the tyrosine ring dynamics. Both Griffin and co-workers[55] and Oldfield and co-workers[56] have studied Phe-d_5-labeled bacteriorhodopsin.

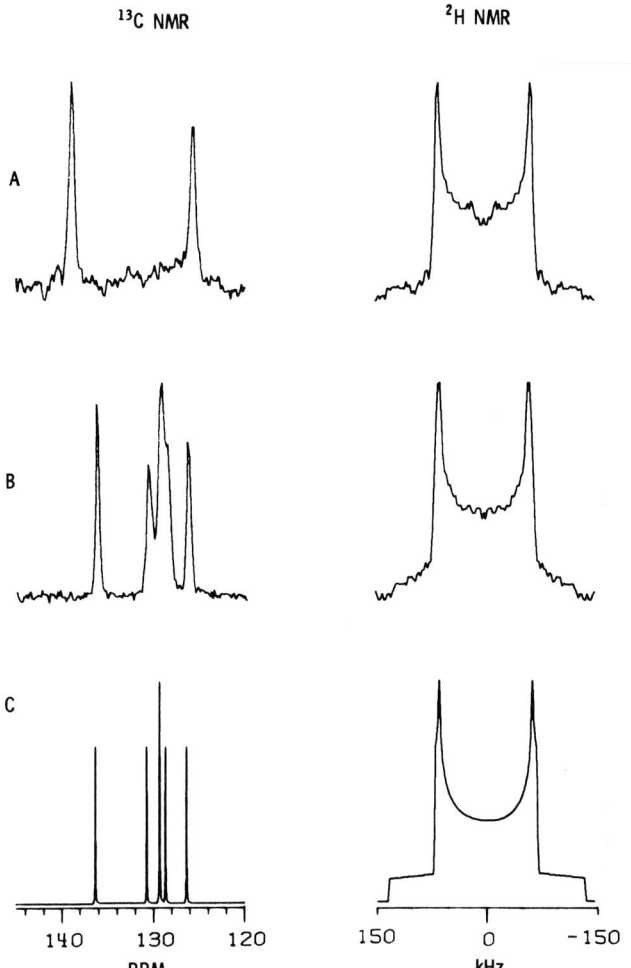

FIG. 22. ^{13}C and ^{2}H NMR spectra of aromatic rings in cyclic peptides. (A) Cyclic (D-Phe-Gly-Ala-Gly-Pro). (B) Cyclic (D-Phe-Pro-Gly-D-Ala-Pro). (C) Calculated spectra for static ring. (From Gierasch et al.[58])

At temperatures near or greater than 37° the phenylalanine rings undergo rapid two site hops in the protein, as demonstrated by the data in Fig. 19.

Aliphatic Sites

The motions of aliphatic side chains can be described by analyzing the motionally averaged powder patterns obtained from specifically labeled sites. Some results have been obtained using ^{13}C NMR.[59a] However, the generally small nonaxially symmetric ^{13}C chemical shielding tensors and dipolar couplings among multiple spins in most aliphatic carbon sites make quantitative work difficult. ^2H NMR provides an alternative approach. The basic experimental protocol is to identify those side chains of interest and label them in specific sites with deuterium. The quadrupolar interaction of the C—D bonds provides the spectroscopic parameters. This interaction is large, axially symmetric, and well characterized. Specific models for motion can be used to calculate readily powder pattern line shapes. The limitation of this approach is the lack of chemical shift resolution among sites. In a protein with more than one of a type of residue, this prevents the analysis of individual side chains. In some cases all of the side chains behave so similarly that the low resolution of the experiment does not destroy its usefulness. In other cases, where there are important dynamic differences among chemically identical sites, there are severe problems.

The type of analysis described for aromatic residues applies equally well to aliphatic sites. All of the motions observed are discrete jumps among several conformations. The situation for longer chain aliphatic side chains (Met, Lys, Arg) is less clear because of the complexities imposed by multiple motions.

The facile reorientation of methyl groups about their C_3 axis has been extensively studied in crystalline alanine, valine, leucine, methionine, threonine, and several small peptides.[43,59a–64] These studies have been extended to several of these residues in proteins. The —CD_3 group is relatively easy to place in the desired locations. The ^2H NMR line shape is sensitive to motions in the 10^{-4}–10^{-7} sec range while the spin-lattice relaxation times reflect motions in the 10^{-7}–10^{-12} sec range. The temperature dependence of these parameters yields the activation energies for the methyl reorientation process.

The ^2H NMR line shapes of labeled methyl groups above about $-135°$ are axially symmetric powder patterns reduced in breadth from the static pattern. Since jumps among three or more equivalent sites and rotations

[59a] L. W. Jelinski and D. A. Torchia, *J. Mol. Biol.* **138,** 255 (1980).

[60] L. W. Jelinski, C. E. Sullivan, and D. A. Torchia, *Nature (London)* **284,** 531 (1980).

[61] R. A. Kinsey, A. Kintanar, M.-D. Tsai, R. L. Smith, N. Jones, and E. Oldfield, *J. Biol. Chem.* **256,** 4146 (1981).

[62] L. S. Batchelder, H. Niu, and D. A. Torchia, *J. Am. Chem. Soc.* **105,** 2228 (1983).

[63] M. A. Keniry, T. M. Rothgeb, R. L. Smith, H. S. Gutowsky, and E. Oldfield, *Biochemistry* **22,** 1917 (1983).

[64] M. A. Keniry, A. Kintanar, R. L. Smith, H. S. Gutowsky, and E. Oldfield, *Biochemistry* **22** (1983).

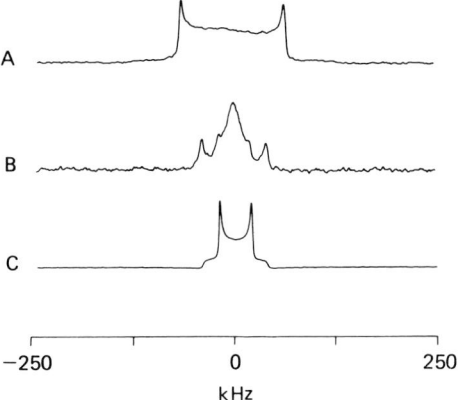

FIG. 23. ^{2}H NMR spectra of polycrystalline L-alanine-d_3. (A) $-150°$; (B) $-96°$; (C) $20°$. (From Batchelder *et al.*[62])

affect powder pattern line shapes in the same way, these characteristic spectra only tell the angle of reorientation, but not the type. By contrast, the analysis of relaxation data enables the discrimination among the various models for methyl reorientation, and it has been shown to be three site jumps.

Figure 23 contains spectra of polycrystalline d_3-labeled alanine at -150, -96, and $20°$.[62] This shows the characteristic powder patterns for a static methyl group, intermediate motion on the time scale of the spin interaction, and fast threefold reorientation. The inversion-recovery measurements of T_1 give spectra where the outer shoulders relax more rapidly than the major singularities as a consequence of the angular dependence of the relaxation across the powder pattern. The correlation times derived from these data utilize the T_1 at the $90°$ (perpendicular) singularities of the powder patterns. For the zwitterionic forms of Ala, Val, Thr, Leu, and Met, the rotational correlation time decreases in going from the methyl group nearest the α carbon (Ala) to the methyl group farthest away (Met). It is likely that the decreased rotational correlation times observed upon increasing the length of the amino acid side chain are due to decreased barriers to methyl rotation.

The interpretation of the line shapes from the other CD_3-labeled aliphatic side chains requires the consideration of additional motions, in particular jump motions around tetrahedral carbon sites in the side chain. This type of motional averaging was first seen in the leucine residues of collagen.[43]

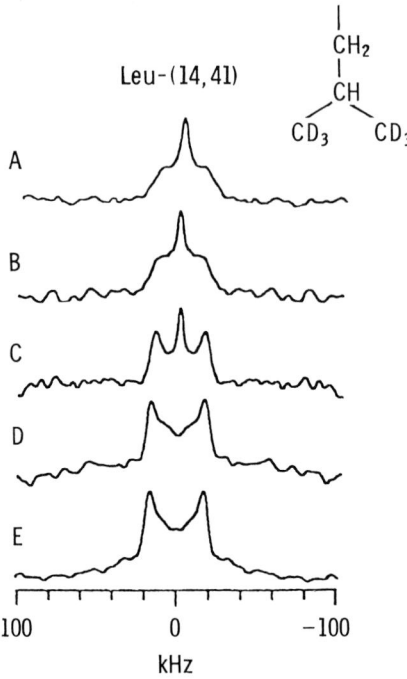

FIG. 24. ^{2}H NMR spectra of CD_3-Leu-28 labeled fd virus as a function of temperature. (A) 25°, (B) 10°, (C) 0°, (D) −10°, (E) −40°. (From Colnago et al.[44])

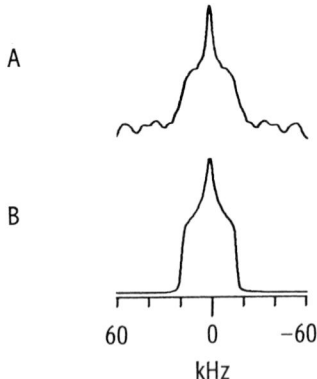

FIG. 25. Comparison of experimental and calculated ^{2}H NMR spectra for CD_3-Leu labeled fd virus. (A) Experimental spectrum at 25° as in Fig. 24A. (B) Calculated spectrum ($\eta = 1$) for tetrahedral jump motions as in Fig. 5B. (From Colnago et al.[44])

fd coat protein has two leucine residues. The ^{15}N NMR powder pattern in Fig. 9C from these residues shows that they both have immobile peptide linkages. The $\eta = 0$ powder pattern for methyl threefold reorientation or the $\eta = 1$ powder pattern with twofold jumps are expected in the fast motional limits for the CD_3-groups of leucine residues. Experimental ^{2}H NMR spectra for CD_3-Leu-labeled coat protein obtained at temperatures between $-40°$ and $25°$ are shown in Fig. 24. Above about $10°$ both of the Leu side chains give powder patterns with $\eta = 1$, indicative of the presence rapid tetrahedral jump motions. The two Leu residues differ in their activation energy for the jump motion, since at $0°$ one of the side chains is immobile and the other undergoes twofold jump motions. Even at temperatures where the jump motions occur in both residues, the backbone is rigid as seen in the ^{15}N chemical shift anisotropy powder pattern and required by the ^{2}H NMR powder pattern line shapes. The experimental line shape is very similar to that calculated for a methyl group undergoing only tetrahedral two-site jump motions, as shown with the comparison in Fig. 25.

fd coat protein has Leu residues at positions 14 and 41. Both have immobile backbone sites. Their side chains undergo tetrahedral two-fold jump motions with somewhat different activation energies. They present a particularly clear-cut case of jump motional averaging in a protein.

Acknowledgments

The research carried out at the University of Pennsylvania has been supported by grants from the N.I.H. (GM-24266 and GM-29754).

[18] Internal Motion of Proteins: Nuclear Magnetic Resonance Measurements and Dynamic Simulations

By C. M. Dobson and M. Karplus

Principles

The accepted view of the structure of a protein is now a time-dependent one with ever present, significant fluctuations about an average structure. There has been considerable discussion of the possible functional importance of such motions.[1] The existence of fluctuations has been demonstrated experimentally using a wide range of different approaches.[2-7] These include the interpretation of hydrogen-exchange measurements, X-ray diffraction temperature factors, fluorescence quenching and depolarization, and nuclear magnetic relaxation rates. A common feature of the experimental approaches is the difficulty of a direct interpretation of the results because of the highly complex and coupled internal motions that occur in protein molecules.

The dynamics of proteins can also be studied using theoretical techniques.[8,9] In order to do this it is necessary to have a knowledge of the potential energy surface, that is of the energy of the system as a function of the atomic coordinates. Given a potential energy function, a variety of approaches may be used to study protein dynamics.[8] In molecular dynamics simulations, Newtonian equations of motions for the atoms of the protein and any surrounding solvent are solved for the atomic positions as a function of time. With currently available computers it is possible to simulate the dynamics of small proteins for up to a few hundred picoseconds. Such periods are long enough to characterize completely the librations of small groups in the protein and to determine the dominant contributions to the atomic fluctuations. To study slower and more complex processes in proteins it is generally necessary to use methods other

[1] M. O'Connor (ed.), *Ciba Found. Symp.* **93** (1983).

[2] I. D. Campbell, C. M. Dobson, and R. J. P. Williams, *Adv. Chem. Phys.* **39,** 55 (1978).

[3] F. R. N. Gurd and M. Rothgeb, *Adv. Protein Chem.* **33,** 74 (1979).

[4] R. E. London, *in* "Magnetic Resonance in Biology" (J. S. Cohen, ed.), Vol. 1, p. 1. Wiley, New York, 1980.

[5] P. G. Debrunner and H. Frauenfelder, *Annu. Rev. Phys. Chem.* **33,** 283 (1982).

[6] E. Clementi and R. H. Sarma (eds.), "Structure and Dynamics: Nucleic Acids and Proteins." Adenine, New York, 1983.

[7] S. W. Englander and N. R. Kallenbach, *Q. Rev. Biophys.* **16,** 521 (1984).

[8] M. Karplus and J. A. McCammon, *Crit. Rev. Biochem.* **9,** 293 (1981); *Annu. Rev. Biochem.* **53,** 262 (1983).

[9] J. A. McCammon and M. Karplus, *Acct. Chem. Res.* **16,** 187 (1983).

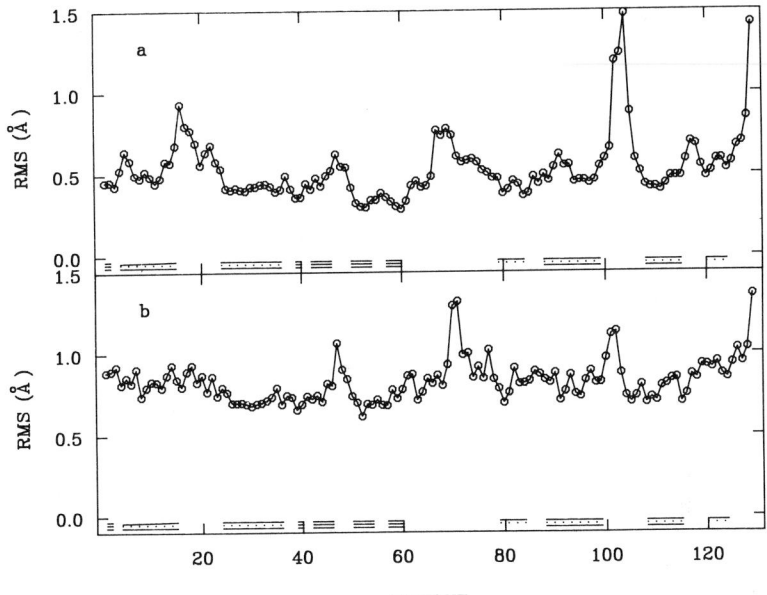

FIG. 1. Calculated and experimental rms fluctuations of lysozyme.[10] Backbone averages are shown as a function of residue number, and were obtained (a) from a molecular dynamics simulation and (b) from X-ray temperature factors without correcting for disorder contributions.

than straightforward molecular dynamics simulations. A variety of approaches have been described, such as stochastic dynamics, harmonic dynamics, and activated dynamics.[8,9]

The molecular dynamics simulation provides detailed descriptions of the fluctuations of individual atoms in the protein. Comparisons of the rms magnitude of these fluctuations with the value of temperature factors from X-ray diffraction studies have now been made for a number of proteins[9-11]; an example is given in Fig. 1.[10] In general, the level of agreement is sufficient to produce confidence in the overall results of the simulation, as well as in the interpretation of the X-ray data. The simulations provide, however, much more information about the internal motion in the protein than the magnitude of atomic fluctuations. They can show, for example, the variation with time of structural parameters such as torsion angles and interatomic distances, or provide correlation functions for

[10] C. N. Post, B. Brooks, P. Artymiuk, J. Stockwell, D. C. Phillips, C. Dobson, and M. Karplus, *J. Mol. Biol.*, in press (1986).
[11] S. H. Northrup, M. R. Pear, J. A. McCammon, M. Karplus, and T. Takano, *Nature (London)* **287,** 659 (1980).

particular types of motion. With this information, it becomes possible to compare the theoretical predictions with a wider range of experimental data, in particular with the parameters obtained from NMR experiments.[12–18]

NMR spectroscopy is now at the point where many resonances of the spectra of several small proteins (MW ≤ 20,000) have been resolved and assigned.[19–22] For larger proteins, resonance assignments are also becoming available under favorable circumstances. NMR parameters such as coupling constants and chemical shifts can then be extracted from these spectra and associated with specific nuclei in the protein (see Fig. 2). In addition, well-defined approaches exist for measuring a variety of relaxation phenomena and for extracting specific relaxation rates from them. Considerable effort has been employed in understanding the relationships between NMR parameters and molecular structure, and in many cases these are very well defined. For molecules of known structure, it is often possible to calculate NMR parameters that agree well with experimental values.[23]

NMR can provide information about the dynamic processes occurring in proteins in several different ways.[24] First, NMR parameters such as spin–spin coupling constants, chemical shift values, and nuclear relaxation rates depend on local interactions within the molecular structure. If different molecular conformations exist and if interconversion among them is rapid the observed parameters will be an average of those for the different structures. Analysis of the experimental values of these parameters can therefore in principle reveal not only the average structure but

[12] R. M. Levy, M. Karplus, and J. A. McCammon, *J. Am. Chem. Soc.* **103**, 994 (1981).

[13] R. M. Levy, M. Karplus, and P. G. Wolynes, *J. Am. Chem. Soc.* **103**, 5998 (1981).

[14] R. M. Levy, C. M. Dobson, and M. Karplus, *Biophys. J.* **39**, 107 (1982).

[15] J. C. Hoch, C. M. Dobson, and M. Karplus, *Biochemistry* **21**, 1118 (1982).

[16] E. T. Olejniczak, C. M. Dobson, M. Karplus, and R. M. Levy, *J. Am. Chem. Soc.* **106**, 1923 (1984).

[17] J. C. Hoch, C. M. Dobson, and M. Karplus, *Biochemistry* **24**, 3831 (1985).

[18] G. Lipari, A. Szabo, and R. M. Levy, *Nature (London)* **300**, 197 (1982).

[19] G. Wagner and K. Wüthrich, *J. Mol. Biol.* **155**, 347 (1982).

[20] M. Delepierre, C. M. Dobson, M. A. Howarth, and F. M. Poulsen, *Eur. J. Biochem.* **145**, 389 (1984).

[21] G. R. Moore, R. G. Ratcliffe, and R. J. P. Williams, *Essays Biochem.* **19**, 142 (1983).

[22] C. Redfield, C. M. Dobson, and F. M. Poulsen, *Eur. J. Biochem.* **128**, 527 (1982).

[23] E. D. Becker, "High Resolution NMR. Theory and Chemical Applications." Academic Press, New York, 1980.

[24] C. M. Dobson, *in* "Intramolecular Dynamics" (J. Jortner and B. Pullman, eds.), p. 481. Reidel, Dordrecht, 1982.

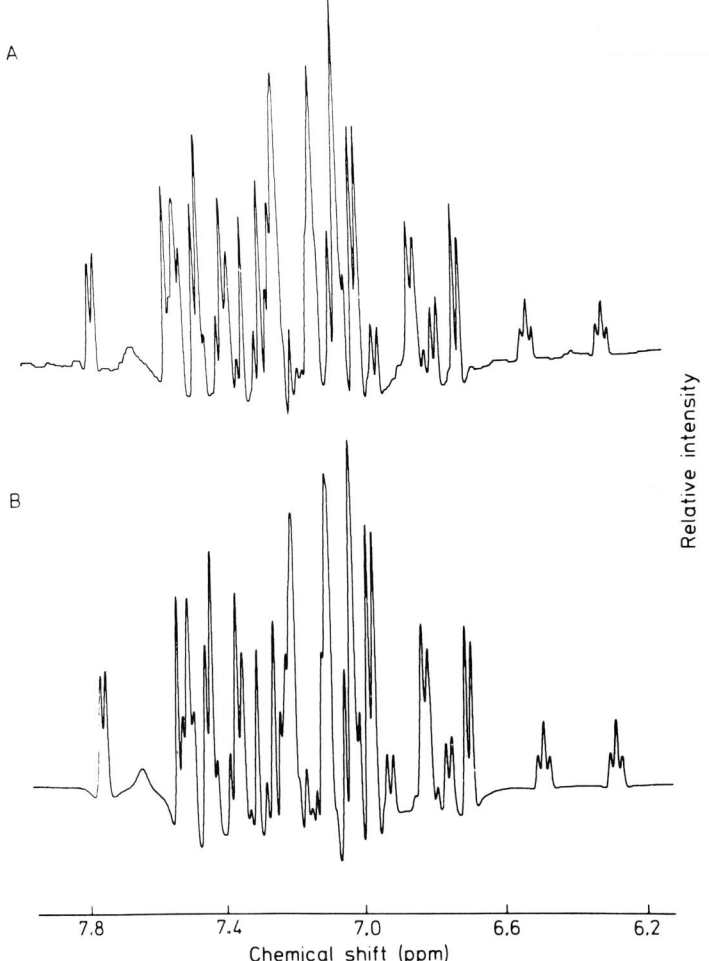

FIG. 2. Aromatic region of the 470 MHz ^{1}H NMR spectrum of hen lysozyme.[22] (A) Experimental spectrum; (B) spectrum simulated by defining NMR parameters for 58 aromatic protons in the protein. All the resonances in this region of the spectrum have been assigned to specific residues in the protein sequence.

the nature of the individual conformations contributing to this average. If the rate of interconversion between different conformations is slow on the NMR time scale, then separate resonances may be observed for the different conformations. In such cases it may be possible to determine the rates of interconversion of the conformations by line shape analysis or magnetization transfer techniques. The dynamics of interconversion of folded and unfolded states of proteins have, for example, been studied in this way.[25] Second, nuclear relaxation processes are dependent on the molecular motions. In solution, overall tumbling of a molecule is often the dominant type of motion contributing to relaxation but internal motions can have a considerable influence. For solids, the internal motions can be of primary importance. The magnitudes and rates of these motions determine their effect on relaxation rates and hence may be investigated. Third, when a system is not at chemical equilibrium, NMR may be used to follow the approach to equilibrium by recording spectra as a function of time. Only relatively slow processes such as hydrogen exchange can, however, be studied directly in this way because the time required to record NMR spectra of proteins is typically several minutes.

The nature of specific motions normally cannot be derived in a direct way from NMR experiments. The procedure adopted is to set up models for the motions and to calculate from these the values of NMR parameters. The models for the dynamic behavior of proteins in solution can range from ones involving only the overall tumbling of rigid molecules to ones in which many internal motions are specifically included. The experimental data can then be used to evaluate to what extent given models are applicable. In practice, whether or not a given NMR experiment can distinguish between different possibilities depends on several factors. These include the accuracy with which the NMR parameters can be measured, the accuracy with which they can be calculated for a given model, and the magnitude of the differences between the predictions of the different models.

The complexity of the internal motions that take place in proteins and other macromolecules is such that the simple models appropriate for small molecules are often of limited value. This is a reason for the need for detailed results from molecular dynamics simulations to supplement the experimental data. From the simulations, more realistic models for internal motions can be formulated and then used in parametrized form to interpret NMR data. Also, the highly detailed description of internal motions provided by molecular dynamics simulations can be used directly to make predictions that can be tested against experimental data. The time scales of the available simulations are too short to permit a study of slow processes such as protein folding or hydrogen exchange. The simulations are of value in examining the dynamic averaging of NMR parameters, and in exploring relaxation phenomena.

[25] C. M. Dobson and P. A. Evans, *Biochemistry* **23**, 4267 (1984).

Methods

In this section, we describe the methods used in the simulation of protein motions and in the application of the results to the interpretation of NMR parameters.

Simulation Procedures

The starting point of a molecular dynamics simulation is a potential energy function for the system from which the forces on the atoms can be determined. An initial set of atomic coordinates and velocities is obtained from the X-ray coordinates of the protein by a preliminary calculation in which the system is equilibrated.[26,27] The X-ray structure is refined by an energy minimization procedure which reduces unfavorable interactions resulting, for example, from overlap of nonbonded atoms or distortion of bond lengths. Random velocities are then assigned to the various protein atoms with a distribution corresponding to a low temperature, and a simulation is performed by integrating the equations of motions for all of the atoms over a period of several picoseconds. The equilibration is continued by alternating new velocity assignments corresponding to successively increased temperatures with intervals of dynamic simulation. The temperature is defined in terms of the mean kinetic energy for the system of atoms and the equilibration period is considered complete when no systematic changes of temperature are evident over a time period of about 10 psec. After the equilibration, the actual simulation is carried out by continuing to integrate the equations of motion for the desired length of time.

Since the computer time required for a simulation increases with the number of atoms involved, a simplified model in which only polar hydrogens are included is often used. In this extended atom approach the presence of nonpolar hydrogen atoms is accounted for by adjusting the mass and radius of the directly bonded heavy atom.[28] For the interpretation of ^{1}H NMR parameters, protons are generated onto the heavy atoms in standard configurations for each coordinate set obtained in the dynamics simulation.[15]

The direct results of a dynamic simulation are the coordinates and velocities of all the atoms as a function of time. The next step is to calculate NMR parameters from the sets of coordinates. In the case of conformationally dependent parameters such as three-bond coupling constants and chemical shifts, the coordinate sets are taken in turn and the

[26] J. A. McCammon, P. G. Wolynes, and M. Karplus, *Biochemistry* **18**, 227 (1979).
[27] W. F. van Gunsteren and M. Karplus, *Macromolecules* **15**, 1528 (1982).
[28] W. F. van Gunsteren and M. Karplus, *Biochemistry* **21**, 2259 (1982).

calculations of the parameters are carried out for each one.[15,17] The variation of the parameters with time is obtained then by examining the calculated values for the coordinate sets corresponding to different times in the simulation. In the case of relaxation properties, in addition to calculations for individual coordinate sets, the correlation of angular and distance-dependent functions of coordinates corresponding to varying time intervals is required. The results of this type of analysis are correlation functions which are used then to predict relaxation behavior.[12–14,16]

A typical dynamic simulation of 100 psec involves some 10^5 coordinate sets because a time step of approximately 10^{-3} psec is appropriate for the numerical integration of the equations of motion. Although the general procedure outlined above can be applied using all of these coordinate sets, in practice some simplification is possible. In particular, because the time scale of the NMR phenomenon is orders of magnitude slower than that of the calculated fluctuations, a selection of coordinate sets can be made.[15] Typically, less than 1 in 10^2 need be analyzed, corresponding to time steps of approximately 10^{-1} psec.

It might be thought that the simplest way to proceed at this point is to compare the predicted NMR parameters with those derived from experiments. In practice this is not usually straightforward. For example, coupling constants and chemical shifts are determined by the average structure of the molecule as well as by the fluctuations about this average. The average structure of a protein in solution is not known directly, although the assumption that it is very similar to the crystal structure may be a good one in many cases. The average structure in the simulation, however, deviates significantly from that of the crystal structure. Typically, the rms differences between C^α atom in the two structures are more than 1 Å (Fig. 3). These differences arise because the structure of the protein adjusts during the equilibration and simulation periods under the influence of the approximate potentials used in the simulation; the simulations are often performed in vacuum and sometimes in dilute solution, while the structural data correspond to a concentrated crystal environment. One approach to this problem is to determine how the NMR parameters calculated from the simulation compare with those calculated for a rigid model of the protein which has the average structure of the simulation.[15] This enables the effects of the fluctuations on the NMR parameters to be examined and analyzed. A further complication arises in the interpretation of relaxation effects because they depend on the overall molecular tumbling as well as the internal motions. The overall tumbling of the protein is not usually modeled in the simulation, because the time scale of the tumbling is considerably longer than that of the simulation. For the relaxation data, therefore, it is necessary to assume separability of tum-

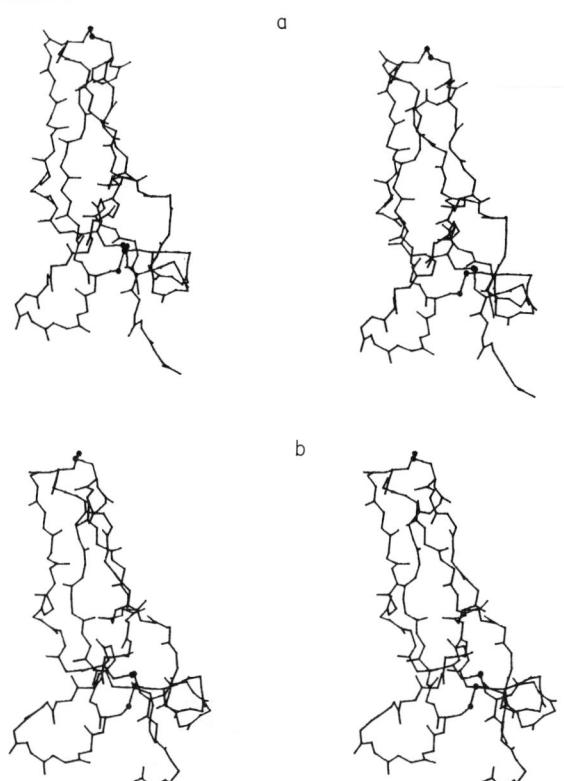

FIG. 3. Drawing of polypeptide backbone of BPTI.[28] The upper (a) stereo pair is the X-ray structure and the lower (b) stereo pair is the average structure from the simulation.

bling and internal motions and to obtain a value for this tumbling rate from experiments or theoretical studies before proceeding with the analysis of the effects of the fluctuations.[12–14,16]

Calculation of NMR Parameters

The simplest NMR parameter to calculate from the dynamics simulation is the spin–spin coupling constant between a pair of protons separated by three bonds. The coupling constants can be defined by an expression of the form,[29,30]

$$^3J_{\text{H–H}} = A \cos^2 \theta + B \cos \theta + C \tag{1}$$

[29] M. Karplus, *J. Am. Chem. Soc.* **85**, 2870 (1963).
[30] V. F. Bystrov, *Prog. NMR Spectrosc.* **10**, 41 (1976).

where $^3J_{H-H}$ is the coupling constant, θ is the H–X–Y–H[1] torsion angle, and A, B, and C are constants. The latter can in principle be obtained from theoretical calculations[31] but in practice have been determined empirically for the various groups of polypeptides.[30,32] For a given protein coordinate set, $^3J_{H-H}$ can be calculated from Eq. (1).

There are a number of contributions to the chemical shift of a nucleus in a protein.[15,33–35] The most fundamental effects depend on the chemical nature of the group containing the nucleus and are therefore relatively constant for nuclei in a given type of amino acid residue. Other contributions to the chemical shifts arise from through space interactions between a nucleus and the residues in its vicinity. These contributions can be estimated by the difference between observed chemical shifts and those appropriate for amino acid residues in unstructured polypeptides, and are of major interest as they are dependent on the conformation of the protein. The dominant through space contribution for side chain protons bound to carbons is often the ring current shift, and this can be calculated rather accurately.[34] The ring current shift arises from the diamagnetic anisotropy of aromatic rings, and depends on the distance between a proton and an aromatic ring and on their relative orientation. Various methods of computing the shift have been described.[34] The simplest is the dipole approximation in which the shift is simply proportioned to $(1 - 3\cos^2\theta)/r^3$; here r is the distance between the proton and the center of the aromatic ring and the angle θ is formed by a vector normal to the ring plane and a vector connecting the center of the ring and the proton.[36] As with the coupling constants, all that is needed for calculation is a coordinate set for the protein.

Under most circumstances, the relaxation behavior of a protein in solution is governed by dipolar interactions between the various nuclear spins.[4,37–39] The dipolar couplings depend on the relative positions of the coupled nuclei and on their relative motions.[38,40] For spin–lattice relaxation, such as observed in T_1 or nuclear Overhauser effect measurements,

[31] M. Barfield and M. Karplus, *J. Am. Chem. Soc.* **91**, 1 (1969).
[32] K. D. Kopple, G. R. Wiley, and R. Tauke, *Biopolymers* **12**, 627 (1973).
[33] H. Sternlicht and D. Wilson, *Biochemistry* **6**, 2881 (1967).
[34] S. J. Perkins, *Biol. Magn. Reson.* **4**, 193 (1982).
[35] A. Pardi, G. Wagner, and K. Wuthrich, *Eur. J. Biochem.* **137**, 445 (1983).
[36] J. A. Pople, *J. Chem. Phys.* **24**, 1111 (1956).
[37] A. Kalk and H. J. C. Berendsen, *J. Magn. Reson.* **24**, 343 (1976).
[38] I. D. Campbell and C. M. Dobson, *Methods Biochem. Anal.* **25**, 1 (1979).
[39] C. M. Dobson, E. T. Olejniczak, F. M. Poulsen, and R. G. Ratcliffe, *J. Magn. Reson.* **48**, 97 (1982).
[40] J. H. Noggle and R. E. Shirmer, "The Nuclear Overhauser Effect." Academic Press, New York, 1971.

it is possible to express the behavior of the magnetization in the form

$$\frac{d[I_z(t) - I_0]_i}{dt} = - \rho_i[I_z(t) - I_0]_i - \sum_{i=j} \sigma_{ij}[I_z(t) - I_0]_j \qquad (2)$$

where $I_z(t)_i$ and I_{0i} are the z components of the magnetization of nucleus i, ρ_i is the direct relaxation rate of nucleus i, and σ_{ij} is the cross-relaxation rate between nuclei i and j. Similar expressions can be derived for spin–spin relaxation such as can be observed in T_2 or line width measurements. The quantities ρ_i and σ_{ij} can be expressed in terms of spectral densities

$$\rho_i = \frac{6\pi}{5} \gamma_i^2\gamma_j^2\hbar^2 \sum_{i \neq j} \left[\frac{1}{3} J_{ij}(\omega_i - \omega_j) + J_{ij}(\omega_i) + 2J_{ij}(\omega_i + \omega_i) \right] \qquad (3)$$

$$\sigma_{ij} = \frac{6\pi}{5} \gamma_i^2\gamma_j^2\hbar^2 \left[2J_{ij}(\omega_i + \omega_j) - \frac{1}{3} J_{ij}(\omega_i - \omega_j) \right] \qquad (4)$$

The spectral density functions can be obtained from the correlation functions for the relative motions of the nuclei with spins i and j[16,41]

$$J_{ij}^n(\omega) = \int_0^a \left\langle \frac{Y_n^2[\theta_{lab}(t)\phi_{lab}(t)] Y_n^{2*}[\theta_{lab}(0)\phi_{lab}(0)]}{r_{ij}^3(0)r_{ij}^3(t)} \right\rangle \cos(\omega t)dt \qquad (5)$$

where $Y_n^2[\theta(t)\phi(t)]$ are second-order spherical harmonics and the angular brackets represent an ensemble average which is approximated by an integral over the molecular dynamics trajectory. The quantities $\theta_{lab}(t)$ and $\phi_{lab}(t)$ are the polar angles at time t of the internuclear vector between protons i and j with respect to the external magnetic field and r_{ij} is the interproton distance. In the simplest case of a rigid molecule undergoing isotropic tumbling with a correlation time, τ_0, this reduces to the familiar expression

$$J_{ij}(\omega) = \frac{1}{4\pi r_{ij}^6} \left[\frac{\tau_0}{1 + (\omega\tau_0)^2} \right] \qquad (6)$$

Because of the large number of magnetic nuclei in proteins the relaxation behavior can be extremely complex. In practice, certain types of relaxation are more straightforward to interpret than others. Two of these are of particular importance. The first is exemplified by measurement of ^{13}C T_1 values for protonated carbons.[42,43] Here, because ^1H spins can be saturated in the experiment, the second term of Eq. (2) vanishes. Furthermore, because the distance from the carbon to its directly bonded proton is short compared with the distance to other protons, the latter can be ignored and Eq. (3) reduces to that describing the motion of the single

[41] I. Solomon, *Phys. Rev.* **99**, 559 (1955).
[42] K. F. Kuhlmann, D. M. Grant, and R. K. Harris, *J. Chem. Phys.* **52**, 3439 (1970).
[43] J. R. Lyerla, Jr. and G. C. Levy, *Top. ^{13}C NMR Spectrosc.* **1**, 79 (1974).

pair. As r_{ij} is essentially fixed for a C—H bond, only the angular variation is important in Eq. (5). For nonprotonated carbons, where the dipolar relaxation is not dominated by protons at fixed distances, an average must be made over many interactions for which the distances as well as the orientations are, in general, time dependent. The second type of behavior that is straightforward to analyze is exemplified by interproton nuclear Overhauser effects.[40] Here it is possible to examine the second term in Eq. (2) and extract cross-relaxation rates σ_{ij} directly for pairs of protons from measurements of the time dependences of nuclear Overhauser effects.[39] Again, if i and j are separated by a fixed distance, for example, the protons of a CH_2 group or an aromatic ring, the relaxation due to the internal motion depends only on the orientational variation of the proton pair.[39,44] Otherwise, averaging over a sum of interproton interactions must be carried out.

The molecular dynamics simulation permits the direct evaluation of the internal motion correlation functions [see Eq. (5)], and hence permits the relaxation behavior described by Eq. (2) to be modeled exactly for a given simulation. The computational problem is much simpler for situations such as the ^{13}C T_1 and ^1H cross-relaxation measurements where r_{ij} does not vary in the simulation. Then only the angular correlation function is required.[12,13,16] In the more general cases where several spin pairs are involved and where r_{ij} values are time dependent the full analysis involving averaging over a number of interactions is needed; details of the procedures are given elsewhere.[14,16]

Examples

Molecular dynamics simulations of two proteins, BPTI and lysozyme, have been examined by the methods discussed in this section. Details of the results are presented elsewhere[12–18]; here we outline the type of behavior that is observed and present some of the consequences of the comparisons between simulation results and experiment.

Coupling Constant Data

In Fig. 4 are shown fluctuations of coupling constant values for two residues of BPTI calculated from a molecular dynamics simulation.[17] The behavior here is typical of that found in BPTI and lysozyme, and demonstrates that very large variations in the coupling constants are predicted. Because of the direct relationship between coupling constants and torsion angles, the origins of these fluctuations are readily understood. In Fig. 5, the variations in dihedral angles are shown for the two residues of Fig. 4.

[44] E. T. Olejniczak, F. M. Poulsen, and C. M. Dobson, *J. Am. Chem. Soc.* **103**, 6574 (1981).

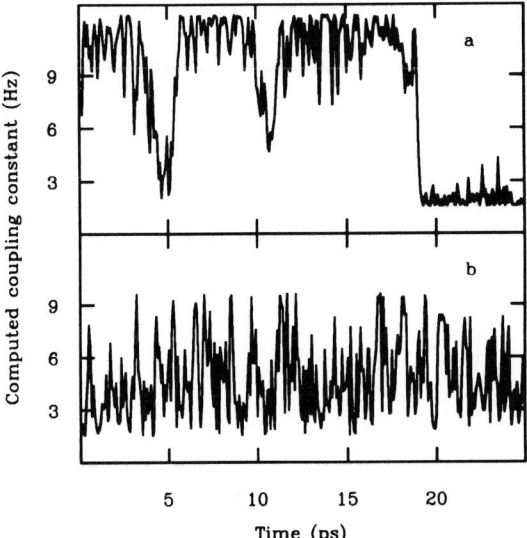

FIG. 4. Fluctuations of three-bond proton coupling constants during a molecular dynamics simulation of BPTI.[17] (a) Glu-7 $^3J_{\beta 1 \gamma 1}$, (b) Arg-42 $^3J_{\beta 1 \gamma 1}$.

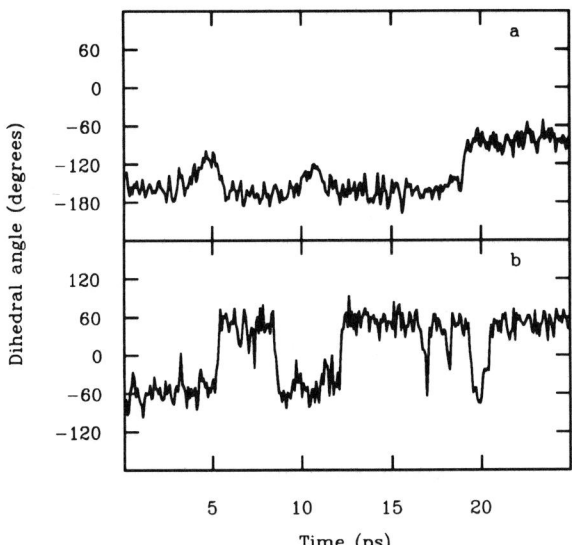

FIG. 5. Fluctuations of interproton dihedral angles during a molecular dynamics simulation of BPTI.[17] The dihedral angles shown are (a) Glu-7 $\theta_{\beta 1 \gamma 1}$ and (b) Arg-42 $\theta_{\beta 1 \gamma 1}$, i.e., those determining the coupling constants shown in Fig. 4.

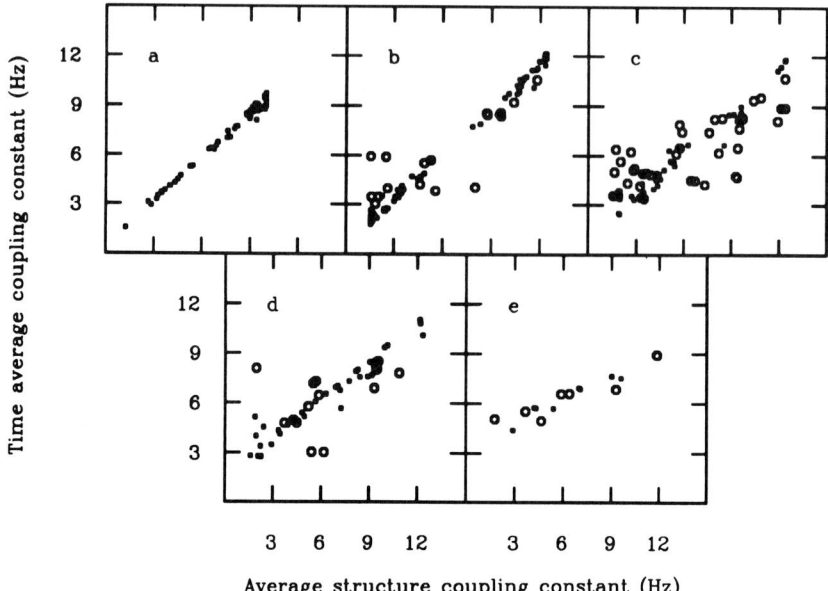

FIG. 6. Comparison of time average and average structure proton coupling constants computed from a molecular dynamics simulation of BPTI.[17] The coupling constants associated with (a) ϕ, (b) χ_1, (c) χ_2, (d) χ_3, and (e) χ_4 torsion angles are shown separately. Small filled circles indicate cases where fluctuations occur about a single torsion angle value, and open circles cases where jumps occur between different torsional states.

Two major types of fluctuations are observed.[17] First, on a very short time scale (1 psec) torsion angle fluctuations with rms amplitudes of ~10° are found, and these represent motions within a single potential well (e.g., fluctuations about a single conformational state). Second, on a longer time scale (1 psec or greater) there occur rare events involving large changes in torsion angles. These represent jumps between potential wells (e.g., fluctuations involving different conformational states).

The time scale of the fluctuations observed in the simulation is such that only values of the coupling constants averaged over time are observable experimentally. To examine the influence of dynamics on the observable values, the average values of coupling constants are computed and compared with values calculated from the average structure of the protein obtained from the simulation[17]; some results for BPTI are shown in Fig. 6. The differences between the two types of averages range from 0 to more than 6 Hz; the values of the coupling constants themselves range between 2 and 14 Hz. Figure 6 shows clearly that all the large differences arise from dihedral fluctuations involving jumps between different potential wells.[17] Fluctuations within a single potential well agree to better than 2 Hz between the two types of averages in all cases.

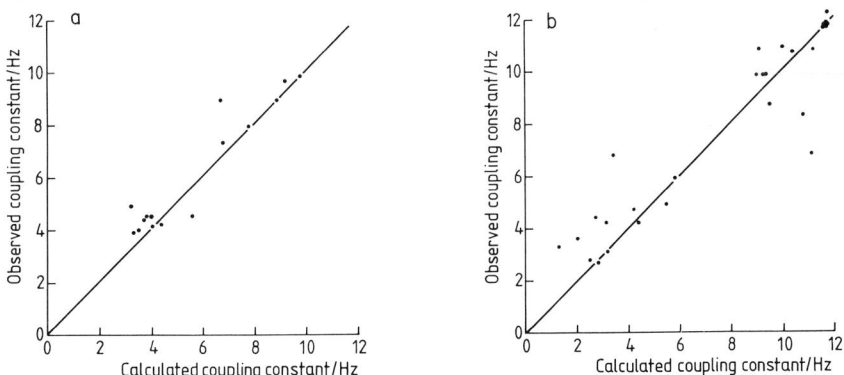

FIG. 7. Comparison of experimental proton coupling constants with values calculated from the crystal structure of lysozyme.[46] (a) $^3J_{HN-H\alpha}$, (b) $^3J_{H\alpha-H\beta}$.

Two trends are apparent from the data of Fig. 6. The first is that for large coupling constants the average structure value tends to exceed the time average value, while the reverse is true for the small coupling constants. This is a result of the geometrical dependence of the coupling constants on the torsion angle. The second trend is that the differences between the two averages are larger for side-chain than for main-chain coupling constants, and they increase as one progresses out along the side chain. This reflects the increasing occurrence of significant populations of more than one rotameric state for bonds further from the main chain.[17]

The comparison of time average coupling constants with average structure values outlined above provides a basis for comparing experimental coupling constants (time average values) with values computed from an X-ray crystal structure (average structure values). Although the data available for proteins are as yet limited,[45] there is evidence that the effects predicted in the simulation are revealed in the experimental values. Some results for lysozyme are shown in Fig. 7[46]; the clear correlation reflects the similarity of the average structure in solution and in the crystal. Significant deviations are observed, however, and it is of particular interest that the two trends apparent from the simulated data of Fig. 6 are detectable experimentally. Overall, the experimental data are consistent with the type of motion predicted to occur in the simulations.

The results obtained in these analyses have consequences for the experimental determination of the structure and dynamics of proteins.[17] In a

[45] K. Nagayama and K. Wüthrich, *Eur. J. Biochem.* **115,** 653 (1981).
[46] M. Delepierre, C. M. Dobson, M. A. Howarth, and F. M. Poulsen, *Eur. J. Biochem.* **145,** 389 (1984).

situation where the structure of a protein is unknown, the results show that motional averaging of the type seen in the simulation does not prevent the use of coupling constants for determining possible values of the ϕ torsion angles. Although in general there will not be a unique value of ϕ defined by the experimental data, one of the allowed values should correspond clearly to the average value of ϕ in the protein. The same conclusion is generally valid for χ_1 but as one moves out along the side chain of a residue the torsion angle fluctuations become larger and significant populations of more than one rotameric state become common. The coupling constant analysis, if used by itself, therefore becomes less effective. Nevertheless, the description of the structure of a protein in solution must be consistent with the coupling constant data and this is an important test of its validity.

The analysis of the simulations also suggests that coupling constants can be used to define experimentally the magnitude of torsion angle fluctuations. One approach of particular interest is applicable where the average structure in solution is known. For small stable proteins such as BPTI or lysozyme there is considerable evidence that the average solution structure is close to the crystal structure, particularly for internal residues. If it is accepted that the two structures are very similar, the analysis (see Fig. 6) suggests that when agreement between experimental coupling constants and those calculated from the crystal structure is poor, jumps occur between different conformers which have significant populations. That experimentally the deviations between the observed coupling constants and those calculated from the average (X-ray) structure are more frequent than found in the simulation suggests that fluctuations over different rotamers are rather common in proteins, more so than is indicated by the available X-ray data. For the cases where agreement is good, fluctuations that do occur are about a single potential well. A fit of the measured coupling constants to the average torsion angle and a Gaussian distribution appears to provide a consistent description of the magnitude of these fluctuations.[17]

Chemical Shift Data

Examination of the explicit time dependence of chemical shifts computed from the individual coordinate sets of the dynamics simulations reveals large high frequency fluctuations for many of the protons. Examples from BPTI are given in Fig. 8. As with the coupling constants, most fluctuations arise from motions on the subpicosecond time scale although longer time variations are also apparent. The large magnitude of the calculated fluctuations (up to ± 6 ppm) is shown by the fact that the rms values

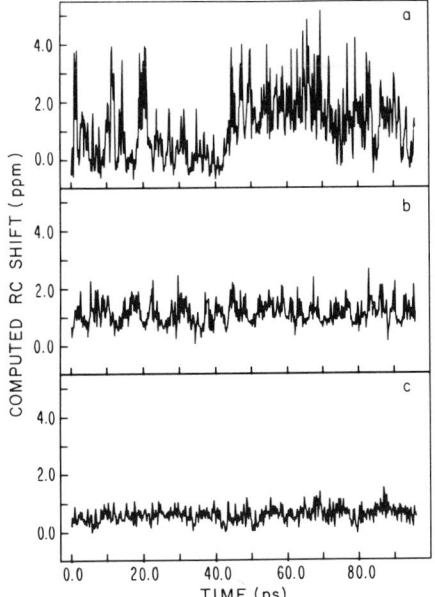

FIG. 8. Fluctuations of ring-current shifts during a molecular dynamics simulation of BPTI.[15] The shifts are calculated for (a) Pro-9 $H^{\beta 2}$, (b) Gly-28 $H^{\alpha 2}$, (c) Ala-48 $H^{\beta 2}$.

are in general comparable to the average shifts. The largest fluctuations tend to arise from the protons that are closest to aromatic rings, as one would anticipate.

In most instances the time evolution of the chemical shift of a proton exhibits random fluctuations about a single average value during the simulation and a nearly normal distribution of shifts is obtained.[15] For a number of protons the calculated shift was found to fluctuate about one average value and then be displaced so that it fluctuates about a different average value. Similar behavior was observed for coupling constants and it reflects conformational transitions which are rare events on the time scale of the simulation. In the case of chemical shifts, transitions involving a 180° flip of aromatic residues are particularly clear examples of this type of behavior (see Fig. 9).

The frequencies of the fluctuations in the chemical shifts obtained from the simulations are so high that again in general only average values need be considered for comparison with experimental behavior. This is true even for most of the rare events such as side chain transitions. In many proteins a single resonance is observed for protons on opposite sides of phenylalanine and tyrosine rings, made equivalent by the rapid

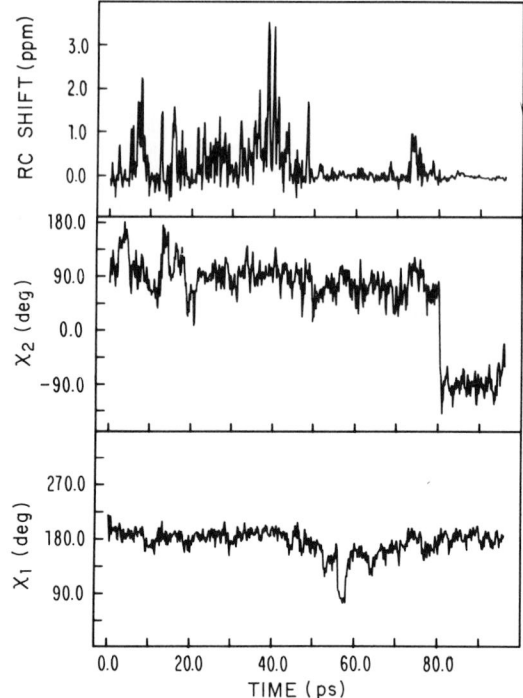

FIG. 9. Comparison between fluctuations of the ring current shift of Phe-22 H[δ1] and of the χ_1 and χ_2 torsion angles of Phe-22 from a molecular dynamics simulation of BPTI.[15]

flips of the aromatic rings.[47–49] Such rapid averaging is not, however, always the case; again the phenylalanine and tyrosine ring protons provide well-studied examples.[50,51] When the rate of ring flipping becomes slower than the chemical shift difference between the protons on opposite sides of the ring, separate resonances can be observed. When the rate of ring flipping is comparable to the chemical shift differences the rate can be determined from exchange effects in the spectrum and examples of this are well documented.[50,51]

For these and other slow processes the standard molecular dynamics simulations are not appropriate to determine the motion. The applications

[47] I. D. Campbell, C. M. Dobson, and R. J. P. Williams, *Proc. R. Soc. London Ser. B* **189,** 503 (1975).

[48] A. Cavé, C. M. Dobson, J. Parello, and R. J. P. Williams, *FEBS Lett.* **65,** 190 (1976).

[49] K. Wüthrich and G. Wagner, *FEBS Lett.* **50,** 265 (1975).

[50] I. D. Campbell, C. M. Dobson, G. R. Moore, S. J. Perkins, and R. J. P. Williams, *FEBS Lett.* **70,** 96 (1976).

[51] G. Wagner, A. DeMarco, and K. Wüthrich, *Biophys. Struct. Mech.* **2,** 139 (1976).

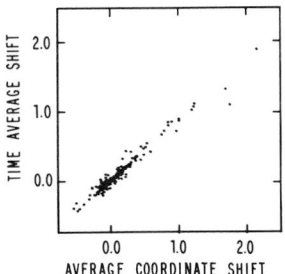

FIG. 10. Comparison of time average ring-current shifts and ring current shifts calculated from the average coordinates computed from a molecular dynamics simulation of BPTI.[15]

of adiabatic mapping and activated dynamics methods have, however, given insight into the processes involved.[52–54] In studies of the aromatic residues of BPTI, the right order of magnitudes for the barriers to rotation have been obtained in these calculations; one of the four tyrosines in BPTI was predicted to have a high barrier and the three others low barriers,[52] in agreement with subsequent experiments.[50] Other studies of ring rotations in BPTI have also considered the phenylalanines.[53,55]

Comparison of the time average ring current shifts with those calculated from the average structure shows very close agreement, although systematic differences are present. In BPTI, for example, (Fig. 10) the shifts calculated from the average structure are typically larger than the average values by some 15%.[15] Analysis of the averaging effects on chemical shifts is more complex than that of coupling constants where only a single conformational parameter is involved. The fluctuations in the chemical shifts have been shown to be the result of both radial and angular contributions. The principal source of the angular fluctuations appears to be torsional motion of the aromatic rings, while the radial fluctuations are often due primarily to motions of the groups containing the shifted protons. The direction, although not the magnitude, of the systematic differences observed between the time average values and values computed from average coordinates is due principally to the angular contribution.

The agreement between the two methods of averaging to obtain the chemical shifts demonstrates the essential role of the average structure in determining the ring current contribution. For only a handful of protons in BPTI is the difference between the time average and average structure

[52] B. R. Gelin and M. Karplus, *Proc. Natl. Acad. Sci. U.S.A.* **72**, 2002 (1975).
[53] R. Hetzel, K. Wüthrich, J. Deisenhofer, and R. Huber, *Biophys. Struct. Mech.* **2**, 159 (1976).
[54] S. H. Northrup, M. R. Pear, C. Y. Lee, J. A. McCammon, and M. Karplus, *Proc. Natl. Acad. Sci. U.S.A.* **79**, 4035 (1982).
[55] G. Wagner, *Q. Rev. Biophys.* **16**, 1 (1983).

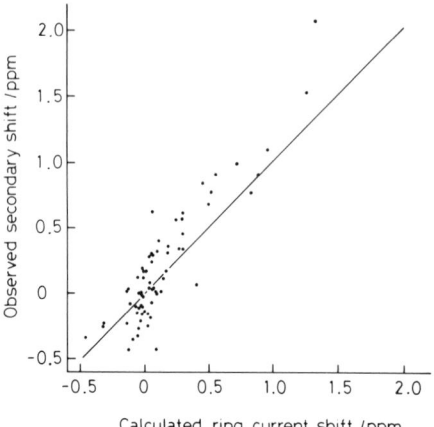

FIG. 11. Comparison of experimental proton chemical shifts of aromatic protons and methyl groups with values calculated from the crystal structure of lysozyme.[56]

ring current shifts calculated to be greater than 0.1 ppm (see Fig. 10). The correlation between the shifts computed from the crystal structure and the experimental shifts for proteins such as BPTI and lysozyme is, however, no better than about ±0.2 ppm[34,35,56] (see Fig. 11). This difference is not due to dynamic effects, but instead arises from differences between the average solution and crystal structures and from contributions to the chemical shift neglected in the ring current calculation.

The chemical shifts, unlike the coupling constants discussed above, are shown not to be a sensitive measure of protein dynamics on the picosecond time scale. Only for a protein which undergoes frequent fluctuation of a much larger magnitude than those examined in this study are the dynamic events likely to be of significance in influencing the observed chemical shift values.

Relaxation Effects

The correlation functions for a rigid molecule tumbling isotropically in solution are simple exponentials; this reflects the decreasing probability, for example, that the orientation of a given vector has remained unchanged as the time period increases. Figure 12 shows a schematic diagram of correlation functions anticipated from molecular dynamics simulations. The rapid loss of correlation in the first few picoseconds results from the effect of the librational potential of the residue containing the nucleus and of collisions between the atoms of the residue and those of the surrounding protein cage. These are analogous to the fast fluctuations

[56] C. M. Dobson, J. C. Hoch, and C. Redfield, *FEBS Lett.* **159,** 132 (1983).

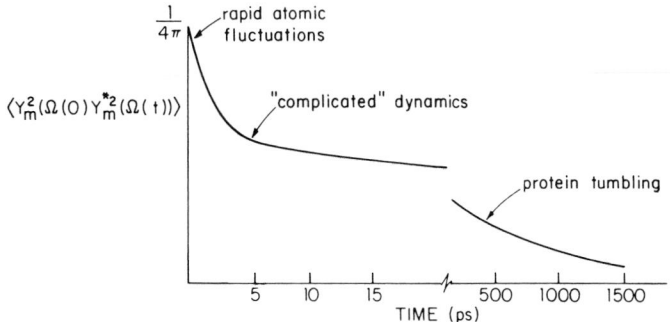

FIG. 12. Schematic diagram of the general behavior of the decay of NMR correlation functions in proteins.[12] The physical origins for the decay in correlation resulting from the different kinds of protein motions are indicated.

seen in the torsion angles and in the calculated coupling constants and chemical shifts in Figs. 4 and 8. A plateau value is then often reached, and the initial decay is followed by a slower loss of correlation over the next few hundred picoseconds. This involves larger scale and more complex motions, such as the rare events seen in Figs. 5 and 9. Finally, the correlation functions will decay to zero approximately exponentially because of the overall tumbling of the molecule referred to above.

In Fig. 13 are shown correlation functions for several protonated carbons of BPTI. In these, as in most cases for BPTI and lysozyme, the motional averaging over the time scale of the simulation is largely complete after 2 psec. This time scale is very much shorter than the NMR resonant frequencies and so the effect is simply to scale the relaxation

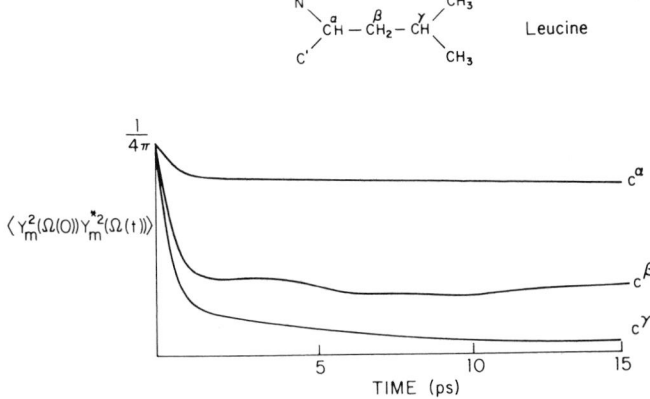

FIG. 13. NMR correlation functions for C^α, C^β, and C^γ of Leu-29 calculated from a BPTI molecular dynamics simulation.[12]

TABLE I
INCREASES IN ^{13}C NMR SPIN–LATTICE
RELAXATION TIMES IN BPTI PREDICTED FROM A
MOLECULAR DYNAMICS SIMULATION[12]

Residue	Motional averaging scale factor	
α-Carbons		
Phe-22	1.10[a]	1.13[b]
Gly-28	1.19	1.19
Leu-29	1.10	1.13
β-Carbons		
Leu-6	1.54	3.48
Ile-18	1.31	2.67
Ile-19	1.13	1.14
Leu-29	2.58	3.20
γ-Carbons		
Leu-6	1.86	4.0
Ile-18	1.78	3.64
Ile-19	1.95	1.95
Leu-29	4.21	11.4

[a] Using values of correlation functions at 2 psec.
[b] Using values of correlation functions at 96 psec.

rates from those expected for a rigidly tumbling molecule by a factor depending on the extent of motional averaging. For ^{13}C T_1 values, the result is an increase in the values over those from the rigid model. Results obtained for different types of carbons are shown in Table I. For the α-carbons of BPTI, the motional averaging effects cause an increase in the T_1 of less than 20% in nearly all the cases examined. As one moves out along the side chains, however, much larger scaling factors are predicted, up to a factor of 4 for the motions contributing to the first 2 psec decay. Further, as Fig. 13 shows, the importance of events on a longer time scale than 2 psec increases, reflecting the greater contribution of lower frequency fluctuations to side chain motion than that of the protein backbone.[12] These slower motions are likely to require more extensive calculations to obtain quantitatively reliable predictions for T_1 values. Nevertheless, the relaxation results correspond well to the extent of mo-

TABLE II

CALCULATED T_1 VALUES OF PHENYLALANINE C^γ
RESONANCES OF BPTI[14]

Residue	T_1 Values (msec)		
	Crystal	Dynamics	
Phe-4	434	599[a]	618[b]
Phe-22	442	633	800
Phe-33	459	502	514
Phe-45	442	491	518
Mean[c]	444	556	613

[a] Using values of correlation functions at 2 psec.

[b] Using values of correlation functions at 96 psec.

[c] Experimental values are between 508 and 553 msec under conditions relevant to the overall rotational correlation time used in the calculations (3.9 nsec).

tional averaging found for coupling constants discussed above. Similar behavior has also been found for the few rather more complex cases of nonprotonated carbons that have been studied.[14]

Rather limited correlation of the predictions for ^{13}C relaxation behavior with experimental results has been possible at this stage because of the lack of data for individual assigned ^{13}C resonances. For protonated carbons in BPTI, there is evidence that greater increases in T_1 values do occur for side chain atoms, e.g., isoleucine C^δ, than for C^α atoms.[57] A detailed comparison for nonprotonated carbons has been carried out[14] involving the C^γ resonances of the four phenylalanine residues of BPTI (Table II). The observed increases over T_1 values estimated for the rigid protein structure are consistent with the increases of 15–25% calculated from the decay of the NMR correlation functions at 2 psec, although this conclusion is dependent upon the assumption of a value for the overall tumbling rate. Calculated values involving the decay of the correlation functions over 96 psec are little different for three of the residues, reflecting the importance of the fast fluctuations. The T_1 value calculated for one of the phenylalanines, Phe-22, is, however, much larger at 96 than 2 psec

[57] R. Richarz and K. Wüthrich, *Biochemistry* **17**, 2263 (1978).

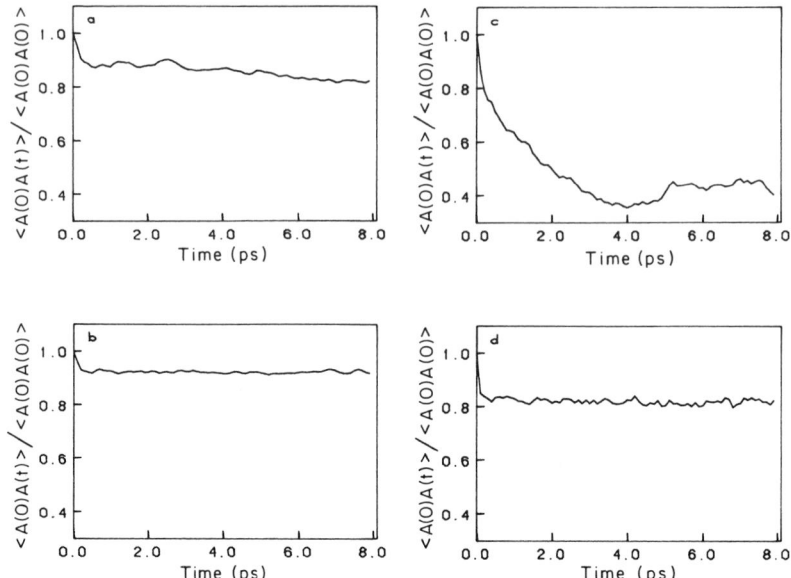

FIG. 14. Decay of the fixed distance interproton internal motion correlation functions for four residues of lysozyme calculated from a molecular dynamics simulation of lysozyme.[16] The intraresidue vectors are (a) Trp-28 $H^{\delta 3}$—$H^{\eta 3}$, (b) Trp-108 $H^{\delta 3}$—$H^{\eta 3}$, (c) Met-105 $H^{\gamma 1}$—$H^{\gamma 2}$, and (d) Ile-98 $H^{\gamma 11}$—$H^{\gamma 12}$.

because the aromatic ring flips once during the course of the 96 psec simulation. That this large value is inconsistent with the experimental data (Table II) suggests that equilibrium rate of aromatic side chain flips is much smaller than would be calculated from even one flip in 96 psec. This is in accord with adiabatic barrier estimates[52–54] and experimental data[50,51] on chemical shifts discussed above. A comparison of experimental order parameters, corresponding to the averaging of the correlation function at the plateau value, with those calculated from a molecular dynamics simulation has been made for 12 of the methyl groups of BPTI.[18] The relative flexibility of the residues studied is reasonably well described by the simulations, although the theoretical order parameters indicate less motion than the experimental ones. This suggests that longer time simulations are needed to allow all accessible bond orientations to be sampled.[18]

Interproton vector correlation functions have been calculated for pairs of protons in lysozyme,[16] and examples are given in Fig. 14. These examples are for proton pairs where the interproton distances are fixed and the correlation functions are similar to those reported for the ^{13}C—H vectors above. For two of the cases shown in Fig. 14 there is a clear plateau region in the correlation function after less than 1 psec. For another case,

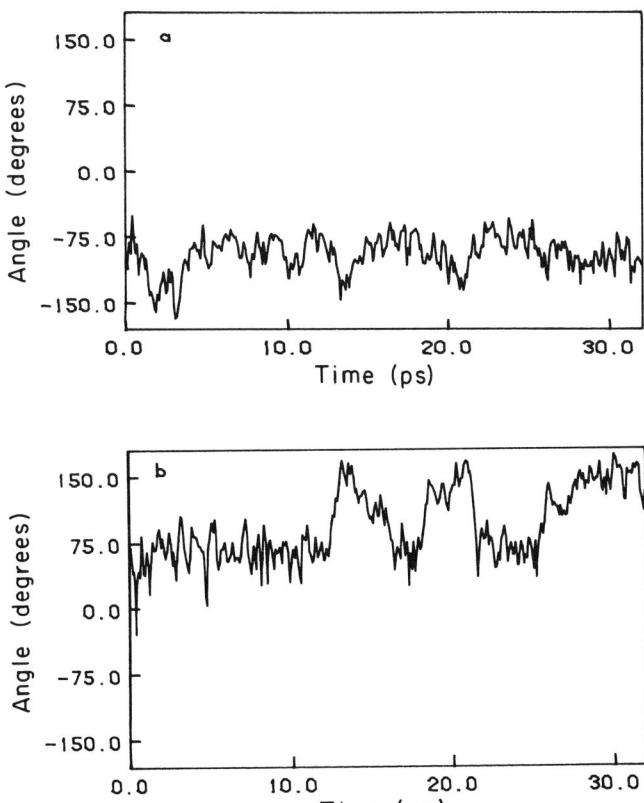

FIG. 15. Torsion angle fluctuations for the Met-105 side chain calculated from a molecular dynamics simulation of lysozyme.[16] (a) χ_1, (b) χ_2.

Trp-28, there is an additional slow loss of correlation over the 8 psec for which the correlation function was calculated. For one case, Met-105, the interproton vectors decay substantially over the 8 psec time scale. Examination of the torsion angles involved (Fig. 15) shows that the reason for this is the occurrence of jumps between different side chain conformations. For the other residues, the fluctuations are essentially about only a single site.[16] The effects on cross-relaxation rates arising from these motions are similar to those on the ^{13}C T_1 values but here the relaxation rates are reduced by scale factors from those anticipated for a rigid model of the protein. The changes in relaxation predicted for the residues of Fig. 14 range from 10% or less for all proton pairs of Trp-108 to as much as a factor of 3 for one of the proton pairs of Met-105.

TABLE III

DECREASES IN PROTON CROSS-RELAXATION RATES IN
LYSOZYME PREDICTED FROM A MOLECULAR
DYNAMICS SIMULATION AND OBTAINED
FROM EXPERIMENT[16]

Residue	Vector	Motional averaging scale factor	
		Dynamics	Experimental
Trp-28	$H^{\epsilon 3}$—$H^{\zeta 3}$	0.86	0.91 ± 0.10
Trp-28	$H^{\zeta 3}$—$H^{\eta 3}$	0.83	0.93 ± 0.07
Ile-98	$H^{\gamma 11}$—$H^{\gamma 12}$	0.82	0.63 ± 0.08
Met-105	$H^{\beta 1}$—$H^{\beta 2}$	0.69	0.49 ± 0.05

Experimental values of cross-relaxation rates are available for the four residues discussed above. Despite the uncertainties introduced by the necessary assumption of a value for the correlation time for overall tumbling, a comparison with the predictions from the simulation is possible (Table III). First, the relative rigidity of the ring of Trp-28 observed in the simulations agrees with the experimental results. Second, the significant degree of motional averaging predicted for proton pairs of Met-105 is observed experimentally. The simulations predict, however, less motional averaging for Ile-98 and Met-105 than is observed. As with the ^{13}C results discussed earlier this can be interpreted as another example of the failure of short time simulations to model well the rare fluctuations involving transitions between different conformational states that are important for relaxation effects. It is also interesting that analysis of the motions observed in protein dynamics simulations shows that they are often highly correlated.[58] This means that interpretation of NMR data in terms of simple models, for example involving independent fluctuations about individual bonds, can give a quite unrealistic view of the true nature of the fluctuations.[16]

The experimental studies of internal motions discussed above make use of fixed distance interactions to simplify the analysis.[44] If nuclear Overhauser effects are measured between protons where distances are not fixed, then the strong distance dependence of the cross-relaxation rates can be used to obtain estimates of the interproton distance.[59] The simplest application of this approach is to assume that proteins are rigid and tumble isotropically. The molecular dynamics simulations can be

[58] S. Swaminathan, T. Ichiye, W. van Gunsteren, and M. Karplus, *Biochemistry* **21**, 5230 (1982).
[59] F. M. Poulsen, J. C. Hoch, and C. M. Dobson, *Biochemistry* **19**, 2597 (1980).

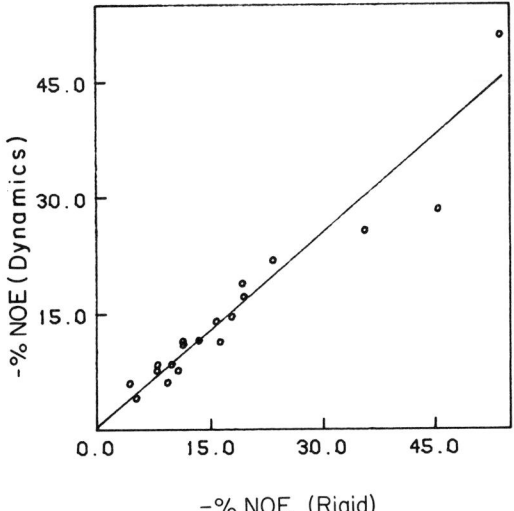

FIG. 16. Comparison of proton nuclear Overhauser effects calculated from a molecular dynamics simulation of lysozyme with values calculated for a rigid model of the protein.[16] The line is a least squares fit with a slope of 0.84.

used to determine whether picosecond motions are likely to introduce important errors into such an analysis. In studies of lysozyme the results from the simulations suggest that the presence of the motions will cause a general decrease in most NOE effects observed in a protein (Fig. 16). The decrease is too small, however, to produce a significant change in the distance dependence of most of the NOE effects. This is consistent with the strong correlations found between experimental NOE values and values calculated from the crystal structure[59,60] (Fig. 17). Specific NOE effects can, however, be altered by the motions to such a degree that the effective distances obtained would be considerably different from those predicted for a static structure. Such possibilities, must, therefore, be considered in any structure determination based on NOE data. This is true particularly for cases involving averaging over larger scale fluctuations.

Conclusions

This chapter has summarized the approaches that can be used to combine molecular dynamics simulations and NMR data to enhance the understanding of internal motions in proteins. Illustrative examples have been given that enable some general conclusions to be drawn.

[60] C. M. Dobson, E. T. Olejniczak, and F. M. Poulsen, *J. Magn. Reson.* **59**, 518 (1984).

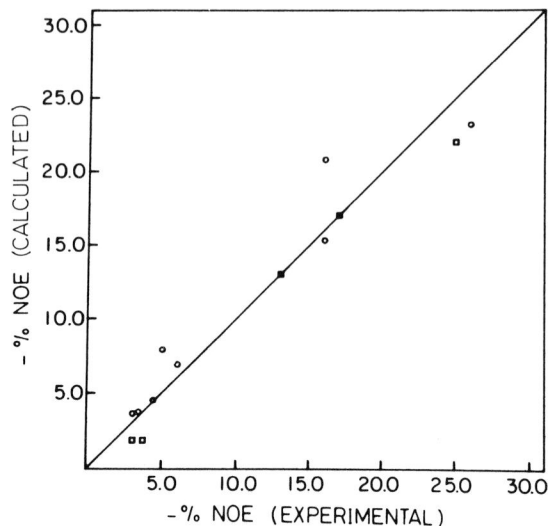

FIG. 17. Comparison of experimental proton nuclear Overhauser effects with values calculated from a rigid model of lysozyme based on the crystal structure.[60]

1. The effects on NMR parameters of both the small and larger scale motions observed in molecular dynamics simulations can be calculated. The effects are found to be important in interpreting NMR spectra, more so for some parameters such as coupling constants and relaxation rates than for others such as chemical shifts.

2. The relatively small fluctuations involving rapid (1 psec or shorter time scale) motions within a given potential well (conformational state) have relatively little effect on NMR parameters. The presently available NMR data are generally consistent with the magnitude of such motions predicted by the simulations. In favorable circumstances it is possible to compare quantitatively the magnitudes of predicted and actual fluctuations for individual residues.

3. The relatively large fluctuations involving rare (much longer than 1 psec time scale) jumps between different potential wells (conformational states) can have substantial effects on NMR parameters. The correlation between the predicted effects and available experimental data is not clear. This reflects the impossibility of predicting the rates of such infrequent events from short time simulations. Much longer simulations or alternative approaches, some of which have already been utilized, are needed here. Nevertheless, it is possible by use of the presently available simulations to demonstrate the types of changes in NMR parameters that are likely to occur as a consequence of such rare events. Currently available results suggest that large discrepancies between experimental NMR ob-

servations and NMR parameters calculated from a rigid model of a protein provide evidence for the occurrence of such motions.

4. On the assumption of a rigid model of a protein, errors in distances or angles calculated from NMR parameters such as NOEs or coupling constants will result from fluctuations of the type found in the molecular dynamics simulations. These errors will, however, be rather small in many cases, particularly for relatively constrained regions of the protein such as the interior. This reflects the significance of the average structure of a protein in determining NMR parameters and implies that the motion can be neglected, at least initially, in the determination of many aspects of protein structure from measured values of NMR parameters. Precise studies will, however, need to take account of the motions, particularly for determining side chain positions.

Overall it is clear that NMR studies can serve as a stringent test of many aspects of molecular dynamics simulations, and that the simulations can enhance the interpretation of NMR experiments. As more NMR data are accumulated, and as the simulations become more sophisticated, the value of the interaction of the two techniques will increase. This would also enable the approaches to be extended from the local fluctuations considered here to more cooperative events such as conformational changes and folding. It is hoped that studies of well-defined systems such as BPTI and lysozyme will provide the basic understanding needed to apply both techniques to more complex or less well characterized macromolecules.

[19] Study of Protein Dynamics by X-Ray Diffraction

By DAGMAR RINGE and GREGORY A. PETSKO

Introduction

It is widely believed that X-ray crystallography is an inherently static technique, incapable of providing information about the dynamic properties of molecules. This notion is false, even for small molecules in tightly packed, solvent-free single crystals. The motions of atoms and groups of atoms leave their traces in the electron density image, much the way footprints remain in the sand long after the beachcomber has walked by. In recent years, the techniques developed by small molecule crystallographers to interpret these "footprints" in terms of intramolecular motion have been applied to crystalline proteins. Such application has been controversial, because the errors and approximations involved in the mea-

METHODS IN ENZYMOLOGY, VOL. 131

surement and analysis of protein crystal diffraction data are much greater than in the case of simple organic and inorganic crystal structures. Nevertheless, a considerable body of evidence now exists that protein crystallography can map, with accuracy, the spatial distribution of the relative motions of the atoms of crystalline proteins. Moreover, it appears that these relative amplitudes of motion describe dynamic behavior very similar to that found for the same molecules in aqueous solution. This chapter summarizes the methods used to extract dynamic information from protein crystal structure data, with particular emphasis on the assumptions made, and considers the relationship between crystallographic results and those obtained by other methods. The possible functional role of the disorder that has been observed is also discussed.

The concept of enzymes as flexible entities is not new. Linderstrøm-Lang and Schellman concluded in the mid-1950s, on the basis of hydrogen exchange data, that protein structures must "breathe" to allow solvent contact with normally buried groups.[1] In 1963, Straub published the astonishingly contemporary view that "the conformation of an enzyme in solution is regarded to be the statistical average of a number of different conformations, the protein structure oscillating between these conformational states."[2] However, it is only recently that the extraordinary variety of motions possible for enzymes has been detailed. The development of many new physical techniques for probing the structure and behavior of macromolecules has allowed a quantitative description of macromolecular dynamics for the first time. Among these techniques X-ray crystallography is unmatched for its high resolution view of spatial properties of the non-hydrogen atoms in a macromolecule. X-Ray diffraction is complementary to spectroscopy and other techniques, which are low in spatial resolution but rich in temporal information. Used together, these techniques provide not only data on how fast overall processes occur for a given biopolymer, but a detailed map of the various motions undergone by different regions of the molecule in different time regimes.

The focus of this chapter, as stated above, is on the application of single crystal X-ray diffraction techniques to the study of protein dynamics. No attempt is made to review either the literature of protein dynamics or protein crystallography exhaustively. This chapter presents a general methodological overview followed by a brief discussion of recent progress in the field.

[1] K. U. Linderstrøm-Lang and J. A. Schellman, in "The Enzymes" (P. D. Boyer, H. Lardy, and K. Myrbäck, eds.), Vol. 1, p. 443. Academic Press, New York, 1959.
[2] F. B. Straub, Adv. Enzymol. 26, 89 (1964).

Overview

Classification of Motions in Proteins

Dynamics is a term that covers all of the intramolecular motions of proteins (commonly used synonyms are flexibility and mobility). These may be the motions of individual atoms, groups of atoms, or whole sections of the molecule. Many decades in time can be encompassed by the full variety of intramolecular protein motions. The terms used to describe protein dynamics are often undefined or imprecise. A simple classification scheme is presented here, together with a recommended nomenclature.

Protein motions can be divided into three broad categories, characterized by the extent and time scale of the motion, and by the method that is used to study it (see the table).

The first category contains atomic fluctuations, such as individual vibrations. These motions are random, very fast, and rarely cover more than 0.5 Å. The energy for these motions comes from the kinetic energy inherent in the protein as a function of temperature. The time scale of these motions is on the order of picoseconds or less, and therefore it is usual to observe atomic fluctuations by vibrational spectroscopy and to model them by molecular dynamics. X-Ray diffraction can give information on the spatial distribution of atomic fluctuations through an analysis of the spreading of atomic electron density produced by such motions.[3]

The second category contains collective motions, such as the movements of groups of atoms that are covalently linked in such a way that the group moves as a unit. Noncovalently interacting groups of atoms may also move collectively. The size of the group ranges from a few atoms to many hundreds of atoms. Entire structural domains may be involved, as in the case of the flexible Fc portion of immunoglobulins, where rigid-body motion of a 50,000 M_r unit may occur.[4] There are two types of rapid collective motions: those that occur infrequently (internal aromatic ring-flips belong to this category), and those that occur with high probability (many collective motions of small groups of neighboring atoms, bonded or nonbonded, are in the picosecond time regime). Collective motions can also be very slow (local unfolding of a polypeptide segment is one example). The energy for collective motions also derives from the thermal energy inherent in a protein as a function of temperature. The time scale of collective motions (from picoseconds to nanoseconds or slower[5]) al-

[3] G. A. Petsko and D. Ringe, *Annu. Rev. Biophys. Bioeng.* **13**, 331 (1984).

[4] P. M. Colman, J. Deisenhofer, R. Huber, and W. Palm, *J. Mol. Biol.* **100**, 257 (1976).

[5] G. Careri, P. Fasella, and E. Gratton, *CRC Crit. Rev. Biochem.* **3**, 141 (1975).

TYPES OF MOTION FOUND IN PROTEINS

Motion	Spatial displacement (Å)	Characteristic time (sec)	Energy source	Method of observation
Fluctuations (e.g., atomic vibrations)	0.01 to 1	10^{-15} to 10^{-11}	$k_B T$	Computer simulation, X-ray diffraction
Collective motions A. Fast, frequent B. Fast, infrequent (e.g., Tyr ring flip) C. Slow (e.g., local unfolding)	0.01 to >5	10^{-12} to 10^{-3}	$k_B T$	NMR, fluorescence, hydrogen exchange, simulation, X-ray
Triggered conformational change	0.5 to >10	10^{-9} to 10^3	Inter- action	X-Ray, spectroscopy

lows some of them to be studied by techniques such as NMR and fluorescence spectroscopy.[6,7]

The third category contains motions which can be described as triggered conformational changes. These are the motions of groups of atoms (i.e., individual side chains) or whole sections of a protein (i.e., loops of chain, domains of secondary and tertiary structure, or subunits) that occur as a response to a specific stimulus. The distance moved can be as much as 10 Å or more. The time scale can be estimated by studies of the rate of binding or turnover reactions. The energy for triggered conformational changes comes from specific interactions, such as electrostatic attractions or hydrogen bonding interactions. The best-known example of a triggered conformational change is the transition in tertiary and quaternary structure that occurs when ligands bind to the iron atoms of hemoglobin.[8,9]

Categories one and two define atomic motions that, whether individual or correlated, involve random excursions about the equilibrium conformation. In contrast, category three is used to classify motions that involve a transition from one equilibrium conformation to another.

[6] G. Wagner, A. Demarco, and K. Wüthrich, *Biophys. Struct. Mech.* **2,** 139 (1976).

[7] I. Munro, I. Pecht, and L. Stryer, *Proc. Natl. Acad. Sci. U.S.A.* **76,** 56 (1979).

[8] M. F. Perutz, *Nature (London)* **228,** 726 (1970).

[9] M. F. Perutz, *Annu. Rev. Biochem.* **48,** 327 (1979).

The Trace of Motion

X-Ray diffraction gives information about all of these motions. Point-atom electron densities are not observed in X-ray crystal structures, even for small molecules at very low temperatures. The electron density of any atom in a structure derived from an X-ray diffraction study is an average of the position of that atom in every unit cell of the crystal (averaged over all possible orientations that the unit cell can take and still maintain the integrity of the crystal, a phenomenon called mosaicity), and also of the possible positions that atom can sample during the diffraction experiment weighted by the amount of time the atom spends in each position. Both effects result in a spread of the electron density of that atom over a range of space. For instance, the vibrations defined in category one would result in a distribution of the electron density about a most probable position. The most probable position is the point of highest electron density and can be used to define the centroid of the amplitude of the vibration. Atomic vibrations are frequent but limited in magnitude. By contrast, the motions defined as category three occur less frequently and cover a wider range of space. The distinguishing feature of these large motions, which makes it possible to observe their effect, is the stability of the starting and ending positions, which are highly populated relative to the intermediate positions. These highly populated positions are the ones observed in an X-ray crystallographic study. The absence of this feature, i.e., discrete states that are highly populated, is the reason that some large collective motions are difficult to observe by X-ray methods. If a motion occurs too infrequently (or too slowly) to produce states of appreciable occupancy, the electron density of the atoms involved is too low to be observed. It is therefore important to realize that when we speak of "the crystal structure of a protein" we really mean the average structure of over 10^{17} molecules. The exact structure a given protein molecule has at any instant in time may be quite different from this average, and may have different properties. Indeed, the average structure may have no physical existence at all. It is merely the best single model that can be fit to the superposition of many slightly different structures. The dynamic information obtainable from X-ray crystallography is simply an estimate of how far each atom deviates from its average position, and in what direction. Because of the averaging over the time of data collection (usually days or weeks), all temporal information about these excursions is lost.

X-Ray crystallography is thus similar to a camera with a very fast lens and film, but whose shutter remains open for a long time. If one had a fast shutter on such a camera and photographed a racehorse in full gallop, one would see a "frozen" picture of the horse in mid-stride. But if the shutter were kept open for several seconds, one would obtain a blurred picture that on close inspection would be seen to be a superposition of many pictures of the horse in different positions. One would be able to estimate the distance traveled by the horse and its principal directions of motion, but it would not be possible to calculate the speed of the animal from the picture alone. The analogy to a crystallographic experiment is a close one: the smearing out of the electron densities for each atom is due to the superposition of many different instantaneous pictures of the structure of the molecule, each with the atoms in slightly different positions. Positions that do not reoccur very often will contribute only weakly and will be lost in the noise, but positions that have a high occupancy will be clearly observed due to the summing of their contributions. The task facing the crystallographer is to deconvolute this blurred image, to model the motions that left this trace.

Motions in Protein Crystals

The Solvent Content of Protein Crystals. Protein crystals are different from crystals of small inorganic and organic molecules such as ammonium sulfate or anthracene. In a protein crystal, the different neighboring molecules make contact with one another at only a few points on their surfaces. There are large open spaces in the lattice filled with the mother liquor of crystallization to prevent the crystal from collapsing. A typical protein crystal is 40–60% solvent by volume.[9a] The average surface amino acid residue in a crystalline protein is in contact, not with another protein molecule, but with solvent. The local environment experienced by these residues is not very different from that in a concentrated solution. Furthermore, there is ample solvent space in the lattice for surface protein atoms to move. Contacts between neighboring molecules influence only the mobility of those few atoms in the contact region.[10]

Both the structure and dynamics of proteins in the ''solid'' state will therefore be relevant to these properties in aqueous solution. There is now abundant evidence that the average protein structure is normally not changed on crystallization.[11,12] Dynamic properties are more difficult to

---

[9a] B. Matthews, *in* ''The Proteins'' (H. Neurath, R. L. Hill, and C. L. Bolder, eds.), Vol. 3, p. 403. Academic Press, New York, 1975.

[10] S. Sheriff, W. A. Hendrickson, R. E. Stenkamp, L. C. Sieker, and L. H. Jensen, *Proc. Natl. Acad. Sci. U.S.A.* **82,** 1104 (1985).

[11] M. W. Makinen and A. L. Fink, *Annu. Rev. Biophys. Bioeng.* **6,** 301 (1977).

[12] J. A. Rupley, *in* ''Structure and Stability of Biological Macromolecules'' (S. N. Timasheff and G. D. Fasman, eds.), p. 291. Dekker, New York, 1969.

assess. It is useful to consider the question of whether such information from a crystalline macromolecule is likely to be pertinent to the flexibility of that same molecule under physiological conditions.

The Flexibility of Proteins in the Crystalline State. Indirect evidence for the flexibility of crystalline proteins has existed for many years. A number of crystalline enzymes possess catalytic activity comparable to that found in solution,[11] so if flexibility is important for function (a generally held but unproven assumption), at least a large measure of flexibility must be retained by the molecules in the lattice. Oxygen-binding proteins such as hemoglobin and myoglobin, which are known to require flexibility of structure for oxygen binding,[13,14] bind ligands reversibly in the crystalline state.[15,16] The weak packing forces in the highly solvent-filled lattice of a protein crystal do not prevent all intramolecular motion, a fact that should not be surprising in view of the fact that aromatic ring flips have been demonstrated in simple crystals of individual amino acids by solid-state NMR.

Direct evidence for the internal mobility of crystalline proteins has come from observations of hydrogen/deuterium and hydrogen/tritium exchange for the amide protons from proteins in the crystalline state.[17–20] Neutron diffraction studies of several crystalline proteins have shown that if these crystals are suspended in deuterated mother liquor for extended periods of time, nearly all of the amide protons in the proteins are exchanged,[21–23] including amides completely inaccessible to solvent in the static crystal structure of the protein.

The rate of exchange of peptide amide hydrogens with solvent reflects the secondary and tertiary structure of the protein. Two extreme models, "local unfolding" and "solvent penetration," have been proposed to explain exchange.[24,25] Both require that the protein be flexible for exchange

[13] M. F. Perutz and F. S. Mathews, *J. Mol. Biol.* **21**, 199 (1966).

[14] D. Ringe, G. A. Petsko, D. Kerr, and P. R. Ortiz de Montellano, *Biochemistry* **23**, 2 (1984).

[15] B. Chance and A. Ravilly, *J. Mol. Biol.* **21**, 195 (1966).

[16] B. Chance, A. Ravilly, and N. Rumen, *J. Mol. Biol.* **17**, 525 (1966).

[17] A. D. Barksdale and A. Rosenberg, *Methods Biochem. Anal.* **28**, 1 (1982).

[18] S. W. Englander and J. J. Englander, this series, Vol. 49, p. 24.

[19] C. K. Woodward and B. D. Hilton, *Annu. Rev. Biophys. Bioeng.* **8**, 99 (1979).

[20] C. K. Woodward, I. Simon, and E. Tüchsen, *Mol. Cell. Biochem.* **48**, 135 (1982).

[21] J. Hanson and B. Schoenborn, *J. Mol. Biol.* **153**, 117 (1981).

[22] A. A. Kossiakoff, *Nature (London)* **296**, 713 (1982).

[23] A. Wlodawer and L. Sjölin, *Proc. Natl. Acad. Sci. U.S.A.* **79**, 1418 (1982).

[24] S. W. Englander, *Ann. N.Y. Acad. Sci.* **244**, 10 (1975).

[25] C. K. Woodward and B. D. Hilton, *Biophys. J.* **32**, 561 (1980).

to take place. If amide hydrogen exchange can be observed in proteins after crystallization, the restrictions caused by the packing of the protein molecules into the crystal lattice cannot have eliminated the flexibility of the molecule.

Direct kinetic evidence for crystalline protein flexibility has been obtained by Tüchsen and Ottesen on the rates of amide hydrogen exchange in protein crystals. The total tritium–hydrogen back-exchange kinetics for hen egg-white lysozyme are nearly identical in cross-linked crystals and in solution.[26,27]

X-Ray Diffraction Techniques

The study of protein dynamics by X-ray diffraction can be divided into two broad categories: qualitative studies at low and medium resolution where the absence of significant electron density for a group of atoms is taken as evidence for disorder in their positions, and studies at high resolution where a quantitative estimate of individual atomic motions is attempted. To interpret the observed electron density distribution in terms of particular types of atomic motion, the application of a physical model is required.

The Debye–Waller Model and Its Assumptions. Small molecule crystallography has developed a number of very sophisticated models for analysis of atomic motions incorporating anharmonicity, libration, and anisotropic vibration. Owing to the scarcity of high-resolution data and the size of the calculations required, these have not yet been applied to proteins. The model being used for macromolecular crystal structure analysis is the simple Debye–Waller model.[28] In the Debye–Waller treatment of atomic motion, the probability of finding an atom a given distance x from its equilibrium position x_0 is considered to be Gaussian. If the motion is also considered to be isotropic, the model states that, in any direction, the motion can be characterized in terms of a mean-square displacement, $\langle x^2 \rangle$. Application of this model is usually done in reciprocal space, the space of the diffraction pattern. It is possible to calculate the expected diffraction from an atom, modified by a Gaussian function that is related to the estimated mean-square displacement of that atom in real space. The form of the Gaussian is

$$\exp(-B \sin^2 \theta/\lambda^2)$$

[26] E. Tüchsen, A. Hvidt, and M. Ottesen, *Biochimie* **62,** 563 (1980).
[27] E. Tüchsen and M. Ottesen, *Carlsberg Res. Commun.* **44,** 1 (1979).
[28] B. T. M. Willis and A. W. Pryor, "Thermal Vibrations in Crystallography." Cambridge Univ. Press, London, 1975.

and B is related to the mean-square displacement by

$$B = 8\pi^2\langle x^2 \rangle$$

B is called the atomic temperature factor or Debye–Waller factor.

It is important to state explicitly the assumptions in the Debye–Waller model of individual atomic motion. The use of the Gaussian means that the potential in which each atom moves is taken to be harmonic, or at least to have a very large harmonic component. It is further assumed that the motion is isotropic (that is, the mean-square radial displacement is simply three times the mean-square displacement in any given direction). The validity of these assumptions in the case of protein dynamics is highly questionable, as indicated below. Nevertheless, they are made because the number of data points is too small to permit the application of more complex models. The assumptions mean that even in the best of cases the mean-square displacements obtained from X-ray crystallography will only be approximations to the actual values.

Determination of Mean-Square Displacements in Crystalline Proteins. Using the Debye–Waller model, each atom in a crystalline protein may be assigned an individual isotropic temperature factor, B. Once a protein crystal structure has been solved and approximate atomic coordinates have been measured for most of the non-hydrogen atoms in the molecule, the preliminary atomic model may be improved by least-squares refinement in which the squared difference between the observed diffraction pattern and that calculated from the model is minimized. If only low-resolution data have been measured for the protein, then it is not appropriate to try to include the mobility of the atoms in the model. But if high-resolution data (usually 2 Å or beyond) are available, then the calculated X-ray scattering contribution from each atom can be modified by the exponential term containing the Debye–Waller B factor, and B becomes an additional variable parameter for each atom in the least-squares refinement. An estimate of the overall B for the entire protein serves to initiate the process, and individual B values will be determined by the iterative least-squares procedure.

In addition to the observed X-ray intensities, known stereochemical information may be inserted in the form of expected bond lengths, bond angles, planarity, and chirality. This structural information is assigned weights relative to that given to the X-ray data, and by appropriate adjustment of these weights a distribution of bond lengths, angles, etc. about the ideal values is obtained in the final refined structure. Because the interatomic distances are not constrained to exact numbers, but are allowed to range about the expected quantities, this modified least-squares process is called restrained refinement. The use of restraints in protein structure refinement is open to criticism on the grounds of insufficient knowledge of the actual distribution of bond lengths and angles in real proteins, but it is only by this method that most crystalline proteins can yield any dynamic information.

Very few protein crystals diffract to high enough resolution (beyond 1.5 Å) to yield enough data to permit unrestrained least-squares refinement with individual atomic temperature factors. Fortunately, it appears that the restraints do not conceal meaningful variations in individual atomic mobilities. However, although mean-square displacements obtained from restrained refinements appear to have considerable relative accuracy, their absolute accuracy is open to doubt. The assumptions of harmonicity and isotropy in the Debye–Waller model are two reasons for this; another is that restraints are applied to the distribution of B factors as well as interatomic distances. The technical aspects of the imposition of relative B factor restraints have been discussed by Konnert and Hendrickson,[29] whose method is the most widely used. To allow refinement of individual isotropic B factors for all atoms in a protein with relatively low resolution data (i.e., 2 Å resolution or lower) the number of independent parameters in the protein model is reduced by restraining the variances of the interatomic distributions to suitably small values. The positional disorder of two covalently bonded atoms is assumed to be highly correlated. In other words, two bonded atoms are restrained to have mean-square displacements that do not differ by a large amount. These restraints may be relaxed later in the refinement procedure when higher resolution data are included. When appropriate weight is given to these restraints in the refinement, it is very difficult for two covalently bonded atoms to differ in B factor by 10 Å^2. Intuition suggests that this is physically reasonable, but it must be kept in mind that restraints represent the imposition of preconceived notions about structure and dynamics on what would otherwise be an objective technique.

Factors Contributing to Observed Mean-Square Displacements. The B factor calculated for each atom as the result of a restrained least-squares refinement of the macromolecular structure against high-resolution X-ray diffraction data in reciprocal space may be viewed in real space as the attempt to fit a Gaussian to the spread of electron density about the average position of that atom. Anything producing a spreading of electron density will contribute to the B factor, and consequently to the estimated mean-square displacement. Atomic motion occurring over the course of the collection of the X-ray data, whether that motion is individual or part of a collective mode, will certainly contribute, as outlined above. It is important to emphasize again that motions producing electron density

---

[29] J. H. Konnert and W. A. Hendrickson, *Acta Crystallogr. A* **36,** 344 (1980).

whose magnitude is at or below the noise level in the structure determination will not be reflected in the final refined value. Experience suggests that this level is about 10–15% occupancy for most high-resolution structures, and is probably 20–25% for structures at worse than 2 Å resolution. However, dynamic disorder in atomic position is not the only possible kind of disorder. The crystal may be heterogeneous with respect to the position of that atom: different molecules in the crystal may have the atom in different places, and these positions may not interconvert because of large potential energy barriers. This is equivalent to saying that the protein has folded into a number of distinct conformations, at least at the site in question. Such static conformational disorder will also cause a spreading of the electron density if the different positions are closer together than the resolution of the structure refinement. If the different positions are far apart, and there are only a few of them so that each one is more than 10% occupied, it is possible to discern the alternate conformations and use as the model the entire static distribution. In this case, alternative positions will be placed into the model for the atom(s) in question, and their occupancies will be nonunity and subject to refinement as well. Several examples of such positional disorder have been observed; one of particular interest concerns an essential catalytic residue His-119 in ribonuclease A.[30] For sperm whale metmyoglobin, about 20 residues out of 153 have been found to have alternate conformations separated by more than 1.5 Å.[31]

It is important to realize that there should be some temperature at which enough kinetic energy exists to convert this static disorder into a dynamic one without denaturing the protein. Thus, the distinction between static disorder and dynamic disorder is only properly made at some specified temperature. The term "conformational substates" has been used to refer to discrete conformations of a protein that differ in some biochemical property, such as ligand binding rate.[32] This term has been misinterpreted as meaning that these substates do not interconvert. Such a statement is only true at the low temperature limit of $T = 0$ K; at any other temperature, interconversion is theoretically possible, although it may be extremely slow and improbable.

Overall lattice vibration of the entire molecule as a rigid body is another possible contributor. Thus far, attempts to identify such vibrations in protein crystal structure analyses have not given any definitive results.[33] It is likely that the solvent-filled interstices in protein crystals

[30] N. Borkakoti, D. S. Moss, and R. A. Palmer, *Acta Crystallogr.* **B38,** 2210 (1982).

[31] J. Kuriyan, M. Karplus, F. Parak, G. A. Petsko, and D. Ringe, submitted (1986).

[32] H. Frauenfelder, G. A. Petsko, and D. Tsernoglou, *Nature* (*London*) **280,** 558 (1979).

[33] M. J. E. Sternberg, D. E. P. Grace, and D. C. Phillips, *J. Mol. Biol.* **130,** 231 (1979).

will not lead to long-range transmission of such vibrations except in certain directions where intermolecular contacts are unusually strong.

The final contribution to the mean-square displacement of every atom is lattice disorder. If the crystal lattice is not perfect (i.e., if every molecule in the crystal is not in exactly the same position relative to the origin of its unit cell as every other molecule) there will be a spreading of the electron density due to this static disorder. Unlike the "frozen-in" conformational disorder discussed earlier, where insufficient energy exists at low temperature for different positions to interconvert, lattice disorder is a temperature-independent property of the particular crystal being measured. It is possible to separate the lattice disorder into two components: translational and rotational. An overall rotational disorder of the molecules should have a magnitude that increases with increasing distance from the axis of rotation; translational lattice disorder should affect all atoms in the protein equally.

Thus, the following expression can be written for every individual atomic mean-square displacement in the protein:

$$\langle x^2 \rangle = \langle x^2 \rangle_v + \langle x^2 \rangle_c + \langle x^2 \rangle_{ld}$$

where $\langle x^2 \rangle_v$ is the contribution due to individual atomic fluctuations (thermal vibrations), $\langle x^2 \rangle_c$ is the contribution due to collective motions, which have a larger potential energy barrier than simple vibrations (and are therefore slower and may be static at some specified temperature), and $\langle x^2 \rangle_{ld}$ is the contribution due to lattice disorder.[32]

Lattice Disorder. Only one detailed attempt has been made to determine the magnitude of the lattice disorder component in a protein structure determination of individual atomic $\langle x^2 \rangle$ values.[32] Frauenfelder *et al.* made use of the insensitivity of the Mössbauer effect to lattice disorder: every Mössbauer-active nucleus absorbs randomly, and the recoilless fraction of the Mössbauer effect (which measures the isotropic mean-square displacement of the atom in question) includes no contribution from stationary disorder. At some temperature above the characteristic temperature of the sample (i.e., the temperature at which the mean time of the relaxation in question is equal to the nuclear lifetime)

$$\langle x^2 \rangle_M = \langle x^2 \rangle_v + \langle x^2 \rangle_c$$

and comparison of this formula with that given above for $\langle x^2 \rangle$ from X-ray diffraction suggests a direct method of determining $\langle x^2 \rangle_{ld}$. One measures $\langle x^2 \rangle$ for some atom in a protein by both Mössbauer absorption and X-ray crystallography, on the same crystal if possible, and subtracts the Mössbauer value from the X-ray value, yielding the lattice disorder at that atom. Unfortunately, only a few nuclei are Mössbauer active, so the number of atoms that can be studied in this way is extremely small. Frauenfelder *et al.* subtracted Parak and Formanek's measurements of the $\langle x^2 \rangle_M$ for the iron atom in the heme of crystalline metmyoglobin[34] (the protein had been doped with ^{57}Fe, an active isotope, at the heme iron position) from $\langle x^2 \rangle$ for the iron from a 1.5 Å metmyoglobin crystallographic refinement.[33] The resulting $\langle x^2 \rangle_{ld}$ at the iron position was 0.045 Å^2. Use of this value to correct every atom in the protein for the effect of lattice disorder required the assumption that the lattice disorder was purely translational, an assumption based on no evidence, but which yielded sensible results. Comparison of this value with the overall $\langle x^2 \rangle$ for the protein, 0.175 Å^2 at room temperature, suggests that in the monoclinic crystal form of sperm whale metmyoglobin lattice disorder contributes less than 30% of the total $\langle x^2 \rangle$. Although this value is reasonable in view of the strong diffraction observed from this crystalline protein, the myoglobin lattice disorder determination has been reevaluated recently due to improved understanding of the Mössbauer effect in crystalline proteins.[35] Any motion having a characteristic time slower than 0.1 μsec will not contribute to $\langle x^2 \rangle_M$. Therefore, $\langle x^2 \rangle_M$ may be systematically too small, leading to an overestimation of the lattice disorder term. Current estimate of $\langle x^2 \rangle_{ld}$ is approximately 0.025 Å^2 for metmyoglobin crystals, or less than 15% of the total mean-square displacement. There is some suspicion that this number may be an upper, rather than a lower, limit. One consequence of this determination is the conclusion that protein crystals that diffract to high resolution (i.e., 1.5 Å and beyond) like myoglobin will have very small lattice disorder contributions.

The problems encountered in the interpretation of the Mössbauer effect in protein crystals, plus the experimental difficulty of the technique and the scarcity of Mössbauer-active nuclei in most proteins, limit the general applicability of this method in crystallographic studies of protein dynamics. Fortunately, it is often unnecessary to measure $\langle x^2 \rangle_{ld}$ accurately. As the foregoing discussion implies, the important question is whether the lattice disorder dominates the measured $\langle x^2 \rangle$ values for the protein of interest. If it does not, the relative $\langle x^2 \rangle$ values will be meaningful, even though they will be raised by some unknown quantity. Phillips and co-workers pointed out that if the same molecule crystallizes in two different crystal forms, comparison of the relative average $\langle x^2 \rangle$ values for the residues in each structure, refined independently, will reveal the effect

[34] F. Parak and H. Formanek, *Acta Crystallogr. A* **27**, 573 (1971).
[35] H. Hartmann, F. Parak, W. Steigemann, G. A. Petsko, D. Ringe Ponzi, and H. Frauenfelder, *Proc. Natl. Acad. Sci. U.S.A.* **79**, 4967 (1982).

of crystal packing.[36] If the structural correlations of $\langle x^2 \rangle$ are the same in the two distinct crystals, lattice disorder cannot be dominant. This was observed for two lysozyme crystals, one of which was in a tetragonal space group and the other in an orthorhombic space group. The only areas of the two lysozyme structures that differed significantly in their relative $\langle x^2 \rangle$ values were those residues that were involved directly in the intermolecular contacts in the two lattices. This result reaffirms that the dynamics of proteins in the crystalline state are not disturbed by crystal contacts, except for very restricted regions of the surface of the molecule. This treatment has been extended to oligomeric proteins by Sheriff *et al.*, who reach the same conclusion.[10] Thus, for most regions of the protein structure in well-diffracting enzyme crystals, the $\langle x^2 \rangle$ values determined by X-ray crystallography will reflect the intrinsic dynamic behavior of the molecule.

The relative magnitude of the lattice disorder contribution is important when two different crystal structures are to be compared for changes in mobility. Restrained least-squares refinement of the crystal structures of a native enzyme and the enzyme-inhibitor complex yields two sets of $\langle x^2 \rangle$ values, but the magnitude of $\langle x^2 \rangle_{ld}$ may not be the same for each of the two data sets. If the inhibited crystal indeed shows lower overall $\langle x^2 \rangle$s than the native, it is possible that the difference arises not from a change in the mobility of the enzyme on binding, but rather from differences in the lattice disorder of the particular crystals chosen for data collection. Therefore, if questions of mobility are paramount in the comparison of two states of a crystalline protein, *one should measure both data sets on the same crystal specimen.* But this is not always possible due to radiation damage. The solution to this problem that we recommend is the minimum function method,[37,38] which involves an assumption about the level of $\langle x^2 \rangle$ values expected for simple molecules. Values of $\langle x^2 \rangle$ smaller than the zero-point vibrational limit are not observed in any structure, and atoms that have $\langle x^2 \rangle$s near to this limit (0.01 Å^2) must have very small lattice disorder contributions. Therefore, if one takes two independently determined sets of mean-square displacements and adjusts the one with the higher overall $\langle x^2 \rangle$ so that its lowest values are equal to the corresponding values in the other structure, the two data sets will have been placed on an approximately identical scale and any lattice disorder difference will have been subtracted out. A corollary of this reasoning is that if two sets of

[36] P. J. Artymiuk, C. C. F. Blake, D. E. P. Grace, S. J. Oatley, D. C. Phillips, and M. J. E. Sternberg, *Nature (London)* **280,** 563 (1979).
[37] H. Frauenfelder and G. A. Petsko, *Biophys. J.* **32,** 465 (1980).
[38] D. Ringe and G. A. Petsko, *Prog. Biophys. Mol. Biol.* **45,** 197 (1985).

$\langle x^2 \rangle$s have roughly equal values for the smallest displacements, the lattice disorder in the two structures may be taken as approximately equal.

It is more important to be aware of the possible effects of lattice disorder and to try to ascertain its relative effect on a given system than to try to measure its magnitude exactly. Experience suggests that if a crystalline protein gives strong diffraction beyond 2 Å resolution it is safe to assume that lattice disorder does not dominate the relative $\langle x^2 \rangle$ values from residue-to-residue.

Temperature Dependence of $\langle x^2 \rangle$. To obtain the maximum amount of information from the $\langle x^2 \rangle$ values, it is desirable to separate the effects of static and dynamic disorder. This can be accomplished by means of a parameter whose variation is sensitive to these quantities in different ways. Temperature is the obvious choice. A static disorder will be temperature independent, while a dynamic disorder will have a temperature dependence related to the shape of the potential well in which the atom moves, and to the height of any barriers it must cross.[32] Simple harmonic thermal vibration decreases linearly with temperature until the Debye temperature T_D; below T_D the mean-square displacement due to vibration is temperature independent and has a value characteristic of the zero-point vibrational $\langle x^2 \rangle$. The high-temperature portion of a curve of $\langle x^2 \rangle$ vs T will therefore extrapolate smoothly to 0 at $T = 0$ K if the sole or dominant contribution to the measured displacement is harmonic motion. In such a plot the low-temperature limb is expected to have values of $\langle x^2 \rangle$ equal to about 0.01 Å^2.[28] Departures from this behavior indicate more complex motion or static disorder.

Proteins are expected to undergo other motions than just vibration in an harmonic potential with weak restoring forces. Examples of a more complicated motion might be the hindered rotation of side chain methyl groups, or the cis–trans isomerization of proline, or large-scale librations of aromatic rings. Any motion with a collective character will have a potential well different from a simple thermal vibration, and will require more energy to occur. The potential for the group of atoms may be temperature dependent, but it is possible that at any given temperature the behavior may appear harmonic. In such cases the high-temperature limb of the plot of displacement vs temperature may still linearly tend toward zero as T decreases, but there will be some temperature of relaxation, T_R, below which the motion will be "frozen-out" and a static distribution of conformations will exist (one point that is often forgotten is that the distribution found in the "frozen" structure will depend on both the relative depth of the potential minima for the various discrete conformations and the time required to reach T_R; rapid cooling is more likely to retain representatives of the entire distribution than slow cooling, which allows the system to "anneal," i.e., allows various higher energy conformations to move into the lowest free energy state). Each member of this distribution is called a conformational substate. If the barriers between substates are small, the distribution can still be dynamic even at temperatures as low as 80 K.[35] If the disorder is still dynamic, then the motion should be thought of as continuous in the absence of other information; conformational substates, in a formal sense, imply a hopping between distinguishable conformations. The existence of such separate structural variants (the alternate conformations of side chains referred to above is an example) is certain only when the distribution has been frozen out.

Below the T_R the mean-square displacement should no longer decrease with decreasing temperature. Evidence from Mössbauer scattering suggests that an important T_R for at least some classes of motion sensed by the iron atom may occur at about 180 K for myoglobin.[39] Thus, measurements of $\langle x^2 \rangle$ at temperatures below this value should show a much less steep temperature dependence than measurements above, if the motion(s) affected form a significant component of the total $\langle x^2 \rangle$. In practice, the relative contributions of vibrations and collective modes will vary from residue to residue within a given protein, and this break in the plot of $\langle x^2 \rangle$ vs temperature has not been observed by X-ray diffraction.

Measurements of $\langle x^2 \rangle$ as a function of temperature should establish conclusively whether or not static disorder dominates the apparent dynamics in a crystalline protein. If the $\langle x^2 \rangle_{ld}$ term or any other static distribution of structures (a conformational disorder that is static at room temperature cannot be distinguished from disorder in crystal packing by lowering the temperature, since both types of disorder will show no decrease in $\langle x^2 \rangle$) is the major contributor to the observed $\langle x^2 \rangle$ there should be little or no temperature dependence for the individual atomic $\langle x^2 \rangle$ values.

It is routine in small molecule X-ray and neutron crystallography to measure data at low temperature to improve precision by reduction of thermal motion, and structures are often done at multiple temperatures to assess the origins of disorder in atomic positions. Albertsson et al. have reported the analysis of the crystal structure of D(+)-tartaric acid at 295, 160, 105, and 35 K. The individual isotropic B factors for the atoms in the structure plotted against temperature show a smooth variation of B with T in the high-temperature regime. Below 105 K B is essentially identical for all atoms and is also temperature independent; the average value of 0.4 Å^2 agrees well with the expected zero-point vibrational value. However,

[39] F. Parak, E. N. Frolov, R. L. Mössbauer, and V. I. Goldanskii, *J. Mol. Biol.* **145**, 825 (1981).

even for this simple structure, not one of the atoms shows B vs T behavior at high temperature that extrapolates to 0 $Å^2$ at 0 K.[40]

Errors in $\langle x^2 \rangle$ *Values.* Before turning to a discussion of the results of crystallographic analyses of protein motion, it is important to establish how precise the observed $\langle x^2 \rangle$ values are, and whether they have any quantitative accuracy. In small-molecule crystallography the precision of isotropic B factors can be calculated from the inverse of the least-squares matrix used in the refinement, but application of restraints in protein refinement makes this impossible. Several small proteins have been refined by unrestrained least-squares at very high resolution.[35,41] Different refinements of the same protein structure have also been compared.[31] It appears that if the resolution of the structure is 2 Å or higher and if the atomic coordinates are well determined (in practice this means a crystallographic R factor of about 20% or less) then the standard deviation of an individual atomic B factor is about 10 to 15% of its magnitude.

Systematic errors must also be considered in this treatment. Small errors in atomic coordinates will cause the B factor for the atom in error to be larger than its true value. Errors in correcting for radiation damage will also cause a systematic shift in B: overcorrection of the high-resolution reflections will make B smaller than it should be, because the high-resolution data will appear to be strong, while undercorrection will cause B to be too large. For this reason, it is recommended that all high-resolution protein data sets should be measured under conditions (such as low temperature) such that radiation damage is minimized. Failure to properly account for absorption of X-rays by the crystal, mother liquor, and capillary tube are likely to affect the refined B factors, but the exact effects have not been characterized. Finally, the resolution of the data set must be taken into consideration. If the analysis is performed at too low a resolution, the B factors will be systematically too low because there will be no relatively weak data at very high resolution to show the effects of large temperature factors. This problem becomes acute when different data sets are being compared. If the two crystals are not measured to the same resolution, the higher resolution data set should be truncated before refinement and comparison of B factors. Without this correction there could be a systematic difference in the overall B factors that does not arise from differences in dynamics or lattice disorder. Experience suggests that 2 Å resolution is the lowest at which this effect may become small; thus, comparison of a 1.9 Å resolution structure with a 1.5 Å resolution structure may be possible without a serious systematic difference due to reso-

---

[40] J. Albertsson, A. Oskarsson, and K. Stahl, *J. Appl. Crystallogr.* **12,** 537 (1979).
[41] K. D. Watenpaugh, L. C. Sieker, and L. H. Jensen, *J. Mol. Biol.* **138,** 615 (1980).

lution (if the two structures are both well refined), while comparison of a 2.5 Å resolution structure with a 1.7 Å resolution structure would be subject to major errors. In the latter case, the minimum-function method should be used to place the two structures on the same scale before any comparisons are attempted.

In conclusion, careful restrained least-squares refinement of protein atomic coordinates combined with the Debye–Waller model of isotropic harmonic thermal motion can yield individual $\langle x^2 \rangle$ values of good precision for every non-hydrogen atom in the molecule. The absolute accuracy of these mean-square displacements is, however, difficult to determine and is likely to be much worse than the precision. Unless some independent measurement of one or more $\langle x^2 \rangle$ values is available, it is unwise to place much faith in the absolute values of individual atomic (or even residue averaged) $\langle x^2 \rangle$s. Examination of the results of such analyses suggests, however, that the relative values of these displacements are meaningful.

Results to Date by Category of Motion

Atomic Fluctuations

Magnitudes. In their studies of the small iron–sulfur protein rubredoxin, Jensen and co-workers demonstrated that it was possible to refine protein crystal structures by least-squares methods if very high-resolution data were available.[42] These investigators were the first to refine individual isotropic B factors for every non-hydrogen atom in a protein. Their analysis has now progressed to 1.2 Å resolution with a crystallographic R factor of 12.8%.[41] The overall B for rubredoxin is 12 Å^2, corresponding to an $\langle x^2 \rangle$, uncorrected for lattice disorder, of 0.15 Å^2. Within the protein, individual B factors vary from 7 Å^2 (0.09 Å^2 for $\langle x^2 \rangle$) to over 50 Å^2 (0.6 Å^2). Atoms or residues with B values larger than 50 Å^2 are nearly invisible in the electron density map even at low contour level, and either occupy many distinct conformations or have large amplitudes of motion. It is important to note that even the smallest of these values is larger than those normally observed for atoms in small organic crystal structures at room temperature, where $B = 3$ to 4 Å^2, $\langle x^2 \rangle = 0.04$ Å^2. Other proteins have given numbers similar to rubredoxin. The overall B for sperm whale metmyoglobin from refinement against 1.5 Å resolution data at $T = 300$ K

[42] K. D. Watenpaugh, L. C. Sieker, J. R. Herriott, and L. H. Jensen, *Acta Crystallogr. B* **29**, 943 (1973).

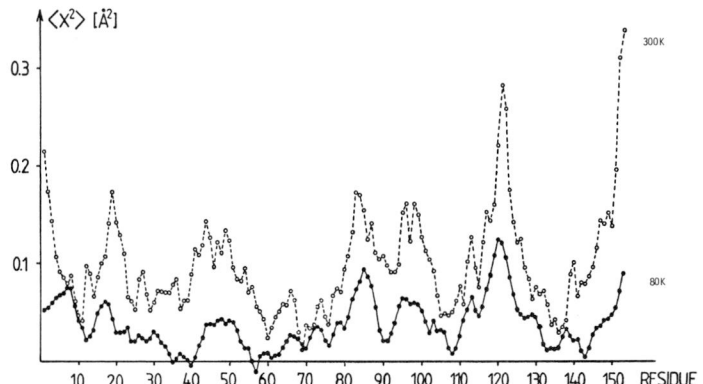

FIG. 1. Plot of the mean main chain $\langle x^2 \rangle$ values against residue number for sperm whale metmyoglobin. Dotted curve, 300 K; solid curve, 80 K.

is 14 $\mathring{A}^2$ (0.18 $\mathring{A}^2$).[35] Human lysozyme,[36] actinidin,[43] and cytochrome c^{44} yield values of the same order. The large magnitudes suggest that, compared to crystals of simple organic substances, protein crystals have either much larger lattice disorder or much larger $\langle x^2 \rangle_c + \langle x^2 \rangle_v$ values, or both. Recently, several larger enzymes have been subjected to restrained least-squares refinement. These proteins have overall B factors of 20 $\mathring{A}^2$ or higher, corresponding to a mean displacement of 0.5 $\mathring{A}$ for the atoms in the molecule.[38]

To compress the enormous amount of information obtained from the isotropic B factor refinement and to reduce variations due to errors, Jensen and associates averaged the temperature factors for the main chain α carbon, nitrogen, and carbonyl carbon atoms for each residue and plotted the averages vs residue number. Such a graph is shown in Fig. 1 for sperm whale metmyoglobin. This type of plot has become a standard way to represent the average mean-square displacement of each amino acid in an enzyme crystal structure.

The extremely high resolution of the rubredoxin structure refinement allowed assignment of individual isotropic B factors to every atom without the imposition of restraints on the difference in B between bonded atoms. Nonetheless, such atoms were found not to differ excessively in temperature factor. This observation suggests that the restraints that have been imposed in nearly all other protein structure refinements are reasonable.

[43] E. N. Baker, *J. Mol. Biol.* **141**, 441 (1980).
[44] T. Takano and R. E. Dickerson, *Proc. Natl. Acad. Sci. U.S.A.* **77**, 6371 (1980).

The overall and individual magnitudes of crystallographically derived $\langle x^2 \rangle$ values imply relatively large and complicated motions for most of the atoms in the protein. The correlation of the magnitudes with protein secondary and tertiary structural features of the protein will now be considered.

Correlations with Structure. The individual atomic $\langle x^2 \rangle$ values determined for a number of crystalline proteins vary widely depending on the position of the residue to which they belong in the tertiary structure of the protein. Their relative magnitudes correlate with intramolecular structural features, with smaller B factors occurring where the atoms appear to be more rigidly bound by covalent and noncovalent interactions. It is this sensible behavior of the $\langle x^2 \rangle$ values that, more than any theoretical consideration, is persuasive of their relative accuracy.

Consider for example the average B factors for some side chains in the refined structure of the sulfhydryl protease actinidin.[43] The average B for the side chain atoms of Lys-17, which is internal, is 6.8 Å^2. The corresponding value for the completely exposed residue Lys-145 is 34.2 Å^2. Similar behavior is observed for Phe-28 and Phe-152: the buried side chain of Phe-28 has averaged $B = 7.3$ Å^2 but the value for Phe-152, which has one edge exposed to solvent, is 11.6 Å^2. Identical correspondence of average B with location in the molecule has been observed in myoglobin[32] and lysozyme,[36] and has now been repeated in numerous other crystalline proteins. Residues involved in strong covalent interactions, such as cysteines in disulfide bridges or ligands to metal atoms, also have significantly smaller B factors than the same residue type when not so involved. Conversely, residues in external loops that project away from the globular mass of the protein tend to have high average B factors. Residues involved in tight secondary structures such as α helices and β-pleated sheets have, on average, lower mean-square displacements than residues in irregular structural features. The termini of the polypeptide chain, unless involved in strong intra- or intermolecular interactions, have very high mean-square displacements, in keeping with the observation that the electron density for these regions is often very weak or absent altogether.[45] Interestingly, active site regions in many enzymes seem to have residues with among the largest mean-square displacements in their respective proteins. In lysozyme, the active site cleft is among the most mobile portions of the structure.

Individual atomic $\langle x^2 \rangle$ values show the same sensible correlation with structure as do the residue averages. The concern has been raised that the

[45] D. Ringe, G. A. Petsko, F. Yamakura, K. Suzuki, and D. Ohmori, *Proc. Natl. Acad. Sci. U.S.A.* **80**, 3879 (1983).

increase in B factor observed as one procedes along an extended side chain reflects not mobility but rather the restraints imposed during the refinement, which themselves decrease as one nears the terminal atoms in the side chain.[37] However, examination of lysine residues in myoglobin (and other proteins) has shown that this is not the case.[46] An example from myoglobin illustrates this point.

Lysine-98 in myoglobin projects away from the molecular surface into the solvent, and its individual atomic $\langle x^2 \rangle$ values, which increase smoothly with distance from the backbone, reflect this fact:

Lys-98	C-1–	C-1–	C-3–	C-4–	C-5–	N-6	
$\langle x^2 \rangle$	0.195	0.21	0.23	0.25	0.26	0.275	(Å^2)

These values also accurately reflect the strength of the observed electron density for this residue: at a contour level of twice the root mean square deviation in the map, there is no electron density beyond the β carbon of the side chain (Fig. 2).

In contrast, Lys-77 refined with identical restraints, but which is salt bridged to the side chain of Glu-18 and consequently lies flat along the surface of the protein, shows B factor behavior indicative of its structural location, not the restraints:

Lys-77	C-1–	C-2–	C-3–	C-4–	C-5–	N-6	
$\langle x^2 \rangle$	0.12	0.11	0.10	0.11	0.11	0.13	(Å^2)

Again, the electron density is consistent with these values, being strong all along the length of the side chain (Fig. 3). Similar behavior has been demonstrated for the B factors of internal and external side chains in many other proteins.

One interesting common feature in all of the careful analyses of protein $\langle x^2 \rangle$ values carried out to date is that the nitrogen, α carbon, and carbonyl carbon atoms of the backbone of a particular residue all have very similar B factors, but the carbonyl oxygen atom of the peptide bond often has a slightly higher B. This indicates that there may be some torsional flexibility of the carbonyl unit. Because of this systematic effect, most workers omit the carbonyl oxygens when computing average main chain B factors.

Plots of the individual $\langle x^2 \rangle$ values vs distance from the center of mass of the protein for a number of different proteins all show a general increase in the magnitude of $\langle x^2 \rangle$ with increasing distance.[32,36] In terms of X-ray determined mobility, proteins can be considered to have a condensed

[46] W. A. Gilbert, J. Kuriyan, G. A. Petsko, and D. Ringe Ponzi, in "Structure and Dynamics: Nucleic Acids and Proteins" (E. Clementi and R. H. Sarma, eds.), p. 405. Adenine, New York, 1983.

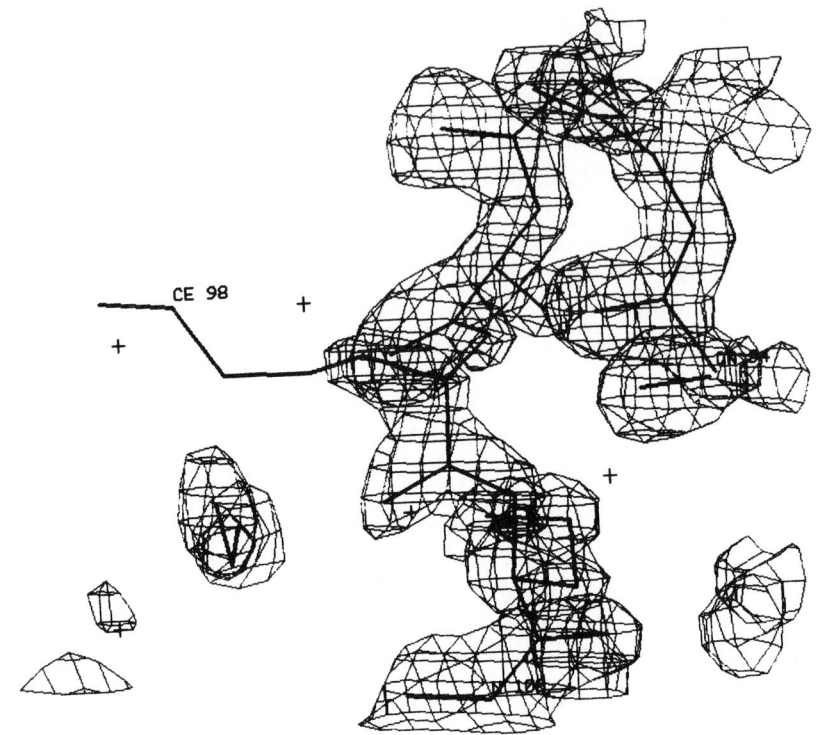

Fig. 2. Electron density and refined crystal structure coordinates for the myoglobin at 300 K in the region of residue Lys-98. Contours are drawn at approximately twice the root mean square deviation of the electron density. The map is a $2F_{obs} - F_{calc}$ map after refinement. The isolated crosses are water molecules; many of these have temperature factors higher than those for the lysine, so no density appears at their positions. The residue has no near-neighboring protein atoms for much of its length.

core and a semiliquid outer shell. However, within each of these regions there is considerable local variation, suggesting an inhomogeneous system. Mobility of the core is therefore likened to an aperiodic solid, while the term semiliquid as applied to the outer shell refers to the large but bounded $\langle x^2 \rangle$ values possible for surface atoms. Even on the surface, however, there are differences in mobility assignable to chemical properties. Charged side chains have larger $\langle x^2 \rangle$ than uncharged side chains, even when the two are exposed to the same degree.

Temperature Dependence. Huber and associates have refined the structure of trypsinogen in methanol–water, initially at 213 K[47] and subse-

[47] T. P. Singh, W. Bode, and R. Huber, *Acta Crystallogr. B* **36,** 621 (1980).

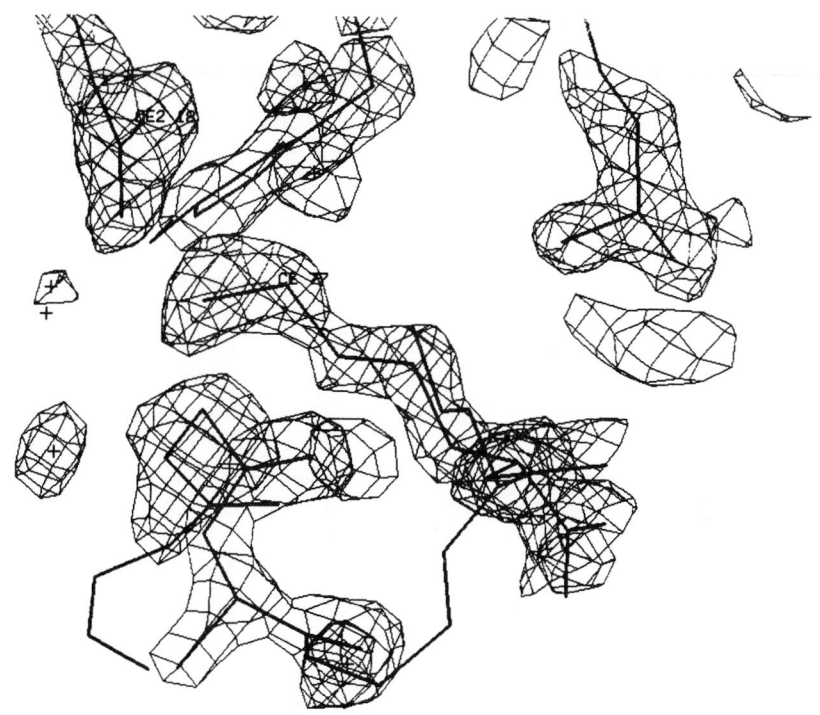

FIG. 3. Electron density ($2F_{obs} - F_{calc}$) and superimposed refined coordinates for the region around Lys-77; contours are drawn at the same level as for Fig. 2. Although this region is also at the surface of the protein, as evidenced by the absence of density for many atoms at this high contour level, the lysine is completely in density. It is stabilized by a salt bridge to the surface residue Glu-18, shown in the upper left.

quently at 173 and 103 K using X-ray data collected with synchrotron radiation.[48] The purpose of this study was to ascertain if the so-called "activation domain," a region of 15% of the enzyme that is disordered in the native crystal structure at room temperature,[49] became ordered at the low temperature. Cooling to 103 K reduced the overall B factor of the trypsinogen structure from 16.1 to 11.5 Å^2, and one or two of the residues at the N-terminal region became more clearly visible, but the majority of the amino acids in the activation domain remained disordered. There are

[48] J. Walter, W. Steigemann, T. P. Singh, H. D. Bartunik, W. Bode, and R. Huber, *Acta Crystallogr. B* **38**, 1462 (1982).
[49] R. Huber, *Trends Biochem. Sci.* **4**, 271 (1979).

three possible explanations for this effect: either the domain is still mobile at 100 K, the disorder is static at all temperatures, or a dynamic disorder has been frozen into a static disorder in which a number of conformations are of nearly equal energy and all are nearly equally populated, but none to such an extent that it is visible above the noise level in the electron density map. A number of elegant chemical labeling[48] and physical measurements[50] have now been carried out by Huber and co-workers, who conclude that the disorder in the activation domain of trypsinogen is dynamic in solution at room temperature with conformational transitions occurring with a reorientation correlation time of 11 nsec.

Stroud and associates, working with essentially the same crystal form of trypsinogen, have reported reasonably clear electron density for much of the activation domain.[51] Differences of this kind have serious implications for the study of protein dynamics by X-ray diffraction, and this difference is particularly serious because it refers to a functionally important part of an enzyme. It is regrettable that no resolution of the dispute has appeared in the literature, and no independent study of the problem appears to have been made. The discrepancy between the results of the two groups may reflect differences in sample preparation, crystallization procedures, and methods of refinement. For example, one group routinely omits low-resolution data in map calculation and refinement, while the other appears not to do so. The effect of omitting low-resolution data, which are sensitive to the presence of highly mobile regions that do not contribute at higher resolution, may explain why one group consistently observes more electron density for the disordered region than does the other. Further study of both the consequences for refinement and Fourier analysis of the omission of regions of data and of the particular situation with trypsinogen is clearly desirable.

In a collaborative study, Hartmann *et al.* have determined the structure of metmyoglobin to 2 Å resolution at 80 K.[35] Data collection at this temperature was achieved by a "flash-freezing" technique that did not involve the use of a cryoprotective mother liquor. The overall B factor decreased from 14 Å^2 at 300 K ($\langle x^2 \rangle = 0.175$ Å^2) to 5 Å^2 at 80 K ($\langle x^2 \rangle = 0.063$ Å^2). The striking drop in $\langle x^2 \rangle$ with cooling observed in this study and that of Huber's group on trypsinogen is the most convincing evidence that proteins are flexible in the crystalline state and that static disorder does not dominate the observed B factors. When the 80 K data were combined with those observed earlier[32] it was seen that not all of the 153 residues in myoglobin had average B factors that extrapolated to near 0 at 0 K. The

[50] T. Butz, A. Lerf, and R. Huber, *Phys. Rev. Lett.* **48,** 890 (1982).
[51] A. A. Kossiakoff, J. L. Chambers, L. M. Kay, and R. M. Stroud, *Biochemistry* **16,** 654 (1977).

temperature dependence of the remainder of the protein was consistent with the notion that conformational substates could be frozen out at sufficiently low temperatures. Fifty-one residues can be modelled with a linear temperature dependence with $\langle x^2 \rangle$ = approximately 0.04 Å^2 at 0 K. The remaining 56 amino acids in myoglobin do not show a good linear fit of $\langle x^2 \rangle$ vs T. The number of temperature points available for this analysis is still small, so these numbers should be regarded as preliminary.

Despite the use of different crystals, data collection methods, absorption corrections and other parameters, the overall correlation of $\langle x^2 \rangle$ with structure is the same at the two temperatures (Fig. 1). Residues with large displacements tend to have large temperature dependence, while residues with small B factors show little change on cooling. Although the observed data do not all fit a simple harmonic motion model, this behavior may be described as quasiharmonic. Huber has noted similar correlation of the temperature dependence of B with its magnitude for trypsinogen.[48]

As mentioned above, Mössbauer data suggest that there is a relaxation temperature (which may reflect changes in the interaction of the surface residues of the protein with the bound solvent) at about 180 K for myoglobin. Unfortunately, the present data for myoglobin could be fit by a single straight line with high temperature scatter or by two lines which intersect at just below 200 K, one line being nearly temperature independent. Although the latter interpretation is in accord with the Mössbauer results, it is important to determine the exact form of the temperature dependence experimentally. Consequently, Parak and associates are currently measuring myoglobin data at a number of additional temperatures (Parak, personal communication).

Comparison of the myoglobin structures at 300 and 80 K revealed a reduction in the volume of the protein at low temperature. The unit cell volume of the crystal is 4.5% smaller at 80 K than at room temperature[36] and this is a reflection of a general shrinkage of the protein by approximately the same amount.[52] The average volume change can be computed by several different techniques, and in all cases the thermal expansion coefficient of myoglobin is found to be between the thermal expansion coefficient of pure water and the value for pure benzene. This fact may reflect a balance between hydrogen bonding and van der Waals forces in the stabilization of the protein. Because of the nonuniformity of these interactions within the molecule, it is expected that the thermal expansion coefficient of myoglobin (or any other globular protein) should be anisotropic. An analysis of the spatial distribution has been carried out, and it

[52] H. Frauenfelder, H. Hartmann, M. Karplus, I. D. Kuntz, Jr., J. Kuriyan, F. Parak, G. A. Petsko, D. Ringe, R. F. Tilton, Jr., M. Conolly, and N. Max, submitted (1986).

is found that some regions of the molecule do move much more than others. In particular, an external loop of charged amino acids, the so-called CD corner, moves away from the center of the protein considerably more than any other region. Taken together with the temperature dependence of the $\langle x^2 \rangle$ values, this study of the thermal expansion of myoglobin provides the most detailed picture yet available of the physical chemistry of a protein. Huber also observed a change in the center of mass of trypsinogen on cooling to 100 K and remarked that this was consistent with an anisotropic contraction of the structure, but no further analysis of these data has appeared.[48] Since myoglobin is an all helical protein with a large prosthetic group, analysis of the thermal expansion of other crystalline proteins of different secondary structure distribution is needed.

The Utility of Molecular Dynamics. Since X-ray diffraction is able to visualize only the highly occupied states of a molecule (averaged over many molecules and many days of data collection), high-frequency, low-amplitude motions are expected to be the major contributors to the observed $\langle x^2 \rangle$. Molecular dynamics can simulate motions on this time scale. Given the atomic coordinates of a protein from an X-ray structure, an empirical potential energy function may be written as a function of the coordinate set. It is then possible to derivatize this function to obtain the force on each atom, and to solve the Newtonian equations of motion for a small time interval, usually a fraction of a picosecond.[53,54] The importance of the potential energy function chosen, the limitations of the method, and general strategies for its use have been discussed by Levitt.[55] Calculations have been carried out for a number of proteins, and the results have been compared with the dynamic picture obtained from X-ray diffraction analysis of $\langle x^2 \rangle$ values.

To compare X-ray-derived mean-square displacements with the radial displacements obtained from the movements of the atoms during the molecular dynamics simulation, one must correct for the assumed isotropy of the Debye–Waller model, and one must remove the lattice disorder term from the X-ray quantity. Thus,

$$\langle r^2 \rangle_{\text{mol dyn}} = 3[(B/8\pi^2) - \langle x^2 \rangle_{\text{ld}}]$$

where the equal sign represents "should equal." The earliest molecular dynamics simulations were performed by McCammon, Karplus, and associates, who studied the basic trypsin inhibitor of bovine pancreas. They found that after 9 psec of simulation the structure had altered in some

[53] J. A. McCammon, B. R. Gelin, and M. Karplus, *Nature (London)* **267**, 585 (1977).
[54] J. A. McCammon, P. G. Wolynes, and M. Karplus, *Biochemistry* **18**, 927 (1979).
[55] M. Levitt, *J. Mol. Biol.* **168**, 595 (1983).

regions, but generally remained close to the X-ray structure.[53] The same conclusions pertained after 96 psec. However, comparison of the mean-square displacements with X-ray $\langle x^2 \rangle$ values was difficult, since the crystallographic refinement method employed for the trypsin inhibitor at that time was not optimum for the determination of individual B factors. McCammon, Karplus, and co-workers were able to perform such a comparison for reduced cytochrome c,[56] where the crystal structure had been refined by restrained least-squares.[44] The lattice disorder correction was made by the minimum-function method, and $\langle x^2 \rangle_{ld}$ was assumed to be 0.07 Å^2. Comparison of the calculated and experimental root mean square (rms) radial fluctuations showed excellent agreement between the two in terms of regions of high and low displacement. This correlation confirms the reliability of both computer simulation and X-ray diffraction for providing detailed information about the internal mobility of proteins.

In the simulations performed to date, the atomic motions have been found to be highly anisotropic and often to produce one or more discrete alternate conformations for certain side chains. The rms fluctuation of an atom in its direction of largest displacement is often twice that observed in its direction of smallest displacement, and may be even larger.[57] There is often very little correlation between the preferred direction and local bonding.[56,58] A quasi-harmonic model in which the potential of mean force for an atomic displacement is itself temperature dependent is consistent with the data.[57,59]

The size of the molecular dynamics calculation necessitates that most simulations are run for an isolated protein in vacuum. Lack of good models for the solvent continuum has also been a contributing factor. Therefore, there should be discrepancies between theory and experiment in those regions of the protein involved in intermolecular contacts in the crystal lattice, and in some of the protruding external side chains. For cytochrome c, the largest differences are found in surface charged residues, particularly lysines.[60] Omission of the solvent and crystal contacts is not found to affect the dynamics of the interior of the protein significantly. Simulation of pancreatic trypsin inhibitor in a Lennard-Jones solvent or with crystal contacts improved the agreement between the aver-

[56] S. H. Northrup, M. R. Pear, J. A. McCammon, M. Karplus, and T. Takano, *Nature (London)* **278**, 659 (1980).

[57] W. F. van Gunsteren and M. Karplus, *Biochemistry* **21**, 2259 (1982).

[58] S. Swaminathan, T. Ichiye, W. F. van Gunsteren, and M. Karplus, *Biochemistry* **21**, 5230 (1982).

[59] B. Mao, M. R. Pear, J. A. McCammon, and S. H. Northrup, *Biopolymers* **21**, 1979 (1982).

[60] M. Karplus and J. A. McCammon, *Annu. Rev. Biochem.* **53**, 263 (1983).

age structure from simulation and the X-ray structure, the rms deviation in atomic positions being reduced from 2.2 to 1.35 Å in the "solvent" and to 1.52 Å in the "crystal".[57,61]

The effect of neglect of solvent and crystal contacts in the dynamics calculations has also been shown in simulations of myoglobin (J. Kuriyan, R. M. Levy, S. Swaminathan, and M. Karplus, unpublished). Although overall agreement between the X-ray $\langle x^2 \rangle$ and the fluctuations from molecular dynamics is good, one region of the protein shows much greater mobility in the simulation than indicated by its B factors. This region is the CD corner, which is also observed to undergo the largest change in conformation on cooling of the crystalline protein to 80 K. As this region is highly charged and contains several lysines and arginines, it is possible that in solution there is substantial coupling between its fluctuations and charge fluctuations in the solvent. Despite these differences, vacuum simulations appear to give an excellent representation of the relative mobilities of internal residues in proteins.

Recently, a 12 psec molecular dynamics simulation has been carried out on the full unit cell of trypsin inhibitor containing 4 protein molecules and 560 water molecules, the latter generated theoretically.[62] The magnitude of the fluctuations of internal residues corresponds to those found in the earlier simulations of the isolated protein, while the structure averaged over the full simulation was found to have an rms coordinate discrepancy of only 1.2 Å from the X-ray structure. The difference between the average structures of the four molecules in the unit cell was found to be of the same magnitude as the difference between each of the molecules and the X-ray structure. Averaging over the four molecules decreased the difference from the X-ray structure, suggesting that no systematic deviation from the X-ray structure occurred. Comparison between the mean square radial fluctuations from this simulation and the $\langle x^2 \rangle$ values from refinement of the trypsin inhibitor crystal structure shows better agreement than seen in earlier simulations of the same protein, but still poorer than observed for myoglobin, lysozyme, or cytochrome c. Forty-seven bound water molecules have been located in the crystal structure of trypsin inhibitor, but only 9 of these are reproduced within 1 Å in the simulation.

van Gunsteren and associates have tried to use the molecular dynamics results from this detailed simulation of a crystalline protein to overcome the limitations of isotropy and harmonicity imposed on the X-

[61] W. F. van Gunsteren and M. Karplus, *Nature (London)* **293**, 677 (1981).

[62] W. F. van Gunsteren, H. J. C. Berendsen, J. Hermans, W. G. J. Hol, and J. P. M. Postma, *Proc. Natl. Acad. Sci. U.S.A.* **80**, 4315 (1983).

ray analysis by the Debye–Waller model. Their reasoning is as follows: molecular dynamics produces a large number of configurations, which together describe the trajectories of all atoms in the unit cell. From each configuration, a model electron density map can be generated and the resulting set of maps may be combined to produce an average map containing electron density distributions for each atom which reflect its motion throughout the simulation, regardless of the complexity of that motion. It should be noted that the same information is generated for the model solvent atoms as well. Fourier transformation of this average map yields structure factors that can be directly compared with the measured X-ray data. The 1 Å discrepancy between the dynamics structure and the X-ray structure is reflected in a 52% R factor for this calculation. However, the R factor for low-resolution data is actually better than that obtained from the crystallographic refinement, indicating better treatment of the bulk solvent in the crystal interstices. Even more encouraging, when the average molecular dynamics positions were shifted back to the X-ray positions, but the fluctuations were taken from the molecular dynamics simulation, the R factor for data between 6.65 and 1.5 Å resolution was 29%, compared with 25.8% when the X-ray temperature factors were used. This approach is likely to be taken by other crystallographers in the future. As confidence in the validity of carefully performed simulations increases, crystallographers will begin to amalgamate their descriptions into the models used in refinement.

The availability of a trypsin inhibitor simulation in the full crystalline unit cell suggests that comparison of the average structure evolved in this environment with that obtained from a molecular dynamics run on an isolated molecule in water would shed light on the effects of crystallization on the protein. van Gunsteren and Berendsen have carried out such a study, using 20 psec simulations.[63] For the solution run, a three-point charge model was used for water.[64a,b] No significant differences were found in the conformations of backbone atoms following the two calculations, except for two flexible loops (residues 24 to 29 and 12 to 15) and the highly flexible C-terminal residues. Those side chains whose time-evolved conformations differed between the "crystal" and "solution" all belonged to polar residues.

To date, only one molecular dynamics simulation has been carried out at low temperatures and compared with corresponding X-ray data. A full molecular dynamics calculation was performed on a decaglycine helix as

[63] W. F. van Gunsteren and H. J. C. Berendsen, *J. Mol. Biol.* **176**, 559 (1984).

[64a] F. M. Richards, *Carlsberg Res. Commun.* **44**, 47 (1979).

[64b] H. J. C. Berendsen, J. P. M. Postma, W. F. van Gunsteren, and J. Hermans, *in* "Intermolecular Forces" (B. Pullman, ed.), p. 331. Reidel, Dordrecht, 1981.

a function of temperature between 5 and 300 K.[65] Below 100 K the harmonic approximation is valid for the motions of this helix, but above that value the average $\langle r^2 \rangle$ for the α carbon atoms is more than twice that found in the harmonic model. These data were compared with the observed $\langle x^2 \rangle$ for the helices in myoglobin at 300 and 80 K.[32] Correcting the X-ray data by an assumed lattice disorder of 0.025 Å² and adjusting the corrected mean square displacements by the factor of three appropriate for the comparison, the root mean square displacements of the α carbon atoms of all the helical regions of myoglobin are found to be 0.48 Å at 300 K and 0.28 Å at 80 K. The simulation yields values of 0.52 Å at 300 K and 0.22 Å at 80 K for the decaglycine helix. Once again, the agreement between theory and experiment is excellent, giving confidence in both approaches for studying protein dynamics.

Implications for Protein Structure Refinement. The results of the crystallographic determination of $\langle x^2 \rangle$ values combined with the molecular dynamics calculations performed on several small proteins have important implications for least-squares refinement of protein structures. It has long been known that the crystallographic R factor, which is an indicator of the residual disagreement between observed reflection amplitudes and those calculated from the current model, is reduced when restraints are loosened or removed altogether in the refinement process. Nevertheless, there has been considerable criticism of those structures that have been refined with relatively large allowed deviations from ideality in bond lengths and angles, or with no restraints on these quantities at all. It is considered a challenge to obtain a reasonably low R factor (say, below 20%) with extremely tight restraints (0.01 Å rms deviation from ideality on bond distances). At the same time, the lowest R factors that have been achieved in this way are still at least 2 to 3 times the extimated error in precision of the data. This discrepancy is often assigned to lack of an adequate model for the bulk solvent, and no doubt this is an important factor. However, studies of protein dynamics suggest an additional answer.

When refinement is performed with loose or no restraints on the internal coordinates of a protein, not only is the bond length distribution broader than if the restraints had been tight, but also the centroid of the distribution is shifted to lower values and the distribution is skewed in that direction. If the results of a molecular dynamics simulation on a protein are analyzed to yield the distribution of bond lengths that evolve during the time of sampling, the same effects are seen. Assuming the simulation is presenting an accurate picture of protein motion, unre-

[65] R. M. Levy, D. Perakia, and M. Karplus, *Proc. Natl. Acad. Sci. U.S.A.* **79**, 1346 (1982).

strained least-squares refinement may be giving results that incorporate the effects of individual and collective fluctuations in a more correct way. In this view, the lower R factor observed when restraints are released means exactly what it seems to: namely, improved agreement between model and data due to a more accurate model.

If this seems surprising, it is well to remember exactly what a protein crystal structure represents. The data obtained reflect the X-ray scattering of an average molecule in the crystal; the average is not only over all unit cells, but also over the many days (and, occasionally, many different crystals) required to collect the full data set. The refined structure that best fits these data will be that of an average molecule. An average molecule is expected to have nonideal bond lengths and angles (and, possibly, some bad nonbonded contacts as well). Such a molecule probably never exists at any instant in time, but is the most accurate reflection of the time- and space-averaged properties of the 10^{17} protein molecules in the crystal.

This argument does not mean that refinement with tight restraints is incorrect. Such a refinement will give a physically reasonable molecule that is, it is hoped, one member of the ensemble that actually exists, and that may even be the most probable member. If a stereochemically reasonable model is desired (for example, in investigations of enzyme-induced strain), restraints should be made tight. But if the object is to compare the crystallographic results with those of molecular dynamics averages, loose restraints or no restraints at all will yield a more appropriate model.

Cavities in Proteins. Given the mobility of atoms inside globular proteins, a pertinent question is where these atoms can move to, since most macromolecules have very tightly packed interiors. Richards pointed out that the packing of atoms in the interior of proteins is not perfect, and that small cavities are found.[64a,b] It is these cavities that provide the room for internal residues to move. Presumably, as the structure fluctuates, new cavities are being formed as atoms move away from their former positions to fill old cavities. This suggests that internal motions might be highly collective in character, even over nonbonded neighbors. The modulation of local displacements by the collective motions and the opening of transient cavities might be expected to show up in types of motion which require a large volume in which to occur. Aromatic ring flipping is one such type.[65] Traditionally, it was assumed that any large cavity inside a globular protein must be occupied by one or more water molecules. Kuntz and Tilton have questioned this assumption, pointing out that there are a number of large cavities inside myoglobin that appear to be highly apolar. Recently, they have carried out an investigation of the binding of xenon in those cavities by both NMR and high-resolution X-ray diffraction.[66,67] Xenon can occupy four of the internal cavities in myoglobin, with only slight alterations in the positions of amino acids in one of the cavities. Filling these cavities has a small effect on the overall $\langle x^2 \rangle$ for the protein, which is reduced by 15% relative to the xenon-free structure at the same temperature (the minimum-function method was used to correct the two data sets for lattice disorder). Larger reductions in $\langle x^2 \rangle$ are observed for some of the side chains in van der Waals contact with the xenon atoms. This study establishes the existence of cavities inside myoglobin, and implies that their existence is correlated with the flexibility of the molecule. It is interesting to note that in the static structure of myoglobin there is no obvious pathway from outside the protein to any of the internal cavities occupied by xenon. The ability of xenon to reach these sites even in the crystalline protein is further evidence that much of the solution flexibility of macromolecules is retained in the crystal state.

The Role of Collective Motions

Evidence that collective motions occur in proteins comes in part from molecular dynamics simulations. Analysis of the time dependence of the atomic fluctuations in the trypsin inhibitor simulations demonstrated that the motions contributing to the temperature factor for any atom could be separated into individual atomic vibrations superimposed on collective motions.[58] Over a 25 psec time period, the high-frequency atomic fluctuations contribute only 40% to the average $\langle x^2 \rangle$ of the main chain atoms. Further, these individual atomic motions tend to be of uniform magnitude over the whole structure, and provide a baseline for the total $\langle x^2 \rangle$. The low-frequency, collective motions provide the variation of $\langle x^2 \rangle$ with structure. Individual atomic fluctuations have a subpicosecond time scale; the collective motions have time scales of at least 1 psec and often longer. Collective motions may involve all the atoms in a residue, an atom plus its nonbonded neighbors, or groups of linked atoms such as an α helix.

Examination of a plot of $\langle x^2 \rangle$ vs residue number from X-ray diffraction such as shown in Fig. 1 is informative in the light of these findings. The observed structural dependence of such a plot must come from collective modes of motion superimposed on a uniform level of atomic fluctuations. However, because isotropic B factors are used to fit the X-ray data, detailed information about the direction of these collective motions is lost. The correlation in B factor between neighboring atoms observed in

[66] R. F. Tilton and I. D. Kuntz, Jr. *Biochemistry* **21**, 6850 (1982).
[67] R. F. Tilton, I. D. Kuntz, Jr., and G. A. Petsko, *Biochemistry* **23**, 2849 (1984).

X-ray determinations of mobility is a reflection of the magnitude of collective motions.[3]

Collective motions destroy the correlation of anisotropy with chemical bonding due to their relatively large amplitudes.[68] As an example, consider a leucine side chain in isolation and in a protein. The predominant directions of motion for the terminal methyl group in the isolated amino acid will tend to correlate with the covalent bond direction, since simple harmonic vibration of the atoms will predominate. When the leucine is placed inside a protein, however, there will be nonbonded contacts with neighboring residues that will restrict motion in certain directions. The motion of the leucine methyl group will depend on the permanence of these restrictions; motions of the surrounding residues away from the methyl will allow the leucine room to move, and it will do so in a direction and with a magnitude that is correlated with the motion of these neighboring atoms. The large contribution of collective motions to the $\langle x^2 \rangle$ values determined from crystallography means that any attempt to fit an anisotropic model to protein fluctuations in an X-ray analysis must not assume that the principal direction of motion will correlate with a covalent bond direction.

Normal mode calculations, though very difficult for large proteins, will be very useful in understanding collective fluctuations in structure. Calculation of the normal modes of low-frequency vibrations in bovine pancreatic trypsin inhibitor revealed a rich variety of motions due to concerted variations in soft variables, i.e., dihedral angles,[69] and in bond lengths and angles as well.[70] Most modes with frequencies above 50 cm^{-1} (time periods of less than 0.7 psec) behaved harmonically at room temperature. As expected, the mean square displacements in atomic positions were found to be determined mainly by collective modes with frequencies below 30 cm^{-1}. The low-frequency modes appear to be delocalized over the protein. Comparison of the main chain average $\langle x^2 \rangle$ computed from the normal mode analysis with values from X-ray crystallography or with values from a full molecular dynamics simulation gave good agreement.

Crystallographic observations on several enzymes confirm that the binding of an inhibitor to the active site of an enzyme often reduces the mean square displacement of residues some distance away, and in some cases appears to affect whole domains of the protein. Because of the close spacing of modes (350 modes between 3 and 216 cm^{-1}, and a similarly

[68] J. D. Morgan, J. A. McCammon, and S. H. Northrup, *Biopolymers* **22,** 1579 (1983).
[69] N. Go, T. Noguti, and T. Nishikawa, *Proc. Natl. Acad. Sci. U.S.A.* **80,** 3696 (1983).
[70] B. Brooks and M. Karplus, *Proc. Natl. Acad. Sci. U.S.A.* **80,** 6571 (1983).

dense distribution up to 1800 cm^{-1}), local effects such as substrate, inhibitor, or allosteric effector binding may, by significant mode mixing, be felt in remote regions of the enzyme.

Not all collective motions contribute significantly to the $\langle x^2 \rangle$ obtained in a protein crystal structure refinement. Any motion that is very slow is unlikely to yield sufficiently strong electron density to be fitted. In other words, the fast collective motions will contribute. If a large-scale collective motion is important, its presence may be inferred from an X-ray diffraction study since the electron density for the particular group of atoms may vanish altogether.[45,71] Quantitative information about such motions cannot be obtained, since there is no electron density to fit, but a qualitative picture which can aid in the interpretation of spectroscopic data is possible. Another class of collective motions that will not be seen by crystallography is that which is fast but infrequent. Tyrosine ring flips are one example. The actual flipping of even a buried ring occurs in less than 1 psec, but the event is very improbable.[6,60] Consequently, the intermediate states in the flip are, averaged over a long period of time, very low in occupancy. X-Ray results confirm this picture: although ring flips are known to occur, the electron density for aromatic rings in protein structures is not spherical.[41,46]

Anisotropic temperature factor analysis should greatly aid in the identification of collective modes. Some protein refinement methods treat certain covalently bonded atoms as a group. This approach has promise but neglects the noncovalent interactions that contribute to the collective motions of atoms in proteins. Although it will be very difficult to extract information about collective motions from a protein crystal structure, it is important to realize that the information is present.

Very Low Frequency Motions. X-Ray diffraction provides information about the spatial distribution of protein dynamics, but not about time scales. Therefore, it is difficult to compare the results of crystallographic studies of B factors with spectroscopic measurements, at least quantitatively. Techniques such as NMR, fluorencense spectroscopy, and Mössbauer spectroscopy should be regarded as complementary to X-ray data. The time domains of most of these methods are much longer than that accessible to molecular dynamics simulations, and it is likely that X-ray diffraction results are also most appropriate to the high-frequency regime. Although one might expect from theoretical considerations that regions of a protein undergoing large atomic fluctuations also demonstrate consider-

[71] T. Alber, D. W. Banner, A. C. Bloomer, G. A. Petsko, D. C. Phillips, P. S. Rivers, and I. A. Wilson, *Philos. Trans. R. Soc. London Ser. B* **293**, 159 (1981).

able mobility on a longer time scale, this fact has never been proven experimentally.

Fluorescence and NMR spectroscopy can identify the presence of a highly mobile region within a protein even if the motion is very slow and involves large structural units. These results can be compared with the absence of strong electron density in a crystal structure determination of the same molecule. The strength of such an approach is the complementary nature of the information provided. Spectroscopy can indicate that a region of the polypeptide chain containing, for example, a tryptophan and a histidine, is unusually mobile. X-Ray diffraction can indicate which segment of the amino acid sequence these residues belong to. One can then return to the spectroscopic approach to examine the time dependence of this motion and its alteration upon substrate binding in the light of the location of the region in the three-dimensional structure.

Hydrogen exchange can be correlated directly with both neutron and X-ray results. Several different classes of peptide amide hydrogens can be classified on the basis of their exchange rates, with the slowest requiring days or longer for complete exchange.[20] Neutron diffraction can indicate which protons in the crystalline molecule are exchanged completely after a given time of soaking the crystal in deuterated mother liquor. Several classes of hydrogens are also observed by this method,[23,72] but even in the crystal nearly all amide protons are able to exchange eventually. These observations indicate that the motions that allow buried protons to exchange with solvent occur even within the crystal lattice. One can then ask whether there is any correlation between the slowest exchanging amide protons and regions of the molecule with very low $\langle x^2 \rangle$. The answer in the case of bovine trypsin is that no correlation is found.[22] Motions permitting solvent to reach protons that are inaccessible in the static structure are apparently too slow (i.e., generate conformations too low in occupancy) to make a major contribution to the individual atomic B factors.

Effects of Binding. Protein crystallography provides an ideal method for determining changes in the mobility of groups of atoms (or the entire molecule) when an enzyme binds substrates, inhibitors, or allosteric effectors. The effects of xenon binding on the $\langle x^2 \rangle$ values of myoglobin have already been noted. Other studies have focused on the changes in B factors that occur when inhibitors bind to ribonuclease A,[73] *Streptomyces griseus* serine protease A,[74] and other enzymes. Reduction in mobility of

[72] A. A. Kossiakoff, *Annu. Rev. Biophys. Bioeng.* **12,** 159 (1983).

[73] A. Wlodawer, M. Miller, and L. Sjölin, *Proc. Natl. Acad. Sci. U.S.A.* **80,** 3628 (1983).

[74] M. N. G. James, A. R. Sielecki, G. D. Brayer, L. T. J. Delbaere, and C. A. Bauer, *J. Mol. Biol.* **144,** 43 (1980).

those residues which make contact with the inhibitor is specific and dramatic. Individual changes in mobility have been seen for the residues of myoglobin on going from the met to the oxy form.[37] Overall and individual changes in $\langle x^2 \rangle$ are observed for ribonuclease A when the inhibitory sulfate ion is removed from the active site.[75]

It is not necessary to carry out complete refinement of a structure to detect changes in the disorder of a residue or group of residues. Huber and associates have solved the structures of several complexes of trypsinogen with peptides and inhibitors, and have noted a marked increase in the clarity of the electron density in the activation domain prior to any refinement.[49,76,77] A 10-residue flexible loop in the structure of triose-phosphate isomerase shows similar behavior: in the native enzyme this loop is almost invisible in the electron density map, apparently due to high thermal motion.[71,78] On binding of substrate, the loop adopts a different average conformation and also becomes highly ordered.[78,79] It is expected that the qualitative assessment of changes in mobility will be one of the most important uses of protein crystallography in the area of enzyme dynamics.

Triggered Conformational Change

When a ligand binds to a protein, not only the mobility but also the average structure of the protein may adjust to accommodate the presence of the new molecule. This new structure has a lifetime as long as the ligand remains bound, or as long as the structure of the ligand remains unchanged. Such changes in protein structure have long been observed by spectroscopic methods, but until recently could not be defined. The availability of structural data has made it possible to observe the end effect of such movements. These motions are not random, but rather the result of a specific stimulus. Motion is inferred because a portion of the protein occupies a different position in the structure with ligand bound compared to the native structure. For instance, when the substrate dihydroxyacetone phosphate binds to the glycolytic enzyme triose-phosphate isomerase, the section of polypeptide chain bounded by Trp-168 and Thr-177 moves more than 10 Å from a relatively external, disordered position to a well-defined one covering the active site pocket. A similar movement of

[75] R. L. Campbell and C. A. Petsko, submitted (1986).
[76] W. Bode, *J. Mol. Biol.* **127,** 357 (1979).
[77] W. Bode, P. Schwager, and R. Huber, *J. Mol. Biol.* **118,** 99 (1978).
[78] T. Alber, Ph.D. thesis, Massachusetts Institute of Technology, Cambridge, 1981.
[79] T. Alber, W. A. Gilbert, D. Ringe Ponzi, and G. A. Petsko, *Ciba Found. Symp.* **93,** 1 (1982).

about 4.5 Å has been observed in the acid protease penicillopepsin when inhibitor is bound.[80]

X-Ray diffraction can observe the initial and final states of such a movement, but gives no information about the path or time of the motion involved. In the case of triose-phosphate isomerase, the mobile loop goes from a relatively disordered state to an ordered one. In terms of dynamics this means that the individual atoms of the loop have a high mobility in the native protein, and a low mobility in the bound form. This is not always necessarily true. Since the energy for triggered conformational change comes from the energy of specific interactions, such as breaking and remaking hydrogen bonds and/or salt bridges, not from thermal energy, it does not follow that a segment involved in such a movement has to be initially mobile in a dynamic sense. In hemoglobin the quaternary structural change on oxygenation is a transformation between two ordered states.[8,9] This particular conformational change is triggered by the binding of a ligand to the sixth coordination position of the octahedral heme iron, with consequent movement of the iron by more than 0.5 Å to a position in the plane of the porphyrin. The rest of the structure, including the intersubunit contacts, alters its conformation in response to the change in structure of the heme.

Any connection between a large-scale triggered conformational change such as hemoglobin undergoes and the small thermal-driven fluctuations experienced by individual atoms is moot at present. Occasionally, it has been found that those regions of the protein whose average conformations change in response to ligand binding also demonstrate large $\langle x^2 \rangle$ values in the absence of ligand but this is not always the case (hemoglobin is one major counterexample; not all of the areas involved in the quaternary structure change are highly mobile originally). Nevertheless, the atomic fluctuations, both individual and collective, are important for the slower, large, ligand-induced movements. Triggered conformational changes must overcome relatively large potential energy barriers caused in part by interatomic contacts. If all protein atoms were completely rigid, such movements would be very difficult. Atomic fluctuations are the "lubrication" that makes it possible for large groups of atoms to rearrange their positions without prohibitively high "friction."

X-Ray diffraction studies cannot give information on the speed of a triggered conformational change, or any other protein movement. Time-resolved protein crystallography may be possible in the future if developments in synchrotron radiation sources and position sensitive area detectors continue.

[80] M. N. G. James and A. R. Sielecki, *J. Mol. Biol.* **163**, 299 (1983).

The types of substances whose binding can trigger a specific conformational change in the average structure of an enzyme include substrates, coenzymes, allosteric effectors, inhibitors, and even protons. Despite its limitations, X-ray diffraction is the only method that can give detailed information on the initial and final states and, possibly, stabilized intermediate states.[81] By combining this information with spectroscopic data and computer simulations (if these can be extended to such large, slow moving systems), it should eventually be possible to define both the pathway and time scale of triggered conformational changes.

The Role of Protein Dynamics in Protein Function

Although there is little direct evidence that the flexibility of enzymes is essential for their biological function, there is considerable indirect evidence. Assuming that flexibility and function are related, one question is whether the entire range of motions that occurs is involved. It seems likely that the answer is yes. Even small amplitude individual atomic motions are likely to be important; they provide the means by which the structure can relax during larger scale movements. Intuition suggests that low-frequency, collective modes are likely to be particularly important, since they could lead to the displacement of individual loops of polypeptide chain or a critical side chain. Triggered conformational changes have been suggested for many years as being of fundamental importance for the catalytic action of many enzymes. Indeed, it is popular to assume a conformational change whenever an unusual, functionally related change in the spectroscopic or chemical behavior of an enzyme has been observed. X-Ray crystallographic studies have now provided evidence for the involvement of flexibility with biological action in a number of proteins.

Access

Over 20 years ago Kendrew, Perutz, Watson, and colleagues noted that there was no obvious pathway in the static structure of myglobin or hemoglobin for oxygen or other ligands to reach the iron atom from outside the protein. Entrance to the heme pocket in metmyoglobin is blocked by the side chains of Val-68, His-64, Arg-45, a bound sulfate ion, and one propionic acid group of the heme itself. The observed on-rate for ligands is sufficiently fast that any process effecting entry must itself be very rapid. The logical conclusion is that rapid fluctuations in the protein structure open a transient channel for ligand penetration. Frauenfelder and associates have studied the ligand binding process in myoglobin by spec-

---

[81] T. Alber, G. A. Petsko, and D. R. Rose, *J. Mol. Biol.,* in press (1986).

troscopy at multiple temperatures, pressures, and viscosities.[82,83] They observe nonexponential time dependence of the binding of both CO and O_2 at temperatures below 160 K. They concluded that the protein must exist in a number of conformational substates at low temperature, each with a different rate constant for rebinding, and that diffusion of the ligand through the protein matrix must occur in a path with multiple energy barriers. Case and Karplus analyzed this problem theoretically. Empirical energy function calculations showed that the rigid protein would have barriers to ligand entrance into the heme pocket on the order of 100 kcal/ mol. Trajectory calculations for the escape of model ligands from the pocket showed a primary exit route between the distal His-64 (E7) and the distal Val-68 (E11).[84] The energy barrier could be reduced to a reasonable magnitude (5 kcal/mol) by small rotational motions of these side chains costing 8.5 kcal/mol in energy.

Inspection of the X-ray derived mean square displacements for myoglobin reveals that the distal side of the heme pocket is, surprisingly, more rigid overall than the proximal side (Fig. 1). Nonetheless, there is a region of relatively high $\langle x^2 \rangle$ values that leads from the binding site at the iron past His-64 and outside the protein near Arg-45, which forms a salt bridge with a propionic acid side chain of the heme.[32] The general direction of this mobile region is in agreement with prediction.

Evidence for the existence of a channel and its probable location has now come from crystallographic studies of the binding of a large ligand, the phenyl group derived from phenylhydrazine, to crystalline myoglobin. Ringe et al. diffused phenylhydrazine into crystals of the met form of the protein and determined the structure of the resulting aryl heme complex at 1.5 Å resolution.[14] The phenyl ring is bound end-on and its steric interactions prevent the closing of the channel which had to open to admit its entrance into the distal pocket. The channel is formed by side chain rotations of Val-68 and His-64; the histidine moves by over 5 Å, folding into a cavity in the protein interior. Movement of the histidine displaces a sulfate ion which occupies part of this cavity in the met structure, and is accompanied by a movement of Arg-45. Rotation of this side chain about two carbon–carbon bonds ruptures the salt bridge to the heme propionic acid, which in turn becomes highly disordered (large increase in $\langle x^2 \rangle$). The net effect of these movements is to create a channel which static accessi-

[82] R. H. Austin, K. W. Belson, L. Eisenstein, H. Frauenfelder, and I. C. Gunsalus, *Biochemistry* **14**, 5355 (1975).
[83] D. Beece, L. Eisenstein, H. Frauenfelder, D. Good, M. C. Marden, L. Reinisch, A. H. Reynolds, L. B. Sorensen, and K. T. Yue, *Biochemistry* **19**, 5147 (1980).
[84] D. A. Case and M. Karplus, *J. Mol. Biol.* **132**, 343 (1979).

bility calculations show is wide enough to accommodate an oxygen molecule. Ringe has referred to the phenyl group as a "molecular doorstop" that holds open a passage normally open only transiently.[14]

Oxygen quenching of aromatic side chain fluorescence is observed with many enzymes, implying similar diffusion through these molecules is possible. The functional correlation of mobility is obvious for the heme proteins: a rigid molecule would be unable to bind or release ligands, and so would be inactive. The heme groups of a number of different enzymes such as cytochrome *P*-450 and horseradish peroxidase are buried within their proteins to protect them from the formation of inactive oxygen-bridged dimers, yet substrates are able to reach the active sites. Presumably, the mechanism by which this happens is similar to that described here for the oxygen storage and transport proteins. Protection of a buried reactive group while permitting rapid binding of the desired ligands is one important functional role for enzyme flexibility.

Regulation

Huber[49] and Alber[78] have discussed the role of protein dynamics in the regulation of enzymatic activity and the recognition of one molecule by another. The trypsinogen–trypsin system is an example of regulation by flexibility. The rigid enzyme is the active species; the proenzyme with its disordered activation domain is unable to bind substrate sufficiently tightly. Covalent modification of trypsinogen produces a triggered conformational change which leads to activation. Thus, mobility is used to reduce the activity of a molecule until it is needed. Crystal structures of other zymogen/active-enzyme pairs should reveal if this principle is a general one.

Recognition

Antibodies provide striking examples of large-scale collective motion in proteins. The structure of the intact IgG molecule Kol showed no significant electron density for the entire 50,000 Da Fc portion of the immunoglobulin. This domain is connected to the Fab portion of the antibody by a hinge region that is thought to be flexible.[4] The Fab "arms" of the protein also have some flexibility as different elbow bend angles are observed between the constant and variable regions in different Fab structures.[85] The functional significance of independent arm and stem movement may lie in the ability to reach antigenic determinants in different

[85] M. Matsushima, M. Marquart, T. A. Jones, P. M. Colman, K. Bartels, R. Huber, and W. Palm, *J. Mol. Biol.* **121,** 441 (1978).

arrangements.[86a,b] This is an example of the importance of a large-scale correlated motion for recognition.

A particularly elegant example of the role of mobility in recognition comes from crystallographic studies of viral proteins and intact viruses: viral proteins, which interact with the viral nucleic acid to build the stable virus particle, are all found to have at least one disordered region. The mobility of this segment of the protein is believed to allow bonding to differently arranged RNA strands.

Allostery

Examples of triggered conformational change providing recognition and regulation are induced fit[86a,b] and allostery. As originally stated, ligand binding stabilized the active form of a conformational equilibrium. For induced fit, inhibitors lacked some essential interactions needed to stabilize this form. Extension to a disordered region in a protein only requires that the correct allosteric effector or substrate order the flexible segment, while inhibitors are unable to do so.

Catalysis

The most controversial role for atomic fluctuations in proteins is in enzymatic catalysis. A side chain correlated motion of the catalytic His-57 in trypsin appears to be essential for serine protease catalysis.[49] The initial structure in which this ring is hydrogen bonded to Ser-195 must rearrange to one in which the imidazole hydrogen bonds to the peptide amide of the leaving group peptide. An aromatic ring flip would accomplish this. Gilbert and Petsko postulate that flexibility of the side-chain of Lys-41 is essential for ribonuclease A catalysis of RNA hydrolysis.[46] The substrate undergoes large changes in stereochemistry and charge configuration during the reaction, and a flexible lysine is needed to follow these changes without interfering with them sterically. Thus one function for flexibility in catalysis is to provide an active site capable of responding to alterations in substrate structure as the reaction proceeds.

An interesting role for a disordered region has been proposed for the E2 component of the multienzyme complex 2-oxoacid dehydrogenase,[87] where a lipoyl-lysine is thought to move rapidly over considerable distances. Limited proteolysis and proton NMR experiments support this

[86a] D. E. Koshland, Jr., *Proc. Natl. Acad. Sci. U.S.A.* **44**, 98 (1985).

[86b] L. M. Amzel and R. J. Poljak, *Annu. Rev. Biochem.* **48**, 961 (1979).

[87] G. C. K. Roberts, H. W. Duckworth, L. C. Packman, and R. N. Perham, in "Mobility and Function in Proteins and Nucleic Acids" (R. Porter, M. O'Connor, and J. Whelan, eds.), p. 47. Pitman, London, 1982.

model. Thus, a large-scale correlated motion of part of an enzyme structure may function to provide a mechanism for active site coupling.

Alber has pointed out that highly mobile regions of an enzyme may play a thermodynamic role in facilitating catalysis by weakening substrate and product binding.[78] Enzymes obtain specificity by offering a number of discrete contact points for interaciton with the substrate, and by having active sites whose size and shape prohibits binding of grossly incorrect molecules. However, if the interactions between enzyme and substrate produce very tight binding of the small molecule, product dissociation is likely to be very difficult. Alber suggests that the observation that several crystalline enzymes contain disordered segments of polypeptide chain that become ordered on substrate or product complexation may be explained by the advantage of entropy loss. If some of the intrinsic binding energy of substrate with enzyme is expended to order a mobile region of the protein, the net binding strength will be weaker than the sum of the interactions would suggest. The enzyme may thus have specificity without excessively tight binding. This argument may also be viewed kinetically: if the rate-limiting step for the reaction is, as often observed, product release, then product release from an enzyme with a disorder-to-disorder transition will be faster than product release from an enzyme where one static conformation is converted to another. Consider the 10 amino acid disordered loop in triose-phosphate isomerase that folds over the substrate and becomes highly ordered. In order for product release to occur, the loop must move out of the way. A thermally driven fluctuation in atomic positions can break transiently the bonds between the loop and the substrate, but the tendency of the loop to resume its open, native conformation when this happens will be increased by the gain in entropy that results from regenerating the multiple conformations of the disordered form. Flexible segments thus may contribute directly to the optimization of enzymatic reaction rates.

Finally, there remains a possibility that the protein structure may favor those modes of motion which lie along the reaction coordinate at the expense of those perpendicular to it. This is the so-called "directed fluctuations" hypothesis.[79] In this view, the intramolecular contacts and arrangement of secondary structural features would lead to a distribution of vibrational modes in the active site which tended to favor the movements of atoms required during catalysis. A review of the various ways in which particular classes of fluctuations in the protein structure might provide means for collimating the thermal energy from solvent bombardment to produce high free-energy events at the active site has been given by Welch et al.[88] There is no experimental evidence in favor of this hypothe-

---

[88] G. R. Welch, B. Somogy, and S. Damjanovich, Prog. Biophys. Mol. Biol. 39, 109 (1982).

sis, or against it. Analysis of the anisotropies of protein $\langle x^2 \rangle$ values in enzyme–substrate complexes stabilized at subzero temperatures may give some evidence.

Transmission of Information

Consideration of proteins as flexible molecules able to fluctuate about an average conformation provides one explanation for the transmission of information from one portion of the structure (say, an allosteric site) to another (the active site). Until recently, the commonly held view was that a signal was always sent from the distant site by disruption or formation of a linked series of noncovalent interactions: a kind of molecular "Rube Goldberg Device," in which rupture of a hydrogen bond leads to weakening of a salt bridge followed by a shift in position of a helix followed by, etc. There is no doubt that such information relays exist; hemoglobin is a well-studied example. However, a protein that contains regions undergoing large, random, collective motions could function without such a network. If the desired new conformation is a member (albeit an infrequent one) of the ensemble of substates that are in equilibrium, then ligand binding may simply shift the equilibrium so that the bound conformation is now the center of a new, possibly narrower distribution.

Summary

Properly carried out, high-resolution X-ray diffraction data collection followed by careful least-squares refinement can give the spatial distribution of the high-frequency mean-square displacements in a protein. These displacements reflect both individual atomic fluctuations in hard variables (bond lengths and bond angles) and collective motions involving soft variables (torsion angles, nonbonded interactions). Lower frequency, large amplitude motions and rapid but improbable motions are not quantifiable, but they may lead to such complete disorder that their existence can at least be inferred from the absence of interpretable electron density for some sections of the structure. Interior residues are more rigid than groups on the surface, and structural constraints are reflected in restricted motion even for surface residues. Amplitudes of motion of 0.5 Å or greater are not uncommon. The temperature dependence of these fast motions varies considerably over the structure. In general, large $\langle x^2 \rangle$ values have large temperature dependence, while small displacements are less affected by temperature; however, exceptions are common. Significant reduction in $\langle x^2 \rangle$ on cooling establishes that proteins are mobile even in the crystalline state, and that static disorder is not the dominant contributor to the individual mean square displacements. Disordered regions in electron density maps are no longer automatically taken as signs of errors in structure determination. It is now recognized that the absence of strong electron density is often an indicator of conformational flexibility. Some of the functional roles for protein dynamics are beginning to be understood.

Missing from these results are the physicochemical details that can be extracted from thermal motion analysis of small molecule crystal structures. Application of these methods to protein data is very difficult, but it is well to remember that just over 10 years ago it was commonly felt that protein structures could not even be refined. Certainly some small, well-diffracting proteins should be amenable to many of the sophisticated small-molecule analyses, as they yield X-ray data to resolutions comparable to simple organic structures.

The most important type of analysis that awaits is anisotropic B factor refinement, which would give the principal directions of motion added to the amplitude information now obtained. Unfortunately, refinement of unrestrained anisotropic thermal elipsoids requires six parameters for each atom instead of a single isotropic B parameter, and even 1.5 Å resolution data do not provide enough overdeterminacy. An alternative approach is to refine selected residues anisotropically while holding the rest of the structure fixed. The only published procedure for restrained anisotropic protein refinement implements restraints based on bond directions[29]; this approach is invalidated by the result from molecular dynamics that collective motions destroy such a correlation. Optimally, very high-resolution data should be coupled with unrestrained anisotropic refinement for at least a few proteins. These studies would provide the information to develop restrained or group-atom methods in the light of unbiased information. Blundell and associates have undertaken precisely this study on avian pancreatic polypeptide, measuring data beyond 1 Å resolution and refining the structure by small-molecule procedures, including full anisotropic thermal elipsoids in the model.[89]

X-Ray diffraction can provide atomic-resolution information about the initial and final states of a process such as ligand binding or triggered conformational change, but the pathway is much more difficult to unravel. Ringe has shown that a bulky ligand can be used to prevent the return of a channel to its resting state after binding, allowing the residues that move to form the channel to be identified.[14] Ligands with chemical "tails" might be used to investigate other channels in proteins. Intermediate states in triggered conformational changes may be trapped at low temper-

[89] I. Glover, I. Haneef, J. E. Pitts, S. Wood, D. S. Moss, I. J. Tickle, and T. L. Blundell, *Biopolymers* **22**, 293 (1983).

atures[90,91] or by careful choice of inhibitors, or by examination of abortive complexes in multisubstrate reactions.

The structure determined by X-ray diffraction studies on a crystalline protein is that of a nonexistent average molecule. It is this structure that will best fit the measured X-ray scattering amplitudes. The average structure contains contributions from all of the different conformational substates that are generated by the thermally driven fluctuations and collective motions of the molecule during the time of data collection,[92] each being weighted according to its probability and lifetime. Any static structure with ideal bond lengths and angles is at best merely one representative from this ensemble of conformations. Most spectroscopic and chemical probes of structure and activity will also reflect the properties of the average molecule, but they may not be sensitive to all members of the conformational distribution and so may see a slightly different average. In certain cases, only one or a few conformations may be seen by a particular technique, or may react with a particular substance. This is the picture of protein structure that has emerged from crystallographic and other studies of protein dynamics.

Acknowledgments

We are grateful to John Kuriyan, Tom Alber, Sir David Phillips, Martin Karplus, Hans Frauenfelder, Robert F. Tilton, Jr., I. D. Kuntz, Jr., Peter Zavodszky, J. Andrew McCammon, and Emil Pai for stimulating discussions.

[90] A. L. Fink and G. A. Petsko, *Adv. Enzymol.* **52,** 177 (1981).
[91] P. Douzou and G. A. Petsko, *Adv. Protein Chem.* **36,** 245 (1984).
[92] A. Cooper, *Sci. Prog. Oxford* **66,** 473 (1981).

[20] Protein Dynamics Investigated by Neutron Diffraction

By A. A. KOSSIAKOFF

Introduction

Characterization of the dynamic properties of protein molecules currently represents one of the most intensely investigated subjects in biophysics. This research activity has been stimulated by a growing body of evidence indicating that in certain cases transient conformational variation plays an important role in biological function.[1-4] In addition, it is apparent that a description of the extent and frequency of various types of dynamic events will be important to provide a better understanding of the forces that confer on a protein its stable, functional conformation. At present, there is little experimentally derived information addressing this issue.

The subject of this chapter deals with the description of two classes of conformational fluctuation that are uniquely suited for investigation by the neutron diffraction technique. The first category of conformational fluctuation will be referred to as "breathing" and is defined here as a set of motions which is involved in a transient redistribution of the packing density in the interior and at the surface of the protein. These deformations by themselves do not involve significant denaturation of the protein structure.

In contrast to protein "breathing" motions, fluctuations described by the term "regional melting" (category II) do involve localized denaturation of the protein's tertiary structure. This usually results from the breaking of some number of hydrogen bonds in the secondary structure. It has been shown that even under physiological conditions, certain segments of the polypeptide chain of a protein molecule undergo conformational excursions, producing deformations extensive enough to solvate groups in the protein interior that in the time-averaged structure are well buried.[5-7] It has not been established experimentally whether these two classes of conformational fluctuation occur through the same type of mechanism differing only in extent, or whether they are the result of different phenomena.

Several types of physicochemical techniques have been exploited in studying protein fluctuations.[1] A major shortcoming with most of these techniques has been that, although they can identify the existence of conformational fluctuations, they cannot readily correlate the motion with the specific region where it occurs. What has been needed are techniques which are not only sensitive to the existence of fluctuations, but can also accurately identify and characterize the regions in which they occur.

[1] F. R. N. Gurd and T. M. Rothgeb, *Adv. Protein Chem.* **33,** 73 (1979).
[2] R. Lumry and R. Biltonen, *in* "Structure and Stability of Biological Macromolecules," p. 165. Dekker, New York, 1969).
[3] R. L. Baldwin, *Annu. Rev. Biochem.* **44,** 453 (1975).
[4] L. A. Blumenfeld, *J. Theor. Biol.* **58,** 269 (1976).
[5] A. A. Kossiakoff, *Nature (London)* **296,** 713 (1982).
[6] A. Hvidt and K. Linderstrøm-Lang, *Biochim. Biophys. Acta* **14,** 574 (1954).
[7] S. W. Englander, N. W. Downer, and H. Teitelbaum, *Annu. Rev. Biochem.* **41,** 903 (1972).

Major advances in identifying specific regions of fluctuation have been made using high-resolution NMR and sophisticated chemical analyses.[8-11] It should be recognized, however, that these methods employ state of the art methodologies that are generally difficult to perform in practice. As an alternative, neutron diffraction in many ways offers a more straightforward approach to the study of certain aspects of protein dynamics. The advantage of neutron diffraction is derived from its capability to locate directly hydrogen atoms in protein molecules, and further, it has the ability to distinguish hydrogens from deuterium atoms.[12,13] Just how these characteristics can be exploited to obtain interpretable experimental data on protein dynamics will be described later in the text.

The developed methodology discussed here is compiled from several crystallographic studies.[14,15] It may at first seem a bit contradictory to attempt to obtain dynamic information from a crystalline protein system; however, there are a number of reasons that suggest that this is quite an appropriate approach. First of all, a significant percentage ($\approx 50\%$ on average) of the volume of a protein crystal is solvent; thus, protein molecules in the crystal form are usually highly solvated systems. Results from a large number of structural analyses indicate that protein crystals are best described as an ordered but open array of molecules held together by a relatively small number of intermolecular contact points.[16,17] Because of the extent of unoccupied volume in the crystal, ions and substrates can readily diffuse through the solvent channels to interact with the protein.

Given the above description of a protein's environment in a typical crystal, protein molecules would be expected to exhibit many of the same chemical and physical properties that they do in solution and this supposition has been supported by a number of investigations.[18] Further, several dynamic studies using H/D exchange have been performed to compare

[8] C. K. Woodward and B. D. Hilton, *Annu. Rev. Biophys. Bioeng.* **8,** 99 (1979).

[9] K. Wüthrich and G. Wagner, *J. Mol. Biol.* **130,** 1 (1979).

[10] A. A. Schreier and R. L. Baldwin, *J. Mol. Biol.* **105,** 409 (1976).

[11] J. J. Rosa and F. M. Richards, *J. Mol. Biol.* **133,** 399 (1979).

[12] B. P. Schoenborn, *Brookhaven Symp. Biol.* **27,** (1976).

[13] A. A. Kossiakoff, *Annu. Rev. Biophys. Bioeng.* **12,** 159 (1983).

[14] A. A. Kossiakoff, *in* "Neutrons in Biology-Basic Life Sciences" (B. P. Schoenborn, ed.), Vol. 27, p. 281. Plenum, New York, 1984.

[15] M. M. Teeter and A. A. Kossiakoff, *in* "Neutrons in Biology-Basic Life Sciences" (B. P. Schoenborn, ed.), Vol. 27, p. 335. Plenum, New York, 1984.

[16] F. M. Richards, *Annu. Rev. Biochem.* **32,** 269 (1963).

[17] M. F. Perutz, *Trans. Faraday Soc.* **B42,** 187 (1946).

[18] J. A. Rupley, *in* "Structure and Stability of Biological Macromolecules," p. 291. Dekker, New York, 1969.

crystal and solution dynamics.[19,20] Interestingly, it was found that a protein's dynamic properties in the crystalline or solvated state are closely comparable. The above evidence supports the validity of using crystallized proteins to study certain dynamic processes.

The following sections are organized to discuss the different experimental approaches used to study protein dynamics by the neutron diffraction technique. In the case of the breathing fluctuations, the conformation and relative degree of order of side chain methyl groups are used as a probe for local motion. Conformational fluctuations involved in the regional melting process are measured by employing the H/D exchange method, data being provided through the ability of neutron diffraction to discriminate between hydrogen and deuterium atoms.

Protein Breathing

Evaluation of the rotational properties of interior side chains can be a useful probe of protein breathing motions.[22-25] It was determined several years ago by NMR spectroscopy that in many proteins bulky ring structures (i.e., Phe and Tyr) can undergo ring "flips" (180° rotations around C_β—C_γ bonds) even when these groups had, in the time-averaged structure, numerous packing close contacts.[26,27] This finding suggested that there operates a set of concerted forces in protein molecules that are of sufficient amplitude to cause rapid redistribution of the molecules' interior packing density.

Neutron diffraction can offer no direct insight into the type of structural fluctuations involved in the flipping of rings. The technique, however, does offer the possibility to study in detail the rotational properties of another interesting class of side chain groups—terminal methyl groups. NMR studies have shown that methyl groups in proteins rotate rapidly (1–15 psec) around their rotor axes.[1,25] However, interpretation of NMR data is limited to giving information about the time frame of reorientation; it cannot address the question as to whether these groups have preferred torsional conformations. The answer to this question has direct bearing

[19] G. H. Haggis, *Biochim. Biophys. Acta* **23**, 494 (1957).
[20] M. Praissman and J. A. Rupley, *Biochemistry* **7**, 2431 (1968).
[21] E. Tuchsen, A. Hvidt, and M. Ottensen, *Biochimie* **62**, 563 (1980).
[22] G. Wagner, A. DeMarco, and K. Wüthrich, *Biophys. Struct. Mech.* **2**, 139 (1976).
[23] E. Oldfield, R. S. Norton, and A. Allerhand, *J. Biol. Chem.* **250**, 6368 (1975).
[24] R. J. Wittebort, T. M. Rothgeb, A. Szabo, and F. R. N. Gurd, *Proc. Natl. Acad. Sci. U.S.A.* **76**, 1059 (1979).
[25] G. Lipari, A. Szabo, and R. M. Levy, *Nature (London)* **300**, 197 (1982).
[26] I. D. Campbell, C. M. Dobson, and R. J. P. Williams, *Proc. R. Soc. London Ser. B* **189**, 503 (1975).
[27] J. A. McCammon and M. Karplus, *Biopolymers* **19**, 1375 (1980).

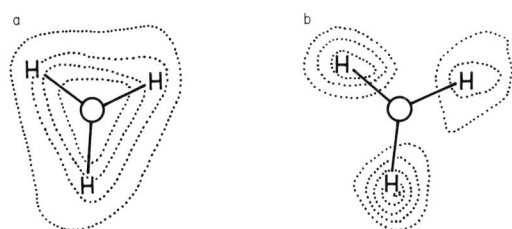

FIG. 1. (a) Neutron density of a typical methyl rotor group in trypsin. The trypsin structure was highly refined at 1.8 Å resolution. (b) Density of a methyl group in the crambin structure refined at 1.4 Å. Note that what seems to be a modest (0.4 Å) increase in resolution actually substantially improves the interpretability of the density map.

on gaining insights into the stereochemical rules applied in protein folding.

Assignments of Methyl Rotor Conformations

Given that in general the rotational order of a methyl group is short while the process of collecting data is quite long (weeks), it might be expected that the scattering from these groups would give a quite disordered pattern in a neutron density map. In actual fact, since the diffraction experiment measures the time-averaged structure, this is not the case. In order for a particular conformation to be identified in the map, what matters is the ratio of the time spent in the preferred conformation versus the total time spent in all other orientations. If this ratio is sufficiently large, the result should be an interpretable set of peaks in the neutron density maps.

High-resolution data and a well-refined phasing model are a prerequisite for any type of detailed methyl rotor analysis.[28] For example, Fig. 1 shows the scattering density of methyl hydrogens at two different resolutions. (The neutron scattering length of a hydrogen is −3.8 fermi and therefore appears as a negative density peak in a neutron Fourier map.) The methyl rotor density in Fig. 1a is taken from the 1.8 Å analysis of trypsin, the density in Fig. 1b is from the 1.4 Å analysis of crambin.

In Fig. 1a, although there is an obvious asymmetric shape to the density that suggests some degree of ordering in a preferred orientation, the map is not of sufficient resolution to distinguish between the orientation shown in the model from one differing by about ±20°. On the other hand, it is apparent from the rotor density shown in Fig. 1b that at a higher resolution (an additional 0.4 Å resolution can make a significant difference) the orientation of the rotor hydrogens can be assigned with high

[28] A. A. Kossiakoff, *Nature (London)* **311**, 582 (1984).

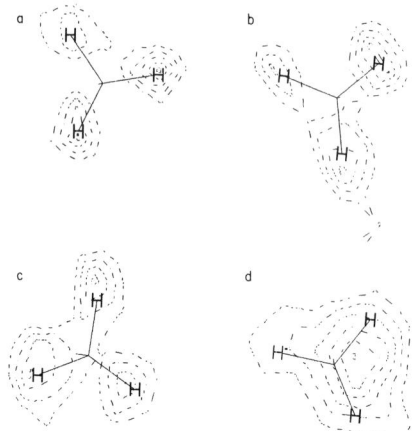

FIG. 2. Sections of a neutron difference Fourier map showing methyl hydrogen densities for several representative methyl groups in crambin. No phasing information about the methyl hydrogens was included in the model (as was also the case in Fig. 1); therefore, hydrogens should appear in the difference map at their true positions but at reduced density. The crambin analysis represents the first time that methyl hydrogens have ever been observed at atomic resolution in a Fourier density map. The groups shown are (a) Ala-24, (b) Thr-21, (c) Thr-28, and (d) Ala-45.

precision. Because of the superior interpretability of the density maps at 1.4 Å resolution, most of the following discussion and examples will be taken from work performed in the crambin analysis.[15]

Developing Criteria of Confidence for Rotor Assignments

Before meaningful interpretations can be made concerning the character of the density distribution around methyl groups, several factors which could possibly lead to misinterpretation have to be evaluated. Of primary concern is phasing bias; this comes from the basic fact that the density model derived from a Fourier map is dominated by the input phases, not the amplitudes. Therefore, it is essential that a method be developed that can allow for interpretation independent of a preconceived model of methyl rotor orientation. Referring to the density drawings in Fig. 2, in order to guard against phasing bias no information about methyl hydrogen locations was incorporated in the phasing model used to calculate the maps. This treatment effectively reduces the overall peak heights in the neutron Fourier map by about 1/2, but should not perturb the peak position or overall shape of the scattering profile (Fig. 3). Note that the density profiles show differing contour patterns and peak intensities. There are three primary factors that will attenuate the scattering density of rotor hydrogens as seen in Fig. 3. Two of the factors, general

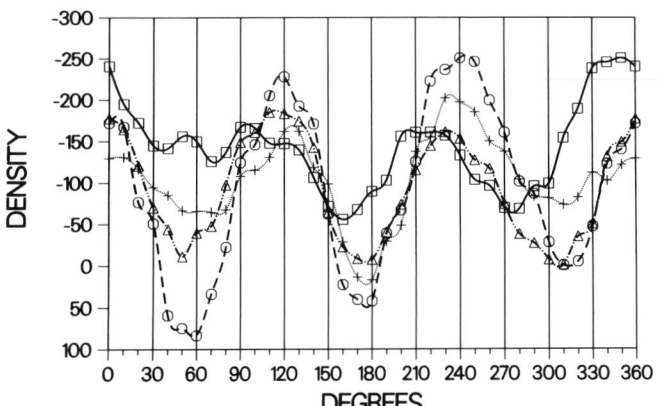

FIG. 3. Examples of representative methyl rotor plots calculated by sampling the map densities in incremental steps around the rotor axes. In practice, these plots are derived by calculating a series of possible methyl hydrogen positions at points differing by a rotational increment of 10°, and then sampling the observed density at each point in the difference map. The fact that many of the rotor plots have a 3-fold symmetrical pattern is due to real structural features, since there is no inherent bias in the phasing to produce the 3-fold character. (□) Val-15 CG1; (△) Thr-21; (○) Ala-27; (+) Ala-45.

rotational disorder and vibrational disorder of the parent methyl carbon, result directly from the physical characteristics of the protein motion. The third factor, incorrect phasing, is associated with refinement errors. These factors are discussed separately below.

1. General rotational disorder involves two types of fluctuations, oscillatory and static disorder. In theory, the effects on methyl densities by these two types of fluctuation can be deconvoluted by integration of the density contained in the immediate vicinity (the so-called density torus) of methyl hydrogens. In the case of oscillatory disorder, the total integrated density contained in a torus defined by the rotor axis would show a decrease as a consequence of the attenuating scattering potential due to high thermal factors of the group atoms. Such an effect is shown in Fig. 4.

The resulting density from statistically disordered methyl groups manifests itself in a different form. For example, if two distinct conformations of equal occupancy exist for a particular methyl rotor, the result would be a density distribution having two maxima of equivalent height separated by the appropriate angular displacement of the two conformers. However, because of the limited resolution (>1.2 Å) of almost all protein structure analyses, the two peaks are not observed as independent features in rotor plots but as an averaged set of density peaks located in a bisecting position between the two actual conformers. To summarize, the

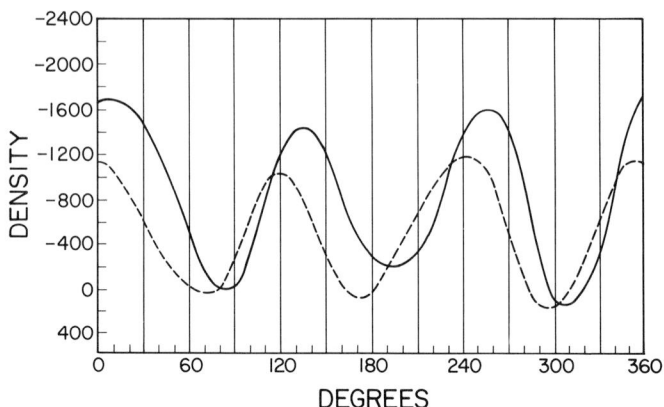

FIG. 4. Comparison of the effects of vibrational and static disorder on methyl rotor plots. The dashed line shows the density profile for a methyl group undergoing the effects of vibrational motion represented by a temperature factor of 15 [a temperature (or B) factor value can be directly related in a positional displacement and an attenuation in scattering intensity]. The solid line depicts the density profile of a statically disordered group where there are two equally occupied positions (with $B = 5$) displaced by 30°. It is clear at the resolution of the plots (1.4 Å) that the two statically disordered positions cannot be identified. The result is a single peak bisecting the two half-occupied positions.

effects of oscillatory and static disorder cannot be differentiated even at high resolution.

2. The effects of vibration disorder of the methyl group are best described by an example. For instance, take a valine residue. A small oscillatory movement around the C_α—C_β bond would effectively displace both γ carbons leading to a smearing of methyl hydrogen densities independent of any rotational disorder of the rotors themselves. This effect can usually be identified in a refined structure by a large temperature factor for the methyl carbons. In point of fact, in the crambin structure there seems to be a definite correlation between side chain length and methyl density. What is observed is that rotors having the highest order by density criteria are those associated with alanines, where side chain motion is strongly coupled to the main chain.

3. Errors in refinement usually involve some misplacement of atom coordinates. Just how much refinement errors will affect assignment of rotor orientations has been investigated by a series of model studies with the following findings. Significant errors in interpretation of rotor orientations appear only at displacements out of the correct rotor plane (rotor plane defined by the plane of the three hydrogens) greater than $\cong 20°$. For displacements of this magnitude to occur there would have to be significant error in the methyl group position. In most reasonably well-refined

structures, errors of this magnitude can be readily identified. Clearly nothing can be said about the true disorder properties of a methyl group unless the methyl carbon position can be accurately defined.

Procedures to Identify Conformationally Preferred Orientations for Methyl Rotors

Rotor densities can be displayed graphically by rotor density plots of the type pictured in Fig. 3. These profiles are simply the density sampled every 10° around the rotation axis at the stereochemically appropriate location for a possible methyl hydrogen position. Most profiles show a pattern for hydrogen positions every 120° around the rotation axis. This indicates a high degree of order. However, previous studies have shown there are a statistically significant number of cases which, by crystallographic criteria, are disordered.[28] As has been discussed, this does not necessarily indicate rotational disorder alone, but could be a function of the inherent vibrational motion of the parent methyl carbon itself and hence of the entire methyl group.

Based on the neutron densities in Fig. 2, Ala-45 CB (c) is the least well ordered methyl group pictured. Its corresponding rotor density plot (Fig. 3), however, shows a rather symmetric profile from which a ''preferred'' orientation can be identified. It was found in several instances that although the fine structures of this type of curve showed varying degrees of asymmetry, the overall shapes of the profiles generally indicated a systematic trend of peaks and valleys. The confidence factor in assigning a specific orientation to a rotor comes from the fact that the peaks and valleys, irrespective of their relative amplitudes, usually appear at intervals of 120°. If no bias has been introduced into the maps, the probability of this occurring by chance is remote.

The identification of the highest populated rotor orientation of each methyl group can be assigned most accurately by a 3-fold averaging procedure. Figure 5 shows rotationally averaged plots for three representative side chain types: alanine, threonine, and valine. Taken together, these plots clearly suggest that the preferred conformation in the vast majority of cases is staggered. Note, however, that there are a few exceptions to this general observation. The clearest example of this is the CG1 rotor of Val-15 which is displaced by about 25° from the staggered orientation. This deviation likely represents a real effect. Referring to Fig. 3, the symmetric nature of the density peaks, each of which is displaced by about 20–30°, is too unmistakably systematic to be due to an artifact. A plausible explanation for this occurrence is that it was found that the CG1 methyl hydrogens make van der Waals contact with a portion of a symme-

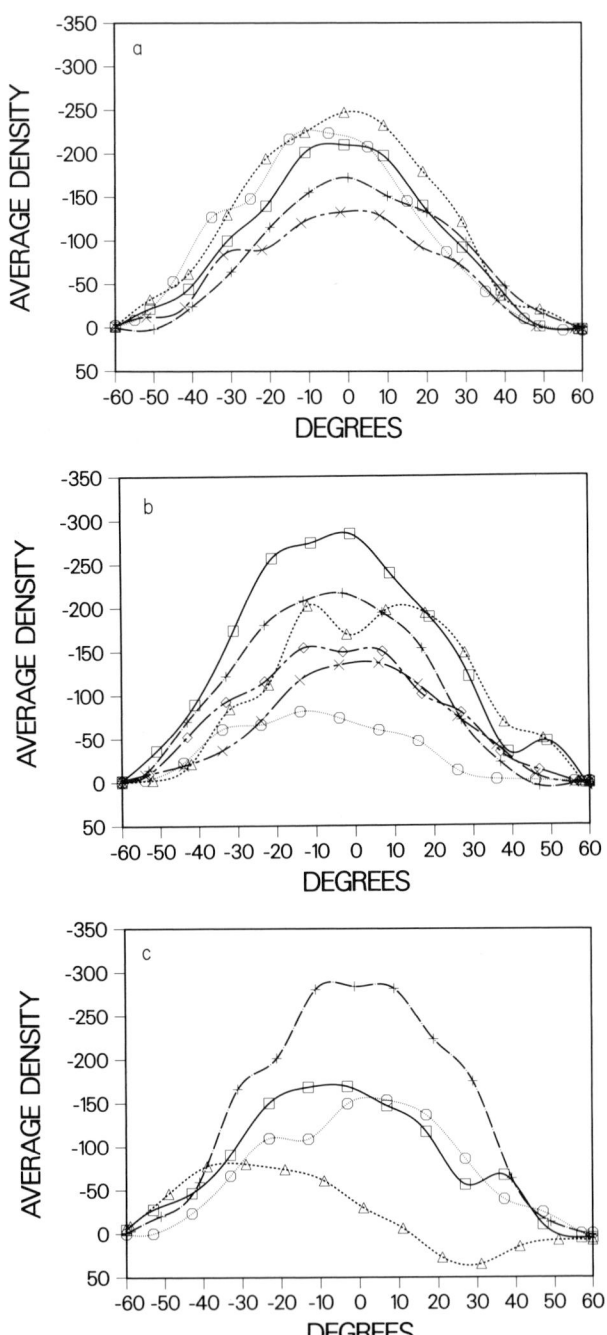

try related molecule in the crystal packing lattice in crambin. On the other hand, at other similar intermolecular contact regions (of which there are several), methyl groups were observed to have staggered conformations; so there is no general rule which can be applied to these types of interactions.

In summary, methyl rotor plots can give quantitative information about the preferred conformation of these groups and can identify cases in which these groups are effected by thermal or static disorder. Interpretation of these rotor data has led to the finding that a majority of rotors adopt a "staggered" conformation in the time averaged structure, suggesting that the principal determinant in conferring conformation is the intrinsic barrier to rotation rather than local structure environment.[28] Experimental observations from NMR (showing rapid rotatory motions) together with these neutron results (indicating highly preferred orientations) depicts methyl rotations as being "quantized" in 120° steps about a position of highest stability.

Regional Melting

The term regional melting describes a class of molecular fluctuations that involves transient denaturation of limited regions of the protein structure. Such a process usually entails the breaking of some number of hydrogen bonds and thus represents as a class a set of conformational changes larger than those associated with protein "breathing" as discussed above. There has been considerable controversy in the literature concerning the extent and nature of the transient denaturation process of proteins at physiological conditions.[7,8,29,30]

Progress toward clarifying the details of the nature of regional melting requires an experimental technique that can identify those segments of the polypeptide chain involved in such transient fluctuations. Since its introduction by Linderstrøm-Lang and colleagues in the 1950s,[6] the hy-

[29] A. Hvidt and S. O. Nielson, *Adv. Protein Chem.* **21**, 287 (1966).
[30] S. W. Englander, D. B. Calhoun, R. K. Englander, N. R. Kallenbach, and R. K. Liem, *Biophys. J.* **32**, 557 (1980).

Fig. 5. Rotationally averaged rotor plots for (a) alanine, (b) threonine, and (c) valine. Plots were calculated by averaging densities at points separated by 120° in the Fig. 2 type plots (i.e., 0, 120, 240°; 10, 130, 250°; etc.). Plots were placed on the same relative scale by adjusting the lowest average density point for each individual group to be at 0. Effects on the methyl densities due to the inherent vibrational motion of the parent carbon atoms were corrected for by an empirical procedure. This was done by developing an empirical function relating the expected attenuation of the methyl peak density to the vibrating motion (B factor) of the parent carbon atom.

drogen exchange (H/D) method has been recognized to offer great potential in this regard. This is because H to D exchange rates can differ by as much as 10 decades in an intact protein,[7] corresponding to the differences in the degree of shielding from the solvent provided at a given site by hydrogen bonding and steric factors in the tertiary structure. H/D exchange has major advantages over other labeling techniques because the D has chemical and physical properties almost identical to those of the proton it replaces in the protein structure, and because potentially labile sites are distributed throughout the protein and thus allow probing of the exchange properties over the whole molecule.

A serious drawback of past H/D exchange experiments has been their inability to relate exchange rates with specific groups or even regions of the polypeptide chain. The absence of such information has limited the utility of exchange methods in clarifying the factors responsible for protein conformational mobility. This limitation has been generally recognized, and recently NMR spectroscopy[31,32] and chemical analyses[10,11] have begun to provide data relating exchange rates to specific structural features in the protein. However, both the NMR and chemical studies involve state-of-the-art techniques that can be difficult to perform in practice. Neutron diffraction is not limited in this way; since it is a crystallographic technique, the precise location of each labile site in the well-ordered segments of the protein is known and can be examined in the neutron density map. The process of assigning a labile site as having either H or D character is a rather straightforward task because the amplitudes of H and D are of opposite sign. The practicality of using neutron diffraction for H/D analysis was first shown by Schoenborn and colleagues[12] and subsequently verified by other investigators.[5,33–35]

Determination of H/D Exchange Ratios

To quantify the H/D exchange ratios for amide peptides several different refinement schemes have been employed.[13,36] In all cases, because of the general noise features observed in the neutron density maps, the obtained ratio values had to be viewed as being more qualitative than quantitative. Consequently, the exchange ratios are most accurately subdivided into three major categories: unexchanged, partially exchanged, and fully exchanged; finer gradation of these categories is of questionable significance at resolutions >1.8 Å. By limiting the number of categories in

[31] G. Wagner and K. Wüthrich, *J. Mol. Biol.* **134,** 75 (1979).
[32] B. D. Hilton and C. K. Woodward, *Biochemistry* **17,** 3325 (1978).
[33] J. C. Hanson and B. P. Schoenborn, *J. Mol. Biol.* **153,** 117 (1981).
[34] A. Wlodawer and L. Sjolin, *Proc. Natl. Acad. Sci. U.S.A.* **78,** 2853 (1981).
[35] S. E. V. Phillips, *Neutrons Biol. Basic Life Sci.* **27,** 305 (1984).
[36] A. Wlodawer, *Prog. Biophys. Mol. Biol.* **40,** 115 (1982).

this manner, one retains the essential informational content of the H/D data while guarding against possible overinterpretation of the relevance of small changes in exchange ratio. The exchange characteristics as a function of secondary structure for the proteins trypsin and crambin are shown in Fig. 6.

The primary obstacle to obtaining accurate H/D occupancies is that, at the resolution of the data used in most protein analyses, any errors in the positional or thermal refinement parameters of the parent amide nitrogen are very likely to affect the density characteristics of the peptide proton. Furthermore, experience has shown that useful H/D information can be obtained only from well-ordered segments of the polypeptide chain. A substantial fraction of the groups classified as partially exchanged on the basis of their zero density is usually found to be located in "loosely ordered" regions of the molecule.[5,14] This finding leaves the interpretation of the exchange status of these groups open to question.

In order to circumvent the above problem, several approaches using a refined nondeuterated structure have been devised.[37] The fact that in the nondeuterated structure all exchangeable sites are occupied by hydrogen atoms allows for a straightforward means of assigning a confidence level on which individual site occupancy factors can be assessed. In practice, this was done by inspecting a difference map for which the peptide protons were not included in the phasing model. For those sites in which the observed density represented a full proton, the errors of the refined structure could be assumed to be small and not likely to interfere with the calculation of the H/D occupancy of the site. Conversely, for sites where the proton density was poorly resolved, it could be concluded that there were problems in the structure in the vicinity of the site and that little confidence would be placed in their observed H/D occupancy.

Another method currently being developed uses the differences in the measured H_2O and D_2O trypsin data to refine the occupancies of the exchangeable sites.[38] A difference map is calculated by using the terms $(F_{D_2O}\phi_{D_2O} - F_{H_2O}\phi_{H_2O})$. The differences in the terms reflect, by and large, differences in solvent scattering (H_2O and D_2O have very different scattering lengths) and H/D exchange; the protein contribution is canceled out since it is effectively the same in both the H_2O and D_2O structures. By dealing properly with the solvent component of the difference vector it is possible to isolate the scattering due to the exchanged sites alone. This type of refinement has several advantages over conventional approaches. First, since it is assumed that there is no protein contribution in the $H_2O - D_2O$ difference vector, effects on the peptide proton due to small errors in the protein model are minimized. Second, the number of param-

---

[37] J. Shpungin and A. Kossiakoff, unpublished results.

[38] A. Kossiakoff, unpublished results.

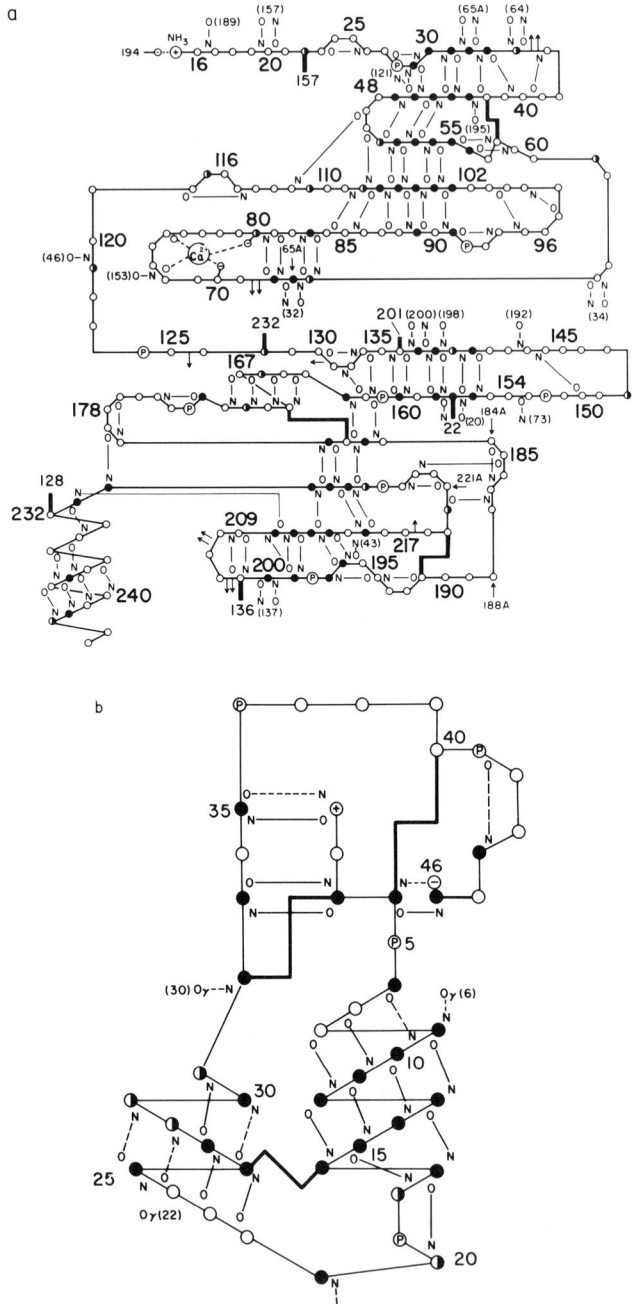

FIG. 6. Schematic representation of H/D exchange at each amide peptide site in (a) trypsin and (b) crambin. Full exchange, ○; partial exchange, ◑; unexchanged, ●; proline, P; insertions, ↓; deletions, ↑; carboxyl side chains, ⊖. Peptide NH and carbonyl oxygens are shown when H bonded. Disulfide bridges are indicated by broad lines.

eters to be refined decreases significantly, from several thousand to several hundred, without a concomitant decrease in the observed number of data, thus greatly increasing the data-to-parameter ratio.

It is tempting to try to quantify the degree of D substitution using some type of least-squares occupancy refinement. Unfortunately, at the resolution of the trypsin analysis ($\cong 2$ Å) (even with the improved data to parameter ratio) attempts to do a full least-squares refinement were not completely satisfactory.[37] However, there are no obvious technical obstacles which should prevent such a refinement strategy from being successful with higher resolution data (i.e., >1.5 Å).

Summary

The most correct model of the molecular structure of a protein molecule is one which describes it as having a number of variable conformational states. These states differ in degree over a large spectrum of structural variation ranging from individual atomic vibrational motion to significant tertiary denaturation. The application of the neutron diffraction techniques discussed above dealt with two classes of conformational fluctuation, "protein breathing" and "regional melting." The utility of the neutron technique stems from the ability to locate hydrogen atoms and to discriminate between hydrogens and deuteriums. This latter attribute allows for performing H/D exchange experiments by identifying individual sites of exchange. With this information it has been possible to discern which regions of the protein molecule undergo regional melting. Protein breathing was explored by analyzing the rotational properties of side chain methyl groups. Such information clearly suggested that most of these groups reside in "staggered" (low energy) conformation in the time averaged structure and are not greatly affected by local structural packing.

Together these two classes of conformational fluctuation span nearly the full range of all motions that might play a role in biological activity. Information about such motions can be obtained from other types of physicochemical methods, but in most cases the interpretation of the data is considerably less definitive than that which can be obtained from a neutron analysis.

Acknowledgments

I would like to thank J. Shpungin and S. Shteyn for technical assistance and to acknowledge the comments of Drs. M. Teeter, J. Hanson, and G. Rose. Some of the work described was carried out at Brookhaven National Laboratory under the auspices of the U.S. Dept. of Energy and NIH Grant GM 29616. The author would also like to thank Genentech, Inc. for support during preparation of the manuscript.

[21] Protein Conformational Dynamics Measured by Hydrogen Isotope Exchange Techniques

By ROGER B. GREGORY and ANDREAS ROSENBERG

Introduction

The hydrogen isotope exchange method was introduced by Hvidt and Linderstrøm-Lang[1] who demonstrated that the rate of the exchange reaction

$$
\begin{array}{c}
R_1 \\
\diagdown \\
\diagup \\
R_2
\end{array}
N{-}H +
\begin{array}{c}
H \\
\diagdown \\
O \\
\diagup \\
{}^2H
\end{array}
\xrightarrow{k}
\begin{array}{c}
R_1 \\
\diagdown \\
\diagup \\
R_2
\end{array}
N{-}{}^2H +
\begin{array}{c}
H \\
\diagdown \\
O \\
\diagup \\
H
\end{array}
\tag{1}
$$

is extremely sensitive to the environment of the exchange site. The method was rapidly established as a useful probe of protein conformational motility and, indeed, for many years provided the only source of evidence to contradict the static pictures of protein structures derived from early interpretations of X-ray crystallographic data. In subsequent years, further evidence of the dynamic nature of protein conformation accumulated, most notably, from the classical studies of fluorescence quenching by Lakowicz and Weber,[2] from NMR studies of aromatic side chain rotation[3,4] and from the analysis of protein crystallographic temperature factors.[5,6] These and other studies of protein conformational dynamics have been reviewed elsewhere.[7-9] The reality of protein conformational fluctuations is now firmly established and much current interest is focused on the implications of dynamic behavior for protein function.[10] Measurement of hydrogen isotope exchange kinetics is one of the principal methods that provides information about protein conformational dynamics and the methodology continues to expand and develop.

[1] A. Hvidt and K. Linderstrøm-Lang, *C. R. Trav. Lab. Carlsberg* **29**, 385 (1954).
[2] J. R. Lakowicz and G. Weber, *Biochemistry* **12**, 4171 (1973).
[3] K. Wüthrich and G. Wagner, *FEBS Lett.* **50**, 265 (1975).
[4] I. D. Campbell, C. M. Dobson, and R. J. P. Williams, *Proc R. Soc. Ser. B* **189**, 503 (1975).
[5] H. Frauenfelder, G. A. Petsko, and D. Tsernoglou, *Nature (London)* **280**, 558 (1979).
[6] P. J. Artymuik, C. C. Blake, D. E. Grace, S. J. Oatley, D. C. Phillips, and M. J. Sternberg, *Nature (London)* **280**, 563 (1979).
[7] G. Careri, P. Fasella, and E. Gratton, *CRC Crit. Rev. Biochem.* **3**, 141 (1975).
[8] F. R. N. Gurd and M. Rothgeb, *Adv. Protein Chem.* **33**, 73 (1979).
[9] M. Karplus and J. A. McCammon, *CRC Crit. Rev. Biochem.* **9**, 293 (1981).
[10] G. R. Welch, B. Somogyi, and S. Damjanovich, *Prog. Biophys. Mol. Biol.* **39**, 109 (1982).

Early exchange studies focused on exchange of peptide group protons but all sufficiently acidic protons can undergo exchange (i.e., those attached to N, O, or S) and a number of techniques have been developed to monitor specific side chain proton exchange reactions of tryptophan indole protons and tyrosine OH protons. Of the carbon-bonded hydrogens, only the imidazole C-2 proton exchanges with measurable rates under conditions appropriate for biological materials. Nevertheless these side chain proton exchange reactions provide additional probes of the internal dynamics of the protein.

There have been major advances in methodology in recent years, particularly in the application of NMR spectroscopic and neutron diffraction techniques to the measurement of single proton exchange rates and in the development of distribution function representations of total protein exchange data. These techniques taken together provide the most powerful approach to the characterization of protein internal motions and yet it must be said that the method still lacks a rigorous mathematical and interpretive framework. There is still no general agreement as to the types of internal motion that mediate exchange or the factors responsible for the attenuated exchange rates observed in proteins and while the recent advances in measurements of single proton exchange rates promise to provide much-needed structural interpretations of the exchange kinetics, it is not yet possible to convert the distributions of exchange rates to distributions of conformational relaxation times. In these respects the full potential of the method has yet to be realized and much of the present research in the field is as concerned with understanding the hydrogen exchange method itself as with its application, as a tool, to the study of biologically important problems.

The purpose of this chapter is to outline the basic principles of the method, its applicability to various biological problems, and to describe the experimental and data analysis procedures that are available for the acquisition and interpretation of hydrogen isotope exchange data. It is not our intention to provide an extensive review of the literature on applications of the method to specific model and protein systems. These have been well reviewed elsewhere.[11–17]

[11] A. Hvidt and S. O. Nielsen, *Adv. Protein Chem.* **21,** 287 (1966).

[12] L. Willumsen, *C. R. Tranv. Lab. Carlsberg* **38,** 223 (1971).

[13] S. W. Englander, N. W. Downer, and H. Teitelbaum, *Annu. Rev. Biochem.* **41,** 903 (1972).

[14] A. Hvidt, *in* "Dynamic Aspects of Conformation-Changes in Biological Macromolecules" (C. Sadron, ed). Reidel, Dordrecht, 1973.

[15] C. K. Woodward and B. D. Hilton, *Annu. Rev. Biophys. Bioeng.* **8,** 99 (1979).

[16] A. D. Barksdale and A. Rosenberg, *Methods Biochem. Anal.* **28,** 1 (1982).

[17] C. K. Woodward, I. Simon, and E. Tüchsen, *Mol. Cell. Biochem.* **48,** 135 (1982).

Principles of Hydrogen Isotope Exchange

Regardless of the method or the type of exchange site (peptide NH or side chain protons) being studied, the experimental quantity of interest is the rate of the isotope exchange reaction measured by monitoring the extent of exchange with solvent as a function of time. The exchange of peptide and most side chain protons is specifically catalyzed by H^+ and OH^- ions. These protons can also undergo direct exchange with water. With the exception of histidine C-2 protons, exchange of labile peptide and side chain protons freely exposed to bulk solvent is extremely rapid. These chemical exchange rates are well established from studies of the free amino acids, small amide model compounds, and model polypeptides. By contrast, peptide and side chain protons in folded proteins exchange with considerably slower rates that are widely distributed and may be attenuated by factors of up to 10^{10}. Many of the labile hydrogens that exchange are well buried within the three-dimensional structure of the protein and with the old concepts of proteins as static, rigid structures the wonder was that they exchanged at all. That all labile hydrogens, given sufficient time, do exchange with solvent indicates that proteins must fluctuate about their time-averaged structures to sample conformations that allow exchange site–catalyst encounters to occur. The attenuated exchange rates therefore reflect the dynamics of transitions among the conformational states available to the protein. The exchange rates are extremely sensitive to the local chemical environment of the exchange sites and to the overall conformational dynamics of the protein and can therefore be employed to monitor perturbations in protein conformational dynamics that may be brought about by ligand binding reactions, chemical modification, subunit association–dissociation reactions, or by changes in pH, ionic strength, temperature, solvent viscosity, and cosolvents.

The method has great versatility since all proteins possess exchangeable hydrogens located in all regions of the protein molecule that are amenable to study by at least one exchange technique or another. Two basic approaches to such studies can be recognized that dictate the experimental technique to be employed: those that provide exchange information for specific single residues within the protein and those that yield the total exchange kinetics of the protein.

The study of single site exchange offers a number of advantages over multiple site methods since the position of the exchange site is known within the three-dimensional structure of the protein and exchange rates may therefore be correlated, unambiguously, with local conformational and chemical features. The approach also introduces considerable analytical simplicity because it yields a single apparent first-order rate constant for each site. Deuterium–hydrogen exchange of single protons can be studied by NMR, neutron diffraction, and, to a limited extent, by UV methods. The UV method, as a single site method, is restricted to studies of tryptophan indole and tyrosine hydroxyl protons and only to those proteins where these residues are few in number and their exchange rates are readily distinguishable. A neutron diffraction exchange study is usually restricted to a single time point and in this sense cannot be regarded as a kinetic technique. Yet it provides unique information about the location of classes of exchangeable protons. NMR techniques are the most generally applicable to single proton exchange studies and have provided extraordinary detail about the pH and temperature dependence of individual protons for a number of proteins.[18–20] However, the study of single proton exchange rates usually provides only a fragmentary view of the dynamics of the whole protein (with the notable exception of BPTI where 58 labile protons have now been assigned and their exchange rates determined[19]). Methodological considerations also restrict the use of the principal single proton exchange technique, NMR, to well-characterized, low-molecular-weight ($<15{,}000–20{,}000$) proteins, which excludes or severely limits its application to many complex biologically interesting systems.

Multiple-site exchange techniques overcome many of these problems and can be employed with a wide variety of physical systems including gels, crystals, precipitates or membranes, and solutions of widely different protein concentration. However, the principal advantage of measuring total exchange kinetics is that it provides information about the conformational dynamics of the protein molecule as a whole, not just local regions of it. There is, however, a price to be paid for such a general view, namely, the difficulty of analyzing the parallel and overlapping exchange reactions of large numbers of hydrogens and the inability of these techniques alone to identify exchange rates with particular labile hydrogens in the protein structure.

The analytical difficulties have, to a large extent, been overcome by the development of Laplace inversion techniques[21,22] which provide exchange rate distribution function representations of the data, although to be successful, this method of analysis requires large quantities of high quality data obtained over an extensive time range.

[18] C. K. Woodward and B. D. Hilton, *Biophys. J.* **32**, 561 (1980).
[19] G. Wagner, *Q. Rev. Biophys.* **16**, 1 (1983).
[20] R. E. Wedin, M. Delepierre, C. M. Dobson, and F. M. Poulsen, *Biochemistry* **21**, 1098 (1982).
[21] D. G. Knox and A. Rosenberg, *Biopolymers* **19**, 1049 (1980).
[22] R. B. Gregory, *Biopolymers* **22**, 895 (1983).

The principal multiple-site techniques involve infrared and ultraviolet spectrophotometric monitoring of deuterium–hydrogen exchange and radiometric monitoring of tritium–hydrogen exchange.

Methods

Exchange experiments involve either (1) the dilution of unlabeled material into a ^2H_2O solution (exchange-in) or (2) initial labeling of the material to isotopic equilibrium by incubation in ^2H_2O or ^3H-enriched H_2O and subsequent transfer of labeled material to an H_2O solution (exchange-out). The loss of isotope from the protein or its accumulation in the solvent is then monitored as a function of time, either continuously (NMR, IR, and UV methods) or after separation of the protein from solvent at discrete time points (solvent density and tritium methods).

Practical considerations such as the quantity of material available, its physical state as a gel or solution, its stability to pH and temperature changes, and its molecular weight inevitably determine the techniques and range of conditions that can be employed and thereby dictate the type and quality of data that may be acquired. Rapidly exchanging peptides and polypeptides must be studied at low pH and temperature in order to bring their exchange rates into a convenient timeframe for measurement. For proteins, the exchange rates of which are greatly attenuated relative to free model peptides, it is often necessary to adopt a method and pH and temperature conditions that will allow the slowest exchanging protons to be observed. This is particularly important if exchange rate distribution functions are to be constructed from the data, since the numerical Laplace inversion methods, described later in this chapter, require the signal to have decayed as close to the noise level of the data as possible in order to provide meaningful distribution functions.

The Use of ^2H_2O

Substitution of ^3H or ^2H for H in proteins and water causes important changes in their physical and chemical properties. Equilibrium and kinetic isotope effects on the exchange reaction itself have been discussed in detail elsewhere[16] and we are concerned here with other problems introduced by the use of isotopically enriched solutions. ^3H isotope is always employed as a trace label and has negligible effects on the properties of water in exchange experiments. However, deuterium–hydrogen exchange-in experiments involve the almost total replacement of H_2O by ^2H_2O which can make comparisons with protein chemical and physical properties determined in H_2O difficult.

TABLE I
COMPARISON OF PROPERTIES OF H_2O AND ^2H_2O

Property	H_2O	^2H_2O
Triple point temperature (°C)	0.01	3.83
Maximum density temperature	3.98	11.19
Dielectric constant (20°)	80.2	79.89
Viscosity (30°)	0.7975	0.969
Density (g ml^{-1})	0.99705	1.10445
pK_w (25°)	14.000	14.869
Dissociation enthalpy (kcal mol^{-1}) (25°)	13.526	14.311

A comparison of some physical and chemical properties of H_2O and ^2H_2O is given in Table I. There are significant differences in the ionization constants of acids and bases in ^2H_2O and H_2O and particularly in the ion product for ^2H_2O and H_2O.[23] The activities of $^2H^+$ and O^2H^- ions can, nevertheless, be established by measurements with a glass electrode calibrated with ordinary aqueous buffers by correcting the observed pH reading, pH*, according to the relationship[24]

$$p^2H = pH^* + 0.4$$

O^2H^- activities can then be established by using appropriate values of the ion product for ^2H_2O.[23] The perturbation of pK_a values of ionizable groups on transfer to ^2H_2O often compensates the error in the glass electrode measurements so that some workers report pH* values directly.

At a more fundamental level, changes in dissociation constants of protein ionizable groups on transfer from H_2O to ^2H_2O may perturb the charge state of the protein. Substitution of ^2H for H also affects the stability and geometry of hydrogen bonds in apparently rather complex ways[11] and may, through changes in the hydrogen bond zero-point vibrational energies, alter the conformational dynamics of hydrogen-bonded structures within the protein. As a result, it is possible that the conformational dynamics of the protein are changing during the course of a deuterium–hydrogen exchange experiment.

Solvent Density Methods

Solvent density methods were employed in the earliest classical attempts to measure protein hydrogen exchange. The protein is allowed to exchange completely in ^2H_2O and is then lyophilized. The dry deuterated

[23] A. K. Covington, R. A. Robinson, and R. G. Bates, *J. Phys. Chem.* **70**, 3820 (1966).
[24] P. K. Glasoe and F. A. Long, *J. Phys. Chem.* **64**, 188 (1960).

protein is redissolved in H_2O and samples of the solution are withdrawn at timed intervals and frozen to $-60°$. Each sample is subsequently lyophilized, the sublimed water collected, and its density determined with density gradient tubes of precisely known composition. The out-exchange of ^2H from the protein into the solvent increases the solvent density, which can be measured to an accuracy of 2 ppm. Detailed discussion of the method is given by Hvidt and Nielsen.[11]

Solvent density methods are rarely employed in exchange studies nowadays because of the effort and time needed to prepare and calibrate the gradient tubes and perform the lyophilization procedures. However, a notable feature of the method is its ability to determine absolute numbers of exchangeable hydrogens, since equilibrium isotope effects can be neglected.

Infrared Methods

The Amide II Band. Hydrogen–deuterium exchange can be followed directly by observation of the very strong, amide II infrared absorption band at 1552 cm^{-1}. On deuteration of the peptide group this band, which is dominated by the N–H bending vibrational mode, is shifted to 1450 cm^{-1}. The amide I band, associated mainly with the carbonyl stretching vibrational mode at 1650 cm^{-1}, is very little, if at all influenced by deuteration of the peptide nitrogen and provides an internal concentration standard that lends the necessary precision to the method.

Since H_2O absorbs strongly in the wavelength region of interest, dilution of small volumes of aqueous protein solution into ^2H_2O is not possible and exchange is initiated instead by dissolving dry protein in appropriately buffered ^2H_2O.[25] Exchange of proteins that are sensitive to freeze drying can be initiated by passage of the protein solution through a Sephadex G-25 column previously equilibrated with ^2H_2O.[26] The solution of protein in ^2H_2O is then rapidly transferred to a thin cell (i.e., 0.1 mm) with CaF_2 windows and the sample spectrum between 1700 and 1400 cm^{-1} recorded as a function of time.

The fraction of exchange at each time point is expressed by

$$x(t) = \frac{A_{II}(t) - A_{II}(\infty)}{\omega A_I} \qquad (2)$$

where $A_{II}(t)$ is the intensity of the amide II band at time t, $A_{II}(\infty)$ is the intensity of the amide II band of the fully deuterated protein, A_I is the

[25] M. Ottesen, *Methods Biochem. Anal.* **20**, 135 (1971).
[26] P. Zvodsky, J. Jahansen, and A. Hvidt, *Eur. J. Biochem.* **56**, 67 (1975).

intensity of the amide I band, and ω the ratio $A_{II}(0)/A_I(0)$ for the fully protonated protein in ^2H_2O.

The intensities of the amide I and II bands can be estimated from the peak heights or peak areas. A_I is measured with reference to a baseline drawn between the absorption minima on either side of the amide I band, while $A_{II}(t)$ may be estimated with $A_{II}(\infty)$ as the baseline. The experimental determination of $A_{II}(\infty)$ is by no means a trivial procedure. Complete deuteration is achieved by performing exchange at high p^2H and/or temperature or by unfolding the protein with denaturants. However, if complete exchange is achieved and the effects of the procedure are fully reversible, subsequent determination of $A_{II}(\infty)$ under the original experimental conditions is quite reliable.

A determination of ω cannot be made directly and its value is generally established by extrapolating the intensity changes to zero time under conditions that minimize exchange. Alternatively, ω can be determined from the spectra of thin films of concentrated protein solutions in H_2O.[27] Values of ω lie between 0.4 and 0.5 for proteins.

Sources of error associated with the method arise from the elevation of the temperature when the thin cell is inserted into the infrared beam, from dissolution of material from the cell windows into the sample and from instrument baseline drift. A more fundamental, but often overlooked source of possible error with the method arises from the implicit, but unestablished, assumption that the change in intensity of the amide II band is directly proportional to the extent of exchange.[11] The infrared oscillator strength of a single oscillator depends on its local molecular environment and the extent and strength of hydrogen bonding, so it is in fact unlikely that each oscillator makes the same contribution to the absorption bands. The method requires relatively high protein concentrations between 3 and 7%. The great advantage of the method is its selectivity for peptide protons.

The Amide A Band. Hydrogen–deuterium exchange can also be followed by measuring the intensity at the N–H stretching frequencies between 3100 and 3400 cm^{-1}. Use of the major amide A band in this region has previously been considered experimentally difficult because of the rapidly growing absorption at the O–H frequency due to protons released from the protein. The O–H band has a maximum at 3400 cm^{-1}, is quite broad, and therefore overlaps and obscures the amide A band. However, the O–H band is symmetric in shape and its contribution to absorption can be estimated.[28] Use of the amide A band has an advantage because it

[27] D. D. Ulmer and J. H. R. Kagi, *Biochemistry* **7**, 2710 (1968).
[28] E. V. Brazhnikov and Y. N. Chirgadze, *J. Mol. Biol.* **122**, 127 (1978).

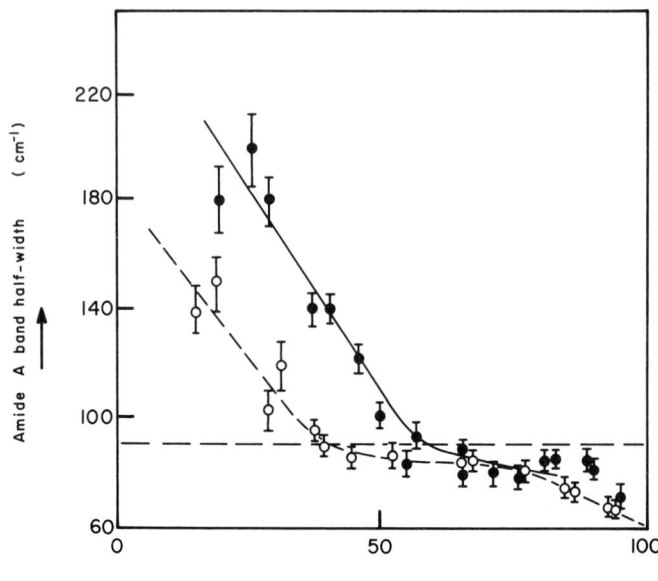

FIG. 1. The dependence of the amide A band half-width as a function of the extent of exchange (data of Brazhnikov and Chirgadze[28]). Bacteriophage T_4 lysozyme, 1 mM KCl, pD 4.4 (O); HEW lysozyme, 1 M KCl, pD 4.4 (●).

contains structural information about the exchange site in addition to accounting for the extent of isotope exchange. Chirgadze *et al.*[29] have shown that the band width of the amide A band varies with the regularity and strength of the secondary structure of which the peptide groups are a part. The variation in band width is substantial, ranging from 59–90 cm^{-1} for α-helical structures to 150–200 cm^{-1} for random structures in proteins[28] and provides an estimate of the degree of geometrical distortion of peptide hydrogen bonds.

Peptide groups in proteins with large amide A band half-widths are found to exchange more rapidly than those with narrow band half-widths indicating that exchange from random structures precedes that from more ordered secondary structures. For example, the dependence of the amide A band half-width on the percentage of exchanged peptide groups (Fig. 1) shows three regions in bacteriophage T_4 lysozyme and hen egg white (HEW) lysozyme, although these are less well defined in the latter. The location of the break in the curves at 90 cm^{-1} correlates well with the proportion of random structure determined from X-ray analysis of these

[29] Y. N. Chirgadze, E. V. Bradhnikov, and N. A. Nevskaya, *J. Mol. Biol.* **102,** 781 (1976).

proteins.[28] It is interesting to note, in this connection, that exchange rate distributions derived for HEW lysozyme show three peaks[22] and that the slowest exchanging protons having amide A band half-widths <80 cm^{-1} also exchange with high activation enthalpies in HEW lysozyme.[30]

The Near-Infrared Method

The near-infrared region between 1.35 and 1.50 μm has been employed to monitor hydrogen–deuterium exchange in polypeptides, amides, and proteins.[16] The extent of exchange is followed by monitoring the increase of the HOD band at 1.4 μm or the decrease of the NH band at 1.5 μm. The method requires very high protein concentrations (10%).

Tritium Methods

Tritium–hydrogen isotope exchange techniques are perhaps the most versatile of all the methods for monitoring total exchange in proteins and generally provide the most precise exchange rate data.

The protein is initially labeled to isotopic equilibrium with a ^{3}H-enriched solution. Exchange-out is initiated by separation of the protein from excess solvent tritium by gel permeation chromatography, rapid dialysis or lyophilization, followed by dissolution into H$_2$O. Samples of the protein solution are then withdrawn at timed intervals, a second separation of the tritium-labeled protein from the solution with which it is exchanging is performed, and the protein and sometimes the solvent are subsequently analyzed for their tritium content by liquid scintillation counting.

Initial tritium labeling of the protein to isotopic equilibrium is usually conducted under conditions of high temperature and pH, typically, 15 to 25 hr incubation at pH 8–9 and 35–45°, to accelerate the exchange reaction. Pilot exchange-out experiments are necessary to determine that maximum labeling of the protein has been achieved. Commercial supplies of ^{3}H-enriched water are available with specific activities of up to 5 Ci ml^{-1}, equivalent to a ^{3}H content of about 0.2%. However, it is seldom necessary to employ solutions with such high specific activities, and more typical isotope contents for initial labeling might vary from 0.0001 to 0.01% depending on the amount of material available and the activity desired in the samples to be analyzed during the out-exchange. The tritium methods therefore involve only a trace labeling of the proteins.

The tritiated material may be lyophilized, provided volatile buffers are employed in the labeling procedure. In this case, out-exchange is initiated

[30] R. B. Gregory, D. G. Knox, A. J. Percy, and A. Rosenberg, *Biochemistry* **24**, 6523 (1982).

by dissolving the dry protein in a solution at the required experimental pH. The speed of transfer depends on the solubility of the material and the quantity that must be dissolved and determines the error introduced in the zero time estimate. The exchange reaction is usually more conveniently initiated by gel filtration on a column preequilibrated with buffer at the required experimental pH.[31] Error in the zero time estimate depends on the rapidity of separation on the column, which in turn is determined by the solvent composition, temperature, and pressure head. Hanson[32] has described an apparatus for rapid high-pressure gel chromatography that can reduce the zero time error to a few seconds. Alternatively, an initial separation can be accomplished by rapid dialysis.[33] The zero time will be poorly defined if several changes of dialyzate are required to effect complete removal of excess tritium. However, rapid dialysis is useful in studying exchange of gels, suspensions, or very dilute solutions. Tüchsen and Ottesen[34] have described a filtration device employing glass microfiber filters for studying the exchange behavior of suspensions of insoluble proteins and cross-linked protein crystals.

Tritium exchange methods are particularly well suited to studies of slow exchanging protons since the out-exchange may be continued for hours or days provided the integrity of the protein can be maintained for such periods. At timed intervals samples of the out-exchange solution are withdrawn for analysis of the protein tritium content. Separation of the protein from solvent may be achieved by cryosublimation,[1,11,35,36] gel filtration,[31] rapid dialysis,[33,37] or by adsorption of the protein onto phosphocellulose paper.[38]

Cryosublimation techniques allow a large number of samples to be collected at well-defined time points but the subsequent lyophilization procedure is time consuming and there is some evidence of exchange occurring between the protein and ice during the cryosublimation step which can introduce errors.[11,39,40] Gel filtration provides a convenient and simple method for separating protein from solvent.[31] The precision of the time point, which depends on the conditions and rapidity of separation on the column, can be improved by forced flow through the column or by

[31] S. W. Englander, *Biochemistry* **2**, 789 (1963).
[32] C. V. Hanson, *Anal. Biochem.* **32**, 303 (1969).
[33] S. W. Englander and D. Crowe, *Anal. Biochem.* **12**, 579 (1965).
[34] E. Tüchsen and M. Ottesen, *Carlsberg Res. Commun.* **44**, 1 (1979).
[35] M. Ottesen, this series, Vol. 11, p. 735.
[36] A. A. Schreier and R. G. Baldwin, *J. Mol. Biol.* **105**, 409 (1976).
[37] A. D. Barksdale and A. Rosenberg, this series, Vol. 48F, p. 321.
[38] A. A. Schreier, *Anal. Biochem.* **83**, 178 (1977).
[39] B. E. Hallaway and E. S. Benson, *Biochim. Biophys. Acta* **105**, 154 (1965).
[40] W. P. Bryan and S. O. Nielsen, *Biochemistry* **8**, 2572 (1969).

minimizing exchange on the column, the latter being achieved by performing the separation at low temperature (0°) on a column preequilibrated with buffer at a pH close to the pH minimum for exchange. The
material eluting from the column is ready for scintillation counting and if
protein concentrations are determined by ^{14}C dual labeling, accurate volumetric transfer of material to the scintillation vials becomes less critical,
although variations in volume and protein concentration may affect
counting efficiency. Alternatively, the separation can be performed by
vacuum filtering the solution through phosphocellulose paper.[38] The protein is adsorbed onto the paper and after several rinses with buffer at a pH
and temperature that minimizes exchange it can be eluted from the paper
with a concentrated salt solution.

Rapid dialysis methods require no second separation of materials. At
timed intervals samples of the protein solution and dialyzate are withdrawn for analysis of tritium activity.

[14]C Dual Labeling

Tritium exchange methods require accurate estimates of the tritium
content and protein concentration of the samples withdrawn at each time
point. Separate determination of the latter can be particularly laborious
and time consuming but can be avoided by trace labeling the protein with
[14]C.[36,37]

Free amino groups (lysine ε-NH$_2$ and NH$_2$-terminal) on the protein
can be converted to their N-methyl or N,N-dimethyl derivatives by re-
ductive methylation with [14C]formaldehyde and a reducing agent such as
NaBH$_4$[41] or NaCNBH$_3$.[42] ^{14}C-labeled protein is added to an excess of
unlabeled protein. Only a small fraction of the total number of protein
molecules is labeled so that the procedure does not affect exchange. The
protein concentration in each sample can then be established from the [14]C
activity of protein standards of known concentration. The tritium content
and protein concentration of samples at each time point can be determined by [14]C—[3]H dual isotope counting with a two-channel counting
technique since the two isotopes have different β-emission energy spectra. However, the energy spectra overlap so it is necessary to determine
the counting efficiency of the two isotopes in each channel to correct for
this "spillover."[43] The number of hydrogens remaining unexchanged per

[41] G. E. Means and R. E. Feeney, *Biochemistry* **7**, 2191 (1968).
[42] N. Jentoft and D. G. Dearborn, *J. Biol. Chem.* **254**, 4359 (1979).
[43] D. L. Horrocks, "Applications of Liquid Scintillation Counting." Academic Press, New
York, 1974.

protein molecule at time t, $H(t)$, is simply derived from the ^{3}H activity of the sample, dpm(t) as

$$H(t) = \text{dpm}(t)/CP \tag{3}$$

where P is the number of moles of protein in the sample and C is the specific ^{3}H activity of the protein exchangeable hydrogen after the initial in-exchange to isotopic equilibrium expressed as dpm per mole of hydrogen. The latter is determined from the measured specific ^{3}H activity per unit volume of the inexchange solution, dpm_0, as

$$C = \text{dpm}_0 I/W \tag{4}$$

where I is the isotopic enrichment factor and W is the number of moles of solvent hydrogen per unit volume. For dilute solutions $W = 111$ mol/liter. However, for concentrated solutions of high-molecular-weight proteins W must be corrected for excluded-volume effects. At isotopic equilibrium, peptide groups are isotopically enriched with tritium in exchange reactions with water. The isotopic enrichment factor for peptide groups in poly (DL-alanine) is 1.24[44] and 1.11 for poly(DL-glutamic acid.[45] However, the generality of such an effect is not at all well established for proteins and will vary from site to site within the protein. The value of I employed in Eq. (3) is therefore somewhat arbitrary and is usually set equal to 1.0 or 1.24, the latter being the value for the model compound poly(DL-alanine). As a consequence, it is not possible to determine absolute numbers of exchangeable hydrogens with tritium exchange methods.

UV Methodology

Recently Takahashi et al.[46] and Englander et al.[47] have described a simple method for determining peptide hydrogen–deuterium exchange rates by monitoring the change in absorbance in the 220–230 nm region. Replacement of hydrogen by deuterium at the peptide nitrogen causes a blue shift of the peptide envelope which produces a small change in absorbance at 220 nm. The exchange reaction is initiated by rapid dilution of an aqueous solution of the material of interest into ^{2}H$_2$O and is monitored by recording the time-dependent decrease in absorbance at the appropriate wavelength with an ultraviolet spectrophotometer. Data can be acquired within a few seconds after manual mixing or within milliseconds

[44] S. W. Englander and A. Poulsen, *Biopolymers* **7**, 370 (1969).
[45] W. H. Welch and G. D. Fasman, *Biochemistry* **13**, 2455 (1974).
[46] T. Takahashi, M. Nakanishi, and M. Tsuboi, *Bull. Chem. Soc. Jpn.* **51**, 1988 (1978).
[47] J. J. Englander, D. B. Calhoun, and S. W. Englander, *Anal. Biochem.* **92**, 517 (1979).

TABLE II
PEPTIDE ABSORBANCE PARAMETERS[a]

Compound	λ (nm)	ε_H (H$_2$O)	ε_H	ε_D	$\Delta\varepsilon$ (kinetic)	$\dfrac{\varepsilon_H - \varepsilon_D}{\varepsilon_H}$ (%)
Poly(DL-alanine)	220	368	349.3	336	16.3	4.43
Poly(DL-lysine)	220	528	477.3	457	20.3	3.84
Oxidized ribonuclease	220	1090	1057.7	1040	17.7	1.62
	230	530	481	70	11.0	

[a] Extinction coefficients are in units of liter mol^{-1} (peptide) cm^{-1}. ε_H (H$_2$O) is the extinction coefficient of the hydrated peptide in H$_2$O. ε_h and ε_D are the extinction coefficients of the hydrated and deuterated peptide in ^{2}H$_2$O, respectively.

with a stopped-flow instrument. The method therefore extends the time range available with other methods by a factor of at least 10 and possibly up to a factor of 10^3 depending on the dead time of the mixing device. The method offers a number of advantages over other methods in terms of the ease and speed with which data can be acquired. High quality data can be collected with very small quantities of material. For example, collection of equivalent exchange data with infrared absorbance methods would require about 100-fold higher concentrations. The method is also suitable for exchange studies of low-molecular-weight materials which cannot be measured with gel permeation chromatographic techniques. The relative change in absorbance on complete deuteration of the peptide is only about 5% in the most favorable cases, which is much smaller than that observed in the infrared, although this is offset to some extent by the greater sensitivity and precision inherent in modern ultraviolet spectrophotometers. Nevertheless, the small absorbance changes, the need to establish accurate extinction coefficients for the fully protiated and deuterated material, and the high background absorbances of interfering protein chromophores place limits on the usefulness of the method for protein exchange studies.

Englander et al.[47] have established the peptide absorbance parameters for several model systems which are given in Table II. Even for poly(DL-alanine), where there is no background absorbance due to side chain chromophores, the total change in absorbance due to the isotope exchange reaction amounts to less than 5%. The absorbance changes are also complicated by solvent perturbations that occur on transfer of the peptide group from H$_2$O to ^{2}H$_2$O. These perturbations are instantaneous and do not contribute to the kinetics of the exchange reaction. As a consequence, the extinction coefficient appropriate for calculations of exchange rates is that observed for the —CONH— species in ^{2}H$_2$O rather than in H$_2$O, which must be determined by extrapolation to zero time. For simple model systems displaying linear semilogarithmic plots, extrapolation to zero time is straightforward, but it may be difficult for proteins and heteropolypeptides where the exchange kinetics deviate from simple first-order behavior. Examples of the application of the UV method to model polypeptide systems are given by Takahashi et al.,[46] Englander et al.,[47] and Gregory et al.[48]

Extension of the method to protein studies is complicated by interference from side chains that absorb in the 200–230 nm region. The side chains with labile hydrogens that may interfere are asparagine, glutamine, arginine, histidine, lysine, serine, threonine, cysteine, tryptophan, and tyrosine. Extensive studies of the effects of deuteration on the ultraviolet spectra of these amino acids by Takahashi et al.[49] established 206 nm as a wavelength suitable for peptide exchange studies. Nevertheless a number of side chains make contributions to the absorbance change at 206 nm that are comparable to that of a peptide group, so the method cannot be considered specific for peptide hydrogens. The extent of interference depends on the exchange rates and relative abundance of the amino acid residues in the protein under study. Exchange rates of arginine, histidine, and tyrosine side chain protons are considerably faster than that of the peptide group but could interfere with estimates of peptide group exchange rates if they are hydrogen bonded or otherwise have reduced accessibility in the protein. Asparagine, glutamine, and tryptophan side chain protons exchange with rates similar to that of the peptide group and therefore cannot be distinguished at 206 nm by the UV method.

Because of the large background absorbance due to side chain chromophores the total change in absorbance on complete deuteration of proteins is very small. Englander et al.[47] observed a ≃4% decrease in absorbance at 220 nm for complete deuteration of oxidized ribonuclease, but of this only 1.6% was attributed to the exchange reaction itself. However, solvent perturbation effects would perhaps be expected to make a relatively smaller contribution to the total absorbance change for folded proteins. Calculation of the fraction of hydrogen remaining unexchanged requires, like the infrared methods, an accurate extinction coefficient for the fully deuterated material and an accurate estimate of the protein concentration. The latter can be achieved spectrophotometrically by measurements between 275 and 280 nm, where time-dependent changes in absorbance due to the exchange reaction are negligible. Conversion of the measured absorbance changes to numbers of hydrogens remaining unex-

[48] R. B. Gregory, L. Crabo, A. J. Percy, and A. Rosenberg, Biochemistry 22, 910 (1983).
[49] T. Takahashi, M. Nakanishi, and M. Tsuboi, Anal. Biochem. 110, 242 (1981).

changed also requires, like infrared methods, an assumption that absorbance changes for each exchangeable proton are all equal.

The UV method is new and we are not aware of any published studies of its use in measuring peptide hydrogen exchange in proteins. The convenience and ease with which data can be obtained undoubtedly makes it the method of choice for rapidly exchanging polypeptides and polyamides. However, for protein systems the small absorbance changes, the probable need to establish extinction coefficients at each experimental condition that is employed (pH, temperature, cosolvents, etc.), and the long-term stability of the ultraviolet spectrophotometer that is used severely limit the usefulness of the method for quantitative studies of slow exchanging peptide protons.

UV spectrophotometric measurements do, nevertheless, provide a convenient method for specifically monitoring the exchange of tryptophan and tyrosine side chain protons. The difference spectrum for L-tryptophan shows peaks at 274, 283, and 291 nm but of these, only the peak at 291 nm is associated with the exchange of indole protons.[50] The exchange reaction can be followed conveniently by monitoring the time-dependent change in absorbance at 291 nm while measurements at 280 nm provide an internal concentration standard. Nakanishi et al.[50] have described the application of the method to an exchange study of tryptophan side chain protons in lysozyme.

The difference spectrum for L-tyrosine shows peaks at 279 and 286 nm.[51] Only the peak at 286 nm is associated with the exchange of the phenol OH proton. Measurements of tyrosine OH exchange rates generally require a stopped-flow instrument. The very fast exchange rates of tyrosine OH protons allow the exchange behavior of tryptophan protons to be readily distinguished from those of tyrosine. A study of the exchange kinetics of tyrosine residues in ribonuclease A by this method has been described by Nakanishi and Tsuboi.[51]

NMR Methodology

Nuclear magnetic resonance techniques are perhaps the most powerful of all methods for providing detailed information about protein internal motions. Chemical shifts, relaxation times, and line width parameters contain, in themselves, a wealth of information about molecular dy-

[50] M. Nakanishi, H. Nakamura, A. Y. Hirakawa, M. Tsuboi, T. Nagamura, and Y. Saijo, *J. Am. Chem. Soc.* **100,** 272 (1978).

[51] M. Nakanishi and M. Tsuboi, *J. Am. Chem. Soc.* **100,** 1273 (1978).

namics.[8,19,52,53] NMR methods have also been extended to studies of hydrogen exchange in proteins. The great advantage of the method is, of course, its ability to resolve and identify individual exchangeable protons within the protein. In principle, it is possible to measure the exchange rates of single protons located in all parts of the macromolecule and thereby avoid the analytical difficulties encountered with tritium tracer, UV, and IR absorption measurements of total exchange kinetics. The main difficulty of the approach is associated not with the measurement of the exchange rates themselves but with the assignment of the resonances that are observed. Without such identification the advantages of the method are lost and the technical difficulties associated with a complex method like NMR favor other techniques for total exchange measurement, although the decrease in total area of the downfield region of the proton NMR spectrum on transfer to ^2H_2O has been employed for total exchange measurements.[54–56]

Proton NMR spectra of proteins are exceedingly complex and it is only in recent years, with the development of high field instruments and Fourier transform methods, that detailed resonance assignments have been possible. Even so, complete or nearly complete resonance assignments of exchangeable protons are available for only a few small proteins and peptides, the most extensively studied of these being bovine pancreatic trypsin inhibitor (BPTI).[19] Detailed assignments of resonances in hen egg white lysozyme have also been reported.[57]

The recent dramatic improvements in NMR spectral resolution have been possible not so much from the use of higher fields but rather from developments in the use of Fourier transform pulse methodology. For example, two-dimensional correlated spectroscopy (COSY) employs a pulse sequence that provides two time variables: the delay time between two consecutive pulses and the free induction decay time following the second pulse. Fourier transformation then provides a two-dimensional frequency domain in which diagonal peaks correspond to the normal one-

[52] T. L. James, "Nuclear Magnetic Resonance in Biochemistry." Academic Press, New York, 1975.

[53] O. Jardetzky and J. C. K. Roberts, "NMR in Molecular Biology." Academic Press, New York, 1981.

[54] K. Akasaka, H. Aoshima, H. Hatauo, S. Sato, and S. Murao, *Biochim. Biophys. Acta* **412**, 120 (1975).

[55] C. Thiery, E. Nabedryk-Viala, A. Menez, P. Fromageot, and J. M. Thiery, *Biochem. Biophys. Res. Commun.* **88**, 950 (1980).

[56] A. Hvidt and E. J. Pedersen, *Eur. J. Biochem.* **48**, 333 (1974).

[57] C. Redfield, F. M. Poulsen, and C. M. Dobson, *Eur. J. Biochem.* **128**, 527 (1982).

dimensional spectrum of chemical shifts while spin–spin coupling effects among protons produce "cross-peaks" located off the diagonal. The method increases resolution by resolving overlapping resonances in the second dimension.[58] Subsequent assignment of well resolved resonances to specific protons involves the use of crystallographic data, comparison with models, chemical modification, isotope exchange nuclear Overhauser effects (NOE), and selective spin decoupling.[57,59]

Exchange rates of single protons with assigned resonances in the NMR spectrum can be measured by observing the H resonances on transfer of the protein to a ^2H_2O solution or, in favorable cases, by line broadening or saturation transfer methods. The first approach takes advantage of the large difference in the magnetic resonance frequencies of ^1H and ^2H for a given applied magnetic field. A sample of dry protein is dissolved in a ^2H_2O solution and the decrease in area of the ^1H resonance peaks is monitored as a function of time. An example of the time-dependent changes in a COSY spectrum obtained for BPTI after transfer to ^2H_2O is shown in Fig. 2.[58] Exchange data may be obtained over a time range from a few minutes to a few months. The observation of the fastest exchanging protons is limited by the time needed to dissolve the protein. However, since the exchange kinetics of individual protons is first order, estimates of the exchange rates can still be obtained from incomplete progress curves.

There are a number of methods for monitoring ^1H—^1H exchange directly. Exchange of protons between two different chemical environments with different NMR chemical shifts will give rise to line broadening if the exchange relaxation time is of the same order of magnitude as the difference in the chemical shifts for the two environments. When this occurs the line width contains an excess component due to chemical exchange in addition to the usual spin–spin relaxation term.[53] Exchange rates of protons between an N—H site and water can be determined directly in this way since methods are now available to suppress the dominating water proton resonances. The method has been successfully used to study N—H exchange in RNA[60,61] but has, as yet, found no major utilization in protein exchange studies.

An alternative to line broadening measurements is the use of saturation transfer and recovery methods in which either the water or amide proton resonance is saturated by a strong, selective radiofrequency field

[58] G. Wagner and K. Wüthrich, *J. Mol. Biol.* **160**, 343 (1982).
[59] A. Dubs, G. Wagner, and K. Wüthrich, *Biochim. Biophys. Acta* **577**, 177 (1979).
[60] H. Fritzsche, L. S. Kan, and P. O. P. Ts'o, *Biochemistry* **22**, 277 (1983).
[61] H. Fritzsche, L. S. Kan, and P. O. P. Ts'o, *Biochemistry* **20**, 6118 (1981).

W = 8.42 ppm

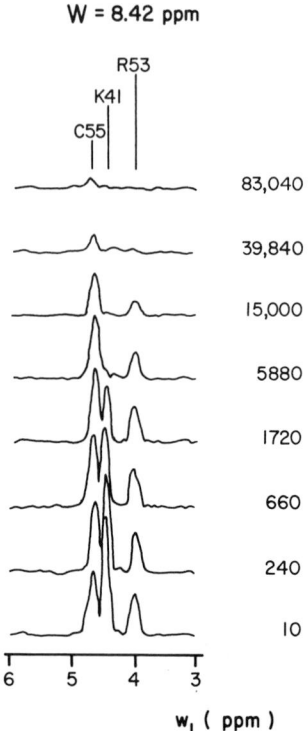

FIG. 2. Cross sections of a COSY spectrum along the ω_1 axis for BPTI. Exchange rates of single protons are determined from the time-dependent decrease in the area of the ^{1}H resonance peaks on exchange with ^{2}H. Data of Wagner and Wüthrich.[58]

and the subsequent recovery, as the saturated spins exchange, is monitored as a function of time.[53,62,63]

A major weakness of NMR techniques is their inherently low sensitivity, which necessitates the use of high protein concentrations (1–20 mM). For example, much of the work on BPTI has been conducted with 5% protein solutions while for lysozyme concentrations between 7 and 10% were required. The size of the protein that can be studied is also limited by the problem of line broadening which increases rapidly with molecular size and leads to poorly resolved resonances.

The exchange of C-2 protons of histidine is particularly well suited to study by NMR since the C-2 proton resonances can be relatively easily assigned. Their exchange is slow with typical half-times for folded proteins between 1 and 10 days at 40° and neutral pH.

[62] J. J. Led and H. J. Gesmar, *J. Magn. Reson.* **49**, 444 (1982).
[63] A. Hvidt, H. J. Gesmar, and J. J. Led, *Acta Chem. Scand.* **B37**, 227 (1983).

Neutron Diffraction

Neutron diffraction methods provide unique protein structural information because of their capacity to locate hydrogen atoms. Furthermore, because the neutron scattering factors for hydrogen and deuterium nuclei have opposite signs it is possible to distinguish these two isotopes and thereby monitor the extent of hydrogen isotope exchange in the crystal. The application of neutron diffraction techniques to protein hydrogen exchange studies was pioneered by Schoenborn.[64,65] Since that time neutron diffraction data have been reported for myoglobin,[66] ribonuclease A,[67] trypsin,[68] lysozyme,[69] and crambin.[70] Detailed descriptions of the methods of data collection and refinement have been given by Kossiakoff[71] and references therein.

Protein crystals employed in neutron diffraction studies are routinely soaked in ^2H_2O solution to reduce the large incoherent background scattering of hydrogen atoms, so, provided conditions are such that exchange of labile protons with deuterium is not carried out to completion, the methods automatically yields exchange information in the form of hydrogen or deuterium occupancy values for each exchange site in the protein. In general, exchange data are confined to a single time point and in this sense the method cannot be regarded as a kinetic one. However, hydrogen occupancies, previously reported as fully, partially, or unexchanged can now be determined to ±15% and with further improvements may provide reliable estimates of exchange rates. Experiments, suggested by Mason et al.,[69] in which a hydrated protein crystal is sealed in a tube with a volume of ^2H_2O, frozen to 258 K for neutron diffraction measurement and then periodically warmed to allow exchange of labile protons with ^2H_2O vapor will provide occupancy data at a number of time points and greatly improve estimates of exchange rates. An example of neutron diffraction exchange data for ribonuclease A[67] is shown in Fig. 3. However,

[64] B. P. Schoenborn, *Cold Spring Harbor Symp. Quant. Biol.* **36**, 569 (1971).
[65] B. P. Schoenborn, *in* "Structure and Function of Oxidation Reduction Enzymes" (A. Akeson and A. Ehreberg, eds.), p. 109. Pergamon, New York, 1972.
[66] J. C. Hanson and B. P. Schoenborn, *J. Mol. Biol.* **153**, 17 (1981).
[67] A. Wlodawer and L. Sjölin, *Proc. Natl. Acad. Sci. U.S.A.* **79**, 1418 (1982).
[68] A. A. Kossiakoff, *Nature (London)* **296**, 713 (1982).
[69] S. A. Mason, G. A. Bentley, and G. J. McIntyre, *in* Neutrons in Biology. Neutron Scattering Analysis for Biological Structures" (B. P. Schoenborn, ed.), p. 323. Plenum, New York, 1983.
[70] M. M. Teeter and A. A. Kossiakoff, *in* "Neutrons in Biology. Neutron Scattering Analysis for Biological Structures" (B. P. Schoenborn, ed.), p. 335. Plenum, New York, 1983.
[71] A. A. Kossiakoff, *Annu. Rev. Biophys. Bioeng.* **12**, 159 (1983).

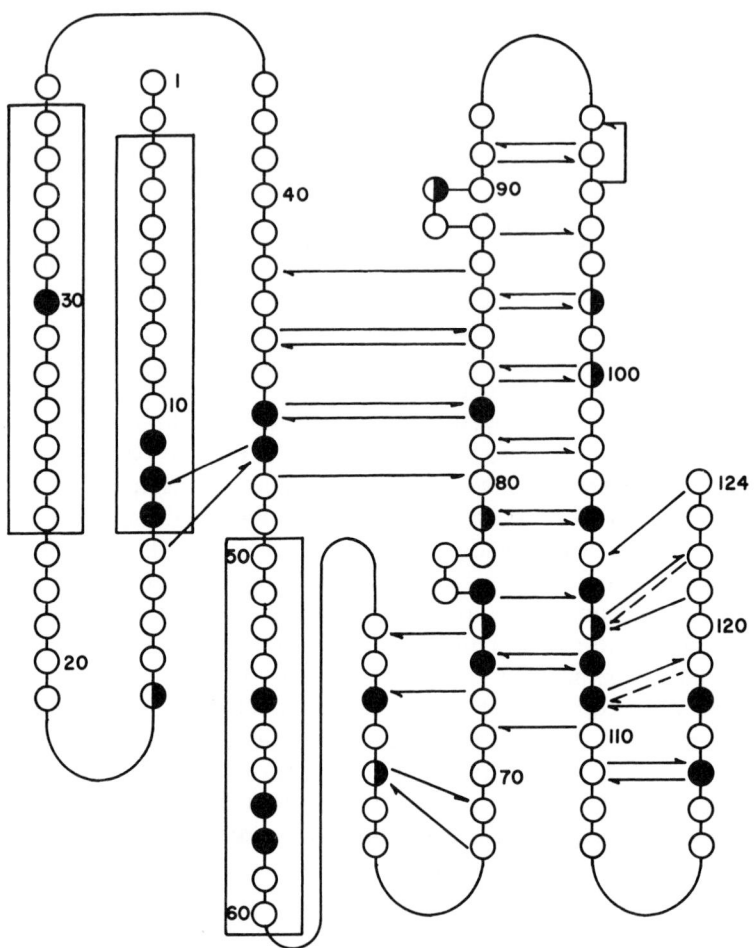

FIG. 3. Location of the slowest exchanging protons (●) in RNase A determined by neutron diffraction. Data of Wlodawer and Sjölin.[67] α-Helix segments are denoted by rectangles.

we defer detailed discussion of the neutron diffraction results to a later section.

Unlike X-ray radiation, the neutron beam causes no radiation damage to the crystal and a single crystal may be employed for all measurements, thereby avoiding the scaling problems associated with X-ray diffraction. Nevertheless, analysis of neutron diffraction data is associated with a severe phasing problem and initial refinement requires a good electron density map derived from X-ray diffraction studies.

At present neutron diffraction studies are confined to just a few research centers. Most protein neutron diffraction work has been performed at Brookhaven National Laboratory in the United States and the Institute Laue-Langevin, at Grenoble, France. The information obtained from neutron diffraction studies of proteins is unique and its ability to correlate exchange behavior with structural features of the protein is of critical importance to an understanding of protein conformational dynamics. It is, therefore, to be hoped that resources at other neutron beam facilities will be expanded to allow protein studies to be performed and thereby extend the application of the method to more systems.

Partial In-Exchange and Difference Exchange Techniques

In most hydrogen exchange experiments all the exchangeable sites of the protein are fully in-exchanged to isotopic equilibrium. However in some situations, for example, where one wants to observe very small changes in exchange behavior with high precision, it is sometimes advantageous to remove the large background of slowly exchanging protons and isolate only a small region of the distribution of exchange rates. This can be done by performing a limited or partial in-exchange in which exchange-in is allowed to proceed only for a limited period of time. Since in-exchange and out-exchange rates for tracers are identical[44] such a procedure labels only the faster exchanging protons, the exact number that is labeled being determined, of course, by the in-exchange period. It is also possible to rid the fully or partially labeled protein of its fastest exchanging protons by initiating exchange-out at conditions of rapid exchange and then "switching" to the conditions of interest. In this way it is possible to isolate a group of protons that would otherwise appear in the middle of a normal out-exchange curve. The approach is useful for improving the signal-to-noise ratio of the data.

The technique can also be employed to isolate exchange sites that are sensitive to ligand binding or protein–protein association reactions from those that are not. This approach, known as difference hydrogen exchange or functional labeling, has been most extensively applied in studies of ligand-induced perturbations of exchange behavior in hemoglobin[72–74] with a view to identifying the regions of the protein that are affected by the allosteric transition. The difference exchange procedure of Englander and Englander[72] is designed to isolate just those protons whose

[72] S. W. Englander and J. J. Englander, *in* "Structure and Dynamics: Nucleic Acids and Proteins" (E. Clementi and R. H. Sarma, eds.). Academic Press, New York, 1983.
[73] E. L. Malin and S. W. Englander, *J. Biol. Chem.* **255,** 10695 (1980).
[74] J. J. Englander, J. R. Rogero, and S. W. Englander, *J. Mol. Biol.* **169,** 325 (1983).

exchange rates are altered in the allosteric transition. Only a third of the slow exchanging hydrogens in hemoglobin is reported to be affected by ligand binding. The protein is only partially in-exchanged so that the slowest protons are not labeled. The allosterically sensitive protons are then isolated by performing the in-exchange in the fast-exchanging oxyHb form, which is later switched to the slow-exchanging deoxyHb form for out-exchange measurements. In principle the method labels only the allosterically responsive sites because any unresponsive sites that are initially rapidly in-exchanged in the oxyHb form also exchange-out rapidly in the deoxyHb form. In practice, some residual contribution from labeled unresponsive sites appears but this is accounted for by performing an in-exchange with the slow deoxyHb form which provides the necessary "background curve."

Englander and co-workers[73] have completely surveyed the allosterically responsive hydrogens in hemoglobin with these techniques. The responsive hydrogens are reported to exchange as a number of discrete classes, each class being characterized by a common exchange rate. Furthermore, the exchange rates of hydrogens within a given class are reported to be accelerated by the same factor on ligand binding. The observation of apparent first-order exchange kinetics for these classes constitutes the major form of evidence supporting the local-unfolding mechanism of exchange in proteins. However, the interpretation of partial in-exchange data in terms of distinct kinetic sets is subject to some criticism. It is very doubtful that simple partial labeling techniques, which rather arbitrarily label a group of protons, can be analyzed in terms of discrete kinetic sets that have any physical meaning. The method amounts to an experimental "curve peeling" procedure. It is always possible to obtain apparent first-order kinetics after partial labeling when the section of the distribution of exchange rates that is selected is sufficiently small. However, we should distinguish simple partial labeling from difference hydrogen exchange approaches, since the latter involve the application of a number of "kinetic filters" that greatly improve selectivity. Nevertheless, the claim can be made that such procedures, rather than fully labeling a small set of ligand-sensitive sites, do, in fact, only fractionally label a large number of sites. Indeed, careful analysis of some apparent first-order plots does reveal telling deviations from first-order behavior. Kinetic sets of 2, 3, or 4 protons probably have little significance but some sets of 12 to 18 protons have been reported, and these, of course, are of great significance to the mechanism, particularly if they are located on a contiguous segment of structure, as has been claimed.[72]

Location of Exchangeable Protons by HPLC

The ability to locate exchangeable protons within the protein structure adds considerable power to the exchange method as evidenced by recent NMR and neutron diffraction studies. To this end, Rosa and Richards[75] developed a medium resolution method to locate slow exchanging protons that involves proteolytic cleavage of the protein and separation of peptides by HPLC. The protein is partially in-exchanged or functionally labeled with tritium under the conditions of interest. The pH of the solution is then rapidly dropped to 2.7 and excess solvent tritium removed by passage through a Sephadex column. The protein is fragmented by a short incubation with pepsin or acid protease and the resulting peptides separated by rapid high-performance liquid chromatography under conditions that minimize further exchange. The peptides are identified by amino acid analysis, Edman-dansyl determinations, and other methods and their tritium content is determined. The separation procedure is critical and must be performed rapidly if loss or redistribution of tritium is to be avoided. At pH $\simeq$ 3.0 and 0° the half-time for exchange of exposed peptide protons is about 70 min. The method has been applied to ribonuclease S[75,76] and hemoglobin.[74] The study of hemoglobin identified a set of 6 to 8 ligand-sensitive protons on a contiguous two-turn segment of the F-FG helix in the β-chain. However, a comparison of the slowest exchanging protons on the S-peptide of ribonuclease determined by the method of Rosa and Richards[75] and by neutron diffraction[67] reveals a large number of discrepancies which at present remain unresolved. The neutron diffraction results must be considered more reliable and less likely to introduce artifacts than the HPLC methods, so at present the latter method remains suspect.

The Chemical Exchange Mechanism

Peptide Exchange

The chemical mechanisms for exchange in model polypeptides and amides have been very extensively studied.[11–17] The exchange reaction is specifically catalyzed by H^+, OH^-, and water. Base-catalyzed exchange proceeds by a rate-limiting proton abstraction to give an imidate anion intermediate which is reprotonated by solvent at diffusion controlled rates.[77]

[75] J. J. Rosa and F. M. Richards, *J. Mol. Biol.* **133**, 399 (1979).

[76] J. J. Rosa and F. M. Richards, *J. Mol. Biol.* **145**, 835 (1981).

[77] A. Berger, A. Loewenstein, and S. Meiboom, *J. Am. Chem. Soc.* **81**, 62 (1959).

$$\underset{R}{\overset{O}{\underset{}{\diagdown}}}C\!-\!N\underset{H^*}{\overset{R}{\diagup}} \quad \underset{OH^-}{\rightleftharpoons} \quad \left[\underset{R}{\overset{O}{\underset{}{\diagdown}}}C\!-\!N\overset{R}{\diagup}\right] \quad \underset{H_2O}{\longrightarrow} \quad \underset{R}{\overset{O}{\underset{}{\diagdown}}}C\!-\!N\underset{H}{\overset{R}{\diagup}}$$

The mechanism of acid-catalyzed exchange in amides has long been thought to involve N-protonation.[11,77]

$$\underset{R}{\overset{O}{\underset{}{\diagdown}}}C\!-\!N\underset{H^*}{\overset{R}{\diagup}} \quad \underset{H^+}{\rightleftharpoons} \quad \underset{R}{\overset{O}{\underset{}{\diagdown}}}C\!-\!\overset{H}{\underset{H^*}{{}^+N}}\!-\!R \quad \underset{H_2O}{\longrightarrow} \quad \underset{R}{\overset{O}{\underset{}{\diagdown}}}C\!-\!N\underset{H}{\overset{R}{\diagup}}$$

However, Martin and Hutton[78] concluded that amide proton exchange occurs predominantly by O-protonation in dilute acids to give an imidic acid intermediate:

$$\underset{R}{\overset{O}{\underset{}{\diagdown}}}C\!-\!N\underset{H^*}{\overset{R}{\diagup}} \quad \underset{H^+}{\rightleftharpoons} \quad \left[\underset{R}{\overset{OH^+}{\underset{}{\diagdown}}}C\!-\!N\underset{H^*}{\overset{R}{\diagup}} \quad \rightleftharpoons \quad \underset{R}{\overset{OH}{\underset{}{\diagdown}}}C\!=\!\overset{+}{N}\underset{H^*}{\overset{R}{\diagup}}\right] \quad \underset{H_2O}{\longrightarrow} \quad \underset{R}{\overset{O}{\underset{}{\diagdown}}}C\!-\!N\underset{H}{\overset{R}{\diagup}}$$

The general situation among secondary amides appears to be somewhat more complex with exchange involving a combination of N- and O-protonation mechanisms, the relative contributions of which are determined by substituent effects, with electron withdrawing groups favoring exchange via imidic acid.[79] Perrin and Arrhenius[79] found that N-acetylglycine methylamide exchanges predominantly via imidic acid with only a 6% contribution from N-protonation and concluded that NH groups of peptides and proteins undergo acid-catalyzed exchange predominantly via the O-protonation mechanism.

Recent NMR studies of BPTI surface NH exchange by Tüchsen and Woodward[80] have demonstrated a good correlation between the acid-catalyzed exchange rate and the carbonyl-oxygen static accessibilities consistent with an O-protonation mechanism.

The rate of the peptide hydrogen exchange reaction is given by

$$\begin{aligned} k_e &= k_0 + k_H[H^+] + k_{OH}[OH^-] \\ &= k_0 + k_H[H^+] + k_{OH} \, K_w/[H^+] \end{aligned} \tag{5}$$

where k_H and k_{OH} are rate constants for specific acid and base catalysis, K_w is the ionization constant of water, and k_0 is the rate constant for direct exchange with water. Until recently, the contribution from direct exchange had been assumed to be negligible and k_0 was often neglected in Eq. (5).

[78] R. B. Martin and W. C. Hutton, *J. Am. Chem. Soc.* **95**, 4752 (1973).
[79] C. L. Perrin and G. M. L. Arrhenius, *J. Am. Chem. Soc.* **104**, 6693 (1982).
[80] E. Tüchsen and C. K. Woodward, *J. Mol. Biol.* **185**, 421 (1985).

TABLE III
ACTIVATION PARAMETERS FOR EXCHANGE OF MODEL POLYPEPTIDE
AND AMIDE SYSTEMS[a]

Compound	Isotope	Catalyst (x)	Temperature (K)	k_x (M^{-1} sec^{-1})	$\Delta H_x^{\ddagger}$ (kcal mol^{-1})	Ref.
Poly(DL- alanine)	^{1}H/^{3}H	H$^+$	273	5.9 E$-$2	15.0	b
			298	5.0 E$-$1	14.4	c
	^{2}H/^{1}H		293	2.75 E$-$1	12.0	d
			298	1.1 E 0	16.0	c
	^{1}H/^{3}H	OH$^-$	273	5.9 E$+$7	3.0	b
			298	7.7 E$+$7	4.0	c
	^{2}H/^{1}H		293	4.0 E$+$8	3.0	d
			298	1.9 E$+$8	1.2	c
	^{1}H/^{3}H	H$_2$O	298	8.0 E$-$6	15.7	c
	^{2}H/^{1}H		298	1.3 E$-$5	21.0	c
	^{2}H/^{1}H		293	8.8 E$-$6	28.0	d
N-Methyl- acetamide	^{2}H/^{1}H	H$^+$	351	2.9 E$+$3	23.0	e
	^{2}H/^{1}H	OH$^-$	351	1.54 E$+$7	6.0	e
	^{2}H/^{1}H	H$_2$O	351	6.0 E$-$3	11.4	e

[a] $\Delta H_{OH}^{\ddagger}$ corrected for ΔH_w°; $\Delta H_{OH}^{\ddagger}$ corrected for ΔH_w°.
[b] Englander and Poulsen.[44]
[c] Gregory et al.[48]
[d] Englander et al.[47]
[e] Hvidt et al.[63]

However, recent studies[47,48,63] indicate that direct exchange with water makes a more significant contribution to exchange than had previously been supposed. Studies of ^{2}H/^{1}H exchange in poly(DL-alanine) indicate relative contributions to the observed exchange rate from acid, base and water-catalyzed exchange at pH$_{min}$ 3.32 at 25° of 31, 31, and 38%, respectively.[48] Activation parameters for exchange reactions in poly(DL-alanine) and N-methylacetamide are listed in Table III. The pH dependence of the chemical exchange reaction can be characterized by the pH value, pH$_{min}$, at which the exchange rate takes its minimum value, k_{min}:

$$pH_{min} = \frac{1}{2}\left(pK_W + \log\frac{k_H}{k_{OH}}\right) \tag{6}$$

$$k_{min} = k_0 + 2k_H[H^+] \tag{7}$$

and by the value of the apparent order of the reaction with respect to [H$^+$] given by

$$\frac{-d\log k_e}{d\,pH} = \frac{k_H[H^+] - k_{OH}[OH^-]}{k_O + k_H[H^+] + k_{OH}[OH]} \tag{8}$$

The relative contribution from direct exchange with water decreases rapidly as the pH is shifted from pH_{min} but because of the large value of $\Delta H_0^{\ddagger}$, k_0 increasingly dominates the observed exchange rate as the temperature is raised, resulting in broad, shallow pH minima and apparent reaction orders that are less than one over an extended range of pH values about pH_{min}.[48] The identification of a large contribution to peptide exchange from direct exchange with water is of particular significance to an understanding of hydrogen exchange mechanisms in proteins. The pH dependence of exchange rates in proteins is often observed to deviate from a simple first-order dependence on $[H^+]$ and $[OH^-]$ which, in many cases, appear in the pH range 2 to 6,[15] exactly the region where deviations would be expected if the water-catalyzed reaction were making significant contributions to the observed exchange. A particularly striking example is provided by the NMR studies of Hilton and Woodward[81] and Richarz et al.[82] on the exchange of single protons in BPTI. The data shown in Fig. 4 consist of the complete pD dependence of exchange rates for the slowest exchanging protons at a number of temperatures. In most cases the p^2H minima are broad and shallow and display apparent reaction orders with respect to $[^2H^+]$ and $[O^2H^-]$ that are less than one.

Wagner and Wüthrich[83] explained the p^2H dependence of exchange rates in terms of a multistate model in which the populations of accessible states of different degrees of ionization vary with p^2H in a manner consistent with the pK values of ionizable groups on the protein, while Hilton and Woodward[81] invoke a two-process model with different p^2H dependences and activation energies for each process and suggest that the broad p^2H minima are due to an accelerated acid-catalyzed exchange rate relative to that of base as protein unfolding is increased with decreasing p^2H. Both models assumed that direct exchange with water is negligible. Reanalysis of the data to include a contribution from direct exchange with water[48] accounts very well for the p^2H dependence and indicates a major contribution from a direct exchange mechanism. Given this interpretation of the data, comparison of k_0 with k_{min} revealed that more than 93% of the observed exchange at p^2H_{min} was attributable to direct exchange with water for protons located in the central portion of the β-sheet structure (Arg-20, Tyr-21, Phe-22, Tyr-23, Glu-31, Phe-33, and Phe-45), while the percentage varied between 64 and 86% for protons located at one end of the β-sheet (Ile-18, Gly-36, Tyr-35). This marked increase in the contribution to exchange by water catalysis in BPTI compared to poly(DL-alanine) is not readily explained in terms of the chemical exchange reaction itself,

[81] B. D. Hilton and C. K. Woodward, *Biochemistry* **18**, 5834 (1979).
[82] R. Richarz, P. Sehr, G. Wagner, and K. Wüthrich, *J. Mol. Biol.* **130**, 19 (1979).
[83] G. Wagner and K. Wüthrich, *J. Mol. Biol.* **134**, 75 (1979).

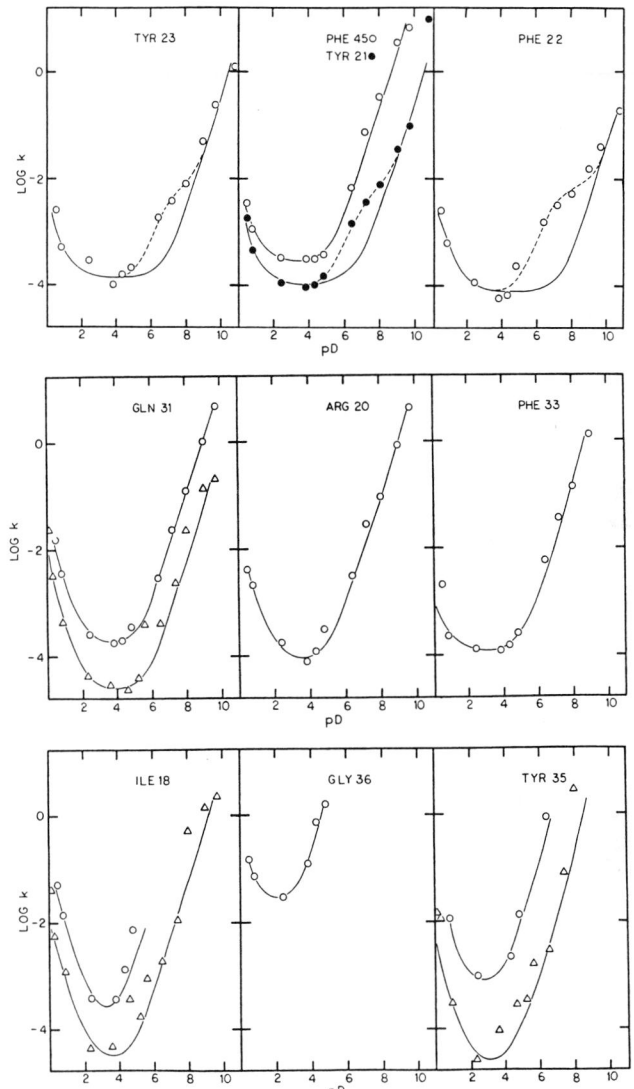

FIG. 4. pD dependence of the exchange rates of 10 protons in BPTI at (△) 22° and (○) 45°. The data are from Richarz *et al*.[82] The calculated curves include a contribution from direct exchange with water.[48] The dashed curves are calculated assuming a discrete charge effect due to ionization of a group with a pK_a 7.0.[48] k in reciprocal hours.

but suggests, instead, a differential accessibility of the BPTI exchange sites to ^2H_2O relative to their accessibility to the charged species, $^2H^+$ and O^2H^-, with the latter being preferentially excluded from the β-sheet region of the protein. By contrast, recent studies of BPTI surface protons at 25° give no evidence of significant direct exchange with water.[80] If these conclusions are correct, they have important implications for the mechanism of exchange of buried hydrogens in proteins since they indicate that exchange occurs from conformations that do not resemble a random coil in bulk solvent.

Klotz et al.[84,85] have observed general catalysis of amide exchange in ^2H_2O–dioxane mixtures. In general, nitrogen bases such as hydroxylamine and imidazole were more effective than carboxylic acids, although in fully aqueous solution general catalysis was found to be insignificant. Nevertheless, the results obtained in dioxane suggest the possibility that polar side chains may act as general acid–base catalysts of peptide exchange in the nonpolar interior of the protein, which would have a profound effect on the pH dependence of exchange.[16] However, we are not aware of any protein systems for which general catalysis has been demonstrated.

The stability of the charged exchange intermediates, the imidic acid and imidate anion, is very sensitive to the immediate chemical environment of the exchange site and this is reflected in the values of k_{OH} and k_H and consequently in the values of pH_{min} and k_{min}. Neighboring groups have a major effect on amide exchange rates. In general electron-withdrawing groups which increase the acidity of the NH proton promote the base-catalyzed reaction and lower pH_{min} while electron-donating groups promote the acid-catalyzed reaction and increase pH_{min}. These nearest-neighbor inductive effects have been quantified by Molday et al.[86] Exchange rates are also sensitive to local and general electrostatic effects. Ionization of charged groups adjacent to an exchange site can have a profound effect on its exchange rate and can give rise to anomalous pH dependencies.[16,48,83] The ionization of a group in the protein with proton dissociation constant K_A gives rise to two species:

$$AH^+ \overset{K_A}{\rightleftharpoons} A + H^+$$

with different hydrogen exchange rates for particular sites within the protein either directly, by affecting the chemical exchange rate, or indirectly by affecting the conformational rearrangements that lead to catalyst-exchange site encounters.

[84] I. M. Klotz and B. H. Frank, *J. Am. Chem. Soc.* **87**, 2721 (1965).
[85] B. H. Leichtling and I. M. Klotz, *Biochemistry* **5**, 4026 (1966).
[86] R. S. Molday, S. W. Englander, and R. G. Kallen, *Biochemistry* **11**, 150 (1972).

The observed exchange rate is given by

$$k_E = \alpha k_{E,A} + (1 - \alpha)K_{E,AH^+} \qquad (9)$$

where α denotes the extent of proton dissociation and $k_{E,A}$ and k_{E,AH^+} represent the intrinsic exchange rates of the two species, A and AH^+, respectively. Examples of exchange behavior in BPTI that are consistent with Eq. (9) are shown in Fig. 4 for Tyr-21, Phe-22, and Tyr-23 in which a pK_A value of 7.0, tentatively assigned to ionization of the amino-terminus, has been assumed.[48]

General electrostatic effects can also influence pH_{min} and $d \log k_e / d pH$ by perturbing the pK value of the exchangeable group[16] or through electrostatic screening and counterion competition among the catalyst species and other ions[87] with consequent perturbation of the local activity coefficients of the catalyst species. The estimation of electrostatic effects by a static accessibility, modified discrete-charge model has been described.[88]

As a consequence of inductive and electrostatic effects the values of pH_{min} for peptide exchange in proteins can vary by about ± 1.0 pH unit from the value of pH_{min} 3.0 for poly(DL-alanine). The recent finding by Tüchsen and Woodward[80] of pH_{min} values for individual BPTI surface protons that extend well beyond the range of values that can be accounted for by such effects is therefore surprising. In some cases pH_{min} values <1 were observed. The results are nevertheless well accounted for by the observed difference in accessibility of the carbonyl oxygen and peptide nitrogen. If an O-protonation mechanism is assumed for acid catalysis and an N-proton abstraction is assumed for base catalysis then the values of k_H and k_{OH} will measure the accessibilities of different atoms in the peptide group and any changes in the relative values of the latter compared to model polypeptides will result in large shifts in pH_{min}. Tüchsen and Woodward[80] also observed k_{min} values for surface NH groups having large static accessibilities that are up to three orders of magnitude lower than those of model polypeptides.

The results described here have a number of important implications for the mechanisms of exchange in proteins and the analytical methods that are employed. It has been assumed that protein NH groups exposed to solvent in the protein–water interface have the same exchange kinetics as model polypeptide NH groups in bulk solvent and it is a prediction of local unfolding models that exchange of buried protons occurs in bulk solvent after exposure of a segment of the protein to solvent by cooperative local unfolding.[16,17,72] We have already noted deviations in the relative

[87] P. S. Kim and R. L. Baldwin, *Biochemistry* **21**, 1 (1982).
[88] J. Matthew and F. M. Richards, *J. Biol. Chem.* **258**, 3039 (1983).

contributions of acid, base, and water-catalyzed exchange for buried protons in BPTI compared to those expected for model polypeptides. These taken together with the wide variation in pH_{min} and k_{min} for the surface protons of BPTI indicate that the exchange mechanism of protein NH groups is considerably more complex than that implied in the simple EX_2 formalism described in a later section. It is clear, for example, that the use of model polypeptide exchange behavior as a means of distinguishing chemical effects from conformational effects in proteins can be very misleading.

Histidine C-2 Proton Exchange

The exchange of histidine C-2 protons is probably the most extensively studied side chain exchange reaction in proteins since such studies not only provide exchange information for specific histidine residues but also yield information about their local chemical environment which is reflected in the histidine pK_a value determined from a study of the pH dependence of exchange rates. Exchange of C-2 protons is very much slower than that of nitrogen or oxygen-bound protons with exchange half-times for free imidazole of 13 hr at pH 8–9, 37° and 2.2 days for free histidine. Exchange proceeds via a rate-limiting OH^- ion attack on the N-3 protonated imidazole with formation of an ylide intermediate which undergoes subsequent rapid reaction with a water molecule:

Rate limiting exchange with water apparently does not occur.[89] The first-order exchange rate is given by

$$k_e = k_{OH} K_W/(K_a + [H^+]) \qquad (10)$$

Studies of the pH dependence of k_e therefore provide estimates of the intrinsic exchange rate, k_{OH} and the dissociation constant, K_a. The exchange behavior of histidine C-2 protons has been very extensively surveyed in a review by Taylor[89] to which the interested reader is directed for further information.

---

[89] S. E. Taylor, *Life Chem. Rep.* **1**, 69 (1982).

Kinetic Formalisms for Proton Exchange

The most commonly applied formal rate equation for exchange of slowly exchanging protons in proteins is the Carlsberg mechanism developed by Hvidt and co-workers.[11,12,14,16] The mechanism in its simplest form assumes that each exchange site, due to fluctuations of the protein conformation about its average native conformation, will at times be "buried" within the protein so that exchange cannot occur and will, at other times, be exposed to solvent where exchange proceeds with rates similar to those of model compounds in bulk solvent.

$$N^* \underset{k_2}{\overset{k_1}{\rightleftharpoons}} I^* \overset{k_e}{\rightarrow} I$$

Buried   Exposed

The availability of the exchange site for exchange is determined by the transconformational reaction characterized by rate constants k_1 and k_2 while k_e is, in principle, given by Eq. (5). Assuming a steady state such that $dI^*/dt = 0$ ($k_2 \gg k_1$), the observed exchange rate, k_{obs}, is given by

$$k_{obs} = \frac{k_1 k_e}{k_2 + k_e} \tag{11}$$

There are two limiting kinetic cases: (1) EX_1 process. If $k_e \gg k_2$, exchange occurs each time the exchange site becomes available and the observed exchange is rate limited by the opening rate constant:

$$k_{obs} = k_1 \tag{12}$$

(2) EX_2 process. If $k_e \ll k_2$, only a small fraction of the transconformational reactions that expose an exchange site to solvent is productive and leads to exchange. The observed rate is given by

$$k_{obs} = \frac{k_1 k_e}{k_2} = K k_e \tag{13}$$

where K is the equilibrium constant for the transconformational reaction, $N^* \rightleftharpoons I$.

The two limiting cases make distinctly different predictions about the pH and temperature dependence of the observed exchange rates. As is clear from inspection of Eqs. (12) and (13), exchange rates for EX_1 processes have no contributions from the chemical exchange step and should display the pH and temperature dependence of k_1 directly. By contrast, exchange rates for EX_2 processes should contain the "V"-shaped pH dependence characteristic of the appropriate model peptide exchanging in bulk solvent in addition to any pH dependence characteristic of the transconformational reaction itself. The EX_1 process generally occurs only at high temperature and pH, or at conditions that favor protein unfolding. EX_1 processes have been proposed to explain the high temperature, high pH data for BPTI[18,19,81] and exchange data for chymotrypsinogen A[90] and lysozyme[91] near their respective unfolding transitions.

Exchange data obtained under less extreme conditions are more commonly interpreted in terms of an EX_2 mechanism. When such a mechanism can be assumed, and k_e is known, the formalism provides estimates of the free energy changes for the transconformational reactions that expose exchange sites to bulk solvent.

The Carlsberg mechanism forms the basis of the contemporary local unfolding mechanism of exchange in proteins. It is not our purpose in this article to discuss the merits of this mechanism, although the basic features of this and other mechanisms are briefly given in a later section. Nevertheless, we hasten to add that several alternative (i.e., catalyst penetration) mechanisms of exchange have been proposed in which exposure of exchange sites to bulk solvent is not necessary for exchange to occur and in which an EX_2 mechanism may not be valid.[92] However, it has proved extremely difficult to formulate exact rate expressions for catalyst penetration mechanisms. As a consequence, the EX_2 formalism (Eq. 13) is also often employed in discussions of these other mechanisms, in which K is taken to represent an equilibrium constant for conformational transitions between catalyst-accessible and catalyst-inaccessible states without implying solvent exposure of the exchange site or is simply employed as an empirical accessibility parameter.

Analysis of Multiple-Site Exchange Data

The hydrogen isotope exchange of a protein with n sites is described by

$$H(t) = \sum_{i=1}^{n} \exp - k_i t \tag{14}$$

where $H(t)$ is the number of hydrogens remaining unexchanged at time t per protein molecule and k_i is the apparent first-order exchange rate for the ith site. In general, the exchange rates for each site are different, although some physical interpretations of the exchange mechanism in proteins suggest that groups of hydrogens exchange as discrete classes.[16,17,72] Equation (14) underlies the immediate attractiveness of the

[90] A. Rosenberg and J. Enberg, *J. Biol. Chem.* **244**, 6153 (1969).
[91] S. Segawa, M. Nakayama, and M. Sakane, *Biopolymers* **20**, 1691 (1981).
[92] R. B. Gregory and A. Rosenberg, *in* "Biophysics of Water" (F. Franks and S. F. Mathias, eds.), p. 238. Wiley (Interscience), New York, 1982.

method, namely, the ability to derive information about the conformational dynamics of the protein as a whole, but it also represents the major restriction to the full exploitation of the approach. For proteins n is usually of the order of 10^2 and the analysis of the parallel exchange of so many hydrogens that is described by Eq. (14) is notoriously difficult. Equation (14) is often approximated by a sum of a small number of exchange sites of class size A_i:

$$H(t) = \sum_{i=1}^{m} A_i \exp - k_i t \qquad m < 5 \tag{15}$$

which is a common representation of many relaxation processes in physics and chemistry. When m is small and is known and good initial estimates of the parameters A_i and k_i are available, damped nonlinear least-squares methods[93] or the method of moments, which only requires an estimate of m,[94,95] can provide physically significant parameter values. However, when m is unknown, as is generally the case with hydrogen exchange data, a poor guess of the number of terms may lead to a completely incorrect set of A_i and k_i parameter values which will approximate the data well enough but have little or no physical meaning. Hamming[96] has illustrated the problems of such an analysis with the simple case of attempting to distinguish between

$$A \exp - kt \tag{16}$$

and

$$\frac{A}{2} \exp - (k + \varepsilon)t + \frac{A}{2} \exp - (k - \varepsilon)t \tag{17}$$

The latter expression is equal to $A \exp - kt \cosh \varepsilon t$. The series expansion of $\cosh \varepsilon t$ is

$$\cosh \varepsilon t = 1 + \frac{\varepsilon^2 t^2}{2} + \frac{\varepsilon^4 t^4}{24} \cdots \tag{18}$$

Only when t is large can we hope to detect a difference between Eqs. (16) and (17) which depends on ε^2. However, when t is large, $\exp - kt$ is small, thus negating our ability to detect the difference. It is just this type of behavior which is responsible for the generally low information content of

[93] S. L. Laiken and M. P. Printz, *Biochemistry* **9**, 1547 (1970).
[94] I. Isenberg, R. D. Dyson, and R. Hanson, *Biophys. J.* **13**, 1090 (1973).
[95] R. D. Dyson and I. Isenberg, *Biochemistry* **10**, 3233 (1971).
[96] R. W. Hamming, "Numerical Methods for Scientists and Engineers." McGraw-Hill, New York, 1962.

sums of exponentials and that makes determination of m so difficult. Equation (15) with m unknown can be solved by using Fourier transform methods[97] which, in principle, automatically determine the value of m. Provencher[98,99] has improved this approach and has also developed an eigenfunction expansion method, available as a Fortran program called DISCRETE, which is both more accurate and more convenient than the Fourier transform methods and appears to have wide applicability. Many examples of protein hydrogen exchange data analyzed with Eq. (15) can be found in the older hydrogen exchange literature and, regrettably, odd examples are still to be found in the current literature where analysis is often performed with a very small number of data points and with inferior methods such as unmodified nonlinear least-squares or "curve-peeling" procedures. Equation (15) may provide meaningful representations of the exchange kinetics of small polypeptides, where there is reasonable justification for assuming a small number of exchange classes but it is generally quite inappropriate for the analysis of protein exchange data since the exchange kinetics of a large number of sites cannot be adequately represented by just a few terms.

A more physically meaningful representation of the data is to replace the sum in Eq. (14) by an integral[21]:

$$H(t) = \int_0^\infty f(k) \exp - kt \, dk = L\{f(k)\} \qquad (19)$$

The experimentally derived function, $H(t)$, is the Laplace transform of the exchange-rate probability density function (pdf), $f(k)$. In principle, the function $f(k)$ can be recovered by an inverse Laplace transformation of $H(t)$. Unfortunately, the comments given above for the discrete class approach also apply with full force to the solution of Eq. (19). $H(t)$ is very insensitive to the underlying exchange rate probability density function. In general, even for data with arbitrarily small noise levels, there exists a large number of solutions that fit the data to within the noise level. In addition, as Provencher[100] has noted, the errors are unbounded, and the solutions may be quite diverse in character. The problem with such an analysis therefore rests principally on the question of solution uniqueness and the development of an appropriate strategy for choosing a physically meaningful solution.

[97] D. G. Gardner, *Ann. N.Y. Acad. Sci.* **108,** 195 (1963).
[98] S. W. Provencher, *J. Chem. Phys.* **64,** 2772 (1976).
[99] S. W. Provencher, *Biophys. J.* **16,** 27 (1976).
[100] S. W. Provencher, *Comput. Phys. Commun.* **27,** 213 (1982).

Closed-Form Laplace Inversion

One approach to the solution of Eq. (19) essentially ignores the question of solution uniqueness, assumes a functional form for $H(t)$ and derives the exchange-rate pdf by closed-form Laplace inversion. Knox and Rosenberg[21] applied this approach to an analysis of exchange in lysozyme where $H(t)$ was approximated by a power law function, modified to account for the exchange of the slowest hydrogens which, at the time, were believed to exchange by a mechanism of cooperative thermal unfolding:

$$H(t) = b(1 + at)^{-n} \exp - ct \tag{20}$$

Laplace inversion of Eq. (20) provides the pdf:

$$L^{-1}\{H(t)\} = f(k) = \begin{bmatrix} \dfrac{ba^{-n}(k - c)^{(n-1)} \exp - (k - c)/a,}{\Gamma(n)} & k > c \\ 0, & k < c \end{bmatrix} \tag{21}$$

where $\Gamma(n)$ is the gamma function and a, b, c and n are adjustable parameters. Equation (21) describes a modified gamma pdf in which the tail of the density has been cut-off and imposed on a delta function at $k = c$. The parent gamma pdf

$$f(k) = \frac{ba^{-n}k^{n-1}}{\Gamma(n)} \exp - k/a \tag{22}$$

with Laplace transform

$$H(t) = b(1 + at)^{-n} \tag{23}$$

has also been employed to describe the exchange of hemoglobin and myoglobin.[16] Examples of modified gamma functions are shown in Fig. 5 as plots of $kf(k)$ against $\log k$. This dimensionless representation of the density is derived from a simple transformation of Eq. (19):

$$H(t) = \int_0^\infty kf(k) \exp - ktd \ln k \tag{24}$$

$kf(k)$ is proportional to the probability density for the activation free energy, $\Delta G^{\ddagger}$. The number of hydrogens exchanging with rates between $\log k$ and $\log k + d \log k$ is then equal to $2.303 \ kf(k)d \log k$. Two other density functions have been employed to describe protein exchange data: the "box-car" function, in which the activation free energy is assumed to be uniformly distributed on some interval $(\Delta G^{\ddagger}_{min}, \Delta G^{\ddagger}_{max})$, a pdf employed by Shenkin[101] to describe the exchange of myoglobin, and the log normal pdf

---

[101] D. Shenkin, Ph.D. thesis, Princeton University, Princeton, New Jersey (1978).

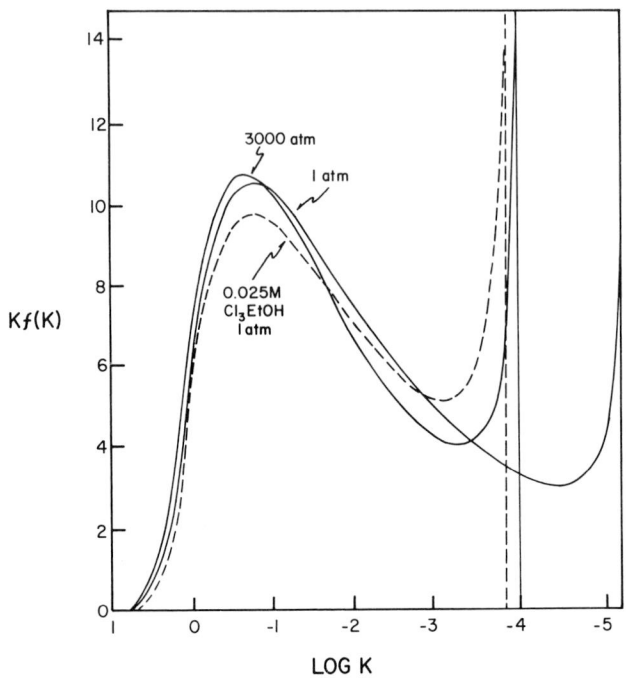

FIG. 5. Examples of closed form exchange rate pdf's for lysozyme [A. D. Barksdale, D. G. Knox, and A. Rosenberg, *Biophys. J.* **32**, 619 (1980)].

in which the activation free energy is assumed to be normally distributed. The latter has been proposed to describe the distribution of exchange rates in soybean trypsin inhibitor, oxidized ribonuclease, and bovine pancreatic trypsin inhibitor.[16] The log normal density function is given by

$$kf(k) = \frac{1}{\sigma \sqrt{2\pi}} \exp - (\ln k - \mu)^2 / 2\sigma^2 \tag{25}$$

The corresponding $H(t)$ function is derived by numerical integration of Eq. (24) after appropriate substitution for $kf(k)$ from Eq. (25). Detailed discussion of the characteristics and application of these functions to hydrogen exchange data has been given by Barksdale and Rosenberg.[16] Despite its simplicity and elegance, closed-form Laplace inversion of hydrogen exchange data can be very misleading. The only criteria for judging the suitability of a particular function that is imposed on the data is the goodness of fit between the calculated and observed $H(t)$ values. However, as we have noted earlier, an enormous number of diverse solutions will approximate the data equally well and this approach provides no guarantee that features of the observed distribution have any physical significance. In fact, distributions derived in this way almost certainly will *not* reflect the true distribution of hydrogen-exchange rates. Fortunately, recent developments in the numerical solution of integral equations now make it possible to perform model-independent Laplace inversions of experimental data and thereby avoid the hazards inherent in fitting data to analytical functions.

Numerical Laplace Inversion

Numerical Laplace inversion of noisy experimental data is a notoriously ill-conditioned problem. Although a number of guides to numerical inversion are available, the difficulties of the problem are such that numerical analysis of multiple relaxation processes is rarely attempted.[102,103] Recently, however, Provencher[100,104] has developed a computer program, CONTIN, for the solution of Fredholm integral equations that have the form

$$H(t_j) = \int_a^b K(k,t_j)s(k)dk + \sum_{i=1}^{N} \beta_i L_i(t_j) \tag{26}$$

Here $H(t_j)$ and t_j are the dependent and independent variable, respectively, $K(k,t_j)$ is a known function, and $s(k)$ is to be estimated. The sum of known $L(t_j)$ terms is optional. For the Laplace inversion of hydrogen exchange data the appropriate kernel in Eq. (26) is

$$K(k,t_j) = k^{-1} \exp - kt$$

which provides the solution $s(k) = kf(k)$. CONTIN computes a constrained, regularized solution of Eq. (26) by incorporating a regularizor term in the function that is to be minimized. The form of the regularizor is such that it penalizes deviations from smoothness. In operation, the program computes a series of constrained solutions in which the size of the regularizor is progressively increased. In accordance with the principle of parsimony, one chooses the smoothest solution that is consistent with the data. The imposition of constraints and the regularization of the solution provide a very effective strategy for dealing with the problem of solution

[102] R. E. Bellman, "Numerical Inversion of the Laplace Transforms." American Elsevier, New York, 1966.

[103] V. I. Krylov and N. S. Skoblya, "Handbook of Numerical Inversion of Laplace Transforms." Israel Program for Scientific Translations I PST Press, Jerusalem, 1969.

[104] S. W. Provencher, *Comput. Phys. Commun.* **27**, 229 (1982).

uniqueness and yields the best compromise between a stable solution and an adequate model. The approach avoids the major hazard of prior specification of the number of degrees of freedom inherent in model-dependent analysis since the number of degrees of freedom is automatically determined by the constraints and the regularizor during the analysis.

The size of the solution set from which a parsimonious solution is chosen can be considerably reduced by incorporating known physicochemical constraints. First, the hydrogen exchange rate pdf is always positive, so the solution, $s(k)$ can be constrained to be positive. Second, it is possible to define a maximum exchange rate, k_{max}, and thereby constrain $s(k) = 0$ for all $k > k_{max}$. Appropriate values for k_{max} may be determined from the well-known exchange rates observed in model random-coil polypeptides such as poly(DL-alanine) and poly(DL-lysine).[44,48] Finally, if the total number of exchangeable hydrogens is known, it is possible to constrain the integral over the solution, $\int_{k_{min}}^{k_{max}} f(k)dk$, to this value. Unfortunately, the total number of exchangeable hydrogens is often difficult to determine in proteins because of contributions from side chain protons and, in the case of tritium methods, because of the uncertainty introduced by equilibrium isotope effects. The application of CONTIN to the analysis of hydrogen exchange data for lysozyme has been described by Gregory[22] and Gregory and Lumry.[105]

Figure 6 shows an example of the exchange rate pdf's obtained at 25°, pH 7.5, for lysozyme. CONTIN estimates moments, standard deviations, number of degrees of freedom, Fisher F distribution values, and random runs probabilities for each solution which together with a visual inspection of the smoothness of the curve provide the basis for the choice of a parsimonious solution from the solution set. The distributions shown in Fig. 6 represent the smoothest three-peak solutions that are consistent with the data. The solution may not have all the detail of the true solution, but the strategy adopted by CONTIN ensures that the features that it does have are demanded by the data and are therefore less likely to be artifacts. CONTIN has very clearly established the exchange rate pdf for lysozyme to be at least bimodal and most probably trimodal.

Some idea of the confidence which may be placed in the analysis can be gained by a comparison of the pdf's obtained from the four independent experiments shown in Fig. 6. The general features of peak 1 and 2 are well reproduced in the four curves, while the density for the fastest exchanging hydrogens is rather poorly defined, a fact which reflects the experimental difficulty of sampling the fastest exchanging hydrogens with the tritium exchange technique. It is apparent that the gamma distribution

[105] R. B. Gregory and R. Lumry, *Biopolymers* **24**, 301 (1985).

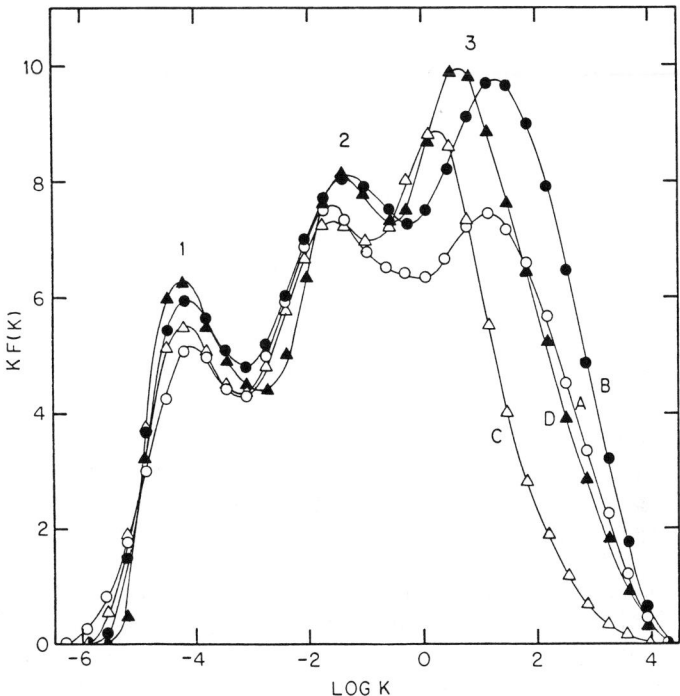

Fig. 6. Exchange-rate pdf's for lysozyme at pH 7.5, 25° determined by numerical Laplace inversion with the program CONTIN.[22,100,104] The curves A–D represent the smoothest three-peak solutions that are consistent with the data for four identical, independent experiments.

derived by closed form Laplace inversion (Fig. 5) is a very oversimplified representation of the true exchange rate pdf which is more properly described by the broad, overlapping three-peak pdf determined by CON-TIN. Comparison of analytically and numerically derived exchange rate pdf's for a number of other systems indicates that oversimplification, and consequent loss of valuable information, is the most general fault of the closed-form Laplace inversion approach which inevitably arises in the attempt to stabilize an ill-posed problem by reducing the number of degrees of freedom. For this reason, and in the absence of any well-established stochastic model for the time course of protein hydrogen exchange reactions, numerical Laplace inversion is always to be preferred.

Data Requirements

The construction of meaningful protein exchange rate distribution functions requires exchange data of high precision collected over a time range that is sufficiently long to allow $H(t)$ to decay to the noise level of the experiment. The latter requirement is essential since it defines the lower limit of the exchange rate pdf, which, unlike the fastest possible exchange rate, is not usually known a priori. Distribution functions derived from exchange data that do not extend to sufficiently long times to sample the slowest exchanging protons can be misleading even though they may approximate very well the limited region of $H(t)$ that is obtained. Collection of an adequate data set for proteins therefore must proceed for long periods of time (i.e., 10^3 to 10^4 min) under pH and temperature conditions that allow exchange of the slowest protons to be monitored. If rank order of exchange is conserved with changes in pH and no pH-dependent conformational changes occur, the effective time range over which data are available can be extended by superimposing exchange curves obtained at different pH values to obtain a single curve in which the time is "normalized" to one particular pH value.[21,26,56,106] Tritium exchange methods are the most suitable for monitoring the slow exchanging protons, but usually cannot provide information about the fastest exchanging peptide and side chain protons which are therefore almost always poorly defined in the exchange rate pdf.

The ability to perform reliable numerical Laplace inversions of noisy experimental data now made possible with Provencher's CONTIN program represents a major advance in the analysis of protein hydrogen exchange data. However, we defer detailed discussion of their interpretation and application to a later section.

"Relaxation Spectra"

An alternative representation of total exchange data developed a number of years ago by Willumsen[12] and Hvidt and co-workers[56,106] involves plotting the fraction of unexchanged hydrogens, X, against $\log(k_e t)$ where k_e is the chemical exchange rate of the model compound poly(DL-alanine). Such plots are called "relaxation spectra," but are essentially a normalized replot of the exchange curve in which the pH dependence of the chemical exchange reaction is accounted for in the calculation of k_e. They are similar to the curve "overlapping" procedure described in the previous section, although in the latter a value of the second-order chemical exchange rate constant is not explicitly included in the calculation. Exchange data determined at different pH values will fall on a continuous curve when recast in this way only if the observed exchange rate is directly proportional to the catalyst concentration. Relaxation spectra are interpreted with reference to an assumed EX_2 mechanism of exchange

---

[106] A. Hvidt and K. Wallevik, *J. Biol. Chem.* **247**, 1530 (1972).

where X is given by

$$X = \frac{H(t)}{n} = \sum_{i=1}^{n} \exp - \beta_i k_e t$$

and β_i is the probability for exposure of the ith site to bulk solvent. Comparison of the "relaxation spectrum" with curves of $X = \exp - \beta k_e t$ calculated at various values of β indicate the distribution of β values while discontinuities in the "spectrum" suggest a pH-dependent conformational change, a change in mechanism, or some deviation in the pH dependence of the chemical exchange rate of sites in the protein from that of poly(DL-alanine). The latter is particularly likely since the pH of minimum exchange and the relative contributions of the acid, base, and water-catalyzed exchange reactions for peptide groups in the protein are known to vary considerably from that observed for poly(DL-alanine).[48] The plot provides a useful visualization of such pH-dependent effects but otherwise its usefulness as a tool in analysis of exchange data is limited since its interpretation and the proposed estimation of exposure probabilities presume the validity of an EX_2 mechanism of exchange.

The Rank-Order of Exchange

Comparison of total hydrogen exchange data obtained under different experimental conditions, i.e., the presence and absence of specific ligands, cosolvents, etc., or the analysis of hydrogen exchange rates as a function of an independent variable such as temperature, pressure, or solvent viscosity is difficult because of the inability of total exchange studies to relate exchange rates to specific sites within the protein. The exchange rate observed at any given time represents an average rate of a number of simultaneously exchanging sites, the exchange order of which is indexed, approximately, by the value of $H(t)$ at that particular time. Further progress in the analysis of these average exchange rates can only be made if the rank order of exchange is conserved at each experimental condition. For example, if at a given temperature, T_1, the order of exchange is $k_n > \ldots > k_i \ldots k_2 > k_1$ then the rank order of exchange is said to be conserved with changes in temperature if, at a different temperature, T_2, the order of exchange remains the same as that at T_1. If rank order is conserved, then values of exchange parameters determined at fixed $H(t)$ values will reflect weighted averages for approximately the same group of sites at each experimental condition. The use of the index, $H(t)$, in this way does not, of course, imply that each site exchanges completely before the next site.

The conservation of rank order of exchange is established by performing "jump" or "switching" experiments. An example of the results of a

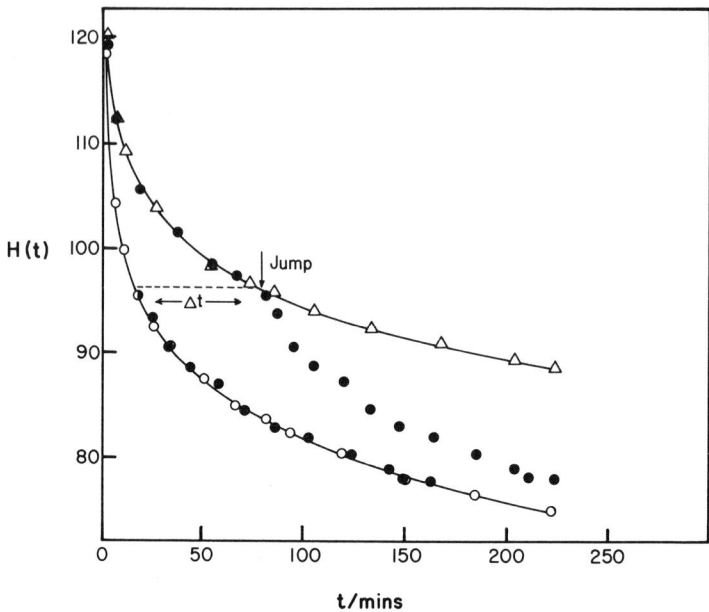

FIG. 7. The conservation of rank order of exchange with changes in glycerol concentration at pH 5.0, 25° exchange-out in the presence (△) and absence (○) of 16 m glycerol. Filled symbols show exchange initially in the presence of 16 m glycerol which was removed at the jump point. The postjump data are also shown replotted against $(t - \Delta t)$. (Data of Gregory et al.[107])

"jump" experiment to establish conservation of rank order with changes in glycerol concentration for lysozyme[107] is shown in Fig. 7. Three samples are prepared. Samples 1 and 2 are maintained at their respective glycerol concentrations of 0 and 16 m throughout the experiment. The third sample is initially prepared with 16 m glycerol and its exchange curve follows that of sample 2. After a period of time, the glycerol is removed from sample 3, which immediately assumes the exchange characteristics of sample 1 but with a time lag, Δt. The post-jump sample 3 data are replotted against $(t - \Delta t)$. If the rank order of exchange is unaffected by the change in glycerol concentration the plot of $H(t)$ against $(t - \Delta t)$ for sample 3 should be superimposable upon the plot of $H(t)$ against t for sample 1 after the concentration jump, as is indeed the case in this example.

[107] R. B. Gregory, D. G. Knox, A. J. Percy, C. Lee, L. Crabo, B. Quibbemann, B. Nelson, and A. Rosenberg, unpublished results (1984).

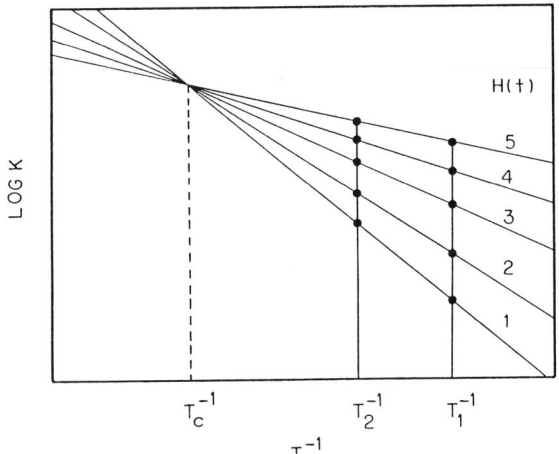

FIG. 8. The conservation of rank order of exchange with changes in temperature for all $T < T_c$ where data show linear enthalpy–entropy compensation behavior.[105]

"Jump" and "switching" experiments have been employed in studies of rank-order conservation on ligand binding to hemoglobin[108,109] and lyso-zyme[110] and the conservation of rank order with changes in temperature is now well established for lysozyme.[110] The latter is particularly interesting because of its relationship to the appearance of enthalpy–entropy compensation in hydrogen exchange[30] and other protein processes. The en-thalpy–entropy compensation relationship for a related series of pro-cesses may be expressed by

$$\Delta G_i(T) = \frac{T}{T_c}\, \alpha(T) + \Delta H_i \frac{(T_c - T)}{T_c} \tag{27}$$

where T_c is the compensation temperature.

The relationship between conservation of rank order and the appear-ance of enthalpy–entropy compensation is illustrated in Fig. 8 where plots of ln k against T^{-1} for different $H(t)$ values have been drawn assuming a monotonic relationship between $\Delta G_i^{\ddagger}$ and $\Delta H_i^{\ddagger}$. The lines intersect to give a common exchange rate when $T = T_c$ as required by Eq. (27). Examina-tion of Fig. 8 shows that if Eq. (27) is valid for activation parameters determined at different values of $H(t)$ then rank order must be conserved with temperature for all values of $T < T_c$.[105]

[108] S. W. Englander and A. Rolfe, *J. Biol. Chem.* **248**, 4852 (1973).
[109] S. W. Englander and C. Mauel, *J. Biol. Chem.* **247**, 2387 (1972).
[110] R. R. Wickett, G. J. Ide, and A. Rosenberg, *Biochemistry* **13**, 3273 (1974).

The sources of linear or nearly linear enthalpy–entropy compensation have been treated in detail elsewhere.[111-114] In proteins compensation behavior appears as a simple consequence of weak coupling of two processes of a linked system. In hydrogen exchange, compensation and hence rank-order conservation with changes in temperature result from coupling between the measured exchange process and the protein conformational fluctuation processes that control access of the catalyst to the exchange site. The compensation temperature itself is characteristic of the type of fluctuation process required to effect exchange and is therefore a very useful parameter for classifying exchange mechanisms.[105]

Analysis of Exchange Rates as a Function of an Independent Variable

If rank order of exchange is established with changes in a given independent variable then further analysis of the average exchange rates, indexed by their $H(t)$ value, as a function of an independent variable is possible. The effect of an independent variable, x, on the average exchange rate is most simply reflected in its effect on the time required to reach a particular $H(t)$ value.[30] Differentiation of Eq. (1) with respect to x yields

$$\frac{dH(t)}{dx} = -t \sum_{i=1}^{n} \frac{dk_i}{dx} \exp - k_i t - \frac{dt}{dx} \sum_{i=1}^{n} k_i \exp - k_i t \tag{28}$$

Since the dependence of t on the variable x is evaluated at constant $H(t)$ values, we define $dH(t)/dx = 0$ and find

$$\frac{d \ln t}{dx} = \left(\sum_{i=1}^{n} \frac{dk_i}{dx} \exp - k_i t \Big/ \sum_{i=1}^{n} k_i \exp - k_i t \right) = \frac{\left\langle \frac{dk}{dx} \right\rangle}{\langle k \rangle} \tag{29}$$

For example, the average activation energy for exchange, $\langle E_A \rangle$, is simply given by this procedure[30] as

$$\langle E_A \rangle = \frac{\langle k E_A \rangle}{\langle k \rangle} = R \left(\frac{d \ln t}{dT^{-1}} \right) \tag{30}$$

Equation (30) is a generalization of the procedure developed by Woodward and Rosenberg.[115] Note, that although it is possible to derive an

[111] R. R. Krug, W. G. Hunter, and R. A. Grieger, *J. Phys. Chem.* **80**, 2335 (1976).
[112] R. R. Krug, W. G. Hunter, and R. A. Grieger, *J. Phys. Chem.* **80**, 2341 (1976).
[113] M. R. Eftink, A. C. Anuseim, and R. L. Biltonen, *Biochemistry* **22**, 3884 (1983).
[114] R. Lumry and R. B. Gregory, *in* "The Fluctuating Enzyme" (G. R. Welch, ed.). Wiley (Interscience), New York, 1986.
[115] C. K. Woodward and A. Rosenberg, *J. Biol. Chem.* **246**, 4105 (1971).

apparent first-order average rate function

$$\langle k \rangle = - \frac{d \ln H(t)}{dt} = \left(\sum_{i=1}^{n} k_i \exp - k_i t \Big/ \sum_{i=1}^{n} \exp - k_i t \right) \qquad (31)$$

the subsequent expression of $\langle k \rangle$ as a function of an independent variable is more complex than Eq. (29) and confuses the interpretation of the dependence of $\langle k \rangle$ on the variable of interest.[16,30] For this reason use of Eq. (29) is always to be preferred. We should also emphasize that comparison of the extent of exchange at a fixed time point under different experimental conditions generally has little meaning since different sites are being compared.

Ligand Binding Equilibria

The exchange of sites influenced by a rapidly established ligand binding reaction can be represented by:

$$P + L \underset{k_2}{\overset{k_2}{\rightleftharpoons}} PL$$
$$\downarrow k_P \qquad\qquad \downarrow k_{PL}$$

where P and PL are the concentrations of the unliganded and liganded protein, respectively, and k_P and k_{PL} the intrinsic exchange rates for these two species. If $k_1, k_2 \gg k_P, k_{PL}$, the apparent exchange rate is given by

$$k = \alpha k_{PL} + (1 - \alpha)k_P \qquad (32)$$

where α represents the fraction of time spent in the liganded state. In favorable cases, hydrogen exchange rates can be utilized to determine the intrinsic exchange rates of the two species. However, care is necessary in the interpretation of ligand binding perturbations on exchange kinetics since it is not always clear that intrinsic exchange behavior for a particular species is being observed. For example, if the system is saturated with ligand L such that the predominant species present is PL, but k_{PL} is perturbed by the binding reaction in such a way that $k_{PL} \ll k_P$ then the preferred exchange pathway will be via ligand dissociation and exchange from the unliganded protein, P. In the limit, the observed exchange rate approaches

$$k \simeq \frac{k_P k_2}{[L]k_1} \qquad (33)$$

The observed difference in exchange rates then simply reflects the ligand binding reaction and not the difference in intrinsic exchange rates.

Mechanisms of Exchange in Proteins

The problem of determining exchange mechanisms in proteins is one of determining just how exchange-site/catalyst encounters occur. Two limiting mechanisms have been proposed: the local unfolding or "breathing" hypothesis[13,72,108,109] and the solvent penetration hypothesis.[81,90,92,115–117] The evidence for and relative merits of these mechanisms have been reviewed in detail elsewhere[15–17,72] and here we only outline their basic features.

The local unfolding model of exchange is essentially an extension of the earlier Carlsberg mechanism.[11] Exchange is suggested to be controlled by cooperative local unfolding reactions of segments of the protein, with consequent disruption of the hydrogen bonds, and movement of the unfolded segment into bulk solvent where exchange takes place. The model emphasizes the importance of hydrogen bonding in determining slow proton exchange. Hydrogens located on an unfolded segment are predicted to exchange as a well-defined class with a common exchange rate. The putative exchange classes have been extensively studied in hemoglobin by partial in-exchange and functional labeling techniques.[72]

By contrast, penetration models suggest that catalyst molecules diffuse to buried exchange sites, the diffusion being mediated by collective small amplitude fluctuations, with exchange occurring within the protein interior. The types of motion envisaged by solvent penetration models are not easily defined. In its simplest form solvent penetration can be considered as a diffusion process within a homogeneous hydrogen-bonded network, although this is now known not to be sufficient to explain exchange behavior in proteins.[21] Nevertheless, penetration mechanisms are still often wrongly suggested to imply a dependence of exchange rate on the depth from the protein surface and to neglect the role of hydrogen bonding. Among the more specific solvent penetration proposals, the "mobile defect" hypothesis[116] suggests that migration of free volume, associated with the redistribution of hydrogen bonds and other secondary structural interactions, provide pathways for the penetration of exchange catalyst molecules to the buried exchange sites. The lability of hydrogen bonds within proteins is therefore considered a major factor in determining exchange rates not only because strong bonding reduces the probability of exchange during a catalyst–peptide encounter but also because of its more general effects on the patterns of internal bond rearrangements.[30,105]

Kossiakoff[68] has suggested a mechanism of "regional melting" based on neutron diffraction results. Exchange is suggested to occur by break-

[116] R. Lumry and A. Rosenberg, *Colloq. Int. CNRS* **246,** 53 (1975).
[117] F. M. Richards, *Carlsberg Res. Commun.* **44,** 47 (1979).

ing a few hydrogen bonds which results in the formation of a solvent-filled cleft in the protein surface. Exchange is assumed to take place *within* the cleft. The mechanism is sometimes viewed as a variant of the local unfolding mechanism, although it specifically excludes the need to extrude a segment of the protein into bulk solvent, considers exchange to require the rupture of only a few hydrogen bonds rather than cooperative disruption of whole protein segments, and acknowledges the exchange environment to have different chemical properties from the bulk phase.

Recent Results from NMR, Neutron Diffraction, and Total Exchange Studies

Recent exchange studies have done a good deal to improve descriptions of exchange mechanisms and, of course, in the process have provided further information about protein dynamics and the types of protein motion that occur. In the following sections we describe some of these studies, especially as they demonstrate the methods of data analysis and the power of NMR, neutron diffraction, and total exchange techniques as complimentary approaches to understanding protein dynamics. In later sections we illustrate, with some selected examples, how the methods have been applied to some specific biochemical problems.

An example of the type of information provided by distribution function representations of total exchange data is illustrated in Fig. 9, which shows out-exchange curves and the corresponding numerically derived exchange rate pdf's for lysozyme between 5 and 45°. These results establish a much more complex pattern of whole molecule dynamic behavior than has previously been supposed which together with results from NMR and neutron diffraction exchange studies suggest not only a need to expand the number of exchange mechanisms that must be considered for proteins but also indicate the existence of dynamically distinct structural classes in proteins.[105]

Contributions from Thermal Unfolding

The description of the exchange mechanism of the slowest exchanging hydrogens in proteins has been a matter of some confusion because the mechanism has not been clearly isolated from thermal unfolding mechanisms. Hilton and Woodward[81] proposed a two-process model of exchange for BPTI in which exchange characterized by low activation energies was associated with a penetration mechanism from the folded state while that characterized by high activation energies was associated with a mechanism involving thermal unfolding of the protein. The relative contributions to exchange from the two processes varied with pH and tem-

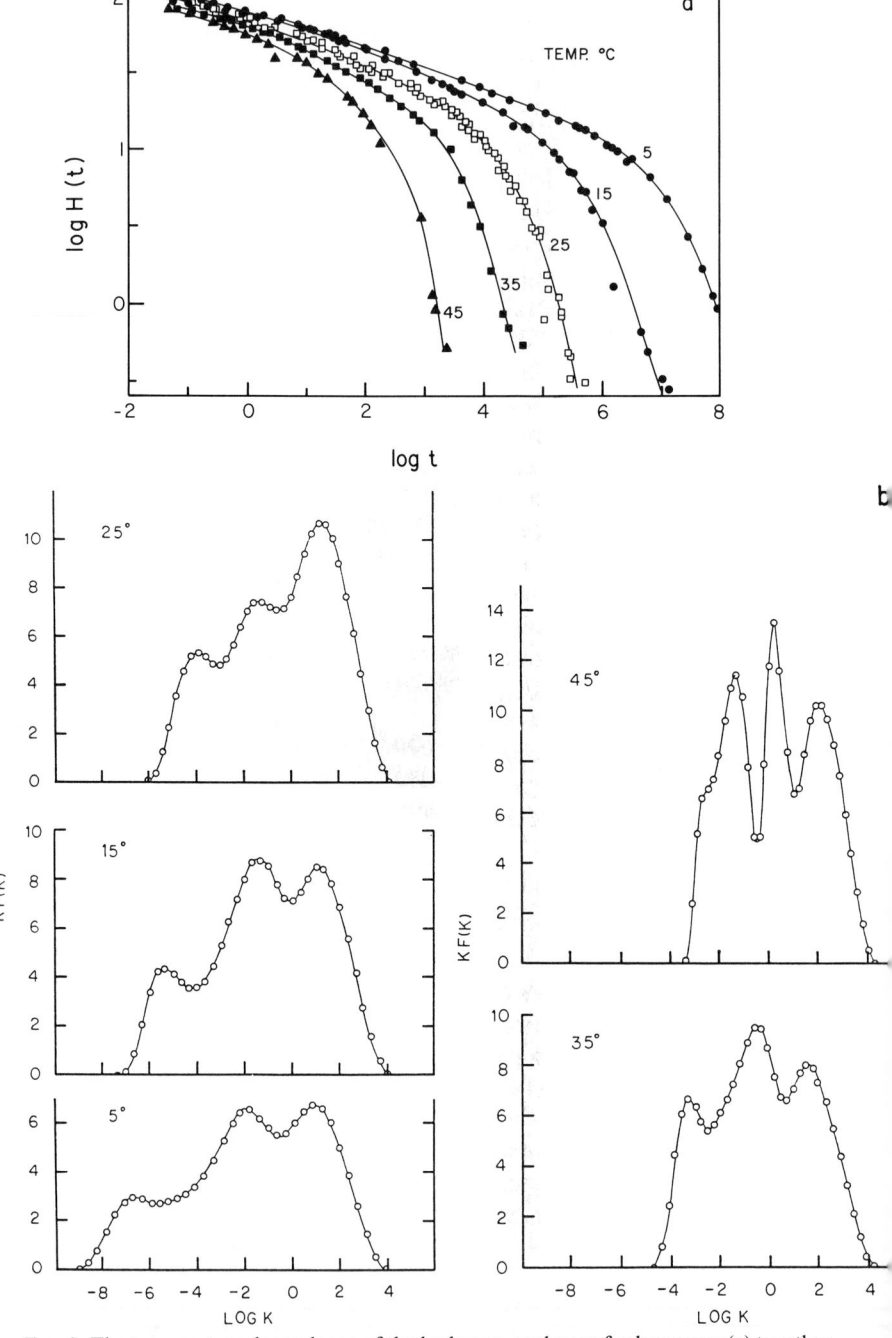

FIG. 9. The temperature dependence of the hydrogen exchange for lysozyme (a) together with the corresponding exchange rate pdf's obtained by numerical Laplace inversion (b).[105] The units of k are min^{-1} normalized to pOH 6.5.

perature. Exchange in lysozyme between 5 and 45° is also characterized by distinctly different activation energies and enthalpy–entropy compensation behavior[30] with exchange of the fast hydrogens characterized by low activation energies and exchange of the slowest hydrogens characterized by high activation energies. The latter was originally suggested to involve thermal unfolding.[21,22,30] However, the work of Wedin *et al.*[20] and Delepierre *et al.*[118] clearly demonstrates that exchange mechanisms involving unfolding processes occur only at temperatures >55° in lysozyme. The activation energies for exchange of tryptophan indole protons in lysozyme, measured in single proton NMR studies, are temperature dependent with a low temperature process ($T \leqslant 55°$) characterized by activation energies between 13 and 40 kcal mol^{-1} and a high temperature process with activation energies between 92 and 120 kcal mol^{-1}. The latter was assigned by Wedin *et al.*[20] to a thermal unfolding mechanism. Interestingly, however, their activation energies and those determined for exchange at high temperatures in BPTI[19] are still somewhat lower than those expected for a mechanism involving the thermally denatured state, suggesting that passage along the unfolding reaction coordinate beyond the activated complex need not be complete for exchange to occur.

Nevertheless these results indicate that all hydrogens exchanging in lysozyme between 5 and 45° do so from the folded state. As further evidence of this, Gregory and Lumry[105] constructed the model-independent activation energy pdf for lysozyme which is shown in Fig. 10. In so doing, use was made of the known conservation of rank order of exchange with changes in temperature. Cumulative exchange rate distributions, $F(k)$, were derived by integration of exchange rate pdf's at each temperature:

$$F(k) = \int_0^k f(k')dk' \tag{34}$$

Since rank order is conserved, $F(k)$ serves as an index of the properties of the same sites at each temperature and it was therefore possible to construct the activation energy cumulative distribution and probability density functions from the temperature dependence of exchange rates determined at fixed $F(k)$ values.

The activation energy pdf shows one broad peak at $E_A \simeq 43$ kcal mol^{-1} that accounts for about 15–17 of the slowest exchanging hydrogens. This overlaps with a second peak that is only partially defined. However, the

[118] M. Delepierre, C. M. Dobson, S. Selvarajah, R. E. Wedin, and F. M. Poulsen, *J. Mol. Biol.* **168,** 687 (1983).

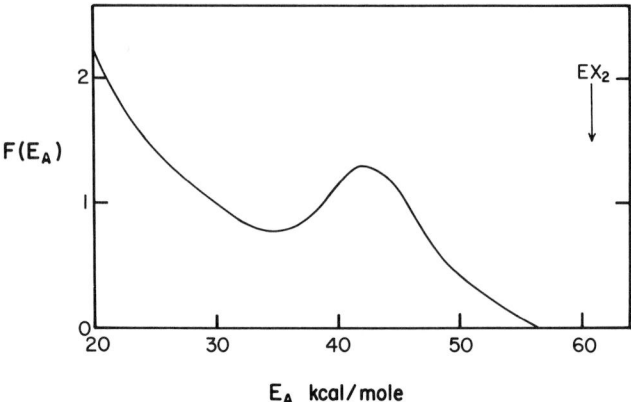

FIG. 10. Activation energy pdf for exchange in lysozyme obtained from the temperature dependence of the cumulative exchange rate distributions.[105] The activation energy expected for an EX_2 mechanism from the unfolded state is indicated.

peak positioned at 43 kcal mol^{-1} is located approximately 18 kcal mol^{-1} lower than that expected for a mechanism involving thermal unfolding.

These results emphasize the need to establish, preferably from independent calorimetric measures, the conditions under which thermal unfolding mechanisms contribute to exchange if exchange data are to be properly interpreted. The general two-process classification of Hilton and Woodward[81] is likely to be valid for most systems but the conditions of temperature and pH under which exchange from the folded state can be assured will, of course, be specific to the system of study.

In lysozyme, at least below 55°, the trimodal distribution function and the appearance of distinct activation energies and enthalpy–entropy compensation behavior must reflect features of the dynamics of the folded state alone.

Exchange of Surface Protons

The exchange behavior of the fastest exchanging hydrogens located on the protein surface has been studied very little presumably because of a belief that exchange of surface protons is similar to that of model compounds and yields very little information about protein dynamics. There is now considerable interest in the dynamic behavior of the protein–solvent interface, particularly with regard to the role of protein surface events, such as surface water relaxation and charged group-water interactions, in enzyme function[7,10] and their general role in coupling the dynamics of the protein to its environment.

In general, studies of whole protein exchange behavior are rarely extended to include the fastest exchanging protons which, like model peptide systems, are amenable to study only at low pH values and temperatures. In exchange rate distributions for lysozyme the fastest exchanging protons are associated with a poorly defined peak of about 50 hydrogens that exchange with activation energies only a few kilocalories greater than that expected for a fully exposed peptide in bulk solvent but with rates that may be attenuated by factors of up to 10^4 relative to model peptides.[105] The only NMR single-proton exchange study of surface protons is that of Tüchsen and Woodward[80] referred to earlier. Their findings of large pH_{min} shifts, the correlation of acid and base-catalyzed exchange rates with carbonyl oxygen and amide nitrogen static accessibilities and exchange rates that are attenuated by factors of up to 10^3 relative to those of exposed peptides indicate a good deal more complexity for this region than had previously been appreciated and suggest that further exchange studies of surface protons would be a very fruitful source of protein–water interface dynamics information.

Exchange of Slow-Protons from the Folded State

Studies of total exchange kinetics cannot relate exchange rates to particular structural features of the protein. However, recent studies of single-proton exchange by neutron diffraction and NMR spectroscopic techniques provide exactly this type of information. To date neutron diffraction exchange data are available for myoglobin,[66] MIP-trypsin,[68] ribonuclease A,[67] and lysozyme[69] and detailed NMR exchange data are available for BPTI.[19,81,82] With the exception of the data for myoglobin which we shall discuss later, the following general pattern emerges. Slow exchanging protons are invariably located within the central regions of antiparallel β-sheet structure and when they appear in α-helices are usually located on the internal surfaces at helix-to-helix and helix-to-sheet contact regions. There is therefore a very strong correlation between the degree of exchange and the extent of hydrogen bonding of an exchange site and sites adjacent to it. However, the location of a site within hydrogen-bonded secondary structure is not sufficient in itself to account for the observed exchange patterns.[67,69] The second factor that appears to be important is the close association of the slow protons with hydrophobic regions of the protein.[105]

Identification of Dynamically Distinct Structural Classes

The differences in exchange behavior of these two classes of protons led Gregory and Lumry[105] to propose the existence of dynamically distinct structural regions in globular proteins. The exchange-stable regions were referred to as "knots" and their special stability against exchange was suggested to be due entirely to electrostatic factors. The favorable peptide dipole–dipole interactions in β-sheet structures, and the possibility for inductive charge displacement through α-carbon atoms enhance hydrogen bond strength in these structures but it was concluded that the principal factor in determining strength is the adjacent layer of aliphatic and aromatic groups which lowers the effective local dielectric constant with significant strengthening of the hydrogen bond network.[105] Shortening of the hydrogen bonds causes the knots to contract essentially to the repulsive limit. Similar electrostatic considerations apply to knots formed on α-helices where nonpolar groups establish a low local dielectric constant, and were suggested to account for the observation of curvature in amphipathic helices reported recently by Blundell et al.[119]

Exchange of protons in the knots is suggested to involve bond rearrangement and free volume redistribution of the more disordered "matrix" regions of the protein to allow catalyst diffusion to the protein interior, which is then supplemented by local distortions that expose one or two exchangeable protons in the knots.[105] Salemme[120] has recently demonstrated that antiparallel β-sheet structures have high conformational flexibility and can respond to fluctuations in their surroundings as cooperative flexible units in which concerted motions of the β-sheet in the form of accordion-like compression along the β-sheet axis and twisting motions about the axis occur without rupturing interchain hydrogen bonds. As a result, the local distortions of β-sheet structure required for proton exchange are rare conformational events below 55°.

The identification of knot structures in proteins has a number of important consequences for whole-protein dynamics. Gregory and Lumry[105] suggest that they are the dominant factor in determining the distribution of bond rearrangements and therefore establish the microstates available to the protein and presumably serve an important role in controlling dynamic behavior critical for protein function. They appear to be critical in establishing kinetic stability and may make some contribution to thermodynamic stability. At a more general level, they illustrate the wide variability of hydrogen bonds and their adaptability to local electrostatic conditions. According to the proposal of Gregory and Lumry[105] the knots must have both a well-ordered hydrogen bond structure and a relatively large number of contact nonpolar groups. Exact geometry appears to be critical and is not easily detected by examination of X-ray diffraction

[119] T. Blundell, D. Barlow, N. Borkakoti, and J. Thornton, *Nature* (*London*) **306,** 281 (1983).
[120] F. R. Salemme, *Nature* (*London*) **299,** 754 (1982).

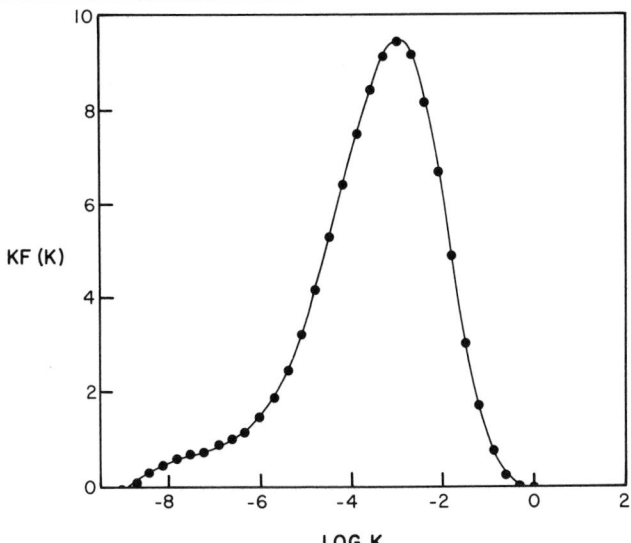

FIG. 11. Numerically derived exchange rate pdf for human carboxyHb A tetramer at 5°. Data of Barksdale and Rosenberg.[16] Units of k are min^{-1} normalized to pH 7.0.

depictions of structure, although even quite subtle variations in hydrogen bond strength are clearly revealed in hydrogen exchange rates.

The existence of distinct structural classes among the globular enzymic proteins that have been mentioned is quite well established but it is apparently not the only strategy adopted by proteins to control their dynamic behavior. Examination of the exchange rate pdf derived from exchange data of Barksdale and Rosenberg[16] for carboxyhemoglobin tetramer shown in Fig. 11 reveals a unimodal distribution with an extended "tail" at slow exchange rates that accounts for about five of the slowest exchanging protons. A few of the latter are probably intrinsically slow-exchanging C-2 histidine protons.[73] There are no other features of the exchange rate pdf that can be attributed to the presence of knot structures. No neutron diffraction data are available for hemoglobin but a neutron diffraction exchange study of myoglobin[66] showed only six unexchanged protons located singly in various regions of the protein with no clear pattern of exchange. These all-helix heme proteins apparently have no significant knot structures and presumably derive kinetic stability in quite a different manner.[105] The most likely candidate appears to be the favorable interaction energy derived from helix–helix dipole interactions, which are known to be large in these proteins.[121] It is interesting to note that the protein from which most of the evidence for local unfolding

[121] W. G. J. Hol, L. M. Halie, and C. Sander, *Nature (London)* **294,** 532 (1981).

mechanisms has been derived is hemoglobin. It is possible, therefore, that its distinctly different dynamic characteristics revealed in exchange rate pdf's may also determine a preference for a different exchange mechanism.

Application of Exchange Methods to Specific Systems

Exchange Behavior of the S-Peptide in Ribonuclease-S

The recent study by Kuwajima and Baldwin[122,123] on the conformation of ribonuclease S is a good example of sophisticated utilization of different exchange methods. RNase is a 14-kDa protein that can be selectively cleaved by subtilisin into two fragments, a 19-residue S-peptide and the S-protein. Neither the S-peptide nor the S-protein alone is enzymatically active. However, when combined, an active reconstituted enzyme molecule results. This property, besides facilitating studies of the active site, allows the study of the conformation of the S-peptide in solution and in the reconstituted molecule. Smaller peptides, representing residues 1–15, also bind with high affinity to S-protein and produce an active reconstituted protein. Kuwajima and Baldwin[122,123] have carried out a detailed study of the motility of the peptide fragments as they associate with S-protein to form the reconstituted molecule. The S-peptide is dominated by a helical segment, consisting of residues 3–13. The fortunate circumstance of rapid recombination of the S-peptide with S-protein allows an ingenious use of the technique of partial labeling. The S-protein is fully exchanged in ^2H_2O, then the peptide, still in the protonated state, is added and exchange of the protons of the S-peptide followed without background interference from exchange of the rest of the protein hydrogens. The exchange behavior of the S-peptide was initially established from studies of the tritium-labeled S-peptide bound to unlabeled S-protein. Eight slow-exchanging protons were observed, with rates attenuated by a factor of about 10^4 relative to those of the free peptide at pH 5.1 and it was established that the observed exchange occurred from the S-peptide/S-protein complex.

The locations of the slow exchanging protons in the 1–15 peptide was established by single-proton NMR exchange studies. Exchangeable proton resonances were assigned to specific residues in the free peptide by COSY two-dimensional NMR spectroscopy, spin decoupling, pH titration, and by comparison of the observed rates with those of model compounds.[86] The slow-exchanging protons were kinetically selected by al-

[122] K. Kuwajima and R. L. Baldwin, *J. Mol. Biol.* **169**, 281 (1983).
[123] K. Kuwajima and R. L. Baldwin, *J. Mol. Biol.* **169**, 299 (1983).

lowing partial out-exchange of the fast-exchanging protons in the complex. After a suitable time interval, the 1–15 peptide with unexchanged slow protons was dissociated from the S-protein by transfer to 3 M urea at pH 2.3 and 0°, conditions that also minimize further exchange, and the NMR spectrum of the free peptide recorded. The eight slowest exchanging protons were located in a single sequence, residues 7 to 14. Seven of these protons are hydrogen bonded in the 3–13 α-helix and the NH of residue 14 is hydrogen bonded to the β sheet.

A study of the pH dependence of the exchange rates demonstrated a transition in exchange behavior between pH 3.0, where the helix is only weakly protected against exchange, and pH 5.0 where the helix is much more stable. A cooperative two-state opening of the helix was observed below pH 3.7 and it was concluded that exchange at pH 3.1 occurs by this opening reaction without dissociation of the peptide from S-protein. However, at pH 5.0 the nature of the exchange reaction changed; protection against exchange increased by two orders of magnitude with exchange occurring preferentially from the NH_2-terminal end. These results are in good agreement with neutron diffraction results[67] in showing greater protection against exchange of the COOH-terminal residues 11, 12 and 13, but disagree with those of Rosa and Richards[75] who found no special protection of residues 7 to 13 in the peptide in studies employing proteolytic cleavage and HPLC.

Inhibitor Binding to Lysozyme

Gregory et al.[124] have recently reported a preliminary study of the effect of tri-N-acetylglucosamine, $(NAG)_3$, inhibitor binding on the total exchange kinetics of lysozyme that illustrates some of the power of complementary tritium total exchange and neutron diffraction exchange techniques in studies of the role of conformational fluctuations in protein function.

The active site cleft of lysozyme can accommodate six units of the N-acetylglucosamine oligosaccharide in subsites labeled A to F. There are two modes of binding: a productive mode in which all six sites are occupied that goes on to form a reactive complex which subsequently undergoes glycoside cleavage between the D and E sites (the so-called β and γ processes), and a nonproductive mode in which only the A, B, and C sites are occupied (the so-called α-process)[125] (Fig. 13). $(NAG)_3$ binds specifically to the A, B, and C sites in a manner that is believed to be very similar to the type of productive (β-process) binding that occurs at these

[124] R. B. Gregory, A. Dinh, and A. Rosenberg, Biophys. J. 45, 16a (1984).
[125] S. K. Banerjee, E. Holler, G. P. Hess, and J. A. Rupley, J. Biol. Chem. 250, 4355 (1975).

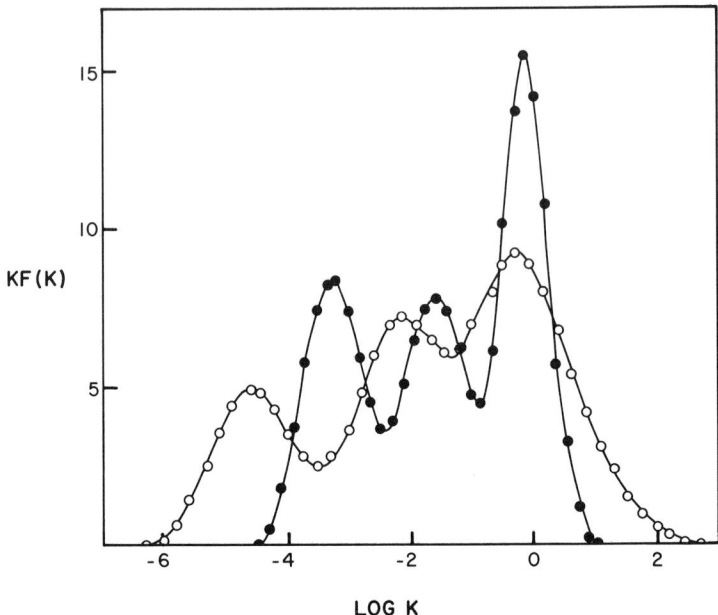

FIG. 12. Exchange rate pdf's for lysozyme at 35° obtained by numerical Laplace inversion of the data of Gregory *et al.*[124] in the presence (○) and absence (●) of saturating concentrations of tri-*N*-acetylglucosamine. Units of k are min^{-1} normalized to pOH 6.5.

sites but does not undergo any significant glycoside cleavage. The (NAG)$_3$–enzyme complex is therefore a very useful model of the dynamic and structural changes that are initially developed as the reaction proceeds along the reaction coordinate. Figure 12 shows the exchange rate pdf's obtained in the presence and absence of saturating concentrations of (NAG)$_3$ at 35°, pH 7.5. The fast exchange peak is very poorly defined by the data and the differences that are observed are not believed to be significant. The other two peaks are more reliability defined and the binding of (NAG)$_3$ is clearly seen to shift both these peaks to slower exchange rates. However, the most significant effect of inhibitor binding is observed for the slowest exchanging protons, the exchange rates of which are decreased by factors of between 10 and 50. With the availability of neutron-diffraction exchange data for lysozyme[69] it was possible to identify the 20 or so slowest exchanging protons within the protein structure. Approximately 24 proton sites were found to have H occupancies $\geq$ 90% after soaking in ^2H_2O at pH 4.2 for 2 months (Fig. 13). These include five protons located in the center of the three-stranded antiparallel β-sheet, 7 protons located on the buried α-helix (residues 28–34), and several pro-

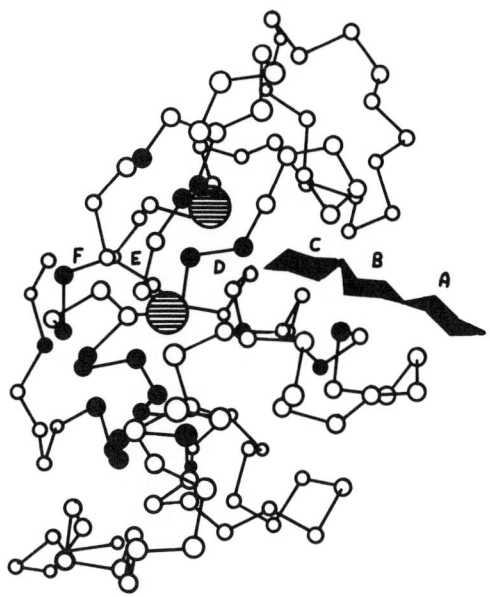

FIG. 13. Structure of the lysozyme-tri-N-acetylglucosamine complex showing the inhibitor binding sites A, B, and C, the position of the two active site carboxylates (shaded circle) and the location of the slowest exchanging protons[69] associated with the slow exchange peak of the exchange rate pdf (●).

tons on the internal surface of the NH$_2$-terminal α-helix that makes contact with the buried α-helix.

The location of the slowest exchanging protons that are affected by (NAG)$_3$ binding is particularly interesting since only three protons are found closely associated with the (NAG)$_3$ binding site itself. Most of the slow protons are located on secondary structure elements at the other end of the active site cleft. Furthermore, these protons are associated with the two structural elements (the β-sheet and buried α-helix) to which the two active-site carboxylates, glutamate-35 and asparate-52, are attached. The results indicate that ligand binding at the A, B, and C sites perturbs the conformational dynamics of those structures that are at some distance from the binding site itself and which are associated with the glycoside cleavage site. Although we cannot be certain of the changes that occur at the catalytic site, it is tempting to speculate that these perturbations caused by ligand binding serve to create a conformation and electrostatic environment at the active site which facilitates the subsequent approach to the transition state as the system moves along the reaction coordinate.

Hydrogen Exchange of Hemoglobin as a Function of Fractional Ligand Saturation

The physical mechanisms that give rise to the well-known cooperative binding of ligands such as oxygen and carbon monoxide to ferrohemoglobin remain a central area of research in protein chemistry. Hemoglobin is certainly the most well studied of all proteins, yet the mechanisms for free energy redistribution among the ligand binding sites and, in particular, the molecular detail of the sequential binding events currently elude us. Recent work by Hallaway et al.[126] designed to follow changes in the conformational dynamics of hemoglobin that take place with increasing occupancy of the ligand binding sites provides a suitable illustration of the use of the hydrogen exchange method in such studies.

Tritium–hydrogen exchange data were acquired as a function of fractional ligand saturation for the cooperative CO-ferrohemoglobin and noncooperative azide-methemoglobin systems. Maintenance of a constant and well-defined degree of ligand saturation is the major source of difficulty in such experiments. For azide binding to methemoglobin, the desired degree of saturation was achieved by dialysis of the fully in-exchanged protein against a buffer containing the required concentration of azide and a tritium activity, identical to that of the in-exchange solution. All subsequent chromatographic separations were performed with columns equilibrated with azide-containing buffers.

For the gaseous ligand CO hemoglobin solutions in-exchanged in either the oxy or deoxy form were divided into two samples. One sample was saturated with CO and the other converted to deoxyhemoglobin with dithionate. Different degrees of saturation were then achieved by mixing appropriate volumes of the two solutions. Subsequent removal of excess tritium and transfer of the out-exchange solutions were performed anaerobically. Second separations were performed with ligand-free buffers to convert the hemoglobin to the slower exchanging deoxy form and hence minimize exchange on the columns.

A typical out-exchange curve is shown in Fig. 14. The linearity of the $\log H(t)$ vs $\log t$ plot over the time range employed is a consequence of the form of the exchange rate distribution function[16] but proves to be particularly fortuitous for the establishment of conservation of rank order of exchange with changes in ligand saturation, on which subsequent analysis rests, since it can be easily shown that the parallel out-exchange curves obtained at different ligand saturations can arise only if conservation of rank order does indeed pertain. The fractional change in exchange rate at

---

[126] P. E. Hallaway, B. E. Hallaway, and A. Rosenberg, *Biochemistry* **23**, 266 (1984).

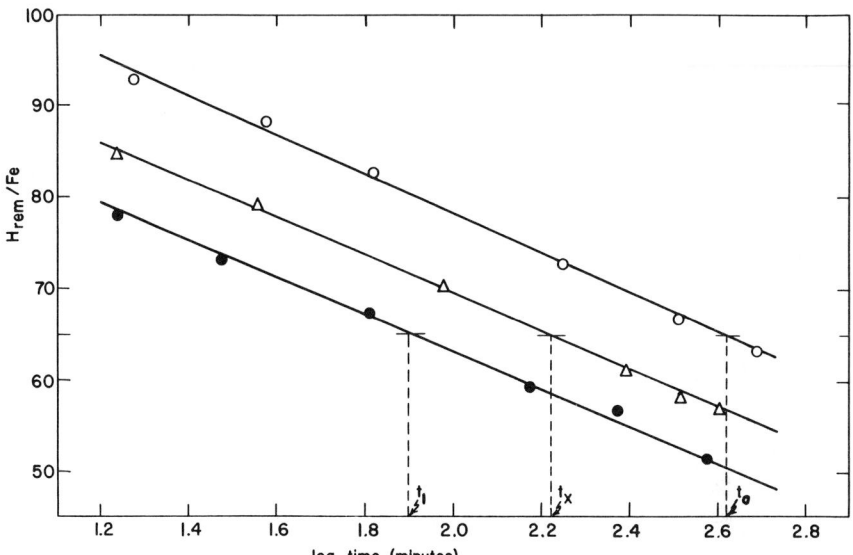

FIG. 14. Typical out-exchange curves for deoxyHb ($\bigcirc$), $Hb(CO)_4$ ($\bullet$), and a partially saturated intermediate ($\triangle$) at pH 7.0, 6°. (Data of Hallaway et al.[126])

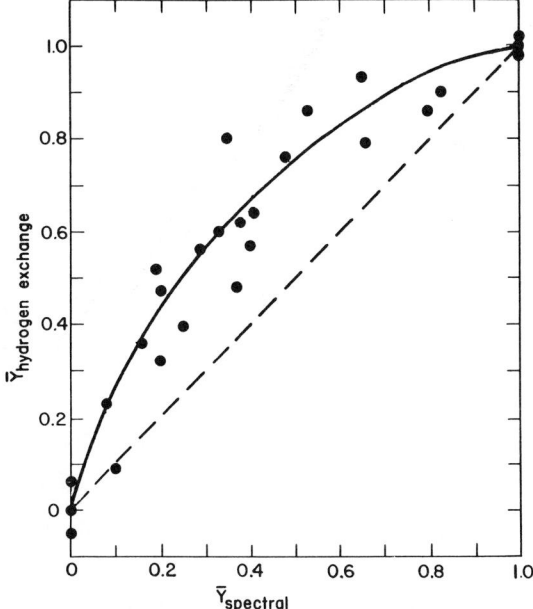

FIG. 15. Fractional hydrogen exchange rate against fractional spectral change for CO binding to Hb A_0 at pH 7.0, 6°. The broken line represents perfect linear correlation. The solid line is the result expected assuming that perturbation of the exchange rate occurs only on binding the first ligand molecule. (Data of Hallaway et al.[126])

a given $H(t)$ value is given by

$$\phi = \frac{t_x - t_1}{t_0 - t_1}$$

where t_x, t_1, and t_0 are the times taken to reach a given $H(t)$ value for the partial liganded intermediate $Hb(CO)_4$ and deoxyHb, respectively. The plot of ϕ as a function of the fractional saturation of metHb by azide, determined spectrophotometrically, demonstrates, as expected, a good linear correlation between these two variables, reflecting the fact that in this noncooperative system the perturbation in hydrogen exchange rates is directly proportional to the number of ligands bound. By contrast, the corresponding plot for CO binding to ferrohemoglobin, shown in Fig. 15, is curved. Further analysis of the data revealed that exchange rates in the cooperative system were perturbed only on addition of the first ligand molecule, indicating that major changes in the conformational dynamics of Hb occur with this step and not with subsequent ligand binding to the other three sites.

Further studies were performed at pH 9.0 where the differences in exchange rates on ligand binding were found to be abolished. Since the cooperativity of CO binding is virtually unchanged at this pH, the perturbations in exchange rates observed at pH 7.0 must reflect only those rearrangements associated with Bohr proton release during the first ligand-binding step.

Acknowledgments

We thank Drs. Erik Tüchsen, Clare Woodward, and Sax Mason for providing results before publication. This work was supported by NSF Grant PCM8303027 and NIH Grant 5-POI-HL-16833.

[22] Individual Breathing Reactions Measured by Functional Labeling and Hydrogen Exchange Methods

By Jose R. Rogero, Joan J. Englander, and S. W. Englander

The exchangeable hydrogens of proteins include the polar side chain protons but more especially the main chain peptide group protons. The latter are distributed uniformly along the polypeptide backbone and form the structurally crucial H-bonds in α-helices, β-strands, and bends. Their exchange with solvent protons depends on transient structural fluctuations and has been widely observed to respond to the binding of small and large molecules and to allosteric and other alterations. Thus the nonperturbing observation of the naturally occurring HX process can provide information about structural fluctuations at sites all through the protein, their dependence on local structure and structure change, and their interplay with function.

To interpret HX measurements in this way one must understand the fluctuational mechanisms underlying the protein HX process. Recent progress in this area suggests that proteins expose their protected, H-bonded hydrogens to exchange with solvent by way of structural unfolding reactions of greater or lesser size. Under conditions of instability—high temperature or extreme pH—the dominant unfolding reaction may involve whole protein denaturation but under normal solution conditions appears to involve only small segments of structure in the cooperative opening. Such local unfolding reactions have now been seen in several small proteins by NMR detection of HX patterns, and have been observed in hemoglobin by the HX methods described below. In these experiments, one observes that a number of peptide group protons, side by side on neighboring residues of an α-helical segment, all exchange at similar rates, even though some of these protons are on the aqueous face of the helix and some are buried on the inner hydrophobic side. Meanwhile other protons on residues immediately adjacent to the cooperatively exchanging segment exchange at very different rates. These observations appear to require as an intermediate in the HX process a structurally distorted form in which the sequential peptide protons are made equally accessible to solvent, even though they are very differently exposed in the native structure. Apparently some region of the protein transiently unfolds, severing H-bonds and reforming them to solvent water. We refer to this process as local unfolding or breathing.

It has been observed very generally that when proteins engage in any functional interaction, some of their exchangeable protons change their exchange rates. The sensitive protons are faster in one functional state and slower in the other. This makes it possible to label selectively with hydrogen isotopes via the HX process just these proton sites and to study their HX rates in the different functional states of the protein. When the hydrogen isotope used is tritium, it is possible to locate the selectively labeled sites by fragmenting the protein and rapidly separating the fragments under conditions that retain much of the tritium label. A brief introduction to selective labeling approaches and a description of the fragmentation and separation method are provided here.

Introduction to Selective Labeling

Kinetic Labeling

The various exchanging peptide NH of a protein span many orders of magnitude in exchange rate, that is, they are greatly spread out on the HX time axis. Proton sites in different time regions of the overall HX curve can be labeled selectively and studied simply by taking advantage of their very different exchange rates. In order to measure the exchange-out of particular NH sites by the loss of tritium label, the label must first be exchanged-in. A brief exchange-in will label only the relatively fast exchanging sites; these can then be studied while the slower unlabeled sites remain transparent to the measurements made. A somewhat longer exchange-in period will label both the faster and intermediate rate protons. Protons that exchange with intermediate rate can be more or less isolated for study, in a kinetic sense, by allowing the faster sites to exchange-out, so that one obtains a sample with only the intermediate rate sites labeled. The behavior of these can then be studied more or less in isolation. For example, by initially preparing samples in this way, one can then ask how the faster sites or the slower sites respond to solvent viscosity or to other solution or ambient variables.

We refer to this simple approach as kinetic labeling to distinguish it from the more sophisticated and considerably more powerful functional labeling approach (see below). In summary, the kinetic labeling approach utilizes a limited exchange-in period (t_{in}) to avoid the labeling of relatively slow sites, and an even shorter exchange-out period ($t_{out} < t_{in}$) to trim away the faster sites. In this way some predetermined segment of the overall multidecade HX curve can be excised, in a general sense, for more detailed study.

Functional Labeling

As with kinetic labeling, a functional labeling experiment starts by exposing the protein to labeling with solvent tritium for a limited time period. Usually this would be done with the protein in the functional form that places the targeted sites in their fast exchanging form. At the conclusion of the limited exchange-in period, the protein is switched to its slower exchanging form. For example, partially labeled oxyhemoglobin would be deoxygenated, oxidized cytochrome c would be reduced, aspartate transcarbamylase could be bound with added substrate analog, pancreatic trypsin inhibitor bound to trypsin, etc. Then the partially labeled protein, now in its slow form, is exchanged-out for a time period longer than the exchange-in time. The fundamental point is that in a functional labeling experiment, unlike kinetic labeling, $t_{out} > t_{in}$. In a kinetic labeling experiment this would cause almost all the bound tritium label to be lost. For example label exchanged in during a 1-hr period would be lost during a subsequent 5-hr exchange-out. Similarly, in a functional labeling experiment that utilizes a 1-hr exchange-in and 5-hr exchange-out, almost all the tritium on nonsensitive sites—that is, on sites that do not become slower in the altered functional state—is lost. Only tritium on sites that are relatively fast in the form used for exchange-in and much slower in the form used for exchange-out will be conserved. These sites are by definition functionally sensitive. Thus sensitive sites that happen to exchange in any particular region of the multidecade HX curve can be targeted, selectively labeled, and studied.

The huge spread in rate of the various protons in a typical protein provides considerable resolution for isolating different sets in this kind of experimentation. The approach is not compromised by the possible presence of sites with compensating rate changes. If some sites are in their fast form in oxyhemoglobin and slow in deoxy, they can be selectively labeled and studied by exchanging-in initially in the oxy form. If other sites are faster in the deoxy form, they can be selected for study by exchanging in initially in the deoxy form. The different sets do not in any way interfere or cancel in the functional labeling mode, though they would do so if one simply measured an overall exchange curve in the usual way.

A Typical Functional Labeling Experiment

A typical functional labeling experiment starts with 0.6 ml of oxyhemoglobin (oxyHb) in solution at 1 mM concentration (50 mg/ml). The actual pH, salt, and temperature to be used are at the investigator's disposal. In addition to protein structural and functional requirements set by the system under investigation, especially pH and temperature can be set to bring the targeted part of the multidecade HX curve into a convenient time window. At pH 5.5 and 0°, even the fastest of the protein's H-bonded peptide NH will exchange on a time scale of minutes; the slowest would require 100 years. At pH 9 and 37°, where exchange chemistry is much faster, the slowest of Hb's NH can be labeled in about 1 day.

Tritium labeling is initiated by adding a small volume of stock tritiated water (THO; 1 Ci/ml) to bring the oxyHb exchange-in solution to a level usually between 2 and 50 mCi/ml, depending on the radioisotope count level expected in the final samples to be counted. After the predetermined labeling period, the protein is switched to its slow exchanging form (deoxygenated by addition of a few mg dithionite), and immediately passed through a Sephadex G-25 column (0.5 ml through a 1×8 cm gel bed). The column has been washed with a solution designed to maintain the deoxy (slow exchange) condition (1 mg ferrous ammonium sulfate/ml with 0.02 M pyrophosphate, or glucose oxidase/glucose/catalase mixture). Thus the column step serves to terminate exchange-in, initiate exchange-out, remove the dithionite and its products, and place the hemoglobin in deoxygenated buffer. The first two-thirds of the effluent Hb peak is collected (the "pool") in a test tube maintained under a flow of argon gas. Exchange-out then proceeds in the pool at any chosen temperature.

An exchange-out curve in the deoxy form is determined by measuring the remaining, still-bound label as a function of exchange-out time. To measure each time point, 0.1 to 0.2 ml of the pool is passed through a deoxygenated Sephadex column (1×4 cm) to remove free tritium, and samples before and through the peak region are collected (2 to 3 drops). These are diluted and analyzed for protein concentration and tritium count level. Protein concentration can be measured by absorbance (dilute Hb into a Drabkin's solution and read absorbance at the cyanometHb Soret band). The level of still-carried tritium is measured by liquid scintillation counting 0.5 ml of the same samples (after measuring absorbance).

Alternative methods for measuring much lower protein concentrations utilize some added ^{14}C-labeled macromolecular marker and double isotope counting.[1,2] At any point in the exchange-out sequence, one can arrange to measure the very same protons in the alternative (fast exchanging) protein form by switching a sample to that form (religanding Hb, e.g., by adding CO-saturated buffer) and continuing to gather exchange-out points.

Further details of experimental manipulation and data computation have been described previously in this series.[3,4]

Data points taken in time as described will portray exchange-out curves in either oxy or deoxyHb. Owing to the selective exchange-in and exchange-out conditions imposed, these curves will portray in the main the exchange of allosterically sensitive sites. The limited exchange-in time acts as a kinetic filter to remove from the measurement sites that exchange (in oxyHb) on a time scale much longer than the exchange-in time used. The longer exchange-out time in the deoxy form acts as a second filter to remove especially allosterically insensitive sites that exchange on a relatively faster time scale. Remaining label on allosterically insensitive sites, that continues to contribute to the curves measured as just de-

[1] A. A. Schrier, *Anal. Biochem.* **83,** 178 (1977).
[2] A. D. Barksdale and A. Rosenberg, *Methods Biochem. Anal.* **28,** 1 (1982).
[3] S. W. Englander and J. J. Englander, this series, Vol. 26, p. 406.
[4] S. W. Englander and J. J. Englander, this series, Vol. 49, p. 24.

scribed, can be independently selectively measured by performing the kinetic selection the other way round, that is by exchanging-in in the deoxy form, where allosterically sensitive sites are in their slow form, thus tend not to become labeled, and then exchanging-out as oxyHb, where the minimal label picked up on the sensitive sites is rapidly exchanged-out. Meanwhile the insensitive sites continue to exchange just as before; by definition their exchange rates are insensitive to the functional form of the protein. The resulting HX curve will therefore display essentially the (background) contribution that the insensitive sites make to the other curves, which are rich in sensitive sites. Subtraction of the background curve then produces a difference curve which portrays the exchange of only sensitive sites. A mathematical description and graphical illustrations of these operations have been provided in Refs. 5 and 6.

General Results

By use of the methods described, the entire Hb HX curve has been surveyed and the allosterically sensitive hydrogens measured. Only one-fourth of the exchanging peptide NH of hemoglobin are allosterically sensitive. These occur in seven fairly distinct kinetics sets, ranging in size from about 6 to 18 protons. The protons in each set exchange at very similar rates, but the various sets have quite different rates. All the sensitive sets are faster in the oxy form and become slower in deoxyHb by factors between 15 and 60, although one set displays a 10,000-fold slowing in deoxyHb.

This behavior has been taken to reveal the underlying behavior of cooperative local unfolding units, visualized as segments of helix that unfold transiently, and exchange as a unit. In this view, the selective labeling methods just described portray for us the HX behavior of the breathing units that happen to be sensitive to hemoglobin's allosteric form. The allosterically indifferent breathing units are removed from the measurement, and this allows the sensitive sets to be seen as kinetically distinct units, since they only account for a fraction of the total NH and are well dispersed over an 8 decade time scale. The insensitive sites, it can then be expected, also exchange in similar units, but these are not easily resolved by the selective labeling methods.

More recent work with hemoglobin has been aimed at defining the location of each of the allosterically sensitive sets.

5 S. W. Englander and A. Rolfe, *J. Biol. Chem.* **248**, 4852 (1973).
6 S. W. Englander, *Ann. N.Y. Acad. Sci.* **244**, 10 (1975).

Locating the Selectively Labeled Sites

Given a sample of protein selectively labeled as just described, it is possible to find the location of the label within the protein. This has been accomplished so far by proteolytic fragmentation of the protein and rapid separation of the fragments by high-performance liquid chromatography (HPLC). This approach was first reported by Rosa and Richards.[7,8] More recent efforts[9,10] have introduced modifications that maintain the protein under normal solution conditions throughout the initial (functional) labeling procedures, minimize loss of exchangeable tritium during the separation, and correct for the tritium losses that inevitably occur. This makes it possible to measure more or less quantitatively the tritium label remaining unexchanged at the time of starting the HPLC analysis and thus to compute the number of NH protons actually involved.

Sample Preparation

In preparing samples for the HPLC analysis, one maximizes initial protein concentration and levels of tritium labeling, since as little as a few nanomoles of oligopeptide fragments may be recovered from the HPLC run. We use as much as 10% initial concentration of Hb, and label with tritium in the 100 mCi/ml range. First column separations may use a 2×6 or a 2×4 cm Sephadex bed, so that volumes of initial sample as large as 1.5 ml can be filtered. Otherwise procedures used are like those described before. It can be noted that such high concentrations of protein in the initial exchange-in mix require a small change in the final data calculation; the factor 111 representing molar concentration of H in H_2O should be reduced by 0.65 for each 1% of protein concentration present in the initial exchange-in solution [see Eq. (1) below].

The protein fragmentation and HPLC separation steps are performed at low temperature (0°) and with pH about 2.7 where HX rate is minimized. At this condition the exchange halftime of even fully exposed peptide NH is over 1 hr, and longer in the partially organic solvent used in HPLC separations. The selectively labeled sample can be quenched to pH 2.7 by passing it through a short Sephadex column buffered at this pH with 0.1 M phosphate ($pK_1 = 2.1$). Alternatively it may sometimes be desirable to attain the low pH by adding a predetermined volume of phosphoric acid to the sample and then passing it through the Sephadex

[7] J. Rosa and F. M. Richards, *J. Mol. Biol.* **105**, 409 (1979).
[8] J. Rosa and F. M. Richards, *J. Mol. Biol.* **145**, 835 (1981).
[9] S. W. Englander, D. B. Calhoun, J. J. Englander, N. R. Kallenbach, E. L. Malin, C. Mandal, and J. R. Rogero, *Biophys. J.* **32**, 577 (1980).
[10] J. J. Englander, J. R. Rogero, and S. W. Englander, *J. Mol. Biol.* **169**, 325 (1983).

column. For example, a good test of the quantitative accuracy of the overall analysis can be obtained by starting with a fully labeled protein sample and measuring the label recovered at the end of the HPLC analysis. If all is well, the number of protons measured on each peptide recovered should be close to the known number of peptide NH on that fragment (1 per peptide NH minus prolines). In this test, if one runs the Sephadex separation at neutral pH instead of pH 2.7, label on all the non-H-bonded NH (exchange halftime at pH 7 ~0.5 sec at 0°) will be lost in this first step.

Subunit Separation

If one is working with a multisubunit protein like hemoglobin, a significant gain can be made in the quality of the ultimate HPLC fragment separation by initially separating the subunits. The α and β chains of Hb can be resolved in a brief HPLC run under conditions that lose little of the exchangeable tritium. This uses a semipreparative μBondapak C_{18} column (7.8 × 30 cm, from Waters Associates). This column, and all subsequent HPLC columns, are run at 0° by packing them in ice along with the injector valve and guard column. For this first column run also the post-column fittings and absorbance flow cell are kept cold.

The α/β subunit separation follows Congote et al.[11] in the use of an acetonitrile–phosphate buffer–trifluoroacetic acid (TFA) solvent system but incorporates certain changes. The run is made nearly isocratically at about 46% acetonitrile; the exact concentration is critical and changes from day to day. For solution A we use 40% acetonitrile, 57% 0.05 M NaH_2PO_4, and 3% TFA (importantly Sequenal grade from Pierce); pH is set to 3.3 (glass electrode) by adding ~25 ml 10 N NaOH per liter. Solution B is 50% acetonitrile, 50% 0.05 M phosphate, set to pH 3.3 with 1 N HCl. The column is prepared by initial washing with 50 ml of 58% solution B. The exact concentration used each day is first established in trial runs with untritiated Hb, and increases with aging of the mixed solvents. The selectively tritiated Hb sample at pH 2.7 is injected into the column after first diluting it by half with the column solvent (0.5 ml Hb ~8 mg + 0.5 ml column solvent). Without this step, a large fraction of the protein elutes at the solvent front. We run a shallow gradient, ~58 to 60% solution B in 10 min, starting 10 min after injection. Hb components elute in the order α chain, β chain, heme (opposite to Congote et al.[11] at room temperature). The heme is washed out of the column with phenyl B solvent (see below), then washed with 50 ml of 70% methanol and stored.

[11] L. Congote, H. Bennett, and S. Solomon, *Biochem. Biophys. Res. Commun.* **81**, 851 (1979).

Tritiated samples are collected through the α and/or β peaks in iced test tubes and can be either used directly or stored overnight or for longer periods frozen in dry ice.

Proteolytic Fragmentation

The selectively labeled Hb or one of the subunit chains isolated therefrom is next subjected to fragmentation by treatment with the acid protease, pepsin. Samples are first freed of the HPLC solvent by rapid dialysis[12] against pH 2.7, 0.05 M phosphate at 0° for 10 min. The protein sample at about 2 mg/ml is then treated with 5 to 10 μl of pepsin at 0° for 10 min. The sample is loaded onto an HPLC column (fatty acid column from Waters Associates) and run at 0° to separate the pepsin fragments.

The pepsin used (Worthington; 12 mg/ml) is prepared by dialysis against pH 4.5, 0.05 M acetate for several hours in a microdialysis stick[12] in the cold, and stored frozen in 50-μl aliqouts. HPLC solution phenyl A is 0.05 M sodium phosphate adjusted to pH 2.7. Solution phenyl B is 10% A, 30% dioxane, 60% acetonitrile, readjusted to pH 2.7 with care, using first 1 N HCl until the solution clears, then 0.1 N. It is important not to overshoot pH 2.7. The water is glass distilled, and solvents are prefiltered through CVWP 04700 Millipore filters.

The α subunit preparation is loaded onto the column, run for 2.5 min with phenyl A solution, then for 2.5 min with 20% phenyl B, then against a linear gradient from 20 to 50% phenyl B for 100 min. β subunit fragments are separated similarly except that the gradient is 15 to 45%. The columns are run at 1 ml/min. Conditions can also be set to deliver rapidly some particular fragment, e.g., by immediately running up to the appropriate solvent strength, then setting a gradient perhaps 4% wide in 20 min. Elution of peptide fragments is followed by absorbance at 230 nm in a 80 μl flow cell (Hellma), with a chart recorder set at 0.2 to 0.6 absorbance units full scale.

Data Analysis

Samples through the peptide peaks are collected by hand in such a way as to maximize their purity; for example, if a peak is overlapped on one side by another peak, collection will emphasize the opposite shoulder. Carried tritium label is measured by counting 0.5 ml samples by liquid scintillation. The organic solvents and low pH do not affect counting efficiency in our counting cocktail (5 g PPO, 100 g naphthalene, 120 ml toluene, 880 ml dioxane). The effective absorbance of each eluant sample

[12] S. W. Englander and D. Crowe, *Anal. Biochem.* **12**, 579 (1968).

is determined from the peak area and sample volume obtained from the recorder trace.

From these data, the number of protons per mole of peptide fragment not yet exchanged-out can be obtained from Eq. (1).

$$H/\text{peptide} = (Ct \times 111 \times E)/(A \times Ct_0 \times 1.19) \qquad (1)$$

Here Ct is the number of counts obtained from 0.5 ml of the peptide sample, and Ct_0 is the count level present in the same volume of the original exchange-in mixture; 111 is the molar concentration of H in water (generally somewhat lower for the exchange-in mixtures used here, see above). The term $Ct \times 111/Ct_0$ then yields the molar concentration of as yet unexchanged H in the peptide sample. A is the absorbance of the collected sample and E the extinction coefficient for that particular fragment; the term A/E then yields the molar concentration of fragment in the sample. We note that in order to translate most reliably measured data into terms of true hydrogens per protein, one should use the factor 1.19 in the data calculation to account for the equilibrium isotope effect.

We have generally found it possible to identify the fragment present in each elution peak by amino acid analysis and recourse to the known amino acid sequence of the hemoglobin chains. The extinction coefficients were obtained by quantitative amino acid analysis of samples having known absorbance. Additionally, extinction coefficients usually can be estimated accurately by summing the coefficients for the absorbing species present. These are, for our solvent conditions, $13.8 \times 10^3 \ M^{-1}$ cm^{-1} for tryptophan residues, 7.3×10^3 for tyrosine, and 500 for each peptide bond (at 230 nm).

The calculation indicated in Eq. (1) utilizes the amount of tritium label carried on the eluted fragments, thus indicates the number of original peptide NH still carried (not yet exchanged) in these fragments. One really wishes to obtain the analogous quantity for each sample before they were subjected to the losses of tritium label that inevitably occur during the fragmentation and separation steps. This can be obtained from information on the exchange rates that occur during these procedures. Fortunately this information is now available. Tritium loss during the HPLC separations runs between 20 and 50%. The details of obtaining these factors are somewhat complex and will be elaborated elsewhere.[13]

[13] J. J. Englander, J. R. Rogero, and S. W. Englander, *Anal. Biochem.* **147**, 234 (1985).

[23] Fluorescence Studies of Structural Fluctuations in Macromolecules as Observed by Fluorescence Spectroscopy in the Time, Lifetime, and Frequency Domains

By JOSEPH R. LAKOWICZ

The phenomenon of fluorescence emission typically occurs on the nanosecond timescale. Consequently, fluorescence spectral parameters are dependent on dynamic processes which occur on this timescale. Kinetic processes on the nanosecond timescale include (1) rotational diffusion and/or segmented motions of fluorophores bound to macromolecules, (2) the diffusion-controlled collision of quenchers with fluorophores, and (3) the reorientation of polar molecules around the dipole moment of the excited state. Experimentally, these processes are revealed by (1) the fluorescence anisotropy, or the time-dependent decays of anisotropy, (2) the reduction in fluorescence lifetime or yield caused by collisional quenching, and (3) time-dependent spectral shifts in the emission spectra.

These dynamic processes can be quantified in several ways. Time-dependent processes are observed most directly using time-domain measurements. In addition to time-domain measurements we also describe how similar information can be obtained in two additional domains, the lifetime domain and the frequency domain. Lifetime-domain experiments require steady-state measurements with varying amounts of collisional quenching. With adequate precautions, reliable information is available using only simple steady-state measurements. Finally, we will describe measurements in the frequency domain. Recent advances in instrumentation have made it possible to reliably obtain fluorescence phase shift and modulation data over a wide range of modulation frequencies. We will describe how frequency-domain measurements can be used to resolve multiexponential decays, and to determine time-resolved decays of anisotropy and time-resolved emission spectra. Frequency-domain fluorometry provides impressive resolution of rapid and/or complex anisotropy decays. More recently the technique of frequency-domain fluorometry has allowed measurements of intensity and anisotropy decays of proteins. This application originally required the use of continuous 280–300 nm lasers, but it is now possible to use pulsed laser sources. Such sources have intrinsic harmonic content extending to several GHz, and the available intensity is about 10^3 higher than possible using a continuous UV source. The results from this method will probably allow comparison of experimental data with molecular dynamics calculations.

Introduction

In recent years there has been considerable interest in the dynamic properties of biological macromolecules. This interest was stimulated in part by the early experiments which indicated that proteins are highly permeable to molecular oxygen,[1] by speculation that structural fluctuations are necessary for biological function,[2] and by molecular dynamics calculations which provide predictions for comparison with experimental results.[3] Among the available techniques to quantify structural fluctuations the fluorescence method is uniquely suitable. This is because of the high inherent sensitivity of this method, and especially because of its appropriate timescale. Fluorescence emission generally occurs 1 to 10 nsec following excitation. Consequently, dynamic processes which occur on this timescale can affect the measured fluorescence quantities. In this chapter we describe three such phenomena, segmental mobility of macromolecule-bound fluorophores, diffusion of quenchers to fluorophores, and time-dependent spectral relaxation.

First, fluorescence anisotropies provide a measure of the average rotational displacement of the fluorophore during the excited state lifetime. Time-resolved anisotropies reveal the time course of these motions. The overall rotational diffusion coefficient of proteins or vesicles can be predicted with reasonable confidence. Hence, one can compare the predicted and observed decays of anisotropy. A faster-than-expected decay of anisotropy is generally taken as evidence for segmental motion of the fluorophore in excess of overall rotational diffusion.

Second, fluorescence lifetimes or quantum yields can be decreased by collisional encounters with quenchers. This process requires diffusion of the quencher to the fluorophore during the lifetime of the excited state. Studies of collisional quenching can reveal the rates at which small molecules diffuse to fluorophores which are bound to macromolecules. For fluorophores buried within a protein or a membrane, this rate reflects permeation of the macromolecule by the quencher. Since proteins are closely packed, quenching of buried residues may reflect volume fluctuations which allow the quencher to penetrate the protein.

A third measure of dynamic behavior is provided by time-resolved emission spectra. Fluorophores which are sensitive to solvent polarity are

[1] J. R. Lakowicz and G. Weber, *Biochemistry* **12,** 4171 (1973).
[2] G. Careri, P. Fasella, and E. Gratton, *Annu. Rev. Biophys. Bioeng.* **8,** 69 (1979).
[3] M. Karplus and J. A. McCammon, *CRC Crit. Rev. Biochem.* **9,** 293 (1981).

expected to show time-dependent spectral shifts. These shifts occur as a result of reorientation of polar groups around the dipole moment of the excited state. If the fluorophore is bound to a macromolecule, then the rate of spectral relaxation may reflect the rate at which the macromolecule can reorient in response to creation of an excited state.

These three time-dependent processes are most directly revealed by measurements in the time domain. However, similar information is available from measurements in the lifetime domain[4] and in the frequency domain.[5,6] The latter measurements are made possible by measurement of fluorescence phase shift and modulation over a wide range of modulation frequencies.[7,7a] In this chapter we attempt to provide a unifying description which reveals the similarities between measurements in each of the three domains, time, lifetime, and frequency.

Time-Domain Measurements

Time-Resolved Fluorescence Anisotropies

The time-resolved decays of fluorescence anisotropy provide considerable detail about the diffusive motions of fluorophores. Such data can reveal whether a fluorophore is free to rotate over all angles, or if the surrounding environment restricts its angular displacement. Time-resolved anisotropies are measured following pulsed excitation. Detailed descriptions of the instrumentation for time-resolved measurements were given previously.[6,8,9] The parallel ($\parallel$) and perpendicular ($\perp$) components of the fluorescence intensity decay as

$$
\begin{aligned}
I_\parallel(t) &= \tfrac{1}{3} I_0(t) [1 + 2r(t)] \\
I_\perp(t) &= \tfrac{1}{3} I_0(t) [1 - r(t)]
\end{aligned}
\tag{1}
$$

where the decay of the total intensity is given by

$$
I_0(t) = I_\parallel(t) + 2I_\perp(t) = I_0 e^{-t/\tau}
\tag{2}
$$

In this equation τ is the fluorescence lifetime. The anisotropy is the ratio of the difference between the parallel and the perpendicular components

[4] J. R. Lakowicz, F. G. Prendergast, and D. Hogen, *Biochemistry* **18**, 520 (1979).

[5] G. Weber, *J. Chem. Phys.* **66**, 4081 (1977).

[6] J. R. Lakowicz, "Principle of Fluorescence Spectroscopy." Plenum, New York, 1983.

[7] E. Gratton and M. Limkeman, *Biophys. J.* **44**, 315 (1983).

[7a] J. R. Lakowicz and B. P. Maliwal, *Biophys. Chem.* **21**, 61 (1985).

[8] M. G. Badea and L. Brand, this series, Vol. 61, p. 378.

[9] T. I. Lin and R. M. Dowbin, *in* "Excited States of Biopolymers" (R. F. Steiner, ed.), p. 59. Plenum, New York, 1983.

to the total intensity,

$$r(t) = \frac{I_\parallel(t) - I_\perp(t)}{I_\parallel(t) + 2I_\perp(t)} \tag{3}$$

It is informative to consider a spherical molecule whose rotational motions are symmetric and unhindered. For such molecules the time-resolved anisotropy, following a δ-pulse excitation, is given by

$$r(t) = r_0 e^{-t/\phi} = r_0 e^{-6Rt} \tag{4}$$

where ϕ is the rotational correlation time of the fluorophore and R is its rate of rotation ($6R = \phi^{-1}$). The anisotropy in the absence of rotational diffusion is r_0.

In the interpretation of time-resolved anisotropies it is important to remember the meaning of r_0. From Eq. (4) one is tempted to assign r_0 to the value of the anisotropy at $t = 0$ [$r(0)$]. While this is correct for an ideal situation, instrumental limitations generally prevent the precise determination of r_0 from the time-resolved data. This is because the time-resolved anisotropies frequently display an initial more rapid decay, followed by a slower decay, presumably due to larger amplitude motions. The faster decay can be missed because of the limited time resolution of most instruments and the difficulties associated with deconvolution of the excitation pulse. In fact, the anisotropies can decay within the lamp profile itself. Hence, the values of $r(0)$ obtained from an attempt to extrapolate to $t = 0$ are frequently less than r_0. However, for most fluorophores, it is not necessary to obtain r_0 by extrapolation to $t = 0$. Instead, the steady-state anisotropy of the fluorophore can be measured at low temperatures and/or high viscosities which are adequate to prevent rotational diffusion. The absorption spectra of most fluorophores are rather insensitive to solvent, so that the r_0 values measured under vitrified conditions (with specific excitation and emission wavelengths) are generally appropriate for the higher temperature studies. The limiting anisotropy (r_0) is a fundamental property of the fluorophore which is independent of its rotational motion. The limiting anisotropy is given by

$$r_0 = \frac{2}{5}\left(\frac{3\cos^2\alpha - 1}{2}\right) \tag{5}$$

where α is the angle between the transition moments for absorption and emission. For parallel moments the angle equals zero and $r_0 = 0.4$. It is important that r_0 be measured under precisely the same experimental conditions which were used to measure $r(t)$. If at all possible, both mea-

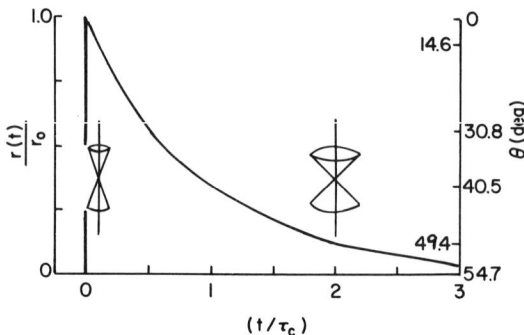

FIG. 1. Time-resolved decay of fluorescence anisotropy. Reprinted with permission from Lakowicz.[6]

surements should be performed in the same instrument. Once r_0 is known then this value can be compared with the anisotropy measured at the earliest observable time. If the apparent anisotropy at $t = 0$ is less than r_0 one may confidently decide that the fluorophore has undergone some displacement prior to the earliest time-resolved measurement of the anisotropy.

An intuitive representation of a time-resolved anisotropy is presented in Fig. 1. In this presentation we assumed that the sample is excited with an infinitely sharp pulse of vertically polarized light. At $t = 0$ the population of excited fluorophores is symmetrically distributed about the z axis. Complete alignment of the excited fluorophores with this axis is not achieved because the probability of light absorption depends upon $\cos^2 \theta$, where θ is the angle between the axis of polarization and the absorption transition moment of the fluorophore. This photoselection is accounted for by the factor of $\frac{2}{5}$ in Eq. (5). Also, at $t = 0$, the anisotropy is not necessarily equal to 0.4, but depends on the angle α between the absorption and emission transition moments.

Subsequent to excitation the distribution of excited fluorophores is broadened due to rotational diffusion. The anisotropy at any time t can be related to an average angular displacement by

$$r(t) = r_0 \left(\frac{3 \cos^2 \gamma - 1}{2} \right) \tag{6}$$

where γ refers to the angular displacement due to rotational diffusion. At $t = 0.1\phi$ one may state that the average fluorophore has rotated by $14.6°$. At $t = 2\phi$ this angle is $49.4°$. Complete depolarization is accomplished when $\gamma = 54.7°$. We stress that the anisotropy is a measure of $\cos^2 \gamma$. These values can be converted to an equivalent angle, as we did above.

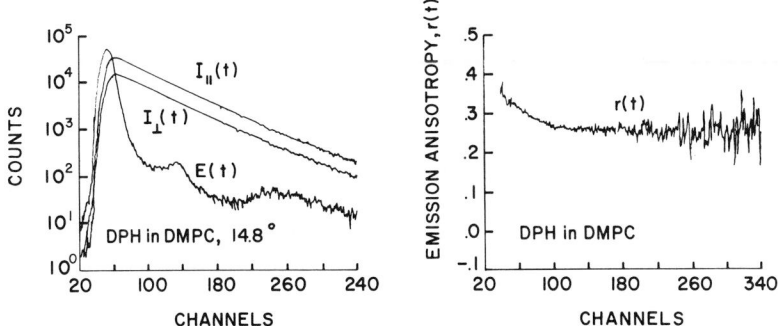

FIG. 2. Time-resolved anisotropies of DPH in DMPC vesicles.[11]

However, this angle is simply a way to visualize the meaning of the anisotropy value, and is not the actual angle through which the individual fluorophores rotate. Indeed, rotation must occur freely over all angles for γ to equal 54.7°. We note that this is a model-independent interpretation of the anisotropy values, and was described by Perrin in 1929[9a] and by Weber in 1952.[9b] In recent years this simple interpretation of the anisotropy values has been recast in more complex formalisms.[9c,9d]

Membrane-Bound Fluorophores

The fluoresence probe diphenylhexatriene (DPH) has been widely used to monitor the properties of lipid bilayers and cell membranes. This is because its steady-state anisotropy is highly sensitive to the phase state of the membranes.[10] Time-resolved anisotropies for DPH-labeled vesicles of dimyristoyl-L-α-phosphatidylcholine (DMPC) are shown in Fig. 2.[11] It is informative to examine the individual polarized components since these are the actual data used to derive $r(t)$. One notices from the left panel of Fig. 2 that $I_{\parallel}(t)$ remain greater than $I_{\perp}(t)$ at all observable times. This is equivalent to an anisotropy which, in contrast to the model data shown in Fig. 1, does not decay to zero. The implication of these results is that the anisotropy decays to some limiting value, called r_{∞}, at times long compared to the lifetime and to the fastest initial decay of anisotropy. These time-resolved decays are generally fit to an equation of

[9a] F. Perrin, *Ann. Phys.* **12**, 169; M. P. Soleillet, *Ann. Phys.* **12**, 23 (1929).
[9b] G. Weber, *Biochem. J.* **51**, 145 (1952).
[9c] A. Szabo, *J. Chem. Phys.* **81**, 150 (1984).
[9d] C. Zannoni, A. Arcioni, and P. Cavatorta, *Chem. Phys. Lipids* **36**, 179 (1983).
[10] B. Lentz, Y. Barenholz, and T. E. Thompson, *Biochemistry* **15**, 4521 (1976).
[11] L. A. Chen, R. E. Dale, S. Roth, and L. Brand, *J. Biol. Chem.* **252**, 2163 (1977).

the form

$$r(t) = r_\infty + (r_0 - r_\infty)e^{-t/\phi} \tag{7}$$

where $\phi = (6R)^{-1}$ now refers to the decay time of the anisotropy to a limiting value r_∞.

Considerable attention has been given to interpretation of the limiting anisotropies (r_∞). These values reflect the degree of hindrance to rotation imposed on the probe by the membrane, and thus should reflect some structural feature of the membranes. In the opinion of this author it is best to interpret the value of r_∞ using

$$\frac{r_\infty}{r_0} = \left\langle \frac{3 \cos^2 \theta - 1}{2} \right\rangle \tag{8}$$

where $\cos^2 \theta$ is the average value of this quantity after the rotational orientation of the probe has randomized to the extent possible within its hindered environment. Using this interpretation $\cos^2 \theta = 1/\sqrt{3}$ for complete depolarization, which can be interpreted as an equivalent angle of 54.7°. This interpretation is independent of any assumed model or assumed energy barrier to probe rotation. In contrast, the "wobbling-in-a-cone" model[12,13] assumes that there is a constant resistance to probe motion within a cone angle θ_c, and infinite resistance at angles greater than θ_c. With this model the experimental observable (r_∞/r_0) is related to the cone angle θ_c by

$$\frac{r_\infty}{r_0} \left[\frac{1}{2} \cos \theta_c (1 + \cos \theta_c) \right]^2 \tag{9}$$

For this model r_∞ can decay to zero only if the cone angle is 90°. The difficulty with this model is that the calculation of θ_c from r_∞/r_0 depends upon the existence of a square-well potential for rotation of the probe in the membranes. The observation of a nonzero r_∞ does not prove that this specific model is appropriate. For this reason it is recommended that limiting anisotropies be presented as the actual experimental values. If angles must be used, than those given from the definition of anisotropy [Eq. (8)] are preferable to those given from an undemonstrated model [Eq. (9)].

Tryptophan Fluorescence from Proteins

Relatively few time-resolved anisotropy decays have been published for the intrinsic tryptophan fluorescence from proteins. This is because most researchers use flash lamps with pulse widths of 1 nsec or longer.[8]

[12] K. Kinosita, S. Kawato, and A. Ikegami, *Biophys. J.* **20**, 289 (1977).
[13] G. Lipari and A. Szabo, *Biophys. J.* **30**, 489 (1980).

The lifetimes of tryptophan fluorescence are rather short (2–5 nsec), and often multiexponential.[14–16] Because of the pulse width of flash lamps and the rapid initial decays of the anisotropy, it has been difficult to obtain reliable time-resolved anisotropies. It seems probable that this situation will change quickly because of the increasing availability of lasers with narrow pulse widths. Nonetheless, even with pulse laser sources, one must consider the possibility of a substantial decay of anisotropy within the light pulse or within the time response of the instrument. Recent simulations by Visser and co-workers illustrate this possibility.[16b] With only a few exceptions[16a,16b] most currently available measurements for tryptophan residues in proteins were obtained using flash lamps or synchrotron radiation.

The time-resolved anisotropy for a protein-bound fluorophore is expected to somewhat more complex than that for a membrane-bound fluorophore. The main additional complication is the need to consider rotational diffusion of the protein, as well as segmental motions of the tryptophan residues. For membrane-bound fluorophores it was not necessary to consider rotational diffusion of the vesicles because vesicle rotation is expected to be slow relative to the decay of fluorescence. Let ϕ_P be the correlation time for overall rotation of the protein, and ϕ_F the correlation time for segmental motion. Then, assuming that the segmental motions occur independently of rotational motion, the anisotropy decays as the product of these separate events. That is

$$r(t) = r_0 e^{-t/\phi_P}[\alpha e^{-t/\phi_F} + (1 - \alpha)] \qquad (10)$$

where α is the fraction of the total depolarization which is lost by the more rapid process. This may be regarded as a somewhat more complex case of the hindered model in which the anisotropy decays rapidly to an apparent limiting value, $r_\infty = r_0(1 - \alpha)$. Subsequently, the remaining anisotropy decays to zero as a result of overall protein rotation (e^{-t/ϕ_P}). It is interesting to notice that the faster motion must be hindered ($\alpha < 1$) in order to observe a multiexponential decay of anisotropy. If the segmental motion is completely unhindered ($\alpha = 1$) then the segmental motions are revealed by an apparent correlation time $[\phi_A = (\phi_P\phi_F)/(\phi_P + \phi_F)]$ which is smaller than that expected for overall rotational diffusion of the protein.

[14] A. Grinvald and I. Z. Steinberg, *Biochim. Biophys. Acta* **427**, 663 (1976).

[15] J. A. B. Ross, C. J. Schmidt, and L. Brand, *Biochemistry* **20**, 4369 (1981).

[16] J. P. Privat, P. Wahl, J. C. Auchet, and R. H. Pain, *Biophys. Chem.* **11**, 239 (1980).

[16a] A. van Hoek, J. Veroort, and A. J. W. G. Visser, *J. Biochem. Biophys. Methods* **7**, 243 (1983).

[16b] A. J. W. G. Visser, T. Yhema, A. V. Hoek, D. J. O'Kane, and J. Lee, *Biochemistry* **24**, 1489 (1985).

A wide range of segmental freedom has been found for tryptophan residues in native proteins. This is illustrated (see Fig. 3) by the time-resolved anisotropies for nuclease B, human serum albumin (HSA), and myelin basic protein.[17] For nuclease the anisotropy decays approximately as a single exponential with ϕ = 9.9 nsec. This value is approximately that predicted for a hydrated sphere

$$\phi_P = \frac{\eta MW}{RT} (\bar{v} + h) \tag{11}$$

where MW is the molecular weight, $\bar{v}$ is the specific volume, h is the degree of protein hydration, and η is the viscosity. Hence, the single tryptophan residue of nuclease appears to be rigidly bound to the protein on the timescale of the fluorescence emission. Contrasting results were observed for myelin basic protein (MBP). The anisotropy decayed with a correlation time of 1 nsec or less. Although subsequent measurements in the lifetime domain indicate somewhat longer correlation times for the single tryptophan residues in MBP,[18] there seems little doubt that this residue displays considerable motional freedom in excess of overall protein rotation.

Segmental freedom intermediate between these extremes was observed for the single tryptophan residue of human serum albumin (HSA). At 4° the anisotropy decayed as a single exponential with ϕ = 31 nsec. This is comparable to the calculated value of 40 nsec, obtained from Eq. 11 with MW = 64,000 and h = 0.2 g/g of protein. Hence, this residue appears immobile at 4°. At higher temperature one notices a rapid component in the anisotropy decay, followed by a slower decay most likely due to overall protein rotation. More extensive studies in the lifetime domain[18,19] gave similar results for HSA. Additionally, the lifetime-domain measurements also indicated that the extent of segmental mobility of both tryptophan and tyrosine residues in protein is highly variable among proteins, and dependent upon their conformation.

Potentially important information is available from the anisotropies observed at t = 0 [$r(0)$]. To compare $r(0)$ with r_0 these values must be measured using the same instrument and using the same conditions for excitation and emission. Such consistency is especially important for tryptophan because its last absorption band displays a complex dependence of r_0 on excitation wavelength.[20] Suppose such carefully measured values

[17] I. Munro, I. Pecht, and L. Stryer, *Proc. Natl. Acad. Sci. U.S.A.* **76,** 56 (1979).
[18] J. R. Lakowicz, B. P. Maliwal, H. Cherek, and A. Balter, *Biochemistry* **22,** 1741 (1983).
[19] J. R. Lakowicz and B. P. Maliwal, *J. Biol. Chem.* **258,** 4794 (1983).
[20] B. Valeur and G. Weber, *Photochem. Photobiol.* **25,** 441 (1977).

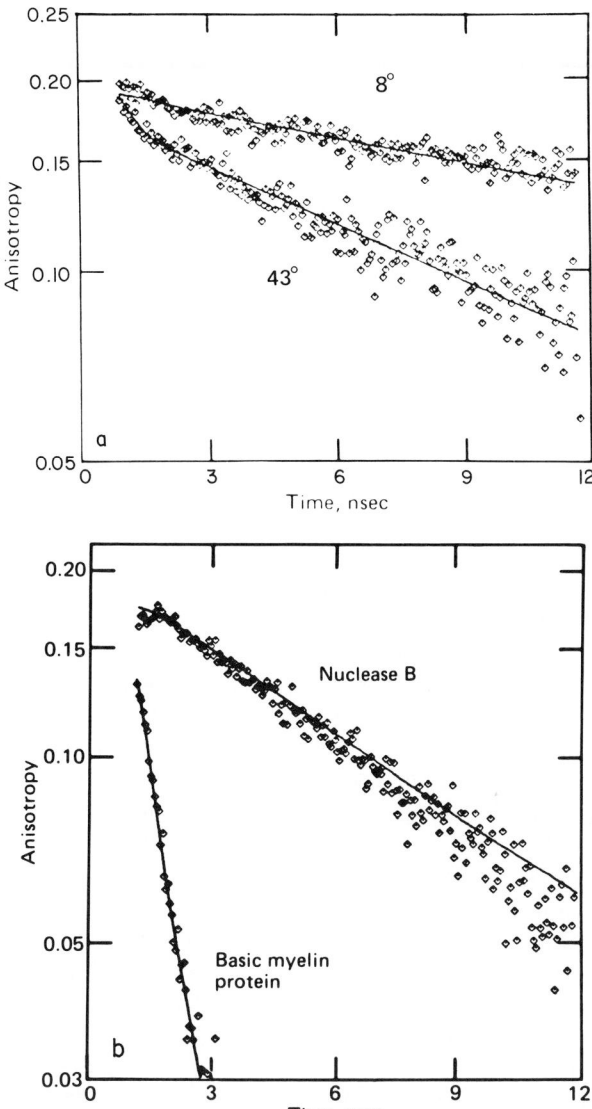

FIG. 3. Time-resolved anisotropies for proteins. (a) HSA; (b) nuclease B and basic myelin protein. Reprinted with permission from Munro *et al.*[17]

of r_0 are available. Then, if $r(0) < r_0$ one could decide that the tryptophan residue displayed segmental mobility with a correlation time too short to be resolved by the experiment, or equivalently, whether the anisotropy has decayed within the time response of the instrument.[16b] Unfortunately,

r_0 values are often not measured, and the potentially interesting comparison of $r(0)$ and r_0 is not possible.

Solvent Relaxation as Quantified by Differential-Wavelength Deconvolution

We now consider a different time-dependent process. Many fluorophores are sensitive to the polarity of their surrounding environment. Generally, the emission spectra of such fluorophores shift to longer wavelength (red shift) as the solvent polarity is increased. A detailed interpretation of solvent-dependent emission spectra requires a considerable understanding of the specific and general interactions between the fluorophore and the solvent.[6] At present most theories of solvent effects regard the solvent of a continuum.[6,20,20a,20b] Consequently, it is difficult to interpret the results in terms of specific molecular events. More recently theories have appeared which treat the solvent effects in terms of specific dipolar interactions.[20c,20d] Such theories will facilitate interpretation of time-resolved spectral data in terms of the structure and dynamics of macromolecules. In this section we will be concerned mainly with observation of the time-dependent spectral shifts.

A schematic representation of solvent relaxation is shown in Fig. 4. Generally, the dipole moment of a fluorophore increases upon excitation. If the solvent is fluid then the rate of solvent relaxation is much more rapid than emission, and a red-shifted emission (R) is observed. If the solvent is cold and/or viscous then the rate of solvent reorientation around the fluorophore is slow, and the emission spectrum shifts to shorter wavelengths (F). At any given wavelength the decay of fluorescence intensity is complex because the emission results from at least two states, and more probably a larger number of intermediate states. Of course the decay may be estimated by a sum of exponential decays, but the actual decay is probably nonexponential. In spite of this complexity it is possible to analyze the time-dependent spectral shifts using data observed in the time domain, the lifetime domain, or the frequency domain (see below). In previous reports time-resolved emission spectra of membrane-bound fluorophores have been described in considerable detail[8,21,22]

[20a] V. E. Lippert, Z. Electrochem. **61,** 962 (1957).

[20b] N. Mataga, Y. Kaifu, and M. Koizumi, Bull. Chem. Soc. Jpn. **29,** 465 (1956).

[20c] R. B. MacGregor and G. Weber, Proc. N.Y. Acad. Sci. **366,** 140 (1981).

[20d] G. Weber, Proc. NATO Adv. Study Inst. Sicily (1984).

[21] M. G. Badea and L. Brand, Biophys. J. **24,** 197 (1978).

[22] J. H. Easter, R. P. DeToma, and L. Brand, Biophys. J. **16,** 571 (1976).

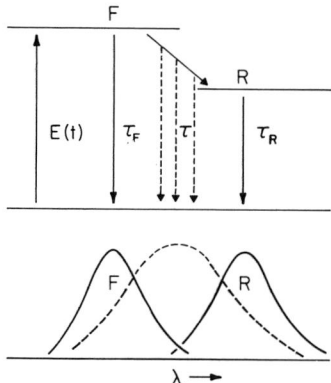

FIG. 4. Schematic representation of solvent relaxation. The spectra for F and R represent the hypothetical spectra for the initially excited and relaxed states, respectively. The steady-state spectrum expected when $\tau = \tau_s$ is also shown (---). From Lakowicz.[6]

These earlier studies demonstrated that the rates of spectral relaxation for membrane-bound fluorophores were dependent upon the chemical composition and phase state of the membranes. In this section we wish to demonstrate how the analysis of time-dependent spectral shifts can be simplified by the technique of differential-wavelength deconvolution.[23,24]

In the presence of spectral relaxation the time-dependent decays of intensity can be complex. This is particularly true for a reversible excited state process, that is, the partial repopulation of the F state from the R state (Fig. 4). For this case the emission may be described as a sum of exponentials, but the decay times and preexponential factors cannot be easily interpreted to yield the decay times or spectra of the individual species.[23] A useful simplification is available which yields the resolved emission spectra of the F and R states.

From Fig. 4 we note that the R state is not excited directly by the light pulse. Rather, the R state is populated by F state molecules, which then decay to the R state. That is, the F state population is in fact the apparent excitation pulse for the R state. This suggests that if the R state emission is deconvolved using the F state emission rather than the lamp profile then the intrinsic properties of the R state would be revealed. That this simplification in fact occurs has been demonstrated theoretically and experimentally.[23,24]

[23] J. R. Lakowicz and A. Balter, *Biophys. Chem.* **16,** 233 (1982).
[24] J. R. Lakowicz and A. Balter, *Biophys. Chem.* **15,** 353 (1982).

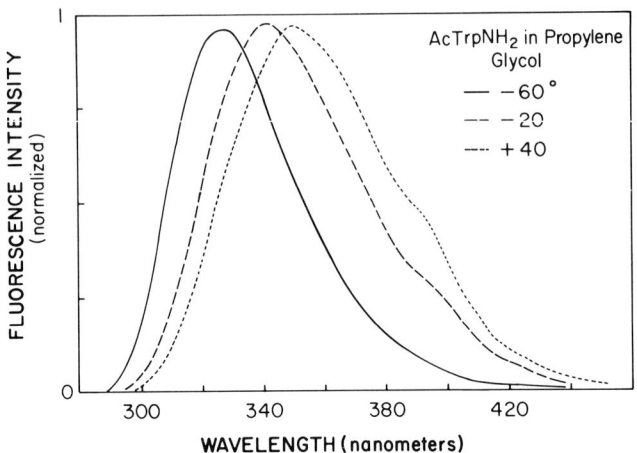

FIG. 5. Steady-state emission spectra of N-acetyl-L-tryptophanamide AcTrpNH$_2$ in pro-pylene glycol. Spectra are shown at $-60°$ (——), $-20°$ (– – –), and $40°$ (- - -). From Lakowicz and Balter.[24]

To take advantage of this simplification there is one requirement. One must be able to select an emission wavelength at which only the F state emits. This is needed so that an F state response can be obtained for use in the deconvolution procedure. Fortunately, this is generally possible. The effect of temperature on the steady state emission spectrum of N-acetyl-L-tryptophanamide (NATA) in propylene glycol is shown in Fig. 5. At low temperature ($-60°$) a blue shifted emission spectrum is observed. As the temperature is increased (-20 and $+40°$) the emission shifts to longer wavelengths. From other evidence[25] it is known that these spectral shifts are the result of time-dependent processes which occur subsequent to excitation. In the case of NATA in propylene glycol (Fig. 5) the emission at 300 nm is due almost exclusively to the F state. Hence, the procedure is as follows. Time-resolved decays of intensity are observed across the emission spectrum. The decay on the blue side of the emission is taken to represent the F state. Then, the decays at longer wavelengths are deconvolved using the 300 nm emission (F state) as the apparent excitation pulse for the R state. Of course, some F state emission may be observed at longer wavelengths. Since the 300 nm emission is defined as the exciting pulse, the F state emission appears to be a zero decay-time component at longer wavelengths. The individual emission spectra of the

[25] J. R. Lakowicz, H. Cherek, and D. R. Bevan, *J. Biol. Chem.* **255**, 4403 (1980).

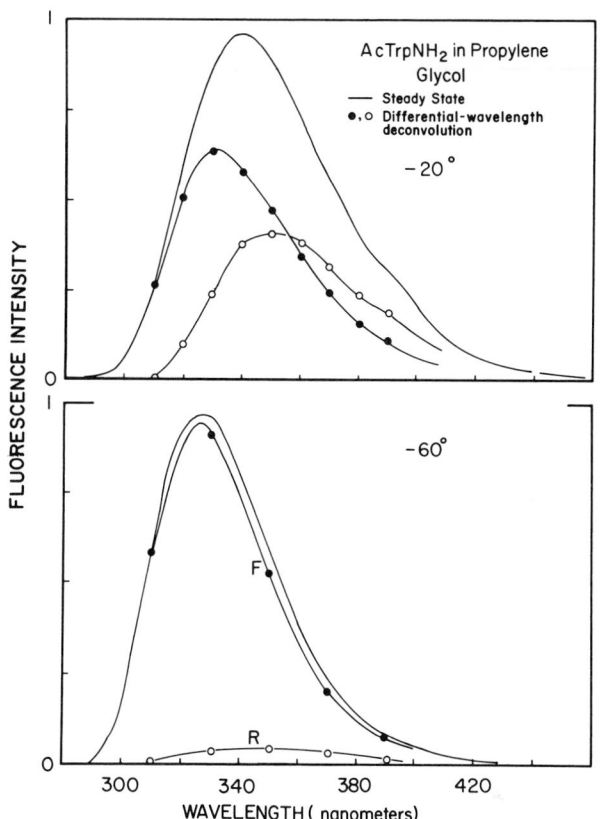

FIG. 6. Resolution of emission spectra of the initially excited and solvent-relaxed states of *N*-acetyl-L-tryptophanamide (AcTrpNH$_2$) in propylene glycol by differential-wavelength deconvolution. Steady state (——), *F* state (●), and *R* state (○). From Lakowicz and Balter.[24]

F and *R* states [$I_F(\lambda)$ and $I_R(\lambda)$] are easily calculated using

$$I_F(\lambda) = \frac{\alpha_0 \tau_0}{\alpha_0 \tau_0 + \alpha_R \tau_R} I(\lambda) \qquad (12)$$

$$I_R(\lambda) = \frac{\alpha_R \tau_R}{\alpha_0 \tau_0 + \alpha_R \tau_R} I(\lambda) \qquad (13)$$

where $I(\lambda)$ is the steady-state emission spectra, α_0 and τ_0 are the amplitude and near-zero decay time of the *F* state, and α_R and τ_R are the amplitude and decay time of the *R* state.

Application of this procedure to the case of NATA in propylene glycol is shown in Fig. 6. At $-20°$, where an intermediate emission spectrum was observed (Fig. 5), the intensities of the F and R states are nearly equal. At lower temperature, where a blue shifted emission was observed, only a small amount of R state emission was found. these results illustrate that, by using this differential procedure, the approximate amounts of F and R state emission can be determined. This procedure should be useful in studies of time-dependent shifts of tryptophan emission from proteins. Of course, it is necessary to determine that the time-dependent shifts are due to relaxation and not ground state heterogeneity.

For completeness we note that Beechem and co-workers have also simplified the analysis of excited state reactions.[25a] They used multiwavelength or global[25b] analysis methods to solve directly for the kinetic constants of the system. Their method is somewhat more complex than differential deconvolution in that it requires computer programs specially designed for this purpose. However, their method is more general than differential deconvolution in that it is not necessary to have a region where only the F state emits.

Lifetime-Domain Measurements

Collisional quenching of fluorescence provides a useful method to obtain time-resolved information, without the exclusive use of time-resolved instrumentation. In the lifetime-domain steady-state measurements are mostly adequate to determine the time-dependent parameters. The principle of the method is that quenching decreases the lifetime of the fluorophore, and thus the relative influence of the time-dependent processes on the steady-state measurements. In principle one can study arbitrarily fast processes by increasing the degree of quenching, but in practice the loss of intensity becomes prohibitive. Since the technique is simple, it has been possible to examine a variety of proteins and membranes. Many small molecules act as collisional quenchers,[6,25c] so that the lifetime-resolved method can be applied in many circumstances.

Lifetime-Resolved Fluorescence Anisotropies

Consider a fluorophore in the presence of a quencher, Q. For purely collisional quenching its lifetime is given by

$$\tau = \tau_0/(1 + K[Q]) \tag{14}$$

where τ_0 and τ are the lifetimes in the absence and presence of quencher, respectively, and K is the Stern–Volmer quenching constant. Collisional quenching provides a means to vary τ, and hence the average time avail-

[25a] J. M. Beechem, M. Ameloot, and L. Brand, *Chem. Phys. Lett.* **120,** 466 (1985).
[25b] J. R. Knutson, J. M. Beechem, and L. Brand, *Chem. Phys. Lett.* **102,** 501 (1983).
[25c] M. Eftink and C. Ghiron, *Anal. Biochem.* **14,** 199 (1981).

able for depolarization. For any given lifetime the steady-state anisotropy $r(\tau)$ is an average of the time-resolved anisotropy over the decay law of the sample,

$$r(\tau) = \frac{\int_0^\infty r(t)e^{-t/\tau}dt}{\int_0^\infty e^{-t/\tau}dt} \tag{15}$$

First we consider the application of this technique to membrane-bound fluorophores. Assume $r(t)$ decays to zero as a single exponential. Then application of Eq. (15) using Eqs. (2) and (4) yields

$$r = \frac{r_0 - r}{6R\tau} \tag{16}$$

A plot of r versus $(r_0 - r)/\tau$ is expected to yield a slope equal to $(6R)^{-1}$ and an intercept of zero. Now consider a hindered fluorophore whose time-resolved anisotropy is given by Eq. (7). In this case application of Eq. (13) yields

$$r = r_\infty + \frac{(r_0 - r)}{6R\tau} \tag{17}$$

A plot of r versus $(r_0 - r)/\tau$ again yields $(6R)^{-1}$ as the slope. The limiting anisotropy is given by the y intercept.

The fact that time-domain information is available in the lifetime domain is illustrated in Fig. 7. The left panel shows lifetime-resolved anisotropies for DPH in mineral oil. In this case the y intercepts are zero, indicating that $r_\infty = 0$. This observation is in agreement with the time-resolved measurements of DPH in mineral oil, which also indicated that $r_\infty = 0$.[26] For DPH bound to DMPC vesicles (right panel) substantially higher y intercepts were observed. Once again, the r_∞ values found by this extrapolation to $\tau = 0$ accurately reflected these same values found from the time-resolved measurements. We note that in the use of the lifetime-resolved method it is essential to correct for any apparent static quenching.[4]

Lifetime-Resolved Studies of Tryptophan and Tyrosine Mobility in Proteins

Lifetime-resolved methods have been especially useful in studies of the segmental motions of aromatic residues in proteins.[18,19] This is because protein fluorescence is easily quenched by molecular oxygen, and this quenching is almost purely collisional. Also, relatively few time-resolved anisotropy decays of proteins have been published. This is because of the limited time resolution and sensitivity of most instruments which use a flash lamp as the source, and the relatively few pulsed laser

[26] R. E. Dale, L. A. Chen, and L. Brand, J. Biol. Chem. **252**, 7500 (1977).

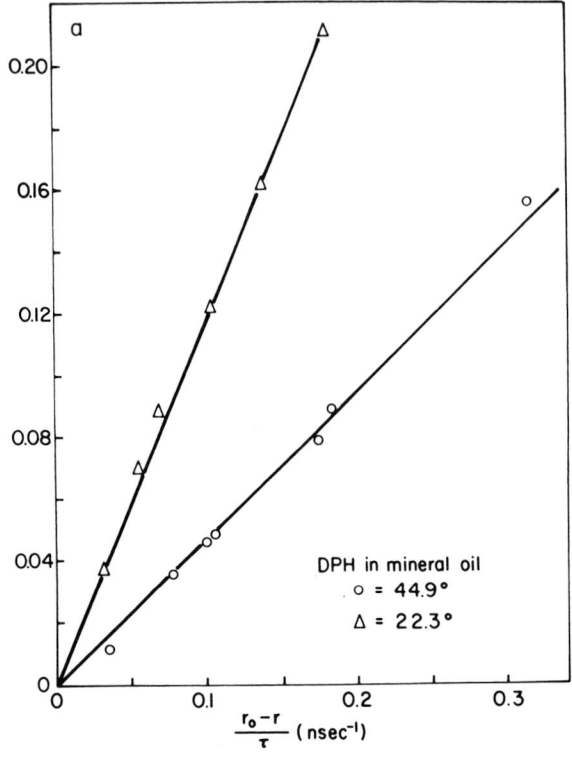

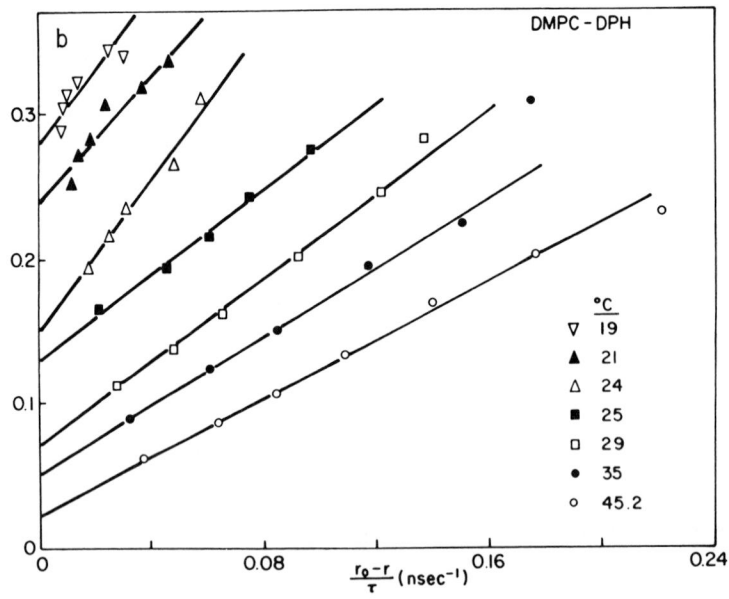

instruments which are in use for this purpose.[16a,26a] When the unquenched lifetime is known, then only steady-state measurements of the anisotropies are required to estimate the anisotropy decay, and the steady-state values can be obtained with high precision. Also, the short lifetimes of tryptophan and tyrosine fluorescence are not a problem because it is not necessary to deconvolute the emission from a relatively wide excitation pulse. Lifetime-resolved measurements can also be performed with quenchers other than oxygen.[27]

For lifetime-resolved anisotropies of proteins one expects the anisotropy to decay to zero because the overall rotational diffusion of proteins in solution is not hindered. In these cases we found the most reliable method to analyze these data is by plotting the reciprocal anisotropy versus the lifetime,

$$\frac{1}{r} = \frac{1}{r_0} + \frac{\tau}{r_0\phi} \tag{18}$$

Then, the correlation time is obtained from the slope, and a estimation of $r(0)$ is obtained by extrapolating to $t = 0$. The extrapolated value of $r(0)$ can be compared with the measured value r_0. If $r(0)$ is substantially less than r_0, one can conclude that the fluorophore has undergone segmental motions which are mostly complete prior to the shortest quenched lifetimes. Additionally, one can compare the measured correlation time with that expected for overall rotational diffusion of the protein [Eq. (11)]. Segmental motions on the nanosecond timescale result in the apparent correlation times which are shorter than that predicted by Eq. (11). Theoretical calculations illustrating these results are described elsewhere.[18,19]

The considerable time resolution of the lifetime-resolved method is illustrated by Fig. 8. Data are shown for small tyrosine peptides. By oxygen quenching it was possible to decrease the fluorescence lifetimes about 3-fold. The steady-state anisotropies measured under these conditions allowed measurement of correlation times as short as 32 psec. Of course, with the lifetime-resolved data it is difficult to determine the precise form of the anisotropy decay, that is, whether the anisotropy decays as a single or a multiexponential. Nonetheless, a 32 psec correlation time at 25° is close to the value measured in the frequency domain at 20° (45 psec). Also these values are faster than what is usually measurable in the

[26a] E. W. Small, J. L. Libertini, and I. Isenberg, *Rev. Sci. Instrum.* **55**, 879 (1984).
[27] M. Eftink, *Biophys. J.* **43**, 323 (1983).

FIG. 7. Lifetime-resolved anisotropies of DPH (a) in mineral oil and (b) in DMPC vesicles. Reprinted with permission from Lakowicz *et al.*[4] Copyright 1979 American Chemical Society.

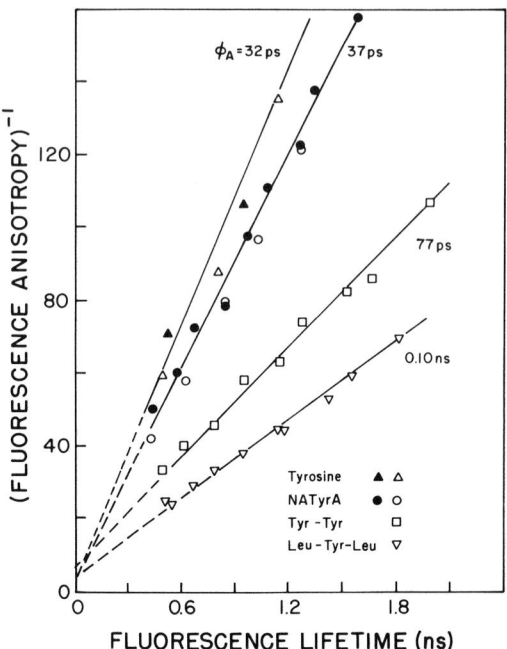

FIG. 8. Lifetime-resolved anisotropies of tyrosine and tyrosine peptides. From Lakowicz and Maliwal.[19]

time domain.[27a,27b] Recently we developed the ability to measure such rapid anisotropy decays for tryptophan fluorescence in the frequency domain. For instance we measured correlation times of 47 and 45 psec for tryptophan and tyrosine derivatives in aqueous buffer at 20°, respectively[27c,27d] (see Fig. 26).

The application of this method to human serum albumin is shown in Fig. 9. At low temperature and without denaturants, the apparent correlation time of 45 nsec is in good agreement with that calculated from Eq. (11) (40 nsec) and with that observed from the time-resolved decay of anisotropy (31 nsec). At higher temperature (44°) the apparent correlation time (8.5 nsec) is less than that expected for overall protein rotation. Such a decrease on the apparent correlation time is consistent with some segmental motion of the tryptophan residue. Such motional freedom was also seen in the time-resolved anisotropy (Fig. 3).

[27a] R. S. Moog, M. D. Ediger, S. G. Boxes, and M. D. Fayer, *J. Phys. Chem.* **86,** 4694 (1982).

[27b] S. A. Rice and G. A. Kenny-Wallace, *Chem. Phys.* **47,** 167 (1980).

[27c] J. R. Lakowicz, G. Laczko, I. Gryczynski, and H. Cherek, *J. Biol. Chem.* **261,** 2240 (1986).

[27d] J. R. Lakowicz, G. Laczko, H. Cherek, and I. Gryczynski, unpublished observations.

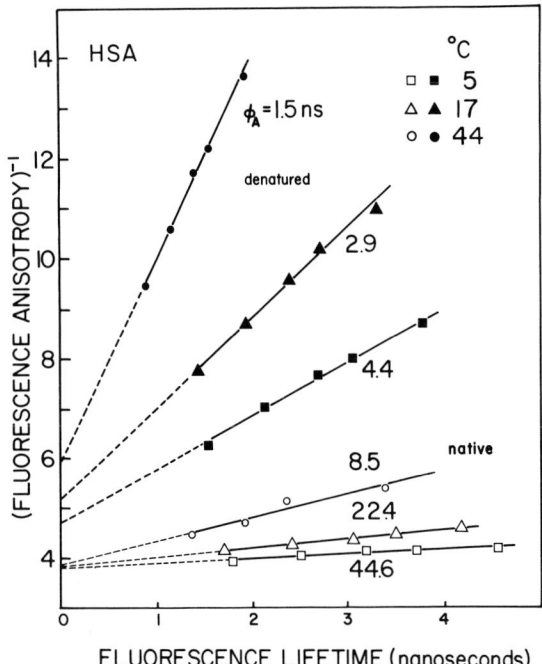

FIG. 9. Lifetime-resolved anisotropies of human serum albumin in the absence and presence of denaturants. Reprinted with permission from Lakowicz *et al.*[18] Copyright 1983 American Chemical Society.

HSA was also examined in the presence of denaturants.[18] The lifetime-resolved anisotropies indicate that the apparent correlation times of the tryptophan residue decreased substantially to values near 3 nsec. Additionally, the y intercepts increased. This result indicate that some fraction of the initial anisotropy is already lost at the shortest quenched lifetimes [Eq. (18)]. The y intercept and the known value of r_0 can be used to calculate an average angular displacement of the residues at times less than 0.5 nsec [Eq. (6)]. For example, for HSA at 44°, in the presence of 6 M guanidine hydrochloride, $r(0) = 0.169$. Using $r_0 = 0.260$ one can readily calculate using Eq. (6) that the fluorophore is mobile through an average angle of 29°. Recalling that complete depolarization is achieved at 54.7°, this angle represents considerable motional freedom.

A final example of lifetime-resolved anisotropies is given in Fig. 10. These data were obtained using oxygen quenching of the single tyrosine residue in histone H1. It is known that this histone exists mostly as a random coil at low pH, and that a more compact folded structure is formed at neutral pH. In the folded state the apparent correlation time is 4.2 nsec. Upon unfolding this value decreases to 1.0 nsec, and the exis-

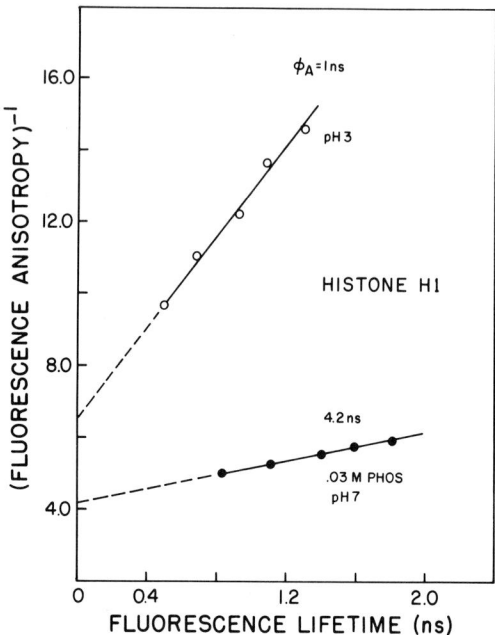

FIG. 10. Lifetime-resolved anisotropies of the singled tyrosine residue in histone H1. From Lakowicz and Maliwal.[19]

tence of faster unresolved motions is indicated by the increased y intercept [decreased $r(0)$]. The results in Figs. 7–10 demonstrate the substantial potential of lifetime-resolved methods for measurement of short correlation times. At the present time, such data are just becoming available from direct time-resolved measurements. Additionally, in the past 2 years we found that the frequency-domain method provides remarkable resolution of complex and/or rapid anisotropy decays (see below and Refs. 7a, 27e, 48).

Lifetime-Resolved Measurements of Time-Dependent Spectral Shifts

To date, time-dependent spectral shifts have been examined almost exclusively in the time domain,[21,22,28] and more recently using data in the frequency domain.[28a,28b] This is because the process is complex, and only the measurements with substantial information content could provide in-

[27e] J. F. Faucon and J. R. Lakowicz, Anisotropy decay of DPH in melittin-lipid complexes by multi-frequency phase-modulation fluorometry. Submitted for publication.

[28] W. R. Ware, S. K. Lee, G. J. Brant, and P. P. Chow, *J. Chem. Phys.* **54,** 4729 (1970).

[28a] J. R. Lakowicz, H. Cherek, G. Laczko, and E. Gratton, *Biochim. Biophys. Acta* **777,** 183 (1984).

[28b] J. R. Lakowicz and H. Cherek, *Chem Phys. Lett.* **122,** 380 (1986).

formation adequate to reveal the details of this process. However, with some reasonable assumptions, the lifetime-domain measurements can provide an estimate of the spectral relaxation times.[29]

Assume that the center of gravity of the emission spectrum $[\bar{\nu}_{cg}(t)]$ or less rigorously, the emission maximum (in cm^{-1}) decays exponential from some initial value $\bar{\nu}_0$ to a final value $\bar{\nu}_\infty$. Then

$$\bar{\nu}_{cg}(t) = \bar{\nu}_\infty + (\bar{\nu}_0 - \bar{\nu}_\infty)e^{-t/\tau_s} \qquad (19)$$

where τ_s is the spectral or solvent relaxation time. For any given lifetime the center of gravity is

$$\bar{\nu}_{cg}(\tau) = \frac{\int_0^\infty \bar{\nu}_{cg}(t)e^{-t/\tau}dt}{\int_0^\infty e^{-t/\tau}dt} \qquad (20)$$

$$= \bar{\nu}_\infty + (\bar{\nu}_0 - \bar{\nu}_\infty)\frac{\tau_s}{\tau_s + \tau} \qquad (21)$$

The observed extent of solvent relaxation depends upon both τ_s and the lifetime τ. Since the lifetime can be reduced by quenching [Eq. (14)], the steady-state spectra at various lifetimes can be used to determine the spectral relaxation time.[29]

Lifetime-resolved emission spectra are shown in Fig. 11. These data are for vesicles of DMPC/cholesterol, labeled with 2-p-toluidinyl-6-naphthylenesulfonic acid (TNS). As the lifetime is decreased from 6.7 to 0.5 nsec the steady-state emission spectrum shifts to shorter wavelengths. Such shifts occur because quenching, like emission, is a random event. Those molecules which emit at longer wavelengths are also those which have remained in the excited state long enough for relaxation to have taken place, and thus whose energy has decayed to lower values. The longer lived fluorophores have a greater probability of encountering a quencher prior to emission. Hence, the longer lived fluorophores are selectively quenched and the emission spectra shift to shorter wavelengths.

Lifetime-resolved emission spectra can be used to estimate the spectral relaxation times. Temperature-dependent data for DMPC/cholesterol vesicles are shown in Fig. 12. The spectral relaxation times are estimated by fitting assumed values of τ_s to the measured values of $\bar{\nu}_{cg}(\tau)$. In such fitting it is difficult to obtain values of τ_s without some knowledge of $\bar{\nu}_0$ or $\bar{\nu}_\infty$.[29] Nonetheless, since only steady-state measurements are used, the measurements permit event short relaxation times to be measured. For example, at 45°, the spectral relaxation times are estimated to be about 0.5 nsec. Once again we note that the information obtained in the lifetime

[29] J. R. Lakowicz and D. Hogen, *Biochemistry* **20**, 1366 (1981).

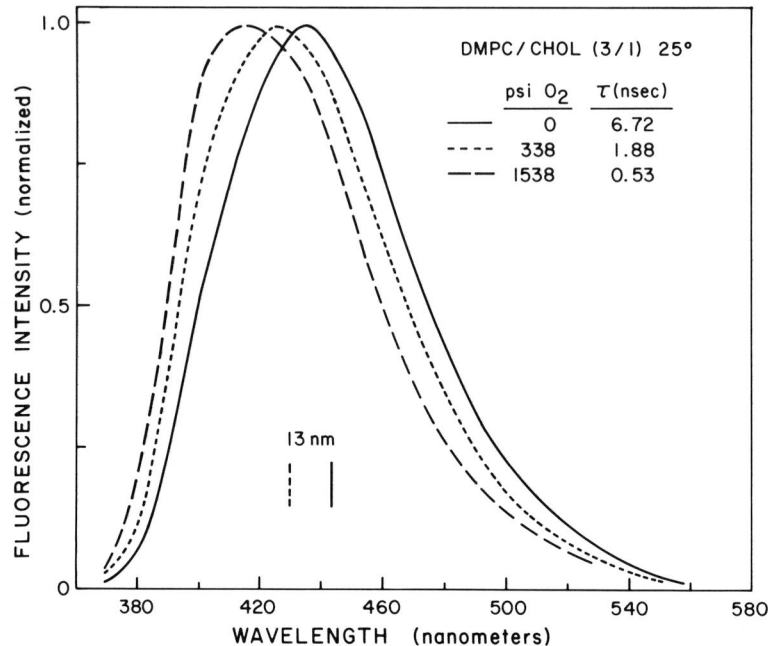

FIG. 11. Effects of oxygen quenching on the emission spectra of TNS-labeled vesicles. The vesicles were DMPC/cholesterol, 3/1 molar ratio. The solid and dashed vertical lines represent the center of gravity of the emission spectra. Reprinted with permission from Lakowicz and Hogen.[29] Copyright 1981 American Chemical Society.

domain is in good agreement with the time-resolved measurements on these same systems[21,22] and with more recent results from the frequency domain.[28a] Additionally, the lifetime-resolved method is potentially useful for estimating even more rapid rates of relaxation. For instance, this method was used to measure the rate of spectral relaxation around indole in ethanol at 25°.[30] The relaxation time was found to be 13 psec. To date, this measurement has not been performed by time-resolved or frequency-domain measurements.

Frequency-Domain Measurements

Until recently resolution of complex decay laws by phase-modulation fluorometry was severely limited by the small number of available modulation frequencies, typically two or three. In the frequency domain the equivalent of a nanosecond excitation pulse are phase and modulation data measured over a wide range of frequencies. During the past 2 years

[30] G. Weber and J. R. Lakowicz, *Chem. Phys. Lett.* **22**, 419 (1973).

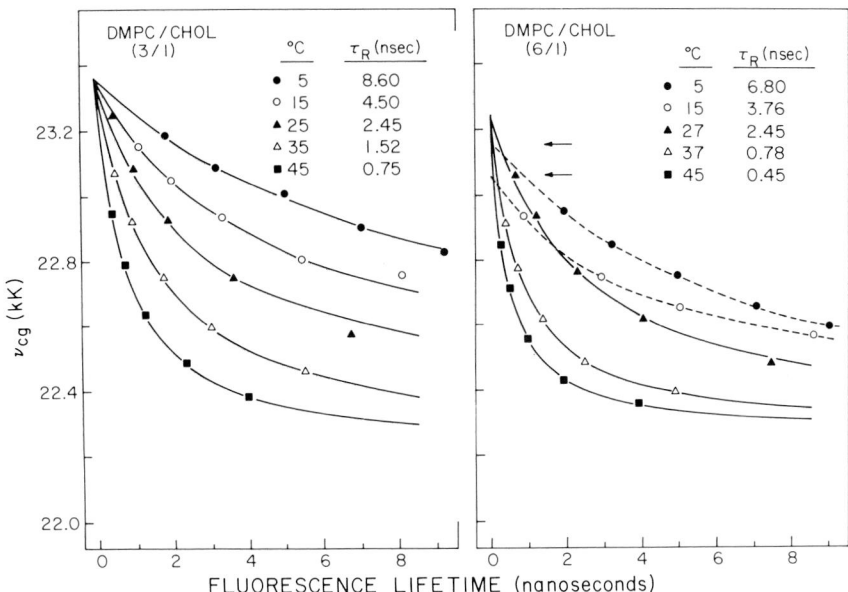

FIG. 12. Lifetime-resolved centers of gravity for TNS-labeled vesicles. Reprinted with permission from Lakowicz and Hogen.[29] Copyright 1981 American Chemical Society.

the potential and practical capabilities of the frequency-domain measurements have been substantially enhanced by the availability of stable and reliable designs for variable-frequency phase-modulation fluorometers.[7,7a] Such instruments are easily assembled using commercially available components, and one such instrument has recently been assembled in this laboratory. With such instruments the resolving power of the frequency-domain measurements is at least equivalent to that available using time-resolved measurements. In the following section we present data which illustrate the resolving power of the frequency-domain measurements. Similar demonstrations for the time-resolved method were presented in earlier reports.[8,9]

The basic difference between time-domain and frequency-domain measurements is the nature of the incident light used to excite the fluorescence, and the parameters which are measured. For harmonic or frequency-domain measurements the light intensity is modulated at frequencies comparable to the decay rates of the sample. The measured quantities are the phase and modulation of the emission, relative to the modulated excitation. During the past 15 years the time-resolved method has been more widely used than the harmonic method. This is for several reasons. First, interpretation of the phase and modulation data has been hindered by lack of a general formalism. Second, there was no strong impetus to develop this formalism because the available phase-modulation fluorometers operated at only a few fixed modulation frequencies. In principle, data at only one or a few frequencies can be used to resolve some complex decay laws.[31,32] Unfortunately, very precise experimental data are required to obtain a reliable resolution of a complex emission based on the phase and modulation data measured at two or three frequencies,[32a] and the available fixed frequencies are usually not the desired values for the resolution of the unknown decay. Additionally, the precision of the phase and modulation measurements has been severely limited by the wavelength and/or geometry-dependent time response of the photomultiplier tubes. This problem can now be easily circumvented.[33] The recent application of phase-sensitive detection provides a practical and reliable method to resolve closely space decay times, especially for those samples which display distinct emission spectra for each emitting species.[34–36]

Instrumentation for a Variable-Frequency Phase-Modulation Fluorometer

The equipment needed for a variable-frequency instrument is similar to that used for most spectrofluorometers, except that it is necessary to modulate the incident light at a range of frequencies (Fig. 13). This can be accomplished with longitudinal or transverse-field electrooptic modulators. In contrast to the generally held view, reasonable modulation can be achieved with relatively low voltage and/or power. Such devices provide useful modulation at frequencies from dc to 250 MHz, depending upon the particular design. Higher modulation frequencies, to 1 GHz, may be possible using advanced electrooptic modulators.[37] In general, modulators with the lowest half-wave voltages are preferable because modulation is accomplished with less RF power, and more importantly, with a higher transmission efficiency for the incident light. The transverse-field electrooptic modulators have low half-wave voltages (100 to 500 V), but these modulators also have small apertures. This requires the use of a laser light source to obtain an adequate throughput of the incident light. In

[31] G. Weber, *J. Phys. Chem.* **85,** 949 (1981).

[32] D. M. Jameson and G. Weber, *J. Phys. Chem.* **85,** 953 (1981).

[32a] D. M. Jameson and E. Gratton, *in* "New Directions in Molecular Luminescence" (D. Eastwood, ed.), p. 67. ASTM STP 822, 1983.

[33] J. R. Lakowicz, H. Cherek, and A. Balter, *J. Biochem. Biophys. Methods* **5,** 131 (1981).

[34] J. R. Lakowicz and H. Cherek, *J. Biochem. Biophys. Methods* **5,** 19 (1981).

[35] J. R. Lakowicz and H. Cherek, *J. Biol. Chem.* **256,** 6348 (1981).

[36] J. R. Lakowicz and S. Keating, *J. Biol. Chem.* **258,** 5519 (1983).

[37] S. R. Adhav, R. S. Adhav, and H. van der Vaart, *Appl. Opt.* **20,** 867 (1981).

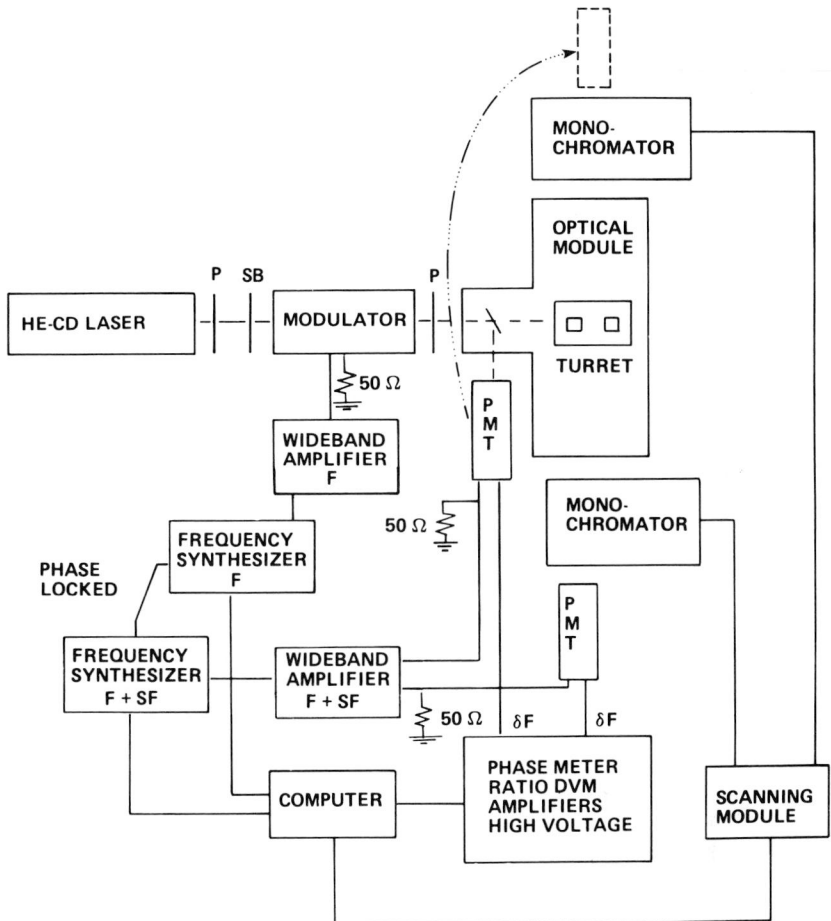

FIG. 13. Schematic diagram of a variable-frequency phase-modulation fluorometer.

contrast, longitudinal-field modulators have higher apertures, 1 cm in our case, but also have higher half-wave voltages (1000 to 2000 V). Nonetheless, these modulators provide useful modulation of laser light to 250 MHz. Recently E. Gratton (University of Illinois, personal communication) has found that a xenon arc lamp could be modulated to 200 MHz with a longitudinal field modulator.

In phase-modulation fluorescence one usually modulates a continuous source. This requirement limits the available excitation wavelengths. An arc lamp is ideal, but its beam diameter and the collimation requirements of E-O modulators makes this a difficult task. There are several convenient ultraviolet CW sources, such as an Ar ion laser at 351 nm, and a He-Cd laser at 325 nm. Unfortunately, it is more difficult to obtain wavelengths for the excitation of protein fluorescence. Such wavelengths (280 to 300 nm) can be obtained using a rhodamine 6G ring dye laser, which is pumped by an argon ion laser. Such a system was just installed in this laboratory, yielding 2 mW of UV from 290 to 300 nm after frequency doubling within the dye laser cavity. Alternatively, one may use a mode locked pulse train to take advantage of the higher doubling efficiency from the high peak powers. This approach provides the additional advantage of using the intrinsic high frequency content of a pulse train.[37a,37b] At this moment the harmonic-content method seems preferable. In this laboratory we obtained 1000-fold more UV, and the frequencies extend well to 2 GHz.[37c]

The signal-to-noise ratio of these instruments is substantially enhanced by the use of cross-correlation detection.[6,38] In this method the photomultiplier gain is modulated at a frequency offset slightly ($\delta\omega = 25$ Hz) from the modulation frequency. This is accomplished, over a wide range of frequencies, using two highly stable frequency synthesizers. The detection electronics are tuned to detect and amplify the 25 Hz cross-correlation signal from the photomultiplier. Conveniently, this low-frequency signal contains all the phase and modulation information which is present in the high-frequency signal. Of course, it is considerably easier to measure the phase and modulation of the low-frequency cross-correlation signal.

Theory of Variable-Frequency Phase-Modulation Fluorometry

The data available from a variable-frequency fluorometer allow one to obtain all the information available using a pulse-fluorometer. Specifically, one can resolve mixtures of exponentially decaying fluorophores, determine time-resolved emission spectra, and determine time-resolved decays of anisotropy. Such measurements are described in the following sections. However, since this technique is relatively new and the theory is not widely available, we will briefly describe how the data are analyzed.

In the frequency domain phase angles and demodulation factors are measured. Conveniently, it is possible to predict the frequency-dependent data for any assumed decay law.[5,6] Suppose the time-resolved decay can

[37a] E. Gratton and R. Lopez-Delgado, *Nuovo Cimento B* **56,** 111 (1980).

[37b] E. Gratton, D. M. Jameson, N. Rosata, and G. Weber, *Rev. Sci. Instruments* **55,** 486 (1984).

[37c] J. R. Lakowicz, I. Gryczynski, and G. Laczko, in press (1986).

[38] R. D. Spencer and G. Weber, *Ann. N.Y. Acad. Sci.* **158,** 361 (1969).

be described by a sum of exponential decays

$$I(t) = \sum_i \alpha_i e^{-t/\tau_i} \tag{22}$$

If the decay is more complex than a sum of exponential it is generally adequate to approximate the decay by such a sum. The frequency-domain data can be calculated from the sine and cosine transforms of $I(t)$

$$N_\omega = \frac{\int_0^\infty I(t) \sin \omega t \, dt}{\int_0^\infty I(t) \, dt} \tag{23}$$

$$D_\omega = \frac{\int_0^\infty I(t) \cos \omega t \, dt}{\int_0^\infty I(t) \, dt} \tag{24}$$

where ω is the circular modulation frequency ($2\pi \times$ frequency in Hz). If needed, nonexponential decay laws can be transformed numerically. For a sum of exponentials these transforms are

$$N_\omega = \sum_i \frac{\alpha_i \omega \tau_i^2}{(1 + \omega^2 \tau_i^2)} \bigg/ \sum_i \alpha_i \tau_i \tag{25}$$

$$D_\omega = \sum_i \frac{\alpha_i \tau_i}{(1 + \omega^2 \tau_i^2)} \bigg/ \sum_i \alpha_i \tau_i \tag{26}$$

The frequency-dependent values of the phase angle (ϕ_{cw}) and the demodulation ($m_{c\omega}$) are

$$\tan \phi_{c\omega} = N_\omega / D_\omega \tag{27}$$
$$m_{c\omega} = [N_\omega^2 + D_\omega^2]^{1/2} \tag{28}$$

The subscript c is used to indicate calculated values based upon the assumed parameters of the decay (α_i and τ_i). The experimental data (ϕ_ω and m_ω) can be analyzed using the method of least-squares.[39,40] The assumed parameter values are varied until χ^2 reaches a minimum

$$\chi^2 = \sum_\omega \frac{1}{\sigma_{\phi\omega}^2} (\phi_\omega - \phi_{\omega c})^2 + \sum_\omega \frac{1}{\sigma_{m\omega}^2} (m_\omega - m_{\omega c})^2 \tag{29}$$

when $\sigma_{\phi\omega}$ and $\sigma_{m\omega}$ are the uncertainties in the measured phase and modulation data. The value of reduced χ^2 is given by $\chi_R^2 = \chi^2/\nu$, where ν is the number of degrees of freedom. This method of analysis is analogous to

[39] J. R. Lakowicz, E. Gratton, G. Laczko, H. Cherek, and M. Limkeman, *Biophys. J.* **46**, 463 (1984).
[40] E. Gratton, J. R. Lakowicz, G. Laczko, M. Limkeman, and B. Maliwal, *Biophys. J.* **46**, 479 (1984).

that used in the analysis of the time-resolved data.[8,41] That is, the measured data are fit to the values predicted for an assumed model. The assumed model is accepted or rejected based upon the goodness of the fit between the measured and calculated data.

We note that the procedure described by Eqs. (22)–(29) is completely general, and can be applied to any assumed decay law. Additionally, these expressions can be extended to permit determination of time-resolved decays of anisotropy. In this case one measures the frequency-dependent phase difference (Δ_ω) between the perpendicular $(\phi_\perp)$ and parallel $(\phi_\parallel)$ components of the modulated emission $(\Delta_\omega = \phi_\perp - \phi_\parallel)$ and the amplitude ratio (Λ_ω) of the parallel $(m_\parallel)$ and the perpendicular $(m_\perp)$ components of the modulated emission $(\Lambda = m_\parallel/m_\perp)$. These values are compared with those expected on the basis of an assumed anisotropy decay law.

Following δ-function excitation the decays of the parallel and perpendicular components of the emission are given by

$$I_\parallel(t) = \tfrac{1}{3}I(t)[1 + 2r(t)] \tag{30}$$
$$I_\perp(t) = \tfrac{1}{3}I(t)[1 - 2r(t)] \tag{31}$$

where $r(t)$ is the time-resolved decay of anisotropy. Generally, $r(t)$ can be described as a multiexponential decay,

$$r(t) = r_0 \sum_i g_i e^{-t/\theta_i} \tag{32}$$

where r_0 is the limiting anisotropy in the absence of rotational diffusion, θ_i the individual correlation times, and g_i the associated amplitudes. The expected values of Δ_ω $(\Delta_{c\omega})$ and Λ_ω $(\Lambda_{c\omega})$ can be calculated from the sine and cosine transforms of the individual polarized decays [Eqs. (23) and (24)]. The frequency-dependent values of Δ_ω and Λ_ω are given by

$$\Delta_{c\omega} = \arctan\left(\frac{D_\parallel N_\perp - N_\parallel D_\perp}{N_\parallel N_\perp - D_\parallel D_\perp}\right) \tag{33}$$

$$\Delta_{c\omega} = \left(\frac{N_\parallel^2 + D_\parallel^2}{N_\perp{}^2 + D_\perp{}^2}\right)^{1/2} \tag{34}$$

where the N_i and D_i are calculated at each frequency. The parameters describing the anisotropy decay are obtained by minimizing the squared

[41] A. Grinvald and I. Z. Steinberg, Anal. Biochem. **59**, 583 (1974).
[42] J. Beechem and L. Brand, Photochem. Photobiol. **37**, S20 (1983).

deviations between measured and calculated values,

$$\chi^2 = \sum_\omega \frac{1}{\sigma_{\Delta\omega}^2} (\Delta_\omega - \Delta_{c\omega})^2 + \sum_\omega \frac{1}{\sigma_{\Lambda\omega}^2} (\Lambda_\omega - \Lambda_{c\omega})^2 \tag{35}$$

In this expression $\sigma_{\Delta\omega}$ and $\sigma_{\Lambda\omega}$ are the estimated experimental uncertainties in the measured quantities (Δ_ω and Λ_ω). The ratio Δ_ω can be converted into a frequency-dependent anisotropy using $r_\omega = (\Delta_\omega - 1)/(\Lambda_\omega \pm 2)$. At low frequency the value of r_ω is equal to the steady-state anisotropy, and at the high frequency limit $r_\omega = r_0$.

We used three models to describe the decays of anisotropy. For an isotropic rotator we used

$$r(t) = r_0 e^{-t/\theta} \tag{36}$$

For asymmetric molecules such as perylene and tryptophan we used anisotropy decay law with two or three correlation times

$$r(t) = r_0 \sum_i g_i e^{-t/\theta_i} \tag{37}$$

with $\Sigma_i g_i = 1$, g_i as floating parameters and r_0 fixed. For probes in membranes such as DPH or perylene in lipid bilayers, we also used the hindered model

$$r(t) = (r_0 - r_\infty) e^{-t/\theta} + r_\infty \tag{38}$$

where r_∞ is the limiting anisotropy observed at times longer than the rotational correlation time of the fluorophore.

Resolution of Multiexponential Decays

The resolution of multiexponential decays is a problem encountered in almost every application of time-resolved fluorescence to a biochemical system. Since frequency-domain fluorometry has only recently become possible, the usefulness of this method for resolving multiexponential decays is not widely appreciated. Hence it is instructive to examine frequency-domain data for single fluorophores and for mixtures of fluorophores, each of which decays exponentially.

The data in Fig. 14 illustrate the use of frequency-domain data for measurement of single exponential decay times. The data are for fluorescein in water (4.3 nsec) and for rose bangal in ethanol (0.83 nsec) and in water (0.13 nsec). At any given frequency larger phase angles and smaller modulations are seen for the longer lifetimes. The phase angles increase and the modulation decreases with increasing modulation frequency. The solid lines (Fig. 14A) represent the theoretical curves calculated for the

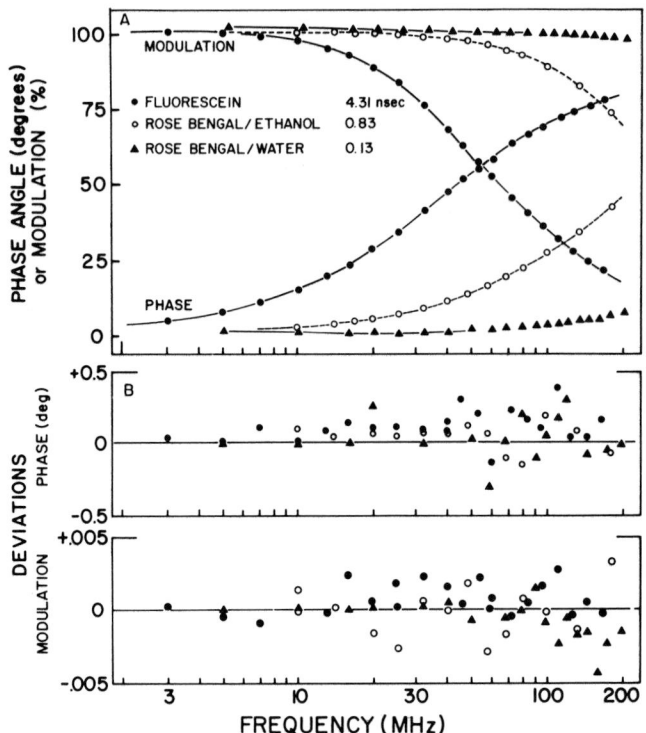

FIG. 14. Phase and modulation data for single-component solutions. (A) Data are shown for fluorescein in water (0.02 M Tris, pH 8) (●), rose bengal in ethanol (○), and in water (▲). (B) The deviations between the measured and calculated values. From Lakowicz and Maliwal.[7a]

lifetimes which yielded minimum value of χ_R^2. Obviously, the calculated values for single exponential decay accurately reproduce the measured values. The agreement between the measured and calculated data is further illustrated in Fig. 14B, which shows the deviations between the measured data and the calculated values. The deviations are randomly distributed around zero. This indicates that the single exponential model is adequate to account for the data, and that the measurements are free of systematic errors. The average deviations are near 0.2° for phase and 0.002 for modulation, using the instrument in this laboratory. These values were used for the calculation of χ_R^2. In contrast to photon counting data, the random errors in phase and modulation cannot be estimated on the basis of counting statistics.[42a] Rather, the level of random error is estimated by comparing the data with the calculated values for cases in which the assumed decay law is appropriate. The fact that the χ_R^2 are often

[42a] A. Grinvald and I. Z. Steinberg, *Anal. Biochem.* **59,** 583 (1974).

less than unity indicates that the actual level of random error is less than 0.2° of phase and less than 0.002 in modulation. If one attempts to fit the data for fluorescein and rose bengal to a double exponential decays of intensity the values of χ_R^2 do not substantially decrease and the amplitude of the assumed second component is negligible. This result demonstrates that inclusion of an additional component is not justified from the data.

In our opinion the major challenge is not measurement of the shortest lifetimes. In biochemical research it is more important to resolve closely spaced decay times. Since the decay times (τ_i) and preexponential factors (α_i) are correlated parameters, one requires precise measurements to obtain reliable estimates of α_i and τ_i. The resolution obtainable from the frequency-domain measurements is illustrated by the examination of two-component mixtures. The data for one such mixture are shown in Fig. 15 for anthracene (4.08 nsec) and 9,10-diphenylanthracene (6.31 nsec). Obviously, a good fit to the measured data (solid circle) is provided by an assumed two-component decay law (solid line).

The lower panels of Fig. 15 show the deviations between the measured and calculated values for the best fit obtainable using a single and double exponential decay law. For this mixture, in which the decay times differ by 40%, the experimental deviations found using a single decay time model are about 10- to 20-fold larger than the random error level of the measurements. For the single-component model the deviations vary systematically with the frequency, indicating that the deviations are not due to random error and are probably due to an inappropriate decay law. Additionally, the value of $\chi_R^2 = 12.7$ is clearly unacceptable. The inclusion of a second component in the decay law results in a 40-fold decrease in χ_R^2 to 0.30, and the deviations become small and randomly distributed around zero. Importantly, the calculated decay times and fractional intensities are in good agreement with the expected values.

We note that lifetimes more closely spaced than 4 and 6 nsec can also be resolved. Using a method analogous to the global method of Brand and co-workers,[42b] we resolved a two-component decay with lifetimes differing by only 8%, 4.1 and 4.4 nsec.[7a] To the best of our knowledge these are the most closely spaced lifetimes which have been resolved to date, excluding those resolved by phase-sensitive detection in which the emission spectra of the components were known. Irrespective of the use of global methods, a 40-fold decrease in χ_R^2 for a two-component fit illustrates the substantial resolution possible with frequency-domain fluorometry. A 40-fold decrease in χ_R^2 is more than is required, which indicates that more closely spaced lifetimes could be resolved.

[42b] J. M. Beechem, M. Ameloot, and L. Brand, *in* "Excited State Probes in Biochemistry and Biology" (A. Szabo and L. Massoti, eds.). Plenum, New York, in press, 1986.

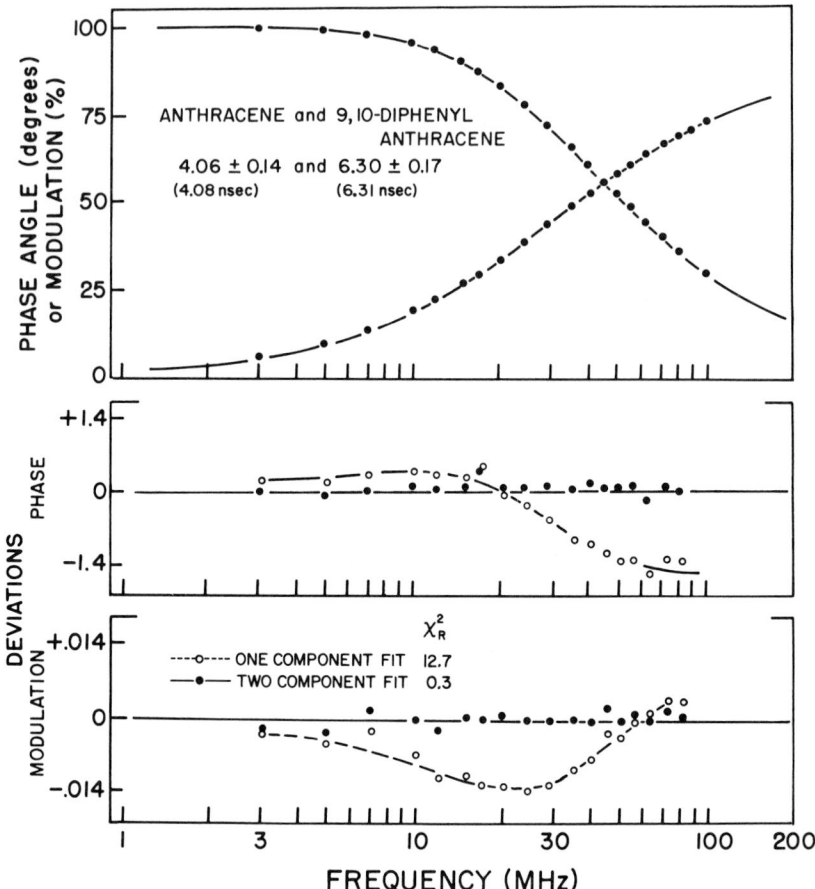

Fig. 15. Phase and modulation data for a mixture of anthracene and 9,10-diphenylanthracene. The dashed line indicates the best one-component fit, and the solid line the best two-component fit. The values not in parentheses are the experimental values and those in parentheses are from measurements on the pure components. From Lakowicz and Maliwal.[7a]

It is well known that the resolution of a three-component decay is considerably more difficult than a two-component decay.[42a] Hence, it is of interest to test the frequency-domain method using a three-component mixture. Data for a three-component mixture of POPOP (1.32 nsec), anthracene (4.1 nsec), and 9-vinylanthracene (7.7 nsec) are shown in Fig. 16. The resolution of a three-component mixture with a 6-fold range of lifetimes is a difficult problem. The two-component fit (Fig. 16) is easily judged to be unsuccessful because of the systematically varying devia-

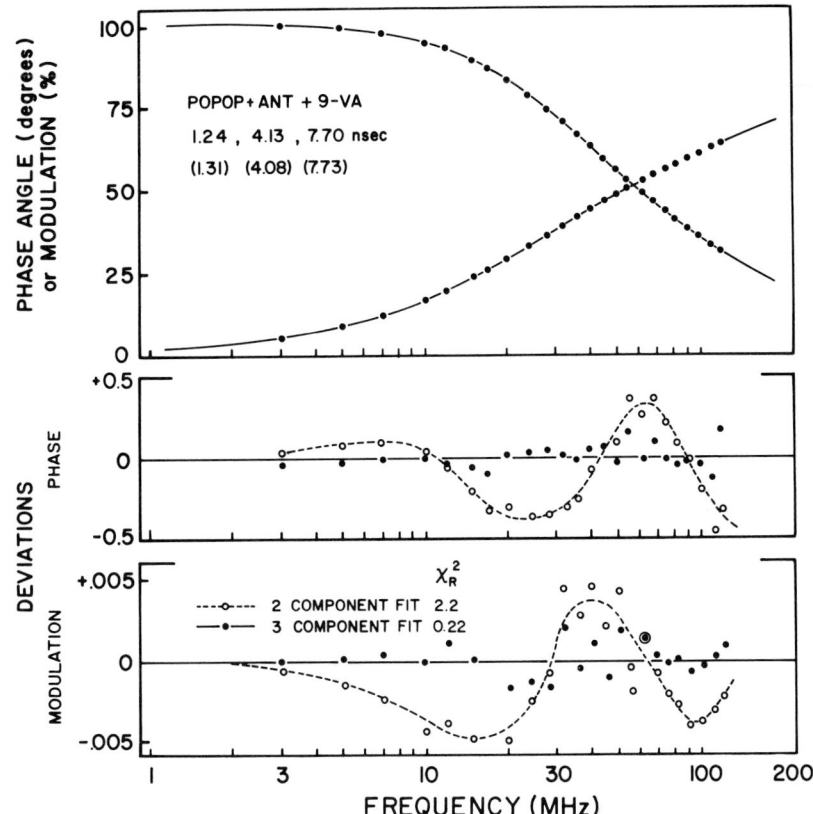

FIG. 16. Phase and modulation data for a three-component mixture of POPOP, anthracene, and 9-vinylanthracene. The fraction intensities of the three components are equal ($f_1 = f_2 = f_2 = 0.33$). From Gratton *et al.*[40]

tions and the large value of χ_R^2. The two-component fit displays deviations which are about 5-fold larger than the expected level of random error. Inclusion of the third component yields a 10-fold decrease in χ_R^2 and more random deviations. From the results in Fig. 15 and 16 it is clear that the frequency-domain method provides good resolution of multiexponential decays of fluorescence. This capability is useful in most biochemical application of fluorescence.

Time-Resolved Emission Spectra from Frequency-Domain Data

Time-resolved emission spectra are the spectra which would be observed at defined times following δ-pulse excitation. The determination of such detailed spectral information is generally regarded as a unique capa-

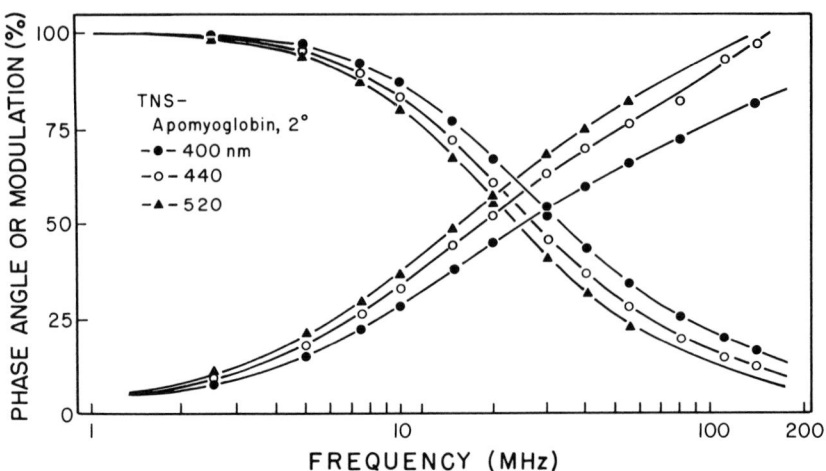

FIG. 17. Variable-frequency phase and modulation data for TNS-labeled apomyoglobin. From Lakowicz *et al.*[43]

bility of the time-resolved method. In fact, Teale[42c] has claimed that phase-modulation methods could not be used to obtain time-resolved emission spectra. If such spectra could be obtained from frequency-domain data, this would demonstrate the essential equivalence of the methods. In fact, this has been accomplished for labeled proteins and membranes.[28a,43] The solvent-sensitive probe *p*-toluidinyl-6-naphthalene-sulfonic acid (TNS) is known to bind to the heme site of apomyoglobin. Phase and modulation data for TNS-labeled apomyoglobin are shown in Fig. 17. From a detailed examination of the data one can conclude that the data could not result from a mixture of directly excited fluorophores, that is, ground-state heterogeneity. Rather, on the long wavelength side of the emission, the data are unambiguous in demonstrating the occurrence of a time-dependent spectral shift. This fact is demonstrated by phase angles in excess of 90°, and by apparent phase lifetimes which are longer than the apparent modulation lifetimes.[44,45] At each wavelength the data were fitted to a multiexponential decay law. To fit the data it was necessary to include a term with a negative preexponential factor. This is the usual criteria used in time-resolved measurements to demonstrate an excited state reaction. The impulse-response functions at each wavelength, when

[42c] F. J. W. Teale, *in* "Time-Resolved Fluorescence Spectroscopy in Biochemistry and Biology" (R. B. Cundall and R. E. Dale, eds.), p. 77. Plenum, New York, 1983.

[43] J. R. Lakowicz, H. Cherek, B. Maliwal, G. Laczko, and E. Gratton, *J. Biol. Chem.* **259**, 10967 (1984).

[44] J. R. Lakowicz and A. Balter, *Biophys. Chem.* **16**, 99 (1982).

[45] J. R. Lakowicz, H. Cherek, and D. R. Bevan, *J. Biol. Chem.* **255**, 4403 (1980).

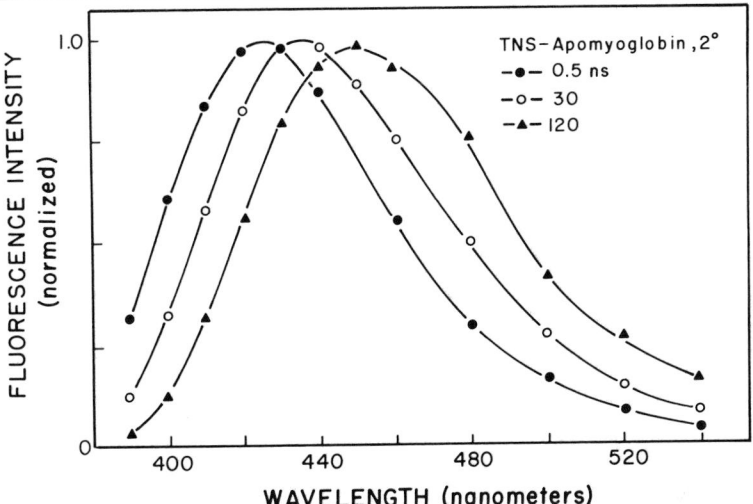

FIG. 18. Time-resolved emission spectra of TNS-labeled apomyoglobin. From Lakowicz
et al.[43]

normalized to the steady-state spectrum, can be used to calculate the
time-resolved emission spectra. Such spectra are shown in Fig. 18. One
notices that time-resolved spectra can be reliably calculated for times less
than 1 nsec, and to times at least five times the average lifetime. Additionally, the time-resolved spectra can be used to calculate the time dependence of the decay of the emission center of gravity (Fig. 19). For TNS-
labeled apomyoglobin the data indicate an initially rapid spectral decay,
followed by a slower decay. The results in Figs. 18 and 19 are in good
agreement with the time-resolved measurements performed on this same
system.[46,47] Such agreement, for the complex phenomena of time-dependent spectral shifts, provides a sound demonstration that equivalent information is currently available in either the time or the frequency domain.

Anisotropy Decay from Frequency-Domain Measurements

Phase-modulation methods are uniquely suited for the measurement of
rapid and/or complex decays of fluorescence anisotropy. This is because
the signal resulting from rotational diffusion is measured directly.[48] One
representative case is shown in Fig. 20[43] for TNS-labeled apomyoglobin
and TNS dissolved in propylene glycol. Also shown in the best fit through
the data obtained using a single rotational correlation time [Eq. (36)]. The

[46] A. Gafni, R. P. DeToma, R. E. Manrow, and L. Brand, *Biophys. J.* **17,** 155 (1977).
[47] R. P. DeToma, J. H. Easter, and L. Brand, *J. Am. Chem. Soc.* **98,** 5001 (1976).
[48] J. R. Lakowicz, H. Cherek, B. Maliwal, and E. Gratton, *Biochemistry* **24,** 376 (1985).

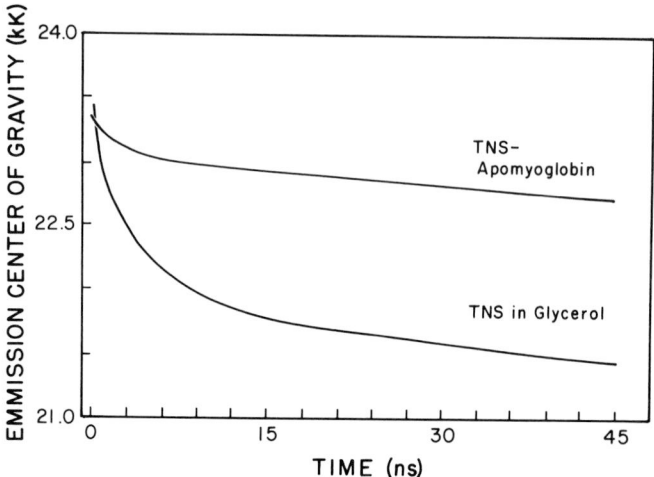

FIG. 19. Time-resolved emission center of gravity for TNS in glycerol and bound to apomyoglobin. From Lakowicz et al.[43]

best fit for TNS-Apo was obtained with a single correlation time of 20.5 nsec at 2°, and with $r_\infty = 0$. These results, once corrected for the difference in temperature and viscosity, are in agreement with the time-resolved measurements of anilinonaphthalene sulfonate-labeled apo-

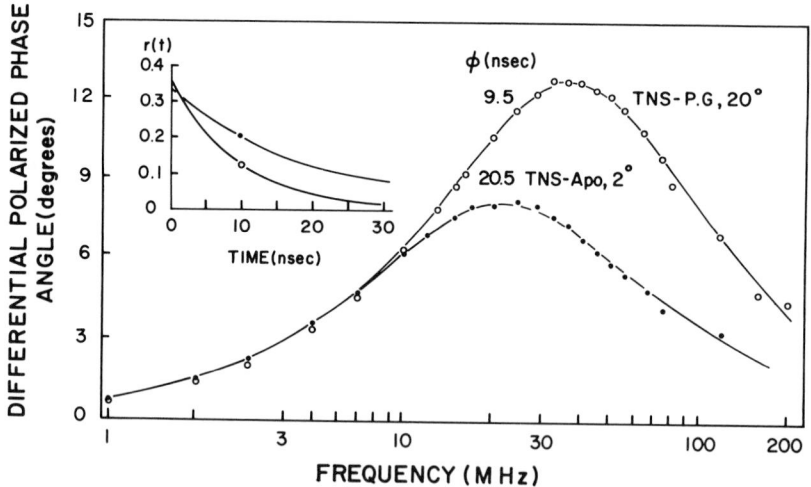

FIG. 20. Frequency-dependent differential polarized phase angles of TNS-labeled apomyoglobin. From Lakowicz et al.[43]

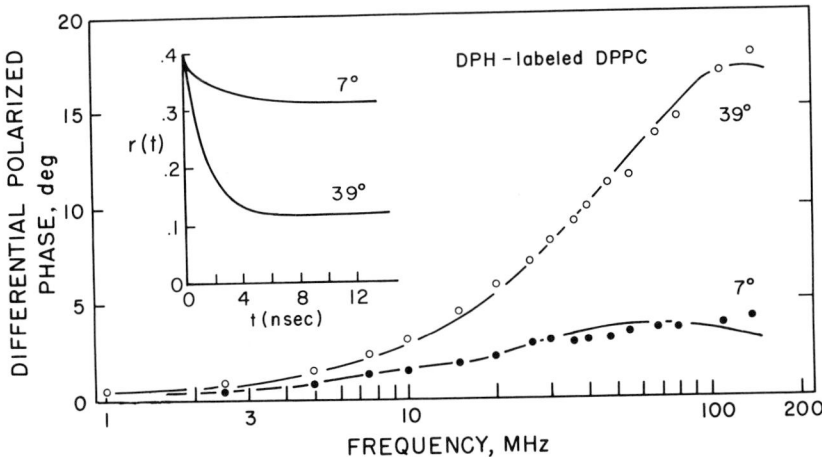

FIG. 21. Frequency-dependent differential polarized phase angles of DPH-labeled DPPC vesicles. The inset shows the time-resolved anisotropies which were derived from the frequency-domain data. From Lakowicz *et al.*[48]

myoglobin, which indicated a rotational correlation time of 9.6 nsec at 25°.[49]

The case of TNS-apomyoglobin may be regarded as rather simple because the rotational diffusion of the protein is unhindered and almost isotropic. Hence it was of interest to see whether the frequency-domain data could reveal a more complex decay of anisotropy. We examined diphenylhexatriene (DPH) in dipalmitoyl-L-α-phosphatidylcholine (DPPC) vesicles above and below their transition temperature (Fig. 21). Substantially smaller differential phase angles were found at low temperature. This was an expected result in consideration of the hindered rotations of DPH in vesicles below the transition temperature. It is known that hindered rotations result in a uniform decrease in the values of tan $(\phi_\perp - \phi_\parallel)$.[50] These data were fit using the hindered rotator model [Eq. (38)]. Also shown in Fig. 21 are the time-resolved anisotropies which were calculated using the frequency-domain data at 7°. The calculated values of the correlation time ($6R^{-1}$) and r_∞ were 3.3 nsec and 0.31. At 39° these values were 1.5 nsec and 0.12. These results are in excellent agreement with the time-resolved data from a number of laboratories.[26,51,52]

Frequency-domain fluorometry also provides reliable measurements of anisotropy decays on the picosecond timescale. This is illustrated by

[49] L. Stryer, *J. Mol. Biol.* **13**, 484 (1965).
[50] J. R. Lakowicz and F. G. Prendergast, *Biophys. J.* **23**, 213 (1978).
[51] S. Kawato, K. Kinosita, and A. Ikegami, *Biochemistry* **16**, 2319 (1977).
[52] J. R. Lakowicz, F. G. Prendergast, and D. Hogen, *Biochemistry* **18**, 508 (1979).

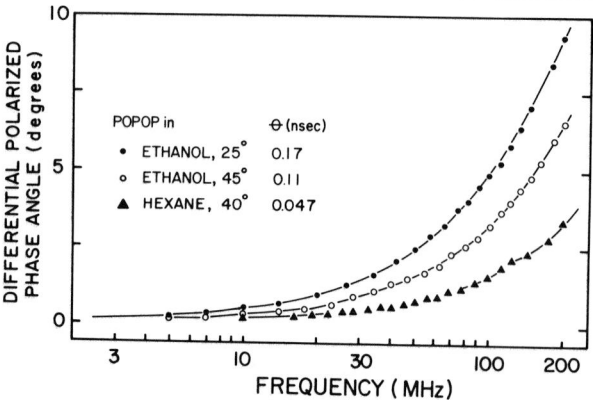

FIG. 22. Determination of picosecond anisotropy decays using frequency-domain fluorometry. Differential polarized phase angles are shown for p-bis[2-(5-phenyloxazolyl)]benzene (POPOP) in fluid solvents. From Lakowicz and Maliwal.[7a]

measurements of the rotational diffusion of POPOP in ethanol at 45° and in hexane at 40° (Fig. 22). The differential polarized phase angle (Δ_ω) is dependent upon the rotational rate. For a single correlation time a plot of Δ_ω versus log ω is nearly Lorentzian in shape. This shape is not evident in Fig. 22 because the rotational rate of POPOP is fast in these solvents, and the maximum value of Δ_ω is displaced beyond the maximum frequency of 200 MHz. The solid lines represent the calculated values of Δ_ω for the correlation times shown on Fig. 22. The frequency-domain method allowed determination of correlation times as short as 0.11 and 0.047 nsec for POPOP in ethanol and hexane at 45°, respectively. For such rapid diffusive motions it is difficult to know if the decay anisotropy is single or double exponential. In these measurements there was no substantial improvement in χ_R^2 upon inclusion of an additional correlation time. The frequency range of such measurements has been extended to 2 GHz. Hence, the measurements may soon be directly comparable with molecular dynamics calculations on proteins.[37c]

Recent data indicate that the frequency-domain fluorometry provides adequate resolution to determine complex anisotropy decays, even when these occur on the subnanosecond timescale. This is illustrated in Fig. 23 for perylene in propylene glycol at 28°. Perylene is a plate-like molecule, and the apparent correlation times are known to be in the ratio of about 7 to 1.[52a] It was not possible to account for the measured data using a single correlation time. This is evident from the mismatch seen in Fig. 23, the systematically varying deviations between the measured and calculated

[52a] M. D. Barkley, A. Kowalczyk, and L. Brand, *J. Chem. Phys.* **75**, 3581 (1978).

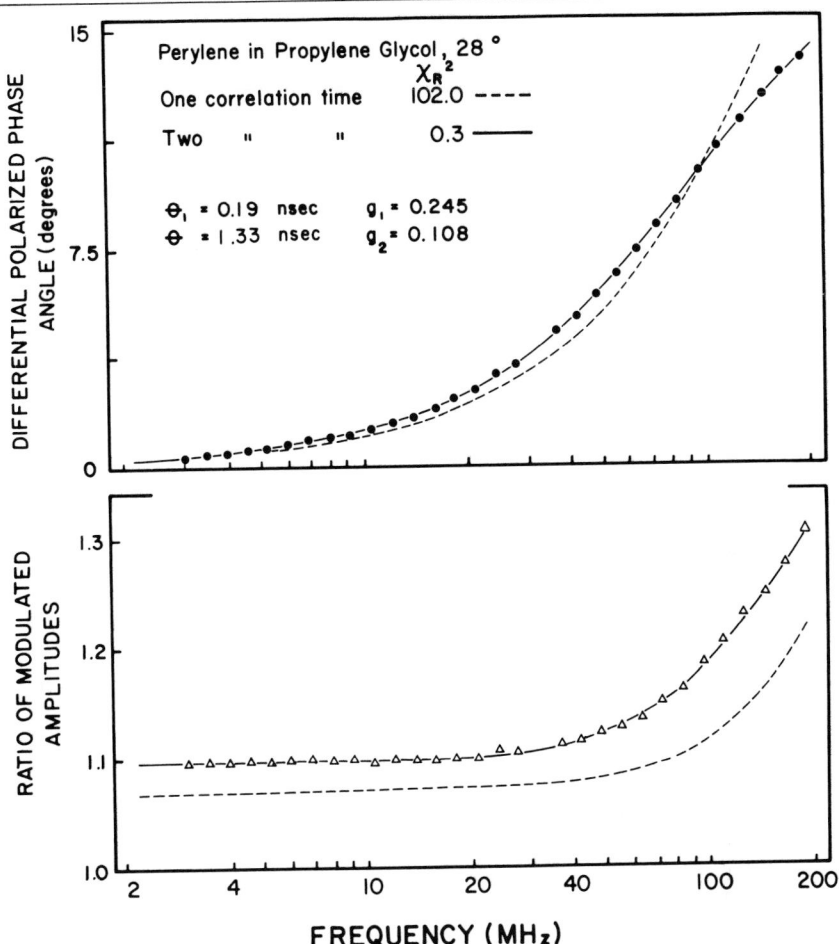

FIG. 23. Resolution of a two-component anisotropy decay on the subnanosecond time-scale (Lacowicz and Maliwal, unpublished observations).

values, and the unacceptable value of $\chi_R^2 = 102$. Inclusion of a second correlation time in the fitting algorithm [Eq. (37)] results in a 340-fold decrease in χ_R^2 and random deviations. (The low value of χ_R^2 is because the actual errors are smaller than estimated values.) It is remarkable that the value of χ_R^2 decreases 340-fold upon inclusion of the second correlation times. This result indicates that we could resolve still more complex anisotropy decays, or a two component decay for which the correlation times are shorter than the values of 0.2 and 1.3 nsec in Fig. 23. This capability is likely to be valuable for interpreting the details of anisotropy

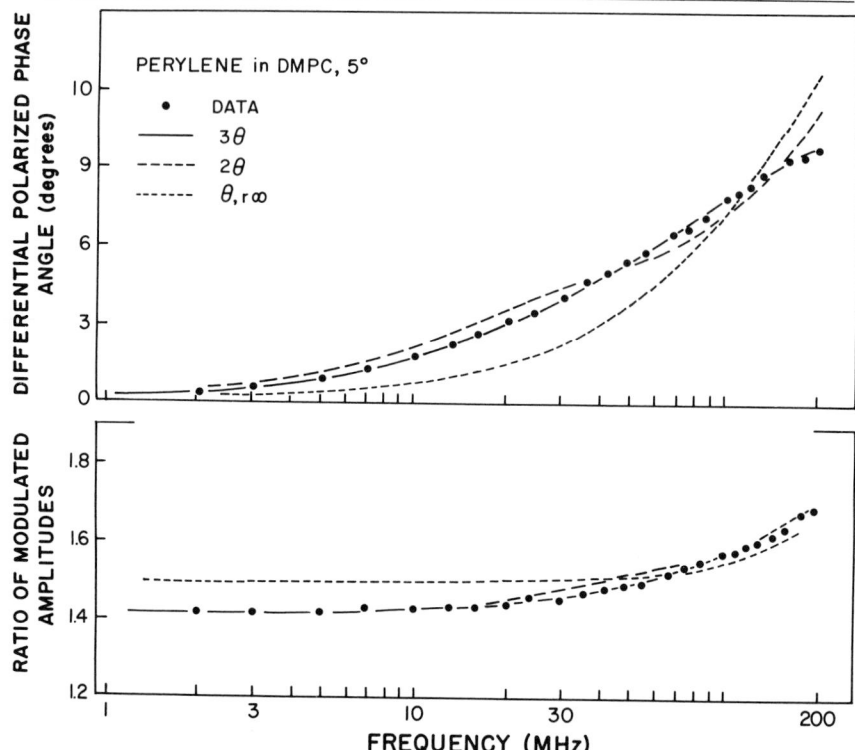

FIG. 24. Three-component anisotropy decay of perylene in DMPC bilayers. The values of χ_R^2 for the three–two correlation time and hindered model fits are 0.5, 14.4, and 106, respectively. From Lakowicz *et al.*[48]

decays of trytophan residues in proteins in terms of the known structures or the calculations of their dynamic properties.

A final example of a complex anisotropy decay is shown in Fig. 24, for the anisotropic molecule perylene in DMPC bilayers. Previous studies[52b] indicated that below the transition temperature ($<25°$) the rotational motions of perylene were hindered by the bilayer. Since perylene is itself asymmetric and displays two correlation times in isotropic solvents, we expected that three correlation times could be present in the membrane. The hindered model [Eq. (38)] was completely inadequate to account for the data ($\chi_R^2 = 106$). Even the fit with two correlation times is easily judged to be inadequate ($\chi_R^2 = 14.4$). The three-correlation time fit provides a reasonable value of $\chi_R^2 = 0.5$, and reasonable correlation times (0.2, 1.4, and 47 nsec). We attribute the two shorter correlation times to

[52b] J. R. Lakowicz and J. R. Knutson, *Biochemistry* **19**, 905 (1980).

rapid motions of perylene within the nonhindered environment, and the 47 nsec correlation time to the hindered motion. It is remarkable that one could obtain a 28-fold decrease in χ_R^2 for distinguishing two versus three rotational correlation times. The results presented in Fig. 20 to 24 demonstrate that the frequency-domain method provides remarkable resolution of complex decays of fluorescence anisotropy.

Protein Fluorescence Intensity and Anisotropy Decays

Until recently it has not been possible to measure protein fluorescence in the frequency domain. This limitation was due to the lack of a continuous laser source with wavelengths from 280 to 300 nm. This limitation has now been overcome using a ring dye laser.[27c] The pump source is a argon ion laser. The dye is rhodamine 6G which is frequency doubled by a crystal placed within the dye laser cavity. Using this source we were able to examine tryptophan and tyrosine decays for a number of proteins.

Figure 25 shows frequency-domain data for monellin. The intensity decay is easily revealed to be more complex than a single exponential. For instance, χ_R^2 for a single decay time fit was 113, which is easily judged to be unacceptable. Even a two decay time fit seems unacceptable given the 4-fold decrease in χ_R^2 when the three-component model is used. At present the resolution provided by the data exceeds our ability to provide an unambiguous interpretation of the results. For instance, perhaps the three decay times found for monellin (Fig. 25) are actually an approximation for a continuous distribution of lifetimes. Alternatively, there may be two or more central lifetimes around which the distribution is clustered. Such analysis is currently in progress.[52c]

The frequency-domain method also allows measurement of rapid and/ or complex anisotropy decays. This is illustrated in Fig. 26 for an uncharged tryptophan derivative (NATA) in propylene glycol and in water at 20°. In the more viscous solvent the data require an anisotropy decay law containing at least two correlation times. It seems probable that the more rapid 0.3 nsec correlation time is due to rapid local motions of the indole ring, whereas the slower motion may be due to overall rotational motion of the NATA molecule. This distinction is ambiguous because indole itself is asymmetric and its anisotropy decay may be expected to be multiexponential.

Our ability to measure rapid tryptophan anisotropy decays is also illustrated in Fig. 26. For NATA in aqueous buffer the correlation time is near 47 psec. For *N*-acetyl-L-tyrosinamide (not shown) at 55° we measured a correlation time of 50 psec. Since the frequency range of these

[52c] J. R. Lakowicz, M. L. Johnson, I. Gryczynski, and H. Cherek, unpublished observations.

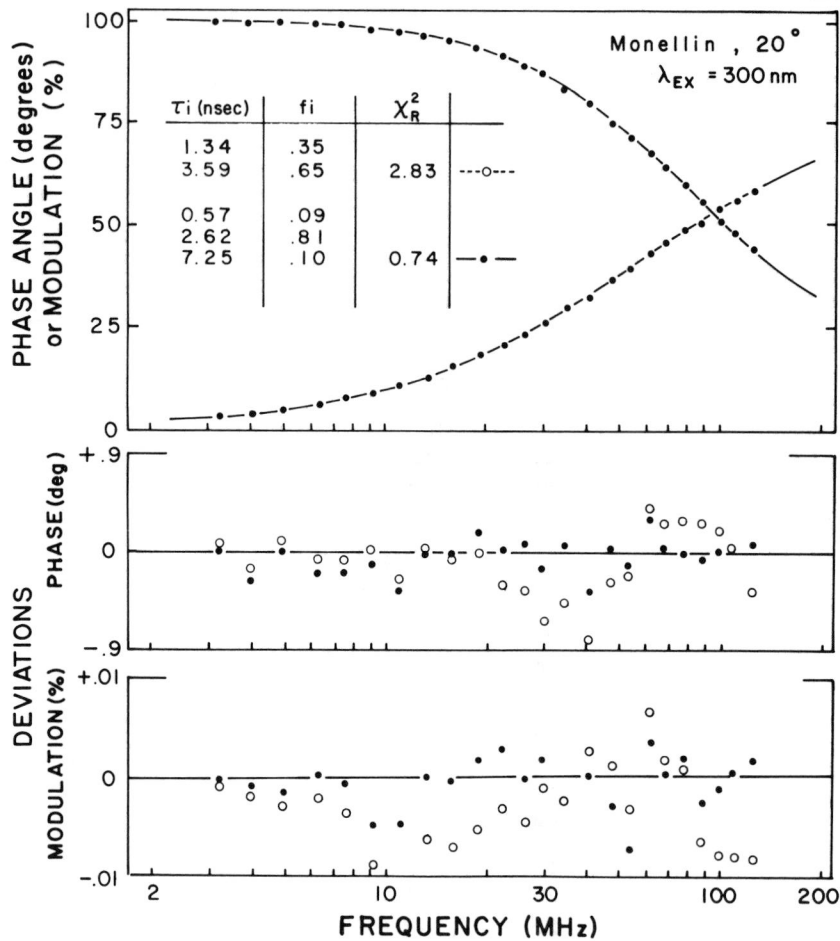

FIG. 25. Intensity decay of monellin from frequency-domain fluorometry.

instruments now exceeds 1 GHz or higher,[37c] we believe it is becoming practical to compare such experimental results with dynamic calculations on proteins.

Finally, the data in Fig. 27 demonstrate that the form of the frequency-domain data is highly sensitive to the nature of the tryptophan anisotropy decay. Data are shown for S nuclease (EC 3.1.31.1) and monellin. The single tryptophan residue of nuclease is held rather rigid by the protein. Hence, one observes the usual bell-shaped dependence of Δ_ω on frequency. (However, we stress that there does appear to be a 12% component with a 0.2 nsec correlation time.) In contrast, the values of Δ_ω for

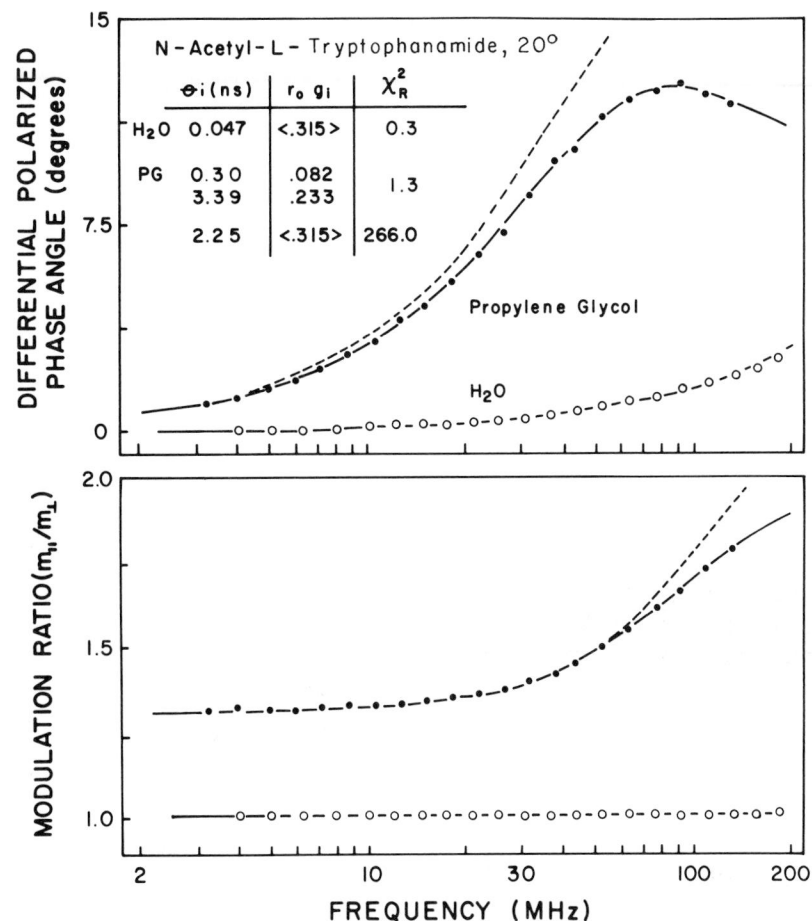

FIG. 26. Anisotropy decays of N-acetyl-L-tryptophanamide from frequency-domain fluorometry. From Lakowicz *et al.*[27c]

monellin increase monotonically with frequency. This high-frequency component of Δ_ω is due primarily to higher amplitude motions on the subnanosecond timescale.

Harmonic Content Fluorometry

During the past 6 months we installed alternative light sources for frequency-domain fluorometry. This source is a frequency-doubled and cavity-dumped dye laser (R6G), pumped by a mode-locked argon ion laser. This system provides 4 psec pulses with a repetition rate near 4 MHz. The advantage of this source is that it can be used directly, without

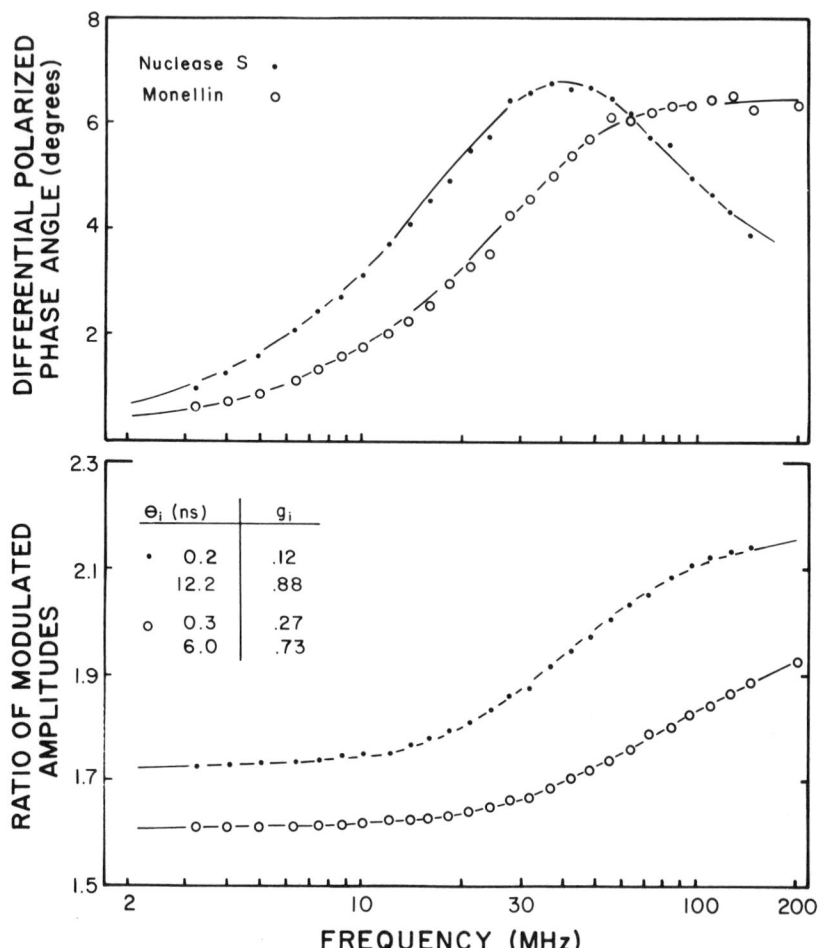

FIG. 27. Anisotropy decays of S nuclease and monellin from frequency-domain fluorometry.

the need for intensity modulation by an electro- or acoustooptic modulator. Importantly, its frequency content extends to many GHz, well beyond the limits of our detector. At present, with a standard photomultiplier tube, we can reliably measure to 400 MHz. With a faster detector, a microchannel plate PMT, this range extends to 2 GHz. Given the resolution obtainable with the previous 200 MHz data, the extended data sets should provide resolution adequate for a direct comparison with calculation on protein dynamics.

Phase-Sensitive Emission Spectra

In the preceding section we described the new possibilities which are provided by frequency-domain fluorescence. Realization of this potential required the construction of variable-frequency phase fluorometers. However, considerable information can be obtained, even at a single modulation frequency, by the use of phase-sensitive detection of fluorescence. Phase-modulation fluorometers have been commercially available for almost 10 years, and in spectroscopy laboratories for a longer period of time. Nonetheless, only in the past several years has the considerable potential of phase-sensitive detection of fluorescence been recognized.[34-36,52d] Many potential applications of this technique are now apparent, such as the identification and resolution of heterogeneous emission and the analysis of excited state processes.[53,54]

The measurement of phase-sensitive emission spectra requires only the addition of a simple phase sensitive detector to a phase fluorometer.[34] Phase-sensitive detection is performed on the low-frequency cross-correlated signals. The principle of this measurement is best illustrated in schematic form (Fig. 28). Assume a sample contains two fluorophores, A and B, and is excited with sinusoidally modulated light. Then, the emission is also sinusoidally modulated. The phase angle and modulation of the emission are moderately complex functions of the fractional intensities ($\alpha_i\tau_i$) and lifetimes (τ_i) of the two emitting species [Eqs. (22)–(28)]. The modulated emission is composed of two sine waves of equal frequencies but different phase shifts and amplitudes. The essential point, which remained unrecognized for a number of years by the practitioners of fluorescence spectroscopy, is that these two sine waves can be simply separated using a phase-sensitive detector. These detectors yield a direct current signal proportional to modulated amplitude and to the cosine of the phase difference between the detector phase ϕ_D and the phase of the sample (ϕ_i). For the two-component mixture illustrated in Fig. 28 the phase sensitive intensity is

$$I(\lambda, \phi_D) = F_A(\lambda)m_A \cos(\phi_D - \phi_A) + F_B(\lambda)m_B \cos(\phi_D - \phi_B) \quad (39)$$

where $F_i(\lambda)$ are the steady-state emission spectra, m_i the demodulation factors of each component, and ϕ_i their phase angles. For single exponential decays the latter two quantities are related to the lifetime by

$$\tan \phi_i = \omega\tau_i \quad (40)$$

[52d] S. Keating-Nakamoto and J. R. Lakowicz, *Anal. Biochem.* **148**, 349 (1985).
[53] J. R. Lakowicz and A. Balter, *Photochem. Photobiol.* **36**, 125 (1982).
[54] J. R. Lakowicz and A. Balter, *Biophys. Chem.* **16**, 117 (1982).

PHASE SUPPRESSION

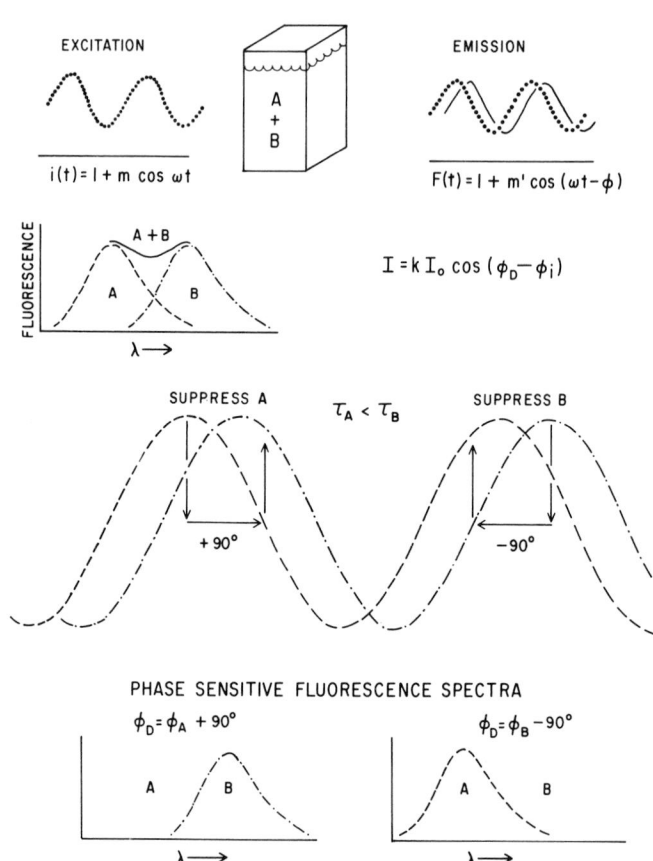

FIG. 28. Schematic description of phase-sensitive detection of fluorescence. From Lakowicz and Cherek.[34]

$$m_i = (1 + \omega^2 \tau_i^2)^{1/2} \tag{41}$$

Equation (39) and Fig. 28 illustrate an important feature of phase-sensitive detection. The detector phase can be selected to be out of phase, $|\phi_D - \phi_i| = 90°$, with any given component and/or lifetime. For a two-component mixture, this allows the direct recording of each emission spectrum from the mixture. For instance, if $\phi = \phi_A + 90°$, then the detector is out of phase with component A. The phase-sensitive emission spectrum is then expected to be superimposable on the steady-state spectrum of component B. Conversely, if $\phi_D = \phi_B - 90°$, then the emission spectrum of component A can be recorded from the mixture. It is impor-

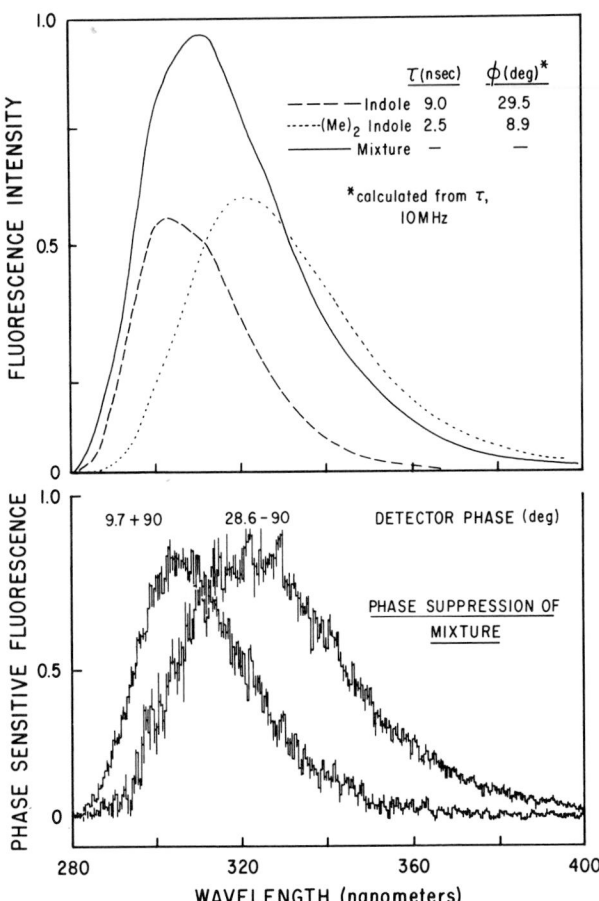

FIG. 29. Steady-state and phase-sensitive emission spectra of a mixture of indole and dimethylindole. From Lakowicz and Cherek.[35]

tant to notice that the phase angle required to suppress the emission from each component is determined by its lifetime. Alternatively, the lifetime can be determined from the phase angle used for suppression, of this component.

At present the use of phase-sensitive detection is just beginning, and only a few results are available. First, we wish to demonstrate that the resolution predicted in Fig. 28 can be realized in practice. The resolution of indole and dimethylindole (DMI) is shown in Fig. 29. Steady-state spectra are shown in the top panel and phase-sensitive spectra in the lower panel. At a modulation frequency of 10 MHz the phase angles of indole and dimethylindole are 8.9° (2.5 nsec) and 28.6° (9 nsec), respec-

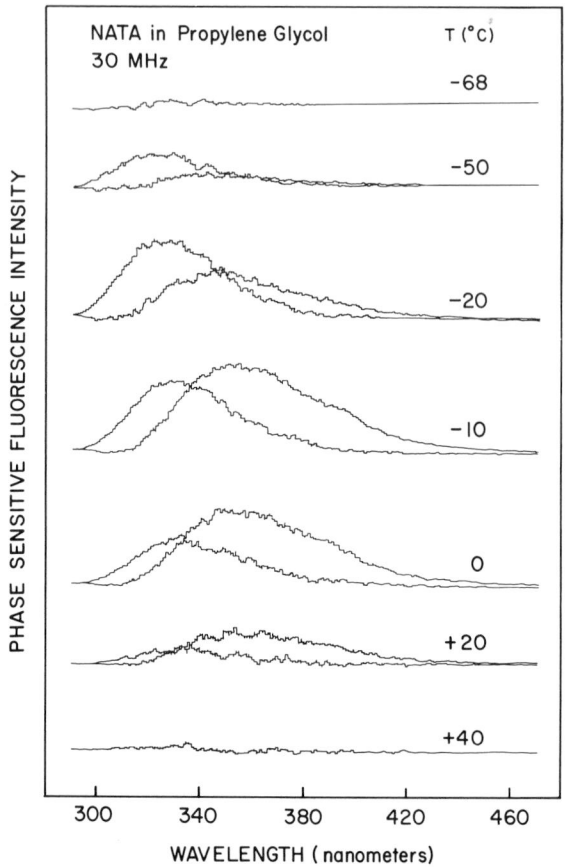

FIG. 30. Phase-sensitive emission spectra of *N*-acetyl-L-tryptophanamide in propylene glycol. The *F* and *R* state emissions were suppressed at 310 and 400 nm, respectively. From Lakowicz and Balter.[53]

tively. When the detector is 90° offset from either of these phase angles, the phase-sensitive spectrum is superimposable on the steady-state spectrum of the other pure compound. This result demonstrates that phase-sensitive detection can be used to resolve a two-component mixture. For accuracy we note that at a single modulation frequency one needs to know either one emission spectrum or one lifetime to determine the other three parameters (one or two lifetimes and two or one emission spectra).

As a final example of the use of fluorescence for analysis of time-dependent processes we describe how phase-sensitive detection was used to study spectral relaxation of *N*-acetyl-L-tryptophanamide (NATA) in propylene glycol. Previously we presented the temperature-dependent spectra of NATA in this solvent (Fig. 5). These spectra suggested that the unrelaxed emission was dominant at 310 nm, and the relaxed emission was dominant at 400 nm. Suppose that this sample is studied by phase-sensitive detection. Then, if the detector phase is 90° offset from the emission at 310 nm, this emission is suppressed, and only the relaxed emission should be observed. Conversely, if the detector phase is 90° offset from the emission at 400 nm, then this emission is suppressed, and only the unrelaxed emission should be observed. The results of this experiment are shown in Fig. 30. Consider the intermediate temperature of $-10°$. On suppression of the emission at 310 and 400 nm two distinct emission spectra are seen. If the temperature is lowered, the blue-shifted emission becomes dominant. At higher temperature the red-shifted emission becomes dominant. The relative intensities of each phase-sensitive spectrum can be used to estimate the spectral relaxation time.[53] At the highest and lowest temperatures the emission is from only the relaxed or the unrelaxed state, respectively, and distinct phase-sensitive spectra are not observable.

More recently this procedure has been generalized to avoid the need to suppress any individual component.[52d,55] Specifically, phase-sensitive spectra are recorded at a number of arbitrary phase angles. These spectra are then fit using a least-squares method to yield the lifetimes and amplitudes of each species in a mixture of fluorophores. One needs to know the steady-state spectra, but three-component resolutions are possible. However, the need to know the steady-state spectra of each component can be eliminated using phase-sensitive spectra obtained at several modulation frequencies.[56]

Summary

Fluorescence spectroscopy provides numerous opportunities for determination of the dynamic properties of macromolecules. Measurements may be performed in the time, lifetime, or the frequency domain. Presently, the highest resolution is provided by the frequency-domain measurements, using the newly available instruments.

[55] S. Keating-Nakamoto and J. R. Lakowicz, *Biophys. Chem.* in press (1986).
[56] S. Keating-Nakamoto, H. Cherek, and J. R. Lakowicz, submitted for publication (1986).

[24] Mössbauer Effect in the Study of Structure Dynamics

By FRITZ PARAK and LOU REINISCH

Introduction

Nuclear resonance absorption of gamma radiation without transfer of recoil energy was discovered by R. L. Mössbauer in 1958. In only a few years this discovery developed into a powerful spectroscopic tool in many fields of science. A comprehensive treatment of Mössbauer spectroscopy can be found in other sources.[1-3]

Mössbauer spectroscopy on the ^{57}Fe nucleus is an experimental method applied to many problems in solid state physics, complex chemistry, and biochemistry. In general, it is used to investigate the electron structure of iron in a compound. The electron level scheme influences the iron nucleus via the hyperfine interactions, yielding slightly different resonance energies. Isomer shift, electric quadrupole splitting, and magnetic hyperfine interactions determine the numbers, positions, and relative intensities of the resonant lines of a Mössbauer spectrum.

In the past several years a quite different parameter of Mössbauer spectroscopy has proven to be of importance for the investigation of iron-containing proteins. The Mössbauer effect in a Mössbauer isotope occurs only with a certain probability, given by the Lamb–Mössbauer factor, f. Simply explained, f is the fraction of gamma radiation absorbed by the nuclei without energy being lost to recoil. For the hypothetical case of an ^{57}Fe atom bound absolutely rigid in space, f in the classical limit is 1. However, an atom is never absolutely fixed. During the characteristic absorption time of the Mössbauer gamma photon (10^{-7} sec), the iron normally moves. A mean square displacement in this time interval, $\langle x^2 \rangle$, corresponds to a reduction of the probability of Mössbauer absorption given by

$$f = \exp(-k^2\langle x^2 \rangle) \tag{1}$$

where $k = 2\pi/\lambda$ with $\lambda = 0.86$ Å in the case of ^{57}Fe. The magnitude of $\langle x^2 \rangle$ depends on the environment of the iron and the coupling to the environment. An ^{57}Fe atom in metallic iron foil is bound quite strongly in the

[1] U. Gonser, ed., "Mössbauer Spectroscopy," Topics in Applied Physics, Vol. 5. Springer-Verlag, Berlin, 1975.

[2] U. Gonser, ed., "Mössbauer Spectroscopy II," Topics in Current Physics, Vol. 25. Springer-Verlag, Berlin, 1981.

[3] N. N. Greenwood and T. E. Gibb, "Mössbauer Spectroscopy." Chapman & Hall, London, 1971.

metal lattice and therefore has a rather rigid environment. Consequently, $\langle x^2 \rangle$ is small, determined only by the lattice vibrations, yielding $f = 0.7$ as a typical value at room temperature. An $^{57}Fe^{2+}$ ion in water marks the other extreme case. The liquid is not at all rigid, and no Mössbauer effect is observed. It is important to note that environment, in the case of ^{57}Fe, means a region of mass of at least 5×10^7 amu. Otherwise the recoil transfer during the absorption of the gamma quantum prevents principally resonance absorption.

An iron atom in a protein represents a situation which is between the two limiting cases previously discussed. Protein molecules are flexible. Large mean squared displacements of the iron can occur. A protein crystal or a frozen protein solution represents, nevertheless, an environment for the iron which is large and rigid enough to permit Mössbauer effect measurements. By measuring f, one finds according to Eq. (1) the $\langle x^2 \rangle$ value of the iron. The structure dynamics of the protein contribute considerably to the $\langle x^2 \rangle$ of the iron. Mössbauer absorption spectroscopy on the iron in proteins is, therefore, a method to examine protein dynamics. The section on Mössbauer Absorption Spectroscopy will give the theoretical and technical details needed to perform and interpret such experiments.

There is, in addition, another quite different application of the Mössbauer effect which allows the study of dynamic properties of condensed matter. Gamma rays from a Mössbauer source can be Rayleigh scattered from the electrons in samples containing no iron. The RSMR (Rayleigh scattering of Mössbauer radiation) technique also allows for the determination of the mean square displacements of the atoms of a molecule. It has already been applied in a few cases to proteins and will be described later in this chapter.

Mössbauer Absorption Spectroscopy

Background

Principle of the Measurement. A Mössbauer spectrometer is similar to an optical spectrophotometer. Radiation with a wavelength λ and intensity I_0 is incident upon a sample and the transmitted intensity is measured. By changing the wavelength of the incoming radiation, absorption maxima can be detected.

In our case the monochromatic incoming radiation stems from the radioactive isotope ^{57}Co which decays to ^{57}Fe. In the last step of this decay, the excited ^{57}Fe goes to its ground state with the emission of radiation with an energy $E_s = 14.4$ keV ($\lambda = 0.86$ Å). This radiation can, under proper conditions, be absorbed by ^{57}Fe nuclei in a sample.

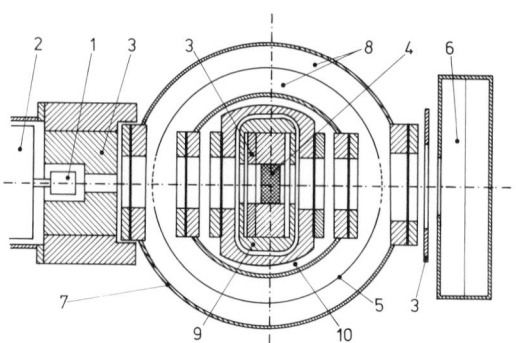

Fig. 1. Mössbauer spectrometer for low temperature investigations. (1) Source; (2) electromagnetic driver; (3) lead pinholes; (4) sample; (5) copper thermal shield; (6) detector; (7) outer wall of the cryostat; (8) thermal isolation vacuum; (9) probe chamber with sample holder; (10) low pressure gas chamber for temperature control.

Figure 1 shows a Mössbauer spectrometer. The ^{57}Co source (1) is mounted on an electromagnetic driving system (2) which by the Doppler shift allows a small energy change $[(v/c)E_s]$ of the emitted ^{57}Fe gamma radiation. This energy shift corresponds to the change of wavelength in an optical spectrophotometer. Here, v is the velocity of the source (up to 30 mm/sec), and c is the velocity of light. The radiation is collimated (3) and is incident upon the sample (4). The number of transmitted gamma quanta is measured by a detector (e.g., a proportional counter) (6). All technical details of the spectrometer are discussed in the section on Experimental Equipment.

A Mössbauer spectrum is the transmitted intensity, or equivalently the number of gamma photons counted, $Z(v)$, plotted as a function of the velocity v of the source. Often a normalization is performed by dividing $Z(v)$ by $Z(\infty)$. Here, ∞ means a high velocity of the source where the resonance condition for Mössbauer absorption is practically destroyed. Figure 2 shows the Mössbauer spectrum of deoxygenated myoglobin.[4] It consists of two absorption lines at -0.2 and 1.52 mm/sec. The distance between the two lines as well as the displacement of the center of symmetry of the two lines with respect to $v = 0$ are determined by the hyperfine interactions, while the area A of the absorption line (hatched in Fig. 2) reflects the dynamic properties of the sample.

The Shape and Area of a Mössbauer Absorption Spectrum. The sample is irradiated with an intensity I_0. Only a fraction of I_0 is absorbed by the Mössbauer nuclei. The number of quanta in the transmitted beam is

[4] F. Parak, E. W. Knapp, and D. Kucheida, *J. Mol. Biol.* **161,** 177 (1982).

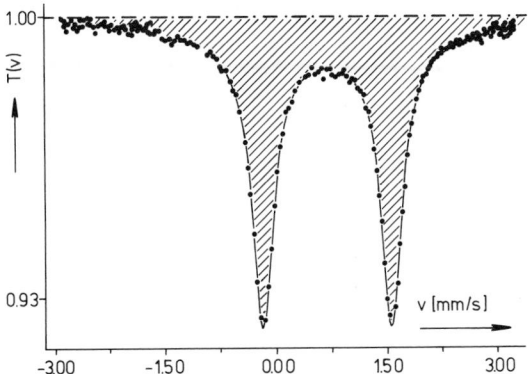

FIG. 2. Mössbauer spectrum of deoxygenated sperm whale myoglobin at 245 K. The sample consist of a large number of small crystals. The logarithm of the hatched area, A_{exp}, is proportional to the mean square displacement of the heme iron atom. From Parak et al.[4]

given by

$$Z(v) = CR \int_{-\infty}^{\infty} Q(v, E_s, E) \exp[-\sigma_a(E_a, E)n_{\text{Fe}}]dE$$

$$+ CR(1 - f_s) + CU \qquad U + R = 1 \qquad (2)$$

C is determined by the activity of the source, R is the fraction of detected radiation coming from the 14.4 keV transition, while U is the fraction of detected radiation with energy other than 14.4 keV. In Eq. (2) the first term represents a decrease of the intensity of the transmitted beam due to Mössbauer absorption. The second term represents photons from the 14.4 keV transition, which lost recoil energy during the emission and, therefore, cannot perform Mössbauer effect. The final term accounts for photons with the improper energy which cannot be discriminated due to the limited energy resolution of the detector. Since the energy resolution of the detector is always much broader than the energy resolution of the Mössbauer effect, U is always nonzero.

In the first term of Eq. (2), $Q(v, E_s, E)$ is the emission spectrum of the source, centered at the energy $[1 + (v/c)]E_s$

$$Q(v, E_s, E) = \frac{f_s(\Gamma_s/2\pi)}{\{E - E_s[1 + (v/c)]\}^2 + (\Gamma_s/2)^2} \qquad (3)$$

and $\sigma_a(E_a, E)$ represents the cross-section for Mössbauer absorption centered at the energy E_a

$$\sigma_a(E_a, E) = \sigma_0 f_a \frac{(\Gamma_a/2)^2}{(E - E_a)^2 + (\Gamma_a/2)^2} \qquad (4)$$

The full-width half max (FWHM) linewidths of the Lorentzian energy distribution emitted by the source and absorbed in the sample are given by Γ_s and Γ_a, respectively. These are usually slightly broader than the theoretical linewidth for several reasons, as discussed in the section on Experimental Equipment. The probability for recoil-free emission or absorption is given by the Lamb–Mössbauer factors f_s and f_a, respectively. For iron, σ_0 equals 2.56×10^{-18} cm^2/^{57}Fe nucleus, and n_{Fe} gives the number of ^{57}Fe nuclei per cm^2 in the sample.

Unfortunately, the integral in Eq. (2) cannot be solved analytically. When the sample does not contain too much iron, which means

$$\sigma_a(E_a, E)n_{Fe} \ll 1 \tag{5}$$

the exponential function can be expanded in a series which is truncated after the linear term of $\sigma_a(E_a, E)n_{Fe}$. Using Eqs. (2)–(5), the Mössbauer spectrum becomes, in the thin absorber approximation

$$T(v) = \frac{Z(v)}{Z(\infty)} = 1 - \frac{R}{(R + U)} f_s t_a \left(\frac{\Gamma_a}{2}\right)$$

$$\frac{(\Gamma_a + \Gamma_s)/2}{\{E_a - E_s[1 + (v/c)]\}^2 + [(\Gamma_a + \Gamma_s)/2]^2} \tag{6}$$

with

$$t_a = n_{Fe}\sigma_0 f_a \tag{7}$$

In practice, Eq. (6) is often used for a phenomenological description of a Mössbauer spectrum, even if the condition of Eq. (5) is not rigidly fulfilled. Corrections which allow the use of Eq. (6) for the determination of dynamic properties of a medium thick absorber are given below.

Here we introduce some other notation which is useful. The Mössbauer absorption is given by

$$\eta(v) = 1 - T(v) \tag{8}$$

and the maximal absorption becomes

$$\eta(0) = \frac{R}{(R + U)} f_s t_a \frac{\Gamma_a}{\Gamma_a + \Gamma_s} \tag{9}$$

The absorption area, A, of the Mössbauer spectrum is defined as

$$A = \int_{-\infty}^{\infty} \eta(v)d\left(\frac{E_s}{c}v\right) \tag{10}$$

which gives

$$A = \frac{R}{(R + u)} f_s t_a \frac{\Gamma_a}{2} \pi \tag{11}$$

Equation (11) is used for the determination of the effective absorber thickness, t_a. Then t_a contains the Lamb–Mössbauer factor, f_a, and likewise information about the protein dynamics of the sample.

For relatively thick absorbers ($t_a \leqq 30$), Eq. (10) together with Eq. (2) was solved by starting the integration over v and developing the results into a series.[5] As a result, one obtains

$$A_{exp} = \frac{R}{(R + U)} f_s L(t_a) \frac{\Gamma_a}{2} \pi \tag{12}$$

A table of L as a function of t_a is made, allowing one to determine t_a from the experimental measurement of $L(t_a)$.[5,6]

Hyperfine Interactions. The hyperfine interactions between the electrons of the iron and the ^{57}Fe nucleus cause a splitting of the Mössbauer absorption described by Eqs. (2) and (6). A number of resonances are observed as seen in Fig. 2. At present the hyperfine splitting cannot be used to determine structure dynamics. Nevertheless, it is necessary to discuss some basic aspects.

The general consequence of the hyperfine interactions is replacing Eq. (6) by

$$T(v) = 1 - \frac{R}{(R + U)} f_s t_a \sum_{\ell=1}^{L} w_{a\ell} \frac{\Gamma_{a\ell}}{2}$$

$$\frac{(\Gamma_{a\ell} + \Gamma_s)/2}{\{E_{a\ell} - E_s[1 + (v/c)]\}^2 + [(\Gamma_{a\ell} + \Gamma_s)/2]^2} \tag{13}$$

The index $\ell = 1$ to L accounts for the L different resonance energies in the absorber. The weighting factor $w_{a\ell}$ gives the relative contribution of the resonance ℓ, with

$$\sum_{\ell=1}^{L} w_{a\ell} = 1 \tag{14}$$

For many applications, the index ℓ can be omitted from $\Gamma_{a\ell}$. We now discuss very briefly some standard characteristics of Mössbauer spectra.

Line shift. In Eq. (6) it was assumed that the energy of the 14.4 keV ^{57}Fe radiation emitted by the source is centered around E_s, while the center of the resonance energy E_a of the absorber is not necessarily equal to E_s. The energy difference primarily comes from the different environ-

[5] G. Lang, *Nucl. Instrum. Methods* **24,** 425 (1963).
[6] D. W. Hafemeister and E. Brooksshera, *Nucl. Instrum. Methods* **41,** 133 (1966).

ment of the iron atoms in the source (nearest neighbor metal ions) and the iron atoms in the absorber (nearest neighbors are typically covalently bound N, C, or S atoms). This energy difference is termed the isomer shift.

Another contribution to the line shift is the temperature shift, or the second-order Doppler shift. The mean square velocity of the iron in an absorber and source can shift the resonant energy. The shift in energy is proportional to the kinetic energy of the Mössbauer atoms in the material, and therefore it is a function of the heat capacity and the chemical composition. Although this parameter, in principle, contains information concerning the dynamics, it has only recently been used in the investigation of myoglobin.[26]

In the following discussion, the line shift, $\delta S_\ell = E_{a\ell} - E_s$ will be treated as an empirical parameter necessary for the simulation of the Mössbauer spectrum.

Electric quadrupole splitting. Such splitting occurs if the nearest neighbor atoms of the iron are not arranged in an octahedron or a tetrahedron. In iron-containing proteins, the symmetry is practically always lower than cubic: electric quadrupole splitting has to be taken into account. For an ^{57}Fe Mössbauer spectrum one obtains two resonance energies. In Eq. (13) L becomes 2. A quadrupole doublet with different absorption areas A_1 and A_2 and consequently different weighting factors w_{a1} and w_{a2} can occur, for instance, in the investigation of a single crystal or if the motion of the iron is considerably nonisotropic (Goldanskii Karjagin effect).

More often, frozen protein solutions or polycrystalline samples with a random orientation of the molecules are investigated. In this case, Eq. (13) has to be used with

$$
\begin{aligned}
\Gamma_{a1} &= \Gamma_{a2} = \Gamma_a \\
w_{a1} &= w_{a2} = 0.5 \\
E_{a1} &= E_a + \tfrac{1}{2}\Delta E_Q \\
E_{a2} &= E_a - \tfrac{1}{2}\Delta E_Q
\end{aligned}
\tag{15}
$$

The quadrupole splitting measured as the distance between the two resonant absorption lines is given as ΔE_Q.

Magnetic hyperfine splitting. In the presence of a magnetic field $\mathbf{B}_{hf}$ at the iron nucleus, resonant absorption occurs at typically six different energies. In samples with randomly oriented molecules the $w_{a\ell}$ values for the six lines are 3/12, 2/12, 1/12, 1/12, 2/12, 3/12, and $\Gamma_{a\ell}$ is equal for all the lines. In practice the ratios are often modified by local order. The magnetic hyperfine field $\mathbf{B}_{hf}$ at the nucleus is mainly due to the polarized d electrons of Fe^{3+}. The spins $\pm 5/2$, $\pm 3/2$, $\pm 1/2$ of the d electrons produce different $\mathbf{B}_{hf}$ values, so that the Mössbauer spectrum becomes a superposition of three spectra according to Eq. (13) ($L = 18$). In proteins the electric quadrupole splitting is always present, in addition to the magnetic hyperfine interactions. As a consequence, the different absorption lines of the Mössbauer spectra are often difficult to resolve.

Due to the coupling of the iron to the thermal bath of the environment another problem arises. Resolved magnetic hyperfine patterns are only found at temperatures around 4.2 K, if at all. At slightly higher temperatures the spins of the $3d$ electrons change their orientations during the time which the nuclear spin needs to precess around the field $\mathbf{B}_{hf}$. This relaxation produces fluctuating $\mathbf{B}_{hf}$ fields and unresolved Mössbauer spectra. At high temperatures (typically 200 K) the spin relaxation becomes so fast, the effective $\mathbf{B}_{hf}$ values becomes zero. Once again, only quadrupole splitting is seen.

In practice, one uses Eq. (13) for all cases where resolved lines are found. At low temperatures L is typically 6, while at high temperatures it becomes 2. The coefficients w_{af} are fitted to the experimental values. For Fe^{3+} proteins the region between $T = 10$ and 240 K normally cannot be simulated by Eq. (13).

Sample Preparation

Isotopic Enrichment. Since 1971[7] the examination of structure dynamics using Mössbauer spectroscopy has been performed only on iron-containing proteins, such as myoglobin,[4,7,8] hemoglobin,[9] or the reaction center of photosynthetic bacteria.[10] In all cases, the natural isotopic mixture of iron containing only 2.2% of the Mössbauer isotope ^{57}Fe had to be replaced by iron highly enriched in ^{57}Fe. Commercially, Fe metal or Fe_2O_3 enriched in ^{57}Fe between 80 and 95% is available.

About 0.1 mg ^{57}Fe per cm^2 of sample area is necessary to measure a Mössbauer spectrum in a reasonable time. [$n_{Fe} = 10^{18}$ ^{57}Fe nuclei/cm^2 in Eq. (7)]. Using the typical sample area of 1 cm^2, this corresponds (in the case of myoglobin) to 30 mg protein for totally enriched material and to 1.36 g of native protein. Since protein dynamics are only present in protein molecules covered by water, the total sample normally contains at least the same amount of water as protein. With unenriched material the sample thickness then becomes about 2 cm. Such a sample would completely absorb the 14.4 keV gamma radiation by electronic X-ray absorption and nearly prevent Mössbauer effect measurements. In our experi-

[7] F. Parak and H. Formanek, *Acta Crystallogr., Sect. A* **27**, 573 (1971).

[8] E. R. Bauminger, S. G. Cohen, I. Nowik, S. Ofer, and J. Yariv, *Proc. Natl. Acad. Sci. U.S.A.* **80**, 736 (1983).

[9] K. H. Mayo, D. Kucheida, F. Parak, and J. C. W. Chien, *Proc. Natl. Acad. Sci. U.S.A.* **80**, 5294 (1983).

[10] F. Parak, E. N. Frolov, A. A. Kononenko, R. L. Mössbauer. V. I. Goldanskii, and A. B. Rubin, *FEBS Lett.* **117**, 368 (1980).

ence, a sample thickness between 3 and 6 mm (protein plus water) can be tolerated with respect to electronic absorption. This example shows that isotopic enrichment of ^{57}Fe cannot be avoided. If enrichment is not possible, one should consider not using Mössbauer absorption spectroscopy.

In this context, it is important to note that the electronic absorption is dramatically increased by salts and metal ions in the solution of the protein. Whenever possible these should be reduced to a minimum.

There are two methods to enrich proteins in ^{57}Fe. In heme-containing proteins, the heme group can often be removed from the molecule and then reconstituted with ^{57}Fe containing heme. The problematic step in this procedure is not the separation and reconstitution, but it is usually the preparation of the ^{57}Fe-containing heme. Several slightly different procedures are described in the literature,[7,11] but in our experience all of them are difficult to reproduce.

A more general method of ^{57}Fe enrichment is the *in vivo* enrichment. It can be applied to enzymes of bacteria and yeast. Cells are grown on a medium containing ^{57}Fe. It is essential to avoid as much contamination as possible from natural iron, for instance from the walls of stainless-steel fermentors. ^{57}Fe is not easily taken up into the cells if it is present in the medium as chloride or sulfate. Better efficiency is obtained with iron citrate. As examples of *in vivo* enrichment we refer to the reaction centers of photosynthetic bacteria.[10]

For the investigation of non-iron enzymes an iron label is necessary. Only a few attempts have been made to develop a technique similar to EPR labeling. In our group, iron *o*-phenanthroline has been used as a nonspecific label for motions on the protein surface.

The Charge State of the Iron. The most information on dynamics is obtained if not only the absorption area but also the line shape of the Mössbauer spectrum can be analyzed. In general, this is only possible for Fe^{2+} compounds. In Fe^{3+} compounds, magnetic relaxations make an analytic description of the Mössbauer spectrum complicated or even impossible. Whenever possible, to ease the analysis, studies of structure dynamics should be performed on Fe^{2+} compounds.

The Aggregate State of the Sample. The Mössbauer effect cannot be observed in a water solution of protein molecules at room temperature. Normally it is necessary to investigate frozen solutions. Some problems have to be taken into account. First of all, investigations can be performed only well below the melting point of the solution. In a range of 10° below the melting point of the solution, diffusion processes may occur which give similar features in the Mössbauer spectrum as protein dynamics. It is therefore difficult to separate these two effects.

---

[11] M. Overkamp, H. Twilfer, and K. Gersonde, *Z. Naturforsch.* **31c**, 524 (1976).

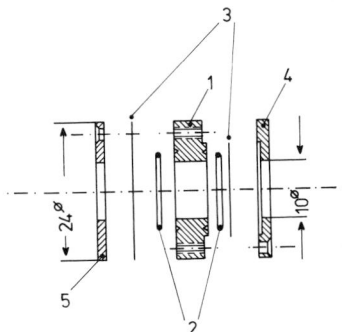

FIG. 3. Sample holder. (1) Plastic ring; (2) indium wire; (3) Mylar windows; (4 and 5) brass rings.

Protein crystals provide a better sample. Experiments can be performed even at room temperature. Comparisons with X-ray investigations are also possible. Crystals which are from the crystallographic point of view unsuitable can be used in a Mössbauer experiment. Even rather small crystals are sufficient since a sample consists, in general, of many crystals. The sample should contain no "excess" mother liquid. On the one hand, drying the crystals prevents structure dynamics in protein molecules. On the other hand, the presence of 3.75 M $(NH_4)_2SO_4$ liquid around the myoglobin crystals drastically increases the electronic absorption of the gamma radiation, making the Mössbauer absorption measurements difficult.

Another sample possibility is protein precipitates. In the case of myoglobin, a precipitate can be obtained by adding $(NH_4)_2SO_4$ to the water solution. The wet precipitate without excess mother liquid can be used for Mössbauer experiments even at room temperature. It should be mentioned, however, here that the environment of each molecule is not as well defined as in a crystal. Part of the information from such samples, in comparison to crystals, may not be resolvable.

Occasionally freeze dried materials are investigated. It should be emphasized once more that proteins with a hydrophilic surface do not exhibit the normal dynamics in the dry state. Such samples yield information only by comparison with water-covered samples.

The Sample Holder. Figure 3 shows the sample holder which is used at present in our laboratory. It consists of a 3–6 mm plastic ring (1) and two Mylar windows (3) pressed together with brass rings (4,5) to make a vacuum-tight container. The indium wire (2) is used as a low-temperature gasket between the Mylar and plastic. When this sort of sample holder is used at room temperature for several months, the mother liquid of the protein crystals can etch small channels through the indium ring. The

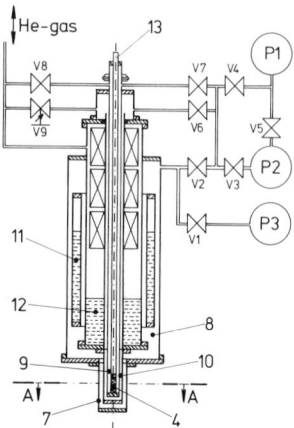

FIG. 4. Cryostat for Mössbauer experiments between 4.2 K and room temperature. (4) Sample; (7) outer wall of the cryostat; (8) thermal isolation vacuum; (9) probe chamber with sample holder; (10) low pressure gas chamber for temperature control; (11) liquid nitrogen reservoir; (12) liquid helium reservoir (13) pole to mount sample in the cryostat; P1, rotation pump; P2, diffusion pump; P3, ion getter pump; V1 to V9, valves. Another view of the same cryostat along cut A–·–A is shown in Fig. 1.

sample then dries out. This can be avoided by initially coating the indium wire in the sample holder with a small amount of silicon grease.

Experimental Equipment

Temperature Control of the Sample. Mössbauer absorption spectroscopy can be performed easily between 4.2 K and room temperature. This opens the possibility of comparing room temperature measurements (where the structure dynamics are active) with low temperature measurements (where protein specific motions are frozen out). It should be pointed out that, in general, only these extended temperature range measurements yield complete information. Measurements at only one temperature or over a small temperature interval are of a limited value.

A cryostat which can maintain the sample at temperatures between 4.2 K and room temperature is shown in Fig. 4. Figure 1 is a cross section along A–·–A of the cryostat tail from Fig. 4. The cryostat is essentially a central probe chamber (9) and two coaxial cryogenic liquid reservoirs (11, 12).

For measurements made below 80 K, reservoir (11) is filled with liquid nitrogen and reservoir (12) with liquid helium. Both chambers are thermally isolated by a vacuum (8) which is established by the rotation pump (P1) together with the diffusion pump (P2). During an experiment, proper adjustment of the valves V1 to V4 allows only the ion getter pump (P3) to maintain the vacuum. The thermal coupling of the probe chamber (9) to the cryogenic liquid in (12) is controlled by the He gas pressure in (10) with the help of the valves V6 and V9. This allows the gross temperature regulation of the sample, while fine regulation is made with a heater mounted on the sample holder. Temperature measurements are made with a simple Si diode. If necessary, a frozen sample can be mounted onto the sample pole (13) and quickly brought into the cold cryostat. To prevent the condensation of water on the sample or cryostat windows, the probe chamber can be evacuated (valve V7) and filed with He gas (valve V8).

The ^{57}Co Source. Mössbauer spectroscopy counts statistically emitted gamma quanta, and the accuracy depends on the absolute number of registered quanta. For the investigation of structure dynamics, a ^{57}Co source with an activity between 50 and 200 mCi is necessary to determine the spectra in a reasonable time. At low temperatures, a Mössbauer spectrum can be collected typically in 1 day or less. Above 200 K and most notably at room temperature, it can take 1 to 3 weeks to collect a good spectrum, even with a strong source.

The choice of the proper source is quite important for several reasons. In order to get a single unsplit emission line, the ^{57}Co has to be diluted by a diamagnetic material which forms a cubic lattice. A very homogeneous matrix material with a small ^{57}Co content (typically 1%, the maximum is about 6%) yields a narrow emission line where Γ_s is close to Γ_{nat}. The matrix material, however, absorbs part of the emitted 14.4 keV radiation. Moreover, highly energetic gamma radiation from the early steps of the ^{57}Co decay gives rise to characteristic X-ray radiation primarily from the K-shell electrons in the matrix. This radiation contributes to the background U in Eq. (13).

The background is also significantly increased by contamination of the ^{57}Co with ^{56}Co and ^{58}Co, even in amounts less than 1%. When buying a source, one should try to get material with the lowest possible contamination. The half-life of these unwanted isotopes is about 70 days in comparison to 270 days for ^{57}Co. Therefore, "older" activity yields a lower background than activity used immediately after the production from a cyclotron.

In our laboratory we generally use ^{57}Co diffused into a rhodium matrix ($^{57}Co\underline{Rh}$). Sources with up to 200 mCi ^{57}Co in a spot of 2×2.5 mm are commercially available. It is, however, preferable to use sources with active areas a bit larger (spots with a 3 to 8 mm diameter). Rh gives characteristic X-ray radiation at about 20.2 keV. In the equipment shown in Figs. 1 and 4, the source is always at room temperature. Problems may

LAMB–MÖSSBAUER FACTOR, f_s, OF A ^{57}Co$\underline{\text{Rh}}$
SOURCE AS A FUNCTION OF TEMPERATURE[a]

Temperature (K)	f_s (± 0.005)
4.2	0.875
10.0	0.874
20.0	0.873
30.0	0.872
40.0	0.870
50.0	0.867
60.0	0.865
70.0	0.862
80.0	0.859
90.0	0.856
100.0	0.852
120.0	0.845
140.0	0.838
160.0	0.830
180.0	0.823
200.0	0.815
220.0	0.808
240.0	0.801
260.0	0.793
280.0	0.786
300.0	0.779

[a] Experimental results from several authors were fitted with a Debye theory, and the values from the fit are given here.

arise in a cryostat with both the source and the absorber at the same temperature. After several cycles of cooling and warming the source, clustering of the ^{57}Co in the rhodium can occur. This leads to line broadening and possible line splitting. The source should be checked often with a reference absorber, such as $K_4Fe(CN)_6$. The line splitting in the source occurs, however, only when it is at low temperatures (4.2 up to 40 K).

In Eq. (13), f_s is together with f_a. So, one must know f_s to find f_a. Frequently this problem can be circumvented. Occasionally, it is convenient to know f_s explicitly, especially if the source and absorber are not at the same temperature. The temperature dependence of f_s for Co$\underline{\text{Rh}}$ is given in the table and can be used as a correction.

In the literature one finds experiments made with ^{57}Co diffused into Cr. Such sources have a low background because of the low energy K-radiation from Cr (about 5.9 keV). These sources are, unfortunately, not always commercially available. The same is true for ^{57}CoO single line sources. Since such a source contains no metal other than Co, these sources have the highest activity for the smallest area and also a very

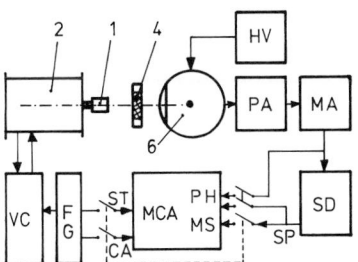

FIG. 5. Block diagram of the electronics of the Mössbauer spectrometer. (1) Source; (2) electromagnetic driver; (4) sample; (6) detector; HV, high voltage supply; PA, preamplifier; MA, main amplifier; SD, single channel discriminator; MCA, multichannel analyzer; FG, function generator; VC, velocity controller.

small background. CoO is, however, antiferromagnetic with a Neel point at 289 K. This limits its use to room temperature (and warm rooms at that). Another slight disadvantage is the large shift $\delta S = 1.1$ mm/sec of the emitted Fe^{2+} line with respect to metallic iron. Mössbauer spectra of iron in proteins are thereby shifted away from the center Doppler velocity, $v = 0$.

Mössbauer Spectrometer. The Mössbauer technique is discussed in a number of articles. Here we describe a standard spectrometer as used in our laboratory (compare Fig. 5). Further details of different Mössbauer spectrometers are described in other sources.[12]

Detection. We use a Reuter Stokes counting tube filled with $Xe(CO_2)$. The tube is biased near -1500 V (HV in Fig. 5), so the output pulse amplitude is proportional to the incoming gamma photon energy. The output pulse is immediately preamplified (PA) and then further amplified (MA) so that pulses corresponding to the 14.4 keV photons are near the middle of single channel discriminator (SD) range. The single channel discriminator is used to select pulses which have an amplitude corresponding to a 14.4 keV photon. For the energy discrimination and for the background correction measurements, the multichannel analyzer (MCA) is working in the pulse height mode (PH). Signals from the amplifier and from the single channel discriminator are used. Measuring a Mössbauer spectrum, the multichannel analyzer operates in the multiscaler mode (MS). The source velocity from $+v_{max}$ to $-v_{max}$ to $+v_{max}$ controls the multiscalar channel selection, and the discriminator pulses are added to the respective channel. After the data collection, the data are read on to magnetic tapes and finally computer analyzed.

The electromagnetic driver. The velocity driver (2) for the Mössbauer

[12] G. M. Kalvius and E. Kankeleit, "Mössbauer Spectroscopy and Its Application." Int. Atomic Energy Agency, Vienna, 1972.

source (1) is oscillated with a constant frequency sine wave. The sine wave is used since it is one of the easiest oscillations for an electromagnetic driver to follow accurately. The amplitude of the wave is directly proportional to the maximum Doppler velocity.

The driving signal is generated in the function generator (FG) and given via the velocity controller (VC) to the driving coil. Motion is monitored by a pickup coil. In the velocity controller the pickup coil signal is used in a feedback circuit to minimize the difference between the driving and pickup signals to less than 0.5%. The function generator also synchronizes the multichannel analyzer operating in the multiscaler mode. At the maximum positive amplitude of the sine wave, the voltage controller opens channel number 1 with a start pulse (ST). As the sine wave progresses, the channel advance pulse (CA) opens the remaining channels consecutively, to count the pulses from the single channel discriminator. The full period of the sine wave corresponds typically to 512 channels.

Data Collection. The pulse height spectrum. In contrast to general Mössbauer spectroscopy, the determination of the Lamb–Mössbauer factor relies on the proper background correction. For this purpose, the factor $R/(R + U)$ has to be determined experimentally.

In order to correct for the background it is necessary to measure the energy distribution of all the counted radiation during the experiment. The electronics operate in the pulse height mode (compare the section on Detection and Fig. 5). As a result, one obtains a spectrum as shown in Fig. 6. In addition to the 14.4 keV line, there is the K-radiation of the Rh at about 20.2 keV, a small amount of K radiation from iron at about 6.5 keV, an escape peak, and a continuous background coming from the Compton effect of high energy quanta produced in the early decay stages of the source. Such a pulse height spectrum needs to be measured before and after each series of Mössbauer spectra. Care must be taken that all conditions are identical during the measurement of the pulse height spectra and the Mössbauer spectra. The value of $R/(R + U)$ can vary due to water or nitrogen condensation on the cryostat windows or temperature dependent misalignment of the absorber. The analysis of the pulse height spectra will be described in the section on Background Correction.

In the next step, the 14.4 keV energy line is separated out with a single channel energy discriminator. The discriminator (SD in Fig. 5) produces a standard pulse (SP in Fig. 5) for every incoming pulse that is within a specified voltage window. Using these standard pulses as coincidence gating pulses in the multichannel analyzer, the pulse height spectra show only the quanta which fall in the voltage window (I_L to I_H) of the single channel discriminator. The normalized pulses are fed later into the multichannel analyzer operating in the multiscaler mode (MS in Fig. 5).

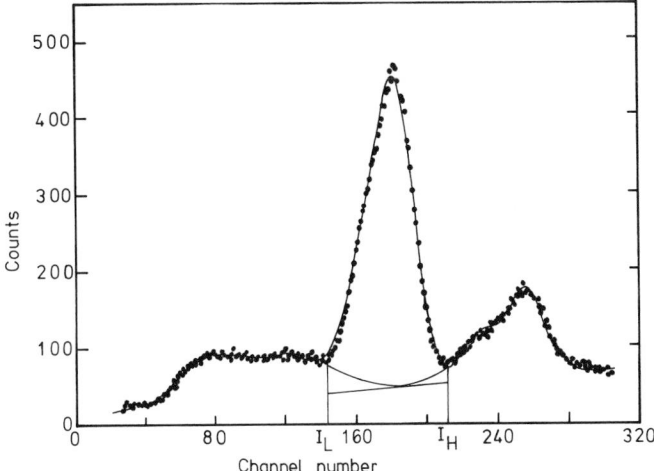

FIG. 6. Pulse height spectrum with a myoglobin sample and a $^{57}CoRh$ source. The spectrum was fitted according to Eq. (17). The parameters are given in the text. The single channel discriminator was set to accept pulses between I_L and I_H for the Mössbauer spectra. The Lorenztian tails of the nearby energy peaks and the linear background present in the "counted" quanta are shown.

The threshold adjustment shown in Fig. 6 is typical. The peak to background ratio cannot be raised significantly by narrowing the energy window. The line width of the 14.4 keV radiation ($\Delta E \approx 2.9$ keV from the detector) is determined by the energy resolution of the detector and not at all by the natural line width of the Mössbauer source ($\Gamma_{nat} = 4.665 \times 10^{-9}$ eV). A very narrow window only results in the loss of many good Mössbauer quanta.

Velocity calibration and linewidth. For our type of Mössbauer investigations, maximal Doppler velocities of the source of about ±4 mm/sec are used very frequently. Maximal velocities of ±10 or even ±30 mm/sec are necessary if magnetic effects occur (Fe^{3+}) or if broad lines from special motions in the protein can be found. As discussed later, such broad lines contain important information concerning protein dynamics. At temperatures above 200 K, high velocity spectra should be collected to ensure that the broad lines are not overlooked.

For the calibration of source velocity in the low velocity region, one uses sodium nitroprusside. This absorber gives two lines separated by 1.7034 ± 0.0014 mm/sec at 298 K. In the high velocity region, one uses a thin iron foil. The foil has six absorption lines with a distance of 10.64 ± 0.015 mm/sec between the first and sixth lines, 6.16 ± 0.008 mm/sec between the second and fifth lines, and 1.678 ± 0.002 mm/sec between the third and fourth lines. As before, the absorber should be at room temperature and in both cases placed as close as possible to the source (never on the detector). An absorber has to be rigidly fixed by a mechanical holder—do not rely on the stability of tape.

The absorption spectra can be fitted by least squares to the theoretical spectrum from Eq. (13) with $L = 2$ or $L = 6$ for the different reference materials, respectively. Then, the channel number of the multichannel analyzer can be correlated with the velocity of the source and, in turn, with the input voltage to the Mössbauer drive system.

For our application it is essential that the broadening of the Mössbauer absorption lines from the equipment not be too large. This broadening normally comes from small vibrations between the driving system and the cryostat. For instance, the vibrations of a vacuum pump can cause artifactually broad lines. Inhomogeneities in the source or absorber also can cause line broadening. In the spectrum of sodium nitroprusside, an experimental linewidth ($\Gamma_{exp} = \Gamma_a + \Gamma_s$, FWHM of the Lorentzian) should not exceed 0.30 mm/sec. Very good equipment gives 0.21 mm/sec for the inner lines of a thin iron foil. The theoretical value of $\Gamma_{exp} = 2 \times \Gamma_{nat} = 0.197$ mm/sec is normally not reached in our equipment.

Collection of the Mössbauer spectrum. To determine f_a, the Mössbauer spectrum must be measured at different temperatures. If the chosen sample contains some liquid (e.g., crystals) and will be measured above and below the freezing point of the liquid, it is advisable to make the high temperature measurements first. Once a sample is frozen, it should not be thawed and refrozen too often. Such cycling normally denatures part of the protein. When using a frozen sample, it is better to avoid changing the temperature systematically. One should first collect a few spectra from very different temperatures and then fill in the gaps. In this way, systematic errors, such as the denaturation of the sample or water slowly condensing on the cryostat windows, will be apparent.

Measurements made near the melting point of the liquid which surrounds the protein must be interpreted with care. Phase transitions in the liquid may produce effects which are not separable from the protein dynamics.

Data Evaluation

Least-Squares Fit. If possible, the measured spectrum is first simulated with a "theory." From the theory, a simulated counting rate, Z^c, is calculated for each point measured, Z^0. The difference between Z^c and Z^0 is then minimized by varying the free parameters of the theory. The Mössbauer spectrum, as the pulse height spectrum, is collected in a multichannel analyzer. The counts, Z_i^0, are obtained then as a function of the channel number. Using I channels, the following expression is then minimized

$$\chi^2 = \frac{\sum_{i=1}^{I} (Z_i^c - Z_i^0)^2}{\sum_{i=1}^{I} (Z_i^c)^2} \tag{16}$$

Many different programs are available to make this minimization efficiently.

Background Correction. The first step of data evaluation is the determination of the background U. First, the pulse height spectrum is least-squares fitted to the function

$$Z_i^c = a_0 i + b + \sum_{j=1}^{J} a_j \exp[-(i - n_j)^2/c_j^2] \tag{17}$$

Here, the spectral lines at different energies are approximated by Gaussians with a linewidth c_j centered at channel n_j. Seven Gaussians ($J = 7$) were used to make the fit in Fig. 6. The parameter $a_0 i + b$ describes a background proportional and independent of energy, respectively. Figure 6 was fit with Gaussians at the following energies: 6.4 keV (iron K-radiation), 9.7 keV, 12.9 keV (Mössbauer line escape peak), 14.4 keV (the Mössbauer line), 18.1 keV (Rh K-radiation escape peak), 20.0 keV (Rh K-radiation), and 21.8 keV.

After the least-squares fit, one has to determine what fraction of each of the seven Gaussians and background are within the energy window (I_L to I_H) of the Mössbauer experiment. When the energy peak of the 14.4 keV radiation can be fit with a single Gaussian, the ratio $R/(R + U)$ is obtained from

$$\frac{R}{(R + U)} = \frac{\sum_{i=I_L}^{I_H} a_{14.4} \exp[-(i - n_{14.4})^2/c_{14.4}^2]}{\sum_{i=I_L}^{I_H} Z_i^c} \tag{18}$$

The summation is performed between channels I_L and I_H, the lower and upper thresholds of the energy window from the single channel discriminator. The index 14.4 defines the Mössbauer line of the pulse height spectrum. In Fig. 6, we obtained $R/(R + U) = 0.82$, which is typical. If this value becomes smaller than 0.5, f_a can be determined with only much less accuracy.

Least-Squares Fit of the Mössbauer Spectra. The first step in the data evaluation of a Mössbauer spectrum is the folding. The multiscalar of a

Mössbauer spectrometer contains the spectrum twice (one time measured from the Doppler velocity $-v_{max}$ to $+v_{max}$ and a second time from $+v_{max}$ to $-v_{max}$). The counted quanta from each velocity can be summed together, ignoring the direction of the acceleration. In order to keep resolution in the velocity domain, the second half of the information from the multiscalar is interpolated between the channels. Hence, the position of the folding point and $+v_{max}$ has to be determined with an accuracy better than one channel.

Together with the folding, the geometry effects in the spectrum are corrected. Geometry effects are simply a change in apparent source intensity due to its position. A Mössbauer spectrum with reasonably well-resolved absorption lines can be fit then with the theoretical function according to Eq. (13).

At this point one has to decide which parameters in Eq. (13) should be varied in order to simulate the experimental data. In general one uses the following three parameters for the description of the absorption line ℓ: the line width $\Gamma_{exp\,\ell}$, which is $\Gamma_{a\ell} + \Gamma_s$ for thin absorbers, the *depth* $\eta_\ell(0)$, which equals $w_{a\ell}\eta(0)$ according to Eqs. (9) and (13), and the energy $E_{a\ell}$ of the center of the absorption line. In addition, $Z(\infty)$ is fit; it is the counting rate in the absence of Mössbauer absorption.

In proteins, the linewidth $\Gamma_{a\ell}$ of the absorber is very often broader than the natural linewidth Γ_{nat}. Inhomogeneities in the iron environment and structural fluctuations cause this broadening. In this instance, $\eta_\ell(0)$ is reduced as a function of $\Gamma_{a\ell}$, while the area, A_ℓ, of the absorption line is kept constant. Since A_ℓ contains the desired information, it is better to fit the uncorrelated parameters $\Gamma_{a\ell}$ and A_ℓ instead of $\Gamma_{a\ell}$ and $\eta_\ell(0)$. It is also convenient to write all parameters in units that are easy to compare to the Mössbauer spectrum. As a result, one can use the following function for the description of the Mössbauer spectrum

$$Z(v) = Z(\infty) \left[1 - \sum_{\ell=1}^{L} \frac{\frac{1}{\pi} A_{exp\,\ell} (\Gamma_{exp\,\ell}/2)}{[\delta S_\ell - v]^2 + (\Gamma_{exp\,\ell}/2)^2} \right] \qquad (19)$$

The number of quanta counted in the channel corresponding to the Doppler velocity v in mm/sec is $Z(v)$. For each absorption line, ℓ, the three parameters $A_{exp\,\ell}$ (mm/sec), $\Gamma_{exp\,\ell}$ (mm/sec), and δS_ℓ (mm/sec) are fitted. The last parameter is $Z(\infty)$. Energy units are generally converted to Mössbauer units by setting 1 mm/sec = 4.80766×10^{-8} eV = 11.6248 MHz.

Each least-squares fit needs starting parameters. One may begin with the following set: $Z^{start}(\infty) = Z(v_{max})$. δS_ℓ is the estimated absorption maximum of line ℓ read from the data plot. Also from the plot, $\Gamma_{exp\,\ell}$ is the estimated FWHM of the absorption line. To find $A_{exp\,\ell}^{start}$, one first finds

$Z_\ell(0)$: the number of counts at the maximum absorption of the ℓth line, and then $A_{\exp\ell}^{\text{start}}$, is given by

$$A_{\exp\ell}^{\text{start}} = \left[\frac{Z(\infty) - Z_\ell(0)}{Z(\infty)} \frac{\Gamma_{\exp\ell} - \Gamma_s}{2}\right] \bigg/ \sum_{\ell=1}^{L} \left[\frac{Z(\infty) - Z_\ell(0)}{Z(\infty)} \frac{\Gamma_{\exp\ell} - \Gamma_s}{2}\right] \quad (20)$$

The linewidth Γ_s has to be determined experimentally. As mention previously, the calibration spectrum with a thin iron foil yields the linewidth of the source. One can assume that

$$\Gamma_s = \Gamma_{\exp}^{\text{cal}} - \Gamma_{\text{nat}} \quad (21)$$

Now using the least-squares fit, one finds the best value for each of the variable parameters. The fitted curve is then plotted together with the experimental Mössbauer spectrum. When the agreement is good, the data interpretation can follow.

As it was briefly discussed in the section on Hyperfine Interactions, the index ℓ runs over the different absorption lines from the hyperfine interactions. Protein structure dynamics may reveal themselves also as broad Lorentzian lines centered about the same energy δS_ℓ as the corresponding narrow line. Figure 7 shows the Mössbauer spectrum of deoxygenated myoglobin crystals at 269 K. In contrast to Fig. 2, $v_{\max}$ was 29.0 mm/sec. It was fit with Eq. (19) assuming quadrupole splitting ($L = 2$) and with $A_{\exp 1} = A_{\exp 2}$ (Fig. 7a). It is obvious that the fitted curve does not fully describe the experimental data. Broad lines at δS_1 and δS_2 are also necessary. To fit such broad lines Eq. (19) has to be modified

$$Z(v) = Z(\infty) \left[1 - \sum_{\ell=1}^{L} \sum_{N=0}^{N_{\max}} \frac{\frac{l}{\pi} A_{\exp\ell}^{(N)} \frac{\Gamma_{\exp\ell}^{(N)}}{2}}{[\delta S_\ell - v]^2 + \left[\frac{\Gamma_{\exp\ell}^{(N)}}{2}\right]^2} \right] \quad (22)$$

The index N in Eq. (22) allows for the existence of $N_{\max} + 1$ lines with different widths and areas $A_{\exp\ell}^{(N)}$ centered at δS_ℓ. Figure 7b gives a fit with $L = 2$ and $N_{\max} = 1$.

It needs to be emphasized, however, that such a fit is not always trivial. Since Eq. (22) contains a large number of variable parameters, local minima in the χ^2 are possible. It is often advantageous to proceed stepwise. For instance, the fit in Fig. 7b is first made with $N_{\max} = 0$, as shown in Fig. 7a. We then start again with $N_{\max} = 1$ and use the parameters obtained from the first fit ($A_{\exp\ell}^{(0)}$ and $\Gamma_{\exp\ell}^{(0)}$) as constants. So in the second step, only $A_{\exp\ell}^{(1)}$ and $\Gamma_{\exp\ell}^{(1)}$ are varied. Iteration proceeds between the two sets of variable parameters until results, such as shown in Fig. 7b, are found.

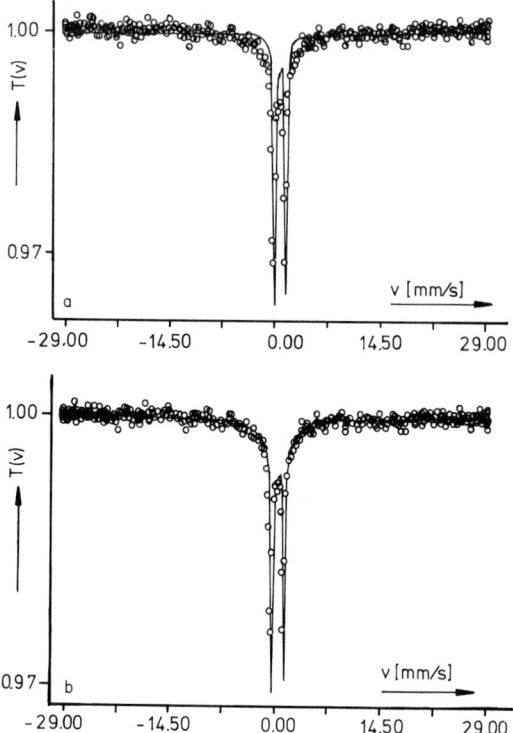

FIG. 7. Mössbauer spectrum of deoxygenated myoglobin crystals at 269 K. (a) A fit with one quadrupole doublet with $\Delta E_Q = 1.65$ mm/sec and $\Gamma_{exp}^{(0)} = 0.37$ mm/sec. (b) A fit with two quadrupole doublets, one as given in part (a) and a second with $\Delta E_Q = 1.65$ mm/sec and $\Gamma_{exp}^{(1)} = 7.19$ mm/sec.

Here we want to add some remarks concerning the interpretation of a Mössbauer spectrum with unresolved absorption lines. In this case, Eq. (19) cannot be applied. Then Γ_{exp} cannot be determined. It is, nonetheless, possible to find A_{exp}. This parameter is

$$A_{exp} = \frac{1}{Z(v_{max})} \sum_v [Z(v_{max}) - Z(v)] \qquad (23)$$

The summation index v runs over all channels of the Mössbauer spectrum. The absorption area obtained in this way also gives the Lamb–Mössbauer factor. It is not very accurate, because thickness corrections cannot be used, and the tails of the Lorenztians are neglected. For thin absorbers containing Fe^{3+}, Eq. (23) yields usable results.

Determination of the Effective Thickness t_a, and Corrections for Thick Absorbers. The experimental absorption area $A_{\exp \ell}$ of a Mössbauer absorption line is determined from a least-squares fit as described in Eq. (19), or by numerical summation according to Eq. (23). Once we have $A_{\exp \ell}^{(N)}$, we can also find $w_{a\ell}^{(N)}$

$$w_{a\ell} = A_{\exp \ell}^{(N)} \Big/ \left[\sum_{\ell=1}^{L} A_{\exp \ell}^{(N)} \right] \tag{24}$$

The ratio $R/(R + U)$ is known from the pulse height spectrum, and the Lamb–Mössbauer factor, f_s, of the source is given in the table. Together with $\Gamma_{a\ell} = \Gamma_{\mathrm{nat}}$ and

$$t_{a\ell} = w_{a\ell} t_a \tag{25}$$

Equation (11) allows the determination of the effective thickness, $t_{a\ell}$ and t_a, in the case of a thin absorber. If no broad lines are found, the index N is omitted.

In the case of thick absorbers, the area $A_{\exp \ell}^{(N)}$ obtained from the least-squares fit with Eq. (22) yields not $t_{a\ell}$, but $L(t_{a\ell})$ according to Eq. (12). Numerical tables are now used to convert $L(t_{a\ell})$ into $t_{a\ell}$.[5,6] Do not mix $t_{a\ell}$ and t_a when making thickness corrections!

It is now possible to correct the linewidth of the Mössbauer spectrum for the sample thicknss effects. From the host of expressions given in the literature, we normally use the following

$$\Gamma_{\exp \ell} - \Gamma_s = \Gamma_{a\ell} [1 + 0.1306 t_{a\ell} + 0.000365 t_{a\ell}^2 \\ - 0.00074 t_{a\ell}^3 + 0.000027 t_{a\ell}^4] \tag{26}$$

One starts with $\Gamma_{a\ell} = \Gamma_{\mathrm{nat}}$. If the values of $t_{a\ell}$ obtained yield the experimentally measured linewidth, $\Gamma_{\exp} - \Gamma_s$, the thickness correction is consistent, and one has the final parameters.

It is also possible that $\Gamma_{a\ell}$ becomes larger than Γ_{nat} due to diffusional line broadening, or sample inhomogeneities. This value should then be used in Eq. (22) to obtain $A_{\exp \ell}$. In turn, one may then obtain a slightly different value for $t_{a\ell}$. At this point one has to iterate between Eq. (22) and Eq. (26), until they give consistent $t_{a\ell}$ and $\Gamma_{a\ell}$ values. Fortunately, usually the iteration is not necessary. The thickness corrections are primarily used only at low temperatures where f_a is large. At such temperatures diffusional broadening is normally absent. In the temperature region where structure dynamics are important, f_a is small and we revert to the thin absorber.

Determination of the Mean Square Displacement of the Iron Label. The value of t_a as just determined contains the f_a factor. In turn, from Eq. (1), we have the mean square displacement $\langle x^2 \rangle$ of the iron. To find f_a from t_a, as seen in Eq. (7), we must know the number of ^{57}Fe atoms per cm^2, n_{Fe}.

In principle, the sample can be destroyed after the Mössbauer measurements and the iron content determined by some analytical technique, such as atomic absorption spectroscopy. In such case, one loses a valuable sample and gives up the possibility of making additional measurements. Moreover, it is not trivial to know the extent of the isotopic enrichment in ^{57}Fe. During the sample preparation, the highly enriched ^{57}Fe might be diluted with natural iron. It becomes convenient to use a normalization procedure.

Eq. (1) and Eq. (7) together yield for $\langle x^2 \rangle$

$$\langle x^2 \rangle = \frac{1}{k^2} \ln(n_{Fe}\sigma_0) - \frac{1}{k^2} \ln(t_a)$$

$$= C_{Fe} - \frac{1}{k^2} \ln(t_a) \tag{27}$$

Plotting $(-1/k^2)\ln(t_a)$ as a function of temperature, one immediately sees the temperature dependence of $\langle x^2 \rangle$. The absolute values of $\langle x^2 \rangle$ are obtained by shifting the plot with the additive constant, C_{Fe}. The easiest way to find this constant is to extrapolate the 50 to 120 K region of the curve to $T = 0$ K (this temperature range represents the normal lattice vibrations as explained in the section below). Then a C_{Fe} is selected so that $\langle x^2 \rangle$ has the classical value of zero at 0 K. Here we clearly neglect quantum mechanical zero point vibrations. The $\langle x^2 \rangle$ values which characterize the structure dynamics of proteins at physiological temperatures are, however, large. So, the small error in C_{Fe} is easily tolerated.

Another method of normalization uses the experimental observation that many iron containing proteins have an $f_a = 0.8$ at 4.2 K. C_{Fe} can be chosen from this point.

From the normalization it is possible to calculate n_{Fe}. The calculated value, of course, can be compared to the estimated value as a control of the normalization.

Figure 8 shows the $\langle x^2 \rangle$ values obtained from the narrow lines of deoxygenated myoglobin crystals as a function of temperature, determined as outlined above.

Interpretation of the Results

Temperature Dependence of the Mean Square Displacement of the Iron. The $\langle x^2 \rangle$ of deoxygenated myoglobin as shown in Fig. 8 exhibits

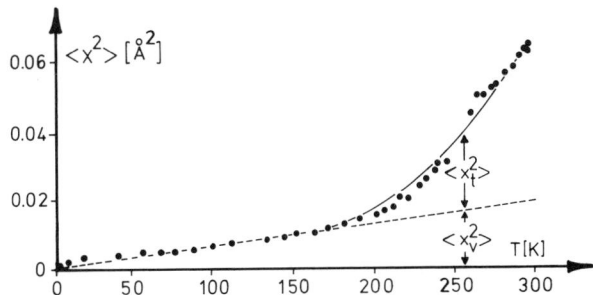

FIG. 8. Mean square displacement of the iron atom in deoxygenated myoglobin crystals. The dashed line accounts for the lattice vibrations. The solid line gives a fit for the added protein specific motion using a restricted overdamped Brownian diffusion model.

different behavior in the low and high temperature regions.

$$\langle x^2 \rangle = \langle x_v^2 \rangle + \langle x_t^2 \rangle \tag{28}$$

Below about 200 K, $\langle x^2 \rangle$ increases linearly with temperature. The absolute values are typical for any solid. One can attribute these motions to lattice or solid-state vibrations. A Debye model can be used to describe the vibrational amplitudes. In Eq. (28) we label such displacements with the index v.

Above about 200 K the $\langle x^2 \rangle$ values increase dramatically with temperature, and the absolute value of $\langle x^2 \rangle$ becomes remarkably large at physiological temperatures. The drastic increase of $\langle x^2 \rangle$ suggests that new modes of motion are present above 200 K. These motions are "frozen out" (thermally prevented) at lower temperatures. The large displacements, given as $\langle x_t^2 \rangle$ in Eq. (28), are believed to be protein specific and are due to the structure dynamics of the protein system.

Even before inferring a model, we already have information concerning the structure dynamics. First, the temperature, T_c, where the protein contribution to the displacement becomes measurable gives a hint of the activation energy of this mode. The absolute $\langle x^2 \rangle$ values may be compared with X-ray diffraction structure investigations. In contrast to the X-ray investigations, which do not tell anything about the time scale for the displacements, Mössbauer spectroscopy gives $\langle x^2 \rangle$ from motions occurring faster than 10^{-7} sec.

In a model independent discussion, one can compare the dynamic properties of different systems, including the investigation of the environment. Membrane bound proteins, for instance, have a lower activation energy for protein motion. The structure dynamics are apparent at $T_c \approx$ 160 K.[10]

For a detailed discussion of the temperature dependence of $\langle x^2 \rangle$, models of the structure dynamics are required. In sections below we discuss models where the intrinsic parameters are determined from the Mössbauer parameters. From such a model we gain a better picture of protein structure dynamics.

Influence of Structure Dynamics on the Shape of a Mössbauer Absorption Spectrum. Line broadening. Slow motions of the iron on the order of 0.1 mm/sec broaden the Mössbauer absorption line. The Mössbauer absorption spectrum can be described by

$$T(v) = 1 - \frac{R}{(R + U)} f_s t_a \sum_{\ell=1}^{L} w_{a\ell} \frac{\Gamma_{a\ell}}{2}$$

$$\frac{\frac{\Gamma_{a\ell} + \Gamma_s}{2} + k^2 \hbar D}{\left[E_{a\ell} - E_s \left(1 + \frac{v}{c} \right) \right]^2 + \left(\frac{\Gamma_{a\ell} + \Gamma_s}{2} + k^2 \hbar D \right)^2} \tag{29}$$

In Eq. (29) $\hbar$ is 0.6582×10^{-15} eV sec [in Mössbauer units, 1.369×10^{-8} (mm/sec) $\times$ sec] and $k = 2 \pi / \lambda$ with $\lambda = 0.86$ Å. D is a diffusion constant (Å²/sec). For the least-squares fit according to Eq. (19), one obtains

$$A_{\exp \ell} = \frac{R}{(R + U)} f_s t_a w_{a\ell} \frac{\Gamma_{a\ell}}{2} \tag{30a}$$

and

$$\frac{\Gamma_{\exp \ell}}{2} = \frac{\Gamma_{a\ell} + \Gamma_s}{2} + k^2 \hbar D \tag{30b}$$

A more instructive value can be obtained from D by

$$\langle x_{qd}^2 \rangle = D \times 10^{-7} \text{ (sec)} \tag{31}$$

Here we have the mean square displacement of the iron in 10^{-7} sec. Information on a longer time scale is not possible since ^{57}Fe can follow motion only during the characteristic time for absorption ($\tau_N = 1.44 \times 10^{-7}$ sec). $\langle x_{qd}^2 \rangle$ may measure the true diffusion of the entire protein molecule. This can occur in frozen solutions at temperatures close to the solvent melting point. However, diffusion processes can also occur in a limited space. As long as the bounded space is not fully explored in 10^{-7}

sec, a quasidiffusive mode is seen. Examples of such quasidiffusion might be the slow motion of large sections of protein molecules or motion of the entire protein in a crystal unit cell about its mean lattice position.

We emphasize that displacements from $\langle x_{qd}^2 \rangle$ do not contribute to the $\langle x^2 \rangle$ value determined from the Lamb–Mössbauer factor in Eq. (1).

Additional broad lines. Figure 7 gives an example with additional broad lines in a Mössbauer spectrum. Such broad absorption lines are extremely interesting in the investigation of protein dynamics. This absorption is due to pronounced modes of motion. The order of magnitude of the time for motion, $\tau^{(b)}$, can be estimated from the line width $\Gamma_{exp}^{(b)}$

$$\tau^{(b)} = 1.411 \times 10^{-7} \text{ (sec)} \frac{\Gamma_{nat}}{\Gamma_{exp}^{(b)} - \Gamma_s} \tag{32}$$

$\Gamma_{nat} = 0.097$ mm/sec. Since Γ_s is normally close to Γ_{nat} and small in comparison to $\Gamma_{exp}^{(b)}$, it may be taken as equal to Γ_{nat}. Times between 10^{-8} and 10^{-9} sec can be resolved in the Mössbauer spectrum.

A Model Describing a Protein Structure as a Discrete Number of Conformational Substates. We start from the picture of one well-defined protein structure as determined from X-ray diffraction investigations. As discussed in the literature,[13-15] such a concept needs to be modified with the assumption of the existence of a number of slightly different structures which are termed conformational substates or microstates of a molecule.

The $\langle x^2 \rangle$ values of the Mössbauer results can be correlated with the spatial mean square distribution of the conformational substates as labeled at the position of the iron. The time threshold of Mössbauer spectroscopy ensures that only transitions between the substates occurring faster than 10^{-7} sec are detected.

The simplest case is a two-state model where the molecule is allowed to jump between the two conformational substates, 1 and 2. Making the transition from 1 to 2 or 2 to 1, the molecule has to overcome an energy barrier ε or $\varepsilon - Q$, respectively. At the iron atom the two conformational substates differ by a displacement $\mathbf{x}_{12}$. The Mössbauer spectrum obtained

[13] R. H. Austin, K. W. Beeson, L. Eisenstein, H. Frauenfelder, and I. C. Gunsalus, *Biochemistry* **14,** 5355 (1975).

[14] W. Doster, D. Beece, S. F. Bowne, E. E. DiIorio, L. Eisenstein, H. Frauenfelder, L. Reinisch, E. Shyamsunder, K. H. Winterhalter, and K. T. Yue, *Biochemistry* **21,** 4831 (1982).

[15] H. Frauenfelder, G. A. Petsko, and D. Tsernoglou, *Nature (London)* **280,** 558 (1979).

from this model is given by

$$T(v) = 1 - \frac{R}{(R + U)} f_s t_a \sum_{\ell=1}^{L} w_{a\ell} \frac{\Gamma_{a\ell}}{2}$$

$$\left[\frac{(1 - B) \dfrac{\Gamma_{a\ell} + \Gamma_s}{2}}{\left[E_{a\ell} - E_s \left(1 + \dfrac{v}{c} \right) \right]^2 + \left(\dfrac{\Gamma_{a\ell} + \Gamma_s}{2} \right)^2} \right.$$

$$\left. + \frac{B \left(\dfrac{\Gamma_{a\ell} + \Gamma_s}{2} + \hbar (k_{12} + k_{21}) \right)}{\left[E_{a\ell} - E_s \left(1 + \dfrac{v}{c} \right) \right]^2 + \left(\dfrac{\Gamma_{a\ell} + \Gamma_s}{2} + \hbar (k_{12} + k_{21}) \right)^2} \right] \tag{33}$$

In comparison to Eq. (19), one has

$$A_{\exp \ell}^{(0)} = \frac{R}{(R + U)} f_s t_a w_{a\ell} \frac{\Gamma_{a\ell}}{2} (1 - B)$$

$$\frac{\Gamma_{\exp \ell}^{(0)}}{2} = \frac{\Gamma_{a\ell} + \Gamma_s}{2}$$

$$A_{\exp \ell}^{(1)} = \frac{R}{(R + U)} f_s t_a w_{a\ell} \frac{\Gamma_{a\ell}}{2} B \tag{34}$$

$$\frac{\Gamma_{\exp \ell}^{(1)}}{2} = \frac{\Gamma_{a\ell} + \Gamma_s}{2} + \hbar (k_{12} + k_{21})$$

The two-state model yields one narrow and one broad line. The broadening is determined by the jump rates k_{12} and k_{21}: the transition rate for substate 1 to 2 and vice versa. From the given barrier heights and the Arrhenius law we can write

$$k_{12} = k_v \exp[-\varepsilon/(R_B T)]$$
$$k_{21} = k_v \exp[-(\varepsilon - Q)/(R_B T)] \tag{35}$$

where R_B equals 8.31 J K^{-1} mol^{-1} and k_v is a frequency, usually taken as 10^{-13} sec^{-1}. In this model, the total absorption area of the Mössbauer spectrum is not changed by transitions between the substates. Only a fraction of the area given by B

$$B = 2P_1 \frac{k_{12} (1 - y)}{(k_{12} + k_{21})} \tag{36}$$

is shifted from the narrow line to the broad line. The distance $\mathbf{x_{12}}$ is contained in y

$$y = \cos(\mathbf{kx_{12}}) \tag{37}$$

where one has to average over all spacial orientations of $\mathbf{x_{12}}$. For random orientation in $\mathbf{x_{12}}$, y is given by

$$y = \frac{\sin(2\pi x_{12}/\lambda)}{(2\pi x_{12}/\lambda)} \tag{38}$$

P_1 and P_2 are the probabilities of finding the molecule in the given state

$$P_2 = P_1 \exp[-Q/(R_B T)]$$
$$P_1 + P_2 = 1 \tag{39}$$

The model can easily be generalized to more states.

Recently the mathematical formulation of structure dynamics based on a model of transitions between discrete conformational substates was greatly improved, allowing the use of complicted potentials.[16]

Model of Restricted Overdamped Brownian Diffusion. Let us again label the protein motion at a certain position within the molecule, for our purpose, the iron atom. We assume that the iron atom performs Brownian diffusion similar to an atom in a viscous liquid. A backdriving force limits the diffusion to a small region, thus preventing the protein from diffusing apart in this model. A protein in a defined conformation will have a diffusional region for the atoms with a radius on the order of 1 Å. The motion is driven stochastically by the thermal bath of the environment. Damping is caused by structural relaxations within the molecule and by the protein environment (e.g., viscosity of water or the lipid bilayer in membrane bound proteins).

An analytical solution is possible for a harmonic backdriving force,[17] and the Mössbauer spectrum is given then by

$$T(\nu) = 1 - \frac{R}{(R + U)} f_s t_a \sum_{\ell=1}^{L} w_{a\ell} \frac{\Gamma_{a\ell}}{2}$$

$$\sum_{N=0}^{\infty} \frac{[k^2 \langle x_t^2 \rangle]^N}{N!} \frac{\frac{\Gamma_{a\ell} + \Gamma_s}{2} + N\hbar\alpha_t}{\left[E_{a\ell} - E_s \left(1 + \frac{\nu}{c}\right)\right]^2 + \left(\frac{\Gamma_{a\ell} + \Gamma_s}{2} + N\hbar\alpha_t\right)^2} \tag{40}$$

[16] W. Nadler and K. Schulten, *Phys. Rev. Lett.* **57**, 1712 (1983).
[17] E. W. Knapp, S. F. Fischer, and F. Parak, *J. Chem. Phys.* **78**, 4701 (1983).

The sum $\ell = 1$ to L accounts again for the hyperfine splitting. In practice Eq. (40) has been used for $L = 2$ with $w_{a\ell} = 0.5$ and $\Gamma_{a1} = \Gamma_{a2} = \Gamma_a$. Due to the bound diffusion, the Mössbauer spectrum is composed of an infinite sum of Lorentzian absorption peaks, each at resonance energy $E_{a\ell}$. The absorption line $N = 0$ corresponds to the narrow line described in Eq. (13). The broader lines ($N \geq 1$) are broadened by $N\hbar\alpha_t$. Here α_t^{-1} characterizes the time the iron atom needs to explore the bounded region. The time can be compared with the model independent quantity $\tau^{(b)}$ of Eq. (32). One may also correlate this characteristic time to the damping, β_t and the frequency of the backdriving force ω_t by

$$\alpha_t = \omega_t^2/2\beta_t \qquad (41)$$

In practice the line with $N = 2$ is often so broad that it cannot be separated from $T(\infty)$. Recently a correction has been tabulated which permits the calculation of the contribution of $N > 1$ if the fit was performed with $N = 1$.[27]

We compare again with Eq. (22) which is used for a practical fit of the Mössbauer spectrum and have

$$A^{(N)}_{\exp\,\ell} = \frac{R}{(R + U)} f_s t_a w_{a\ell} \frac{\Gamma_{a\ell}}{2} \frac{[k^2 \langle x_t^2 \rangle]^N}{N!} \qquad (42a)$$

and

$$\Gamma^{(N)}_{\exp\,\ell} = \Gamma_{a\ell} + \Gamma_s + 2N\hbar\alpha_t \qquad (42b)$$

Note that according to Eqs. (42) α_t is determined by $\Gamma_{\exp}$. It is also noted that t_a according Eq. (40) gives the Lamb–Mössbauer factor with $\langle x^2 \rangle = \langle x_v^2 \rangle + \langle x_t^2 \rangle$ according to Eq. (28).

Equation (40) describes the shape and area of the Mössbauer spectrum at a single temperature. In order to get a quantitative description of the temperature dependence of the protein specific displacement $\langle x_t^2 \rangle$, the following picture can be used.[18]

The protein is either frozen in one of the possible conformational substates characterized by the distribution $\langle x_c^2 \rangle$, or it is in a transition state. In the transition state the protein is flexible and performs Brownian oscillations until it is again trapped in a conformational substate. If $p_t(T)$ is the probability of the protein being in the transition state, one obtains

$$\langle x_t^2 \rangle = p_t(T)\langle x_c^2 \rangle \qquad (43)$$

where the temperature dependence of structure dynamics is present in

[18] E. W. Knapp, S. F. Fischer, and F. Parak, *J. Phys. Chem.* **86**, 5042 (1982).

$p_t(T)$. Taking $\langle x_c^2 \rangle$ from X-ray diffraction data or treating it as a free constant, together with $\langle x_t^2 \rangle$ from Mössbauer spectra, $p_t(T)$ is determined. It can be correlated with the energy and entropy change of a protein going from a conformational substate to the transition state

$$\Delta\varepsilon = \varepsilon_{ct} - \varepsilon_{tc}$$
$$\Delta s = s_{ct} - s_{tc} \tag{44}$$

where the indices ct and tc refer to the change from the conformational substate to the transition state and vice versa. Accordingly, $p_t(T)$ can be written now as

$$p_t(T) = \{1 + \exp[(\Delta\varepsilon - \Delta sT)/(R_B T)]\}^{-1} \tag{45}$$

The model allows the determination of the energy and entropy changes involved in the structure dynamics from the Mössbauer data.

Results on myoglobin indicate that structure dynamics of proteins are driven by the gain in entropy.[19]

Conclusion

Mössbauer spectroscopy of ^{57}Fe in the temperature range of 4.2 K to room temperature is a well established experimental technique. It has been proven to be a useful tool for the investigation of structure dynamics of iron-containing proteins. Samples enriched in ^{57}Fe produce Mössbauer spectra where the area and line shape reflect protein motions at the iron atom. The Mössbauer spectra are sensitive not only to the amplitude of motion, but also to the time dependence. The time threshold of 10^{-7} sec is a unique feature of ^{57}Fe Mössbauer spectroscopy. The most detailed information is obtained from motion with a characteristic time of 10^{-8} to 10^{-9} sec. The investigations with deoxygenated myoglobin show that such motion is the dominant mode in this protein.

Although a detailed understanding of the experimental results depends (as in nearly every investigation) on the model used in the data evaluation, Mössbauer spectroscopy also gives some model independent results. Since results such as protein motion are rather slow, they are not sensitive to the applied model.

Mössbauer absorption spectroscopy gives a very detailed picture of the physical nature of the structure dynamics seen at the position of the iron. Comparisons with other techniques (e.g., X-ray diffraction[15,20] or

[19] F. Parak and E. W. Knapp, *Proc. Natl. Acad. Sci. U.S.A.* **81,** 7088 (1984).
[20] H. Hartmann, F. Parak, W. Steigemann, G. A. Petsko, D. Ringe Ponzi, and H. Frauenfelder, *Proc. Natl. Acad. Sci. U.S.A.* **79,** 4967 (1982).

dielectric relaxation measurements[21]) help to generalize the picture and therefore dramatically increase the efficiency.

Rayleigh Scattering of Mössbauer Radiation (RSMR)

Principle of the Method

Figure 9 shows the experimental arrangement. The Mössbauer source (1) is mounted on an electromagnetic driving system (2) as in Fig. 1. The radiation is collimated (3) and incident upon the sample (4). The nonscattered radiation is absorbed by a beam stop (5). Radiation scattered with an angle 2ϑ is counted by the detector (6). The sample does not need to contain ^{57}Fe. The Rayleigh scattering occurs by the electrons of all atoms in the sample, similar to normal X-ray diffraction experiments. The Mössbauer effect is introduced with the use of an analyzer, A, containing ^{57}Fe. With the analyzer, we are able to determine what fraction of the radiation is scattered from the sample recoil free. In the simplest case, the analyzer is used at position 1 and position 2 combined with two different Doppler velocities of the source. At the velocity v, Mössbauer absorption occurs in the analyzer. At a large velocity ($v = \infty$) the Mössbauer effect is destroyed. With the analyzer in position 1 ($2\vartheta = 0$), one measures the counting rates $Z_0(v)$ and $Z_0(\infty)$, where the number in parentheses refers to the Doppler velocity and the index to the analyzer position. At position 2 (scattered beam at 2ϑ) one measures $Z_{2\vartheta}(v)$ and $Z_{2\vartheta}(\infty)$. According to Eq. (8) one finds $\eta_0(v)$: a measure of the amount of recoil free radiation emitted by the source. Similarly, $\eta_{2\vartheta}(v)$ according to Eq. (8) measures the probability of Mössbauer effect in the scattered beam. Without energy transfer during the scattering, $\eta_0(v)$ would equal $\eta_{2\vartheta}(v)$. Rayleigh scattering in the sample is, however, not completely elastic. Coupling to the motions of the scatterer, energy transfer is possible. In this way, the dynamic properties of the sample can enter and one writes

$$f_R = \gamma_R \exp(-Q^2\overline{\langle x^2 \rangle^R}) = \frac{\eta_{2\vartheta}(v)}{\eta_0(v)} \qquad (46)$$

with

$$\gamma_R = I^R / (I^R + I^C) \qquad (47)$$

and

$$Q = 4\pi \sin\vartheta/\lambda \qquad (48)$$

[21] G. P. Singh, F. Parak, S. Hunklinger, and K. Dransfeld, *Phys. Rev. Lett.* **47**, 685 (1981).

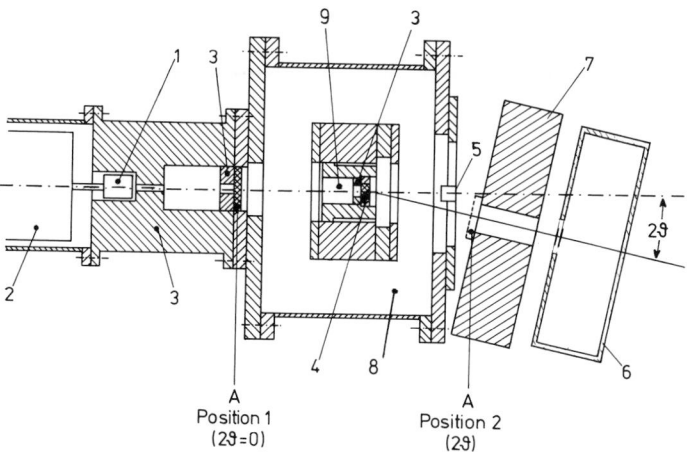

FIG. 9. Equipment for RSMR experiments. (1) Source; (2) electromagnetic driver; (3) lead shielding and pinholes; (4) sample; (5) beam stop; (6) detector; (7) lead shielding; (8) isolation vacuum; (9) probe chamber; A, analyzer at position 1 or position 2.

In Eq. (46) γ_R corrects for the Compton effect which converts the incoming gamma radiation into quanta with lower energy. In Eq. (47), I^R and I^C are the intensities which come from Rayleigh scattering and Compton effect, respectively. The ratio can be calculated as shown in the section on Experiments and Data Evaluation.

The crucial factor containing the protein dynamics is $\overline{\langle x^2 \rangle^R}$. It gives the mean square displacement of the scattering atoms, averaged over all the atoms in the sample. When investigating a protein solution, the average involves all atoms of the protein, of the solvent molecules, and the windows of the sample holder. It is obvious that contributions from the protein alone have to be separated by difference measurements. Since Rayleigh scattering in the sample occurs from the radiation emitted by the ^{57}Fe transition ($\tau_N \approx 10^{-7}$ sec), this characteristic transition time becomes the threshold of the RSMR technique. Only displacements occurring with a time faster than 10^{-7} sec contribute to $\overline{\langle x^2 \rangle^R}$.

It is emphasized once more that this technique can be applied to iron free samples. The only requirement is a large enough protein concentration to have a measureable effect. Protein precipitates or samples composed of many small crystals are preferable because of their high protein concentration.

A general introduction to RSMR is given by Albanese and Deriu.[22]

[22] G. Albanese and A. Deriu, *Riv. Nuovo Cimento* **2**, 1 (1979).

Experimental Details

In many respects we refer the reader back to the section on Experimental Equipment. Only a few pecularities of the scattering geometry need to be mentioned here.

The Analyzer. An important component of the RSMR equipment is the analyzer. It is composed of $K_4{}^{57}Fe(CN_6)$, ^{57}Fe enriched stainless steel, or $Li_3{}^{57}FeF_6 + (NH_4)F_3(^{57}FeF_3)_2$. The last absorber is especially useful if sources with a large lineshift δS (e.g., CoO) are used. This combination of Fe complexes yields, as a result of the different hyperfine splittings, an effective cross section for Mössbauer absorption which is nearly constant over more than 1 mm/sec source velocity. Analyzers with $n_{Fe} = 7 \times 10^{19}$ ^{57}Fe nuclei/cm² yield good results. An increase in the ^{57}Fe content increases both Mössbauer and X-ray electronic absorption so that the total transmitted radiation becomes too small.

Position 1 and position 2 of the analyzer are crucial for the accuracy of the results. For position 2 the analyzer should not be hit by the direct beam in order to avoid scattering from it.

Comparing Eqs. (9) and (46), it is obvious that the background, U, cancels only if the ratio $R/(R + U)$ is identical for the two position of the analyzer. If position 2 is too close to the detector, the apparent acceptance angle for scattered radiation is much larger than when the analyzer is in position 1. High energy radiation from the ^{57}Co source Compton scattered from the analyzer can give different background levels. Sometimes $\eta_{2\vartheta}(v)/\eta_{2\vartheta}(\infty)$ is found to be larger than $\eta_0(v)/\eta_0(\infty)$, which is clearly physically nonsense. One way to avoid artifacts in the experiment is to check carefully the entire system with a test absorber. At room temperature a polyacrylic disk about 2 mm in thickness should give $f^R \approx 0.7$. Position 1 and position 2 of the analyzer have to be varied until a minimum of scattered radiation from the analyzer reaches the detector. We shall come back to this problem in the next section.

The Spectrometer. We begin by mentioning the beam profile of the primary and scattered radiation. The size of the active area of the source together with the pinhole size introduce an uncertainty in the beam direction. The result is a smearing of the scattering angle 2ϑ. As seen from Eq. (46), 2ϑ enters into the data evaluation. Thus the divergence of the primary beam should be kept small. RSMR experiments suffer from low counting rates due to the limited specific activity of a Mössbauer source. The geometrical arrangement is, therefore, critical. We typically use sources where the activity is concentrated on a spot no larger than 3 mm in diameter. The distance between source and sample is 17 cm. The aperature [(3) in Fig. 9] is constructed so that a beam with a 7 mm diameter is incident upon the sample [(4) in Fig. 9]. It is advisable to coat the lead pinholes with aluminum foil in order to reduce the characteristic K-radiation from the lead. The scattered beam is restricted to a 2 cm profile with a pinhole (7) built from a lead brick 4 cm thick. In fig. 9, the scattering angle 2ϑ is 11.3°. The accuracy of this angle is determined by the ability to measure the geometry. The acceptance angle of the counter is ±5°.

Typical counting rates with a 100 mCi ^{57}CoRh source are about 1 count per second or less. The development of stronger point sources (^{57}CoO and ^{57}CoSb$_3$[23]) will give higher counting rates, or offer the alternative to reduce the acceptance angle of the counter and thereby increase the accuracy.

The Mössbauer source is mounted on an electromagnetic driving system as explained before in absorption spectroscopy. All the electronics, including the detector (we use a Reuter Stokes proportional counter filled with Xe) can be identical to the absorption spectroscopy method. In the most general case, one measures $\eta_0(v)$ and $\eta_{2\vartheta}(v)$ over a large velocity range. The results yield not only f^R but also information of line broadening from the scatterer which can be interpreted as explained in the section on Interpretation of the Results. The low counting rates make such experiments extremely time consuming. Therefore, the typical RSMR measurement is performed very often only at two velocities as discussed in the section on Principle of the Method. One velocity, v, is zero if δS between the source and analyzer is negligible. The other velocity (infinity—30 to 100 mm/sec) completely destroys resonance. For such experiments no multichannel analyzer is necessary. Two counters are sufficient. A simple pulse generator generates the maximum velocity amplitude. A flip-flop gates either the maximum velocity or the zero velocity detector signal through the Mössbauer electronics. The two counters are then synchronized with this flip-flop. A better choice is four counters, each counting at the velocities $+v$, $-v$, $+\infty$, and $-\infty$. The comparison of the counting rates at $+\infty$ and $-\infty$ allows for the correction of geometry effects. To destroy Mössbauer absorption, we normally use a sine velocity profile with a maximum velocity of approximately 100 mm/sec. In reality, the Doppler shifted energy is not always out of resonance. The velocity is varied continuously between $+v_{max}$ and $-v_{max}$. As a result, the analyzer is in resonant conditions for a very short time. The small absorption can be neglected. The advantage of the sine velocity profile is that the mechanical system follows it easily. Experiments have shown that a constant velocity drive with a large velocity has a tendency to make uncontrolled

[23] R. L. Mössbauer, F. Parak, and W. Hoppe, p. 5 in Ref. 2.

motions. These motions allow resonance absorption, often stronger than those present with the sine velocity profile.

Temperature Control. As in the case of Mössbauer absorption spectroscopy, most information is obtained by following the temperature dependence of $\overline{\langle x^2 \rangle^R}$. In Fig. 9 a simple cryostat is shown which consist of only a sample chamber (9) thermally isolated with a vacuum system (8). Cold vapor from liquid N_2 is pulled through the chamber to obtain a well-defined temperature between 80 K and room temperature. In principle, one can use the cryostat and temperature controller according to Fig. 1 and 4. One needs large windows on the cryostat so that the scattering angle is not too restricted. Moreover, the number and thickness of the mylar windows should be carefully minimized to avoid additional scattering of the primary beam.

Experiments and Data Evaluation

Background Correction. One should never trust implicity that the ratio $R/(R + U)$ cancels in Eq. (46). A determination of $R/(R + U)$ according to the section on Background Detection should be done for both positions of the analyzer. If $R/(R + U)$ changes with the analyzer in either position, the reason for the additional scattering should be found. The use of a detector with high energy resolution (e.g., a Ge solid state counter) can be helpful in finding the origin of the additional scattering (L-lines of lead and so on). The determination of $R/(R + U)$ should be made frequently during the measurement, since the high energy radiation of the sources decreases with time. It is especially important to repeat the background measurements whenever the source is exchanged. The ratio $R/(R + U)$ should also be large for optimal results. A weaker source yielding lower counting rates but a larger $R/(R + U)$ value is preferable.

Correction for Compton Scattering. Equation (47) gives the relative probability that scattered quanta are Rayleigh scattered by the sample electrons. The intensity, I^R, coming from Rayleigh scattering from atom of type i is proportional to f_i^2. The atomic form factors, f_i, are tabulated in the International X-ray Tables[24] as a function of the scattering angle, 2ϑ, for all atoms. Also listed in the tables is the Compton scattering from all the atoms as a function of scattering angle 2ϑ ($F_i = I^C$). For a sample

---

[24] "International Tables for X-Ray Crystallography, Vol. III." Kynoch Press, Birmingham, England, 1968.

containing three types of atoms, Eqs. (46) and (47) become

$$
f_R = \frac{\left[\sum_{i=1}^{3} n_i f_i^2\right] \exp(-Q^2 \langle x^2 \rangle^R)}{\sum_{i=1}^{3} n_i (f_i^2 + F_i)} \tag{49}
$$

The three types of atoms, in the case of proteins, are normally C, N, and O. The relative number of the atoms in the sample is given by n_i. It is assumed that all atoms have the same average mean square displacement $\langle x^2 \rangle^R$.

Contributions from Different Components of the Sample. At present the analysis of RSMR of proteins is normally based on the assumption that the various components of the sample scatter incoherently although coherence effects can be taken into account.[28] One has to deal with the intensities scattered from the different atoms. Let us take, as an example, the investigation of a concentrated solution of myoglobin in a solvent. In this case one can distinguish three components in the sample which contribute to the total scattering: the mylar windows, the solvent, and the myoglobin molecules. Equation (49) can be rewritten

$$
f_R^{cal} = \frac{\sum_c \left[u_c \sum_i n_i^{(c)} f_i^2 \exp(-Q^2 \langle x^2 \rangle_c^R)\right]}{\sum_c \left[u_c \sum_i n_i^{(c)} (f_i^2 + f_i)\right]} \tag{50}
$$

The index c runs over the scattering components. In the above example, these components are the windows (index corr), the solvent (index s), and the myoglobin (index Mb). The factors u_c giving the relative fraction of the component c in the beam are normalized to

$$
\sum_c u_c = 1 \tag{51}
$$

The value $n_i^{(c)}$ gives the relative number of atoms of type i in component c. The first control measurement is performed with the empty sample holder, yeilding $f_{R\,corr}^{exp}$. Comparing this value with f_R^{cal} calculated from Eq. (50) with $u_{corr} = 1$, $u_s = 0$, and $u_{Mb} = 0$ yields $\langle x^2 \rangle^R$ from all the mylar windows in the beam. The index i runs over all the atoms in mylar. The chemical composition of the windows may be estimated.

The second measurement is made with the solvent in the sample holder, yielding f_{Rs}^{exp}. This value has to be compared with Eq. (50) where

$u_{corr} + u_s = 1$ and $u_{Mb} = 0$. The value $\overline{\langle x^2 \rangle^R_{corr}}$ has to be taken from the first experiment. The weighting factors u_{corr} and u_s need to be estimated from the thickness of the windows and of the sample using the specific density of the different materials. One is now able to determine $\langle x^2 \rangle^R_s$. With very thin windows in the equipment f^{exp}_{RS} is not measurable influenced by the windows and the first experiment can be omitted ($u_{corr} = 0$).

The third measurement is performed with the myoglobin solution, yielding $f^{exp}_{R\cdot Mb}$. The calculated $f^{cal}_{R\cdot Mb}$ value now has to account for the contributions of the myoglobin, the surrounding solvent, and the correction for the windows. All components contribute with the weight u_{corr}, u_s, and u_{Mb} where $u_{corr} + u_s + u_{Mb} = 1$. The values $\overline{\langle x^2 \rangle^R_{corr}}$ and $\langle x^2 \rangle^R_s$ have to be taken from the first two experiments. Equation (50) now allows the determination of $\overline{\langle x^2 \rangle^R_{Mb}}$, the average mean square displacement of the atoms in myoglobin occurring at times faster than 10^{-7} sec.

The described procedure shows the strategy to decompose the different contributions to f_R. In the above example, the present limits of RSMR are seen. For example, it is well known that water binds on the surface of proteins. This water has a reduced mobility in comparison to bulk water and gives a measurable $\exp(-Q^2 \langle x^2 \rangle^R_s)$ value. For unbound water this value becomes zero. This distinction was neglected in the above treatment. In a first approach one can model the bound water with a viscous liquid (e.g., glycerol), measure its $\langle x^2 \rangle^R$, and introduce a fourth component. Such procedures are only as good as the assumed model.

At the present state, absolute values of $\langle x^2 \rangle^R_{protein}$ determined from RSMR measurements have to be taken with care. The main information of the method comes from the temperature dependence of $\overline{\langle x^2 \rangle^R}$.

The Average Mean Square Displacement in Myoglobin

The first study on the investigation of large intramolecular movements by RSMR have been described for metmyoglobin by Krupyanskii et al.[25] They will be used here as an example. Figure 10 gives the mean square displacements averaged over all the atoms of myoglobin for 4 samples measured between 100 and 300 K. In all cases a correction for the water content was used. Sample "a" contained freeze-dried myoglobin which was stored for about 3 days in air with 37% relative humidity. Within the error bars, the $\langle x^2 \rangle^R$ values vary linearly with temperature. There is no indication for the onset of protein specific motions at higher temperatures. The dried protein sample shows no structural dynamics.

[25] Yu. F. Krupyanskii, F. Parak, V. I. Goldanskii, R. L. Mössbauer, E. E. Gaubman, H. Engelmann, and I. P. Suzdalev, Z. Naturforsch. 37c, 57 (1982).

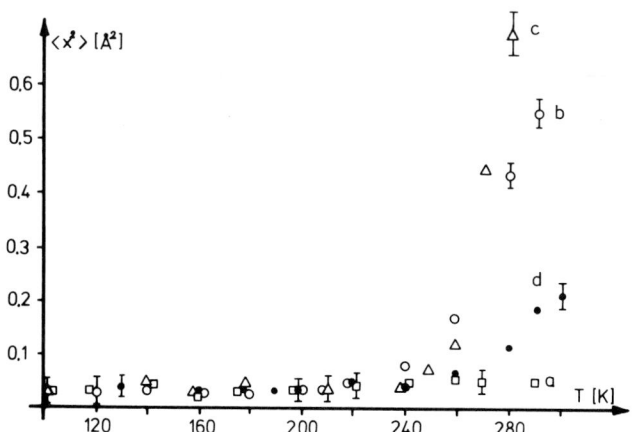

FIG. 10. Average mean square displacements in myoglobin measured with RSMR. (a) (□) Freeze-dried sample kept at 37% relative humidity; (b) (○) freeze-dried sample kept at 94% relative humidity; (c) (△) myoglobin in solution; (d) (●) polycrystalline sample.

Sample "d" contained a large number of small crystals of myoglobin. The sample is comparable with that used for the Mössbauer absorption spectroscropy (see Figs. 2 and 7). The temperature dependences of $\overline{\langle x^2 \rangle^R}$ and $\langle x^2 \rangle$ at the iron atom are clearly similar. In this context, we remember once more that the time threshold for motion sensitivity for Mössbauer spectroscopy and RSMR is the same. The results indicate that the motions labeled at the iron atom are indeed representative of the average motion of all atoms in the molecule, suggesting that the picture of a restricted overdamped Brownian diffusion may characterize the structure dynamics. There are also, however, clear differences. The onset of protein specific motion occurs at a temperature near 240 K for RSMR (it was near 200 K at the position of the iron atom from Mössbauer spectroscopy). Averaged over the entire protein molecule, the activation energies needed to reach the transition state are higher than those at the iron atom. To estimate the accuracy of the absolute value of $\overline{\langle x^2 \rangle^R}$, one has to compare with the average values obtained from X-ray structure analysis. For metmyoglobin, $\overline{\langle x^2 \rangle^X}$ is about 0.14 Å² at room temperature, while $\overline{\langle x^2 \rangle^R}$ is 0.21 Å². This difference was recently explained. The assumption of incoherent scattering is not strictly valid for investigations of a polycrystalline sample. New investigations were performed at different scattering angles.

[26] L. Reinisch, J. Heidemeier, and F. Parak, *Eur. Biophys. J.* **12**, 167 (1985).
[27] I. Nowik, E. R. Bauminger, S. G. Cohen, and S. Ofer, *Phys. Rev.* **31A** (1985).
[28] D. C. Champeney and G. W. Dean, *J. Phys. C: Solid State Phys.* **8**, 1276 (1975).

As a result the contribution of the crystal water was determined with higher precision yielding consistent values of $\overline{\langle x^2 \rangle^X}$ and $\overline{\langle x^2 \rangle^R}$.[29]

Sample "b" contained freeze-dried myoglobin kept for 3 days in air with 94% relative humidity. As shown in Fig. 10, structure dynamics are possible if the molecule is covered by a sufficiently thick water layer. Comparing with the polycrystalline sample, we see that protein dynamics are slightly hindered by the close packing of a crystal.

Extremely large displacements are found when myoglobin is in solution (sample "c"). The interpretation here is, however, difficult since diffusion of the entire molecule may occur. This cannot be easily separated from structural dynamics with the limited measurements made.

Future Perspectives

The application of the RSMR technique to the study of structure dynamics is now only in its infancy. At present, Mössbauer absorption spectroscopy gives a more detailed answer than RSMR. We are, nonetheless, convinced that in the future, RSMR will be of considerable value for the investigation of biomolecules. The principal advantage is clear: there exist no limitations to iron-containing molecules. Still, in iron-containing molecules, Mössbauer absorption spectroscopy will give valuable information and provide a method to correlate results.

In principle, RSMR can be considered as a combination of X-ray structure investigation and Mössbauer spectroscopy. Using a single crystal as a sample, the intensity scattered into each of the Bragg reflections can be analyzed with respect to elastic and inelastic coherent contributions. For inorganic crystals, such investigations have already given interesting results.[22] With better technology, such experiments will also become possible on protein crystals. In this way, the $\langle x^2 \rangle^R$ values of each atom in the molecule could be determined.

New technology necessary for such experiments has been in the developmental stage for several years. Details of some of the advances can be found in Ref. 23. Three areas presently being developed include (1) very intense Mössbauer sources, namely $^{57}CoSb_3$ and ^{57}CoO, (2) technology for investigating protein crystals at very low temperatures,[20] and (3) an area sensitive proportional counter for 14.4 keV radiation, which will permit the performance of experiments in a reasonable time, even with low counting rates. This counter permits the collection of several hundred

[29] G. U. Nienhaus and F. Parak, *Proc. Mössbauer Conf. Leuven, 1985.* Hyperfine Interactions, **29**, 1451 (1986).

Bragg reflections at the same time, and it has the high sensitivity and energy discrimination of a proportional counter.

This technology, which was developed originally for the solution of the phase problem in protein X-ray structure analysis by Mössbauer scattering, has already given first results as discussed by Nienhaus and Parak.[29]

Acknowledgments

This work was supported in part by the Deutsche Forschungsgemeinschaft and the Bundesministerium für Forschung und Technologie. One of us (L.R.) is grateful to the Alexander von Humboldt Foundation for a stipendium to work in the Federal Republic of Germany.

Author Index

Numbers in parentheses are footnote reference numbers and indicate that an author's work is referred to although the name is not cited in the text.

Y

Subject Index

Human serum albumin
  fluorescence lifetime-resolved anisotro-
      pies, in absence and presence of
      denaturants, 536–537
  time-resolved anisotropies for, 526–
      527
Hybridization, as tool for analysis of
    association, 240–242
Hydrogen–deuterium exchange
  amide A band, 455–457
  amide II band observation, 454–455
  near-infrared region used to monitor,
      457
  UV methodology, 460–463
Hydrogen exchange, 189, 190–191, 443–
    444, 508–509
  analysis of exchange rates, as function
      of independent variable, 492–493
  correlated with neutron and x-ray
      results, 423
  and environment of exchange site, 448
  EX$_2$ mechanism, 488–489
  experiments
    infrared methods, 454–457
    methods, 452
    solvent density methods, 453–454
    use of ^2H_2O, 452–453
  fluctuational mechanisms underlying,
      509
  Laplace inversion. See Laplace inver-
      sion
  ligand binding equilibria, 493
  measurement of, and protein conforma-
      tional dynamics, 448–508
  mechanisms in proteins, 494–495
    breathing hypothesis, 494
    mobile defect hypothesis, 494
    regional melting, 494–495
    solvent penetration hypothesis, 494
  multiple site, study of, 450–451
  multiple-site data
    analysis, 480–493
    requirements, 487–488
  nuclear magnetic resonance methods,
      464–466
  principles of, 450–452
  rank-order of exchange, 489–492
  rate, 448
    in different functional states of
        protein, 509

ratios, determination of, 444–447
single site, study of, 450–451
studies, approaches to, 450
total exchange techniques, 495–496
Hydrostatic pressure, effect on proteins,
    231

I

Immunoglobulin
  Fc portion, 391
  hinge-bending, 301–303
Immunoglobulin light chain
  constant fragment, refolding kinetics,
      134
  fast-folding and slow-folding forms, 71
  unfolding and refolding, study, 66–69
Inclusion bodies, 224
Indole, and dimethylindole, steady-state
    and phase-sensitive emission spectra
    of mixture of, 565–566
Induced fit, role of protein dynamics, 429
In-exchange, 452, 510
  partial, 469
Infrared spectroscopy, 306
Insulin, 267
Iron-containing protein. See also Heme
    protein
  Mössbauer spectroscopy, 568
Isomeric specific proteolysis
  advantages, 117
  application to specific proline residue in
      large protein, 111
  future prospects, 124–126
  general strategy for protein substrates,
      111–112
  methods, 112–116
  principles, 109–111
Isomeric specific proteolysis hydrolysis
  assay for products, 115–116
  buffer solutions, 113
  denaturants, 113
  enzyme solution, 113–115
  inhibition and precipitation of protease,
      115
  irreversible substrate modifications, 115
  pH, 112–113
  temperature, 113
Isotope exchange, through dissociation,
    209

of protein motility
    manifestations, 308
    sequence-specific resonance assign-
        ments, 324–326
    in solution, 307–326
protein resonance assignment, 364
relaxation effects, 380–387
single-proton exchange studies, 451,
    499, 502
    of surface protons, 499
solid state, 328–330
    of aliphatic sites, 340–343, 357–361
    of aromatic sites, 337–340, 352–357
    of biological systems, 328–335
    goals of, 328–329
    of immobile sites, 335–337
    motional averaging in, 335–342
    of motions of phenylalanine and
        tyrosine side chains, 352–357
    protein dynamics study, 327–361
study of protein dynamics, 364–367,
    495
as test of molecular dynamics simula-
    tions, 389
time scales, 309–311
two-dimensional, 3–4, 51, 138, 143,
    324
Nuclear magnetic resonance spectra,
    powder patterns, 335
Nuclear Overhauser effect, 311, 316, 465
    between adjacent labile protons, 316–
        317
    experimental, comparison with values
        calculated from rigid model, 387–
        388
    measurements, 370
        between protons where distances
            are not fixed, 386–389
Nuclease-(1–126), preparation, 194–196
Nuclease-(6–48), preparation, 193–194
Nuclease-(49–126). *See* Nuclease-(50–126)
Nuclease-(49–149), preparation, 193–194
Nuclease-(50–126), preparation, 196–197
Nuclease-(50–149), preparation, 193–194
Nuclease-(99–149), preparation, 196
Nuclease-(111–149), preparation, 196
Nuclease A, fragment complexes, fragment
    exchange, 211–213
Nuclease B, time-resolved anisotropies
    for, 526–527

Nuclease S, frequency-domain fluorome-
    try, anisotropy decays, 560–562
Nuclease T, 211
    bound with ligands, unfolding, 217
    fragment exchange, 207

O

Oligomer
    aggregation, at intermediate denaturant
        concentrations, 249
    association, analysis of intermediates
        by proteolysis, 242–243
    chemical modification during dissocia-
        tion or reassociation, 249
    correct subunit folding and aggregation,
        kinetic competition of, 247–248
    dissociation
        by changes in physical parameters,
            230–232
        by chemical denaturants, 225–230
        by dilution, 232–233
        by incubation with concentrated
            guanidine-HCl, 226
        by ligand effects, 232–233
        by pH variation, 227–228
        temperature and ligand effects,
            230
    dissociation/association
        circular dichroism, 243
        fluorescence emission, 243
        light scattering, 243
        UV difference spectra, 243
    dissociation/reassociation
        superimposed by side-reaction
            (aggregation) in transition
            range, 228–229
        transition, 225
    hybridization
        with chemically modified subunits,
            240–243
        experiments with isoenzymes, 240–
            243
    irreversible aggregation, 230
    mechanism of association, cross-linking
        with bifunctional reagents, 235–
        240
    as model for complex assemblies, 222–
        224